Methods of Enzymatic Analysis

Volume III Errata

page 212, line 6 from top

read: acetate
instead of: urate

page 315, lines 1 and 2 are missing. They should read:

dissolve 1.47 mg FH_4 with 0.9 ml $KHCO_3$ solution (6), add 0.1 ml 2-mercaptoethanol solution (2).

page 498, solution 9.

read: 60 kU/l
instead of: 600 U/l

page 500, pipetting scheme, column "concentration in assay mixture", for LDH

read: 6 kU/l
instead of: 60 U/l

page 600, right-hand column, line 19 from bottom

read: Glutamate-oxaloacetate transaminase . . .
instead of: Glutamate
– oxaloacetate transaminase . . .

verlag chemie

Editor-in-Chief:
Prof. Dr. rer. nat. Hans Ulrich Bergmeyer
Hauptstraße 88
D-8132 Tutzing
Federal Republic of Germany

Editorial Consultant and Language Editor:
Prof. Donald W. Moss, Ph. D., D. Sc.
Appletrees 42
Greenways Hinchlay Wood
Surrey, KT 10 OQD, United Kingdom

Editors:
Dr. rer. nat. H. Jürgen Bergmeyer
In der Neckarhelle 168
D-6900 Heidelberg
Federal Republic of Germany

Dr. rer. nat. Marianne Graßl
Frauenchiemsee-Straße 20
D-8000 München 80
Federal Republic of Germany

Note

The methods published in this book have not been checked experimentally by the editors. Sole responsibility for the accuracy of the contents of the contributions and the literature cited rests with the authors. Readers are therefore requested to direct all enquiries to the authors (addresses are listed on pp. XIX – XXVI).

Previous editions of "Methods of Enzymatic Analysis":
1st Edition 1963, one volume
2nd printing, revised, 1965
3rd printing, 1968
4th printing, 1971
2nd Edition 1974, four volumes
2nd printing, 1977
3rd printing, 1981

Previous editions of "Methoden der enzymatischen Analyse":
1. Auflage 1962, one volume
2. neubearbeitete und erweiterte Auflage 1970, two volumes
3. neubearbeitete und erweiterte Auflage 1974, two volumes

Deutsche Bibliothek, Cataloguing-in-Publication Data
Methods of enzymatic analysis / editor-in-chief: Hans Ulrich Bergmeyer. editors: Jürgen Bergmeyer and Marianne Graßl.
– Weinheim; Deerfield Beach, Florida; Basel: Verlag Chemie
Dt. Ausg. u. d. T.: Methoden der enzymatischen Analyse
NE: Bergmeyer, Hans Ulrich [Hrsg.]
Vol. III. Enzymes 1: oxidoreductases, transferases / ed. consultant: Donald W. Moss. – 3. ed. – 1983
ISBN 3-527-26043-9 (Weinheim, Basel)
ISBN 0-89573-233-5 (Deerfield Beach)
NE: Moos, Donald W. [Hrsg.]

Production manager: Heidi Lenz
Composition: Krebs-Gehlen Druckerei, D-6944 Hemsbach
Printing: Hans Rappold Offsetdruck GmbH, D-6720 Speyer
Bookbinding: Klambt-Druck GmbH, D-6720 Speyer
Printed in the Federal Republic of Germany

Preface to the Series

"Methods of Enzymatic Analysis" appeared for the first time in 1962 as a one-volume treatise in German. Several updated and improved editions in English and German have been published since then. The latest English edition appeared in 1974.

In the meantime, enzymatic analysis has continued to find new applications, refinements and extensions at a pace that justifies – indeed, demands – the preparation of a new and completely revised edition. However, the field has grown so enormously that it can no longer be surveyed adequately by one person. Fortunately, therefore, I am supported in this new enterprise by Dr. M. Graßl, who is highly experienced in biochemical analysis, and Dr. J. Bergmeyer, who represents the younger generation of biochemists.

With the 1974 edition of "Methods of Enzymatic Analysis" as a starting point for our work towards the new edition, it soon became obvious that many chapters had to be eliminated, re-written or added. Moreover, the increased number of analytes that can now be determined enzymatically and of enzymes regularly requiring analysis, especially in the clinical laboratory, together with the emergence of an entirely new field of application through the technique of the enzyme-immunoassay, demanded a new arrangement and subdivision of the contents, if the vast range of material was to be dealt with properly and lucidly.

The result is the plan of the work printed on the page opposite the title page of this volume. Of course, it would be impossible to publish a whole series such as this at one moment and still maintain an equal degree of topicality for all contributions. Therefore, we decided to produce the series at a pace of several volumes per year. The volumes will not necessarily appear in their numerical order, but will be made available as they can be planned and completed.

As before, the purpose of the work is to provide reliable descriptions of well-developed procedures of enzymatic analysis in the broadest sense of the term. Special efforts are being made to arrange every chapter, and to coordinate the contents of all chapters, in such a way that the volumes are useful as laboratory manuals for daily work.

Internationally-agreed enzyme nomenclature as well as quantities and units correlating with the "Système International d'Unités" are used wherever possible in order to make statements and data unambiguous and comparable over time and space.

All contributions are and will be written in English: however, contributors come from all over the world and their manuscripts naturally show various versions of English. These have to be harmonized in style and spelling in order to achieve uniformity throughout the series without, we hope, entirely eliminating each author's personal approach. Professor Donald W. Moss has kindly agreed to undertake this task. We agreed with him to use modern English spelling, but to try to minimize differences between British and American practice. We hope that this will be considered as a fair solution and one which will make the series accessible to as wide a readership as possible.

Thanks are due to the authors in the first place for responding so readily to our invitations, for writing their chapters so diligently within a short time and for communicating their experience and expertise. We are also indebted to all colleagues who gave their advice and to Professor Moss for accepting the task of language editor. Finally I wish to record my gratitude to Verlag Chemie for the fruitful and excellent cooperation during all stages from planning to production.

Tutzing, February 1983 Hans Ulrich Bergmeyer

Preface to Volume III

The number of enzymes used (in whatever sense) in science and technology has grown to such an extent that it is necessary to devote three volumes to the determination of their catalytic activities.

A general section introduces this volume. It examines the question why enzymes are still generally measured by means of their catalytic activities, and considers the possibilities and limitations of other approaches such as active site titration or determination of enzyme mass by immunochemical methods. The measurement of enzymatic activities in body fluids other than blood, serum or plasma has gained importance in the clinical laboratory. The special features of such specimens with regard to measurement techniques are pointed out. Further chapters deal with the principles of isoenzyme analysis.

The main part of the volume comprises working instructions for oxidoreductases and transferases. We have attempted to structure all chapters so as to distinguish clearly between the assay and its validation, and general considerations describing applications of the method, enzyme properties relevant in analysis, alternative methods of determination, international reference methods, standards, and enzyme effectors. Thus, the short paragraphs "Methods of Determination" introduce the reader to methods other than those described in this volume, so that he can if necessary choose another method more suited to his needs. Also, as far as they exist, international reference methods are pointed out.

No methods for automated instruments are described since the variety of analyzers has become so great that the user must be referred to manufacturers' manuals. The same is true for closed systems which do not employ conventional or tranferable chemistries in every case. Also, details of determinations of the catalytic activities of enzymes with solid phase devices are not described here, but a short survey of the prinicples involved is given in Volume II.

Since the very beginning it has been the philosophy of "Methods of Enzymatic Analysis" that experienced authors should report their methods which have proved successful at the bench. Consequently, methods described in this volume operate at various measuring temperatures, e.g. 25 °C, 30 °C or 37 °C. This is in agreement with the concept that the analyst should be free to choose the conditions of the analytical procedure, including measuring temperature. On the other hand, the internationally agreed measuring temperature for reference methods is 30 °C. When a result is gained at a temperature other than 30 °C, it should be related to that of a reference method. This can be done by using temperature conversion factors or, better, by using standard enzyme preparations. The editors have asked the authors to provide temperature conversion factors if they were not in a position to give values for 30 °C. Nevertheless, it should be kept in mind that the methods described have been optimized for the temperatures given.

In general, results of measurements are given in international units per litre of sample volume (U/l) *and* in 10^{-9} Katal per litre (nkat/l) as internationally recommended. For the sake of clarity, the editors have made a major effort to have all calculations clearly and uniformly described throughout the book.

The editors hope to have selected those methods which are most generally related to the needs of biochemists and other users.

Thanks are due to the authors, to the editorial consultant of this volume, Professor D. W. Moss, as well as to all colleagues and the authors of former editions for giving advice and participating in fruitful discussions.

Tutzing, August 1983 Hans Ulrich Bergmeyer

Contents

Contents of Volumes I – X

(Chapter Headings only)

Volume V
Enzymes 3: Peptidases, Proteinases and Their Inhibitors

Volume VI
Metabolites 1: Carbohydrates

Volume VII
Metabolites 2: Tri- and Dicarbonic Acids, Purine and Pyrimidine Derivatives, Coenzymes, Inorganic Compounds

Volume VIII
Metabolites 3: Lipids, Steroids, Drugs

Volume IX
Metabolites 4: Proteins, Peptides, Amino Acids

Volume X
Molecular Biologicals (tentative)

Contributors

Aebi, Hugo E. †
Medizinisch-Chemisches Institut
Universität Bern
Bühlstr. 28
CH-3000 Bern
Switzerland

p. 273

Assmann, Gerd
Westfälische Wilhelms-Universität
Zentral-Laboratorium
Westring 3
D-4400 Münster
Federal Republic of Germany

p. 342

Banholzer, Peter
Medizinische Poliklinik der
Universität München
Pettenkoferstr. 8a
D-8000 München 2
Federal Republic of Germany

p. 393

Becker, Reinhild
Bayer AG Leverkusen
Institute of Pharmacokinetics
POB 101709
D-5600 Wuppertal 1
Federal Republic of Germany

p. 286

Berge, Paul-Gerhard
Westfälische Wilhelms-Universität
Zentral-Laboratorium
Westring 3
D-4400 Münster
Federal Republic of Germany

p. 374, 407

Berger, Eric G.
Medizinisch-Chemisches Institut
der Universität Bern
Postfach
CH-3000 Bern 9
Switzerland

p. 374

Bergmeyer, Hans Ulrich
Boehringer Mannheim GmbH
Biochemical Research Center
Bahnhofstr. 9 – 15
D-8132 Tutzing
Federal Republic of Germany

p. 171, 352

Bernt, Erich
Boehringer Mannheim GmbH
Nonnenwald 2
D-8122 Penzberg
Federal Republic of Germany

p. 171

Brand, Karl
Institut für Physiologische Chemie
der Universität Erlangen-Nürnberg
Fahrstr. 17
D-8520 Erlangen
Federal Republic of Germany

p. 334

† deceased

Brolin, Sven E.
Dept. of Medical Cell Biology
Box 571
S-751 23 Uppsala
Sweden

p. 540

Butterworth, Peter J.
Dept. of Biochemistry
Chelsea College
University of London
London SW 3 6LX
United Kingdom

p. 2

Ceriotti, Giovanni
Ospedale Civile di Padova
Laboratorio Centrale di Analisi
I-35100 Padova
Italy

p. 319

Cohen, Philip
Biochemistry Department
Dundee University
Medical Sciences Institute
Dundee DD1 4HN
Scotland

p. 487

Deneke, Ulfert
Boehringer Mannheim GmbH
Postfach 310120
D-6800 Mannheim
Federal Republic of Germany

p. 133

Deutsch, Jack
Clinical Chemistry Laboratory
Dept. of Pathology
State University
Downstate Medical Center
Brooklyn, NY 11203
U.S.A.

p. 190

Diamondstone, Thomas I.
330 West Johnson Street
ABT H4
Philadelphia, PA 19144
U.S.A.

p. 456

Dwenger, Alexander
Medizinische Hochschule Hannover
Abteilung für Klinische Biochemie
Konstanty-Gutschow-Str. 8
D-3000 Hannover 61
Federal Republic of Germany

p. 382

Ellis, Graham
Dept. of Biochemistry
The Hospital of Sick Children
and Dept. of Clinical Biochemistry
The University of Toronto
Toronto, Ontario M5G IX8
Canada

p. 183, 560

Elstner, Erich F.
Technische Universität München
Lehrstuhl für Botanik
Arcisstr. 21
D-8000 München
Federal Republic of Germany

p. 293

Fujii, Hisaichi
Dept. of Internal Medicine
Institute of Medical Science
University of Tokyo
4-6-1, Shirokanedai, Minato-ku
Tokyo 108
Japan

p. 496

Gerhardt, Willie
Munkavaegen
Lasarettet
S-25187 Helsingborg
Sweden

p. 508, 510, 518, 530

Gerlach, Ulrich
Medizinische Klinik und Poliklinik
Westfälische Wilhelms-Universität
Westring 3
D-4400 Münster
Federal Republic of Germany

p. 112

Goldberg, David M.
Dept. of Biochemistry
The Hospital for Sick Children
and Dept. of Clinical Biochemistry
The University of Toronto
Toronto, Ontario M5G IX8
Canada

p. 183, 258, 560

Gröbner, Wolfgang
Medizinische Poliklinik
der Universität München
Pettenkoferstr. 8a
D-8000 München 2
Federal Republic of Germany

p. 393, 399

Grötsch, Horst
Hoechst AG, Frankfurt/M
Pharma Biochemie Süd H821
D-6230 Frankfurt 80
Federal Republic of Germany

p. 42

Hallinan, Terence
Dept. of Biochemistry & Chemistry
Royal Free Hospital School of Medicine
8 Hunter Street
London WC1N 1BP
United Kingdom

p. 365

Hardie, D. Grahame
Dundee University
Dept. of Biochemistry
Medical Sciences Institute
Dundee DD1 4HN
United Kingdom

p. 474, 481, 487

Heinz, Fritz
Medizinische Hochschule Hannover
Institut für Klin. Biochemie
und Physiol. Chemie
Roderbruchstr. 101
D-3000 Hannover
Federal Republic of Germany

p. 210

Henderson, A. Ralph
University Hospital
P.O. Box 5339
Postal Stn., A,
London, Ontario N6A 5A5
Canada

p. 138

Henslee, Jerry G.
The University of North Carolina
School of Medicine
Dept. of Biochemistry and Nutrition
Chapel Hill, NC 27514
U.S.A.

p. 200

van Hoof, François
Université Catholique de Louvain
International Institute of
Cellular and Molecular Pathology
UCL 7539
Avenue Hippocrate 75
B-1200 Bruxelles
Belgium

p. 38

Hørder, Mogens
Odense University Hospital
Dept. of Clinical Chemistry
DK-5000 Odense
Denmark

p. 416, 434, 444

Jaeken, Jaek
Dept. of Medicine
Universitaire Instelling Antwerpen
Universiteitplein 1
B-2610 Wilrijk
Belgium

p. 35

Jakobs, Karl H.
Pharmakologisches Institut
der Universität Heidelberg
Im Neuenheimer Feld 366
D-6900 Heidelberg
Federal Republic of Germany

p. 6

Jones, Mary Ellen
The University of North Carolina
School of Medicine
Dept. of Biochemistry and Nutrition
Chapel Hill, NC 27514
U.S.A.

p. 200

Kramer, James J.
The University of North Carolina
School of Medicine
Dept. of Biochemistry and Nutrition
Chapel Hill, NC 27514
U.S.A.

p. 200

Kusche, Jürgen
II. Chirurgische Universitäts-Klinik
Biochem. u. Experim. Abt.
Ostmerheimer Str. 200
D-5000 Köln 91
Federal Republic of Germany

p. 227, 237

Lindena, Joachim
Zentrum Biochemie
Medizinische Hochschule Hannover
Abteilung für Klinische Biochemie
Karl-Wiechert-Allee 9
D-3000 Hannover 61
Federal Republic of Germany

p. 20

Löffler, Werner
Medizinische Poliklinik
der Universität München
Pettenkoferstr. 8a
D-8000 München 2
Federal Republic of Germany

p. 399

Lorenz, Wilfried
Abt. Exper. Chirurgie u. Pathol.
Biochemie Zentrum für
Operative Medizin I
Robert-Koch-Str. 8
D-3550 Marburg
Federal Republic of Germany

p. 237

Mattenheimer, Hermann
Rush-Presbyterian
St. Luke's Med. Center
1753 W. Congress Parkway
Chicago, IL 60612
U.S.A.

p. 19, 42

Menzel, Hans-Jürgen
Westfälische Wilhelms-Universität
Zentral-Laboratorium
Westring 3
D-4400 Münster
Federal Republic of Germany

p. 342

Miwa, Shiro
Dept. of Internal Medicine
Institute of Medical Science
University of Tokyo
4-6-1, Shirokanedai
Minato-ku,
Tokyo 108
Japan

p. 496

Moss, Donald W.
Royal Postgraduate Medical School
University of London
Hammersmith Hospital
London, W12
United Kingdom

p. 49, 85, 101

Nealon, Daniel A.
Center for Laboratories
and Research
New York State Dept. of Health
Albany, NY 12201
U.S.A.

p. 434

Oßwald, Wolfgang
Technische Universität München
Lehrstuhl für Botanik
Arcisstr. 21
D-8000 München
Federal Republic of Germany

p. 293

Outlaw, William H.
Dept. of Biological Science
The Florida State University
Tallahassee, FL 32396
U.S.A.

p. 176

Pütter, Johann
Bayer A.G.
Pharmakokinetik
Postfach 101709
D-5600 Wuppertal 1
Federal Republic of Germany

p. 286

Rafael, Johannes
Institut für Biochemie I
der Universität Heidelberg
Im Neuenheimer Feld 328
D-6900 Heidelberg 1
Federal Republic of Germany

p. 266

Raijman, Luisa
University of Southern California
School of Medicine
2025 Zonal Ave.
Los Angeles, CA 90033
U.S.A.

p. 326

Reckel, Sylvia
Medizinische Hochschule Hannover
Institut für Klin. Biochemie
und Physiol. Chemie
Roderbruchstr. 101
D-3000 Hannover
Federal Republic of Germany

p. 210

Rej, Robert
Center for Laboratories and Research
New York State Dept. of Health
Albany, NY 12201
U.S.A.

p. 416, 434, 444

Sauer, Hansjörg
Gesellschaft für Strahlen- und
Umweltforschung mbH
Landwehrstr. 81
D-8000 München 2
Federal Republic of Germany

p. 304, 312, 468

Schalhorn, Andreas
Medizinische Klinik III
Klinikum Großhadern der
Ludwigs-Maximilians-Universität
Marchioninistr. 15
D-8000 München 70
Federal Republic of Germany

p. 251

Schmidt, Ellen
Medizinische Hochschule Hannover
Medizinische Klinik
Abteilung für Gastroenterologie
Karl-Wiechert-Allee 9
D-3000 Hannover-Kleefeld
Federal Republic of Germany

p. 216

Schmidt, Friedrich Werner
Medizinische Hochschule Hannover
Medizinische Klinik
Abteilung für Gastroenterologie
Karl-Wiechert-Allee 9
D-3000 Hannover-Kleefeld
Federal Republic of Germany

p. 216

Schriewer, Hilko
Westfäl. Wilhelms-Universität
Zentral-Laboratorium
Westring 3
D-4400 Münster
Federal Republic of Germany

p. 407

Shaw, Leslie M.
Hospital of the University
of Pennsylvania
Dept. of Pathology
3400 Spruce Street/G1
Philadelphia, PA 19104
U.S.A.

p. 349, 357

Siedel, Joachim
Boehringer Mannheim GmbH
Biochemical Research Center
Bahnhofstr. 9–15
D-8132 Tutzing
Federal Republic of Germany

p. 156

Smith, Alistair F.
Dept. of Clinical Chemistry
Royal Infirmary
Edinburgh, EH3 3YW
Scotland

p. 163, 166

Spooner, Richard J.
Dept. of Biochemistry, Royal Infirmary
Glasgow
Scotland

p. 258

Springer, Stefan A.
Dept. of Biological Science (Unit 1)
Florida State University
Tallahassee, FL 32306
U.S.A.

p. 176

Strømme, Johan H.
Central Laboratory
Ulleval Hospital
Oslo 1
Norway

p. 357

Den Tandt, Wilfried Richard
Dept. of Medicine
Universitaire Instelling Antwerpen
Universiteitplein 1
B-2610 Wilrijk
Belgium

p. 35

Thonar, Eugene J.-M. A.
Dept. of Biochemistry
Rush-Presbyterian-St. Luke's
Medical Center
1753 West Congress Parkway
Chicago, IL 60612
U.S.A.

p. 30

Trautschold, Ivar
Medizinische Hochschule Hannover
Abteilung für Klinische Biochemie
Konstanty-Gutschow-Str. 8
D-3000 Hannover 61
Federal Republic of Germany

p. 20, 382

Vassault, Anne
Hôpital Necker – Enfant Malades
Laboratoire de Biochimie A
149 rue de Sèvres
F-75743 Paris Cedex 15
France

p. 118

Verdon, Bernadette
Medizinisch-Chemisches Institut
der Universität Bern
Postfach
CH-3000 Bern 9
Switzerland

p. 374

Wahlefeld, August Wilhelm
Boehringer Mannheim GmbH
Biochemical Research Center
Bahnhofstr. 9 – 15
D-8132 Tutzing
Federal Republic of Germany

p. 126, 352

Wakabayashi, Yasuo
The University of North Carolina
School of Medicine
Dept. of Biochemistry and Nutrition
Chapel Hill, NC 27514
U.S.A.

p. 200

Whisler, Kenneth E.
Dept. of Biochemistry
Rush-Presbyterian-St. Luke's
Medical Center
1753 West Congress Parkway
Chicago, IL 60612
U.S.A.

p. 26

Wieme, Roger J.
University Hospital Gent
Dept. of Clinical Pathology
De Pintelaan 185
B-9000 Gent
Belgium

p. 49, 54

Wilhelm, Andreas
Westfälische Wilhelms-Universität
Zentral-Laboratorium
Westring 3
D-4400 Münster
Federal Republic of Germany

p. 407

Wilmanns, Wolfgang
Medizinische Klinik III
Klinikum Großhadern der
Ludwigs-Maximilians-Universität
Marchioninistr. 15
D-8000 München 70
Federal Republic of Germany

p. 251, 312, 468

Wulff, Karl
Boehringer Mannheim GmbH
Biochemical Research Center
Bahnhofstr. 9 – 15
D-8132 Tutzing
Federal Republic of Germany

p. 525

Youngman, Richard J.
Technische Universität München
Lehrstuhl für Botanik
Arcisstr. 21
D-8000 München
Federal Republic of Germany

p. 293

1 General

1.1 Different Kinds of Enzyme Determination

Peter J. Butterworth

1.1.1 The Problem

Understanding how metabolism operates *in vivo* is largely a matter of understanding the properties of the individual chemical steps in the metabolic pathway, each catalyzed by a particular enzyme. To reach this goal it is helpful to know the relative (and absolute) cellular concentrations of the enzyme and its substrate and to be in possession of considerable knowledge of the kinetic properties of the enzyme; in particular, the response of catalytic activity to alterations in the concentrations of substrate, and of substances or metabolites that stimulate or inhibit the activity.

Usually, the relevant kinetic properties can be ascertained relatively easily, but studies with this aim are performed *in vitro* under conditions of pH, ionic strength and buffer composition that differ markedly from those prevailing *in vivo*. It is possible in some cases to extrapolate from *in vitro* to *in vivo* conditions, but a major difference between *in vitro* experiments and the cellular state is always likely to be found in the very low concentration of the enzyme protein used in the kinetic study compared with that existing in the cell or organelle. The difference may be of one to two orders of magnitude [1]. With relatively high enzyme concentration, a significant fraction of the total endogenous substrate, coenzyme, activators and inhibitors may be enzyme-bound at any instant. This condition is usually deliberately avoided in *in vitro* kinetic studies and therefore the properties of the enzyme established in the laboratory may have little relevance to the cellular state. To make sense of any cellular 'kinetic' control systems that may operate, the biochemist needs to know the cellular concentration of the enzyme so that its behaviour may be studied under conditions that mimic those present in the cell. If a hormonal stimulus, for example, is found to be associated with a change in the steady-state cellular concentration of the enzyme, it may be possible to interpret this change meaningfully in terms of altered metabolic flux.

Cellular metabolism is not the only area of biochemistry in which the absolute concentration of enzyme protein needs to be determined in order to interpret experimental results. An enzymologist working with purified enzyme preparations and studying the mechanism of enzyme-substrate or enzyme-inhibitor interactions needs a measure of concentration in order to calculate the stoichiometry of the enzyme-substrate interaction, i.e. the number of catalytic or binding sites that exist on each molecule of the enzyme. This information, together with details of the subunit structure of the enzyme, is an important element in the understanding of structure-function relationships in the enzyme protein.

A particular area in which the question may arise as to whether the activity of a particular enzyme can be assumed to represent its concentration is in the use of measurements of enzyme activities in human body fluids, especially blood serum, as an aid to clinical diagnosis. With relatively few exceptions, the intracellular concentration of an enzyme greatly exceeds its extracellular value. A leakage of enzymes from damaged or dead cells therefore leads to an increase of enzyme activity in the plasma or serum. Also, it is possible that the proliferation of a type of cell which produces a particular enzyme may result in increased extracellular enzyme activity, as may an increase in the rate of enzyme synthesis.

The great sensitivity and specificity with which the catalytic activity of enzymes can be measured (see below) allows minimal tissue damage or small changes in enzyme production to be detected. However, the measurement of only those enzyme molecules that are catalytically active leaves unanswered the question of whether inactive molecules are also released from damaged tissues, and indeed whether, in some cases of tissue damage, such inactive molecules may predominate. Immunochemical methods provide the opportunity to attack this problem so that it becomes possible to measure all enzyme molecules, inactive as well as active. Such measurements offer a potentially great increase in sensitivity and diagnostic usefulness. The results obtained so far suggest that in some instances of acute tissue damage, such as myocardial infarction, both catalytic activity measurements and immunochemical methods give similar information; however, this may not always be the case in more chronic conditions.

Although there is a general need to determine enzyme concentration, the means by which this can be achieved is not straightforward. The most obvious method, and that which is used most frequently, is to measure the protein concentration. Methods for protein assay, e.g. the absorbance at 280 nm, the biuret reaction, or that of *Lowry et al.* [2] (see Vol. II, chapter 1.3.4), are all based on the general properties of proteins, but because every protein is unique in its composition, the UV absorbance or the colour yield in a spectrophotometric assay are unlikely to be identical for different proteins. If sufficient quantity of the enzyme under study is available in purified form, a suitable standard curve can be constructed for any of the popular methods. This situation is realized very seldom, however, and bovine serum albumin is usually preferred as the standard or reference protein.

If the enzyme sample is impure or it is a component of a tissue homogenate or body fluid, non-specific methods of protein assay are of little use for determining its concentration. Nevertheless, a measure of the total protein present is often used as a reference for comparison purposes; e.g. enzyme concentration in a tissue homogenate is frequently expressed in terms of units of activity per mg of total protein (so called specific activity – see below). The goal of any enzyme isolation scheme is the attainment of a large and constant specific activity. A constant value suggests that all non-enzymic protein has probably been removed [3].

If a solution of pure enzyme is available, a dry-weight analysis may be possible. This should provide an exact value for the enzyme protein content, but it is useless for impure samples or for tissue extracts or body fluids. Even with solutions of pure enzyme protein, the amount of the protein that has to be sacrificed to obtain a reliable dry weight estimate may be a considerable deterrent, considering the great expense in labour and materials that usually accompanies enzyme purification.

When dealing with a mixture of proteins it is necessary to use methods which are capable of selecting the protein of interest in the mixture. Methods based on the catalytic or immunological specificity provide the only real means of doing this.

1.1.2 Catalytic Activity

1.1.2.1 Measurement of Catalytic Activity

Under the conditions in which the concentration of the enzyme is much less than that of the substrate, the majority of enzymes obey the familiar rate law of *Michaelis* and *Menten* which relates the velocity of the catalyzed reaction to the molar concentration of the enzyme $[E]_0$ and substrate [S]

$$v = \frac{k\,[E]_0 \times [S]}{K_m + [S]}$$

or

$$v = \frac{V \times [S]}{K_m + [S]}$$ (when the maximum velocity is substituted for $k \times [E]_0$).

Under a defined set of physical conditions of temperature, pH etc., constant values can be assumed for k, the rate constant for the formation of products at the steady state, and for K_m, the *Michaelis* constant. Hence the activity of the enzyme is directly proportional to the concentration, i.e. mass per unit volume, of the enzyme. In addition, the equation predicts that when [S] is greatly in excess of the K_m value, the catalytic rate becomes independent of [S] and the enzyme is said to be saturated with respect to substrate [3].

The behaviour of some enzymes is not well described by the *Michaelis-Menten* model, yet knowledge of their activities in tissue homogenates or cellular extracts is often of great interest. Several enzymes which display anomalous kinetics are now believed to be of great significance in controlling the rate and direction of metabolic reactions. For a multi-subunit enzyme that displays co-operative interactions between identical catalytic sites situated on different subunits, the rate law at relatively high substrate concentrations may be represented by an equation of the form

$$\frac{v}{V} = \frac{\psi\,[S]^n}{1 + [S]} \quad \text{or} \quad v = \frac{V\,[S]^n}{1/\psi + [S]^n}.$$

(where V is the maximum velocity and ψ is a constant).

This is a form of the well-known *Hill* equation [4]. Provided that $[S]^n$ is large compared with $1/\psi$, the measured velocity approximates to the maximum velocity, i.e. it is apparently independent of [S] but is directly proportional to the concentration of enzyme. Thus, if similar assumptions are made, the activities of multi-subunit, co-operative enzymes can be determined using the same strategies as for '*Michaelis-Menten*' enzymes.

Briefly stated, these principles form the rational basis for the determination of relative or absolute 'amounts' of enzymes in terms of their catalytic activities. This is the most specific, sensitive and rapid type of enzyme determination available. Its theory and practice are discussed at length in Vol. I.

Analytical procedures based on spectrophotometry in visible or ultraviolet light routinely detect chromophores at concentrations of approximately 1 μmol per ml of solution. If we assume an average protein molecular weight of the enzyme of 100000 and a turnover number (see below) of 100 per min, then calculation shows that only 10^{-10} mole of enzyme is required in order to form 1 μmole of product in 1 minute in a 1 ml volume of reaction mixture. Thus, the enzyme concentration need be only 10^{-7} mol/l in this case in order for measurable amounts of product to be formed. Since the assumed turnover number of 100 is very modest by the standard of most enzymes, and because it is possible to prolong the period of the incubation beyond 1 minute to allow more product formation, detection of even smaller concentrations of enzyme is possible. In addition, the greatly improved performance of modern laboratory spectrophotometers enables reliable measurement of relatively low absorbance in the visible and ultraviolet range. Other techniques of detection, such as fluorimetry (Vol. I, chapter 3.4) offer still greater detectability. The combination of all these factors allows the detection of many enzymes at concentrations as low as 10^{-10} mol/l.

The relatively large turnover numbers of enzymes result in a great amplification of their signal; a single molecule of enzyme transforms several hundred molecules of substrate per minute, and this amplifying action has been suggested as the reason for the natural selection of certain multi-enzyme cascade systems in cellular function; e.g. the blood clotting mechanism, hormonal stimulation of glycogen breakdown and the complement reaction.

The greatest disadvantage of methods based on activity determination is that the result is entirely method dependent. Even minor alterations in any of the assay parameters such as temperature, pH, buffer strength and composition, nature of the substrate (in the case of non-specific enzymes, e.g. hydrolases) and its concentration have profound effects on the measured catalytic activity. With care, such effects are unlikely to present problems to investigators looking for relative differences in activity between samples, but the direct comparison of absolute values between laboratories is made difficult by the use of different assay conditions in different laboratories. It is also possible for false impressions of the reactivity of the enzyme in the cell to be obtained because, for reasons of analytical precision and convenience, artificial substrates are frequently used in a laboratory assay. The specific activity of the enzyme in the tissue extract towards its true (physiological) substrate may not necessarily be identical to that determined with the artificial one. Fortunately, with esterases (e.g. alkaline phosphatase) for which the true substrates are generally unknown and for which artificial substrates are particularly useful, their mechanism of action normally dictates an equal rate of reaction at saturating substrate concentrations whatever the identity of the substrate. Nevertheless, whenever the true substrate is unknown it is wise to regard specific activity measurements with a certain amount of caution.

Implicit in the assumption that a change in activity is directly related to a change in the number of molecules of enzyme present, is the assumption that every molecule of a particular enzyme has the same *intrinsic* activity as any other molecule. If the activity of the enzyme in a tissue is found to increase as a consequence of some perturbation, activity measurements alone are unable to distinguish whether the increase represents a net increase in the number of molecules of enzyme present in the tissue, or whether new enzyme molecules of a higher intrinsic specific activity have appeared. In the latter case, the steady-state concentration of the enzyme in the cell may not have undergone any increase; it is possible even for the steady-state level to have decreased. Claims have been made from time to time that enzymes of altered specific activity can be induced [5] but further investigation has usually cast doubt on this. However, since it is known that many proteins are subjected to post-translational modification as part of a natural maturation process, it is possible that molecules at different stages of processing or destruction may differ in their specific activities. Under the influence of an agent that promotes the synthesis of an enzyme, it is possible that a wider spectrum of enzyme molecules at various stages of maturation may exist than in a more quiescent period of enzyme production and turnover.

1.1.2.2 Conversion of Catalytic Activity to Substance Concentration of Enzyme

If the molar catalytic activity (kat/mol, U/mol), formerly "turnover number", of the enzyme under the conditions of the assay is known, the actual mass concentration of the enzyme in a sample can be calculated. Molar catalytic activity is the number of moles of substrate transformed per mole of enzyme per unit time. Hence, for a measured activity z (kat, U) and a known molar catalytic activity (kat/mol, U/mol), the amount of enzyme present in the reaction mixture is given by the quotient of both.

Obviously, the calculation requires an accurate value for the molar catalytic activity. This may be obtained from literature values in some cases, or from the results of determinations by the experimenter if pure samples of enzyme of known concentration are available. It must be remembered that *all* measurements of catalytic activity are method-dependent.

Therefore, if an experimenter obtains a precise estimate of the molar catalytic activity of a given enzyme under defined conditions, these conditions must be adhered to rigidly in a subsequent assay, if the actual enzyme concentration is to be calculated from the activity measurement. *Roth* [6] recommended the use of this approach as a way of overcoming the difficult problem of the standardization of enzyme activity measurements in clinical chemistry. If the recommendation were adopted, different laboratories would be encouraged to record their results in terms of substance concentration instead of activity concentration (units of activity per litre). Thus a direct comparison between the results from various laboratories would be possible irrespective of whether the laboratories employ identical or dissimilar sets of conditions for their assays, provided that each used the appropriate value of the molar catalytic activity.

Roth used spectrophotometric titration (see below) to determine the turnover number of pig heart aminotransferase [6]. He successfully applied the value so obtained to the calculation of the absolute concentration of the enzyme in samples of pig serum. He looks forward to the commercial availability of human enzymes standardized by active-site titration.

It is probably true to state that reliable values for the molar catalytic activity are unavailable in most instances and this is the reason why enzyme 'concentration' is frequently expressed in terms of the measured catalytic rate, rather than in units of 'amount of substance' per litre such as mol/l. However, for comparison purposes, e.g. between tissues, or for monitoring changes in enzyme concentration in the cell associated with development, hormonal induction or repression, and drug treatment, activity measurements are perfectly adequate if expressed as specific activities, i.e. units/mg total protein.

1.1.3 Active-site Titration

Because of uncertainties in knowledge of molar catalytic activities (see above) determination of the concentration of enzyme molecules in solution may in some cases be achieved by direct methods of estimation rather than by calculation from figures obtained in an assay of catalytic activity. Many of the direct methods are based on titrations of the active site of the enzymes and exploit the specificity and biological activity of the protein. In most instances the method adopted is specific to the particular enzyme under study and is not necessarily of general applicability. This is because the methods usually rely on certain features of mechanism of enzyme action that are not possessed by all enzymes.

1.1.3.1 Pre-steady State Burst

Some enzymes, particularly those belonging to the class of hydrolases, have a mechanism that involves the formation of a covalent intermediate between the enzyme and substrate during the course of catalytic turnover. For example, the protease chymotrypsin hydrolyzes 4-nitrophenyl acetate by what is, in effect, a two-step process. The first step is the formation of acetyl-chymotrypsin with the release of 4-nitrophenol and the second is the hydrolysis of the acyl-enzyme intermediate with the release of acetate and the regeneration of active chymotrypsin

(a)

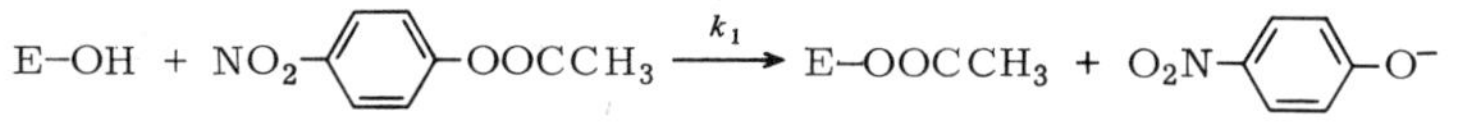

(b) $$E{-}OOCCH_3 + H_2O \xrightarrow{k_2} E{-}OH + CH_3COOH$$

The hydrolysis of the covalent intermediate is relatively slow; i.e., the rate constant k_2 is much less than k_1, the rate constant for the formation of the intermediate. Therefore, in the steady state when the enzyme is in a condition in which the formation and destruction of the intermediate are balanced, the rate of the reaction, i.e. the rate of release of the 4-nitrophenolate ion, is governed by the relatively slow step with rate constant k_2. However, in the brief interval of time that elapses between the initial mixing of the enzyme and substrates and the establishment of the steady state, there will be a release of 4-nitrophenol that is relatively rapid since it is governed by the 'fast' step, k_1. This rapid initial rate of product formation is the basis of a method introduced by *Hartley & Kilby* [7] for active-site quantitation. The rapid release of product will slow exponentially to the steady-state rate as the reaction proceeds. Thus a plot of the release of 4-nitrophenol against time has the appearance of Fig. 1.

In the pre-steady state period one mole of 4-nitrophenol is released for each mole of active site of chymotrypsin present. Therefore, extrapolation of the plot of the steady state portion of the plot to zero time gives a positive intercept on the vertical axis. The magnitude of the intercept, the 'burst height', is taken as a direct measure of the concentration of the enzyme active site.

The method has been widely applied to proteases such as chymotrypsin and to non-specific phosphatases. It was used as an 'all or none' assay during a study of chemical modification of chymotrypsin to decide whether the fall in activity accompanying modification represented destruction of the catalytic site, or was a consequence of impaired catalytic efficiency brought about by modification of some other portion of the enzyme molecule [8]. In order to allow correct interpretation of the data, the experimenter needs to know the number of subunits which make up a molecule of the enzyme. Clearly, a dimer of identical subunits would be expected to release two moles of product per mole of enzyme during the burst. Even here however, there may be a degree of uncertainty in interpretation because of the possible existence of 'half-of-the-sites reactivity' (also known as extreme negative co-operativity) claimed for some enzymes. In such a condition, only half of the possible number of active sites is

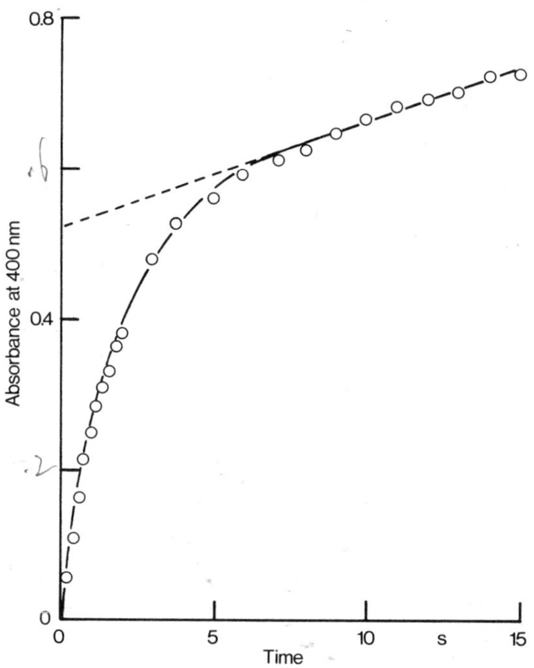

Fig. 1. Time course of the hydrolysis of 4-nitrophenol acetate catalyzed by chymotrypsin. For a mechanism that involves formation of a covalent intermediate, provided that the rate limiting step is the decomposition of the intermediate, extrapolation of the steady-state portion of the curve to the zero time yields a 'burst' height that equals the concentration of the active sites of the enzyme. (Redrawn with permission from *Spencer & Sturtevant,* J. Amer. Chem. Soc. *81*, 1874 (1959). Copyright 1982 American Chemical Society).

expressed during a catalytic turnover. When there is considerable disagreement amongst investigators about whether the 'half-of-the-sites reactivity' of a particular enzyme is a real phenomenon or not, as is the case for alkaline phosphatase for example, interpretation of the stoichiometry revealed by the burst height becomes a matter for individual judgment. With monomeric enzymes such as trypsin and chymotrypsin, the result should be unequivocal.

A major drawback to the method is the need for special instrumentation to allow measurements during the short time interval before attainment of the steady state. A stopped-flow spectrophotometer is usually used so that readings can be taken within milliseconds after the very rapid mixing of the enzyme with suitably-buffered substrate. The response time of chart recorders is too long for them to be of use for data acquisition over the very short reaction periods and so an oscilloscope is used for this purpose. In order to obtain measurable signals, relatively high concentrations of enzyme are needed and the sample must be free from contamination by enzymes of similar and overlapping specificity. These requirements render the method of limited direct applicability to tissue extracts or body fluids, but the method remains one of the most useful for estimating enzyme concentration in purified samples.

It is sometimes possible to arrange the conditions of the assay in order to overcome the need for the special equipment designed for rapid mixing and recording of data, so that conventional laboratory spectrophotometers can be used instead. Selection of an artificial substrate that releases a fluorescent product during the pre-steady state burst increases the sensitivity of the detection of the product by one to two orders of magnitude over absorbance measurements, enabling more dilute enzyme samples to be used. The requirements of rapid mixing and short-time observation can be at least partially overcome by the use of 'poor' substrates. A poor substrate is one that is hydrolyzed at a detectable rate but a rate that is much slower than of a typical normal or 'good' substrate. With a poor substrate the pre-steady state period may be extended to several minutes. When the poor substrate is combined with sub-optimal conditions of pH or temperature the approach to the steady state may be further slowed. Hence conventional methods of mixing enzyme and substrate are sufficiently rapid relative to the time needed to establish the steady state to remove the need for elaborate systems to achieve virtually instantaneous mixing.

1.1.3.2 Active-site Labelling

By taking advantage of the specificity of an enzyme for its substrate or substrate analogue it is possible to devise strategies that result in the incorporation of a suitable labelled substrate or analogue into the active site region of the protein. This may be done even when the enzyme is part of a complex mixture of proteins. For the method to be useful it is essential that the adduct formed between the enzyme and substrate is stable, so that it can be isolated. This usually means that the combination of the label with the enzyme must be covalent. Also, since a substrate is converted by the enzyme, a potent competitive inhibitor or substrate analogue is to be preferred. With some enzymes, the conditions of the experiment can be adjusted so that any covalent intermediate that is formed during the normal catalytic sequence is made resistant to further reaction; i.e. it is trapped and can therefore be isolated. Trapping may be brought about by sudden denaturation of the enzyme with acid, for example.

The enzyme solution is mixed with a suitably radiolabelled analogue. The specific radioactivity of the analogue must be known and its concentration in the mixture must be great enough to ensure that the enzyme active sites are saturated by the analogue. The enzyme-analogue complex is then separated from the free, uncomplexed label by a suitable method such as by precipitation of the labelled protein with perchloric acid or by gel filtration on small columns. Radioactivity in the precipitated material or the column fraction is then determined by scintillation spectrometry or by a suitable alternative method. The concentration of the labelled protein in the mixture can then be calculated. With a known stoichiometry for the enzyme-analogue reaction, the number of moles of enzyme in a known volume of solution may be determined. As with pre-steady state burst titrations, the existence of half-of-the-sites reactivity may complicate the interpretation of experimental results (see above).

The essential requirements for the method are, first, the availability of a suitable analogue that can be obtained with a high specific radioactivity and, second, the formation of a covalent adduct between the enzyme and the analogue. A disadvantage that may have been considered important at one time was the need for special facilities for handling radioactive material and for counting equipment. However, since nearly all biochemistry laboratories have facilities for radiochemistry, with scintillation and/or γ-counters as standard items of equipment, this disadvantage need hardly be considered any longer. The method can sometimes be extended to enzymes that do not normally form covalent intermediates by performing the experiment in the presence of an additional chemical reagent which converts an unstable complex (that would dissociate during any isolation procedure) into a stable covalent one. An example of this is provided by the reduction with sodium borohydride of adducts formed either between free amino groups and pyridoxal phosphate or between a carboxyl group on the substrate and a free protein amino group [9].

1.1.3.3 Spectrophotometric and Fluorescence Titration

Many coenzymes, e.g. NAD(H) and pyridoxal phosphate, and some substrates (e.g. nucleotides such as ATP) have characteristic absorption spectra in free solution. When the coenzyme or substrate becomes bound to an enzyme there is sometimes a change in the spectrum which can be detected in a spectrophotometer. When NAD binds to glyceraldehyde-3-phosphate dehydrogenase an absorption band at 360 nm appears [10]. If such a spectral change occurs, difference spectrophotometry can be used to monitor the binding of the ligand to the enzyme [11]. As the protein is titrated with the ligand, a limiting absorption difference is eventually reached signifying the end point of the titration, i.e. the enzyme is fully saturated with the ligand. From knowledge of the absorption coefficient of the ligand-enzyme complex, the concentration of the complex, and therefore the concentration of the enzyme, may be calculated from the end point.

A protein possessing several identical subunits will bind several molecules of ligand and therefore the number of binding sites needs to be known if the enzyme concentration, rather than active-site concentration, is required.

Chemical modification of certain amino acid residues is accompanied by a change in absorption. Perhaps the best known example is the increase in absorption at 250 nm associated with mercaptide formation between 4-chloromercuribenzoate (PCMB) and thiol groups of cysteine [12]. Therefore, if the total number of thiol groups in an enzyme is known, it is possible to calculate the enzyme concentration from the change in absorption at 250 nm following reaction with PCMB. The need for information on the amino acid composition tends to limit the applicability of this kind of method, but frequently it is found that a particular residue located at the catalytic site is considerably more reactive than similar residues located elsewhere in the enzyme. If this condition is met, it may be possible to do the titration under conditions in which only the active-site residue becomes modified and so obtain a value for the enzyme concentration.

It is not unusual for a fluorescent compound to undergo a change in fluorescence intensity and a shift in the wavelength of maximum emission when the compound becomes bound to protein. Many studies have been made of this phenomenon using the fluorescent probe ANS*. The change in signal has been interpreted as indicating a

* ANS anilinonaphthalene sulphonic acid.

change in the environment of the probe, from the polar environment of the free solution to an apolar environment when sequestered by a hydrophobic region of the protein [13]. The fluorescence enhancement could in principle provide the basis of a method for determining enzyme concentration but it seems not to have been widely used so far. ANS is probably of limited use as a general reagent for the titration of enzymes because it lacks the specificity that is needed in order to titrate selectively a particular enzyme protein. ANS would be suitable for a pure protein. It is possible, however, to apply fluorescence titration to a number of enzymes which act on a fluorescent substrate that binds specifically at the active site region. Dehydrogenases may be titrated with the coenzyme NAD [11].

If the fluorescence of a solution containing the enzyme is monitored during incremental additions of the fluorescent probe, an increase in fluorescence is observed until all of the protein molecules in the solution have become fully liganded with the probe (Fig. 2).

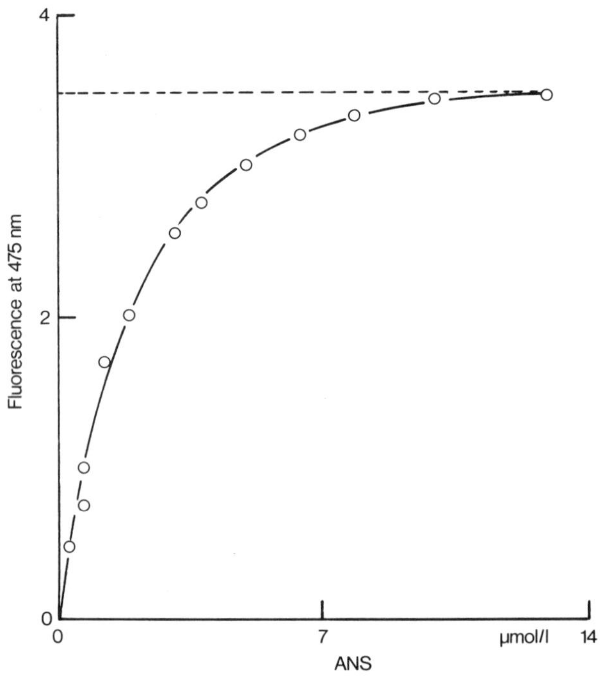

Fig. 2. Fluorescence titration of calf intestinal alkaline phosphatase with ANS. The fluorescence was excited at 370 nm and monitored at 475 nm. ANS was added sequentially to the enzyme solution until a constant fluorescence indicated that the equivalence point was reached. The experimental points have been corrected for any fluorescence emanating from unbound ANS. The concentration of the enzyme may be calculated from the limiting fluorescence, if the fluorescence yield of a standard enzyme-ANS complex is known (see text).

Any increase in fluorescence emanating from the free probe or enzyme can be corrected for by appropriate control titrations. If the fluorescence yield of the probe bound to the enzyme is known, the concentration of enzyme in the sample may be calculated from the limiting value of the fluorescence reached in the titration.

In principle, the method is a good one but its usefulness is limited by the need to quantitate the limiting fluorescence by reference to a known standard solution of the enzyme. Since, unlike spectrophotometry, the signal recorded by a fluorimeter is not usually directly comparable from experiment to experiment, literature values for the maximum fluorescence yield (comparable to absorption coefficients) are not available. Thus a 'standard' solution of pure enzyme must be titrated alongside the 'unknown' solution.

The applicability of the titration methods is limited to relatively pure samples and, as indicated, interpretation of the results requires some knowledge of the number of probe- or ligand-binding sites present in the molecule, i.e. the binding stoichiometry. It seems, in fact, that titrations have mainly been applied to the determination of the binding stoichiometry of a protein solution of known concentration rather than to the measurement of concentration *per se* [11].

1.1.4 Determination of Enzymes as Antigens

All of the methods discussed so far exploit the catalytic properties of enzyme proteins and the special relationship that exists between the active site of an enzyme and its substrate or substrate analogue. A special relationship of a different kind exists between a protein and a specific antibody that has been raised against the protein (cf. Vol. I, chapter 2.7, p. 233). The precise recognition of a protein by an antibody can provide the basis for an estimation of the concentration of the protein antigen. The development of such methods owes much to the needs of immunologists to identify and estimate proteins which have no obvious catalytic activity and therefore lack the relatively straightforward means of identification possessed by enzymes. The advantages of immunoassay lie in its low detection limits and high specificity. The specificity is an immunological one and is independent of the biological function or activity of the antigenic protein. Therefore, the method measures inactive enzyme molecules provided that they have unaltered antigenic properties, as well as active molecules. This ability can be of great use in the investigation of inborn errors of metabolism in which mutant enzymes of altered or even zero catalytic activity may by synthesized by cells. Also, detection of precursor (zymogen) forms of enzymes may be possible, as well as protein fragments produced during degradative turnover.

Immunochemical methods depend on the availability of specific antisera directed against the particular enzymes under study. The antibody is raised in an experimental animal, such as a rabbit, by injection of a sample of the protein antigen. Any impurity in the antigen will result in an antibody of imperfect specificity that can be the source of a number of interfering reactions in the assay itself. One difficulty of the method is that, even with injection of pure antigen, each new batch of antiserum will be found to differ in detectability and degree of specificity for the antigen. Cell hybrids formed by fusion of a tumour cell with a lymphocyte that secretes antibody may be cloned to provide a long-term source of relatively large amounts of a highly specific antibody. Exploitation of such *monoclonal antibodies* [14] will greatly increase the ease and accuracy of performance of immunoassay. A monoclonal antibody is a well-defined chemical reagent, in contrast to a conventional antiserum, which is a heterogenous mixture that varies in composition from batch to batch.

It is possible to select various hybrid clones that all secrete antibody to a common antigen, but the antibodies differ with respect to the structural features of the antigen that are recognized. Therefore use of the various monoclonal antibodies, either singly or in combination, provides the investigator with an almost limitless degree of specificity or selectivity in an assay system.

1.1.4.1 Selective Precipitation

The principle of this method is that excess antibody is added to a solution of the enzyme to precipitate it as an insoluble antigen-antibody complex. The complex may be separated by centrifugation and analyzed for its enzyme content. The analysis is not difficult if, as is usually the case, enzymic activity is retained in the antigen-antibody complex. The activity of the precipitated complex is compared with a standard curve obtained by treating, under the same conditions, known amounts of pure enzyme. The selectivity of a suitable antibody can permit the precipitation of a particular enzyme from a mixture of enzymes of overlapping catalytic specificity. This enables an estimate to be made of its concentration in the mixture, which would be impossible by activity measurements alone.

If the enzyme-antibody complex remains in solution it may be necessary to aid precipitation by the addition of a second antibody that reacts with the first. For example, in an assay of placental alkaline phosphatase, antiserum to the enzyme was raised in a rabbit and an antibody against rabbit γ-globulin was produced in a goat [15]. Addition of the goat antiserum to the enzyme-antibody complex resulted in an insoluble 'sandwich' of enzyme-antibody$_1$-antibody$_2$. The assay allowed the detection and reliable measurement of 5 ng of placental alkaline phosphatase, i.e approximately 6×10^{-14} moles of enzyme.

When the activity of the enzyme is masked in the complex, radiolabelling of the enzyme or antibody, for example by iodination with ^{125}I, allows the complex to be measured by the radioactivity of the separated precipitate.

1.1.4.2 Competition Assays

Radioimmunoassay (RIA) is based on the principle of competition between radiolabelled and unlabelled antigen for a specific antibody present in limited concentration in the assay mixture. The labelled antigen (Ag*) and antibody (Ab) form a reversible Ag*Ab complex. Unlabelled antigen (Ag) present in the mixture displaces Ag* from the complex as AgAb is formed. The radioactivity of the complex is thereby decreased and the magnitude of the decrease is a measure of the concentration of antigen in the test mixture.

The labelled antigen is most frequently produced by iodination with the isotope ^{125}I and the radioactive Ag*Ab complex is measured in a γ-counter after separation by, for example, precipitation or gel filtration. Formation of an enzyme-antibody$_1$-antibody$_2$ sandwich (see above) may be necessary to effect precipitation. For example *Veneziale et al.* [16] have described a method that reliably measures approx. 10 ng of phosphofructokinase (PFK) in extracts of skeletal- or heart muscle. In their method anti-PFK was produced in goats and the radioactive complex of iodinated-PFK and antibody was precipitated with an antigoat γ-globulin of rabbit serum.

An adaptation of the method has been developed that removes the costs and dangers associated with radiochemistry. This is the so-called enzyme linked immunosorbant assay (ELISA) technique (cf. Vol. I, chapter 2.7). In this adaptation the label is an enzyme that is attached, usually covalently, to the antigen. The Ag*Ab complex is then detectable by assay of the linked enzyme. The enzyme chosen for the labelling is one that can be assayed easily and sensitively by spectrophotometric techniques and is reasonably resistant to the procedures required to form the covalent enzyme-antigen adduct so that the enzyme retains its catalytic properties even when linked to the antigen. Alkaline phosphatase has been used by several investigators. ELISA exhibits as low detection limits as the conventional RIA method for the detection of the antibody. ELISA uses the same principles as RIA and thus an essential requirement remains a pure sample of antigen from which suitable standards may be prepared. So far, the ELISA technique has not been applied extensively to the determination of enzymes.

When an enzyme exists as multiple forms (i.e. isoenzymes) the differences in structure between the proteins making up the isoenzyme family may be very small. Therefore a considerable overlap in the immunospecific properties of the isoenzymes is probable. Hence, an antibody raised against a particular variant is likely to react to a greater or lesser degree with the other isoenzymes. This cross reactivity will result in competition for antibody binding sites when the antibody is added to a mixture of isoenzymes. If there is a significant cross reaction, the measured AgAb complex will not provide a reliable estimation of the concentration of the particular variant. In order to obtain a better result it will be necessary first to separate the isoenzymes in some way before immunoassay. Immunodiffusion methods that combine electrophoresis with immunoassay are particularly suitable for estimating the relative amounts of cross-reacting isoenzymes (see below). However, the increasing availability and use of monoclonal antibodies is likely to overcome the difficulties of cross reactivity among related, but non-identical, proteins and allow reliable estimation of the relative concentration of each of the several isoenzymes that may be present in the sample.

1.1.4.3 Immunodiffusion

In this technique, the antigen is allowed to diffuse into a medium containing antibody. Precipitation occurs as a result of AgAb complex formation and, provided that Ab is in excess, the extent of the precipitation is a function of the concentration of the antigen. The diffusion is usually carried out in agarose gel. This material acts as an anti-convectant so that the AgAb precipitate remains concetrated at the site of interaction. Any spreading caused by local thermal currents renders the interaction zones diffuse and difficult to quantitate. Agarose has an exclusion limit for molecular weights that is greater than 10^7 so that almost any protein or enzyme is able to move through the gel. The degree of precipitation is determined either by non-specific staining with a stain such as Coomassie Blue, or by adoption of a specific enzyme stain for examples in which the insoluble AgAb complex retains catalytic activity. Protein stains allow approx. 50 ng of protein to be detected reliably but if the catalytic activity can be exploited the sensitivity can often be increased up to 100-fold. Thus, 0.5 ng of enzymes may be detectable.

The disadvantages of the method are that the rate of diffusion through the agar is slow so that several days may be needed for complete reaction, and relatively large amounts of antibody are required which should be distributed uniformly throughout the gel. The process can be speeded up considerably by the application of an electric field in order to promote electrophoretic migration of the antigen into the antibody containing gel. This principle is the basis of many electro-immunoassay methods [17].

The antigen whose concentration is to be measured is placed in a well in the gel and molecules move under the electrophoretic field into regions of the gel containing antibody. Migration of molecules of antigen continues until they are precipitated as AgAb complexes. Therefore, the greater the concentration of antigen in the well, the further a fraction of the molecules will penetrate before precipitation; hence the distance travelled will be proportional to the antigen concentration. In this electro-immunodiffusion technique, the shape of the precipitation areas revealed by protein or enzyme staining resembles rocket traces and therefore the method is often called 'rocket' electrophoresis.

The technique is relatively easy to apply because antigen samples of known concentration can be used as standards and run alongside the unknowns. From a curve obtained by plotting the distance travelled by the rocket against antigen concentration, the concentration in the unknown solution may be estimated. The measured variable is the height of the rocket trace and not the intensity of the staining reaction. This is an important advantage because of the uncertainty in the relationship between the intensity of a stained zone on a gel and the concentration of the protein at the location of the band.

When dealing with a mixture of proteins of overlapping immunospecificity, it is possible to separate the mixture by conventional electrophoresis in one direction before applying the electric field to cause migration into the antibody gel at right angles to the first direction. This two-dimensional approach allows an estimation of the concentrations of each of the reacting components in the mixture.

For successful application of the technique, the antibody must not migrate in the same direction as the antigen during the electrophoresis. In practice, any migration may cause a problem by disturbing the uniform distribution of the antibody in the gel. It may be possible in some cases to ensure the appropriate relative mobilities of the antigen and antibody by selective chemical modification. For example, carbamoylation or acetylation of protein amino groups abolishes their positive ionic charge and increases the anodal mobilities of proteins in which these groups occur. Obviously such reactions need to be performed with care in order to avoid a loss of immunospecific properties.

1.1.5 Conclusion

The determination of the absolute concentration of an enzyme is straightforward in principle but is usually made difficult in practice because of the need for purified samples to serve as standards or because of the requirement for special items of equipment. Specific activity measurements allow relative comparisons of enzyme concentration to be made between samples which may contain many different kinds of protein. Knowledge of relative concentrations may be perfectly adequate for a great many experimental situations, but if an absolute value is sought then either the turnover number must be known or certain of the methods described above must be applied.

Immunoassays are especially valuable because their specificity and low detection limits are comparable to methods involving the measurement of catalytic activity. Because they are not dependent on the biological activity of the protein under investigation, they can measure inert forms or fragments of the enzyme and can distinguish whether a change in catalytic activity is a consequence of an absolute change in protein concentration or the production of enzyme forms of altered catalytic efficiency. As ELISA and immunodiffusion methods are developed further (cf. also Vol. I, pp. 65, 254) and monoclonal antibodies become readily available, immunological assays are likely to become more widespread for mass determination of enzymes and other proteins.

Table 1 compares the detection limits of several of the methods described in this chapter when applied to the determination of human placental alkaline phosphatase. The figures in the table have been taken or calculated from reports published in the literature. For the calculations, an enzyme molecular weight of 90500 [18] and a specific activity of 900 units per mg protein [19] were assumed.

Table 1. Comparison of methods available for the measurement of human placental alkaline phosphatase

Method	Smallest amount detected (ng)
Spectrophotometric assay of hydrolysis of substrate	1
Active-site labelling [19]	8×10^4
Pre-steady state burst [20]	10^4
Immunodiffusion [21]	5
Rocket electrophoresis [22]	1

It is clear from the table that measurements of catalytic activity and immunoassays show considerably lower detection limits than any of the other methods described. When the specificity of both these kinds of assay is also taken into account, it is obvious that they cannot be easily matched by any of the alternative techniques available.

References

[1] *P. A. Srere,* Enzyme Concentrations in Tissues, Science *158*, 936 – 937 (1967).

[2] *O. H. Lowry, N. J. Rosebrough, A. L. Farr, R. J. Randall,* Protein Measurement with the Folin Phenol Reagent, J. Biol. Chem. *193*, 265 – 275 (1951).

[3] *M. Dixon, E. C. Webb,* Enzymes, 2nd edit., Longmans, Green & Co., London, 1964, pp. 63 – 67.

[4] *A. V. Hill,* The Possible Effects of the Aggregation of the Molecules of Haemoglobin on its Dissociation Curves, J. Physiol. *40*, iv – vii (1910).

[5] *R. P. Cox, N. A. Elson, S. Tu, M. J. Griffin,* Hormonal Induction of Alkaline Phosphatase Activity by an Increased Catalytic Efficiency of the Enzyme, J. Mol. Biol. *58*, 197 – 215 (1971).

[6] *M. Roth,* Molecular Titration of Aspartate Aminotransferase as an Approach to Standardization of Enzyme Assay, in: *A. Burlina* (ed.), Proc. 6th Int. Symp. on Clin. Enzymology, Kurtis, Milano, 1976, pp. 163 – 168.

[7] *B. S. Hartley, B. A. Kilby,* The Reaction of p-Nitrophenyl Esters with Chymotrypsin and Trypsin, Biochem J. *56*, 288 – 297 (1954).

[8] *F. Wold,* Macromolecules Structure and Function, Prentice-Hall Inc., New Jersey, 1971, p. 80.

[9] *H. R. Mahler, E. H. Cordes,* Biological Chemistry, 2nd edit., Harper & Row, New York, Evanston, San Francisco & London, 1971, pp. 336 – 337.

[10] *E. Racker, I. Krimsky,* The Mechanism of Oxidation of Aldehydes by Glyceraldehyde-3-Phosphate Dehydrogenase, J. Biol. Chem. *198*, 731 – 743 (1952).

[11] *J. E. Bell, K. Dalziel,* Studies of Coenzyme Binding to Rabbit Muscle Glyceraldehyde-3-Phosphate Dehydrogenase, Biochim. Biophys. Acta *391*, 249 – 258 (1975).

[12] *P. D. Boyer,* Spectrophotometric Study of the Reaction of Protein Sulfhydryl Groups with Organic Mercurials, J. Amer. Chem. Soc. *76*, 4331 – 4337 (1954).

[13] *L. Brand, J. R. Gohlke,* Fluorescent Probes for Structure, Annu. Rev. Biochem. *41*, 843 – 868 (1972).

[14] *G. Köhler, C. Milstein,* Derivation of Specific Antibody-producing Tissue Culture and Tumour Lines by Cell Fusion, Eur. J. Immunol. *6*, 511 – 519 (1976).

[15] *S. Iino, A. Koaru, O. Toshitugu, H. Suzuki,* A New Method of Radioimmunoassay for Human Placental Alkaline Phosphatase, Clin. Chim. Acta *42*, 161 – 165 (1972).

[16] *C. M. Veneziale, J. B. Hansen, M. L. Johnson,* Specific Radioimmunoassays for Skeletal and Cardiac Muscle 6-Phosphofructokinase and Pyruvate Kinase, in: *J. J. Langone, H. van Vunakis* (eds.), Methods in Enzymology, Vol. *74*, Academic Press, New York 1981, pp. 232 – 244.

[17] *C.-B. Laurell,* Electroimmunoassay, in: *J. J. Langone, H. van Vunakis* (eds.), Methods in Enzymology, Vol. *73*, Academic Press, New York 1981, pp. 339 – 360.

[18] *A. J. Gottlieb, H. H. Sussman,* Human Placental Alkaline Phosphatase: Molecular Weight and Subunit Structure, Biochim. Biophys. Acta *160*, 167 – 171 (1968).

[19] *K. B. Whitaker, D. W. Moss,* Titration of Human Placental Alkaline Phosphatase with Radioactive Orthophosphate, Clin. Chim. Acta *71*, 277 – 284 (1976).

[20] *D. A. Byers, H. N. Fernley, P. G. Walker,* Studies on Alkaline Phosphatase. Catalytic Activity of Human Placental Alkaline Phosphatase Variants, Eur. J. Biochem. *29*, 205 – 209 (1972).

[21] *G. J. Doellgast,* Immunoquantitation of Human Placental Alkaline Phosphatase using Radial Immunodiffusion, Anal. Biochem. *82*, 278 – 288 (1977).

[22] *D. T. Forman, D. W. Moss, K. B. Whitaker,* Determination of Human Placental Alkaline Phosphatase by an Electroimmunoassay ('Rocket') Technique, Clin. Chim. Acta *68*, 287 – 290 (1976).

1.2 Measurement of Catalytic Activity in Body Fluids other than Blood

1.2.1 Introduction

Hermann Mattenheimer*

This chapter reviews some aspects of the enzymology of lymph, cerebrospinal fluid (CSF), synovial fluid, saliva, tears, and urine. It is intended a) to provide the reader with the background in regard to the origin of the enzymes and their physiological and pathological presence in these fluids, b) to review diagnostic applications, and c) to describe practical aspects of sampling, sample preparation, and techniques of enzyme analysis. Measurement of the catalytic activity of enzymes in these specialized fluids is relevant in physiological and pathophysiological research. Measurement is also applied for diagnostic purposes in CSF, saliva, tears, and urine.

Lymph is recognized as a transport vehicle for cellular enzymes after their release from tissues into the extracellular space under physiological and pathological conditions. Investigations of the enzyme content of lymph in various parts of the body and of the regionally different permeability of blood capillaries for proteins have enhanced the understanding of the intricate system by which enzymes are distributed throughout the extracellular space, including the intravascular compartment.

The diagnostic potential of measurement of the activity of enzymes in CSF may not have been fully explored for at least three obvious reasons: a) the prevalence of brain diseases and neurological disorders that warrant investigation of CSF is low, b) lumbar puncture carries a certain risk for the patient, c) the sample volume is small.

Enzyme measurements in the synovial fluid as a research tool support the efforts being made towards the elucidation of the pathogenesis of acute and chronic inflammation of synovial joints.

Saliva and tears can easily be obtained by harmless sampling techniques. Measurement of changes of their enzyme content has recently been used to detect genetically determined diseases, primarily lysosomal storage diseases. Since the activities of many of the enzymes are comparatively low in these fluids, substrates have been introduced for sensitive fluorimetric methods.

Enzymes are physiological constitutions of urine. Pathological enzymuria is observed in diseases of the kidney and in a variety of primary extrarenal diseases. The

* co-ordinating author of chapter 1.2

majority of enzymes in the urine originates from renal tissue. Urine contains enzyme inhibitors and activators that must be eliminated before measurement of activity. Standardization of the sample preparation has not yet been accomplished but appears to be mandatory before the determination of urinary enzymes becomes generally accepted as a routine diagnostic technique.

Citation of the literature in this chapter is by no means comprehensive. However, the inclusion of references to review articles and conference proceedings should enable the reader to find original publications. Enzyme abbreviations that appear in the following sections are listed in Table 1.

Table 1. Enzyme abbreviations

AAP	Alanine aminopeptidase	EC 3.4.11.-	GlDH	Glutamate dehydrogenase	EC 1.4.1.2
AK	Adenylate kinase	EC 2.7.4.3	β-Gluc	β-Glucuronidase	EC 3.2.1.31
ALAT	Alanine aminotransferase	EC 2.6.1.2	γ-GT	γ-Glutamyl transferase	EC 2.3.2.2
ALD	Aldolase	EC 4.1.2.13	LAP	Leucine aminopeptidase	EC 3.4.11.1
AP	Alkaline phosphatase	EC 3.1.3.1	LDH	Lactate dehydrogenase	EC 1.1.1.27
ASA	Arylsulphatase	EC 3.1.6.1	Lys	Lysozyme (Muraminidase)	EC 3.2.1.17
ASAT	Aspartate aminotransferase	EC 2.6.1.1	MDH	Malate dehydrogenase	EC 1.1.1.27
ChE	Cholinesterase	EC 3.1.1.8	NAG	β-N-Acetylglucosaminidase	EC 3.2.1.30
CK	Creatine kinase	EC 2.7.3.2	SDH	Sorbitol dehydrogenase	EC 1.1.1.14

1.2.2 Lymph

Joachim Lindena and Ivar Trautschold

Enzymes in the lymph, including lymph formation and flow, methods for sampling from regional lymphatics, and sample preparation for enzyme assays have recently been comprehensively reviewed by the authors [1].

1.2.2.1 Lymph Formation and Flow, Collecting Vessels

According to the classical *Starling-Landis* concept, lymph is the minute net volume of the capillary fluid balance, which is maintained by a dynamic equilibrium between hydrostatic and osmotic forces acting across the capillary wall. These forces are considered to be almost equal in magnitude, therefore, cancelling each other out [2].

Lymph from any tissue, except renal medulla and choroid plexus, which lack lymph vessels, contains all the proteins that can be detected in blood plasma. Lymphatic vessels are essential for the continual movement of these proteins in one direction throughout the extracellular fluid of the body, from plasma to tissue fluid to lymph and back to the plasma [3].

It has been proposed that molecules are transported across capillary walls by different mechanisms: convection, permeation (diffusion), pinocytosis, filtration, escape through small and large pores (leaks) or accompanying the passage of recirculating lymphocytes through the post-capillary venules in the lymph nodes; which mechanism predominates depends on the barrier characteristics of each organ.

The thoracic duct centres lymph from the abdominal viscera and lower extremities, proteins. The extravascular return rate reflects the lymphatic protein transport and is the reciprocal of the extravascular transit time, which covers a wide range: short in liver, kidney and lungs, and long (up to 5 – 6 days) in muscle and skin [4].

The thoracic duct centres lymph from the abdominal viscera and lower extremeities, the left upper limb, left side of the head and neck, and parts of the heart, while the right lymphatic duct drains most of the lungs, heart, serous cavities, right upper extremities and right side of the head and neck. Compared to these two lymph systems in the body, lymph flow is small from the cervical and subclavian lymph ducts, which also finally enter the intravasal compartments at the root of the neck. Thoracic duct lymph flow in anaesthetized mammals varies between 1 – 2 $ml \times kg^{-1} \times h^{-1}$ and 3 – 5 $ml \times kg^{-1} \times h^{-1}$, [5], whereas right duct lymph flow amounts only to about 5 – 20% of this value. Hepatic and intestinal lymph contribute probably in equal parts to the total lymph in the thoracic duct. In conscious or exercising animals, however, lymph flow exceeds these values by 50 and nearly 300%, respectively. This increase is probably due in part to enhanced lymph flow from exercising muscles by action of the tissue pump and in part the increased liver lymph flow due to the increased pressure and massaging effect to the abdominal lymphatics by muscular movement. During anaesthesia and at rest an extravascular fluid pool would seem to be accumulated and emptied during exercise.

1.2.2.2 Distribution and Transport of Enzymes in the Extracellular Space

Lymph as transport vehicle

Lymph has been recognized as a transport vehicle for enzymes. Enzymes released from cells, by physiological cell-turnover or due to a pathological event, can reach the circulating blood by three different routes. If the cells of origin are in direct contact with the blood, as are blood cells or the endothelial cells of the blood vessels, the released enzymes are rapidly distributed throughout the intravasal space, and the extent of the initial dilution depends on the plasma volume. If the cells of origin have direct contact with the interstitial as well as the intravascular space and/or if the capillary permeability of that region is high (e.g. liver and spleen), the released enzymes are not only rapidly distributed throughout the intravascular compartment but also through parts of the interstitial compartment and the lymphatic system. If the cells of origin are in direct contact with the interstitial space only and/or the permeability of the capillary membranes is low (e.g. skeletal muscle), the released enzymes reach the intravasal compartment preferably by lymphatic transport and rarely by a direct interstitial-venous entry. Because of the time consuming lymphatic transport the time course of appearance of enzymes in plasma lags behind the interstitial-venous increase.

Lymph-plasma ratios of enzyme activities are useful in estimating origin and fate of particular enzymes in lymph and plasma. Plasma protein [6] and cholinesterase [7] have been used as reference proteins. The latter is exclusively synthesized by the liver and directly released into the blood. The transcapillary escape of cholinesterase is limited because of its molecular mass of 350000. A lymph-plasma enzyme activity ratio exceeding unity with the reference protein indicates that the enzyme is derived primarily from the interstitial space.

Measurement of enzymes in thoracic duct lymph and plasma of the dog and the lymph-plasma ratios strongly suggests that SDH*, MDH, ASAT and CK are transported predominantly via the lymphatics, whereas a partial lymphatic transport can be assumed for LDH, GlDH, ALAT, γ-GT, LAP and ALD [6]. From enzyme measurements in plasma and truncus lumbalis lymph of the dog it was concluded that all of the enzymes studied (CK, MDH, LDH, AP, ASAT, ALD, AK, ALAT, and LAP) are of interstitial origin and reach the intravascular compartment via the lymphatic system. Their lymph-plasma ratios exceed the ratio of 0.21 for ChE and decrease in the order from 6.5 for CK to 0.53 for LAP [7].

It is almost impossible to accomplish experimentally an isolated and controlled elevation of enzyme activities in the interstitial compartment, which would be necessary for the quantitative investigation of extracellular enzyme transport. All one can do is to produce an injury to a limited region of a tissue in which the cells have no direct contact with the intravascular space.

Skeletal muscle and myocardium

Peripheral skeletal muscle is a tissue with a particularly low capillary permeability [8]. Furthermore, skeletal muscle offers the great advantage in that, in contrast to other organs, the rate of outflowing lymph can easily be influenced just by changes in its physical activity [9]. Lymph flow from muscle is mainly a function of muscular activity: it stagnates completely with immobility and is stimulated by active as well as passive motion [10]. Consequently, an injury to cells of skeletal muscle, kept immobile, should not result in an increase of enzyme activities in blood plasma. We were able to confirm this assumption experimentally and no doubt remains that enzymes from skeletal muscle are transported from the interstitial into the intravascular compartment via the lymphatic system. Moreover, interruption of blood

* Abbreviations see Table 1, page 20.

flow to skeletal muscle for one hour did not cause enzyme release from muscle cells [7, 11, 12].

The practical importance of lymph flow and its impairment was stressed [13] by comparison of CK activities in plasma after myocardial infarction and lesions of skeletal muscles. Although the latter contains nearly ten times more CK than the myocardial muscle, even severe lesions do not result in such distinct elevations of CK as one usually sees after a myocardial infarction. The explanation is simple: the injured muscle will be kept reflexively immobile resulting in a stagnation of the lymph flow. The enzymes released from injured cells remain in the interstitial space, because there is no means of transportation. After myocardial infarction, in contrast, lymph flow does not cease, because most parts of the muscle are still working, and enzyme transport continues. The lymphatic transport accounts for 15 – 50% of the total of the released enzymes [14], whereas the greater part can reach the blood directly due to the much higher capillary permeability within the myocardium as compared to skeletal muscle [15].

Liver

The significance of diagnostic enzymology in hepatic diseases is probably due to the extremely high capillary permeability in the liver. Hepatic alkaline phosphatase molecules, for example, probably pass freely through the openings in the sinusoidal endothelial lining and are transported out of the liver chiefly in sinusoidal blood, which flows more rapidly than liver lymph [16]. Hepatic lymph apparently derives the bulk of its proteins from the plasma [17].

Physical exercise

The well-known increases in activities of enzymes in plasma during strenuous physical exercise are mainly due to an increased lymph flow. Release of enzymes from muscle cells appears to be minor, if it occurs at all. The following concept of enzyme activity changes during and after physical exercise is based on experimental findings [6, 7, 11, 18, 19, 20] as well as on clinical observations.

During physical exertion the system of flux equilibria between the interstitial and the intravascular compartment becomes disturbed. The first acute effect is a decrease in plasma volume, because of an increased systemic blood pressure. This leads to an intravascular concentration of macromolecules of about 10 – 15%. In skeletal muscle, the elevated hydrostatic pressure increases the filtration of solution into the interstitium. The increased filtration, combined with the propulsion by the pumping action of the working muscles, enhances the lymph flow from muscles. The lymph entering the intravascular compartment contains enzymes, which become concentrated in the plasma because the capillary membranes within skeletal muscle are almost impermeable to proteins, whereas fluid is filtered back at nearly the same enhanced rate as it flows into the plasma.

The accumulated proteins and enzymes slowly get back into the interstitial space, but preferentially in regions where the capillary permeability is high, as in liver. Thereby they induce an increase of the lymph flow from liver, because the organism tries to keep the colloid osmotic equilibrium in balance. In other organs the intravascular concentration of macromolecules leads, with an increasing colloid-osmotic pressure, to a reduction of fluid filtration into the interstitium, and thus to a reduction in lymph flow from these organs. To what extent these physiological mechanisms affect the plasma activity of a particular enzyme in plasma depends on its lymph-plasma activity ratio, especially for lymph from skeletal muscle. Since CK in plasma originates mainly from skeletal muscle and the CK lymph-plasma ratio is considerable higher than the ratio for other enzymes, (see above) it is not surprising that the increase in activity of CK in plasma during and after exercise exceeds the activity increases of other enzymes.

It should be mentioned that due to the ratio of the interstitial to intravasal space of 4:1, macromolecules with a lymph-plasma ratio of less than one can also rise considerably in plasma during physical activity.

It is highly probable that the effect of physical exertion on enzyme activities in plasma is caused predominantly, if not totally, by an alteration of the extracellular distribution of cellular enzymes, and that the total extracellular activities do not change at all.

1.2.2.3 Lymph for Enzyme Analysis

Sampling methods

Detailed descriptions have been published of lymph sampling methods in man and various animals [5]. References to improved and refined techniques can be found in [1].

Sample preparation

An anticoagulant should be added, because lymph contains all coagulation factors (but no platelets) and components for fibrinolysis. Centrifugation is essential in order to remove the relatively numerous leukocytes and any contaminating erythrocytes.

Lipaemic lymph should be clarified by high force centrifugation (e.g. *Beckman* Airfuge) or by adding Frigen (*Hoechst*).

Storage of lymph at $-70\,°C$ caused no loss of activity of most enzymes [6, 7, 11, 12].

Techniques of measurement

Assay techniques usually applied to serum and other body fluids are applicable to measuring enzyme activities in lymph.

References

[1] *J. Lindena, I. Trautschold,* Enzymes in Lymph: A Review, J. Clin. Chem. Clin. Biochem. *21,* 327 – 346 (1983).

[2] *G. Rutili, K.-E. Arfors,* Protein Concentration in Interstitial and Lymphatic Fluids from the Subcutaneous Tissue, Acta Physiol. Scand. *99,* 1 – 8 (1977).

[3] *F. C. Courtice,* Lymph and Plasma Proteins: Barriers to their Movement Throughout the Extracellular Fluid, Lymphology *1,* 9 – 17 (1971).

[4] *N. Rossing,* Intra- and Extravascular Distribution of Albumin and Immunoglobulin in Man, Lymphology *11,* 138 – 142 (1978).

[5] *J. M. Yoffey, F. C. Courtice,* Lymphatics, Lymph and the Lymphomyeloid Complex. Academic Press, London – New York 1970.

[6] *R. Friedel, R. Bode, I. Trautschold, H. Mattenheimer,* Die Lymphe als Verteilungsraum für Zellenzyme. II. Verteilung und Transport von Zellenzymen im extrazellulären Raum, J. Clin. Chem. Clin. Biochem. *14,* 119 – 128 (1976).

[7] *J. Lindena, W. Küpper, I. Trautschold,* Effect of Transient Hypoxia in Skeletal Muscle on Enzyme Activities in Lymph and Plasma, J. Clin. Chem. Clin. Biochem. *20,* 95 – 102 (1982).

[8] *W. P. Paaske,* Capillary Permeability in Skeletal Muscle, Acta Physiol. Scand. *101,* 1 – 14 (1977).

[9] *H. Schad, H. Brechtelsbauer,* Thoracic Duct Lymph Flow and Composition in Conscious Dogs and the Influence of Anaesthesia and Passive Limb Movement, Pflügers Arch. *371,* 25 – 31 (1977).

[10] *H. Schad, H. Brechtelsbauer,* Thoracic Duct Lymph in Conscious Dogs at Rest and During Changes of Physical Activity, Pflügers Arch. *367,* 235 – 240 (1977).

[11] *J. Lindena, W. Küpper, R. Friedel, I. Trautschold,* Lymphatic Transport of Cellular Enzymes from Muscle into the Intravascular Compartment, Enzyme *24,* 120 – 131 (1979).

[12] *J. Lindena, W. Küpper, I. Trautschold,* Enzymatic Composition of Canine Leg Lymph, Enzyme, *28,* 18 – 27 (1982).

[13] *R. Friedel,* Der Mechanismus des Übertritts von Enzymen in das Blut, Medizin in unserer Zeit *6,* 177 – 184 (1978).

[14] *H. Nordbeck, H. Kahles, C.-J. Preusse, G. Spieckermann,* Enzymes in Cardiac Lymph and Coronary Blood under Normal and Pathophysiological Conditions, J. Mol. Med. *2,* 255 – 263 (1977).

[15] *A. E. Taylor, W. H. Gibson, H. J. Granger, A. C. Guyton,* The Interaction Between Intracapillary and Tissue Forces in the Overall Regulation of Interstitial Fluid Volume, Lymphology *6,* 192 – 208 (1973).

[16] *A. L. Baker, S. C. Hauser,* Alkaline Phosphatase Activity in Lymph and Serum of Bile Duct Ligated Rats, Digestion *18,* 103 – 109 (1978).

[17] *Ch. C. Dive, A. C. Nadalini, J. F. Heremans,* Origin and Composition of Hepatic Lymph Proteins in the Dog, Lymphology *4,* 133 – 139 (1971).

[18] *R. Friedel, H. Mattenheimer, I. Trautschold, G. Forster,* Der vorgetäuschte Enzymaustritt. Verteilung und Transport von Zellenzymen im extrazellulären Raum, I. Mitteilung, J. Clin. Chem. Clin. Biochem. *14,* 109 – 117 (1976).

[19] *R. Friedel, F. Diederichs, J. Lindena,* Release and Extracellular Turnover of Cellular Enzymes, in: Advances in Clinical Enzymology. *E. Schmidt, F. W. Schmidt, I. Trautschold, R. Friedel* (eds.), *Karger*, Basel, 1979, pp. 70 – 105.

[20] *J. Lindena, R. Friedel,* Effect of Physical Exercise on Enzyme Activities in Extracellular Fluid Compartments, J. Clin. Chem. Clin. Biochem. *18,* 649 – 651 (1980).

1.2.3 Cerebrospinal Fluid

Kenneth Whisler

Measurement of enzymatic activity in cerebrospinal fluid (CSF) has been marked by uncertainty and has not gained wide acceptance. This stands in contrast to the importance placed on enzymatic activity in blood as a diagnostic tool.

The major drawbacks to CSF studies have been inadequate sample volume and an insufficient patient population. CSF is usually obtained through lumbar puncture which carries a certain amount of risk to the patient.

The studies of CSF become important when one considers the source of the fluid. The fluid which is formed in the choroid plexuses eventually bathes the surface of the subarachnoid spaces around the brain and spinal cord. The fluid is isolated from the blood by the blood-brain barrier. Because of the source and isolation of this fluid and its relationship to the brain, its enzymatic activity must be related to neurological diseases. Investigators have tried to correlate enzyme levels to brain dysfunction and to determine the pragmatic value of each enzyme assay.

It may become necessary to measure different enzymes simultaneously on a single specimen and to establish a characteristic enzyme pattern for the particular neurological disease.

1.2.3.1 CSF Enzymes in Various Diseases

Brain tumours

ASAT* and LDH including the isoenzymes have been extensively studied. *Savory* and others have summarized a number of papers on ASAT measurements [1, 2]. These studies are inconclusive.

LDH increases in brain tumours, particularly in leptomeningeal metastases and highly malignant brain tumours. The level of LDH-5 has also been shown to increase in certain metastasized tumours [3, 4].

β-Glucuronidase has been reported to be increased in leptomeningeal metastases. Marked elevations of β-glucuronidase with chronic meningitis are presumptive evidence of neoplastic disease [5]. AK shows non-specific changes in certain malignancies [6, 7].

Cerebral infarction and ischaemia

CK, AK, ASAT and LDH have been the enzymes most studied in these conditions. ASAT and LDH seem to show no correlation between the disorder and its severity,

* Abbreviations see Table 1, page 20.

although there are some reports of increased ASAT being related to cerebral ischaemia. CK and AK both show increases in these brain disorders [6, 7]. Some studies have been done on CK isoenzymes. CK-BB has been found to be elevated in certain patients with brain damage, the assumption being that CSF reflects the isoenzyme pattern of the brain. The remaining isoenzymes have also been found to be elevated. This could reflect changes in the permeability of the blood-brain barrier. The interpretation of CK isoenzyme levels is difficult. CK in CSF is reviewed by *Savory* [2] and *Seidenfeld* [8].

Cerebral injury

The level of increased ASAT and LDH shows significant correlation to the degree of injury, thus these two enzymes may be of diagnostic use in assessing brain damage [1, 2, 7]. ALAT has been reported to be elevated in cerebral injury [9].

Bacterial infection

LDH levels are increased in convulsions of toxic origin and significantly elevated in acute bacterial meningitis. The enzyme tests in CSF may be of diagnostic value in distinguishing bacterial infection from aseptic meningitis. LDH is presumed to be non-bacterial in origin. The isoenzyme determination of LDH is also helpful [1, 2, 8, 10]. CK and CK-BB have been measured and have little importance [11]. Increased activity of AK and β-glucuronidase have been measured [4, 12].

Imanishi reports that patients with infectious diseases have increases in lysosomal enzymes [13]. *Venable* found that patients with meningitis and neurosyphilis had increased lysozomal activity of varying degrees [14].

Neurological disorders

Total LDH has been measured in a wide variety of neurological disorders. LDH levels are increased in convulsions of toxic origin [7]. There is a suggestion of LDH change in some progressive neurological diseases [15]. The pragmatic value of LDH determinations has been summarized by several authors [1, 2, 8, 10]. One study measured AP in CSF and found no increase in cases of delirium tremens [9]. CK has been found to increase in schizophrenic patients and convulsions of toxic origins [16]. AK increases in schizophrenia whereas acetylcholinesterase shows a significant lowering in senile dementia of *Alzheimer* type [17]. *Davis* found no change in acetylcholinesterase activity in depression and schizophrenia [18]. In studying psychiatric patients and measuring levels of dopamine-β-hydroxylase (E.C. 1.14.17.1), it was shown that the enzyme had a positive correlation with age, was not altered in seizure conditions and alcoholism, but had a lower activity in manic depression [19].

Miscellaneous diseases

There is a wide variety of enzymes measured in CSF that have at this time no pragmatic value, but listing them shows the wide range of the number and type of enzymes found. β-Lactamase I (E.C. 3.5.2.6), a bacterial enzyme, is found only in bacterial infection [20]. An isoenzyme of aldolase (aldolase C) is particularly concentrated in the brain and appears in CSF [21]. Reports show that so called CSF hydroxylase (E.C. 1.14.16.1) requires a folic acid derivative as a co-factor. This co-factor is reduced in *Parkinson*'s disease [22]. There are reports of two cholesterol esterases (E.C. 3.1.1.13) [23] and a neuron-specific enolase (E.C. 4.2.1.11) to be present in CSF [24]. X-prolyl dipeptidyl-aminopeptidase (E.C. 3.4.13.8) activity has been measured. There is no know physiological significance for this enzyme [25]. 2′,3′-Cyclic-AMP phosphodiesterase (E.C. 3.1.4.16) is an enzyme found in very low concentrations in CSF, and it is believed to be derived from myelinic tissue [26].

Measurement of patterns of enzyme activity increases rather than of single enzymes is recommended [1]. The following constellations were found:

- Inflammatory processes: LDH > MDH > ALD > ALAT > ASAT;
- Vascular diseases: ALAT > ALD = ASAT > LDH > MDH;
- Tumours: LDH > MDH > ALD > ALAT = ASAT;
- Degenerative diseases, myelopathies, or diseases of the motor-nerve system: ALD > ALAT > ASAT > LDH > CK > MDH;
- Toxic or metabolic peripheral neuropathies: ALD > MDH = ASAT > CK > ALAT > LDH.

1.2.3.2 Cerebrospinal Fluid for Enzyme Analysis

Sampling and sample preparation

CSF is usually obtained by lumbar puncture at L4 – L5 to avoid damage to the spinal cord. This technique is always done under aseptic conditions and is contra-indicated if there is any infection in the region of the puncture site. If the puncture can be performed as an elective prodecure the patients should be fasting. Before any fluid is removed the pressure has to be evaluated. CSF pressure is directly related to the pressure in the jugular and vertebral veins and varies between 50 and 180 mm H_2O. The specimen may be safely obtained from the patient if the pressure is normal and there is no marked decrease in pressure when fluid is removed. Normal CSF is clear and colourless. If the specimen is bloody one must distinguish whether the specimen is from a traumatic puncture or if there is pathological bleeding. CSF may be cloudy, xanthochromic or have increased viscosity. These conditions have to be evaluated as to the appropriateness of the specimen. Usually most assays are done on a fresh specimen but if delays occur then the specimen should be frozen. The specimen may have to be centrifuged prior to performing the assays.

Techniques of measurement

A wide variety of laboratory methods and techniques has been used on CSF without difficulty. In general, the methods used for the determination of enzymes in CSF are the available serum enzyme methods. There may be some important modifications for certain enzyme assays. For example, volume or assay time may differ from serum assays. CSF specimens are hard to obtain and so the volume needed may be critical and repetition may be impossible. Micro methods are recommended. Refrigeration should be utilized to maintain the stability of most CSF enzymes for a short time period.

Most of the commonly measured enzymes have levels that are lower in CSF than in serum. Because of this any CSF containing blood will give falsely elevated results. The levels of some enzymes are so low that they can only be measured if the specimen is concentrated. This can be accomplished by using an apparatus like the Minicon-B Macro-Concentrator (Amicon Corp., Lexington, MA) which utilizes a semi-permeable membrane between the fluid and an absorbent in a special chamber. Ultrafiltration through a collodion membrane by pressure is an alternative method.

References

[1] *I. R. Navarro, J. M. U. Vales, S. C. DelPozo, O. O. Manchado,* Die Bedeutung der Liquorenzyme für die klinische Praxis, Münch. Med. Wochenschr. *115*, 2293 – 2299 (1973).

[2] *J. Savory, J. P. Brody,* Measurement and Diagnostic Value of Cerebrospinal Fluid Enzymes, Ann. Clin. Lab. Sci. *9*, 68 – 79 (1979).

[3] *B. Breebaart, H. Becker, F. A. Jonsebloed,* Investigation of Reference Values of Components of Cerebrospinal Fluid, J. Clin. Chem. Biochem. *16*, 561 – 566 (1978).

[4] *W. R. Wasserstron, M. K. Schwartz, M. Felisher, M. B. Posner,* Cerebrospinal Fluid Biochemical Markers in Central Nervous System Tumors: a Review, Ann. Clin. Lab. Sci. *11*, 239 – 251 (1981).

[5] *E. Shuttleworth, N. Allen,* CSF Beta-Glucuronidase Assay in the Diagnosis of Neoplastic Meningitis, Arch. Neurol. *37*, 684 – 687 (1980).

[6] *A. Terent, G. Ronquist,* Cerebrospinal Fluid Markers of Disturbed Brain Cell Metabolism in Patients with Stroke and Global Cerebral Ischemia, Acta Neurol. Scand. *62*, 327 – 377 (1980).

[7] *J. L. Viallard, J. Gaulme, B. Dalens, B. Dastugue,* Cerebrospinal Fluid Enzymology: Creatine Kinase, Lactate Dehydrogenase Activity and Isoenzyme Pattern as a Brain Damage Index, Clin. Chim. Acta *89*, 405 – 409 (1978).

[8] *J. Seidenfeld, L. J. Marton,* Biochemical Markers of Central Nervous System Tumors Measured in Cerebrospinal Fluid and their Potential Use in Diagnosis and Patient Management: a Review, J. Natl. Cancer Inst. *63*, 919 – 931 (1979).

[9] *R. Hemminssen, P. Kramp, J. Dissins,* Delirium Tremens: Some Clinico-Chemical Features. A study of Alanine Aminotransferase, Alkaline Phosphatase, Prothrombine and Enolase, Acta Psychiatr. Scand. *62*, 503 – 510 (1980).

[10] *E. Stensland, S. Sandberg, R. Berge, J. A. Aarli, I. Romslo,* Adenylate Kinase in Neurological Diseases (letter), Lancet *13*, 792 – 793 (1979).

[11] *H. Briem, L. Rombo,* Creatine Kinase Isoenzyme Activity in Cerebrospinal Fluid and Plasma from a Patient with a Fatal Bacterial Meningitis, Clin. Chim. Acta *112*, 361 – 364 (1981).

[12] *I. Johnson, H. Briem, G. Ronquist,* Cerebrospinal Fluid Content of Adenylate Kinase, Lactate and Glutathione in Patients with Meningitis, Scand. J. Infect. Dis. *11*, 11 – 15 (1979).

[13] *K. Imanishi, M. Ando, T. Ideta, H. Tokuomi,* Histochemical Studies of Lysosomal Enzymes and Nitroblue Tetrazolium Reduction of Phagocytes in the Cerebrospinal Fluid of Patients with Infectious Diseases of the Central Nervous System, Acta Neurol. Scand. *64*, 54 – 65 (1981).

[14] *G. S. Venables, S. J. Proctor, D. Bates, N. E. Cartlidge, D. A. Shaw,* Intracranial Disease in non-Hodgkin's Lymphoma, Q. J. Me. *49,* 111 – 131 (1980).
[15] *C. Wikkels, C. Blomstrand, P. Nordquist,* Cerebrospinal Fluid Investigations in Multi-infarct Dementia and Senile Dementia, Acta Neurol. Scand. *61,* 1 – 11 (1981).
[16] *M. H. Zweis, A. C. Van-Steirteghem, E. F. Torrey,* Creatine Kinase Isoenzymes in the Serum and CSF of Schizophrenic Patients, Arch. Gen. Psychiatry *38,* 1107 – 1109 (1981).
[17] *H. Soininen, T. Halonen, P. J. Rickinen,* Acetylcholinesterase Activities in Cerebrospinal Fluid of Patients with Senile Dementia of Alzheimer Type, Acta Neurol. Scand. *64,* 217 – 224 (1981).
[18] *K. L. Davis, L. E. Hollister, J. Livesey, P. A. Berger,* Cerebrospinal Fluid Acetylcholinesterase in Neuropsychiatric Disorders, Psychopharmacology *63,* 155 – 159 (1979).
[19] *P. Lerner, F. K. Goodwin, D. P. van-Kammen, R. M. Post, L. F. Major, J. C. Ballenger, W. Lovenberg,* Dopamine-beta-Hydroxylase in the Cerebrospinal Fluid of Psychiatric Patients, Biol. Psychiatry *13,* 685 – 694 (1978).
[20] *R. H. Yolken, W. T. Huges,* Rapid Diagnosis of Infections Caused by Beta-Lactamase-Producing Bacteria by Means of an Enzyme Radioisotopic Assay, J. Pedriatr. *97,* 715 – 720 (1980).
[21] *V. J. Willson, J. G. Graham, I. N. McQueen, R. J. Thompson,* Immunoreactive Aldolase C in Cerebrospinal Fluid of Patients with Neurological Disorders, Ann. Clin. Biochem. *17,* 110 (1980).
[22] *A. C. Williams, R. A. Levine, T. N. Chase, W. Lovenberg, D. J. B. Calne,* CSF Hydroxylase Co-factor Levels in some Neurological Diseases, J. Neurol. Neurosurg. Psychiatry *43,* 735 – 738 (1980).
[23] *R. C. Johnson, S. N. Shah,* Evidence for Two Cholesterol Ester Hydrolases in Human Cerebrospinal Fluid, J. Neurochem. *37,* 594 – 596 (1981).
[24] *A. M. Parma, R. J. Marangos, F. K. Goodwin,* A More Sensitive Radioimmunoassay for Neuron-Specific Enolase Suitable for Cerebrospinal Fluid Determinations, J. Neurochem. *36,* 1093 – 1096 (1981).
[25] *T. Kato, K. Iwase, T. Nagatsu, S. Sakakibara, K. Fujita,* Comparison of X-Prolyl Dipeptidyl-Aminopeptidase Activity in Human Cerebrospinal Fluid with that in Serum, Experientia *35,* 20 – 21 (1979).
[26] *K. Eickhoff, R. Heipertz,* The Activity of 2′,3′-Cyclic Nucleotide 3′-Phosphohydrolase in Human Cerebrospinal Fluid, Clin. Chim. Acta *92,* 303 – 305 (1979).

1.2.4 Synovial Fluid

Eugene J.-M. A. Thonar

The primary function of the diarthrodial joint (synovial joint) is to facilitate the execution of the basic mechanical task of motion. Synovial fluid, by definition an inherent and essential component of such joints, has two main functions: the lubrication and nutrition of the joint tissues. Although enzyme-induced joint destruction is prominent in rheumatoid arthritis and related diseases in which inflammation is a leading feature, a great deal of disagreement exists with respect to the role enzymes play in non-inflammatory conditions such as osteoarthritis.

While enzyme-induced degradation of intra-articular macromolecules or components is thought to play a leading role in the pathogenesis of inflammatory joint diseases, there is as yet no clear evidence that any enzyme analyses of synovial fluids will prove useful diagnostic tools. Indeed, it is thought that synovial joints respond to antigenic challenges in a limited manner. The majority of the present studies on synovial fluid enzymes in disease have concentrated on the condition that affects the largest number of individuals, i.e. rheumatoid arthritis. It is likely that many of the enzymes which, as discussed below, have been detected in unusual amounts in rheumatoid synovial effusions will, in the future, be detected in other inflammatory conditions that affect synovial joints. More likely, enzyme analyses will prove useful in improving our understanding of the pathogenesis of the processes of inflammation, both chronic and acute, which cause great damage in and impair function of synovial joints. They will further help in delineating the pathogenetic processes in inflammatory and non-inflammatory conditions.

Every year new enzymes are added to what is already an impressive list of enzymes which have been found elevated in rheumatoid joint effusions. Examples that illustrate the diversity of enzymes so far detected are given in Table 1.

Table 1. Enzymes in synovial fluid of patients with rheumatoid arthritis

Enzyme		Reference
Lactate dehydrogenase	(EC 1.1.1.27)	19
Pyruvate kinase	(EC 2.7.1.40)	19
Aspartate aminotransferase	(EC 2.6.1.1)	19
5′-Nucleotidase*	(EC 3.1.3.5)	20
Leucine aminopeptidase*	(EC 3.4.11.1)	20, 21
Pepsin	(EC 3.4.23.1)	22
Trypsin	(EC 3.4.21.4)	22
Acid phosphatase	(EC 3.1.3.2)	23
Alkaline phosphatase	(EC 3.1.3.1)	14
Cathepsin D	(EC 3.4.23.5)	4
Cathepsin G	(EC 3.4.21.20)	24
Muramidase (Lysozyme)	(EC 3.2.1.17)	24
α-Mannosidase	(EC 3.2.1.24)	25
N-Acetyl-β-D-glucosaminidase	(EC 3.2.1.30)	26, 27
β-Glucuronidase	(EC 3.2.1.31)	28
Myeloperoxidase	(EC 1.11.1.7)	24
Collagenase	(EC 3.4.24.3)	29

* So-called plasma membrane marker enzymes [20].

1.2.4.1 Origin

There is indirect support for the concept that the enzymes are not derived from the circulation for, in most cases, the enzyme activities in rheumatoid sera are within normal limits [1, 2]. It has been shown that synovial cells [3], leukocytes infiltrating the synovial tissue [4] and cartilage cells [5] all play a role in the elaboration of the various enzymes which, in inflammatory diseases, make their way into the synovial fluid. *Swann* [6] classified, on the basis of their origin, the macromolecular constituents of synovial fluid as a) constituents derived from the blood; b) substances secreted by the joint tissues; and c) products derived from the catabolism of the joint. Enzymes and regulators of enzymatic activity have been shown to originate from each of these sources and while it is obvious that many of the enzymes that feature prominently in diseased joints are derived from infiltrating leukocytes, more recent work has made it clear that in some cases at least they are derived from non-vascularized intra-articular tissues such as synovium and cartilage [4].

1.2.4.2 Pragmatic Value

It is not clear whether these enzymes play a role in the aetiology of rheumatoid arthritis. Most studies have been based upon data from patients with rheumatoid arthritis, ignoring less prominent inflammatory diseases. It is possible that many of the elevated enzyme activities simply represent the common response during intra-articular inflammation or during insult to intra-articular components that occurs in non-inflammatory conditions such as osteoarthritis. As yet, there is no clear evidence that enzyme determinations will prove useful in either a diagnostic capacity or as a correlate of disease activity.

Proteoglycan and collagen, the two major components of articular cartilage matrix, are responsible for many of the physiological and physicochemical properties of this aneural, alymphatic and avascular tissue [7]. It is clear from studies on both human osteoarthritic articular cartilage [8] and experimental models of osteoarthritis [9] that minor changes in the concentration, integrity or organization of these matrix components will adversely affect the capacity of the tissue to function optimally as a shock absorber. Inter- and intramolecular collagen cross links can be destroyed by elastase (EC 3.4.21.11) and cathepsins G isolated from polymorphonuclear leukocytes [10]. The soluble monomeric collagen that results from such an attack has no structural function and can be degraded by specific collagenases [11]. It has been postulated that in osteoarthritis, a non-inflammatory joint disease, the subsequent reorganization of the existing as well as newly synthesized matrix constituents will yield, even in the absence of further enzymatic degradation, a cartilage that will become progressively eroded because it is functionally inadequate for load bearing [8].

1.2.4.3 Synovial Fluid for Enzyme Analysis

Sampling and sample preparation

Isolation and preparation of synovial fluid are usually performed under conditions that will not interfere with the enzyme determinations that will be carried out. Fluid, obtained by aspiration, should be examined and discarded when contaminated with blood [12]. Often, little attention is paid to the amount of fluid aspirated. The amount of fluid that can be aspirated from the knee, the joint most commonly studied, is usually much higher in rheumatoid-arthritis patients than in controls [13]. Consequently, elevated levels of enzymes (see Table 1) which are usually calculated per gram of protein or millilitre of fluid do not necessarily reflect what will be even more significant elevations when calculated in terms of absolute amount of enzyme in the total volume of joint fluid.

The fluid is sometimes [14] but not always [2, 12] defibrinated by shaking with glass balls prior to brief centrifugation to remove cells, fibrin and particulate matter. In some cases, the pellet becomes the subject of further analysis, for example in the study of intracellular phosphatases [14]. This pellet, which is sometimes discarded [2, 15], contains 53% and 58% of the total caseinolytic and cathepsin D activity of rheumatoid synovial fluid and significant amounts of enzymes capable of degrading proteoglycans [12].

Techniques of measurement

A number of factors must be taken into careful consideration if one is to ascribe any role to enzymes in intra-articular tissue destruction. Enzyme determinations do not always reflect that activity which the enzyme exhibits within the fluid. For example, assays for acid phosphatases and alkaline phosphatase are best performed at approx. pH 5 and pH 10 respectively [14]. The pH of rheumatoid synovial fluid is stated to be 7.43 (range 7.31 to 7.64) [16]. Consequently, a significant increase in the concentration of these and other enzymes which have optimum activity at acid or alkaline pH may have little or no significance in the study of intra-articular degradation of connective tissue elements. Recent studies [4, 11] have focused on enzymes such as collagenase and neutral proteinase all of which are active at pH 7.5. *Peltonen et al.* [4], studying neutral proteinases activity in rheumatoid and control synovial fluid, have modified their original assay, using azocaseinolysis at pH 7.5 to get a clear indication of what proteolysis may be like within the joint.

The presence of enzyme inhibitors within the fluid is another factor that should be considered. Rheumatoid synovial fluid contains higher levels of some proteinase inhibitors than control synovial fluid [17]. For example, trypsin and chymotrypsin inhibitors capacities are greatly increased [4]. Further, it should be borne in mind that cartilage, the major target of attack, is renowned for its unusually high content of proteinase inhibitor [18] (cf. Vol. V, chapter 2). It is possible that in disease a selective loss of these small molecular weight components may render cartilage more susceptible to degradation by enzymes circulating in synovial fluid or, more likely, by enzymes secreted by cells in the pannus that overlies the cartilage.

In summary, it may be said that studies of synovial fluid enzymes in joint disease have offered little to suggest that such determinations will one day become routine, either as diagnostic tools or as indicators of the level of activity of disease processes. Before that can happen, more work is necessary to elucidate the mechanisms that lead to the degradation of intra-articular tissues such as cartilage.

References

[1] *T. Vanha-Perttula,* Chromatographic Fractionation and Characterization of Rat Testicular Acid Phosphatases, Biochim. Biophys. Acta *227*, 390 – 401 (1970).

[2] *J. L. Kalliomaki, T. Vanha-Perttula,* Activity of Four Hydrolytic Enzymes in Serum and Synovial Fluids of Patients with Rheumatoid Arthritis, Scand. J. Rheumatol. *1*, 21 – 26 (1972).

[3] *K. D. Muirden,* Lysosomal Enzymes in Synovial Membrane in Rheumatoid Arthritis. Relationship to Joint Damage, Ann. Rheum. Dis. *31*, 265 – 271 (1972).

[4] *L. Peltonen, J. Puranen, K. Lehtinen, L. K. Korhonen,* Proteolytic Enzymes in Joint Destruction, Scand. J. Rheumatol. *10*, 107 – 114 (1981).

[5] *J. F. Woessner, Jr.,* Cartilage Cathepsin D and its Action on Matrix Components. Fed. Proc. *32*, 1485 – 1488 (1973).

[6] *D. A. Swann,* Macromolecules of Synovial Fluid, in: *L. Sokoloff* (ed.), The Joints and Synovial Fluid, Vol. I, Academic Press, New York 1981, pp. 407 – 435.

[7] *G. E. Kempson, H. Muir, S. A. V. Swanson, M. A. R. Freeman,* Correlations between Stiffness and the Chemical Constituents of Cartilage on the Human Femoral Head, Biochim. Biophys. Acta *215*, 70 – 77 (1970).

[8] *M. B. E. Sweet, E. J.-M. A. Thonar, A. R. Immelman, L. Solomon,* Biochemical Changes in Progressive Osteoarthrosis, Ann. Rheum. Dis. *36*, 387 – 398 (1977).

[9] *C. A. McDevitt, E. M. M. Gilbertson, H. Muir,* An Experimental Model of Osteoarthrosis. Early Morphological and Biochemical Changes, J. Bone Jt. Surg. *59*, 24 – 35 (1977).

[10] *P. M. Starkey, A. J. Barrett, M. C. Burleigh,* The Degradation of Articular Collagen by Neutrophil Proteinase, Biochim. Biophys. Acta *483*, 386 – 397 (1977).

[11] *E. D. Harris, Jr.,* Role of Collagenases in Joint Destruction in: *L. Sokoloff* (ed.), The Joints and Synovial Fluid, Vol. I, Academic Press, New York 1978, pp. 243 – 272.

[12] *G. C. Wood, R. H. Pryce-Jones, D. D. White,* Chondromucoprotein-Degrading Neutral Protease Activity in Rheumatoid Synovial Fluid, Ann. Rheum. Dis. *30*, 73 – 77 (1971).

[13] *M. W. Ropes, W. Bauer,* Synovial Fluid Changes in Joint Disease. Harvard University Press, Cambridge, Mass., 1953.

[14] *P. S. Mach, C. Auscher, A. Kahan, A. LeGo, F. Delbarre,* Biochemical and Cytochemical Studies of Acid and Alkaline Phosphatases in Synovial Fluid, Biomedicine *20*, 61 – 67 (1974).

[15] *K. D. Muirden, P. Deutschmann, M. Phillips,* Articular Cartilage in Rheumatoid Arthritis: Ultrastructure and Enzymology, J. Rheumatol. *1*, 24 – 33 (1974).

[16] *D. L. Gardner,* Pathology of rheumatoid arthritis, in: *J. T. Scott* (ed.), *Copeman*'s Textbook of the Rheumatic Diseases, 5th ed., Churchill Livingstone, Edinburgh and London 1978, pp. 273 – 317.

[17] *D. Brackerts, J. Hagmann, F. Kueppers,* Proteinase Inhibitors in Rheumatoid Arthritis, Ann. Rheum. Dis. *34*, 225 – 230 (1975).

[18] *K. E. Kuettner, J. Hiti, R. Eisenstein, E. Harper,* Collagenase Inhibition by Cationic Proteins Derived from Cartilage and Aorta, Biochem. Biophys. Res. Commun. *72*, 40 – 46 (1976).

[19] *G. P. Kerby, S. M. Taylor,* Enzymatic Activity in Human Synovial Fluid from Rheumatoid and Non-Rheumatoid Patients, Proc. Soc. Exp. Biol. Med. *126*, 865 – 868 (1967).

[20] *I. Takala, P. Makisara, E. Kulonen,* The Activities of Plasma Membrane Marker Enzymes in Rheumatoid Synovial Tissues and Fluids, Scand. J. Rheumatol. *6*, 33 – 36 (1977).

[21] *A. Michajlik, J. Gietka,* Changes of some Chemical Constituents of Synovial Fluid in Rheumatoid Arthritis and Osteoarthritis, Abstr. VII Eur. Rheum. Congr. 16.11 (1971).

[22] *T. Vartio,* Proteolytic Enzymes of Synovial Fluid with Special Reference to Intra-Articularly Applied Hydrocortisone, Ann. Med. Exp. Fenn. *38*, 94 – 99 (1960).

[23] *G. Beckman, L. Beckman, R. Lemberg,* Acid Phosphatase Activity in the Synovial Fluid of Patients with Rheumatoid Arthritis and other Joint-Disorders, Acta Rheumatol. Scand. *17*, 47 – 56 (1971).

[24] *N. M. Hadler, J. K. Spitznagel, R. J. Quinet,* Lysosomal Enzymes in Inflammatory Synovial Effusions, J. Immunol. *123*, 572 – 577 (1979).

[25] *D. Pugh, P. G. Walker,* The Localization of N-Acetyl-β-Glucosaminidases in Tissues, J. Histochem. Cytochem. *9*, 242 – 250 (1961).

[26] *B. A. Bartholomew, A. L. Perry,* Alpha-Mannosidase Activity in Synovial Fluid, Acta Rheum. Scand. *16,* 304–310 (1970).
[27] *K. D. Muirden, K. Mills,* Lysosomal Enzymes, Synovial Proliferations, Lymphocytes and Joint Destruction in Rheumatoid Arthritis, Abstr. VII Eur. Rheum. Congr. 16.10 (1971).
[28] *R. F. Jacox, A. Feldmahn,* Variations of Beta Glucuronidase Concentration in Abnormal Synovial Fluid, J. Clin. Invest. *34,* 263–267 (1955).
[29] *D. E. Wooley, M. J. Crossley, J. M. Evanson,* Collagenase at Sites of Cartilage Evasion in the Rheumatoid Joint, Arthr. Rheum. *20,* 1231–1239 (1977).

1.2.5 Saliva

Wilfried Richard Den Tandt and Jaek Jaeken

Enzyme measurements in saliva appear useful in the diagnosis of genetically determined diseases. Enzymes that have been identified in saliva are shown in Table 1. This list is not exhaustive but includes the most commonly measured enzymes.

Table 1. Enzymes in Saliva

Enzyme		Reference
α-D-Galactosidase	(EC 3.2.1.22)	3, 4
β-D-Galactosidase	(EC 3.2.1.23)	3, 4
α-D-Mannosidase	(EC 3.2.1.24)	3, 4
β-D-Glucuronidase	(EC 3.2.1.31)	3, 4
N-Acetyl-β-D-glucosaminidase	(EC 3.2.1.30)	1, 3, 4
N-Acetyl-β-D-galactosaminidase	(EC 3.2.1.53)	3
α-D-Glucosidase	(EC 3.2.1.20)	3, 4
α-L-Fucosidase	(EC 3.2.1.51)	3, 4, 5
β-D-Glucosidase	(EC 3.2.1.21)	3
Acid Phosphatase	(EC 3.1.3.2)	2, 3
Arylsulphatase A	(EC 3.1.6.1)	3
Arylsulphatase B	(EC 3.1.6.1)	2
N-Acetyl-α-D-glucosaminidase	(EC 3.2.1.50)	3
α-Amylase	(EC 3.2.1.1)	6, 7, 8
Lysozyme	(EC 3.2.1.17)	6, 9
Ribonuclease	(EC 3.1.4.22)	6, 10
Neuraminidase	(EC 3.2.1.18)	4

1.2.5.1 Applications

The procedure of obtaining saliva is totally harmless. The determination of hexosaminidase A allows homozygotes and heterozygotes of *Tay-Sachs* disease to be detected [1]. Apart from this report, research of the applicability of saliva enzymes for detection of inborn errors of metabolism has been neglected although arylsulphatases A and B have been determined using a fluorimetric method [2]. With this method it would theoretically be possible to detect metachromatic leukodystrophy, known to be caused by a deficiency of arylsulphatase A, and mucopolysaccharidosis VI (*Maroteaux-Lamy*), which is characterized by a deficiency of arylsulphatase B. We have confirmed the deficiency of arylsulphatase A in metachromatic leukodystrophy using the more classical colorimetric assay [3].

α-L-Fucosidase was measured previously [3, 4, 5] in parotid, submaxillary-sublingual and whole saliva. From our work it appears that this enzyme was also deficient in the saliva of 2 patients with fucosidosis [3].

Of several glucosidases that were detected in whole saliva, only N-acetyl-β-glucosaminidase and α-L-fucosidase were found in sterile parotid and submaxillary-sublingual saliva. The others, including neuraminidase, were found to have originated from oral micro-organisms [4]. We were likewise unable to detect neuraminidase in saliva, even with a highly sensitive fluorimetric technique using methylumbelliferyl-N-acetyl-neuraminic acid as substrate. No influence of bacterial contamination was observed under the experimental conditions for measuring arylsulphatase A and α-L-fucosidase, since the respective enzyme deficiencies were present in two patients with fucosidosis and 2 patients with metachromatic leukodystrophy [3].

We have no experience with enzyme measurements in saliva related to other lysosomal storage diseases, although on a theoretical basis it is likely that specific enzyme deficiencies would also be reflected in this fluid. We would like to mention, however, that if lysosomal enzyme determinations are to be used for clinical application, some caution is needed in the case of the enzymes α-D-glucosidase, α-L-fucosidase and β-D-glucosidase.

Acid α-D-glucosidase is known to be deficient in *Pompe's* disease (glycogenosis type II). However, enzyme determinations in leukocytes do not allow the disease to be detected reliably and, therefore, no speculation about the diagnostic value of determinations in saliva can be made.

A polymorphism of α-L-fucosidase in plasma of the human population is known to exist. About 10 – 15% of the normal population has low activity of this enzyme and, therefore, determination in plasma cannot be used for diagnosis. It has not yet been determined whether salivary fucosidase is also polymorphic.

The biochemical diagnosis of *Gaucher's* disease can be made by the determination of β-D-glucosidase. However, this enzyme is absent from the supernatant of saliva. We have not tested whether this is due to specific absorption or because the enzyme activity in whole saliva reflects only the intracellular activity of the epithelial cells and (or) leukocytes present in the saliva.

Salivary lysozyme is increased in patients with sarcoidosis [6].

1.2.5.2 Saliva for Enzyme Analysis

Sampling and sample preparation

Buccal hygiene is essential before sampling, but teeth should be brushed without a dentifrice. Sampled saliva should be chilled in ice. Sampling is done by spitting into a test tube. In young children or patients with mental retardation saliva is obtained with a suction device, made of a micro centrifuge tube with the bottom cut off, or a plastic tubing (not glass!) with an inner diameter of 1 – 2 mm. Either device is attached to a hand suction bulb. Parotid saliva can be obtained with a special suction cup. We have not used this method in the study of lysosomal enzymes, and it would probably be impractical with unco-operative or mentally retarded patients.

In order to decrease the viscosity and to facilitate pipetting, a metal pin is swirled in the test tube for about 15 seconds. The saliva is then centrifuged at 10000 *g* for 30 minutes in a refrigerated centrifuge. Uncentrifuged saliva is recommended for β-glucosidase (see above). Saliva can be stored at –20 °C for one day apparently without noticeable loss of enzyme activities. Longer storage is not recommended, because loss of activity occurs.

Techniques of measurement

Methylumbelliferyl substrates and fluorimetric measurement of the enzymatically liberated methylumbelliferone are highly recommended. 4-Nitrophenyl-N-acetyl-α-D-glucosamine and 4-nitrocatechol-sulphate are suitable substrates for N-acetyl-α-D-glucosaminidase and arylsulphatase A, respectively. Enzyme activity is expressed per ml saliva or per mg protein, determined by the *Lowry* method (cf. Vol. II, chapter 1.3.4). There are no further pecularities in comparison with blood plasma or serum.

References

[1] *J. D. Singer,* Hexosaminidase in Tears and Saliva for Rapid Identification of Tay-Sachs Disease and its Carriers, Lancet II, 1116 – 1119 (1973).

[2] *H. Rinderknecht, M. C. Geokas, C. Carmack, B. J. Haverback,* The Determination of Arylsulfatases in Biological Fluids, Clin. Chim. Acta *29*, 481 – 491 (1970).

[3] *W. R. Den Tandt, J. Jaeken,* Determination of Lysosomal Enzymes in Saliva. Confirmation of the Diagnosis of Metachromatic Leukodystrophy and Fucosidosis by Enzyme Analysis, Clin. Chim. Acta *97*, 19 – 25 (1979).

[4] *R. Menguy, Y. F. Masters, L. Desbaillets,* Human Salivary Glycosidases, Proc. Soc. Exp. Biol. Med. *134*, 1020 – 1025 (1970).

[5] *S. Shizukuishi, R. Nakamura, A. Tsunemitsu, Y. Uesugi,* α-L-Fucosidase Activity in Human Saliva and Dental Plaque, Arch. Oral Biol. *12*, 721 – 722 (1977).

[6] *J. A. Beeley, D. M. Chisholm,* Sarcoidosis with Salivary Gland Involvement: Biochemical Studies on Parotid Saliva, J. Lab. Clin. Med. *28*, 276 – 281 (1976).

[7] *M. Boehm-Truitt, E. Harrison, R. O. Wolf, A. L. Notkins,* Radioimmunoassay for Human Salivary Amylase, Anal. Biochem. *85*, 476 – 487 (1978).

[8] *B. Gillard, H. C. Markman, S. Feig,* Differences Between Cystic Fibrosis and Normal Saliva α-Amylase as a Function of Age and Sex, Pediatr. Res. *12*, 868 – 872 (1978).
[9] *G. Virella, J. Goudswaard, R. J. Boackle,* Preparation of Saliva Samples for Immunochemical Determinations of Immunoglobulins and for Assay of Lysozyme, Clin. Chem. *24*, 1421 – 1422 (1978).
[10] *A. Bardon, D. Shugar,* Properties of Purified Salivary Ribonuclease and Salivary Ribonuclease Levels in Children with Cystic Fibrosis and Heterozygous Carriers, Clin. Chim. Acta *101*, 17 – 24 (1980).

1.2.6 Tears

François Van Hoof

The clinical biochemistry of tears has been recently reviewed [1]. Among other enzymes tears contain lysosomal acid hydrolases, and measurement of their activity is successfully used for the diagnosis of several inborn lysosomal storage diseases.

1.2.6.1 Origin of Tears and Enzymes in Tears

The origin and composition of tears is complex. A three-layered film covers the anterior part of the eye. The oily surface of this film contains mainly cholesteryl esters and is secreted by the glands of *Meibom, Zeiss* and *Moll.* The aqueous middle layer is the product of the lacrimal glands. The inner layer contains glycoproteins and originates from the goblet cells of the conjunctival epithelium. The normal secretion of tears amounts to about 1 – 2 ml per day with a protein content of 2 – 6 mg per ml. Lysozyme, a bacteriolytic enzyme, constitutes about one fourth of the protein.

The lacrimal glands are the main source of lysozyme and other lysosomal acid hydrolases. Smaller amounts may originate from the conjunctiva, which is also the source of enzymes of the energy producing metabolism [1, 2]. Tears also contain α-amylase produced in the lacrimal glands [2, 3]. The corneal epithelium is the source of collagenase [4] and lactate dehydrogenase [5], which appear in tears in measurable amounts only after mechanical irritation of the eye or in patients with corneal ulcerations.

1.2.6.2 Applications

Acid hydrolases that have been identified in tears are listed in Table 1 together with their pragmatic values in the diagnosis of lysosomal storage disorders. Measurement of alkaline phosphatase may be useful in the detection of hypophosphatasia [11].

Table 1. Acid Hydrolases in Tears

Enzyme	E.C. number	Substrate[a] (pH)	Normal activity (mean ± SD) milliunits/g protein 37 °C	Extreme normal values	Pragmatic value in diagnosis	References
α-L-Arabinosidase	3.2.1.55	MU (4.4)	about 1		Deficient in G_{M1}-gangliosidosis	
α-L-Fucosidase	3.2.1.51	MU (4.4)	514 ± 20	101 – 1650	Deficient in α-fucosidosis	7
α-D-Galactosidase	3.2.1.22	MU (4.4)	76 ± 4	22 – 168	Thermolabile activity (60 to 80% of total) is lacking in *Fabry* disease	8, 9
β-D-Galactosidase	3.2.1.23	MU (4.4)	202 ± 20	42 – 627	Deficient in G_{M1}-gangliosidosis	10
α-D-Glucosidase	3.2.1.20	MU (4.4)	60 ± 3	12 – 130	Deficient in Type II glycogenosis infantile form	6
β-D-Glucosidase	3.2.1.21	MU (5.0)	0.75 ± 0.18	0.11 – 1.86	Deficient in some patients with *Gaucher* disease	11
β-D-Glucuronidase	3.2.1.31	MU or PhP (5.0)	99 ± 23 75 ± 20	23 – 298 25 – 127	Possible use in diagnosis of β-D-glucuronidase deficiency	
α-L-Iduronidase	3.2.1.76	PhI (4.4)	157 ± 16	55 – 439	Deficient in *Hurler* and *Scheie* diseases	6
α-D-Mannosidase	3.2.1.24	MU (4.4)	256 ± 9	101 – 627	Deficient in α-L-mannosidosis	12
N-Acetyl-α-D-glucosaminidase	3.2.1.50	4-NP (4.4)	38 ± 4.5	11 – 94	Deficient in *Sanfilippo* B disease	11

Table 1 (continued)

Enzyme	E.C. number	Substrate[a] (pH)	Normal activity (mean ± SD) milliunits/g protein 37 °C	Extreme normal values	Pragmatic value in diagnosis	References
N-Acetyl-β-D-glucosaminidase	3.2.1.30	MU (4.4)	4828 ± 130	1072 – 9932	Deficient in *Sandhoff* disease. Thermolabile isoenzyme deficient in *Tay-Sachs* disease. Hyperactivity in Mucolipidosis II	6, 13, 14
Acid phosphatase	3.2.3.2	MU (5.0)	2980 ± 297	500 – 8371	Possible use in acid phosphatase deficiency	6
Acid phosphodiesterase	3.1.4.1	MU (5.3)	55 ± 8	32 – 74	Possible use in *Niemann-Pick* disease	
Acid shingomyelinase	3.1.4.12	HNP (4.4)	89 ± 21	32 – 309	Deficient in *Niemann-Pick* disease types A and B	11
Arylsulphatase A	3.1.6.1	4-NC (5.0)	42.4 ± 6	11 – 102	Deficient in metachromatic leucodystrophy and mucosulphatidosis	15
Arylsulphatase B	3.1.6.1	4-NC (6.0)	51.6 ± 9.9	18 – 111	Deficient in *Maroteaux-Lamy* disease and mucosulphatidosis	
Acid proteases	3.4.22.1 + ?	GAMN (4.4)	about 1000		?	4
Lysozyme	3.2.1.17	ML (6.2)	514 ± 32[b] (20 °C)	146 – 1113	Less active in keratitis sicca	2, 16

a Substrates: 4-methylumbelliferyl- (MU), phenyl- (PhI), 4-nitrophenyl- (4-NP), 4-nitrocatechol- (4-NC), phenolphthalein- (PhP) derivatives; hexadecanoyl-4-nitrophenyl phosphorylcholine (HNP), micrococcus lysodeictus (ML), Gly-Arg-4-methoxy-β-naphthyl-amide (GAMN)
b Activity expressed in micrograms of purified egg lysozyme per mg tear protein.

1.2.6.3 Tears for Enzyme Analysis

Sampling

Tears may be collected directly with a capillary tube or a syringe, after stimulation with a crystal of NaCl or with ammonia vapour. Alternatively, a strip of filter paper may be inserted into the conjuntival sac. The strip, when wetted with tears, may simply be allowed to dry without significant loss of enzyme activity. Dried strips may be mailed to the reference laboratory [6]. With this technique the main source of error is contamination with the patient's eye cosmetics.

Techniques of measurement

Enzymes are eluted from the filter paper strips by vigorous agitation in NaCl, 0.15 mol/l, or directly in the assay buffer. The recommended enzyme substrates are included in Table 1.

References

[1] *N. J. Van Haeringen,* Clinical Biochemistry of Tears, Survey Ophthalmol. *26,* 84 – 96 (1981).

[2] *N. J. Van Haeringen, E. Glasius,* Enzymatic Studies in Lacrimal Secretion, Exp. Eye Res. *19,* 135 – 139 (1974).

[3] *N. J. Van Haeringen, F. Ensink, E. Glasius,* Amylase in Human Tear Fluid: Origin and Characteristics, Compared with Salivary and Urinary Amylases, Exp. Eye Res. *21,* 395 – 403 (1975).

[4] *Pi-Kwang Tsung, F. J. Holly,* Protease Activities in Human Tears, Curr. Eye Res. *1,* 351 – 355 (1981).

[5] *C. MacKay, D. H. Abramson, R. M. Ellsworth, F. D. Kitchin, M. Pesche,* Lactate Dehydrogenase in Tears, Am. J. Ophthalmol. *90,* 385 – 387 (1980).

[6] *F. Van Hoof, J. Libert, G. Aubert-Tulfkens, M. V. Serra,* The Assay of Lacrymal Enzymes and the Ultrastructural Analysis of Conjunctival Biopsies: New Techniques for the Study of Inborn Lysosomal Diseases, Metabolic Ophthalmol. *1,* 165 – 171 (1977).

[7] *J. Libert, F. Van Hoof, M. Tondeur,* Fucosidosis: Ultrastructural Study of Conjunctiva and Skin and Enzyme Analysis of Tears. Invest. Ophthalmol. *15,* 626 – 639 (1976).

[8] *D. L. Johnson, M. A. DelMonte, E. Cotlier, R. J. Desnick,* Fabry Disease: Diagnosis by α-Galactosidase Activities in Tears, Clin. Chim. Acta *63,* 81 – 90 (1975).

[9] *J. Libert, M. Tondeur, F. Van Hoff,* The Use of Conjunctival Biopsy and Enzyme Analysis in Tears for the Diagnosis of Homozygotes and Heterozygotes With Fabry Disease, in Birth Defects: Original Article Series 12(3), pp. 221 – 239 (1976).

[10] *A. Tsuboyama, F. Miki, M. Yoshida, Y. Ogura, T. Miyatake,* The Use of Tears for Diagnosis of GM_1 Gangliosidosis, Clin. Chim. Acta *80,* 237 – 242 (1977).

[11] *F. Van Hoof,* Unpublished Data (1982).

[12] *J. Libert, F. Van Hoof, J.-P. Farriaux, M. Tondeur, E. Vamos-Hurwitz, I. Mandelbaum, P. Danis,* La Mannosidose: Diagnostic par biopsie conjonctivale et analyse enzymatique des larmes, Bull. Mém. Soc. Franc. Ophthalmol. *89,* 291 – 298 (1977).

[13] *J. D. Singer,* Hexosaminidase A in Tears and Saliva for Rapid Identification of Tay-Sachs Disease and its Carriers, Lancet II, 1116 – 1119 (1973).

[14] *P. J. Carmody, M. C. Rattazzi, R. G. Davidson,* Tay-Sachs Disease: The Use of Tears for the Detection of Heterozygotes, New Engl. J. Med. *289*, 1072 – 1074 (1973).

[15] *J. Libert, F. Van Hoof, D. Toussaint, H. Roozitalab, K. R. Kenyon, R. Green,* Ocular Findings in Metachromatic Leukodystrophy. An Electron Microscopic and Enzyme Study in Different Clinical and Genetic Variants, Arch. Ophthalmol. *97*, 1495 – 1504 (1979).

[16] *R. Avisar, R. Menaché, P. Shaked, J. Rubinstein, I. Machtey, H. Savir,* Lysozyme Content of Tears in Patients with Sjögren's Syndrome and Rheumatoid Arthritis, Am. J. Ophthalmol. *87*, 148 – 151 (1979).

1.2.7 Urine

Horst Grötsch and Hermann Mattenheimer

The diagnostic value of hyper-enzymuria in renal diseases has been extensively reviewed [1 – 7]. In addition reference is made to the proceedings of two international conferences on the subject [8, 9]. Despite intensive research in many laboratories, enzyme measurements in the urine as diagnostic parameters in renal diseases have not yet been generally accepted as routine procedures. *Burchard et al.*, who are in the forefront in this field, have expressed their reservations as follows (p. 111 in [9]): "at present, it is not justified to recommend the general application of urinary enzyme determinations in basic medical care. Due to difficulties in interpreting the results obtained, these methods should be used for the present only in nephrological and endocrinological centres".

It has been recognized for some time that certain drugs cause hyper-enzymuria [2]. Enzyme measurements in the urine are, therefore, now more frequently included in the assessment of possible nephrotoxic effects of drugs.

1.2.7.1 The Origin of Enzymes in Urine

Several sources may contribute enzymes to the urine: renal tissue and the epithelium of the urinary tract, blood plasma (glomerular filtration and/or impaired tubular re-absorption) bloods cells (erythrocytes and leikocytes), and bacteria. Of these, blood plasma and blood cells have been overestimated as significant sources.

Micro-organisms apparently do not contribute to the activities of the commonly measured enzymes [1].

Renal tissue, and particularly the proximal convoluted tubules, must be regarded as the main sources of enzymes in the urine [1], (p. 119 in [8]). There are, however, some exceptions. Enzymes of low molecular mass, such as α-amylase (45000), pepsinogen (42500), ribonuclease (14000), and lysozyme (14500), are partially filtered and to a large extent re-absorbed in the proximal tubules. An increased excretion of lysozyme is an indicator of atrophy or degeneration of proximal tubules [3].

More than 30 enzymes have been identified in the urine, and many have been studied as to their diagnostic usefulness in renal diseases. However, only a few have survived this "screening". Of diagnostic value are enzymes of the brush border of the proximal tubuli, AAP*, AP, γ-GT, and lysosomal enzymes, β-Gluc, NAG, ASA, and lysozyme. Among these enzymes lysozyme occupies an exceptional position, because it is also filtered from plasma by the glomeruli (see above). AAP, AP and γ-GT are integral parts of the brush border membrane. More than 90% of AAP resides peripherally; γ-GT is a transmembranous protein, 30% belonging to the peripheral and about 70% to the inner and basal parts of the membrane; AP is part of the inner membrane core [10].

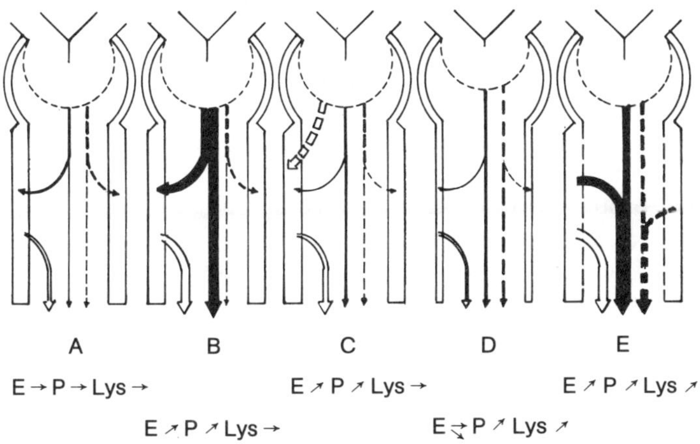

Fig. 1. Concept of the correlation between the reabsorption of lysosomotropic agents by the tubular epithelium and urinary excretion of proteins and enzymes [3]

E = tubular enzymes
P = Protein
Lys = Lysozome
Lysosomotropic agents
normal →, increased ↗, decreased ↘

A Physiological condition: tubulur re-absorption of protein filtered in small amounts by the glomeruli; consecutive release of small amounts of cell enzymes; lysozyme filtered by the glomeruli is almost completely re-absorbed.

B Increased glomerular filtration of protein with increased re-absorption and increased release of cell enzymes.

C Filtration of lysosomotropic agents which compete with protein in re-absorption; the excretion of protein and cell enzymes is increased.

D Nephrocirrhosis of various genesis: the protein concentration is elevated in the reduced volume of the glomerular filtrate; the re-absorption of protein is diminished; depending on the degree of the tubular epithelium, the release of cell enzymes is decreased; the re-absorption of lysozyme is diminished.

E As a result of the disintegration of the tubular cells, serum and cell enzymes, and serum and structural proteins appear in large amounts in the final urine.

* Abbreviations see Table 1, page 20.

Thoughts on the hitherto poorly understood mechanism of enzyme release from tubular cells into the urine were stimulated by an interesting working hypothesis [3], which links enzymuria to lysosomotropism. Among lysosomotropic substances which were found to provoke enzymuria are mannitol, dextran, certain x-ray contrast media, aminoglycoside antibiotics, bile acids, and serum proteins. In brief, the enzyme release mechanism is thought to be as follows: the lysosomotropic agent in the glomerular filtrate, including physiologically filtered serum proteins, is taken up into the cells of the proximal convoluted tubule by endocytosis (pinocytosis). The endocytotic vesicles combine with lysosomes, which contain the acid hydrolases (digestive enzymes). Substances digested slowly or not at all remain contained in a "residual body" together with digestive enzymes. The residual body is than extruded into the tubular lumen. By this mechanism lysosomal enzymes (e.g. β-Gluc) and brush border enzymes (e.g. AAP) are released into the urine. The concept of the correlation between the re-absorption of lysosomotropic agents by the tubular epithelium and urinary excretion of protein and enzymes is shown in Fig. 1.

In reviewing this working hypothesis, there are no difficulties in understanding the proposed mechanism under pathological conditions, i.e., increased protein filtration and/or the presence of other lysosomotropic agents in the primary urine. Considering the physiological excretion of brushborder and lysosomal enzymes, however, one may wonder whether the normal protein re-absorption can actually be held responsible. Could the normal amounts of these enzymes in the urine not be explained by physiological cell turnover? The theory says nothing about cytoplasmic enzymes. How is their appearance in the urine to be explained?

1.2.7.2 Examples of Application

Since it is not within the scope of this chapter to review comprehensively the application of enzyme measurements in the urine, only a few examples will be cited.

Out of a larger spectrum of enzymes which are elevated in the urine in various kidney diseases, *Burchardt et al.* [3, 11] selected AAP, β-Gluc and lysozyme. The activity pattern of these enzymes in the urine, together with the degree of protein excretion, allows the assessment of the type of glomerular and tubular lesions in renal and primary extrarenal diseases (Table 1).

Another field of application is the monitoring of the function of renal transplants. Again, a combination of brush border membrane and lysosomal enzymes was successfully applied [5]. AAP, γ-GT and NAG behaved similarly during graft rejection. Increases of AAP activity were the most sensitive indicator of a reversible rejection. Irreversible rejection was characterized by significantly elevated excretion of all the enzymes, with γ-GT giving the highest absolute values. Reduced enzyme excretion was a sign of a functionless graft.

In 1980, *Meyler* [12] listed about sixty drugs with nephrotoxic side effects. It has also been recognized for some time that certain drugs cause hyper-enzymuria [2, 3],

Table 1. Protein and enzyme constellation in renal and primary extrarenal diseases

Disease Group	Proteinuria	Findings in urine: Alanine amino-peptidase	β-Glucu-ronidase	Lysozyme
Acute renal failure	↑	↑	↑↑	↗
Pyelonephritis, acute exacerbation	↑	↑	↗	↑
Chronic pyelonephritis without retention	↗	↗	→	→
Chronic pyelonephritis with retention	↑	↗	↗	↑
Chronic glomerulonephritis with nephrotic syndrome	↑↑	↑	↗	→
Chronic glomerulonephritis without retention	↗	↗	→	→
Chronic glomerulonephritis with retention	↑	↗	↗	↑
Hyperthyreosis	↗	↗	→	→
Virus hepatitis (one week)	↗	↗	↗	→
Benign bile duct occlusion	↗	↗	↗	→
Malignant bile duct occlusion	↗	↑↑	↗	→

and enzyme measurements in the urine are applied increasingly in assessing nephrotoxicity. Hyper-enzymuria is often a more sensitive indicator and is observed earlier than changes of other biochemical parameters, such as serum creatinine and creatinine clearance.

In the early phase of drug trials (animal experiments) histological studies should also be performed, both on the light- and electron microscopical level. This is essential in order to understand the nature and severity of the renal lesion, which may certainly vary from drug to drug and may be doses dependent. Measurements of urinary enzymes alone in studying nephrotoxicity cannot answer questions such as: what change in the kidney can still be accepted as physiological or tolerable? What quantitative changes in urinary enzyme activities and/or what patterns of changes will predict an imminent transition from perhaps clinically insignificant tubular changes to renal insufficiency? These and other questions pertinent to the definition of drug nephrotoxicity were raised by speakers and discussants at an international symposium ([9], chapter on toxicology).

1.2.7.3 Urine for Enzyme Analysis

Collection

No agreement exists on what is the optimal urine collection period for enzyme measurements. Circadian excretion rhythms have been observed for several enzymes and they appear to be different for individual enzymes [3, 5, 13]. Some investigators, therefore, prefer the 24 h collection in order to obtain an average value, but shorter periods down to 3 h have also been recommended. Standardization of collection periods within a laboratory is, of course, necessary. 24 h urines should be collected under toluene. If urines have to be stored, they should be kept in the refrigerator, and sodium azide should be added.

The normal urine volume varies within relatively wide limits. It is not possible, therefore, to establish a consistent reference volume for expressing enzyme activity, and enzyme mesurements in random urine samples are not to be recommended. The activity should be measured in timed specimens and expressed in units per collection period. Another option is to use urinary creatinine or creatinine clearance as reference quantities [4].

Preparation for analysis

Urine constitutes a very unfavourable environment for enzymes, which may alter the molecular structure of enzyme proteins and thereby affect enzyme activity. For example, the hydrogen ion concentration varies considerably, and an enzyme which is stable at neutral pH may be inactivated at acid pH.

Urine contains inhibitors, which must be removed before enzyme analysis [4], (p. 22 in [9]). In some instances inhibition may be overcome by diluting the urine. But the concentration of inhibitors appears to vary from specimen to specimen [13], and dilution may fail if their concentration is high.

Gel filtration

The most commonly used procedures for inhibitor removal are dialysis and gel filtration [14]. The latter is more practicable and should be recommended. Frequently one finds that the activity of some enzymes decreases following gel filtration, and the question has been raised of enzyme activators in the urine, but none has so far been identified.

Recently we investigated the effect of gel filtration (Sephadex G25, medium) on the activity of AAP, γ-GT, NAG, and β-Gluc in thirty 8 h and thirty 24 h urines randomly selected from the clinical chemistry laboratory of our hospital [15]. Compared to the activity in crude urine (= 100%), the activity of AAP increased to an average of 180% with a range of 90 – 450%; the activity of γ-GT decreased slightly to an average of 95% with a range of 75 – 108%; about the same average change was found for NAG, but the range was 27 – 158%; the widest scattering was observed with β-Gluc, with an average increase to 125% and a range of zero to 800%. There were no apparent differences between the 8 and 24 h urines.

Sephadex G50 fine was used in the originally described method [14] and in a scaled-down adaptation (p. 25 in [9]). Sephadex G25 medium, of which small ready to use columns with 9 ml bed volume are available (*Pharmacia*, PD10), was recommended by *Grötsch* (unpublished).

Recommended procedure

- Prepare column according to the manufacturer's instructions.
- Add 4 ml NaCl, 0.15 mol/l, to column and drain completely.
- Centrifuge urine for 10 min at 1000 *g*.
- Add 2.5 ml of the supernatant to the column and drain completely.
- Add 4 ml NaCl, 0.15 mol/l.
- Collect the total eluate for enzyme assays (dilution factor 1.6).
- Regenerate column with distilled water.

Techniques of measurement

Analytical methods may be used which are described elsewhere in this book. An agreement has yet to be reached on the standardization of techniques specifically for urine.

Imprecision

The overall imprecision of enzyme assays includes errors introduction by gel filtration. The data in tables 2–4 were obtained by *Grötsch et al.* (unpublished) using Sephadex G-25 M (PD10) columns. Although gel filtration increased the imprecision, the variation stayed well within acceptable limits. An exception is β-Gluc with very low activity.

Table 2. Within-column imprecision of gel filtration and enzyme assays.
Ten aliquots of a human urine were passed through one column. The gel was regenerated after passage of each aliquot. Enzyme activity concentration in U/l.

	AAP	NAG	β-Gluc		γ-GT
$\bar{x}$	2.65	3.98	0.76	5.03	18.8
s	0.16	0.18	0.12	0.13	0.79
CV %	6.03	4.52	15.8	2.58	4.20

Table 3. Between-columns imprecision of gel filtration and enzyme assays.
Ten aliquots of a human urine were passed through 10 individual columns. Enzyme activity concentration in U/l.

	AAP	NAG	β-Gluc		γ-GT
$\bar{x}$	2.69	3.78	0.84	4.58	18.2
s	0.09	0.22	0.12	0.15	0.8
CV %	3.35	5.82	14.29	3.28	4.4

Table 4. Within-run imprecision of enzyme assays. Repeated analysis of rat urine following gel filtration. Enzyme activity concentration in U/l.

	AAP	NAG	β-Gluc
$\bar{x}$	6.56	4.90	18.0
s	0.08	0.15	0.3
CV %	1.22	3.06	1.66
n	15	10	10

Quality control

Quality control material for urinary enzymes is not commercially available. *Maruhn* (p. 26 in [9]) prepares pooled urine by concentration, Sephadex gel filtration, addition of bovine serum albumin and division into aliquots, each sufficient for a day's work. The aliquots are stored at −25 °C. No procedure has yet been proposed for the quality assurance of the gel filtration part of urinary enzyme analysis.

Reference values

We desist from listing reference values for enzymes in the urine. Lack of standardization of assay techniques and whether or not gel filtration (or dialysis) was applied has led to differences between reported values. It was also recognized that reference values often did not follow a normal distribution [11, 16]. Reference values for eleven human urinary enzymes have been published, with the application of the appropriate transforming functions [16].

Note for experiments with rats

It has bee reported that rats develop hyper-enzymuria, when isolated singly in metabolic cages. This "stress" effect was most pronounced 1 – 3 days after isolation. A relatively stable and near normal state was reached after 6 – 10 days (p. 88 in [9]).

References

[1] *H. Mattenheimer,* Enzymes in the Urine, Med. Clinics of North America *55,* 1493 – 1508 (1971).
[2] *W. P. Raab,* Diagnostic Value of Urinary Enzyme Determinations, Clin. Chem. *18,* 5 – 25 (1972).
[3] *U. Burchardt, J. E. Peters, L. Neef, H. Thulin, C. A. Grundig, R. J. Haschen,* Der diagnostische Wert von Enzymbestimmungen im Harn, Z. Med. Labor.-Diagn. *18,* 190 – 212 (1977).
[4] *H. Mattenheimer,* Enzymes in Renal Diseases, Ann. Clin. Lab. Sci. *7,* 422 – 432 (1977).
[5] *G. Horpácsy,* Die Anwendung der Enzymdiagnostik in der Praxis der experimentellen und klinischen Nierentransplantation. F. K. Schattauer Verlag, Stuttgart-New York 1979.

[6] *R. G. Price,* Urinary Enzymes, Nephrotoxicity and Renal Disease, Toxicology *23*, 99 – 134 (1982).
[7] *R. E. VanderLinde,* Urinary Enzyme Measurements in the Diagnosis of Renal Disorders, Ann. Clin. Lab. Sci. *11*, 189 – 201 (1981).
[8] *U. C. Dubach* (ed.), Enzymes in Urine and Kidney. Proceedings of the International Conference on Enzymes in the Urine, Rheinfelden, Switzerland, 1967. Hans Huber Publishers, Bern-Stuttgart-Vienna 1968.
[9] *U. C. Dubach, U. Schmidt* (eds.), Diagnostic Significance of Enzymes and Proteins in the Urine. Proceedings of the International Symposium on the Biochemical Aspects of the Diagnostic Value of Urine. Kandersteg, Switzerland, 1978. Hans Huber Publishers, Bern-Stuttgart-Vienna 1979.
[10] *J. E. Scherberich, C. Gauhl, W. Mondorf,* Biochemical, Immunological and Ultrastructural Studies on Brush-Border Membranes of Human Kidney, in: Biochemical Nephrology, *W. G. Guder, U. Schmidt* (eds.). Hans Huber Publishers, Bern-Stuttgart-Vienna 1978, pp. 85 – 94.
[11] *U. Burchardt, J. E. Peters, L. Neef, R. J. Haschen,* Harnenzyme in der nephrologischen Diagnostik, Dtsch. Gesundheitswesen *30*, 2408 – 2412 (1975).
[12] *L. Meyler,* Side Effects of Drugs, 9rh edition, Exerpta Medica, Amsterdam 1980.
[13] *G. Szasz,* γ-Glutamyltranspeptidase-Aktivität im Urin, Z. klin. Chem. klin. Biochem. *8*, 1 – 8 (1970).
[14] *M. Werner, D. Maruhn, M. Atoba,* Use of Gel Filtration in the Assay of Urinary Enzymes, J. Chromatogr. *40*, 254 – 263 (1969).
[15] *H. DeBruin, H. Grötsch, V. Malerczyk, H. Mattenheimer,* to be published in J. Clin. Chem. Clin. Biochem. (1983).
[16] *D. Maruhn, I. Fuchs, G. Mues, D. Bock,* Normal Limits of Urinary Excretion of Eleven Enzymes, Clin. Chem. *22*, 1567 – 1574 (1976).

1.3 Methods for Isoenzyme Analysis

1.3.1 Isoenzymes and Other Multiple Forms of Enzymes

Roger J. Wieme and Donald W. Moss

The early history of enzymology provided many instances in which catalytically-similar enzymes from different sources could be shown to differ in other respects. Analogous enzymes from frequently used sources such as yeast and animal tissues (e.g., yeast and liver alcohol dehydrogenases) were known to differ. It was also realized that even enzymes with similar catalytic actions obtained from various tissues of a single species were not always identical, as in the case of non-specific acid phosphatases from human prostatic and other tissues. However, observations such as these continued to be regarded as rather exceptional. Therefore, it was not until newer methods of separating mixtures of proteins, such as gel electrophoresis and ion-exchange chromatography, began to be applied to enzyme preparations, from the late 1950's onwards, that the existence of multiple molecular forms of an enzyme with a particular type of catalytic activity was generally accepted.

1.3.1.1 Definitions and Nomenclature

In the years 1957–1958 several authors, working independently, reported that very similar enzymatic activities can be carried by different molecular species. *Markert* and *Møeller* [1] proposed to use the term "isozyme" to describe such related enzymes occurring within a single species. Other workers, pointing to the fact that this occurrence is nothing more than a special case of protein multiplicity, continued to use the non-committal terminology of "enzyme multiplicity" [2] or referred to "multiple molecular forms of enzymes" [3].

In 1977 an international agreement was reached, followed by a recommendation by the IUPAC-IUB Commission on Biochemical Nomenclature [4]. The term *isoenzyme* is now restricted to those forms with similar enzymatic activity occurring within a single species, as a result of the presence of more than one structural gene. The original spelling "isozyme" is still preferred by many authors and is recognized as an acceptable alternative by the Commission.

According to this definition, the following categories are regarded as isoenzymes: (I) genetically-independent proteins arising from the presence of multiple gene loci (such as the mitochondrial and cytoplasmic forms of malate dehydrogenase or other enzymes); (II) enzyme variants deriving from the occurrence of allelic genes at a particular gene locus – these isoenzymes are also called allelozymes (numerous examples exist among human enzymes; e.g. the allelozymes of glucose-6-phosphate dehydrogenase); (III) heteropolymers (non-covalent hybrid molecules of two or more different polypeptide chains) such as the "intermediate" isoenzymes of lactate dehydrogenase.

All other causes of enzyme multiplicity are excluded from this definition, such as forms due to conjugation or degradation (e.g. γ-glutamyl transferase or chymotrypsin, respectively), polymers of a single subunit (e.g. glutamate dehydrogenase), naturally-occurring complexes, such as those which may form *in vivo* between various enzymes and immunoglobulins, and allosteric modifications of enzymes.

No recommendations were given by the Commission as how to interpret "similar enzymatic activity", but from the context it can be inferred that a documented isoenzymic relationship can only exist between well-categorized enzymes, i.e. those which have been endowed with a systematic enzyme name.

A proposal concerning the nomenclature of isoenzymes has also been made by the Commission on Biological Nomenclature. An isoenzyme should be designated by its normal name (systematic or trivial) followed by an arabic number, preferably corresponding to its electrophoretic mobility, the lowest figure being given to the most anodal isoenzyme. If an enzyme fraction can be resolved into subfractions (e.g. by isoelectric focusing) these can be designated by appending lower-case subscript letters (e.g. LDH-1a). Rules were also given for the designation of interconvertible enzymes [4].

1.3.1.2 Significance of Isoenzyme Analysis

The enzymatic activities of the members of a family of isoenzymes are similar, but by no means identical in all cases. As a rule, slight differences are observed in pH optimum, or in sensitivity to inhibitors or to denaturing agents and also in affinity constants for the substrate. The differences in properties between isoenzymes determined by multiple gene loci presumably account for their emergence during evolution and their functional significance in present species. Therefore, from their very discovery, isoenzymes have been linked with cell differentiation and development, suggesting a progressive adaptation to changing environment. Yet, in many cases, no clear reasons for the existence of isoenzymes have been found. These are considered remnants of biological evolution. Isoenzymes have also been studied in relation to tumour growth since, in a few instances, tumour cells have been found to produce "new" or embryonal isoenzymes [5, 6].

The allelozymes which result from the existence of allelic genes may or may not differ in their catalytic properties. In some cases the mutant allele may produce no active enzyme at all, although a gene product recognizable by immunological means may be produced [8]. If the affected enzyme occupies a key position in metabolism, failure to produce an active enzyme results in severe, or even fatal, disease. Even if active enzyme is produced by an allelic gene, it may be so altered in its properties that normal metabolism is impossible. The result may be an obvious disease, or the effect of the altered enzyme may only be seen when the affected individual is challenged by some unusual stress, such as the administration of a particular drug.

The existence of tissue-specific forms of human enzymes (whether due to genetic or non-genetic causes) increases the potential specificity and sensitivity of enzyme measurements carried out for diagnostic purposes. These potential advantages can be realized with the aid of appropriate analytical methods.

The study of isoenzymes and other multiple forms of enzymes is thus of interest to biochemists, geneticists, clinicians and biologists of many disciplines. Therefore, isoenzyme analysis (with the term isoenzyme being interpreted in a wide, operational sense) has become routine in many laboratories [9].

1.3.1.3 General Principles of Isoenzyme Analysis

The isoenzymes (or other multiple forms) of an enzyme may differ in some or all of a variety of properties. These differences form the basis of methods of analysis of isoenzyme mixtures.

Separative methods

By definition, the multiple forms of an enzyme differ in their molecular structures; therefore, they are ultimately separable from one another. Separative techniques can be regarded as the definitive techniques of isoenzyme analysis.

Zone electrophoresis is the most useful single technique in the analysis of mixtures of multiple forms of enzymes of all kinds. It requires only small samples, when combined with suitably sensitive methods of detecting the separated isoenzyme zones. A further advantage of the technique is that the separation achieved is generally unaffected by the degree of purity of the sample being separated. Electrophoresis is therefore particularly valuable for investigating the isoenzyme composition of unpurified extracts of cells or tissues, as well as serum or other body fluids, and in consequence it has been used extensively for genetic and clinical studies.

Differences in net molecular charge between multiple forms of enzymes also open up the possibility of applying other analytical techniques such as isoelectric focusing.

The changes in one, or at most a few, amino acid residues that account for the structural differences between isoenzymes produced by allelic genes typically do not result in marked differences in molecular size. Therefore, unless relatively large sections of the amino-acid sequence have been deleted by mutation, separation methods which depend on differences in molecular size are unlikely to be useful in the resolution of mixtures of allelozymes. Similarly, isoenzymic molecules determined by multiple gene loci generally differ little in size so that size-dependent separation methods are not helpful. On the other hand, aggregation of enzyme molecules, or the attachment of enzyme molecules to other proteins or to non-protein constituents such as lipids is a frequent cause of heterogeneity of enzymes in biological fluids and tissue extracts and separations based on size differences are correspondingly valuable.

Variations in carbohydrate side-chains appear to account for the existence of non-isoenzymic multiple forms of some enzymes. Affinity chromatography, which exploits the specific affinity of lectins for some carbohydrate residues, is a useful separation procedure in some of these instances.

Immunological analysis

Analytical methods based on antigenic differences between multiple forms of enzymes can be regarded as bridging the gap between separative and non-separative methods. Precipitation with an antiserum raised against a particular enzyme form can be used to remove that form from a mixture of multiple forms. Alternatively, reaction with a specific antiserum can improve electrophoretic separation, while immobilized antibodies, used as the stationary phase in affinity chromatography, provide specific and highly-resolving separation systems. Immunological methods can also be employed in non-separative analysis, e.g. by means of inhibiting antisera.

Multiple forms of enzymes which arise by post-translational modifications such as aggregation are unlikely to differ in their antigenic determinants. Many of the carbohydrate residues which occur in glycoproteins are of such wide distribution that multiple forms of enzymes which differ only in their carbohydrate moieties are likely to be immunologically similar. Similarly, minor modifications to the polypeptide chain, such as the removal of one or more amide groups, are unlikely to affect the antigenic properties of the molecule. Therefore, immunological methods find their major applications in the analysis of mixtures of multi-locus isoenzymes.

Non-separative methods

Differences in such catalytic properties as relative rates of reaction with alternative substrates or substrate analogues (for enzymes of less than absolute specificity), *Michaelis* constants, and response to inhibitors, almost invariably exist between isoenzymes determined by multiple gene loci. These differences can form the basis of non-separative methods of analysis. Since the differences between isoenzymes may be relatively small, careful standardization of the conditions of measurement is essential for reliable analysis (cf. Vol. I, chapter 2.2.1, pp. 86 – 97; 2.2.2, pp. 97 – 104). The selective binding of certain inhibitors to particular isoenzymes has recently been made the basis of separative methods, by using the immobilized inhibitor as the stationary phase in affinity chromatography.

Significant alterations in catalytic properties often result from the mutations which give rise to allelozymes. Analytical methods based on catalytic differences are useful in these cases also. On the other hand, the various post-translational modifications which may give rise to multiple forms of enzymes usually do not produce alterations in catalytic properties.

The catalytic activity of enzyme molecules depends on the integrity of their three-dimensional structures. These structures are mainly stabilized by numerous hydrophobic interactions between amino-acid side-chains in the interiors of the enzyme molecules. Differences in resistance to inactivation by denaturing agents such as heat or urea therefore indicate the existence of structural differences between enzyme molecules, and can be made the basis of qualitative and quantitative isoenzyme analysis.

As would be expected from the fact that they differ in primary structures, true isoenzymes typically differ in their resistance to denaturation. This is true both for isoenzymes derived from multiple gene loci and for those originating from allelic genes. Multiple enzyme forms due to post-translational modifications generally do not differ greatly in stability, although minor stabilizing effects may result from association of enzyme molecules with non-enzyme moieties.

Non-separative methods of analysis are most useful when the differences in properties between the isoenzymes are clear-cut, and when the isoenzyme mixture contains only two components. They are less successful when the mixture to be analyzed contains several enzyme forms with rather similar properties; e.g. the hybrid isoenzymes of lactate dehydrogenase.

References

[1] *C. L. Markert, F. Møller,* Multiple Forms of Enzymes: Tissue, Oncogentic, and Species Specific Patterns, Proc. Natl. Acad. Sci. USA *45*, 753 – 763 (1959).

[2] *R. D. Cahn, N. O. Kaplan, L. Levine, E. Zwilling,* Nature and Development of Lactic Dehydrogenases, Science *136*, 962 – 969 (1962).

[3] *R. J. Wieme, An Assessment of the Isoenzyme Concept, in: N. van Thoai, J. Roche* (eds.), Homologous Enzymes and Biochemical Evolution, Gordon and Breach, New York 1968, pp. 19 – 40.
[4] IUPAC-IUB Commission on Biochemical Nomenclature, Nomenclature of Multiple Forms of Enzymes. Recommendations (1976), J. Biol. Chem. *252*, 5939 – 5941 (1977).
[5] *W. H. Fishman,* Carcinoplacental Isoenzyme Antigens, Adv. in Enzyme Regulation *11*, 293 – 321 (1973).
[6] *W. E. Criss,* A Review of Isoenzymes in Cancer, Cancer Res. *31*, 1523 – 1542 (1971).
[7] *F. Schapira, A. Hatzfeld, A. Weber,* Resurgence of Some Fetal Isozymes in Hepatoma, in: *C. L. Markert* (ed.), Isozymes, Vol. *3*, Academic Press, New York 1975, pp. 987 – 1003.
[8] *T. A. Tedesco, W. J. Mellman,* Galactosemia: Evidence for a Structural Gene Mutation, Science *172*, 727 – 728 (1971).
[9] *D. W. Moss,* Isoenzyme Analysis, The Chemical Society, London 1979.

1.3.2 Separation Techniques

Roger J. Wieme

1.3.2.1 Introduction

Isoenzymes differ as proteins and therefore can be separated by the techniques of protein chemistry such as salting-out, precipitation with organic solvents or by specific adsorption, e.g. to immunoglobulins or lectins. However, the most powerful separation procedures are designated by the generic term "differential migration techniques", in which physical separation is achieved by displacement along a separation track. In chromatography the migrants are carried by a non-selective flow and are selectively retarded by interaction with the stationary phase. In electrophoresis the driving force itself is selective (related to electric charge of the migrant) but the separation effect can be increased by interaction with the stationary phase. The present trend is to treat both techniques together (e.g. [1, 2]).

The main chromatographic procedures available for isoenzyme separation are: gel chromatography, ion-exchange chromatography (including chromatofocusing), affinity chromatography and high performance liquid chromatography (HPLC). Zone electrophoresis can be classified according to the stabilizer used: cellulose acetate membranes, starch gel, agarose gel, polyacrylamide gel (PAGE), beaded gels such as dextran granules. Several separation-enhancing procedures can be applied, mainly isoelectric focusing (IEF) and the use of pore gradients.

Choice of the procedure or combination of procedures that will achieve the desired separation is still a matter of intuition and experience, but can be guided by experiment. Useful information can also be gathered from product catalogues that are commercially available, mainly from *Biorad, LKB* and *Pharmacia.*

Separation parameters

Mobility

Velocity of migration is the basic characterization parameter. It is expressed in various ways.

In chromatography the dimensionless parameter R is used: the ratio of the velocity of the zone under consideration to that of the non-selective flow. If one refers to the front of the solvent, the parameter becomes the well-known R_f value. R_f is also the ratio of the time a migrant spends in the flow medium to the total duration of the experiment and reflects distribution between the two phases. If the migrant does not interact with the stationary phase, R_f tends to 1. If it is irreversibly fixed, R_f becomes zero.

In electrophoresis m (mobility) is used, i.e. the distance covered by a zone in unit time at unit field strength.

(a) $$m = \frac{d}{t \times E}$$

d = distance (cm), t = time (s), E = field strength (E/cm).

The value of E is not often known; therefore, m_z, the observed velocity of a zone, is used.

(b) $$m_z = \frac{d}{t}$$

The term mobility is used very loosely, often without considering the fact that the experimental conditions (temperature, ionic strength, electric field) are not always homogeneously distributed over the migration field nor time-independent. Some of these difficulties disappear if the distance covered is referred to the distance covered by a reference substance. This is called relative mobility m_r.

(c) $$m_r = \frac{d_x}{d_r}$$

d_x = distance covered by zone x (signed)
d_r = distance covered by reference zone (signed)
(distances measured from point of zero mobility)

A well-characterized protein, such as human serum albumin, can serve as a reference; alternatively, a dye which moves at high speed, preferably following a buffer front, can be used.

The ratio then becomes

(d) $$m_f = \frac{d_x}{d_f}$$

d_f = distance covered by the front (signed)

m_r and m_f are dimensionless ratios that characterize a molecular species within a certain experimental framework. The analogy between m_f and R_f is obvious.

It is important that clear agreement is obtained on these terms since some confusion exists (e.g. [3]). Yet measurement of migration velocity is important for characterization of a molecular species, and for the description of a zymogram (i.e. the pattern of separated isoenzyme zones).

Zone width and zone spreading

A general definition of the width of a zone with bell-shape distribution is the distance between inflexion points of the distribution curve. If the zones deviate markedly from this distribution, as in 2-dimensional experiments, other criteria can be employed for contour mapping (4). Here we consider the migration along a linear track. Generally, the spreading factors act symmetrically and in a large number of small repetitive steps. The distribution is then expected to be *Gaussian* and the width of the zone is equal to $2s$ with s^2 the variance of the distribution. Only such distributions are considered further.

In the course of migration zones broaden essentially by three mechanisms:

1. Thermal diffusion (related to the thermal diffusion coefficient).
2. Eddy diffusion (refering to the variability of migration path for individual molecules).
3. Non-equilibrium spreading.

Non-equilibrium spreading can be the dominant spreading factor. It occurs whenever a zone moves relative to a stationary phase with which it interacts by fixation. Consider a point on the front of a zone; a point follows where concentration is higher; the stationary phase has to adjust to this higher concentration but does this only after a given (short) time. In that time the front has migrated without interaction. Therefore some of the molecules in the front move at greater than mean velocity (i.e. that of the zone itself). Now consider the descending slope of the curve describing the zone distribution. Here the converse is true: part of the molecules move at a slower speed. At the peak of the concentration curve, however, the slope is zero and migration occurs at exactly the expected value. Therefore, the centre of the zone migrates at the exact R_f value, but the front and the back move away from each other. The faster the exchange with the medium, the less is the influence of non-equilibrium; the faster the fluid moves, the greater is the spreading.

The power of a separation procedure is ideally described by objective parameters such as resolution and peak capacity rather than by the subjective assessment of the zymogram. However, definitions and mathematical treatments of these concepts are outside the scope of this essentially practical chapter, but they may be consulted in refs. 5 and 6.

Zone sharpening

Althouth not always clearly realized, this is a most useful procedure that can markedly increase separation. It is used in chromatography as well as in electrophoresis when a sample is to be applied as a thin starting zone, or for collecting broadened zones at the end of an experiment. It can also be employed during the separation process itself by running the zones into a region of decreasing mobility (pore gradient, pH gradient). Zone-front braking is also encountered in chromatofocusing.

A zone narrows if migrating into a region of lower mobility (sample application in affinity chromatography) or if the back of the zone is accelerated (elution in ion exchange). The effect occurs as long as the front and back of the zone migrate at different velocities. If m_a is mobility after the factor is acting and m_b mobility before, and if Z_a zone-width after, and Z_b zone-width before, one has

$$\frac{Z_a}{Z_b} = \frac{m_a}{m_b}$$

with sharpening or broadening according to the ratio m_a/m_b.

In *electrophoresis*, additional procedures are available based on application of *Kohlrausch* moving boundaries. Such a boundary (separating a more rapid ion from a slower one) tends to remain sharp because the field strength distribution is such that the "deviating" migrant molecules are restored to the boundary. In its simplest form, first proposed by *Poulik* [7], a slower anion (borate) follows a rapid one (citrate). This boundary overruns the separated enzymes at the end of the experiment and re-sharpens these zones because the field strength is higher in the medium with the slower anion (acceleration of the back of the zone). The much-used *Poulik* discontinuous buffer system has inspired many variants [8]. In disc-electrophoresis this has culminated in a combination of zone sharpening procedures [9].

Isoelectric focusing (IEF) derives its virtues from a very particular situation. The focused zones do not migrate and stay narrow because they are compressed on both sides due to the restoring force arising when the migrant tends to drift (see further).

Estimation of molecular weight from chromatographic experiments

Several chromatographic and electrophoretic media exhibit size-selective interactions with the solute.

Gel chromatography

It explicitly exploits these interactions. The migration parameter is K_{av} (compare K_{av} with m_f).

(e) $$K_{av} = \frac{V_e - V_v}{V_t - V_v}$$

V_e = elution volume of migrant zone
V_v = void volume of column
V_t = total column volume

A nearly linear relationship exists between K_{av} and $\log M_r$ of the migrant [10].

Gel electrophoresis

The basic studies of *Ferguson* [11] with *starch gel* have demonstrated that the mobility of the migrant is related to the constitution of the medium and also to the migrant size. A linear relationship exists between $\log m_r$ and T according to the classical equation

(f) $$\log m_T = \log m_0 - K_r T$$

m_T = mobility at gel concentration T
m_0 = mobility at gel concentration zero
T = concentration of gel matrix
K_r = retardation coefficient

The interaction between the migrant and the pores of the medium has received several different theoretical treatments (cf. [12]), from which it is clear that K_{av}, as well as K_r, has a definite relationship to molecular weight. However, it is difficult to decide upon its exact nature on the basis of experimental evidence. An empirical approach is therefore preferred, with calibration by several well-characterized proteins which bracket the unknown migrant. Calibration can be made with enzymes but then only non-denaturating conditions can be employed.

For gel chromatography, K_{av} is generally plotted against $\log M_r$, giving a slightly S-shaped *selectivity curve.* In gel electrophoresis, several possibilities exist. The classical *Ferguson* procedure requires the measurement of mobility at several gel concentrations to obtain a K_r value that is then plotted against $\log M_r$. It mainly remains of interest when checking for an anomalous behaviour occurring in SDS electrophoresis (see further).

More often an electrophoretic run is made in a pore gradient (linear or concave). As shown by *Slater* [13] different molecular species will come to a near-standstill in a pore gradient at a specific value of T which is independent of the electric charge. The pore size at that T value is called *limiting pore size.* The log of the migration distance to near-stop is plotted against $\log M_r$. This should obtain for a linear gradient, but even a concave gradient retains the relationship.

An alternative has been proposed by *Manwell* [14] for isoenzymes. He uses commercially available polyacrylamide gels with an exponential (concave) pore gradient. Velocity of migration is found to decrease progressively in the course of the run and is considered to be proportional to the distance separating the migrant from its point of limiting pore size. Theoretically this occurs after infinite time. In practice a plot is made of $\frac{1}{d}$ versus $\frac{1}{t}$ (where d is distance already covered in time t). This plot is linear and extrapolation for infinite time gives a position d_{limit}. This position corresponds to a known gel concentration (G) and pore size. Thus $\frac{1}{G}$ is plotted against $\left(\sqrt[3]{M_r}\right)$ of calibrating migrants.

A *Ferguson* experiment can be shortened by running the separation in a transverse pore gradient. All migrants are then subjected, within a single run, to a broad range of pore sizes [15].

SDS electrophoresis

Several other procedures exist for M_r estimation but they are denaturating so that enzyme activity disappears. In the much used *SDS-pore gradient* system [16], the protein molecule is completely dissociated and the subunits unfolded by heating in a reducing medium in the presence of a high concentration of the chaotropic agent SDS. This highly charged anion binds in a nearly constant ratio (1.4 g SDS per g of polypeptide chain) and confers a large electric charge that overrides all those existing on the polypeptide chains; these acquire identical electric charge density and migrate at the same speed. If this occurs in a pore-interacting system, migration velocity becomes exclusively related to chain length. A gel of constant pore sizes can also be used. In SDS-electrophoresis zones are narrower than expected even in the absence of pore gradients. The explanation is unclear.

1.3.2.2 The Zymogram

Enzymes can be stained as proteins after electrophoresis. However, their concentration in crude extracts is several orders of magnitude lower than can be revealed by conventional procedures, except, in some cases, by the more recent silver stains (see further). They can still be detected by their enzymatic activity. In a preparative mode, the separation medium is sliced and the enzymes eluted (perhaps aided by electrophoresis), or the samples are collected from a chromatographic column and assayed for enzyme activity. However, an enzymatic staining reaction can be applied to the separation medium itself. This results in a zymogram [17] which directly reveals the enzyme multiplicity, allows side-by-side comparisons of mobility and even yields quantitative data (percentages between fractions). The zymogram technique is mainly applied to electrophoretic or chromatographic slabs. Its simplicity commends itself in routine clinical chemistry and in all large-scale investigations of repetitive character (mainly forensic and taxonomic applications).

Development of a zymogram

In the most common and also simplest form, the substrate solution, containing the system for linking the enzymatic activity to the development of a "stain", is poured on top of the separation slab as a thin fluid layer and left undisturbed for 1/2 hour to 1 hour, generally at 37 °C and protected from light. It is preferable to stabilize the reaction mixture against convective mixing. This can be effected by soaking the solution into a paper strip or cellulose acetate membrane that is layered on top. Much more effective is the gelification of the medium with 1% agarose. When the agarose solution at 2% in distilled water is brought to the boil, then cooled down until lukewarm (ca. 45 °C) and mixed in equal amounts with the substrate solution, there is little danger of denaturation. This mixture is poured gently on top of the separation slab, being retained within the area of interest by a rubber frame.

The direct agarose-overlay technique has the following advantages: all reaction products are trapped; the medium is optically clear and allows direct photometry; it can be dried into a thin transparent film; it consumes minimal amounts of expensive reagents, and gelification tends to stabilize labile enzymes or reaction intermediates such as phenazine methosulphate. The gel assures practically anaerobic conditions, a point that has to be considered if the stain develops by oxidation in air. In a recent comparative evaluation, the agarose-overlay procedure was found the most satisfactory [18]. When the separation medium is cellulose acetate strips they are best layered onto the gelified substrate mixture. Only when the enzymes are firmly entrapped, as in starch gel or in polyacrylamide gel (especially in the disc type using small cylinders) can the incubation occur in a free medium that may even be stirred.

Often it is desirable to develop zymograms for different enzymatic activities on a single pherogram. In starch gel that has been sliced, two identical surfaces are obtained. For other media, and also for the slab type of gel chromatography, replication of the pherogram can be made onto filter paper or on a thin-layer of 1% agarose. With the latter up to six copies can be obtained [19].

An interesting variant procedure is of increasing interest. Here the substrate is already included in the separation medium. The first application was for the detection of plasminogen. Fibrin is included in the agar gel and, after separation, plasminogen is activated with urokinase. Proteolysis occurs around the active zones which appear as clarifications when stained for protein. This procedure has been extended to other hydrolytic enzymes [20]. If the enzyme is able to bind to the gelified substrate, a rather complex combination of binding/digestion occurs during the run, but if the conditions are such that the enzyme is not captured (pH, chelating agent, chaotropic agent) migration is unhindered. Activity is restored by incubation with a suitable medium. The high sensitivity of the procedure is claimed as an asset (down to the pg level for most hydrolases [21]).

Enzymatic staining reactions

Useful information can be gathered from histochemistry text books (see also the important contribution by the *Harris* team [22]). Generally the conditions can be more relaxed for zymograms because there are fewer problems with substantivity or with diffusion artefacts. Even perfectly soluble stains can be employed. Nevertheless, an insoluble terminal product is an asset, e.g. by allowing the support to be dried before photometric scanning.

Staining procedures can be described as direct or indirect. In the direct method the enzyme itself generates the coloured product. In the indirect mode, the substance produced by the enzymatic activity constitutes the substrate for a linked exogeneous enzyme that is added together with a convenient indicator (enzyme linked stain). Staining reactions can also be distinguished as positive or negative, according to the appearance of a stained zone or of a "hole" in a stained background.

If NAD(P) is involved, direct photometry can be employed with transparent media at 339 or Hg 365 nm in the absorption mode, or at Hg 405 nm in the fluorimetric mode. The latter has the advantage of higher sensitivity, of direct linear response and of the integrating character of the measurement, but poses instrumental requirements that are not available, as a rule, in the routine laboratory. Therefore, preference goes to the development of visible stains that are submitted to axial scanning in the absorption mode. Good commercial equipment is available and several laboratory-made adaptations have been described [23].

The two main staining reactions are those concerned with NAD(P) linked dehydrogenases and those with the α-naphthol- or β-naphthol-linked specific substrates for hydrolases (e.g. alkaline phosphatases).

Dehydrogenase-linked reactions

The electrons of reduced NAD(P) are channelled through phenazine methosulphate to a tetrazolium salt that is reduced to a highly-coloured, insoluble formazan. Phenazine methosulphate is very light-sensitive so that the reaction should occur strictly in the dark. Meldola blue has been proposed as a more stable substitute and has been tested on over 30 enzymes [24]. Aldehyde dehydrogenase has also been revealed in this way [25]. In our hands it was found more stable, but also somewhat less sensitive, although this depends on the enzyme under study. We found no real advantages for LDH isoenzymes. The most used tetrazolium salt is MTT (methyl thiazolyl tetrazolium). Next comes NBT (nitro blue tetrazolium), and we still also use INT (iodonitrotetrazolium) for LDH detection.

A problem is that the formazan undergoes condensation to a diformazan; the monoformazan tends to a red shade, the diformazan to a deep purple. This transition does not always occur at the same speed for all isoenzyme fractions and tends to lag markedly with massive formazan production. This presents difficulties in spectrophotometric quantitation of fractions. A remarkable improvement can be obtained by adding Triton X-100 at 0.05% to the reaction medium [26].

Whereas the enzymatic reduction of NAD(P) is generally combined with the transfer of the electrons to formazan, *Burlina* [27] has separated these two reactions in the detection of CK isoenzymes on cellulose acetate. The cellulose acetate strip is layered on an agar-gel film containing the conventional initial substrates. Then the strip is removed and dipped for 1 to 2 min in a solution containing phenazine methosulphate and a tetrazolium salt.

No coupling agent is necessary for some electron transfer reactions to tetrazolium salts; e.g with monoamine oxidase acting on tryptamine.

Hydrolase-linked reactions

The second most important group of staining reactions concerns hydrolytic activity (esterases, peptidases). Here an α- or β-naphthol is coupled by an ester, amide or peptide link to a suitable group (e.g. α-naphthyl phosphate). The α- or β-naphthol freed by enzymatic activity couples to a diazonium salt, resulting in a highly coloured (generally red) azo dye. Direct coupling is generally preferable to post-coupling. Although diazonium compounds are available commercially, these products are stable only in the dry state. Double salts, e.g. with zinc chloride, are preferred. Diazonium compounds are particularly unstable in alkaline media. α-Naphthol is more critical in this respect than β-naphthol. A pH-value of 8.2 may be regarded as the upper limit for rapid coupling. On the other hand, α-compounds are generally better substrates for the enzyme reaction, since they are hydrolyzed more rapidly. Bis-diazonium compounds tend to couple only at one end in acidic media, whereas both ends react in alkaline media. A very satisfactory monodiazonium compound is Fast Garnett GBS salt (acid sulphate) produced by *ICI*. It is very stable, dissolves easily in water, and couples rapidly. It gives a brown colour with α-naphthol and a red colour with β-naphthol.

Alkaline phosphatase also cleaves p-toluidinium-5-bromo-4-chloro-3-indoxyl phosphate; indoxyl acetate has similarly been used as a substrate for esterases. The formation of the dye occurs only in the presence of oxygen, and is activated by Cu^{2+}.

Hydrolases can also be detected by fluorigenic stains using 4-methylumbelliferone combined in a specific bond type. However, there is some discrepancy between the enzymatic optimum (generally slightly acid) and the alkaline pH which is optimal for fluorescence yield. Therefore, a two-step procedure is often necessary with a short final stage because the umbelliferone derivatives tend to hydrolyze spontaneously at alkaline pH. Umbelliferone is an example of a highly diffusible stain.

Oxidase-linked reactions

Oxidases and peroxidases have a place of their own. Oxidases reduce molecular oxygen; peroxidases reduce peroxides. Cytochrome oxidase is an example of the former, converting cytochrome c to the reduced state. This can be regenerated by oxidation with dimethyl-4-phenylenediamine; the oxidized 4-phenylenediamine reacts with α-naphthol to form the blue dye, indophenol. Peroxidases are readily detected because they can oxidize many substances, but dianisidine and benzidine are now superseded by the less carcinogenic o-toluidine, tetramethylbenzidine and 3-amino-9-ethyl-carbazole. Fluorigenic substrates for peroxidase have been described and could be of value for isoenzyme detection [28].

A very large number of staining methods are specific for a single enzyme type (e.g. acid phosphatase from red cells, amylase). They are described where appropriate. Indirect ("linked") systems are also numerous and much ingenuity has been manifested in their development. Experimental skill is required because optimization is a multi-parameter affair, often needing expensive and less-stable reagents.

Note on "nothing dehydrogenase"

All new procedures for enzyme detection using an enzymatic staining reaction should be subjected to several blank tests. The zymogram should be developed without any substrate and also in various incomplete reagent combinations. Especially with the widely used tetrazolium system for dehydrogenase detection, coloured zones may appear with incomplete systems and the term "nothing dehydrogenase" (NoDH) has been suggested for those stains. This term refers to real enzymatic activity or to the development of spots by non-substrate reagents. Especially tetrazolium salts can be reduced by thiol groups of lipoproteins and of serum albumin [29], a point that tends seriously to interfere with LDH analysis in human serum. As for real enzymatic activity, alcohol dehydrogenase has found to be involved or other enzymes with firmly bound substrates.

Quantitative aspects and quality control

Remarkably few investigations have been carried out on this important aspect and the physicochemistry of development of a zymogram has been little studied. Generally one is satisfied if a linear range is found in the dose/response curve. However, if one is measuring the comparative percentages of fractions, these ratios are only meaningful if the determination is made at a moment when all fractions are still in the linear range; a highly improbable situation when weak fractions are measured together with strong ones (cf. serum LDH). In addition, the spectral properties should be identical for all zones being compared. Finally, the photometry itself should be effected under correct conditions on spots that do not always have the same shape; the stain should be evenly distributed over the measuring slit, which requires very small slits, and the sampling should be representative for all zones. These photometric conditions are only realized in more elaborate photometers, in which the whole of each spot is scanned with a very small slit and the value integrated [30].

Routine applications are usually limited to applying the sample as a narrow streak, followed by migration and photometric sampling of a representative section. Under these conditions good reproducibility can be obtained and the results employed for comparative purposes with reasonable reliability, but it is not to be expected that in concentration ranges so vastly different as in the serum LDH system, the "real" values can be obtained. Even more, the very essence of isoenzymes means that "real values" as determined from activities do not have "real meaning".

The fact that percentage values depend so much on the electrophoretic medium is an added complication in quantitative analysis (cf. CK isoenzymes in agarose and cellulose acetate). Furthermore, simply switching to another tetrazolium salt will result in systematic differences. Here the chromatographic procedures offer more promise.

Under these conditions, quality control is a formidable task. Efforts are being made, especially for LDH analysis [31, 32]. The variability in the detection of LDH-5 in serum is one of the major problems. Correct estimation of the LDH-1/LDH-2 ratio remains a challenge. The accuracy of isoamylase determinations, an apparently simple goal, has also been investigated [33, 34].

1.3.2.3 Differential Migration Techniques

Gel filtration

Theoretical aspects

This separation method relies on partition in a column or slab between two aqueous phases of identical composition, one freely flowing, the other entrapped by a gel maze in a beaded form. A molecule which, due to its size, cannot diffuse into the gel, follows the interstitial water flow and elutes with the void volume. All the water is available to a very small molecule which therefore elutes only when a volume equal to the total water content has passed through the column. Migrants of intermediate size elute in an intermediary position, according to their K_{av} value. A migrant will never become permanently trapped in the column if no other factors intervene. Separation occurs according to migrant size, and the molecular weight can be estimated if calibrating proteins are run as references. Generally K_{av} (preferred to $V_{elution}/V_{void}$) is plotted against $\log M_r$ giving a selectivity curve valid for that particular gel and column [10] with, however, some restrictions related to the shape of the migrant [35] or possible dissociation [36].

The main dispersing factor is the granularity of the gel beads, which produces a large eddy dispersion. If high resolution is aimed at, these beads should be as small as possible but the water flow then requires a higher pressure head. They should also be strictly spherical and of narrow size distribution. It is here that HPLC brings marked improvement.

The procedure can be applied in a thin layer arrangement with downwards perfusion. The fractions are not eluted but the experiment is stopped when a reference substance that follows the void volume has reached the end of the track. Elution volume is replaced by migration distance when estimating M_r.

Practical aspects

Resolution increases with the square root of column length but remains poor, due partly to the absence of zone sharpening procedures. The procedure is therefore primarily intended for group separations (e.g. desalting of a protein solution, isolation of polymers or isolation of enzyme/immunoglobulin complexes) and for estimation of molecular weight.

A large variety of beaded gels is available: cross-linked dextran (Sephadex), polyacrylamide gel (Sephacryl, Bio-Gel P), agarose (Sepharose, Bio-Gel A) and cross-linked agarose (Sepharose CL). Together they cover a migrant M_r range of 500 to well over 10^6. They are all of great chemical and physical stability and withstand SDS, urea, 8 mol/l, or guanidine hydrochloride, 6 mol/l. The rigidity of the beads is important because it limits the hydrostatic pressure that can be applied in order of inducing a satisfactory water flow. The most rigid are the polyacrylamide gels and the cross-linked agarose gels. The finest beads (diameter ~100 μm) give the sharpest separations.

In all of these gels the void volume is approximately 1/3 of the column volume. It is determined by noting the elution volume of some excluded substance such as Blue Dextran 2000.

Sample volume should be nearly 1% of column volume, but for simple separations this can be made much larger. The viscosity of the sample is very important and should not be greater than twice that of the column buffer, a point that is especially important when applying concentrated protein solution or using a density gradient in layering the sample on top of the column.

Upwards elution can be applied and recycling is easily performed. Increasing the hydrostatic pressure initially increases the flow and then decreases it, as can be seen from the commercially available graphs of flow rate versus pressure.

Some residual charged groups (carboxyl groups) may be present on the gel matrix and therefore a minimal ionic strength (> 0.02) is necessary to reduce ion/ion interactions.

In thin layer experiments, sample volumes as used in similar electrophoretic techniques are employed. Layer thickness is 0.5 to 1 mm. The finest bead grade in a very thick slurry is chosen. The development is descending, with the slanting plate inclined at 10° to 20° (front rate 3 cm/h). After the run, an imprint is made on filter paper (Whatman 3 MM) and developed according to conventional zymogram techniques.

Calibration is effected for M_r determinations with commercially available proteins (enzymes). In thin-layer experiments one should remember that R_f is practically the inverse of K_{av} and that therefore $1/R_f$ against $\log M_r$ should be plotted for comparable selectivity curves.

Applications

Perhaps the most important routine use is for detection of complexes between serum enzymes and immunoglobulins or lipoproteins [37] or for separating the koinozymic alkaline phosphatase [38]. α-Amylases, that would bind to Sephadex, can nevertheless be studied when an inhibitor is added to the sample [39].

Ion-exchange chromatography (including chromatofocusing)

Theoretical aspects

Charged groups that are carried, e.g., by proteins can displace ions of small size that are counter-ions for a charged, solid phase. The following equilibria exist:

for anion exchange, $p^- + Cl^-(B)^+ \rightleftharpoons Cl^- + p^-(B)^+$

for cation exchange, $p^+ + H^+(B)^- \rightleftharpoons H^+ + p^+(B)^-$

where p refers to a protein, (B) to an electrically charged binder: $(B)^+$ is an anion exchanger and $(B)^-$ a cation exchanger.

The equilibrium constant K for these systems is of the order of 10^{-2}. No other types of interaction should occur in pure ion exchange. The main experimental parameters are those of electrophoresis: temperature, pH (determining the charge on the binder) and μ, ionic strength (determining the thickness of the electric double layer). At low ionic strength the counter-ions are scarce, so the p binds firmly; at high ionic strength the binding is weakened and small counter-ions displace p (elution).

An ion exchanger is characterized by its capacity (corresponding to the charged groups in equivalents per ml), bead porosity and, finally, by the nature and strength of the charged group(s).

In protein analysis only a few types of charged groups are used: for anion exchange, aminoethyl or diethylaminoethyl, for cation exchange, mainly carboxymethyl. All are of the weak type, i.e. they are charged only in a relatively narrow pH range. A typical example is diethylaminoethyl that loses its charge above pH 9. If separation at higher pH is needed, the stronger quaternary aminoethyl group is required. Carboxymethyl is negatively charged only above pH 4.5, while sulphopropyl is satisfactory down to pH 2.

These groups are covalently attached (mostly by an ether link) to a beaded matrix of the type used in gel filtration, with such a pore structure that no size-selective interaction occurs and a maximum of surface is exposed. Cellulose has also been much used, especially the treated microcrystalline variety with high porosity which offers granule sizes smaller than those of agarose matrices and hence a slightly better resolution. Hydroxylapatite has a place of its own: rather than ion exchange, absorption comes into play. This sometimes gives separations that are otherwise not obtained.

Column ion exchange is a versatile procedure with the following advantages: zone sharpening tends to occur spontaneously during the initial fixation phase and also upon elution; dilution of the sample is small; flow is relatively fast and for some applications columns can be short; finally, it is a high-capacity technique.

Practical aspects

Many of the considerations which apply in gel chromatography apply also here, because accessibility and flow rates are mainly determined by the beaded matrix. Interactions of the gel-filtration type curtail resolution.

Experimental variables are pH, ionic strength and temperature. If the pI of the investigated substance is known, the starting pH should be chosen 1 pH unit above or below this value in order to obtain well-ionized migrants. In addition, ionic strength should be as high as possible (at least 0.1). Variation in ionic strength has a more marked effect on the structure of the column at low values and can give problems with the liquid flow. All solutions should be de-gassed before use and their temperature should be close to that of the experiment. NaCl is nearly always employed as the natural counter-ion. Often elution is obtained by progressively changing conditions (gradient profile, especially in ionic strength). Ampholytes have been suggested as eluting agents (ampholyte displacement chromatography) and extremely good separations have been obtained of the isoenzymes of neuraminidase [40]. The mechanism is not quite clear but is probably related to chromatofocusing.

Chromatofocusing

Sluyterman [41] described a highly resolving modification of ion exchange. He used the stationary phase as a buffer that is titrated by the addition of the eluting buffer. In this way a pH gradient develops. This gradient is initially narrow and situated at the top of the column but, as elution proceeds, broadens while advancing over the column. After the passage of a buffer volume equal to the void volume, the front of the gradient will reach the exit port; the tail, however, will still be far behind. If the pH of the effluent is then monitored, it will be observed to shift progressively towards the final eluting pH. The profile should be linear for optimal utilization of the column. Only after passing several void volumes will the tail reach the exit port.

The pH gradient, when moving over the column, improves protein separation. Sharpening factors are shallowness of the pH gradient and low ionic strength. Elution velocity, however, does not intervene directly. The ion exchange capacity should be as high as possible and equilibration fast [42].

The separation principle is related to that used in isoelectric focusing, but there is no heat generation and hence no cooling problems. The procedure therefore can easily be scaled up for preparative applications. Simple buffers can be used, but better separations are obtained if the ion exchanger and the eluting buffer have been especially adjusted for generation of a flat and smooth pH gradient.

Applications

Ion exchange has played an important role from the very beginning of isoenzyme studies. It is still of value but is more promising in the HPLC version.

It has found limited application in routine work because it is relatively time consuming, even when use is made of micro-columns and the number of eluted fractions is reduced to a minimum. Nevertheless, the procedure of *Mercer* for the separation of CK isoenzymes on mini-columns [43] has met some success and has been combined with the separation of LDH isoenzymes [44]. LDH chromatography was improved for better separation of LDH-1 from LDH-2 in order to establish the LDH-1/LDH-2 ratio [45] and to determine simultaneously CK-MB [46]. A discussion on different procedures for the clinically important MB determination is enlightening [47]. Improvements of *Mercer's* method have also been described [48]. Ion-exchange procedures for routine separation of alkaline phosphatase isoenzymes exist [49]. Mitochondrial GOT has been separated from its cytoplasmatic counterpart in high purity and good yield [50].

High performance liquid chromatography (HPLC)

Theoretical aspects

This procedure does not employ any new separation principle but depends on instrumental refinements that increase separation speed and resolution by optimizing the spreading parameters. The main problem is to increase the fluid flow without increasing non-equilibrium spreading. To this end a rapid exchange is obtained by the use of very porous spherical beads of diameter of about 10 μm. In addition these beads should withstand high pressure gradients and their size distribution should be as narrow as possible. The development of very homogeneous silicon or glass beads with rigidly controlled pore sizes in a range of 6 – 400 nm was the first requirement. However, these cannot be employed for protein separation in this form, even for simple exchange such as gel chromatography, because their surface is hydrophobic and carries charged groups that block the entrance to the pore channels. Coating with a neutral hydrophilic layer, such as glycerylpropyl bonded as an organosilane phase to the beads, provides a simple and extremely efficient solution in gel chromatography. Very linear relationships are then found to link $\log M_r$ and K_{av}. The void volume of these columns is larger than usual (1 : 2). For anion exchange, charged groups are introduced by copolymerization with epoxy monomers at the surface and linking to the glycerylpropyl silane-bonded phase carriers.

More recently rigid spherical polymeric methacrylate ester matrices have been developed [51] that are sufficiently hydrophilic to be used directly in gel chromatography (Spheron, Skodex). They can also be substituted with the desired ionic species.

An interesting variant consists in the physical coating of the bead surface with a pellicle. First a layer of low-molecular weight polyethylene imine is absorbed onto the carrier beads; this film is then stabilized by cross-linked adjacent amino groups.

Practical aspects

In isoenzyme separation, involving large molecules, non-equilibrium spreading is still a limiting factor to an increase of fluid flow. Nevertheless remarkable improvements in resolution can be obtained, especially in gel chromatography where it was most lacking until now. The gain in separation time remains appreciable so that routine chromatographic separations can be considered. A drawback is the high cost of the columns and of the auxiliary equipment, but good supports can be re-utilized many times.

The detection system should be able to keep pace with the rapid separation. For enzyme activity detection, two basic procedures are available [52]. In the first, monitoring with a kinetic procedure is effected at the entrance and the exit of the reaction coil. The result is not simply the difference between the two curves since there is peak broadening, but this can be compensated for by an algorithm. In the second, a parallel system, direct comparison is possible between a non-reactive system (at low temperature) and an active system (at 37 °C). Monitoring occurs in UV or in fluorimetry of high dynamic range ($1:10^3$) [53].

Affinity chromatography

Theoretical aspects

This chromatographic procedure exploits the occurrence of binding reactions of high affinity, with the enzyme/substrate interaction as a typical example. A high degree of irreversibility is obtained by multiplication of the attachment points, favoured by the co-operative character of the fixation. This also includes a high specificity, i.e. adequate attachment points are to be present in well-determined positions. The ligands are bonded to an insoluble matrix and can be employed in batch procedures. However, if the material is used in a column, an important factor, described by *Cuatrecasas* [54] as "The concept of progressively perpetuating effectiveness", is introduced. In the migrating type of chromatography, the front of the zone being bound is retarded by interaction with the stationary phase and so increases its concentration, giving a shift towards fixation on the matrix. The decay of the complex, on the other hand, is independent of concentration. Therefore a column procedure is much more effective than a batch process. The affinity interaction can also be coupled to electrophoresis (e.g. using lectins) [55].

The equilibrium constant in this type of chromatography should lie between 10^{-4} and 10^{-8}. If higher than this, the binding is ineffective; if lower, it becomes difficult to recover the bound material without fundamentally changing its structure.

Affinity chromatography is not intended to be a polyvalent separation procedure, but is rather directed to a single substance, or group of substances that can be very effectively isolated in one step, even from very dilute and crude extracts. The problem is to find and attach the proper ligand to a support. Initially use was made of naturally occurring ligands (antibodies, enzyme inhibitors, coenzymes, protein A, lectins). Artificial ligands have also been employed such as Cibacron Blue F3G-A and AMP analogues.

The ligands are covalently bound to an insoluble matrix that should exhibit high ligand density, and be easily accessible to the substance to be bound. It should also allow a good water flow. Practically all matrices used in gel chromatography can be used. Agarose beads (cross-linked or not) are often favoured. They are easily derivatized, are very stable and practically inert. A special problem arises when small ligands are used to capture large binding molecules. This ligand should then be "presented" to the binding molecule by interposition of a spacer arm, some 6 carbon atoms long, between ligand and matrix. This results in a marked increase of binding efficiency. If the spacer arm is too long, hydrophobic interactions decrease efficiency.

Recovery of the bound material remains a problem. When attachment is strong it is often sufficient to scoop out the top of the column. Suspending the matrix in a small volume of buffer shifts the equilibrium towards dissociation. One can also elute by imposing a drastic change in pH, or any other specific procedure such as the application of a competing binder. Acid elution is a classical procedure for dissociating antigen/antibody complexes. The point of concern is retention of biological activity. A gentle procedure is electrophoretic desoption with a re-usable electrophoretic cell [56].

Practical aspects

Affinity chromatography is a simple and extremely rapid procedure. Since in most instance no real migration of the fixed zone is needed for separation, the column can be short. If a suitable dissociating system is available the column can be re-used many times; however, crude extracts tend to block the water flow and therefore the sample should be extensively centrifuged. Lipoproteins can be a problem.

A great variety of suitable immobilizing agents are commercially available, ranging from beaded matrices already activated with CNBr and thus ready for coupling, to ready-made specific binding matrices, such as Blue-Sepharose Cl-6B which binds NAD- and NADP-dependent enzymes. In between come matrices already provided with spacer arms ready to bind with the chosen ligand.

Applications

The value of affinity chromatography for the purification of an entire class of enzymes has clearly been stressed in 1974 by *Kaplan* [57], who also showed how isoenzymes can be differentiated by competitive elution with cofactors. Equilibrium constants can be determined [58].

Cibacron Blue F3G-A, a triazine dye closely related to Blue Dextran, has been much used as a ligand for nucleotide-dependent enzymes [59 – 61]. Arylsulphatases A and B can also be separated in this way [62]. Oxamate and AMP-analogues have been used especially in relation to LDH isoenzymes [63 – 65].

Further typical applications are found in the detection of enzyme-immunoglobulin complexes with protein-A as ligand in a batch procedure [66], for the study of plasminogen and plasmin isoenzymes using lysine adducts [67], for unravelling with concanavalin the multiplicity of glycoprotein enzymes such as adenosine deaminase [68], γ-glutamyltransferase [69], alkaline phosphatase [70]. Pyridoxal-dependent enzymes have also been separated [71]. Alkaline phosphatases have been studied using phosphonic acids as ligands [72, 73].

Electrophoresis

Electrophoresis is by far the most useful separation procedure for the study of isoenzymes. Several recent monographs have covered the field of protein electrophoresis [74 – 78], but none has recently been dedicated to isoenzymes in particular. The following remain of interest: 79 – 82.

The value of electrophoresis resides in the fact that it is a gentle, flexible and rapid method of separation that does not require expensive instrumentation. The resolving power can easily be increased by adding pore-size interactions or a pH gradient. The 2-dimensional combination of two different systems increases this resolving power even further. A much used combination of this type has been called ISODALT separation (isoelectric focusing + Dalton units-related separation). It consists of IEF followed by SDS-pore gradient electrophoresis. Finally, extremely sensitive detection procedures (silver stain of polypeptides, zymogram technique) are available.

Specific problems

It is generally not clearly realized that electrophoresis primarily separates only by charge differences and not on a base of migrant size. Intuitively one expects that frictional forces should differentiate between large and small molecules, according to *Stokes'* law. However, this is compensated for by the driving force being proportional to the electric potential or charge density of the migrant. In this way the influence of the *Stokes'* radius is eliminated, as is beautifully illustrated by the behaviour of proteins in SDS-polyacrylamide gel electrophoresis.

Another feature that sometimes deceives the newcomer is electro-osmosis. This term designates water flow, generally directed towards the cathode, that occurs when an electric field is applied to an electrolyte enclosed by walls. It is reversed electrophoresis, provoked by the electric charge of the container (carrier) which, of course, is prevented from moving, so that the water phase flows in the opposite direction. The intensity of this flow is found by relating the distance covered by an uncharged molecule (e.g. Blue Dextran 2000) relative to the application slot to that of human serum albumin used as a reference.

Electro-osmosis is important on cellulose acetate, in starch gel and in agarose gel. For ordinary separations this is immaterial and can even present advantages if in this way some fractions can be removed from the application slot or from interfering migrants, as in LDH electrophoresis. For IEF it should be as low as possible. Modified carriers are available with practically zero electro-osmosis.

A specific feature of electrophoresis is the production of heat in the course of the separation. The *Joule* heat (expressed in Joule $\times$ s^{-1} or watt per cm^3) is equal to the product of the conductivity coefficient and the square of the field strength. Heat production leads to an increase of temperature which in turn accelerates migration. The zones are distorted if the temperature becomes unevenly distributed and special cooling procedures are necessary. This is one of the major difficulties in high voltage electrophoresis of the analytical type and also in preparative applications when the dimensions are scaled up, rendering cooling less efficient.

Electrophoresis is practically always of the zone type, which means that migration of independent zones is obtained, generally in a stabilizing medium. In contrast with chromatography the flat bed ("slab method") is more popular than the tube (or column) set-up, especially for isoenzyme studies because the access to slabs is more direct, cooling is more efficient and samples are more easily compared side-by-side on a single slab. For two-dimensional separations, however, small tubes are often used in the first (focusing) separation since they are easily manipulated for application in the second run. Gel tubes of the *Ornstein-Davis* type [9] are also prefered for full exploitation of the advantages of discontinuous buffer systems (disc-electrophoresis and isotachophoresis) since here the sample zones completely fill the cross-section, a point that favours stacking.

The relative merits of horizontal and vertical arrangements for slab electrophoresis are controversial. The vertical (descending) type of run has advantages in sample application, especially if this is made in fluid form, since density gradients tend to develop that favour penetration of the sample into the slab. On the other hand, the adherence of the medium to the walls is more critical, especially if local volume changes occur that tend to detach the bed. The horizontal set-up is always preferred for IEF separations, since here the medium is markedly heterogeneous due to the pH gradient.

Sample application generally presents no difficulty, but for the highly resolving procedures all particulate material should carefully be removed by centrifugation and painstaking care is needed in preventing proteolytic cleavage of the sample proteins. In certain applications the sample is equilibrated with the electrophoresis buffer in order to adjust the salt concentration.

Electrophoretic media

Only the following media have great practical importance: cellulose acetate (membrane), starch gel, agarose gel and polyacrylamide gel. Agarose and polyacrylamide gels are also used in granular form, together with some other beaded media (Sephadex, Pevikon).

Cellulose acetate membranes

These are mostly used in routine clinical chemistry laboratories. With nearly the same arrangement as for electrophoresis of serum proteins, LDH, CK and AP can be separated and revealed by any of the zymogram techniques.

Fixation, staining and drying are very rapid and a relatively high staining sensitivity with nigrosine is obtained. Typical of this medium is very low protein adsorption. Clarification is perfect but the loading capacity is low and fluid stabilization precarious. Electro-osmosis flow is marked so that for IEF the film must be pre-treated (methylation of the carboxyl groups [83]). However, it has the advantage that only small amounts of ampholytes are needed. Resolution in gelified media is nevertheless better, so that this medium commends itself mainly as a medium-quality all-round electrophoretic support.

Starch gel

This was the first medium combining molecular-sieve effects with electrophoretic separation. It is still preferred by several groups for routine zymogram development, especially in genetic studies [22]. Resolving power is outstanding, but it is not easy to adjust the pore size to definite values. In fact, pore constitution is very different from that of polyacrylamide gel and comes closer to that of agarose gel (thick fibre bundles). However, the dry-matter content being larger, pores are smaller. The notion of limiting pore size is ill-defined in starch gel [84]. Electro-osmosis is rather high, a point that renders the medium unsuitable for IEF, but little research has been done on this point. The medium is extremely gentle to enzymes, diffusion is severely restricted and the fact that one needs to slice the gel, together with the opacity of the slab, exposes perfectly clean migration surfaces.

Gelification is obtained by boiling the slightly hydrolyzed starch (a commercial product) with the buffer (at a concentration of ca. 12 g/100 ml) for a few seconds with swirling, followed by de-gassing and pouring into a mould. The slab technique is used exclusively. Starch gel is very well suited as the second-dimension medium with agarose gel or cellulose acetate as the first, especially if in the second run a buffer discontinuity is applied.

Agarose gel

Agar is a complex carbohydrate substance that can be purified by washing (removing soluble impurities) and physicochemical treatment (e.g. solubility in petroleum ether). The term agarose does not refer to a single, pure gelling substance found in agar, but to a collective of polysaccharides, all of similar structure. The class agarose is best defined as those galactans having the basic repeating unit: 1 – 3 linked β-D-galactopyranose and 1 – 4 linked 3,6 anhydro α-L galactopyranose [85].

An important parameter when chosing an agarose is its electro-osmosis flow. Generally, three grades of decreasing osmotic flow are available (range: $0.1 < M_r < 0.3$). Not only sulphate and pyruvate contents but also other factors intervene. Mixtures can be made for adapting the separation to good positioning of the application slot relative to the interesting fractions. In addition, a very high quality product is generally catalogued, characterized by an extremely low electro-osmosis flow and good gel strength (gelification even at 0.2%). Electro-osmosis can still further be reduced by adding 10% sorbitol (or 5% glycerol or methylcellulose) [86].

Special low-gelling agaroses are available commercially. They solidify at ca. 25 °C and melt again at 50 – 60 °C. Pore interaction with much smaller molecules than is the rule with agarose gels has been observed [87]. Low-gelling agarose finds a place in some preparative applications, for the study of heat-sensitive material and in sample application.

Agarose gel has the advantage of allowing the free migration of most proteins and protein complexes. Therefore, the zero-m_r variety is ideal for IEF [88]. If use is made of the high-gelling variety very large particles such as viruses can be studied. It is also the basic medium for all kinds of immunological combinations with electrophoresis. It can be poured in a great variety of shapes, but the tube method is less satisfactory because the gel is difficult to remove. On the other hand, it is suitable for very thin-layer modifications (thickness < 1 mm). In the zymogram technique it offers a good compromise between stabilization and rapidity of diffusion of the reagents. Staining and de-staining is very rapid and the layer can be dried into a transparent film. The gel has good optical properties. For improving adherence to the support, use can be made of the commercial support called Gelbound film.

Agarose is generally used at a concentration of 1% and accomodates nearly all buffers. It can be combined in composite gels with polyacrylamide, improving the rigidity of low concentration polyacrylamide gels.

Polyacrylamide gel

This is the most used gel medium, except in routine applications, although the disadvantages are well-known [88]. Its preparation entails the manipulation of rather toxic monomers, but the gel itself is harmless. There still remains some irreproducibility in the polymerization and manipulations are not so simple as with agarose gel.

On the other hand the judicious combination of electrophoretic and size-selective migration offers additional separation power. The medium has good optical properties down to 300 nm, and even to 250 nm. It has good elasticity, but is sticky and becomes easily stretched, distorting running dimensions. Adherence to glass can be increased by silane treatment. Its main assets are, however, absence of electro-osmosis and the possibility of chosing the pore size by a combination of monomer and dimer concentration, and even of grading the pore size at will in a gradient profile.

The gel is obtained by polymerizing the monomer acrylamide with some bisacrylamide in the presence of the catalysts ammonium persulphate or TEMED (tetramethylethylenediamine).

In a typical gel monomer concentration T is 7.5 g/100 ml buffer with C (ratio of cross-linker to total monomer) set at 5%. Generally C is kept constant with T chosen between 3 and 30 g/100 ml. Pore sizes then range from 3 nm to 0.5 nm and interactions occur with migrants of M_r ranging from 10^6 to 10^3. Paradoxically, when the proportion of cross-linker is increased above 5%, pore size increases: an increase of cross-linking causes bundling of the fibres. For pore sizes allowing the movement of migrants of $M_r > 10^6$, a 1 to 2 T gel ($C = 5\%$) is fortified with 0.5% agarose.

Polymerization is never complete. Catalysts also remain and can interfere with the separation of substances sensitive to oxidation, but pre-electrophoresis easily eliminates persulphate. The medium accomodates practically all buffer systems, all types of reducing agents and all detergents and combinations of these. An often-used chaotropic agent is sodium dodecyl sulphate (SDS) [89]. As a rule enzymatic activity is then lost, but occasionally it can be restored [90].

Polyacrylamide gel is perfectly suited for filling small glass tubes (in fact the gel polymerizes best when in contact with glass) and the small cylinders can easily be removed by teasing with a needle after electrophoresis. Slabs are best used *in situ*.

Sample application for vertical slab electrophoresis is made generally directly on top of the gel, beneath the electrode buffer, with the sample dissolved in 10% sucrose to increase density. In the horizontal system, slots are provided that are filled in the conventional way and sealed with warm agarose solution.

Staining for protein is not so rapid. The most used stain of high sensitivity is Coomassie Blue R 250. It has a sensitivity of the order of 0.1 µg protein. Bands generally intensify in the destaining solution in the course of a few days. Other protein stains have been described [91] but none attains the sensitivity of a silver stain [92, 93]. In zymogram development it is best to stabilize the reagents since colour development is rather slow. Many applications have been described and have been reviewed several times (e.g. [94]).

Special electrophoretic procedures

Discontinuous buffer systems

Buffer discontinuities are used for compressing (concentrating) large sample volumes, to reclaim zones that have been broadened by long runs, to stack pre-separated fractions and to sweep fraction out of an application slot. This is an application of the *Kohlrausch* theory on moving boundaries (cf. p. 57).

A large number of discontinuous buffer systems based on these principles for anodic or cathodic protein migration have been described [8]. The most complex application is that of the *Ornstein* and *Davis* type of electrophoresis in polyacrylamide gel (disc electrophoresis) where several types of discontinuities in pore size, pH and buffer compositions are simultaneously employed.

Isotachophoresis

The *Kohlrausch* theory is applied here in another way. If finite amounts of proteins of mobilities intermediate between those of the rapid and slow ions are introduced, an additional phenomenon occurs: they "stack" in an order of decreasing mobility and adjust their concentration so as to keep pace by an adjustment of field strength. A new "system" develops where, between a rapid and slow electrolyte, the protein ions are stacked in a succession of zones whose width is proportional to the initial amount. Generally these widths are very small and no usable separation is obtained until they are de-stacked by changing the conditions; e.g. overrun by the previously slow ion that now becomes rapid due to a change in pH, as in the *Ornstein-Davis* disc electrophoresis, or an adequate amount of an electrolyte (e.g. an amino acid [95]) of intermediate mobility is intentionally added to separate the two zones of interest. Isotachophoresis has yet found little application in enzyme separations except for preparative purposes [96].

Pore gradient electrophoresis

The resolution of an electrophoretic run can be improved if pore-selective interactions are added. The effect is enhanced when the pore sizes are distributed over a gradient. Proteins are then migrating into zones of decreasing pore sizes. Polyacrylamide is unrivalled here. By carefully chosing the range of monomer concentration (T = 3 to 30 g/100 ml solution) at a constant proportion of cross-linker (C = 5%) the resolution of complex mixture can be optimized [97]. Two stages occur in the separation. In a first, the fractions continuously shift relatively to each other in accordance with differences in M_r as well as in electric charge. In the second step, the final positions (limiting pore sizes) have been acquired and the pattern is stabilized. In between, for each separation couple a specific, optimal separation time exists. The final stage is used for M_r determination (cf. p. 57).

Isoelectric Focusing (IEF) (ISOF)

Theory

This is actually the most important electrophoretic variant for isoenzyme studies. IEF has been mainly developed by *Kolin, Rilbe* (alias *Svensson*) and *Vesterberg* (cf. [98]). A protein is an ampholytic substance with electric charge depending on pH. At a given pH (its isoelectric point, IP) it has zero time-average electric charge. At higher pH it is positively, at lower pH negatively charged. Therefore, if one succeeds in creating a stable, continuous pH gradient over the separation field, a protein – wherever applied – acquires a charge according to the local pH and migrates towards the pI point where it comes to a standstill. In a shallow pH gradient (e.g. 1 pH unit over 10 cm of migration track) many fractions can find a place, if the zones can be kept narrow. This occurs automatically because, if for some reason the migrant tends to drift from the point of pI, it acquires a charge that pushes it back to the starting point (focusing effect). The focused zone has still a finite width because some thermal diffusion always remains. If it is true that IEF ends up with a steady state where all fractions stay at their pI, there still remains some global displacement of the superimposed pH gradient (cathodic drift of the gradient). Its nature is not entirely explained although mainly electro-osmosis seems to intervene [99].

Zone width is independent of time. In addition the apparent diffusion coefficient D is reduced by a factor composed of two parameters; the first is the change in electric charge with pH, primarily a molecular constant; the second is the change of pH with distance, an experimental parameter.

Optimal conditions for IEF are a high field strength and a shallow pH gradient. In addition the protein should exhibit a marked change in electric charge around its pI. Under these optimal conditions, a Δ pI of 0.01 can yield well-resolved zones, so that 100 fractions will fit into the 10 cm track described above, a performance that makes IEF unequalled in resolution.

Practical aspects

If the separation obtained by IEF is compared with other electrophoretic separations, many more fractions are revealed and, as a rule, an homogeneous fraction is resolved into a series of closely spaced band (e.g. amylase). It is now clear that, generally, these are not artefacts, but constitute charge isomers of a given protein species [100]. They are not real isoenzymes, but enzyme variants that arise by secondary modification of the polypeptide chain (isomerization, amidation, addition of carbohydrate groups etc.). The possibility of artefact is still to be considered, especially because, at the low ionic strength proper to IEF, charge interaction with the ampholytes can occur [101]. This can be verified by applying the sample alternatively at the anodal and cathodal side of the pI point and checking for correspondence; or one of the fraction can be eluted and re-applied, looking for the appearance of heterogeneity.

IEF is well-suited for the development of a zymogram because at the end the medium contains only small amounts of salts (ampholytes) and this favours the overriding of the local pH by the pH of the substrate solution used for the development of the zymogram.

Agarose has received much attention since pore size effects are to be avoided for rapid and complete focusing. The agarose should be of the type with very low electroosmotic flow. Thin films (less than 1 mm) are economical, give good resolution due to efficient cooling, and allow rapid fixation or zymogram development. IEF is sometimes effected in granulated beds (Sephadex G-75 or Pevikon) or on treated cellulose acetate films, the former mainly for small-scale preparative purposes. Alternatively, composite acrylamide/agarose gels can be used [102].

IEF can also be applied to dissociated or solubilized proteins in urea, 8 mol/l, and/or reduced with dithiothreitol. If detergents are necessary for solubilization they should be of the non-ionic or zwitter-ion type [103].

Zone distortion can be a problem and is mainly due to an irregular distribution of conductivity (conductivity gaps creating hot spots) or to fluid creeping between the gel and the support [104]. This can be minimized for agarose gel using Gelbond film as a carrier, and for polyacrylamide gel using silane coated glass that is covalently bonded in the course of the gel polymerization.

Sample application is not critical, but should not introduce too much foreign salt. The sample is incorporated into the gel when it is prepared, or applied, after a pre-focusing run, in conventional ways.

Ampholytes interfere with most protein stains. Therefore their removal is advocated, using a rapid fixation step, followed by a conventional Coomassie Brilliant Blue R 250 stain [105]. Fast Green does not interact with ampholytes and can be used directly [106]. For agarose gels special staining conditions prevail [88].

A note on ampholytes

The development of IEF has been made possible by inventing simple means for creating the desired pH gradient. A variety of procedures are available [107], ranging from the mixing of two buffers to the use of special ampholytes. These are low molecular weight amphoteric substances of marked multiplicity (in amino and carboxylic groups). They focus like proteins, establish the pH at their pI and ensure some residual conductivity (ionic strength of a focused ampholyte is of the order of 0.01). These ampholytes are available from LKB under the brand name Ampholine. Later, sulphate and phosphate groups have been introduced in addition (Biolytes, Pharmalytes, Servalytes). Ampholytes should have a low M_r and not interfere with protein separation and characterization. They are available in a large choice of pH-gradient ranges. They replace the conventional buffer in a final concentration of ca. 1 g/10 ml. Sucrose, sorbitol and/or glycerol are often added to a concentration of 10% in order to increase viscosity and to reduce the global gradient drift. The electrode vessels are protected against invasion by ampholytes by a strong base (e.g. NaOH, 0.5 mol/l) for the cathode, and a strong acid (e.g. H_3PO_4, 0.5 mol/l), for the anode.

The measurement of pI

An important characterization parameter in IEF is the specification of the pI. To this end the gel can be sliced, chopped in distilled water and the pH measured. pH can also directly be measured at the surface of the matrix, but precision is rather poor as far as localization is concerned: although special electrodes are available, it is difficult to confine the measurement to a small area. The result is also temperature-dependent (pI increases with temperature) and, if neglected, this introduces errors of the order of 0.01 pH. Therefore pI should be measured at the temperature of the separation because correction factors are very uncertain. The pH gradient is always positive from anode to cathode (law of pH monotony). Calibrating substances (carbamylation trains) have been proposed as internal markers for the pH axis [108].

Two important variants of IEF

Immobilized pH gradient: incorporation of the ampholyte chemically in a matrix of polyacrylamide of large pore has been suggested [109, 110]. The pH is established by derivatives of acrylamide (Immobiline®) that can cover a pH range of 3.8 to 9.5. This immobilized pH gradient is insensitive to salt effects and has a linearity that depends exclusively on the quality of the gradient established in pouring the monomer solution. It has at least five times the loading capacity of conventional IEF and, of course, there is no gradient drift. The gradient can be made very shallow with only 0.01 pH difference over a 1 cm track, so that differences of 0.001 pH unit can be revealed. In preparative work it is interesting that soluble ampholytes do not contaminate the preparation.

Combined electrofocusing and transverse electrophoresis

Because the focused ampholytes are at their pI, they carry zero net charge. Therefore if such a focused system is submitted to a transverse electric field, little displacement will occur within a given short time. However, a sample applied over the entire width can be separated electrophoretically during that interval. In fact each application point corresponds to a run at a given pH. The pattern thus obtained reveals the titration curve of the different proteins: zero mobility at the zone of their pI, anodal and cathodal migration above and below that point [111].

The procedure is valuable for studying enzyme/ligand interactions [112] and also for chosing optimal electrophoretic conditions [113].

2D-electrophoresis

The two-dimensional combination of two different electrophoretic separation systems increases the resolving power markedly. The number of resolvable fractions can be set equal to the product of each step; thus a maximum of ca. 10.000 fractions can be separated in conventional set-ups. The sequential use of IEF and of pore gradient separation is especially meaningful. In the first dimension separation rests on charge, in the second on size, so that the overall separation reveals charge as well as size isomerism. The ISODALT combination (SDS electrophoresis in the second dimension) proposed by *O'Farrell* [114], has met great popularity, but is not directly applicable to enzyme studies since enzymatic activity is lost. Here more conventional size selective procedures have been used [115, 116].

The description of 2D zymograms poses some problems [4] and also the comparison of complex patterns can be difficult [3, 117]. A simple procedure for making a photographic record can be useful [118]. Note that the silver stain methods [92, 93] or the radioactive label of polypeptide chains can come quite close to the sensitivity obtained by using enzymatic activity. Zymograms are generally developed under scaled-down dimensions and excellent micro-slab procedures are available [119].

References

[1] *E. Heftmann,* Chromatography. A Laboratory Handbook of Chromatographic and Electrophoretic Methods, third edition, Van Nostrand Reinhold, New York 1975.

[2] *B. L. Karger, L. R. Snyder, C. Horvath,* An Introduction to Separation Science, Wiley-Interscience, New York 1973.

[3] *M. Capel, B. Redman, D. P. Bourque,* Quantitative Comparative Analysis of Complex Two-dimensional Electropherograms, Analyt. Biochem. *97*, 210 – 228 (1979).

[4] *P. F. Lemkin, L. E. Lipkin,* GELLAB: A Computer System for 2D Gel Electrophoresis Analysis I. Segmentation of Spots and System Preliminaries, Computers and Biomedical Research *14*, 272 – 297 (1981).

[5] *J. C. Giddings,* Dynamics of Chromatography, part 1, Principles and Theory, in: *J. C. Giddings, R. A. Keller* (eds.), Chromatographic Science, Vol. 1, Dakka, New York 1965.

[6] *R. J. Wieme,* Theory of Electrophoresis in Chromatography, in: *E. Heftmann* (ed.), Chromatography, Van Nostrand Reinhold, New York 1975, pp. 228 – 281.

[7] *M. D. Poulik,* Starch Gel Electrophoresis in Discontinuous System of Buffers, Nature *180*, 1477 – 1479 (1957).

[8] *T. M. Jovin,* Multiphasic Zone Electrophoresis, IV. Design and Analysis of Discontinuous Buffer Systems with a Digital Computer, Ann. N.Y. Acad. Sci. *209*, 477 – 496 (1973).

[9] *L. Ornstein,* Disc Electrophoresis – I. Background and Theory, Ann. N.Y. Acad. Sci. *121*, 321 – 349 (1964).

[10] *G. K. Ackers,* Analytical Gel Chromatography of Proteins, Adv. Prot. Chem. *24*, 343 – 446 (1970).

[11] *K. A. Ferguson,* Starch Gel Electrophoresis – Application to the Classification of Pituitary Proteins and Polypeptides, Metabolism *13*, 985 – 1002 (1964).

[12] *P. G. Righetti,* On the Pore Size and Shape of Hydrophilic Gels for Electrophoretic Analysis, in: *R. C. Allen, P. Arnaud* (eds.), Electrophoresis '81, W. de Gruyter, Berlin – New York 1981, pp. 3 – 16.

[13] *G. G. Slater,* Stable Pattern Formation and Determination of Molecular Size by Pore-Limit Electrophoresis, Analyt. Chem. *41*, 1039 – 1041 (1969).

[14] *C. Manwell,* A Simplified Electrophoretic System for Determining Molecular Weights of Proteins, Bioch. J. *165*, 487 – 495 (1977).

[15] *J. Margolis, K. G. Kenrick,* Polyacrylamide Gel Electrophoresis in a Continuous Molecular Sieve Gradient, Anal. Biochem. *25*, 347 – 362 (1968).

[16] *L. M. Hjelmeland, A. Chrambach,* Electrophoresis and Electrofocusing in Detergent Containing Media: A Discussion of Basic Concepts, Electrophoresis *2*, 1 – 11 (1981).

[17] *R. L. Hunter, C. L. Markert,* Histochemical Demonstration of Enzymes Separated by Zone Electrophoresis in Starch Gels, Science *125*, 1294 – 1295 (1957).

[18] *M. Gemeiner, I. Miller, H. Czikeli,* Ultrathin-layer Isoelectric Focusing of Enzymes in Liver Samples of Wagtails, Electrophoresis *3*, 146 – 151 (1982).

[19] *W. Pretsch, D. J. Charles, K. R. Narayanan,* The Agar Contact Replice Technique after Isoelectric Focusing as a Screening Method for the Detection of Enzyme Variants, Electrophoresis *3*, 142 – 145 (1982).

[20] *R. H. Cruickshank, G. C. Wade,* Detection of Pectic Enzymes in Pectin-Acrylamide Gels, Anal. Biochem. *107*, 177 – 181 (1980).

[21] *T. L. Brown, M. G. Yet, F. Wold,* Substrate-Containing Gel Electrophoresis: Sensitive Detection of Amylolytic, Nucleolytic and Proteolytic Enzymes, Analyt. Biochem. *122*, 164 – 172 (1982).

[22] *H. Harris, D. A. Hopkinson,* Handbook of Enzyme Electrophoresis in Human Genetics, North-Holland Publishing Co., Amsterdam 1976 (with some yearly supplements).

[23] *D. K. Shortess,* A Low-Cost Polyacrylamide Gel Scanning Densitometer, Anal. Biochem. *100*, 15 – 19 (1979).

[24] *V. S. Turner, D. A. Hopkinson,* The Use of Meldolablue in Isozyme Stains after Electrophoresis, FEBS Letters *105*, 376 – 378 (1979).

[25] *S. Harada, D. P. Agarwal, H. W. Goedde,* Isoelectric Focusing in Cellulose Acetate Membrane and Flatbed Granulated Gel: Application to the Study of Human Aldehyde Dehydrogenase Isozymes, in: *B. J. Radola* (ed.), Electrophoresis '79, W. de Gruyter, Berlin – New York 1980, pp. 687 – 693.
[26] *H. H. Gustke, V. Neuhoff,* Kritischer Vergleich der Lactat-Dehydrogenase-Isoenzym-Fraktionierung mit Isoelektrischer Fokussierung und Diskelektrophorese in Polyacrylamid-Mikrogelen, Hoppe-Seyler's Z. Physiol. Chem. *359*, 1481 – 1489 (1978).
[27] *A. Burlina,* Improved Electrophoretic Separation of Creatine Kinase Isoenzymes, Clin. Chem. *26*, 317 – 320 (1980).
[28] *K. Zaitsu, Y. Ohkura,* New Fluorogenic Substrates for Horseradish Peroxidase: Rapid and Sensitive Assays for Hydrogen Peroxide and the Peroxidase, Anal. Biochem. *109*, 109 – 113 (1980).
[29] *K. S. Sri Venugopal, P. R. Adiga,* Artifactual Staining of Proteins on Polyacrylamide Gels by Nitrobluetetrazolium Chloride and Phenazine Methosulfate, Anal. Biochem. *101*, 215 – 220 (1980).
[30] *R. J. Wieme,* Description of a Densitometer for Direct Colorimetry of Irregularly Shaped Chromatographic Spots, J. Chromatography *1*, 166 – 171 (1958).
[31] *D. McKenzie, A. R. Henderson,* The Quality Control of Serum Lactate Dehydrogenase Isoenzyme Separations Using Commercial Quality Control Sera, Clin. Chim. Acta *51*, 53 – 66 (1974).
[32] *F. Y. Leung, A. R. Henderson,* Is the Method You Use for Lactate Dehydrogenase Isoenzyme Determination Accurate? Clin. Chem. *27*, 1623 – 1624 (1981).
[33] *P. Leclerc, J.-C. Forest,* Electrophoretic Determination of Isoamylases in Serum with Commercially Available Reagents, Clin. Chem. *28*, 37 – 40 (1982).
[34] *T. Rammeloo, P. M. M. van Haard, M. H. Beunis,* Two Serum Pancreatic Isoamylase Determinations Compared, Clin. Chem. *28*, 145 – 149 (1982).
[35] *S. C. Meredith, G. R. Nathans,* Gel Permeation Chromatography of Asymmetric Proteins, Anal. Biochem. *121*, 234 – 243 (1982).
[36] *M. Le Maire, E. Rivas, J. V. Moller,* Use of Gel Chromatography for Determination of Size and Molecular Weight of Proteins: Further Caution, Analyt. Biochem. *106*, 12 – 21 (1980).
[37] *P. J. Trocha,* Lactate Dehydrogenase Isoenzymes Linked to Beta-lipoproteins and Immunoglobulin A, Clin. Chem. *23*, 1780 – 1783 (1977).
[38] *M. E. de Broe, R. J. Wieme, G. N. Logghe, F. Roels,* Spontaneous Shedding of Plasma Membrane Fragments by Human Cells in Vivo and in Vitro, Clin. Chim. Acta *81*, 237 – 245 (1977).
[39] *J. Kamaryt, J. Zemek, L. Kuniak,* Determination of the Molecular Weight of α-Amylase, as the Enzyme-Inhibitor Complex, using Thin Layer Gel Filtration on Sephadex, J. Clin. Chem. Clin. Biochem. *20*, 451 – 455 (1982).
[40] *D. H. Leaback, H. K. Robinson,* Ampholyte Displacement Chromatography – A New Technique for the Separation of Proteins Illustrated by the Resolution of β-N-Acetyl-D-Hexosaminidase Isoenzymes Unresolvable by Isoelectric Focusing or Conventional Ion-Exchange Chromatography, Biochem. Biophys. Res. Commun. *67*, 248 – 254 (1975).
[41] *L. A. AE. Sluyterman, O. Elgersma,* Chromatofocusing: Isoelectric Focusing on Ion-Exchange Columns. I. General Principles, J. Chromatography *150*, 17 – 30 (1978).
[42] *L. A. AE. Sluyterman, J. Wijdenes,* Chromatofocusing. IV. Properties of an Agarose Polyethyleneimine Ion Exchanger and its Suitability for Protein Separations, J. Chromatography *206*, 441 – 447 (1981).
[43] *D. W. Mercer,* Separation of Tissue and Serum Creatine Kinase Isoenzymes by Ion-Exchange Column Chromatography, Clin. Chem. *20*, 36 – 40 (1974).
[44] *D. W. Mercer,* Simultaneous Separation of Serum Creatine Kinase and Lactate Dehydrogenase Isoenzymes by Ion-Exchange Column Chromatography, Clin. Chem. *21*, 1102 – 1106 (1975).
[45] *D. W. Mercer,* Improved Column Method for Separating Lactate Dehydrogenase Isoenzymes 1 and 2, Clin. Chem. *24*, 480 – 482 (1978).
[46] *D. W. Mercer,* Immunochemical Determination of LD-1 and LD-2 in CK-MB Fraction of Column Eluates, Clin. Chem. *27*, 779 – 780 (1981).

[47] *P. M. Bayer, M. Boehm, P. Hajdusich, H. Hotschek, H. Koehn, W. Unger, G. Wider,* Immunoinhibition and Automated Column Chromatography Compared for Assay of Creatine Kinase Isoenzyme MB in Serum, Clin. Chem. *28*, 166–169 (1982).
[48] *M.-Y. Hsu, M. M. Kohler, L. Barolia, R. J. L. Bondar,* Separation of Five Isoenzymes of Serum Lactate Dehydrogenase by Discontinuous Gradient Elution from a Miniature Ion-Exchange Column, Clin. Chem. *25*, 1453–1458 (1979).
[49] *L. Kahan, V. L. W. Go, F. C. Larson,* Increased Activity in Serum of an Alkaline Phosphatase Isoenzyme in Cancer: Analytical Method and Preliminary Clinical Studies, Clin. Chem. *27*, 104–107 (1981).
[50] *F. Y. Leung, A. R. Henderson,* Isolation and Purification of Aspartate Aminotransferase Isoenzymes from Human Liver by Chromatography and Isoelectric Focusing, Clin. Chem. *27*, 232–238 (1981).
[51] *F. E. Regnier, K. M. Gooding,* Review. High-Performance Liquid Chromatography of Proteins, Anal. Biochem. *103*, 1–25 (1980).
[52] *R. R. Schroeder, P. J. Kudirka, E. C. Toren,* Enzyme-Selective Detector Systems for High-Pressure Liquid Chromatography, J. Chromatography *134*, 83–90 (1977).
[53] *T. D. Schlabach, J. A. Fulton, P. B. Mockridge, E. C. Toren,* Determination of Serum Isoenzyme Activity Profiles by High Performance Liquid Chromatography, Clin. Chem. *52*, 729–733 (1980).
[54] *P. Cuatrecasas,* Affinity Chromatography of Macromolecules, in: *A. Meister* (ed.), Advances in Enzymology, Vol. 36, Interscience, New York 1972, pp. 29–89.
[55] *T. C. Bøg-Hansen,* Affinity Electrophoresis. Quantification and Characterization of Glycoproteins with Lectins, in: *B. J. Radola* (ed.), Electrophoresis '79, W. de Gruyter, Berlin–New York 1980, pp. 193–202.
[56] *M. R. A. Morgan, N. A. Slater, P. D. G. Dean,* The Electrophoretic Desorption of Human Serum Albumin from an Affinity Chromatography Matrix Using a Preparative Desorption Cell, Anal. Biochem. *92*, 144–146 (1979).
[57] *N. O. Kaplan, J. Everse, J. E. Dixon, F. E. Stolzenbach, Ch.-Y. Lee, Ch.-L. T. Lee, S. S. Taylor, K. Mosbach,* Purification and Separation of Pyridine Nucleotide-Linked Dehydrogenases by Affinity Chromatography Techniques, Proc. Natl. Acad. Sci. USA *71*, 3450–3454 (1974).
[58] *S. J. Johnson, E. C. Metcalf, P. D. G. Dean,* The Determination of Dissociation Constants by Affinity Electrophoresis on Cibacron Blue F3G A-Agarose-Polyacrylamide Gels, Anal. Biochem. *109*, 63–66 (1980).
[59] *Y. D. Clonis, C. R. Lowe,* Triazine Dyes, a New Class of Affinity Labels for Nucleotide-Dependent Enzymes, Biochem. J. *191*, 247–251 (1980).
[60] *A. Adinolfi, D. A. Hopkinson,* Blue Sepharose Chromatography of Human Alcohol Dehydrogenase: Evidence for Interlocus and Interallelic Differences in Affinity Characteristics, Ann. Hum. Genet. Lond. *41*, 399–407 (1978).
[61] *S. E. Fisher, G. S. Whitt,* Purification of the Creatine Kinase Isozymes of the Green Sunfish (Lepomis cyanellus) with Blue Sepharose CL-6B, Anal. Biochem. *94*, 89–95 (1979).
[62] *A. Ahmad, A. Surolia, B. K. Bachhawat,* Affinity Chromatographic Separation of Arylsulfatase A and B Using Cibacron Blue-Sepharose, Biochim. Biophys. Acta *481*, 542–548 (1977).
[63] *A. H. J. Kolk, L. Van Kuyk, B. Boettcher,* Isolation of Human Lactate Dehydrogenase Isoenzyme X by Affinity Chromatography, Biochem. J. *173*, 767–771 (1978).
[64] *B. K. Bachman, C.-Y. Lee,* Purification of Human Lactate Dehydrogenases by General Ligand Affinity Chromatography, Anal. Biochem. *72*, 153–160 (1976).
[65] *S. M. Pettit, D. A. Nealon, A. R. Henderson,* Purification of Lactate Dehydrogenase Isoenzyme-5 from Human Liver, Clin. Chem. *27*, 88–93 (1981).
[66] *K. Bauer, P. M. Bayer, E. Deutsch, F. Gabl,* Binding of Enzyme-IgG Complexes in Human Serum to Protein-A Sepharose CL-4B, Clin. Chem. *26*, 297–300 (1980).
[67] *L. Summaria, F. Spitz, L. Arzadon, I. G. Boreisha, K. C. Robbins,* Isolation and Characterization of the Affinity Chromatography Forms of Human Glu- and Lys-Plasminogens and Plasmins, J. Biol. Chem. *251*, 3693–3699 (1976).
[68] *D. M. Swallow, L. Evans, D. A. Hopkinson,* Several of the Adenosine Deaminase Isozymes are Glycoproteins, Nature *269*, 261–262 (1977).

[69] *E. Köttgen, W. Reutter, W. Gerok,* Two Different Gamma-Glutamyltransferases during Development of Liver and Small Intestine: a Fetal (Sialo-) and an Adult (Asialo-) Glycoprotein, Biochem. Biophys. Res. Commun. *72,* 61 – 66 (1976).

[70] *A. L. Latner, A. W. Hodson,* Differential Precipitation with Concanavalin A as a Method for the Purification of Glycoproteins: Human Alkaline Phosphatase, Anal. Biochem. *101,* 483 – 487 (1980).

[71] *R. Collier, G. Kohlhaw,* Affinity Chromatography of Transaminases, Anal. Biochem. *42,* 48 – 53 (1971).

[72] *M. Landt, S. C. Boltz, L. G. Butler,* Alkaline Phosphatase: Affinity Chromatography and Inhibition by Phosphonic Acids, Biochemistry *17,* 915 – 919 (1978).

[73] *R. A. Stinson, L. E. Seargeant,* Comparative Studies of Pure Alkaline Phosphatases from Five Human Tissues, Clin. Chim. Acta *110,* 261 – 272 (1981).

[74] *A. H. Gordon,* Electrophoresis of Proteins in Polyacrylamide and Starch Gels, North-Holland Publ. C^0, Amsterdam – London 1969.

[75] *I. Smith,* Chromatographic and Electrophoretic Techniques. Vol. II. Zone Electrophoresis, W. Heinemann Medical Books Ltd., London 1969.

[76] *B. H. Hames, D. Rickwood, Gel Electrophoresis of Proteins. A Practical Approach, IRL Press Limited, London and Washington DC 1981.*

[77] *A. T. Andrews,* Electrophoresis, Theory, Techniques and Biochemical and Clinical Applications, Clarendon Press, Oxford 1981.

[78] *P. G. Righetti, C. J. Van Oss, J. W. Vanderhoff,* Electrokinetic Separation Methods, Elsevier/North-Holland 1979.

[79] *J. H. Wilkinson,* Isoenzymes, E. & F. N. Spon Ltd., London 1965.

[80] *A. L. Latner, A. W. Skillen,* Isoenzymes in Biology and Medicine, Academic Press, London and New York 1968.

[81] *N. van Thoai, J. Roche,* Homologous Enzymes and Biochemical Evolution, A Colloquium Organized under the Auspices of the Advanced Study Group of NATO, Gordon and Breach, New York 1968.

[82] *C. L. Markert,* Isozymes. III. Developmental Biology, Academic Press, New York – London 1975.

[83] *J. Ambler, G. Walker,* Isoelectric Focusing of Serum Proteins on Modified Thin Cellulose-Acetate Membranes, Clin. Chem. *25,* 1320 – 1322 (1979).

[84] *O. Smithies,* Molecular Size and Starch-gel Electrophoresis, Arch. Biochem. suppl. *1,* 125 – 131 (1962).

[85] *S. Arnott, A. Fulmer, W. E. Scott, I. C. M. Dea, R. Moorhouse, D. A. Rees,* The Agarose Double Helix and Its Function in Agarose Gel Structure, J. Mol. Biol. *90,* 269 – 284 (1974).

[86] *R. B. Cook,* Agarose Gel Electroendosmosis: its Enhancement, Reduction and Practical Ramifications, in: *R. C. Allen, P. Arnaud* (eds.), Electrophoresis '81, W. de Gruyter & Co, Berlin – New York 1981, pp. 219 – 228.

[87] *Z. Buzas, A. Chrambach,* Un-supercoiled agarose with a degree of molecular sieving similar to that of cross-linked polyacrylamide, Electrophoresis *3,* 130 – 134 (1982).

[88] *O. Vesterberg,* Isoelectric Focusing of Proteins in Agarose Gels and Detection of the Proteins by Different Staining Procedures, in: *B. J. Radola* (ed.), Electrophoresis '79, W. de Gruyter & Co, Berlin – New York 1980, pp. 95 – 104.

[89] *L. M. Hjelmeland, A. Chrambach,* Electrophoresis and Electrofocusing in Detergent Containing Media: A Discussion of Basic Concepts, Electrophoresis *2,* 1 – 11 (1981).

[90] *P. Babczinski,* Fractionation of Yeast Invertase Isozymes and Determination of Enzymatic Activity in Sodium DodecylSulfate-Polyacrylamide Gels, Anal. Biochem. *105,* 328 – 333 (1980).

[91] *C. M. Wilson,* Studies and Critique of Amido Black 10B, Coomassie Blue R, and Fast Green FCF as Stains for Proteins after Polyacrylamide Gel Electrophoresis, Anal. Biochem. *96,* 263 – 278 (1979).

[92] *C. R. Merril, M. L. Dunau, D. Goldman,* A Rapid Sensitive Silver Stain for Polypeptides in Polyacrylamide Gels, Anal. Biochem. *110,* 201 – 207 (1981).

[93] *L. D. Adams, D. W. Sammons,* A Unique Silver Staining Procedure for Color Characterization of Polypeptides, in: *R. C. Allen, P. Arnaud* (eds.), Electrophoresis '81, W. de Gruyter & Co, Berlin – New York 1981, pp. 155 – 165.

[94] *R. C. Allen,* Past, Present, and Future Uses of Electrophoresis in Clinical Medicine, in: *B. J. Radola* (ed.), Electrophoresis '79, W. de Gruyter & Co, Berlin – New York 1980, pp. 631 – 645.

[95] *N. Y. Nguyen, A. Chrambach,* Amino Acid Spacing in Isotachophoresis on Polyacrylamide Gel: a Critical Evaluation, Anal. Biochem. *94*, 202 – 210 (1979).

[96] *B. K. Mesrob, R. A. Vassileva,* Preparation and Some Chemical Characteristics of Milk-Clotting Protease from Bacillus Mesentericus 76. Int. J. Peptide Prot. Res. *17*, 89 – 92 (1981).

[97] *D. Rodbard, A. Chrambach,* Estimation of Molecular Radius, Free Mobility, and Valence Using Polyacrylamide Gel Electrophoresis, Anal. Biochem. *40*, 95 – 134 (1971).

[98] *H. Rilbe,* Historical and Theoretical Aspects of Isoelectric Focusing, Ann. N.Y. Acad. Sci. *209*, 11 – 22 (1973).

[99] *H. Rilbe,* Stable pH Gradients – A Key Problem in Isoelectric Focusing, in: *B. J. Radola, D. Graesslin* (eds.), Electrofocusing and Isotachophoresis, Proceedings of the International Symposium, August 2 – 4, 1976, Hamburg, Germany, W. de Gruyter, Berlin – New York 1977, pp. 35 – 50.

[100] *E. Gianazza, P. G. Righetti,* Facts and Artefacts in Isoelectric Focusing, in: *B. J. Radola* (ed.), Electrophoresis '79, W. de Gruyter & Co, Berlin – New York 1980, pp. 129 – 140.

[101] *D. L. Hare, D. I. Stimpson, J. R. Cann,* Multiple Bands Produced by Interaction of a Single Macromolecular with Carrier Ampholytes during Isoelectric Focusing, Arch. Biochem. Biophys. *187*, 274 – 275 (1978).

[102] *D. E. Jackson, C. A. Skandera, J. Owen, E. T. Lally, P. C. Montgomery,* Isoelectric Focusing of IgA and IgM in Composite Acrylamide-Agarose Gels, J. Immunol. Methods *36*, 315 – 324 (1980).

[103] *L. M. Hjelmeland, D. W. Nebert, A. Chrambach,* Electrofocusing of Integral Membrane Proteins in Mixtures of Zwitterionic and Nonionic Detergents, Anal. Biochem. *95*, 201 – 208 (1979).

[104] *M. Jonsson,* Distorted Zones in Isoelectric-Focusing Systems, Electrophoresis *1*, 141 – 149 (1980).

[105] *R. E. Allen, K. C. Masak, P. K. McAllister,* Staining Protein in Isoelectric Focusing Gels with Fast Green, Anal. Biochem. *104*, 494 – 498 (1980).

[106] *O. Vesterberg, L. Hansen, A. Sjösten,* Staining of Proteins after Isoelectric Focusing in Gels by New Procedures, Biochim. Biophys. Acta *491*, 160 – 166 (1977).

[107] *H. Rilbe,* Steady-State Rheoelectrolysis – A Method for Isoelectric Focusing without Carrier Ampholytes. I. The pH Course to be Expected on Rheoelectrolysis of a Buffer Solution Composed of a Weak Acid and its Salt with a Strong Base. II. The pH Course to be expected on Rheoelectrolysis of a Buffer Solution Composed of a Weak Acid and a Weak Base, Electrophoresis *2*, 261 – 272 (1981).

[108] *S. L. Tollaksen, J. J. Edwards, N. G. Anderson,* The Use of Carbamylated Charge Standards for Testing Batches of Ampholytes Used in Two-Dimensional Electrophoresis, Electrophoresis *2*, 155 – 160 (1981).

[109] *B. Bjellqvist, K. Ek, P. G. Righetti, E. Gianazza, A. Görg, W. Postel,* Immobilized pH Gradients for Isoelectric Focusing. I: Theory and Methodology, in: *H. Peeters* (ed.), Protides of the Biological Fluids. Proceedings of the Thirtieth Colloquium, 1982, Vol. 30, Pergamon Press, Oxford 1983, pp. 599 – 602.

[110] *A. Görg, W. Postel, R. Westermeier, B. Bjellqvist, K. Ek. E. Gianazza, P. G. Righetti,* Isoelectric Focusing in Immobilized pH-Gradients. III. Application to Two-Dimensional Separations, in: *H. Peeters* (ed.), Protides of the Biological Fluids. Proceedings of the Thirtieth Colloquium, 1982, Vol. 30, Pergamon Press, Oxford 1983, pp. 607 – 610.

[111] *P. G. Righetti, E. Gianazza,* Recent Developments in Titration Curves of Proteins by Isoelectric Focusing-Electrophoresis, in: *R. C. Allen, P. Arnaud* (eds.), Electrophoresis '81, W. de Gruyter & Co, Berlin – New York 1981, pp. 655 – 665.

[112] *D. Lostanlen, G. Gacon, J.-Cl. Kaplan,* Direct Enzyme Titration Curve of NADH: Cytochrome b_5 Reductase by Combined Isoelectric Focusing/Electrophoresis, Eur. J. Biochem. *112*, 179 – 183 (1980).

[113] *A. Rosengren, B. Bjellqvist, V. Gasparic,* A Simple Method of Choosing Optimum pH-Conditions for Electrophoresis, in: *B. J. Radola, D. Graesslin* (eds.), Electrofocusing and Isotachophoresis. Proceedings of the International Symposium, August 2 – 4, 1976, Hamburg, Germany, W. de Gruyter, Berlin – New York 1977, pp. 165 – 171.

[114] *P. H. O'Farrell,* High Resolution Two-Dimensional Electrophoresis of Proteins, J. Biol. Chem. *250*, 4007–4021 (1975).
[115] *T. Manabe, K. Tachi, K. Kojima, T. Okuyama,* Two-Dimensional Electrophoresis of Plasma Proteins without Denaturing Agents, J. Biochem. *85*, 649–659 (1979).
[116] *J. Klose, J. Nowak, W. Kade,* Two-Dimensional Electrophoresis: High Resolution of Proteins and Automatic Evaluation of the Patterns under Different Methodical Conditions, in: *B. J. Radola* (ed.), Electrophoresis '79, W. de Gruyter & Co, Berlin – New York 1980, pp. 297–312.
[117] *M. M. Skolnick,* An Approach to Completely Automatic Comparison of Two-Dimensional Electrophoresis Gels, Clin. Chem. *28*, 979–986 (1982).
[118] *A. C. Terranova,* Diazo Transparencies of Polyacrylamide Slab Gels, Anal. Biochem. *107,* 443–445 (1980).
[119] *H.-M. Poehling, V. Neuhoff,* One- and Two-Dimensional Electrophoresis in Micro-Slab Gels, Electrophoresis *1*, 90–102 (1980).

1.3.3 Immunological Techniques in Isoenzyme Analysis

Donald W. Moss

The antigenic determinants of protein molecules consist of relatively small regions, in which both the nature of the amino-acid side-chains and their three-dimensional arrangement are important in conferring antigenic specificity. In view of their different molecular structures, it is not surprising that isoenzymes determined by multiple gene loci are almost always found to be antigenically distinct. Antigenic differences also exist in some cases between isoenzymes (allelozymes) determined by allelic genes. However, since allelozymes often differ by only a single amino-acid residue, they may be immunologically indistinguishable, at least when polyclonal antisera are used. Indeed, mutations that abolish catalytic activity (presumably by the elimination of a catalytically essential amino acid) may nevertheless leave some antigenic determinants unaltered, so that an immunologically recognizable gene product (a "cross-reacting material") may be detectable (e.g., [1]). However, the greater specificity of monoclonal antibodies seems likely to extend the range of immunological methods in isoenzyme analysis. For example, it has recently been shown that monoclonal antibodies can distinguish between the closely similar allelozymes of human placental alkaline phosphatase [2].

The ability to prepare antisera which are specific for individual members of a family of isoenzymes opens up a wide range of analytical techniques of potentially great sensitivity and specificity. In many cases, interaction between the isoenzyme antigen and its specific antibody does not involve the active centre of the enzyme, which remains accessible to the substrate. Therefore, the enzyme-antibody complex is catalytically active. The formation of catalytically active isoenzyme-antibody complexes is an advantage in various methods of immunochemical analysis which depend on the location of the complexes precipitated in gels. In some instances, the reaction between isoenzyme and antibody abolishes or greatly reduces catalytic activity, probably because of steric hindrance. Inhibiting antibodies are useful in some forms of analysis in solution. Radioimmunoassay of isoenzymes is entirely independent of catalytic activity.

The application of immunochemical methods to isoenzyme analysis depends on the production of antisera specific for particular isoenzymes; this in turn requires the availability of suitably purified isoenzyme preparations to serve as antigens. Antibodies against a particular antigen may vary considerably in their specificity and affinity for the antigen from one preparation of antiserum to another. Variations in specificity can have a marked effect on the extent to which the antiserum cross-reacts with molecules with a close structural relationship to the immunizing antigen, such as other members of a family of isoenzymes. All batches of antiserum should therefore be checked for their cross-reactivity towards potentially interfering antigens, e.g., by immunodiffusion techniques. The specificity of the antiserum for the chosen isoenzyme can be improved by selecting the required antibodies from the rest by a separation process such as affinity chromatography, or by adsorbing cross-reacting antibodies with the potential antigens for which they have an affinity. An example of the latter approach is the treatment of antiserum to human intestinal alkaline phosphatase with the antigenically similar placental phosphatase, to increase its specificity for the intestinal isoenzyme [3].

A great increase in the specificity of antibodies has become available recently through the monoclonal antibody technique. By selecting a single clone of cells, an antiserum can be prepared which contains only a single specific type of antibody. Cross-reacting antibodies present in antisera raised in whole animals are eliminated.

1.3.3.1 Immunochemical Reactions in Solution

In this type of analysis, the isoenzyme antigen is allowed to react with a specific antiserum and the amount of bound isoenzyme is then estimated. The rate of reaction is rapid, depending mainly on the concentrations of the reactants. It depends much less on factors such as ionic strength (provided that the reactants are soluble) and pH within the range of approximately pH 5 – 8, provided that the ionization of an essential binding-group is not significantly altered. Increasing temperature accelerates the reaction. However, combination is usually so rapid at room temperature that it is not necessary to increase the risk of denaturation of antigen or antibody by the use of higher temperatures.

Decrease in enzyme activity is a convenient index of the extent of antibody-antigen reaction, if combination with antibody inhibits catalytic activity. The use of inhibiting antisera is valuable in routine analysis of large numbers of samples, since the process of combination may be essentially complete in a period of minutes and no separation of bound and free isoenzyme (e.g., by centrifugation) is necessary.

Enzyme inhibition is not invariably a consequence of combination with antibody, and indeed complete inhibition probably occurs in only a minority of cases. However, conditions can usually be found by varying the relative proportions of antigen and antibody under which the antigen-antibody complex is insoluble and so can be removed. The reaction of enzyme and antibody can thus be measured by the partition of catalytic activity between the sediment and the supernatant.

The insolubility of the antigen-antibody complex depends on the formation of polymeric structures in which several antibody molecules are linked by divalent molecules of antigen. The relationship between antibody and antigen concentrations is critical in promoting precipitation. Therefore, a titration of antigen with antibody is needed to select the most favourable ratio of these reactants for precipitation to take place. Unlike the initial reaction between antibody and antigen, the formation of insoluble complexes is usually slow, so that a period of incubation at 37 °C of up to one hour, or storage at lower temperatures overnight or for as long as several days, or both, may be necessary for complete precipitation.

Precipitation of soluble antigen-antibody complexes can be brought about by the addition of a second antibody, directed against the immunoglobulin with which the primary antigen (the isoenzyme) has combined. Precipitation with a second antiserum is also useful in searching for complexes which may form between the anti-isoenzyme antiserum and cross-reacting antigens.

Immunoprecipitation methods

Numerous reports have been made of the use of antigen-antibody reactions in solution as an aid to the differentiation of isoenzymes, beginning with their application to lactate dehydrogenase [4], creatine kinase [5], aspartate aminotransferase [6], alkaline phosphatase [7], among many other isoenzyme systems. Precipitation of an isoenzyme-antibody complex with a second antibody has been used to measure prostatic acid phosphatase in serum [8] and isoenzyme 1 of lactate dehydrogenase [9] in human serum.

Immunochemical methods with increased sensitivity have been devised in order to detect small amounts of carcinoplacental isoenzymes of alkaline phosphatase in the sera of patients with cancer. These isoenzymes are produced by some tumours, apparently as a result of re-expression of the normal placental phosphatase gene. Although the carcinoplacental isoenzymes may differ in some respects from the normal placental isoenzyme, they retain the ability to react with antiserum to normal placental alkaline phosphatase. In a particularly sensitive method, a monospecific anti-placental phosphatase antiserum was polymerized with ethyl chloroformate, rendering the antibody insoluble [10]. Incubation of heat-treated serum samples with the antiserum for one hour was followed by recovery of the now insoluble, but still catalytically-active, placental alkaline phosphatase by centrifugation. The enzymic activity of the pellet was determined with a fluorigenic substrate to increase sensitivity still further. Carcinoplacental isoenzyme contributing less than 1% of the total alkaline phosphatase activity of the serum could be determined in this way. The phosphatase activity bound to anti-placental phosphatase antibodies which had been polymerized by reaction with glutaraldehyde was determined in a similar technique [11]. High sensitivity was also obtained in an immunoassay for human prostatic acid phosphatase, in which the enzyme in serum was bound to a specific antibody coupled to Sepharose 4B [12]. The activity of the bound isoenzyme recovered by centrifugation, was determined fluorimetrically with α-naphthyl phosphate as substrate.

The determination of isoenzymes of creatine kinase in serum constitutes the most widespread application of immunochemical reactions in solution in the field of clinical biochemistry at the present time [13]. Rabbit antisera to either purified MM- or BB-creatine kinase isoenzymes were each found to be specific for their respective homologous antigens, with no cross-reaction with the heterologous isoenzyme [14]. Both antisera reacted with the hybrid MB-creatine kinase, as is generally the case with hybrid isoenzymes. The extent of combination of creatine kinase isoenzymes in tissue extracts or serum samples with the antisera was estimated from the residual enzymic activity, either after incubation for successive periods of 30 min at room temperature and 4 °C (immunoinhibition), or after centrifugation following these incubations (immunoprecipitation). Activity of the MM isoenzyme was completely inhibited by its specific antiserum. However, BB-creatine kinase activity was inhibited to the extent of about 88% by anti-BB antiserum, although the residual activity was completely removed from solution by centrifugation. The MB isoenzyme was inhibited to the extent of about 80% by either antiserum, with 10% of the original activity remaining in the supernatant after centrifugation; i.e., the hybrid isoenzyme reacted incompletely with a single antiserum. Addition of the second antiserum was needed to precipitate the MB isoenzyme completely. The isoenzyme composition of samples containing creatine kinase was calculated from the fractional residual activities, a and b, measured after precipitation with anti-MM and anti-BB antisera in the following way:

$$a = BB + 0.1\,MB\,; \qquad BB = a - 0.1\,MB$$

$$b = MM + 0.1\,MB\,; \qquad MM = b - 0.1\,MB$$

$$1 = BB + MM + MB = a - 0.1\,MB + b - 0.1\,MB + MB$$

therefore,

$$MB = 1.25 \times (1 - a - b) \times 100 \quad \%\text{ of total CK.}$$

The results of this type of analysis agreed well with the known compositions of mixtures of the three isoenzymes.

Immunoinhibition methods

Immunoprecipitation procedures are time-consuming compared with methods based on inhibition alone, and are effective over a less wide range of enzyme activity. Most analyses of creatine kinase isoenzymes in serum are carried out in order to determine the amount of the MB isoenzyme present. The method can be further simplified by using only one inhibiting antiserum, if it is assumed that BB isoenzyme is absent and that any B subunits are present only as components of the MB isoenzyme.

Two similar methods have been described which employ a human MM-creatine kinase-inhibiting antiserum prepared in goats [15, 16]. Inhibition of MM-creatine kinase was shown to be essentially complete within 3 min of adding this antiserum, with 50% inhibition of the MB dimer and no effect on the BB isoenzyme. The degree of inhibition of MM-creatine kinase was found to be 99%. There is some inhibition of residual activity of the B subunit of MB-creatine kinase as a result of combination of the M protomers with anti-MM antibodies, amounting to about 5%. Therefore, doubling the measured residual activity after immunoinhibition of a mixture of MM- and MB-creatine kinases underestimates the original activity of the latter isoenzyme by up to 10%.

Methods of this type are designed to detect and measure an isoenzyme which is a minor component of the sample; therefore, methods of assay of enzyme activity are needed which are capable of determining low residual activities reliably. A sensitive ATP-dependent luciferase reaction has been used in one immunoinhibition method for creatine kinase isoenzymes [17]. An effective activator of creatine kinase (e.g., N-acetyl cysteine) is also required. Furthermore, blank reactions which can be neglected when measuring total enzyme activity may be comparable with the remaining, uninhibited isoenzyme activity, so that the residual reaction rate should be corrected. For example, suppression of the interfering activity of adenylate kinase by inhibitors is adequate when measuring total creatine kinase activity, but not for all samples when immunoinhibition methods are used to determine the MB isoenzyme. Coefficients of variation of replicate analyses by these methods are of the order of $\pm 5-10\%$, depending on the level of activity of B subunits, and MB-creatine kinase activities as low as $3-4$ U/l at 25 °C can be measured.

1.3.3.2 Immunodiffusion in Gels

Gels provide a favourable environment for the precipitation of antigen-antibody complexes, by reducing convection currents and thermal agitation which disrupt the growth of the lattice of antigen and antibody molecules. When antigen and antibody diffuse towards each other in gels, concentration gradients are generated which intersect to provide the relative concentrations appropriate for precipitation. When precipitation has begun, migration of further molecules of antigen and antibody into the precipitation zone increases the amount of precipitated complex. However, the passive diffusion of large molecules through gels is slow, so that a long period is usually needed to accumulate significant amounts of precipitate. Low temperatures (e.g. 4 °C) are often chosen for diffusion in gels: although rates of diffusion are reduced compared with those at higher temperatures, the insoluble complex is more stable. Diffusion of antigen or antibody through the gel is accelerated in several methods by applying an electric potential.

Inactive enzyme-antibody precipitates can be located after diffusion in gels by staining with dyes which are taken up by proteins, or, if the concentration of the antigen is high enough, by the appearance of opalescent precipitin lines in the clearer gel matrix. However, the advantages of immunodiffusion methods of all kinds in isoenzyme analysis are realized most clearly when the antigen-antibody precipitate retains its catalytic activity. It is then possible to locate and identify isoenzyme-antibody precipitates by specific staining reactions similar to those used to locate isoenzyme zones after electrophoresis [18].

Classification of immunodiffusion methods

A large variety of methods of immunodiffusion analysis have been described, many of which have been used in isoenzyme analysis. They can be classified in several ways. A useful system (Fig. 1) is to divide them into methods in which diffusion through the gel takes place passively, or is accelerated by an electric field. Within each of these categories are "single" or "double" diffusion techniques. Only one component migrates in single diffusion methods: typically, the antigen migrates through a gel which contains the antibody. Both antigen and antibody migrate in double immunodiffusion methods.

The various procedures have certain practical features in common. Agarose is now almost invariably chosen to prepare the gels, because of the clarity, purity, and mechanical strength of its gels at low concentrations, and the absence of negatively

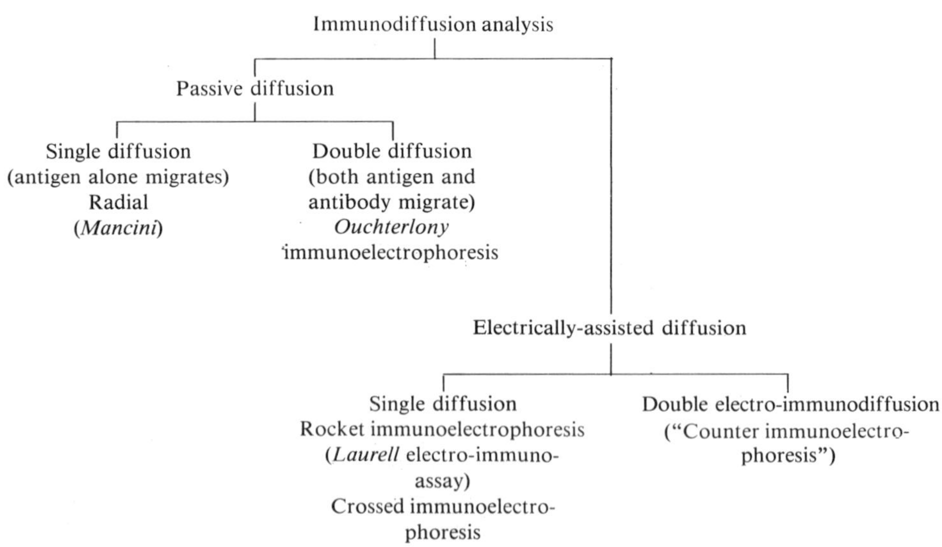

Fig. 1. A classification of methods of immunodiffusion analysis

charged groups which may retard the migration of basic proteins. Concentrations of agarose of approximately 8 g/l are usually used, giving pore sizes which do not form a barrier to larger molecules. Gels can be prepared in saline or in dilute buffer solutions. The choice of a buffer will depend to some extent on the nature of the enzyme reaction subsequently to be carried out in the gel. If the nature of the buffer required for demonstration of enzyme activity is such that it may interfere with formation of the gel, or with combination of antibody and antigen, it can be introduced after immunodiffusion by soaking the gel in it. Agarose solutions are prepared by heating the polysaccharide to about 100 °C in the solvent and become gels at a temperature of 40 – 45 °C. The gelation temperature varies with different agarose preparations. Those gelling at lower temperatures have the advantage that antibodies can be incorporated just before the gel sets with less risk of denaturation.

Although now largely supplanted by agarose, agar gels have been extensively used in immunodiffusion. The electro-osmosis which occurs in these gels can be turned to advantage in some electro-immunodiffusion techniques. Various other gels which have been employed as supporting media in zone electrophoresis, such as starch or polyacrylamide gels, have also been applied to immunodiffusion methods, as has cellulose acetate, but without acquiring the popularity of agarose or agar. Immunodiffusion is now almost invariably carried out in a layer of gel 1 – 2 mm thick on glass plates, older techniques in which the gel is formed in a small test-tube being seldom used.

Passive immunodiffusion techniques

Radial diffusion of antigen from wells 1 – 2 mm in diameter into an agar or agarose gel layer containing antiserum to form precipitin rings (the *Mancini* technique) is a simple method of estimating antigen concentration, and one which has been used for the determination of isoenzymes. However, the *Ouchterlony* double-immunodiffusion test has been used much more extensively in characterizing antisera raised against specific isoenzymes and in investigating the antigenic relationships between isoenzymes, e.g., [1, 3, 18 – 26]. Several antigen wells are arranged in a circle with a single antiserum well at the centre in the *Ouchterlony* technique; this allows the reactivity of several potential antigens (e.g., isoenzymes) towards a single antiserum to be compared. Undisturbed crossing of precipitin lines between the antiserum and different antigens indicates non-identity. Incomplete fusion with spur formation shows partial cross-reaction, and complete fusion of the precipitin lines indicates that the different antigens are identical in this respect.

Immunoelectrophoresis

The most important variant of double immunodiffusion is immunoelectrophoresis. In this technique, separation of a mixture of antigens by electrophoresis in one dimension is followed by diffusion of the separated zones, in a second dimension at right angles to the first, to meet and react with antibody diffusing from a trough cut parallel to the direction of electrophoretic migration. In contrast to electrically-assisted immunodiffusion methods, the application of the electric field is a preliminary to the conjunction of antigen and antibody, which takes place by passive diffusion. The advantage of immunoelectrophoresis which has made it so valuable in isoenzyme analysis is that several isoenzymes can be separated and characterized by their electrophoretic mobilities, and their antigenic relationships can be clarified, by the use of a single technique, e.g. [18, 19, 21 – 27]. Since diffusion from the separated zones takes place in both lateral directions, two different antisera can be used after each separation, one on each side of the zones, further facilitating immunochemical comparisons.

A single gel (typically agarose) is frequently used for both the initial electrophoretic separation and the subsequent immunodiffusion. However, combinations of different gels for the stages of electrophoresis and immunodiffusion may be useful, particularly when more concentrated gels of starch or polyacrylamide are needed to separate enzyme zones of different molecular sizes by molecular sieving during electrophoresis [28, 29]. Passive diffusion of large molecules in these small-pore gels is so slow that they are unsuitable as media for immunodiffusion processes. Therefore, a longitudinal slice of the electrophoresis gel, or the whole gel in the case of electrophoresis in small cylinders of polyacrylamide gel, is placed on a glass plate on which molten agar or agarose gel is poured to form a gel layer surrounding and in contact with the electrophoresis gel. After the agarose gel has solidified, a trough is cut parallel to the electrophoresis strip to receive the antiserum.

The practical considerations in the first stage of immunoelectrophoresis are similar to those applying generally to zone electrophoresis in gels (chapter 1.3.2.3, p. 71). In the subsequent immunodiffusion phase, important factors are the spacing between the centre line of the separated enzyme zones and the antiserum trough, and the dilutions of the antiserum and antigen solutions to be used. If these factors are not optimal, precipitation may take place too close to the antiserum trough to allow significant irregularities, crossing or fusion of precipitin arcs to be seen. Similarly, leakage of antisera from the trough can blur precipitin arcs, and irregularities or splits in the walls of the troughs may result in distortions of precipitin arcs which may simulate various types of immunochemical interactions. Given carefully controlled and appropriate conditions, however, the shapes of precipitin arcs developed in immunoelectrophoresis can provide much information about the antigenic relationships between enzyme zones.

Electrically assisted immunodiffusion

Immunoprecipitation resulting from passive diffusion in gels is a slow process, hours or even days being required to achieve detectable precipitation. Diffusion can be accelerated by applying an electric field. These methods are not only more rapid, but also are more suitable for quantitative analysis than passive-diffusion procedures. Both single and double electro-immunodiffusion techniques have been applied to the study of multiple forms of enzymes. In the former, as in passive single immunodiffusion, the antigen migrates through a gel containing the stationary complementary antibody. More rarely, antibody molecules travel through a field of antigen. In double electro-immunodiffusion (often called counter immunoelectrophoresis), both antigen and antibody molecules migrate under the influence of the electric field.

Rocket immunoelectrophoresis [30], also known as *Laurell* electro-immunoassay, can be regarded as a variant of the *Mancini* radial immunodiffusion technique in which antigen molecules (i.e. isoenzyme molecules) are drawn from their wells through the antibody-containing gel under the influence of an electric field. As in passive diffusion, migration continues until antigen-antibody complexes are precipitated. The distance into the gel at which precipitation occurs depends on the antigen concentration, since the concentration of antibody is initially constant throughout the gel; thus, the height of the rocket-shaped precipitin peak is a measure of antigen concentration. The peak heights given by antigen solutions of unknown concentration are compared with those of standard antigen solutions run simultaneously. Relative concentrations of different antigen solutions are more accurately reflected by the areas enclosed by their respective rockets, rather than simply by their heights, but any advantage thus obtained is more than offset by the practical difficulties of measurement of the areas. Even with simple methods, peak heights of the order of 1 – 5 cm can be measured with an accuracy of about 0.2 mm. The overall precision of the method depends on various additional factors. However, coefficients of variation of less than ±7% have been found for replicate determinations of placental alkaline phosphatase by rocket electro-immunoassay [31].

The sensitivity of the method is also the resultant of several factors. The distance of migration should be great enough for the peak height to be accurately measurable, and can be increased for dilute antigen solutions by reducing the concentration of antiserum in the gel. Enough antigen-antibody complex must be present at the tip of the rocket to be located by staining or other means, and low concentrations of the two components means that the concentration of their complex at the point of precipitation will in turn be low. Although the detection of weak rockets by conventional protein stains may be impossible, catalytically active enzyme-antibody precipitates can be detected with great sensitivity, as in other forms of immunodiffusion analysis.

The antiserum used in rocket immunoassay ideally should not cross-react with isoenzymes related to the one under study, since identification of a particular isoenzyme by its electrophoretic mobility is not possible in this one-dimensional technique, as it is in classical immunoelectrophoresis or "crossed" immunoelectrophoresis. Some degree of cross-reaction which involves a minor isoenzyme component and which produces small, faint rockets may not interfere with the identification and measurement of the selected isoenzyme. However, if the proportions of cross-reacting antigen is considerable, it may not be possible to identify the isoenzyme peaks correctly.

The application of a uniform electrical field is necessary in electro-immunoassay to prevent distortion of the peaks. Good electrical connections to the buffer reservoirs are required across the whole width of the anode and cathode ends of the gel, and the gel layer should be of even thickness (typically about 1 mm). The gel (usually 8 g/l agarose) is made in a suitable buffer solution, e.g., sodium barbitone (0.2 – 1 mol/l)-barbituric acid, pH 8.6. The more dilute buffer solutions reduce heat production during electrophoresis and aid precipitation, but have lower buffering capacity. Voltage gradients of 2 – 8 volts/cm are usually employed, development of the rockets taking from more than 8 h to less than 2 h, according to voltage. Unprecipitated proteins are removed by washing with saline solution or water before staining.

The need to keep the antibody molecules essentially stationary during the migration of antigen in electrically-assisted single immunodiffusion imposes limitations on the pH at which the process is carried out. Although a small movement of antibody molecules under the influence of the electric field may not affect the results, a significant degree of electrophoretic migration will result in an uneven distribution of antibody molecules in the gel so that quantitative assessment is impaired. However, antibody molecules generally have little net charge at alkaline pH values, such as pH 8.6, in contrast to many isoenzyme antigens, which exhibit appreciable electrophoretic mobility under these conditions. The effects of small inequalities in antibody distribution resulting from movement of antibodies during electrophoresis can be overcome in quantitative analysis by the inclusion of standard solutions of antigen amongst the unknown samples.

When the antigen and antibody have similar ionization characteristics, it becomes difficult or impossible to select pH conditions under which their electrophoretic mobilities are sufficiently different for electro-immunodiffusion to be practicable. In such cases, useful differences in net charge at a given pH can sometimes be created by chemical modification of one or other component of the system (usually the antigen). Needless to say, such modification must not materially alter the activity of an enzyme antigen, if this property is to be used to locate the antigen-antibody precipitate. Acetylation or carbamoylation of the free amino groups of proteins prevents their ionization to positively-charged NH_3^+ residues, thus increasing the net negative charge of protein molecules and their electrophoretic mobility towards the anode. Carbamoylation with cyanate has been used to improve the migration of isoenzyme C (isoenzyme II) of human carbonic anhydrase during electro-immunoassay [32].

Rocket electro-immunoassay has been applied to the analysis of several enzymes and isoenzymes, including carbonic anhydrase [32], placental alkaline phosphatase [31], prostatic acid phosphatase [33], glucose-6-phosphate dehydrogenase in human leukaemias [34], and several hydrolytic enzymes in a human diploid cell line [35].

Crossed immunoelectrophoresis [36] combines the separation and identification of isoenzyme zones given by electrophoresis with the rapid and quantitative immunochemical analysis characteristic of rocket electro-immunoassay. A preliminary electrophoretic separation of the mixture of antigens (i.e., the isoenzymes) is followed by a second migration of the zones in a direction at right angles to their original movement into an antibody-containing gel, again under the influence of an electric field. Agarose gel (ca. 10 g/l) is usually chosen as the medium for both migrations, with conditions of buffer composition and pH, voltages, and antibody concentration in the second gel, selected as in one-dimensional immunoelectrophoresis and rocket electro-immunoassay. In earlier descriptions of crossed immunoelectrophoresis, the original sample was applied into a slit in the gel cut transversely to the direction of electrophoretic migration, and a narrow strip was cut longitudinally from the centre of the strip after electrophoresis to serve as the sample gel from which the second migration originates. However, since such a strip contains variable proportions of the several antigens, this procedure is less suitable for quantitative analysis than later versions in which electrophoresis originates from a small, circular well containing the sample, the full width of the electrophoresis zones being included in the second migration. In either case the gel strip is transferred to a second glass plate on to which the molten antibody-containing gel is poured so as to make contact with one of the long edges of the electrophoresis gel.

After migration is completed, unprecipitated proteins are removed by washing and the precipitin peaks are located, preferably by specific reactions dependent on enzymic activity. The relative amounts of different antigens are proportional to the areas under their respective peaks. Absolute amounts of antigens can be determined if solutions of known concentration are available for calibration purposes. Although its quantitative aspect is probably the greatest single advantage of crossed immunoelectrophoresis over conventional immunoelectrophoresis, interactions of various kinds between the precipitin lines produced by different antigens are also more easily studied in the well-developed peaks of the crossed technique than in the arcs of immunoelectrophoresis.

Crossed immunoelectrophoresis has so far been used to a relatively limited extent in isoenzyme analysis. It was used to determine the amounts of two α-amylase isoenzymes during germination of barley seeds [37]. In a study of the reactions of a variety of enzymes in microsomes and plasma membranes from rat liver with antisera to each of these cell fractions, the greater resolving power of crossed immunoelectrophoresis was demonstrated by the recognition of ten antigens with uridine-diphosphatase activity in the microsomal fraction by this technique, compared with only three which would be resolved by conventional immunoelectrophoresis [38, 39]. Specific enzymic staining reactions were used to identify the antigen-antibody precipitates in these investigations.

Counter immunoelectrophoresis is an electrically-assisted variant of double immunodiffusion in which both antigen and antibody migrate in the electric field. To ensure that they meet and precipitate, conditions must be chosen under which the antigen and antibody move in opposite directions. Ideally, therefore, the pH of the medium should be between the respective isoelectric pH values of these two components, so that one moves towards the cathode and the other towards the anode. This condition is difficult to achieve in practice. Nevertheless, the principle of counter immunoelectrophoresis can be applied under circumstances in which the antigen migrates electrophoretically towards the anode while the antibody moves in the reverse direction largely by electro-endosmosis. A medium such as agar gel, which promotes electro-endosmotic flow, is therefore an advantage in this technique.

The antiserum and antigen (isoenzyme) solution are introduced into wells (about 3 mm dia.) cut in the gel layer and separated by a distance of about 1 cm in the direction of migration, with the antiserum in the more-anodal well and the antigen solution in the well nearer to the cathode. After electrophoresis for 1 – 2 h, depending on voltage, pH and buffer concentration, the antigen-antibody complex precipitated between the wells is made visible by staining for protein or enzyme activity. The time of electrophoresis is critical since over-running may cause the precipitate to re-dissolve in excess of antibody or antigen. The method is generally used for rapid quantitation of an antigen and this can be achieved by determining the greatest dilution of the antigen solution which produces a detectable precipitate.

Two counter immunoelectrophoresis assays have been described for the prostatic isoenzyme of acid phosphatase in human serum [40, 41].

Many ingenious variations and extensions of electrically-assisted immunodiffusion techniques have been described, designed generally to improve the recognition of specific antigens and to facilitate the study of cross-reactions between antigens. For example, close spacing of sample wells in rocket electro-immunoassay or crossed immunoelectrophoresis allows fusion of the precipitates formed by neighbouring cross-reacting antigens to take place. A similar effect be obtained in counter immunoelectrophoresis by the use of two antigen wells side by side, while a preliminary electrophoretic separation of the antigen solution followed by counter immunoelectrophoresis of antiserum from a trough parallel to the original direction of migration converts the counter immunoelectrophoretic technique into an electrically accelerated version of conventional immunoelectrophoresis.

1.3.3.3 Radioimmunoassay and Related Methods

Radioimmunoassay occupies an important place in the determination of a wide range of antigens, haptens, and antibodies, and a significant number of radioimmunoassays of isoenzymes have now been described. For the application of radioimmunoassay to isoenzyme analysis, a quantity of the purified isoenzyme is required, not only for the production of a specific antiserum, but also for radioactive labelling and to serve as a calibration standard in the assay. The specificity of the assay is determined by the specificity of the antibodies. Sensitivity is dependent on the titre of the antiserum and the affinity of its antibodies for the antigen. With an antiserum which is satisfactory in these respects, radioimmunoassays are able to determine amounts of antigen in the nanogram range or below.

The ability to distinguish between bound and free antigen is an essential requirement of competitive-binding assays. The two phases must be separated physically in radioimmunoassay, although this is not necessary in certain other competitive-binding methods. Antigen-antibody complexes do not precipitate at the low concentrations typical of radioimmunoassay, but can be precipitated by addition of a second antiserum, directed against the first antibody. Alternatively, the isoenzyme-specific antibodies may be attached to a solid matrix such as a cross-linked dextran or cellulose, either covalently, or by non-specific adsorption, e.g. to the walls of a plastic test-tube. The antibodies themselves may be chemically polymerized to render them insoluble. Various separation methods used in other applications of radioimmunoassay, such as filtration, or precipitation with salts or organic solvents, are less able to separate bound and unbound isoenzymes satisfactorily because of the general similarities in properties of the bound and free forms.

Radioactive iodine, in the form of the isotope ^{125}I (half-life 60 days), is usually used to label enzyme and isoenzyme antigens for radioimmunoassay. The labelled isoenzyme antigen must not only retain those features of its structure which are recognized by the antibodies, but also should have an affinity for the antibodies similar to that of the unlabelled molecule or the sensitivity of the assay may be reduced. A study of radioactively labelled MM and BB isoenzymes of creatine kinase showed that the extent of immunoprecipitation of the isoenzymes by their complementary antisera was not affected by a fourfold increase in the degree of labelling, demonstrating that, in this case at least, the process does not markedly affect the antigen-antibody reaction [42]. In addition, the labelled preparation must be free from unlabelled antigen and from labelled or unlabelled degradation products. After labelling, therefore, the purity of the antigen is usually checked and increased if necessary by electrophoretic or various chromatographic procedures. Labelling may or may not affect the catalytic activity of an enzyme antigen, depending on the relationship between the labelled side-chains and the active centre. However, although persistence of catalytic activity may provide evidence that labelling has not damaged the enzyme molecule, the sensitivity of radioimmunoassay does not depend on catalytic activity.

Radioimmunoassays have been described for carbonic anhydrase isoenzymes [43], type II carboxypeptidase B from human pancreas [44], porcine pancreatic amylase [45], human hexosaminidases A and B [46], human prostatic acid phosphatase [20, 47], human placental and carcinoplacental (*Regan*) alkaline phosphatases [29], and human creatine kinase isoenzymes [48]. All these methods make use of ^{125}I as the radioactive label and in most cases the double antibody technique, in which the antibody-bound isoenzyme is precipitated with an anti-γ-globulin antiserum and recovered by centrifugation, is employed to separate the bound and unbound phases.

These methods display the great sensitivity characteristic of radioimmunoassay. However, incubation periods of several hours are often required, either for the equilibration of the isoenzyme present in the sample and its labelled counterpart with the specific antiserum, or for the subsequent separation reactions, or both. Although this does not represent a serious drawback in assays carried out for non-clinical purposes, or in clinical assays for isoenzymes such as prostatic acid phosphatase or placental alkaline phosphatase, it can represent a marked disadvantage of radioimmunoassay compared with other methods of determining amounts of the isoenzymes of creatine kinase in human serum, when the clinical need is often for a rapid confirmation of the occurrence of myocardial infarction.

Conventional radioimmunoassays have an inherent disadvantage where hybrid isoenzymes are concerned. Hybrid isoenzymes typically possess the antigenic determinants of both types of subunits of which they are composed; e.g., the MB isoenzyme of creatine kinase reacts with antisera to either the M or the B subunit. Therefore, a radioimmunoassay which uses an anti-B antiserum will measure both MB and BB isoenzymes present in the sample: i.e. the assay will lack specificity for the

MB isoenzyme, if this is the object of the assay. Specificity can be increased by two-site immunoradiometric assay. Two such assays, specific for human CK-MB, have been described [49, 50].

In the first procedure [49], serum is added to tubes coated with anti-CK-BB antibody. Isoenzymes containing B subunits (i.e. CK-MB and CK-BB) remain adsorbed to the tubes when the serum is discarded. After washing, anti-CK-MM antibodies labelled with ^{125}I are added. These become bound only to those sites where molecules of the MB isoenzyme are already bound. After removal of the unbound antibody by washing, the bound radioactivity is a measure of the MB-CK of the original sample.

Centrifugation is used in the second method [50] to separate soluble and insoluble phases. Serum is incubated with a mixture of rabbit antiserum to CK-M subunits and ^{125}I-labelled sheep antiserum to CK-B. Goat anti-rabbit immunoglobulin antiserum is then added. The anti-immunoglobulin antibodies are previously rendered insoluble by conjugation with polyvinylidene and amylose particles. Sedimentation of the particles by centrifugation brings down the two isoenzymes which contain M subunits, leaving excess labelled anti-B antibodies, free or bound to CK-BB, in the supernatant. The precipitated fraction of the labelled antibodies represents those bound to CK-MB; therefore, the radioactivity of the washed precipitate is proportional to the CK-MB content of the serum.

1.3.3.4 Conclusions

It is probable that immunological methods of analysis will play an increasing part in isoenzyme analysis, particularly where multi-locus isoenzymes and allelozymes are concerned. There is no reason to suppose that the methods outlined in this chapter exhaust the possible ways of determining the extent of the reaction between the isoenzyme antigen and the corresponding antibody. For example, one recently-described assay for isoenzymes of aspartate aminotransferase depends on the alteration of the scattering of light by a layer of indium particles which occurs when they are coated with dielectric layers, such as layers of antigens and antibody molecules [51]. Spots of solutions of purified cytoplasmic or mitochondrial isoenzymes of aspartate aminotransferase are placed on a layer of indium deposited on a glass slide. After incubation, unabsorbed antigen is removed by washing. Antiserum to the isoenzyme being determined and the serum sample are incubated together, then the prepared slides are incubated in the mixture. Serum proteins are adsorbed to the surface of the slide as a monolayer, except at the isoenzyme spots. Free specific antibodies in the serum will bind to the spots forming a visible double layer, which can be quantitated by densitometry. Free isoenzymes in the serum specimen compete with bound isoenzymes for their specific antibodies; thus affecting the absorbance of the double layer.

References

[1] *T. A. Tedesco, W. J. Mellman,* Galactosemia: Evidence for a Structural Gene Mutation, Science *172*, 727 – 728 (1971).

[2] *C. A. Slaughter, M. C. Coseo, M. P. Cancro, H. Harris,* Detection of Enzyme Polymorphism by Using Monoclonal Antibodies, Proc. Natl. Acad. Sci. USA *78*, 1124 – 1128 (1981).

[3] *F. G. Lehmann,* Immunological Relationship between Human Placental and Intestinal Alkaline Phosphatase, Clin. Chim. Acta *65*, 257 – 269 (1975).

[4] *J. S. Nisselbaum, O. Bodanksy,* Reactions of Lactic Dehydrogenase from Various Rabbit Organs with Antirabbit-muscle Lactic Dehydrogenase, J. Biol. Chem. *234*, 3276 – 3280 (1959).

[5] *H. M. Eppenberger, D. M. Dawson, N. O. Kaplan,* The Comparative Enzymology of Creatine Kinases. I. Isolation and Characterization from Chicken and Rabbit Tissues, J. Biol. Chem. *242*, 204 – 209 (1967).

[6] *N. Lang, S. Massarat,* Immunologie und Enzymkinetik der Glutamat-Oxaloacetat-Transaminase (GOT) und der Glutamat-Pyruvat-Transaminase (GPT), Klin. Wochenschr. *43*, 597 – 602 (1965).

[7] *M. Schlamowitz, O. Bodansky,* Tissue Sources of Human Serum Alkaline Phosphatase, as Determined by Immunochemical Procedures, J. Biol. Chem. *234*, 1433 – 1437 (1959).

[8] *B. K. Choe, E. J. Pontes, M. K. Dong, N. R. Rose,* Double-antibody Immunoenzyme Assay for Human Prostatic Acid Phosphatase, Clin. Chem. *26*, 1854 – 1859 (1980).

[9] *M. Usategui-Gomez, R. W. Wicks, M. Warshaw,* Immunochemical Determination of the Heart Isoenzyme of Lactate Dehydrogenase (LDH_1) in Human Serum, Clin. Chem. *25*, 729 – 734 (1979).

[10] *M. Usategui-Gomez, F. M. Yeager, A. Fernandez de Castro,* A Sensitive Immunochemical Method for the Determination of the Regan Isoenzyme in Serum, Cancer Res. *33*, 1574 – 1577 (1973).

[11] *F. G. Lehmann,* Immunological Methods for Human Placental Alkaline Phosphatase (Regan Isoenzyme), Clin. Chim. Acta *65*, 271 – 282 (1975).

[12] *C. Lee, M. C. Wang, G. P. Murphy, T. M. Chu,* A Solid-phase Fluorescent Immunoassay for Human Prostatic Acid Phosphatase, Cancer Res. *38*, 2871 – 2878 (1978).

[13] *H. Lang,* Creatine Kinase Isoenzymes. Pathophysiology and Clinical Applications, Springer Verlag, Berlin, 1981.

[14] *E. Jockers-Wretou, G. Pfleiderer,* Quantitation of Creatine Kinase Isoenzymes in Human Tissues and Sera by an Immunological Method, Clin. Chim. Acta *58*, 223 – 232 (1975).

[15] *D. Neumeier, W. Prellwitz, W. Würzburg, M. Brundobler, M. Olbermann, H. J. Just, M. Knedel, H. Lang,* Determination of Creatine Kinase Isoenzyme MB Activity in Serum Using Immunological Inhibition of Creatine Kinase M Subunit Activity. Activity, Kinetics and Diagnostic Significance in Myocardial Infarction, Clin. Chim. Acta *73*, 445 – 451 (1976).

[16] *W. Gerhardt, L. Ljungdahl, J. Börjesson, S. Hofvendahl, B. Hedenäs,* Creatine Kinase B-subunit Activity in Human Serum. I. Development of an Immunoinhibition Method for Routine Determination of S-creatine Kinase B-subunit Activity, Clin. Chim. Acta *78*, 29 – 41 (1977).

[17] *A. Lundin, I. Styrelius,* Sensitive Assay of Creatine Kinase Isoenzymes in Human Serum Using M Subunit Inhibiting Antibody and Firefly Luciferase, Clin. Chim. Acta *87*, 199 – 209 (1978).

[18] *J. Uriel,* Characterization of Enzymes in Specific Immune-Precipitates, Ann. N.Y. Acad. Sci. *103*, 956 – 979 (1963).

[19] *L. Korngold,* The Antigenic Properties of Alkaline Phosphatase of Human Kidney, Kidney Tumors and Urine, Int. Arch. Allergy *37*, 366 – 375 (1970).

[20] *A. G. Foti, H. Herschman, J. F. Cooper,* A Solid-phase Radioimmunoassay for Human Prostatic Acid Phosphatase, Cancer Res. *35*, 2446 – 2452 (1975).

[21] *C. L. Markert, E. Appella,* Immunochemical Properties of Lactate Dehydrogenase Isozymes, Ann. N.Y. Acad. Sci. *103*, 915 – 928 (1963).

[22] *F. C. Grimm, D. G. Doherty,* Properties of two Forms of Malic Dehydrogenase from Beef Heart, J. Biol. Chem. *236*, 1980 – 1985 (1961).

[23] *H. W. Goedde, K. Altland,* Evidence for Different "Silent Genes" in the Human Serum Pseudocholinesterase Polymorphism, Ann. N.Y. Acad. Sci. *151*, 540 – 544 (1968).

[24] *W. J. Arnold, J. C. Meade, W. N. Kelly,* Hypoxanthine-guanine Phosphoribosyltransferase: Characteristics of the Mutant Enzyme in Erythrocytes from Patients with the Lesch-Nyhan Syndrome, J. Clin. Invest. *51*, 1805 – 1812 (1972).

[25] *Y. Shibuta, T. Higashi, H. Hirai, H. B. Hamilton,* Immunochemical Studies on Catalase. II. An Anticatalase Reducing Component in Normal, Hypocatalasic and Acatalasic Human Erythrocytes, Arch. Biochem. Biophys. *118*, 200 – 209 (1967).

[26] *D. Stumpf, E. Neuwelt, J. Austin, P. Kohler,* Metachromatic Leukodystrophy $(MLD)_x$. Immunological Studies of the Abnormal Sulfatase A, Arch. Neurol. *25*, 427 – 431 (1971).

[27] *M. Kaminski,* Immunoelectrophoretic Examination of Avian Esterases, Nature *209*, 723 – 725 (1966).

[28] *D. W. Moss,* Immunochemical Similarities Between Major and Minor Components of Human Placental Alkaline Phosphatase, Clin. Chim. Acta *43*, 447 – 449 (1973).

[29] *B. Jacoby, K. D. Bagshawe,* A Radioimmunoassay for Placental-type Alkaline Phosphatase, Cancer Res. *32*, 2413 – 2420 (1972).

[30] *C. B. Laurell,* Electroimmuno Assay, Scand. J. Clin. Lab. Invest. *29*, Suppl. 124, 21 – 37 (1972).

[31] *D. T. Forman, D. W. Moss, K. B. Whitaker,* Determination of Human Placental Alkaline Phosphatase by an Electroimmunoassay ("rocket") Technique, Clin. Chim. Acta *68*, 287 – 290 (1976).

[32] *B. Nørgaard-Pedersen,* Red Cell Carbonic Anhydrases B and C (CA B and CA C), Scand. J. Immunol., Suppl. 1, 125 – 128 (1973).

[33] *V. Milisauskas, N. R. Rose,* Immunochemical Quantitation of Prostatic Phosphatase, Clin. Chem. *18*, 1529 – 1531 (1972).

[34] *A. Kahn, D. Cottreau, J. F. Bernard, P. Boivin,* Post Translational Modifications of Glucose-6-Phosphate Dehydrogenase in Human Leukemias, Biomedicine *22*, 539 – 594 (1975).

[35] *V. Milisauskas, N. R. Rose,* Immunochemical Quantitation of Enzymes in Human Diploid Cell Line WI-38, Exp. Cell Res. *81*, 279 – 284 (1973).

[36] *C. B. Laurell,* Antigen-antibody Crossed Electrophoresis, Anal. Biochem. *10*, 358 – 361 (1974).

[37] *T. C. Bøg-Hansen, J. Daussant,* Immunochemical Quantitation of Isoenzymes. α-Amylase Isoenzymes in Barley Malt, Anal. Biochem. *61*, 522 – 572 (1974).

[38] *F. Blomberg, M. Raftell,* Enzyme Polymorphism in Rat-liver Microsomes and Plasma Membranes. 1. An Immunochemical Study of Multi-enzyme Complexes and other Enzyme-active Antigens. Eur. J. Biochem. *49*, 21 – 29 (1974).

[39] *M. Raftell, F. Blomberg,* Enzyme Polymorphism in Rat-liver Microsomes and Plasma Membranes. 2. An Immunochemical Comparison of Enzyme-active Antigens Solubilized by Detergents, Papain or Phospholipases, Eur. J. Biochem. *49*, 31 – 39 (1974).

[40] *A. G. Foti, J. F. Cooper, H. Herschman,* Counterimmunoelectrophoresis in Determination of Prostatic Acid Phosphatase in Human Serum, Clin. Chem. *24*, 140 – 142 (1978).

[41] *T. M. Chu, M. C. Wang, W. W. Scott, R. P. Gibbons, D. E. Johnson, J. D. Schmidt, S. A. Loening, G. R. Prout, G. P. Murphy,* Immunochemical Detection of Serum Prostatic Acid Phosphatase. Methodology and Clinical Evaluation, Invest. Urol. *15*, 319 – 323 (1978).

[42] *V. S. Fang, H. W. Chu, H. Y. Meltzer,* Enzymic, Electrophoretic and Immunoreactive Properties of Labeled MM and MB Isoenzymes of Human Creatine Kinase, Enzyme *23*, 210 – 214 (1978).

[43] *V. E. Headings, R. E. Tashian,* A radioimmunoassay for Quantifying Carbonic Anhydrase Isozymes in Crude Lysates, Biochem. Genet. *4*, 285 – 295 (1970).

[44] *M. C. Geokas, F. Wollesen, H. Rinderknecht, F. Martinson,* Radioimmunoassay for Pancreatic Carboxypeptidase B in Human Serum, J. Lab. Clin. Med. *84*, 574 – 583 (1974).

[45] *J. P. Ryan, H. E. Appert, J. Carballo, R. H. Davies,* Radioimmunoassay of Serum Pancreatic Amylase in Normal and Pancreatectomized Pigs, Proc. Soc. Exp. Biol. Med. *148*, 194 – 197 (1975).

[46] *B. Geiger, R. Navon, Y. Ben-Joseph, R. Arnon,* Specific Determination of N-acetyl-β-D-hexosaminidase Isozymes A and B by Radioimmunoassay and Radial Immunodiffusion, Eur. J. Biochem. *56*, 311 – 318 (1975).

[47] *P. Vikko, E. Sajanti, O. Janne, L. Peltonen, R. Vihko,* Serum Prostate-specific Acid Phosphatase: Development and Validation of a Specific Radioimmunoassay, Clin. Chem. *24*, 1915 – 1919 (1978).

[48] *R. Roberts, B. E. Sobel, C. W. Parker,* Radioimmunoassay for Creatine Kinase Isoenzymes, Science *194*, 855 – 857 (1976).
[49] *V. J. C. Wilson, H. M. Jones, R. J. Thompson,* A Two-site Immunoradiometric Assay for the MB Isoenzyme of Creatine Kinase, Clin. Chim. Acta *113*, 153 – 163 (1981).
[50] *M. Usategui-Gomez, R. W. Wicks, B. Farrenkopf, H. Hager, M. Warshaw,* Immunochemical Determination of CK-MB Isoenzyme in Human Serum: A Radiometric Approach, Clin. Chem. *27*, 823 – 827 (1981).
[51] *R. Rej, C. R. Keese, I. Giaever,* Direct Immunological Determination of Aspartate Aminotransferase Isoenzymes, Clin. Chem. *27*, 1597 – 1601 (1981).

1.3.4 Selective Inactivation of Multiple Forms of Enzymes

Donald W. Moss

The catalytic activity of an enzyme depends on the specific three-dimensional structure of its molecule. This is stabilized mainly by numerous hydrophobic interactions between the amino acids of the polypeptide chains composing the enzyme molecule. The conformations of the individual polypeptides and the number and strength of the stabilizing bonds are determined by the primary structures of the polypeptide chains themselves. It is not surprising, therefore, that enzyme molecules of different structures differ in the rates at which they are denatured (i.e., at which the higher levels of molecular structure are disrupted) with an accompanying loss of catalytic activity, when exposed to agents which weaken the intramolecular bonds. Even the minor changes in amino-acid sequence, often involving only the replacement of a single amino acid by another, which result from allelic mutation are almost invariably accompanied by alterations in enzyme stability compared with that of the common form of the enzyme. Similarly, isoenzymes derived from multiple gene loci frequently differ in resistance to inactivation. Differences in stability are less marked between multiple forms of enzymes due to post-translational modification, but may occur in some cases.

Measurements of differences in rates of inactivation under controlled conditions are used extensively in distinguishing between isoenzymes, and can form the basis of methods of quantitative analysis of isoenzyme mixtures. Exposure to elevated temperatures or to concentrated solutions of urea or guanidine are most frequently chosen as the denaturing agents. Alternative procedures which may be useful include inactivation by oxidizing or reducing reagents, or by acid or alkaline conditions. However, whatever method is chosen, careful control of experimental conditions is essential to ensure that the small differences in stability which may be associated with minor variations in structure between isoenzymes are detected and measured reproducibly.

1.3.4.1 Inactivation by Heat

The heat-stabilities of individual members of a family of isoenzymes may differ only slightly; on the other hand, marked differences may be observed. Both close similarities and wide differences in stability to heat can co-exist amongst multiple forms of a single enzyme, e.g., human alkaline phosphatase. Placental alkaline phosphatase is completely stable to heating at temperatures up to 70 °C, whereas the enzyme prepared from bone loses half its activity in 8 min at 55 °C at pH 7 [1, 2]. In contrast, the difference in heat-stability between alkaline phosphatases from bone and liver is small (half-inactivation times at 55 °C are 8 and 18 min, respectively). However, it constitutes one of the few differences in properties by which these tissue-specific forms of the enzyme can be distinguished.

Isoenzymes which originate from multiple gene loci frequently differ in resistance to inactivation by heat. Differences in heat-stability between mitochondrial and cytoplasmic isoenzymes have been observed for aspartate aminotransferase [3] and NADP-dependent isocitrate dehydrogenase [4, 5] as well as for other mitochondrial and extra-mitochondrial isoenzymes. Isoenzymes with different stabilities to heat determined by multiple gene loci include the A and B forms of human hexosaminidase [6] and the MM and BB isoenzymes of creatine kinase from several species [7], as well as the well-studied isoenzymes of lactate dehydrogenase [8].

When hybrid isoenzymes consisting of unlike subunits can form, they usually show stability properties intermediate between those of the respective homopolymers; e.g., the isoenzymes of lactate dehydrogenase exhibit a progressive reduction in heat-stability from the most-stable H_4 tetramer (LDH-1) through the mixed tetramers H_3M_1 (LDH-2), H_2M_2 (LDH-3), and HM_3 (LDH-4) to the least-stable M_4 tetramer (LDH-5).

1.3.4.2 Practical Considerations in Analysis of Isoenzyme Mixtures by Heat-inactivation

Temperature-coefficients for the inactivation of enzymes by heat are high; e.g. values of 40 and 44, respectively, have been reported for human liver and bone alkaline phosphatases [9]. Therefore, slight variations in temperature between inactivation experiments can have the effect of obscuring the differences between isoenzymes in qualitative analysis and can introduce large errors into quantitative analysis by heat-inactivation.

Precise control of temperature is aided by the use of a large volume water-bath equipped with vigorous mechanical stirring, and with thermostatic control through a mercury-in-glass contact thermometer. Small volumes of isoenzyme samples should be introduced rapidly into thin-walled glass tubes, previously placed in the bath to attain temperature equilibrium. (Glass tubes are preferable to plastic tubes because of the greater thermal conductivity of the former material). At the end of the timed incubation period the tubes are removed rapidly to ice.

Rates of inactivation of isoenzymes by heat are generally affected by changes in factors such as pH, protein concentration, and the concentrations of substrates and co-factors. For example, the heat-stabilities of isoenzymes of lactate dehydrogenase and alkaline phosphatase are both pH-dependent [10, 11]. The effects of pH on stability can be controlled by buffer solutions in the examination of tissue-extracts or purified isoenzymes. However, a considerable dilution with buffer solutions is needed to overcome the strong buffering capacity of serum samples and this adversely affects the accuracy of assays of residual enzyme activity. Measurements of the heat-stabilities of isoenzymes in serum are usually made without addition of buffer solutions, therefore, although variations in pH undoubtedly contribute to the scatter of values of heat-stability which is observed between serum samples which contain mainly a single type of isoenzyme [11]. Although increases in protein concentration generally have a stabilizing effect on enzyme solutions, the BB dimer of human creatine kinase has been shown to be less stable when incubated at 37 °C in solutions containing human or bovine serum albumin [12]. The increased stability of enzymes in the presence of their substrates is well known and in some cases a differential effect on isoenzymes is found. For example, the heat-stability of the LDH-5 isoenzyme is increased in the presence of NADH [10, 13] and this substance has been incorporated in a heat-stability method for estimating the proportions of lactate dehydrogenase isoenzymes in serum [13].

The shorter analysis times required for selective inactivation methods compared with separative procedures such as electrophoresis or chromatography are of considerable advantage in routine clinical diagnostic analysis. Selective inactivation methods are particularly useful when separation between the isoenzymes of diagnostic importance is not clear-cut, as is the case for the liver and bone forms of alkaline phosphatase, for example, and these methods can be adapted to automated methods of enzyme analysis. However, some of the many variables which can affect the reliability and reproducibility of heat-inactivation measurements have already been mentioned, while the need to measure low residual activities calls for sensitive methods of assay of enzyme activity. These considerations have not always been given due weight in devising or applying heat-inactivation methods for clinical use. Nevertheless, such methods, in which residual activity is measured after heating serum samples for a single, fixed period, are widely employed to assess the relative proportions of the different isoenzymes of lactate dehydrogenase (e.g., [13]) or alkaline phosphatase (e.g., [14]). Measurements of the lability of amylase in serum at 65 °C have been used as an aid to the investigation of pancreatic disease [15].

Results are usually expressed and interpreted as the percentage of the enzymic activity of the unheated specimen remaining after heating; e.g. , less than 20% residual alkaline phosphatase activity after 10 min incubation at 56 °C suggests that the specimen contains predominantly bone phosphatase, while after 60 min at 60 °C, less than 30% residual lactate dehydrogenase activity indicates a shift towards the less-anodal isoenzymes and more than 60% a relative increase in the more anodal forms of this enzyme.

A disadvantage of these simple methods is their insensitivity to small changes in isoenzyme composition, because of the wide range of heat-stabilities observed with normal specimens. Although a small increase in the proportion of, say, a heat-labile isoenzyme will result in an abnormally low heat-stability in a serum in which the value is already towards the lower end of the normal range, a considerably greater increase will be needed if the heat-stability is initially at the higher end of the normal range. Consequently, total activity of the enzyme in question typically must be more than twice the upper limit of normal before a clear segregation of a series of samples into high or low heat-stability categories is observed [11].

Single-period heat-inactivation methods are also insensitive to changes in the relative proportions of isoenzymes when more than two forms are present in the sample. For example, the heat-stabilities of the hybrid isoenzymes of lactate dehydrogenase are not widely different. An approximate division of the activity of this enzyme among most-labile, most-stable, and intermediate-stability forms in serum can be achieved by heating samples for 30 min at each of two temperatures: the most labile isoenzymes are inactivated at 57 °C; the most stable survive incubation at 65 °C, while the difference between the residual activities at these two temperatures corresponds to the activity of the isoenzymes of intermediate stability [13].

A separate incubation at a different temperature can be used to detect an isoenzyme with distinctive heat stability: e.g. placental or carcinoplacental (*Regan*) isoenzymes of human alkaline phosphatase are the only forms of this enzyme which survive heating at 65 °C for 30 min or more.

Parallel measurements of heat-stability and of some other distinctive property can also be used to improve the analysis of multi-component mixtures. In the case of serum alkaline phosphatase, inactivation by heat has been combined with measurements of the activity in the presence of L-phenylalanine, an uncompetitive inhibitor of the intestinal isoenzyme, to correct for the effects of the latter isoenzyme on estimates of liver phosphatase, since these two isoenzymes have rather similar heat-stabilities [16, 17].

Electrophoretic discrimination of enzyme zones may be improved by comparing the patterns obtained with heated and unheated samples. Alternatively, selective inactivation of isoenzyme zones can be achieved *in situ* after electrophoretic separation in supporting media such as starch or polyacrylamide gels, by incubating the gels in sealed containers or by placing them between heated metal plates [18] before carrying out the appropriate reaction to demonstrate residual enzyme activity.

1.3.4.3 Quantitative Heat-inactivation Analysis

Heat-inactivation analysis in which a single period of heating is used is a qualitative or semi-quantitative procedure, except in those rare cases in which all but one of the isoenzymes are completely inactivated under the chosen conditions, or, even more rarely, one is completely inactivated while the others are unaffected. However, heat-inactivation can be made the basis of quantitative analysis, at least when two isoenzymes of different heat-stabilities are present in the sample.

The progress-curve of residual activity of a mixture of two such isoenzymes with respect to time of incubation at a fixed temperature consists of the sum of two exponential functions. Therefore, a plot of logarithm of residual activity against time can be resolved into two straight lines by "curve stripping", each representing the decay of the activity of one of the isoenzymes. The original activity of each isoenzyme is obtained from the intercept of its decay curve on the ordinate at zero time. Although in theory the curve-stripping technique can be applied to multi-component mixtures, the method works best in practice when only two, or in a few cases three, isoenzymes are present.

The method has been extensively studied in the quantitative analysis of bone and liver alkaline phosphatases in human serum, for which alternative methods such as densitometric scanning of electrophoretic patterns are less suitable [9].

In the analysis of progress-curves of heat-inactivation it is not necessary to assume that the heat-stabilities of specific isoenzymes are quantitatively identical in each serum specimen, provided that the qualitative differences in stabilitites between isoenzymes are maintained. Also, slight variations in incubation temperature from one determination to another do not affect the comparability of results, though constancy of temperature during each estimation is obviously essential.

Construction of heat-inactivation curves by multiple estimations of residual activity (e.g. as in [9]) is impracticable for routine clinical analysis of serum specimens. A method has been described by which residual alkaline phosphatase activity during incubation of buffered serum at 60 °C is recorded automatically with the *Technicon* AutoAnalyzer [19]. Subsequent conversion of the data to semilogarithmic plots of residual activity versus duration of incubation is required.

One simplified method for the quantitative analysis of mixtures of liver and bone alkaline phosphatases in serum is based on measurements of initial activity and activities remaining after incubation for periods of 15 and 25 min at 56 °C [20]. This procedure, which retains the advantage that the results are independent of between-sample variations in isoenzyme heat-stabilities possessed by complete curves, assumes that the portion of the decay curve between 15 and 25 min essentially corresponds to inactivation of liver alkaline phosphatase alone. The validity of this assumption under various circumstances and possible sources of error, as well as analytical details of the method and some clinical applications, have been discussed extensively [21]. Coefficients of variation of estimates of liver and bone alkaline phosphatases in serum are of the order of ±6% and ±10%, respectively, for elevated activities of the two isoenzymes. A mean half-life of liver alkaline phosphatase at 56 °C of 452 s ± 55 SD was found by analysis of 110 normal sera and 461 sera from patients with various diseases by the three-point method [21]. This value is close to that of 456 s ± 90 SD determined from full heat-inactivation curves [9], providing evidence of the validity of the simplified procedure.

1.3.4.4 Inactivation by Urea and Related Compounds

Concentrated solutions of urea or derivatives of urea such as guanidine and methylurea have a profound disruptive effect on higher levels of protein structure.

Therefore, differential actions of these compounds on the structures and catalytic activities of multiple forms of enzymes are much used in isoenzyme analysis and characterization. The reagents may have several effects on enzyme molecules, although these are not always distinguished in reports of their use.

Exposure to high concentrations of urea (6 – 12 mol/l) or rather lower concentrations (of the order of 5 mol/l) of guanidine may cause dissociation of polymeric isoenzymes into their component monomers, i.e., quaternary structure is disturbed. In some cases, re-combination of monomers to form active enzyme molecules occurs when the urea concentration is reduced, e.g., by dialysis. On the other hand, recovery of catalytic activity may be incomplete or absent, and in these instances the tertiary structure of the monomers has presumably also been disrupted.

Reversible, inhibitory effects of low concentrations of urea have been used in distinguishing between isoenzymes. However, these seem to be related to differences in catalytic properties rather than in resistance to denaturation, and the differential effects of urea and related compounds on higher levels of molecular structure are more useful in isoenzyme analysis. Plots of residual activity after treatment with urea as a function of urea concentration typically show a marked discontinuity, corresponding presumably to the concentration at which unfolding of the protein molecule takes place.

Inactivation of isoenzymes by concentrated solutions of urea and its derivatives is markedly dependent on temperature and on time of exposure to the reagents. Other variables, such as pH and the presence and concentration of substrates, also influence rates of denaturation; consequently, all these factors must be controlled to ensure reproducible results. Cyanate, itself an inhibitor of some enzymes, forms readily in solutions of urea, which should therefore be freshly prepared.

As with denaturation by heat, selective inactivation in the presence of urea has been made the basis of quantitative analysis of mixtures of isoenzymes of lactate dehydrogenase or alkaline phosphatase in serum [22, 23]. The principles of such analytical methods are similar to those used in heat-inactivation, i.e., measurement of residual activity after a period of exposure to the denaturing agent. In general, the results of the two approaches are qualitatively similar, in that those isoenzymes which are more heat-resistant are also less readily inactivated by urea. However, the correlation is not complete, suggesting that the modes of action of the two inactivating agents are not identical.

Because inactivation by urea can be carried out at the same temperature at which initial and residual enzymic activity is measured, urea-inactivation methods can be incorporated into fully automated systems of analysis more easily than is the case for inactivation by heat. Automated analysis has the advantage that, compared with manual methods, the conditions of exposure to the denaturing agent, which have such a pronounced effect on the extent of inactivation, can be standardized more completely for each sample. Several automated methods of isoenzyme analysis which make use of selective inactivation by urea have been described, particularly for determination of alkaline phosphatase isoenzymes in serum [24 – 26]. The scope of these methods is increased by the inclusion of L-phenylalanine, to provide an estimate of the contribution of intestinal alkaline phosphatase to the total activity. As with single-period heat-inactivation techniques, however, the methods based on denaturation by urea assume that the stability properties of specific isoenzymes are quantitatively identical in each serum.

If the rate of an enzyme-catalyzed reaction is monitored continuously in the presence of a denaturing agent, the residual activity at any instant can be derived from the slope of the reaction progress curve, and as many of such instantaneous rates as desired can be used in computing the curve of progress of inactivation. This approach has been used with urea as the denaturant and with a computer program to reduce the data to activities of bone and liver phosphatase [27].

1.3.4.5 Other Selective Inactivation Procedures

Many enzymes are metalloproteins and are therefore inactivated by metal-chelating agents, such as EDTA, which remove the constituent metal atoms. As with urea, the irreversible, time-dependent inactivation of metallo-enzymes by EDTA should be distinguished from reversible inhibition. In the case of EDTA, the latter is due to binding of soluble metal ions which may be activators of the catalytic process. These effects have been studied for alkaline phosphatase isoenzymes, which are zinc-containing proteins, and differences between human bone, intestinal and placental phosphatases exposed to EDTA have been demonstrated [28].

Several other agents with effects on the properties of proteins in general have been shown selectively to affect the activities of particular isoenzymes under certain circumstances. Some of these, such as alcohols or other organic solvents which are protein precipitants, were used in early attempts to demonstrate organ-specific characteristics of enzymes. Prostatic acid phosphatase is more susceptible to inactivation by ethanol (40% by vol.) than the enzyme from some other tissues [29, 30] while formaldehyde strongly inhibits erythrocyte acid phosphatase but not the prostatic isoenzyme [31]. Exposure to extremes of pH causes denaturation of proteins. Comparison of rates of inactivation of the liver and small-intestinal isoenzymes of human alkaline phosphatase at pH 2 – 4 at 0 °C shows the latter isoenzyme to be more resistant to this treatment [32].

References

[1] *D. W. Moss, E. J. King,* Properties of Alkaline Phosphatase Fractions Separated by Starch Gel Electrophoresis, Biochem. J. *84*, 192 – 195 (1962).

[2] *F. C. Neale, J. S. Clubb, D. Hotchkis, S. Posen,* Heat Stability of Human Placental Alkaline Phosphatase, J. Clin. Path. *18*, 359 – 363 (1965).

[3] *H. Wada, Y. Morino,* Comparative Studies on the Glutamic-Oxaloacetic Transaminases from the Mitochondrial and Soluble Fractions of Mammalian Tissues, Vitam. Horm. (N.Y.) *22*, 411 – 444 (1964).

[4] *D. M. Campbell, D. W. Moss,* Some Properties of Human Isocitric Dehydrogenase Isoenzymes, Proc. Assoc. Clin. Biochem. *2*, 10 – 11 (1962).
[5] *M. Islam, J. L. Bell, D. N. Baron,* Purification and Comparative Properties of Isoenzymes of Nicotinamide-Adenine Dinucleotide Phosphate-Isocitrate Dehydrogenase from Rat Heart and Liver, Biochem. J. *129*, 1003 – 1011 (1972).
[6] *D. Robinson, J. L. Stirling,* N-acetyl-β-glucosaminidases in Human Spleen, Biochem. J. *107*, 321 – 327 (1968).
[7] *D. M. Dawson, H. M. Eppenberger, N. O. Kaplan,* The Comparative Enzymology of Creatine Kinases. II. Physical and Chemical Properties, J. Biol. Chem. *242*, 210 – 217 (1967).
[8] *P. G. W. Plagemann, K. F. Gregory, F. Wroblewski,* Die elektrophoretisch trennbaren Lactat-dehydrogenasen des Säugetieres. III. Einfluß der Temperatur auf die Lactatdehydrogenasen des Kaninchens, Biochem. Z. *334*, 37 – 48 (1961).
[9] *L. G. Whitby, D. W. Moss,* Analysis of Heat Inactivation Curves of Alkaline Phosphatase Isoenzymes in Serum, Clin. Chim. Acta *59*, 361 – 367 (1975).
[10] *E. S. Vesell, K. L. Yielding,* Protection of Lactate Dehydrogenase Isozymes from Heat Inactivation and Enzymatic Degradation, Ann. N.Y. Acad. Sci. *151*, 678 – 689 (1968).
[11] *D. W. Moss, M. J. Shakespeare, D. M. Thomas,* Observations on the Heat-Stability of Alkaline Phosphatase Isoenzymes in Serum, Clin. Chim. Acta *40*, 35 – 41 (1972).
[12] *D. A. Nealon, A. R. Henderson,* Lability of Human Creatine Kinase Isoenzymes at 37 °C: a Complication of Electrophoretic Separation, J. Clin. Path. *28*, 834 – 836 (1975).
[13] *F. Wroblewski, K. F. Gregory,* Lactic Dehydrogenase Isozymes and their Distribution in Normal Tissues and Plasma and in Disease States, Ann. N.Y. Acad. Sci. *94*, 912 – 932 (1961).
[14] *S. Posen, F. C. Neale, J. S. Clubb,* Heat Inactivation in the Study of Human Alkaline Phosphatases, Ann. Intern. Med. *62*, 1234 – 1243 (1965).
[15] *L. A. Donaldson, W. B. McIntosh, M. J. Brodie,* Amylase Thermolability in Body Fluids, Scand. J. Gastroenterol. *12*, 637 – 639 (1977).
[16] *J. Winkelman, S. Nadler, J. Demetriou, V. J. Pileggi,* The Clinical Usefulness of Alkaline Phosphatase Isoenzyme Determinations, Am. J. Clin. Path. *57*, 625 – 634 (1972).
[17] *J. Stepan, V. Volek, J. Kolar,* A Modified Activation-Inhibition Method for Determining the Serum Activity of Alkaline Phosphatase Isoenzymes, Clin. Chim. Acta *69*, 1 – 9 (1976).
[18] *P. J. McAlpine, D. A. Hopkinson, H. Harris,* Thermostability Studies on the Isozymes of Human Phosphoglucomutase, Ann. Hum. Genet. *34*, 61 – 71 (1970).
[19] *C. PetitClerc,* Quantitative Fractionation of Alkaline Phosphatase Isoenzymes According to their Thermolability, Clin. Chem. *22*, 42 – 48 (1976).
[20] *D. W. Moss, L. G. Whitby,* A Simplified Heat-Inactivation Method for Investigating Alkaline Phosphatase Isoenzymes in Serum, Clin. Chim. Acta *61*, 63 – 71 (1975).
[21] *K. B. Whitaker, L. G. Whitby, D. W. Moss,* Activities of Bone and Liver Alkaline Phosphatases in Serum in Health and Disease, Clin. Chim. Acta *80*, 209 – 220 (1977).
[22] *W. G. Brydon, A. F. Smith,* An Appraisal of Routine Methods for the Determination of the Anodal Isoenzymes of Lactate Dehydrogenase, Clin. Chim. Acta *43*, 361 – 369 (1973).
[23] *J. J. Fennelly, J. Dunne, K. McGeeney, L. Chong, M. Fitzgerald,* The Importance of Varying Molecular Size, Differential Heat and Urea Inactivation of Phosphatase in the Identification of Disease Patterns, Ann. N.Y. Acad. Sci. *166*, 794 – 807 (1969).
[24] *B. E. Statland, H. H. Nishi, D. S. Young,* Serum Alkaline Phosphatase: Total Activity and Isoenzyme Determinations made by use of the Centrifugal Fast Analyzer, Clin. Chem. *18*, 1468 – 1474 (1972).
[25] *W. Gerhardt, M. L. Nielsen, O. V. Nielsen, J. S. Olsen, B. E. Statland,* Routine Measurements of Liver and Bone Alkaline Phosphatase in Human Serum. Differential Inhibition by L-Phenylalanine and Carbamide (urea) on the LKB 8600 Reaction Rate Analyzer, Clin. Chim. Acta *53*, 281 – 290.
[26] *J. C. Kim, M. Statsny, J. P. Manning,* Use of a "Three-Isoenzyme Diagram" in Assessing Activity of Alkaline Phosphatase Isoenzymes, Clin. Chem. *20*, 816 – 819 (1974).
[27] *P. B. Brown, K. O. Lewis,* Reaction Rate Retardation as a Method for Serum Alkaline Phosphatase Isoenzyme Measurement. Ann. Clin. Biochem. *17*, 192 – 198 (1980).
[28] *R. A. J. Conyers, D. J. Birkett, F. C. Neale, S. Posen, J. Brudenell-Woods,* The Action of EDTA on Human Alkaline Phosphatases. Biochim. Biophys. Acta *139*, 363 – 371 (1967).

[29] *W. Kutscher, W. Wörner,* Prostataphosphatase. 2. Mitteilung, Hoppe-Seyler's Z. Physiol. Chem. *239*, 109 – 126 (1936).
[30] *F. K. Herbert,* The Differentiation Between Prostatic Phosphatase and other acid Phosphatases in Pathological Human Sera, Biochem. J. *38*, xxiii-xxiv (1944).
[31] *M. A. M. Abul-Fadl, E. J. King,* The Inhibition of Acid Phosphatases by Formaldehyde and its Clinical Application for the Determination of Serum Phosphatases, J. Clin. Path. *1*, 80 – 90 (1948).
[32] *P. B. Scutt, D. W. Moss,* Reversible Inactivation of Alkaline Phosphatase in Acid Solution, Enzymologia *35*, 157 – 167 (1968).

2 Oxidoreductases: Dehydrogenases Acting on CHOH Groups

2.1 Sorbitol Dehydrogenase

L-Iditol : NAD^+ 5-oxidoreductase, EC 1.1.1.14

Ulrich Gerlach

General

Sorbitol dehydrogenase, SDH, was first partially purified in 1951 by *Blakley* [1] from rat liver. Since then SDH has been demonstrated in various human, animal and plant tissues [2 – 5]. *Smith* crystallized the enzyme in 1962 [6]. It is mainly located in the cytoplasm and in the mitochondria [7] of liver, kidney and seminal vesicles [8, 9].

Comparative studies have shown that SDH from *Acetobacter suboxydans* [5] and from liver can be differentiated by their substrate specificities. Depending on the type of coenzyme used, the dehydrogenases from *Acetobacter suboxydans* yield either D-fructose or L-sorbose as the oxidation product of D-sorbitol [10].

The concentration of SDH in the cell is several orders of magnitude higher than that in the extracellular space. This large concentration gradient leads to an increase of SDH activity in the serum of patients when the membrane of liver cells is damaged.

Application of method: in clinical diagnosis and in biochemistry. The clinical importance [8, 9, 11 – 13] of the assay of SDH is based on the observation that SDH is virtually absent from serum of healthy subjects, whereas it is detectable in liver damage (infectious, toxic or hypoxic). SDH activity in serum can therefore serve as a virtually organ-specific indicator of liver cell damage [8, 9]. The physiological function [14, 15] of cytoplasmic sorbitol dehydrogenase (equation a) is possibly associated with an NADP-specific aldose reductase which converts glucose into sorbitol.

Enzyme properties relevant in analysis: the catalytic activity of human serum SDH is of hepatocellular origin. Substrate conversion proceeds linearly with time. The molecular weight of SDH is not known. The *Michaelis-Menten* constant for sorbitol of the enzyme purified from liver is 0.7 mmol/l at pH 8.7 [1]. The pH optimum depends on the source of the enzyme and is between 6.1 and 8.1 [1, 16].

Methods of determination: the catalytic activity [17, 18] can be measured spectrophotometrically. Estimation of enzyme mass, of isoenzymes or active-site titration is not in use.

International reference methods and standards: neither standardization at the international level nor the existence of reference standard materials is known.

Enzyme effectors: special activators of SDH which need to be added to the assay mixture are not known. SDH activity is inhibited by borate, cyanide, monoiodoacetate, EDTA, $AgNO_3$, $HgCl_2$ and 2,3-dimercaptoethanol.

Assay

Method Design

Principle

(a) D-Sorbitol + NAD^+ $\underset{}{\overset{SDH}{\rightleftharpoons}}$ D(−)-fructose + NADH + H^+

The amount of fructose reacting per unit time, measured by the decrease in absorbance due to the oxidation of NADH, is a measure of the catalytic activity of SDH. SDH activity can be measured in either direction. For measurement in serum the use of D-fructose as substrate is preferable.

Blanks are not necessary for routine assays. The method described below can be adapted to automated equipment.

Optimized conditions for measurement: the optimum concentration is 0.4 mol/l fructose in triethanolamine buffer, pH 7.4 [13]. The pH optimum of the reaction with fructose is pH 6.1 in triethanolamine buffer and in tris(hydroxymethyl)aminomethane buffer. The reaction in Tris buffer is somewhat faster than in triethanolamine buffer, but it requires a higher concentration of fructose to saturate the enzyme. The SDH reaction in various buffers is about 1 − 2 orders of magnitude faster with NADH than with NADPH.

In the following method conditions are described which are most satisfactory for the measurement of SDH activity in serum in several states of liver disease.

Temperature conversion factors: conversion factors cannot be applied generally to the individual case. The value of the following relative reaction rates thus lies in indicating the temperature dependence of the reaction and, at the most, providing a qualitative comparison of values up to the upper limit of normal.

°C	25	30	37	40
F	1.00	1.23	1.63	1.74

Equipment

Spectrophotometer or spectral-line photometer capable of exact measurement at 339 nm, Hg 334 or Hg 365 nm, preferably with a thermostatted cuvette holder; water-bath; recorder or stopwatch.

Reagents and Solutions

Purity of reagents: triethanolamine hydrochloride, sodium hydroxide (2 mol/l), reduced nicotinamide-adenine dinucleotide, NADH, D(–)-fructose; suitable commercial preparations are available.

Preparation of solutions (for about 12 determinations): all solutions in re-purified water (see Vol. II, chapter 2.1.3.2).

1. Triethanolamine buffer (0.2 mol/l; pH 7.4):

   dissolve 37.2 g triethanolamine hydrochloride (TEA) in water, adjust to pH 7.4 with NaOH, 2 mol/l, and dilute to 1000 ml with water.

2. Reduced nicotinamide-adenine dinucleotide (β-NADH, 12 mmol/l):

   dissolve 15 mg NADH, sodium salt, with 1.5 ml $NaHCO_3$ solution, 1% (w/v).

3. D(–)-fructose (4.0 mol/l):

   dissolve 72.06 g D(–)-fructose in water and dilute to 100 ml with water.

Stability of solutions: the triethanolamine buffer is stable for about a year at room temperature and the D-fructose solution at 4 °C for about four weeks, providing that it is not contaminated with micro-organisms. The NADH solution should be prepared freshly each week and stored at 4 °C.

Procedure

Collection and treatment of specimen: collect blood from the vein. Avoid the addition of anticoagulants, even though heparin does not affect the enzyme activity appreciably [12, 13]. Centrifuge after coagulation. Serum from haemolysed blood can be analyzed without any appreciable error because erythrocytes are low in SDH.

Stability of enzyme in the sample: the enzyme is thermolabile in serum. Table 1 shows the stability at different temperatures [12].

Table 1. Stability of SDH in serum at different temperatures, the initial value is the activity of fresh serum

Period of storage	Loss of activity (% of initial value) after storage of serum at 21 °C	4 °C	–18 °C
24 hours	20.5	5.5	0
48 hours	49	10.6	0

Details of measurement in tissues: for the determination of SDH activity in tissue [2, 8, 9, 15] the samples are homogenized and centrifuged in the cold at high speed. Care must be taken that the sample is processed quickly, because under anaerobic and post-mortem conditions the concentration of sorbitol in tissues rises [19]. The clear supernatant fluid can be used directly for the assay, but the optimum conditions for human serum do not necessarily apply to tissue preparations; in particular, different conditions apply to purified SDH preparations [1, 15, 20]. With sorbitol as substrate, the most favourable conditions for crude rat liver extracts are Tris buffer (pH 9) and a sorbitol concentration of 0.15 mol/l. Proportionality between the activity and the amount of enzyme added as well as linear progress curves are obtained. In the reverse direction the reaction (with fructose, 0.4 mol/l) in triethanolamine buffer (pH 7.4) is about 8 times faster; in Tris buffer (pH 7.4) it is about 5 times faster.

It should be noted that the determination of liver SDH usually includes the cytoplasmic SDH and the mitochondrial xylitol dehydrogenase. These enzymes are virtually homo-specific, but not identical [14].

Assay conditions: wavelength 339, Hg 334 or Hg 365 nm; light path 10 mm; final volume 3.0 ml; 25 °C (thermostatted cuvette holder). Adjust temperature of solutions to 25 °C before starting the assay; read against air or water. Prepare a reagent blank (containing water instead of serum) for each series of measurements.

Measurement

Pipette successively into the cuvette:			concentration in assay mixture	
TEA buffer	(1)	1.6 ml	TEA	0.321 mmol/l
NADH solution	(2)	0.1 ml	NADH	0.0012 mmol/l
serum		1.0 ml	volume fraction	0.333
mix and incubate for about 30 min at 25 °C until the absorbance is constant				
fructose solution	(3)	0.3 ml	fructose	1.19 mmol/l
mix and read the absorbance at 60 s intervals for 5 – 8 min.				

The values $\Delta A/\Delta t$ must be <0.030/min at 365 nm. Otherwise dilute sample 5 to 10 fold or measure in shorter intervals.

Calculation: use calculation formula (k) or (k_1), (cf. 'Formulae', Appendix 3). Values for ε: cf. Appendix 4. The following relations are valid (Δt in minutes):

	Hg 334 nm	339 nm	Hg 365 nm	
$b =$	$486 \times \Delta A/\Delta t$	$476 \times \Delta A/\Delta t$	$883 \times \Delta A/\Delta t$	U/l
$b =$	$8099 \times \Delta A/\Delta t$	$7944 \times \Delta A/\Delta t$	$14720 \times \Delta A/\Delta t$	nkat/l

Validation of Method

Precision, accuracy, detection limits and sensitivity: for values of about 30 U/l in serum there is an imprecision of about ±10 U/l for manual performance. The standard deviation is $s = 0.14$ U/l. Data about accuracy are not available since standard reference material is not established yet. For measurements at Hg 365 nm with sufficiently sensitive photometers the lowest possible value in serum to be measured is approximately 1 U/l with an imprecision $< \pm 25\%$. Sensitivity is approximately $\Delta A/\Delta t = 0.005/10$ min.

Sources of error: false results occur with insufficient pre-incubation before the start of the enzyme reaction because metabolites in serum, especially ketoses, can react with dehydrogenases in serum with resultant oxidation of NADH. Influences of therapeutic substances on the measured activity are unknown. With regard to the influence of enzyme inhibitors in NADH preparations of insufficient quality, cf. Vol. II, chapter 2.1.2.2.

Specificity: SDH catalyzes the following reaction in the presence of NAD or NADH: ribitol ⇌ D-ribulose; xylitol ⇌ D-xylulose; L-iditol ⇌ L-sorbose; allitol ⇌ D-allulose; L-gala-D-glucoheptitol ⇌ perseulose (L-gala-heptulose) and D-altro-D-glucoheptitol ⇌ sedoheptulose (D-altroheptulose). After sufficiently long pre-incubation and stabilization of the absorbance, SDH activity is measured specifically after the start of the reaction with D(−)-fructose.

Reference ranges: the catalytic concentration of SDH in the serum of healthy subjects is less than 1.5 U/l. The SDH activity content in human organs is given in Table 2.

Table 2. SDH activity content in human organs obtained at autopsy

Organ	SDH activity (25 °C) U/g fresh wt	U/g dry wt
Liver	5.216	20.57
Prostate	1.243	7.70
Kidney	1.108	5.27
Spleen	0.382	1.90
Testes	0.173	1.16
Lymph nodes	0.193	0.85
Heart, left ventricle	0.091	0.51
Skeletal muscle	0.121	0.45

References

[1] *R. L. Blakley,* The Metabolism and Antiketogenic Effects of Sorbitol. Sorbitol Dehydrogenase, Biochem. J. *49*, 257 – 271 (1951).

[2] *H. Holzer, J. Haan, S. Schneider,* Zum Mechanismus des anaeroben Glucose- und Fruktose-Abbaues im Mäuse-Ascites-Carcinom, Biochem. Z. *326*, 432 – 463 (1955).

[3] *H. G. Williams-Ashman, J. Banks, S. K. Wolfson,* Oxidation of Polyhydric Alcohols by the Prostate Gland and Seminal Vesicle, Arch. Biochem. *72*, 485 – 494 (1957).

[4] *J. McCorcindale, N. L. Edson,* Polyol Dehydrogenases, The Specificity of Rat Liver Polyol Dehydrogenase, Biochem. J. *57*, 518 – 523 (1954).

[5] *A. C. Arcus, N. L. Edson,* Polyol Dehydrogenases, The Polyol Dehydrogenases of Acetobacter Suboxydans and Candida Utilis, Biochem. J. *64*, 385 – 394 (1956).

[6] *M. G. Smith,* Polyol Dehydrogenases, Crystallization of the L-Iditol Dehydrogenase of Sheep Liver, Biochem. J. *83*, 135 – 144 (1962).

[7] *S. Hollmann, O. Touster,* The L-Xylulose-Xylitol Enzyme and Other Polyol Dehydrogenases of Guinea Pig Liver Mitochondria, J. Biol. Chem. *225*, 87 – 102 (1957).

[8] *U. Gerlach,* Pathologischer Übertritt von Sorbitdehydrogenase ins Blut bei Lebererkrankungen, Klin. Wochenschr. *35*, 1144 – 1145 (1957).

[9] *U. Gerlach,* Zur klinischen Bedeutung der Aktivitätsmessung von Sorbitdehydrogenase im menschlichen Blutserum, Klin. Wochenschr. *37*, 93 – 98 (1959).

[10] *J. T. Cummins, T. E. King, V. H. Cheldelin,* The Biological Oxidation of Sorbitol, J. Biol. Chem. *224*, 323 – 329 (1957).

[11] *U. Gerlach,* Enzymaktivität im Serum bei Krankheiten der Leber und Gallenwege, in: *F. W. Schmidt* (eds.), Praktische Enzymologie, Hans Huber Verlag, Bern und Stuttgart 1968, pp. 165 – 196.

[12] *E. Schmidt, F. W. Schmidt, E. Wildhirt,* Ferment-Aktivitäts-Bestimmung in der menschlichen Leber, VI. Klin. Wochenschr. *37*, 1221 – 1229 (1959).

[13] *H. Schön, H. Wüst,* Untersuchungen über die Sorbitdehydrogenase des Serums und ihre Bedeutung für die Leberdiagnostik, Klin. Wochenschr. *38*, 497 – 500 (1960).

[14] *H. G. Hers,* Le Mécanisme de la Transformation de Glucose et Fructose par les Vesicules Séminales, Biochim. Biophys. Acta *22*, 202 – 203 (1956).

[15] *S. Hollmann,* Trennung, Reinigung und Eigenschaften der mitochondrialen Xylit-Dehydrogenasen der Meerschweinchenleber, Hoppe-Seylers Z. physiol. Chem. *317*, 193 – 216 (1959).

[16] *S. Hollmann,* Polyol-Dehydrogenasen, Dehydrogenasen von On- und Leronsäuren, Glucose-Dehydrogenase, in: *Hoppe-Seyler/Thierfelder,* Handbuch der physiologischen und pathologisch-chemischen Analyse, Vol. VI, part A, Springer Verlag, Berlin – Heidelberg – New York 1964, pp. 704 – 732.

[17] *W. R. Todd, J. Vreeland, J. Myers, E. S. West,* A Rapid Micromethod for the Quantitative Estimation of Sugar Alcohols, J. Biol. Chem. *127*, 269 – 273 (1939).

[18] *J. D. Anderson, P. Andrews, L. Hough,* The Incorporation of ^{14}C into the Sorbitol of Plum Leaves, Biochem. J. *72*, 9P-10P (1959).

[19] *H. G. Britton,* Sorbitol in Animal Tissues, Biochem. J. *69*, 5P (1958).

[20] *J. B. Wolf,* Sorbitol Dehydrogenase from Liver, in: *S. P. Colowick, N. O. Kaplan* (eds.), Methods in Enzymology, Vol. *I*, Academic Press, New York 1955, pp. 348 – 350.

2.2 Lactate Dehydrogenase

L-lactate: NAD oxidoreductase, EC 1.1.1.27

2.2.1 UV-method with Pyruvate and NADH

Anne Vassault

General

Lactate dehydrogenase, LDH, the "reduzierende Gärungsferment" of *Warburg* [1] has been purified and crystallized from various animal muscles as well as from yeast and bacteria [2]. In human tissue extracts, the LDH activity (per g wet weight) decreases in the following order [3, 4]: kidney > heart > skeletal muscle > pancreas > spleen > liver > lung > placenta. LDH activity is present in erythrocytes, thrombocytes and leukocytes. The enzyme is located in the cytoplasma of the cells.

The enzyme catalyzing the reversible reduction of pyruvate has been shown [5, 6] to exist in five genetically distinct molecular forms in mammals (isoenzymes). They differ in their catalytic and immunological properties, amino acid composition and physical characteristics, though they have similar molecular weights. Each isoenzyme is a tetramer composed of two types of monomeric subunits, usually referred to as the M (muscle) or A form, and the H (heart) or B form. As the names indicate, the M subunit is produced mainly in tissues capable of anaerobic metabolism, such as skeletal muscle, while the H subunit is mainly found in tissues with aerobic metabolism, such as cardiac muscle [7].

Application of method: in biochemistry and in clinical biochemistry.

Enzyme properties relevant in analysis: the catalytic activity of serum LDH is due to the presence of several enzyme proteins which may originate from liver, heart, skeletal muscle, erythrocytes, tumours, etc. Isoenzymes have been shown to display different kinetic properties towards, e.g., pH, substrates and cofactors.

Methods of determination: a number of procedures have been developed for measuring total LDH activity. The most direct is a spectrophotometric measurement by monitoring the increase or decrease of absorbance of NADH at 339 (Hg 334 or 365) nm. Thus, in the pyruvate to lactate reaction [8, 9], the rate of decrease is measured, whereas in the lactate to pyruvate reaction [10] the rate of increase is measured (eqn.

(a) on p. 120). Colorimetric measurements have been employed using phenazine methosulphate [11 – 13] or diaphorase* [14] as electron carriers, and tetrazolium salts which are colourless before they have been reduced by NADH. These procedures have been largely given up because non-specific reactions occur; therefore the procedures currently recommended by several national scientific societies are direct spectrophotometric measurements using the pyruvate to lactate reaction [15 – 17] (Table 1).

Table 1. Recommended methods for total LDH measurement

References	Temperature °C	Buffer mmol/l	Buffer pH	Pyruvate mmol/l	NADH mmol/l	EDTA mmol/l	NaCl mmol/l	Volume fraction
1972 DGKC [15]	25	Phosphate 50	7.5	0.6	0.18	–	–	0.033
1974 SCE [16]	37	Tris 50	7.4	1.2	0.15	5	–	0.022
1982 SFBC [17]	30	Tris 80	7.2	1.6	0.20	–	200	0.0164

DGKC Deutsche Gesellschaft für Klinische Chemie
SCE Scandinavian Committee on Enzymes
SFBC Société Française de Biologie Clinique

These methods differ in the temperature, the buffers and the concentration of pyruvate used. Up to now, the optimal concentration of pyruvate has been difficult to select, because of the higher optimal concentration required for the M subunit than for the H subunit. Recently, a method at 30 °C has been proposed [18] in which Na^+ and Cl^- ions are added. The presence of these ions in the reaction mixture allows differences in the kinetic properties of the different isoenzymes towards the pyruvate concentration to be minimized (Fig. 1).

International reference methods and standards: standardization at the international level is not available so far; for available standard reference material cf. Vol. II, chapter 2.3.

Enzyme effectors: formation of a competitive inhibitor can occur during storage of NADH solutions, producing large variations of the maximal rate of LDH measurement. The formation of these inhibitors is known to occur especially when NADH is dissolved in phosphate buffer. It is now well documented [19] that Tris buffer/NADH solution can be stored at +4 °C during 5 days and at −20 °C during several months without any variation in measured LDH activity.

* NADH: lipoamide oxidoreductase, EC 1.6.4.3.

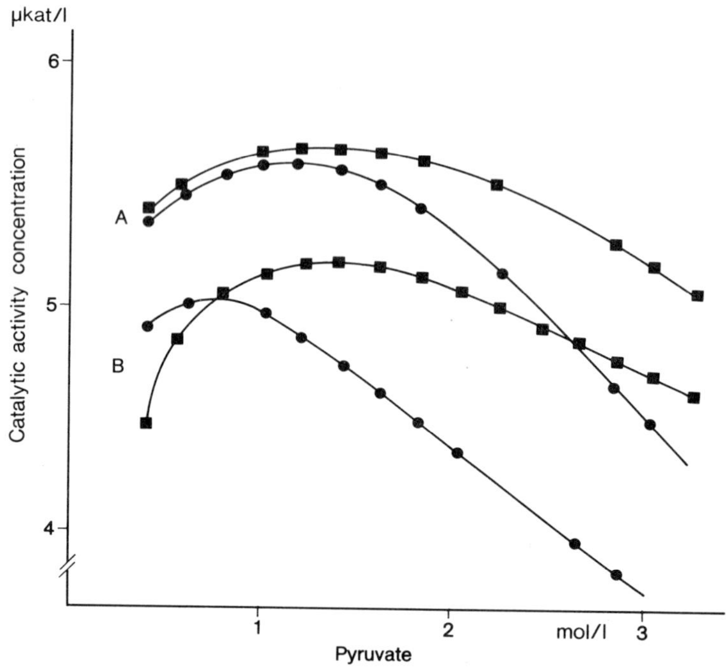

Fig. 1. Choice of the optimal concentration of pyruvate. Influence of Na^+ and Cl^- ions on measured catalytic activity of total LDH in serum at 30 °C.
A: serum from a patient with acute hepatitis
B: serum from a patient with acute myocardial infarction.
● Without NaCl
■ With NaCl

Assay

Method Design

Principle

(a) Pyruvate + NADH + H^+ $\xrightleftharpoons{\text{LDH}}$ L(+)-lactate + NAD^+.

The LDH activity is measured at 30 °C as the amount of pyruvate consumed, by continuously monitoring the decrease in absorbance due to oxidation of NADH at 339 (Hg 334 or Hg 365) nm. The equilibrium is far on the side of lactate and NAD. For routine measurement, blanks are not necessary. The method described below can easily be adapted to automated equipment.

Optimized conditions for measurement: the conditions of the reaction have been selected in order to obtain, at 30 °C, an equal relationship between activity measurement and catalytic concentration, whatever the isoenzyme pattern of the sample might be (i.e. to minimize the differences in the behaviour of the isoenzymes), to obtain a maximal velocity with linear reaction rate over a wide range of activity, and to obtain as little sensitivity as possible to pH variations or to variations in molarity or volume

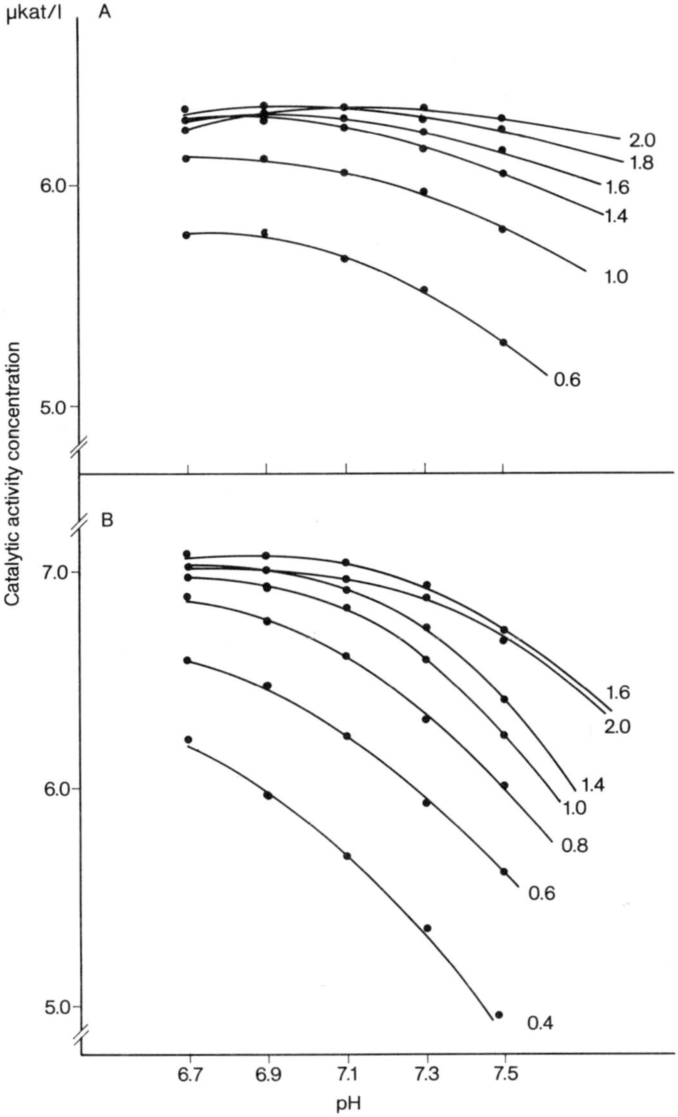

Fig. 2. Dependence of the activity on pH at several pyruvate concentrations for samples A and B. (Tris buffer, 80 mmol/l; pH 7.2; NaCl, 200 mmol/l; NADH, 0.2 mmol/l; 30 °C)
A: serum from a patient with acute hepatitis
B: serum from a patient with acute myocardial infarction.

fraction. The choice of the pyruvate to lactate direction is advantageous [20] especially because V is about three times greater and reagent solutions are more stable than for the lactate to pyruvate reaction.

At high concentrations of pyruvate, substrate inhibition can be observed, especially for the H isoenzyme, so that the optimal concentration of pyruvate is quite different for H and M subunits [21 – 23]. This inhibition can be modified by a change of the electrolyte concentration in the reaction mixture [24]. Addition of Na^+ and Cl^- ions (200 mmol × l^{-1}) increases the apparent K_m for pyruvate of the H subunits so that it becomes close to the apparent K_m of the M subunits. Thus, the optimal concentration of pyruvate for the H subunits is close to that for the M subunits (Fig. 1). At 30 °C with the selected conditions, the optimal pH is about 7.2 for the different isoenzymes and allows the use of tris(hydroxymethyl)aminomethane (Tris) buffer (Fig. 2).

In the following method, pyruvate, 1.6 mmol/l; NADH, 0.2 mmol/l; NaCl, 200 mmol/l; Tris buffer, 80 mmol/l; pH 7.2 are used as these conditions are the most satisfactory for the measurement of total LDH catalytic activity in serum after myocardial infarction, in acute hepatitis, leukaemias, anaemias or malignant diseases.

Temperature conversion factors: they are dependent on the nature of the isoenzymes and cannot be recommended. Nevertheless a mean correction factor of 0.66 has been statistically established for the conversion of results obtained at 37 °C related to the values measured at 30 °C.

Equipment

Spectrophotometer or spectral-line photometer capable of exact measurement at 339 nm, Hg 334 or Hg 365 nm and with temperature control within ± 0.1 °C by means of a thermostatted cuvette holder; recorder or stopwatch.

Reagents and Solutions

Purity of reagents: NADH and pyruvate must be free from inhibitors (cf. Vol. II, chapter 2.1.2.2); suitable commercial preparations are available.

Preparation of solutions (for about 30 determinations): all solutions in re-purified water (cf. Vol. II, chapter 2.1.3.2).

1. Tris/NaCl solution (Tris, 81.3 mmol/l; NaCl, 203.3 mmol/l; pH 7.2):

   dissolve 4.92 g Tris and 5.95 g NaCl in 400 ml water. Adjust to pH 7.2 at 30 °C with 5 mol/l HCl. Make up to 500 ml with water.

2. Tris/NaCl/NADH solution (Tris, 81.3 mmol/l; NaCl, 203.2 mmol/l; NADH, 0.244 mmol/l; pH 7.2):
   dissolve 0.017 g NADH, disodium salt, in 100 ml solution (1).

3. Tris/NaCl/pyruvate solution (Tris, 81.3 mmol/l; NaCl, 203.3 mmol/l; pH 7.2; pyruvate, 9.76 mmol/l):
   dissolve 0.107 g pyruvate, crystallized monosodium salt, in 100 ml solution (1).

Stability of solutions: store solution (1) at 0 – 4 °C. This solution is stable as long as no bacterial contamination occurs. Solution (2) can be frozen and stored for at least 6 months. Solution (3) is stable at 4 °C or – 20 °C for not more than 2 months.

Procedure

Collection and treatment of specimen: collect blood from the vein without stasis. Addition of lithium heparinate (15 U/ml) does not affect the results of the measurement of LDH. Addition of EDTA (1 mg/ml), oxalate (3 mg/ml), citrate (0.1 mg/ml), fluoride (2 mg/ml), iodoacetate (0.5 mg/ml) is not suitable [25].

Centrifuge for 10 min at about 3000 *g* in order to obtain serum or plasma. Use only fresh serum or plasma free from haemolysis.

Stability of the enzyme in the sample: according to [25, 26] the enzyme in serum is stable for a 48 h period at +20 °C or +4 °C. After freezing, activity decreases according to the isoenzymic pattern (10 to 20 percent in 48 h).

Details for measurement in tissues: simple homogenization in a *Potter-Elvehjem* homogenizer (cf. Vol. II, chapter 1.2.2) is sufficient to extract all the enzyme completely because it is located in the cytoplasm of the cells. The activity in tissues is determined in the supernatant fluid obtained after centrifugation of the homogenate at high speed. For LDH measurement in urine, dialysis or treatment with an anion exchange resin are necessary before the determination [27].

Assay conditions: wavelength 339 (Hg 334, Hg 365) nm; light path 10 mm; final volume 3.05 ml; temperature 30 °C (thermostatted cuvette holder). Measure against air. Before starting the assay, adjust temperature of solutions to 30 °C.

Calculation: use calculation formula (k) or (k_1), resp. (cf. "Formulae", Appendix 3). Values for ε: cf. Appendix 4. The following relations are valid (Δt in minutes):

	Hg 334 nm	339 nm	Hg 365 nm	
$b =$	$165 \times \Delta A/\Delta t$	$161 \times \Delta A/\Delta t$	$299 \times \Delta A/\Delta t$	μkat/l
$b =$	$9918 \times \Delta A/\Delta t$	$9682 \times \Delta A/\Delta t$	$17941 \times \Delta A/\Delta t$	U/l

Measurement

Pipette successively into the cuvette:			concentration in assay mixture	
Tris/NaCl/NADH	(2)	2.5 ml	Tris NaCl NADH	80 mmol/l 200 mmol/l 0.2 mmol/l
sample (serum or other sample solutions)		0.05 ml	volume fraction	0.0164
mix thoroughly, bring the mixture at 30 °C				
Tris/NaCl/pyruvate	(3)	0.5 ml	pyruvate	1.6 mmol/l
mix, read absorbance and start stopwatch exactly 30 s after mixing. Repeat the reading after exactly 1 min, 2 min or monitor the reaction on a recorder.				

The value $\Delta A/\Delta t$ must be <0.16 for 60 s at 339 or Hg 334 nm or <0.08 for 60 s at Hg 365 nm.

Validation of Method

Precision, accuracy, detection limits, and sensitivity: for a value of 4 μkat/l(240 U/l) in serum a within-run imprecision of + 0.17 μkat/l (10 U/l) has been found and a run to run imprecision of +0.22 μkat/l (13 U/l) for manual performance (level of significance defined as risk, $\alpha = 0.05$). The relative standard deviation of interlaboratory measurement is about 7% ($n = 8$ laboratories).

Data about accuracy are not available since standard reference material is not yet established.

For measurements at 339 nm with sufficiently sensitive photometers the lowest possible value which can be measured is 0.1 μkat/l (6 U/l). Sensitivity is found to be $\Delta A/\Delta t = 0.005/10$ min (Gilford Stasar III photometer).

Linearity limit is found to be 25 μkat/l (1500 U/l) corresponding to ΔA of 0.16 (for a measuring time of 2 minutes at 339 nm).

Sources of error: influences of therapeutics on the measured activity are unknown. Interference due to the presence of the LDH released from thrombocytes, leukocytes or erythrocytes is possible with plasma samples obtained by low-speed centrifugation [28]. Endogenous compounds such as pyruvate cannot affect LDH measurements. With regard to the influence of enzyme inhibitors in NADH preparations of insufficient quality, see p. 119.

Specificity: apart from pyruvate, 2-hydroxy- and 2-oxo-acids such as hydroxy pyruvate, phenyl pyruvate and 4-hydroxylphenyl pyruvate can be only slowly reduced. Oxaloacetate and 2-oxoglutarate are not significantly reduced. Nevertheless, the relative reaction rates of various 2-hydroxy- and oxo-acids differ between the various LDH isoenzymes.

Reference ranges: in human serum at 30°C, usual values have been found to be 2.3 μkat/l (140 U/l) to 4.6 μkat/l (280 U/l) for adults [25] with sera free from haemolysis.

In the neonatal period values range from 6.9 μkat/l (415 U/l) to 11.5 μkat/l (690 U/l). These values decrease with age to reach the usual adult values at 12 – 14 years.

References

[1] *O. Warburg,* Wasserstoffübertragende Fermente. Verlag Dr. W. Sänger, Berlin 1948, p. 40.

[2] *M. Dixon, E. C. Webb,* Enzymes, Academic Press, New York 1958.

[3] *F. Wroblewski,* The clinical significance of lactic dehydrogenase activity in the milieu intérieur. Scand. J. Clin. & Lab. Invest. *10,* Suppl. 31, 230 – 253 (1957).

[4] *S. B. Rosalki,* "α-Hydroxybutyrate dehydrogenase" Activity of Human Tissue Homogenates, Clin. Chim. Acta *8,* 415 – 417 (1963).

[5] *J. B. Neilands,* Studies on Lactic Dehydrogenase of Heart. I. Purity, Kinetics and Equilibria, J. Biol. Chem. *199,* 373 – 381 (1952).

[6] *A. L. Latner, A. W. Skillen,* Isoenzymes in Biology and Medicine, Academic Press 1968, p. 92.

[7] *G. Pfleiderer, E. G. Wachsmuth,* Alters- und funktionsabhängige Differenzierung der Laktatdehydrogenase menschlicher Organe, Biochem. Z. *334,* 185 – 198 (1961).

[8] *F. Wroblewski, J. S. LaDue,* Lactic Dehydrogenase Activities in Blood, Proc. Soc. Exp. Biol. Med. *90,* 210 – 213 (1955).

[9] *R. J. Henry, N. Chiamori, O. J. Golub, S. Berkman,* Revised Spectrophotometric Methods for the Determination of Glutamic-Oxalacetic Transaminase, Glutamic-Pyruvic Transaminase, and Lactic Acid Dehydrogenase, Am. J. Clin. Pathol. *34,* 381 – 398 (1960).

[10] *E. Amador, L. E. Dorfman, W. E. C. Wacker,* Serum Lactic Dehydrogenase (L.D.H.) Activity – Analytical Assessment of Current Assays, Clin. Chem. *9,* 391 – 399 (1963).

[11] *M. M. Nachlas, S. I. Margulies, J. D. Goldberg, A. M. Seligman,* The Determination of L.D.H. with a Tetrazolium Salt, Anal. Biochem. *1,* 317 – 326 (1960).

[12] *A. L. Babson, G. E. Phillips,* A Rapid Colorimetric Assay for Serum Lactic Dehydrogenase, Clin. Chim. Acta *12,* 210 – 215 (1965).

[13] *A. L. Babson,* The Chemical Differentiation of Tissue Lactic Dehydrogenase, Clin. Chim. Acta *16,* 121 – 125 (1967).

[14] *C. C. Allain, C. P. Henson, M. K. Nadel, A. J. Knobelsdorff,* Rapid Single-step Kinetic Colorimetric Assay for Lactate Dehydrogenase in Serum, Clin. Chem. *19,* 223 – 227 (1973).

[15] Deutsche Gesellschaft für Klinische Chemie: Recommendations of the German Society for Clinical Chemistry, Z. Klin. Chem. Klin. Biochem. *10,* 6, 281 – 291 (1972).

[16] Committee on Enzymes, the Scandinavian Society for Clinical Chemistry and Clinical Physiology: Recommended Methods for the Determination of Four Enzymes in Blood, Scand. J. Clin. Lab. Invest. *33,* 291 – 306 (1974).

[17] Commission Enzymologie de la Société Française de Biologie Clinique: Recommendations for Determining the Catalytic Concentration of Lactate Dehydrogenase in Human Serum at 30°C, Ann. Biol. Clin. *40,* 160 – 162 (1982).

[18] *A. Vassault, I. Maire, L. Sebille, D. Bozon, P. Lalègerie, M. Bailly,* An Improvement for the Determination of Total Lactate Dehydrogenase Catalytic Activity whatever the Isoenzymes. 3rd International Congress for Clinical Enzymology, Salzburg (1981).

[19] *W. Gerhardt, B. Kofoed, L. Westlund, B. Pavlu,* Quality Control of NADH – Evaluation of Methods for Detection of Inhibitors and Specifications for NADH Quality, Scand. J. Clin. Lab. Invest. *33*, 3 – 51 (1974).

[20] *F. B. Howell, S. Mc Cune, R. Schaffer,* Lactate-to-Pyruvate or Pyruvate-to-Lactate Assay for Lactate Dehydrogenase: A Reexamination, Clin. Chem. *25*, 2, 269 – 272 (1979).

[21] *W. B. Novoa, A. D. Winer, A. J. Glaid, G. W. Schwert,* Lactic Dehydrogenase V. Inhibition by Oxamate and Oxalate, J. Biol. Chem. *234*, 1143 – 1148 (1959).

[22] *J. W. Burgner II, G. R. Ainslie, Jr., W. W. W. Cleland, W. J. Ray Jr.,* Bimodal Substrate Inhibition of Lactate Dehydrogenase. Factors Affecting the Enzyme in vivo, Biochemistry *17*, 9, 1646 – 1653 (1978).

[23] *W. E. Trommer, H. Huth, H. R. Wenzel,* The nature of the Substrate Inhibition in Lactate Dehydrogenases as Studied by a Spin-labeled Derivative of NAD, Biochim. Biophys. Acta *567*, 49 – 59 (1979).

[24] *E. Rivedal, T. Sanner,* Role of Ions in the Regulation of Porcine Lactate Dehydrogenase, Biochim. Biophys. Acta *567*, 60 – 65 (1979).

[25] Commission Enzymologie de la Société Française de Biologie Clinique: Additional Information Regarding the Recommended Method for Measuring the Catalytic Concentration of Lactate Dehydrogenase in Human Serum at + 30 °C, Ann. Biol. Clin. *40*, 163 – 164 (1982).

[26] *B. Lacour, B. Cherruau, J. B. Roullet, M. Bailly,* Etude des isoenzymes de la lactate deshydrogenase: valeurs fréquentes en fonction de l'âge chez l'enfant, étude de la conservation, Ann. Biol. Clin. *36*, 425 – 432 (1978).

[27] *N. J. Hochella, S. Weinhouse,* Automated Assay of Lactate Dehydrogenase in Urine, Anal. Biochem. *13*, 322 – 335 (1965).

[29] *L. Hollaar, A. Van Der Laarse,* Interference of the Measurement of Lactate Dehydrogenase (LDH) Activity in Human Serum and Plasma by LDH from Blood Cells, Clin. Chim. Acta *99*, 135 – 142 (1979).

2.2.2 UV-method with L-Lactate and NAD

August W. Wahlefeld

General

Enzyme properties relevant in analysis: to determine the catalytic activity of LDH by either method one has to consider the inherent chemical characteristics of the individual isoenzymes as they relate to temperature, pH, substrate concentration, coenzyme concentration, buffer ions etc. All investigations compromise with respect to optimal reaction conditions for LDH-1 and LDH-5: the assay described below is especially sensitive in this respect concerning the substrate concentration of lactate. *Buhl* [1] reports an optimal lactate concentration for LDH-1 at about 30 mmol/l at 30 °C in diethanolamine and 2-amino-2-methyl-1,3-propandiol buffer, whereas for LDH-5 a plateau seems to be reached at 120 mmol/l at 30 °C.

Another analytical problem, and one which is not yet clearly understood, is the lag phase of the reaction with lactate and NAD as reactants: *Howell* [2] describes lag phases of at least 20 – 30 s after starting with the sample and we found even longer lag phases of up to 3 min at 30 °C, not correlated with catalytic activity but rather with sample-inherent factors (3 samples out of 78 in our study).

On the other hand, the reaction rate deviates from linearity during the time course: it is attributed to the formation of "dead-end" complexes of the enzyme with NAD and pyruvate produced in the lactate-to-pyruvate reaction [2, 3].

International reference methods and standards: no efforts have been undertaken yet to establish a reference method. For standard reference material cf. Vol. II, chapter 2.3.

Enzyme effectors: the formation of LDH-inhibitors from NAD has been observed for the assay described here and these have been characterized as non-competitive inhibitors with high affinity ($K_I = 3.5 \times 10^{-6}$ mmol/l) [5].

The formation of the inhibitors is increased in strongly alkaline solution (pH about 10 and higher) by inorganic phosphate ions, carbonate and oxalate ions. In sodium tetraborate/NaOH, pH 9.73, no inhibitor is formed under comparable conditions.

Assay

Method Design

Principle

$$\text{L-Lactate} + \text{NAD}^+ \xrightleftharpoons{\text{LDH}} \text{pyruvate} + \text{NADH} + \text{H}^+ .$$

The amount of L-lactate oxidized at 30 °C (or 25 °C; 37 °C) per unit time, measured by the continuous increase in absorbance at 339 nm (Hg 334; Hg 365) due to the reduction of NAD, is a measure of the catalytic activity of LDH. The equilibrium constant of the LDH reaction is $K = 2.7 \times 10^{11}$ mol/l (30 °C); it strongly favours the formation of lactate and NAD instead of the oxidation of lactate by NAD.

Therefore, the measurement has to be carried out in an alkaline medium and at relatively high concentrations of lactate and NAD to obtain reasonable reaction velocities for accurate photometric readings.

Assay blanks are not necessary for routine measurements. The method described below can easily be adapted to automated equipment by using the sample itself to start the reaction. NAD can be used as a starting reagent without significant difference [6]. However, with the human isoenzymes LDH-1 and LDH-2 in either pyrophosphate or Tris buffer, 10 to 20% lower results are obtained by starting the reaction with lactate.

Optimized conditions for measurement: conditions for measurement of catalytic activity of LDH in serum must minimize errors which result from variations in the isoenzyme composition of serum. Also, the influence of buffer ions has to be considered. These influences have been studied recently in detail by *Buhl et al.* [1, 7, 8], *Howell* [2] and *Adolf & Koko* [10]. For the non-isoenzyme-biased reaction *Adolf et al.* claim to afford 90 ± 2% of maximum relative activity for any LDH isoenzyme mixture. These authors made use of the activating influence of potassium chloride at 150 mmol/l on overall velocity (personal communication).

A survey of optimized conditions is given in Table 1.

Table 1. Optimized conditions for measurement of LDH; 30 °C

Authors	Buffer	pH	L-Lactate mmol/l	NAD mmol/l	Volume fraction of sample (φ)
Buhl et al. [1, 7, 8]	diethanolamine/ AMP-diol*	8.7	50	5.0	0.050
Howell [2]	AMP**	9.0	72	5.5	0.032
Adolf & Koko [11]	Tris	8.9	118	5 to 10	not stated
this method	Tris	8.9	52	6.5	0.0385

Although discussion of the above-mentioned points is still in progress, the concentration for NAD is not critical and should be above 5 mmol/l. The pH values at 30 °C concentrate at 8.9, with a range between 8.7 and 9.2. The choice of buffer is still in question: *Buhl* considers the pK_a-values of buffer ions, incompatibility between pH optima and NAD stability, inhibitory (pyrophosphate buffer) and stimulatory (AMP, diethanolamine) effects and, in addition, the overlapping pH optima of LDH-1 and LDH-5 isoenzymes.

Only diethanolamine, AMP-diol and Tris buffer ions are to be considered. In the following we describe the method using Tris buffer, despite the fact that the pK_a of Tris (8.0 at 30 °C) is less satisfactory at the optimal pH of 8.9. However, the smallest dependence of activity on buffer concentration is observed with Tris ions, the formation of LDH-inhibitor from NAD has been investigated for Tris buffer [5] but not yet for the other buffer ions in question, and *Bails & Edwards* [9] report considerable sample blank reactions in DEA and AMP-diol buffers due to alcohol dehydrogenase in pathological sera.

The reduced buffer capacity with Tris seems not to be a problem, since volume fractions of only 0.038 or even lower are used.

* 2-amino 2-methyl 1,3-propanediol.

** 2-amino 2-methyl 1-propanol.

The low volume fractions are of importance for two reasons:

- to obtain sufficient dynamic range for the measurement, i.e. 10 fold range of the upper reference value without diluting the sample;
- to assure proper zero-order kinetics during the course of monitoring the conversion rate. It is recommended to use the time interval of 30 s after start for 2 min to record the changes in absorption at 339 nm, (Hg 334 nm or Hg 365 nm).

Temperature conversion factors: the reliability of using temperature conversion factors for the LDH reaction with lactate as substrate has been investigated in detail by *Buhl* [1] for the DEA and AMP-diol buffer systems. He found differences between isoenzymes, but concludes that, for practical purposes, one factor can be used for both buffer systems and for all isoenzyme mixtures. The individual errors are less than 10%.

The temperature conversion factors in Tris buffer and with the assay described here are preliminary, but agree with the figures of *Buhl* [1].

measured	factor for conversion to	
°C	30 °C	37 °C
30	1	1.8
37	0.56	1

The standard deviation is about 10%; therefore, we recommend that these factors should be verified before use.

Equipment

Spectrophotometers or spectral-line photometers capable of exact measurement at 339 nm, Hg 334 nm or Hg 365 nm can be used. A thermostatted cuvette holder is required, and a recorder or a calculator programmed to calculate catalytic activity concentration, *b*, should be available.

Reagents and Solutions

Purity of reagents: NAD preparations must be free from inhibitors of LDH: suitable commercial preparations are available.

L-lactate should be clearly specified; highly concentrated solutions of L-lactic acid, especially, contain considerable amounts of lactides, inter-lactic acid esters. The concentration of L-lactate should be determined enzymatically after alkaline hydrolysis at pH = 10 in dilute solutions for about 24 h at ambient temperature (cf. Vol. V). Tris and KCl should be of the highest analytical purity available.

Preparation of solutions (for about 40 determinations): all solutions are made with re-purified water (cf. Vol. II, chapter 2.1.3.2).

1. β-Nicotinamide-adenine dinucleotide (β-NAD, 172 mmol/l):

   dissolve 114 mg NAD, free acid, *Boehringer Mannheim* grade I, in 4.0 ml water, containing NaN_3, 1 mg/ml, as a preservative.

2. Tris/L-lactate (Tris, 112 mmol/l, pH 9.30, 25 °C; L-lactate, 56 mmol/l; KCl, 170 mmol/l):

   dissolve 1.36 g Tris base in 75 ml water, add 1.28 g KCl and 508 mg L-lactic acid, stir and adjust to pH 9.30 (25 °C) with NaOH, 2 mmol/l. Sodium azide (100 mg) should be added to obtain sufficient antimicrobial preservation. Dilute with water to 100 ml. The exact pH should be checked again.

Stability of solutions: store all solutions at 2 °C to 8 °C. The NAD-solution (1) is stable for 1 week. The Tris/L-lactate solution (2) is stable as long as no bacterial contamination occurs.

Procedure

Collection and treatment of specimen: collect blood from vein without stasis; avoid haemolysis strictly. Citrate and EDTA plasma can be used as well as serum. Centrifuge for 10 min at about 3000 *g* to separate plasma or serum. Fluoride- and heparin-plasma cause activation, whereas oxalate plasma inhibits LDH activity and cannot be used.

Stability of enzyme in the sample: according to our experience, the enzyme in serum loses about 2–3% of its activity within 3 days at room temperature, and 8% at +4 °C. It is not possible to freeze the sample at −20 °C because of dramatic loss of activity.

Details for measurement in tissue: no recommendations can be given.

Assay conditions: wavelength 339 (Hg 334; Hg 365) nm; light path 10 mm; final volume 2.60 ml; 30 °C (thermostatted cuvette holder). Measure against air. Before starting the assay, wait until the temperature of solutions has reached 30 °C.

Measurement

Pipette successively into the cuvette:			concentration in assay mixture	
β-NAD	(1)	0.10 ml	NAD	6.5 mmol/l
Tris/L-lactate	(2)	2.40 ml	L-lactate	52 mmol/l
			KCl	155 mmol/l
			Tris	100 mmol/l
mix thoroughly with plastic spatula, wait for 5 min, until temperature of 30 °C is reached				
sample (serum or other sample)		0.10 ml	volume fraction	0.0385
mix, start stopwatch and read absorbance after 30 s and repeat readings after 1 and 2 min or monitor the reaction on a recorder. Use increase of absorbance between 30 and 150 s for calculations.				

When the values $\Delta A/\Delta t$ exceed 0.260/min at 339 nm or 0.130/min at Hg 365 nm, dilute sample 1 + 3 with 0.9% sodium chloride solution, or measure in shorter intervals, e.g. between 30 and 60 s.

Calculation: use calculation formula (k) or (k_1), respectively (cf. "Formulae", Appendix 3). Values for ε: Appendix 4. The following relations are valid for catalytic concentration in the sample (Δt = 1 min):

	339 nm	Hg 365 nm	
$b =$	$4.21 \times 10^3 \times \Delta A/\Delta t$	$7.65 \times 10^3 \times \Delta A/\Delta t$	U/l
$b =$	$7.01 \times 10^4 \times \Delta A/\Delta t$	$1.27 \times 10^5 \times \Delta A/\Delta t$	nkat/l

Validation of Method

Precision, accuracy, detection limits and sensitivity: the following values of imprecision from day to day were found for manual performance:

$n = 12$;	65 U/l	RSD	5.8%,
	149 U/l	RSD	3.2%.

Data about accuracy are not available since a reference method is not yet established. Studies on method comparison with the optimized method of the German Society of

Clinical Chemistry [11] (pyruvate as substrate) result in the following regression equations:

x = optimized method; y = method described here

sera from patients with heart disease, n = 21	$y = -29.7 + 0.503 \times x$; $r = 0.989$
sera from patients with liver diseases; n = 21	$y = -5.9 + 0.493 \times x$; $r = 0.990$
all samples, n = 42	$y = -6.7 + 0.464 \times x$; $r = 0.989$

For measurements at 339 nm with sufficiently sensitive photometers the lowest value which can be measured is approximately 15 U/l with an imprecision of about 12%. Analytical sensitivity is $\Delta A/\Delta t$ = 0.005/5 min with the Eppendorf photometer.

Sources of error: interference by LDH-inhibitors from NAD preparations [4, 5] has been described above, but with NAD of adequate purity and a limited period of use of NAD solutions it should not be an analytical problem. LDH activities which are dramatically and falsely elevated will be found in samples with haemolysis (haemoglobin above 50 mg/100 ml cannot be tolerated), since erythrocytes contain high concentrations of LDH.

Influences of drugs on the measured activity are almost non-existant; triamterene is reported spuriously to elevate activity [12].

Specificity: the LDH activity is measured specifically by the method described; other NAD oxidoreductases do not interfere because of the high sample dilution and the low endogeneous concentrations of necessary substrates in serum or plasma.

Reference ranges: in human serum reference values for adults of 35 – 100 U/l at 30 °C have been established.

References

[1] *S. N. Buhl, K. Y. Jackson, R. Lubinski, R. E. Vanderlinde,* Optimal Conditions for Assaying Human Lactate Dehydrogenase by the Lactate-to-Pyruvate Reaction: Arrhenius Relationships for Lactate Dehydrogenase Isoenzymes 1 and 5, Clin. Chem. *23*, 1289 – 1295 (1977).

[2] *B. F. Howell, S. McCune, R. Schaffer,* Lactate-to-Pyruvate or Pyruvate-to-Lactate, Assay for Lactate Dehydrogenase: A Re-Examination, Clin. Chem. *25*, 269 – 272 (1979).

[3] *N. O. Kaplan, T. Everse, T. Admiraal,* Significance of substrate inhibition of dehydrogenases, Ann. N. Y. Acad. Sci. *151*, 400 – 408 (1968).

[4] *H. Gallati,* Bildung und Reinigung des Oxidoreductase Inhibitors aus NAD, J. Clin. Chem. Clin. Biochem. *14*, 3 – 8 (1976).

[5] *H. Gallati,* Beiträge zur Charakterisierung des Oxidoreduktase Inhibitors, J. Clin. Chem. Clin. Biochem. *14*, 15 – 22 (1976).

[6] *S. N. Buhl, K. Y. Jackson, R. Lubinski, R. E. Vanderlinde,* Effect of Reaction Initiator on Human Lactate Dehydrogenase Assay, Clin. Chem. *22*, 1098 – 1099 (1976).
[7] *S. N. Buhl, K. Y. Jackson, R. Lubinski, R. E. Vanderlinde,* A Search for the Best Buffer to Use in Assaying Human Lactate Dehydrogenase with Lactate-to-Pyruvate Reaction, Clin. Chem. *22*, 1872 – 1875 (1976).
[8] *S. N. Buhl, K. Y. Jackson,* Optimal Conditions and Comparison of Lactate Dehydrogenase Catalysis of the Lactate-to-Pyruvate and Pyruvate-to-Lactate Reactions in Human Serum at 25, 30 and 31 °C, Clin. Chem. *24*, 828 – 883 (1978).
[9] *R. Bails, J. B. Edwards,* Alcohol Dehydrogenase Interference in Lactate Dehydrogenase Assay, Clin. Chem. *26*, 525 – 526 (1980).
[10] *P. Adolf, I. S. Koko,* A Multi-Parameter Study on Non-Isoenzyme Biased Lactate Dehydrogenase Determinations, Clin. Chem. *23*, 1161 (1977), Abstract.
[11] Empfehlungen der Deutschen Gesellschaft für Klinische Chemie. Standardisierung von Methoden zur Bestimmung von Enzymaktivitäten in biologischen Flüssigkeiten, J. Clin. Chem. Clin. Biochem. *10*, 182 – 192 (1972).
[12] Effects of Drugs on Clinical Laboratory Tests. Clin. Chem. *21*, 323 D-324 D (1975).

2.2.3 Colorimetric Method

Ulfert Deneke

Assay

Method Design

Principle

(a) $\text{L-(+)-Lactate} + \text{NAD}^+ \xrightleftharpoons{\text{LDH}} \text{pyruvate} + \text{NADH} + \text{H}^+$

(b) $\text{NADH} + \text{MTT}^* \xrightarrow{\text{NADH:dye-DH}^{**}} \text{formazan} + \text{NAD}^+$

The amount of L-(+)-lactate reacting per unit time, measured by the increase in absorbance due to the formation of the formazan, is a measure of the catalytic activity of LDH. The equilibrium constant of reaction (a) is far in favour of lactate formation. At pH 7 according to [1], $K' = 2.76 \times 10^{-6}$ mol/l at 25 °C in phosphate buffer.

* MTT = 3-[4,5-dimethylthiazolyl-(2)]-2,5-diphenyl-2H-tetrazolium-bromide.
** NADH: dye-DH = NADH: dye oxidoreductase, EC No. not yet assigned; this enzyme is different from diaphorase, i.e. NADH: lipoamide oxidoreductase, EC 1.6.4.3, but is often also called diaphorase.

Although the indicator reaction with MTT and NADH: dye-DH is essentially irreversible, maximum velocity of the LDH reaction at pH 8, with a lag phase reduced to less than one minute, is only obtained if glutamate and ALT* are present to convert pyruvate to alanine. Blanks are not necessary for routine assay. The method described below can be performed either as a continuous monitoring assay or as a fixed-time test (cf. Vol. I, chapter 2.3).

The mixture of the two detergents octylphenylpolyethyleneglycol ether and coconut-pentaethoxymethylammonium methylsulphate alone results in good linearity in this method. Other detergents or mixtures thereof are not as effective. In the fixed-time test, the third detergent isotridecyl-octaethyleneglycol ether also clears highly lipaemic samples spontaneously.

Optimized conditions for measurement: the characteristics of the LDH proteins that are most important in measuring their activities are their different substrate and pH optima. In the following method lactate, 40 mmol/l, NAD 2.0 mmol/l and pH 8.0 are used as conditions which approximate to those most satisfactory for measurement LDH in serum for the diagnosis of heart and liver diseases.

The NADH: dye-DH is preferred with respect to diaphorase [2] and PMS** [3, 4] because it provides better linearity and a shorter lag phase in the indicator reaction. It is also free of blank reactions.

Temperature conversion factors: temperature conversion factors of the isoenzymes LDH-1 and LDH-5 from rabbit heart and muscle are 0.38 and 0.57, respectively, related to the value measured at 25 °C with the method described here. Human isoenzymes LDH-1 and LDH-5 are similar in their temperature conversion factors. Although [5] gives temperature conversion factors related to the forward reaction of LDH it seems inappropriate to quote them in this test because of the major variability of the isoenzymes in human serum and the large differences in the temperature conversion factors.

Equipment

Spectrophotometer or spectral-line photometer permitting exact measurement at 578 nm, preferably with a thermostatted cuvette holder; water-bath; recorder or stopwatch.

Reagents and Solutions

Purity of reagents: the contamination of the ALT and NADH: dye-DH with LDH should be <0.001%. NAD must be free from inhibitors (cf. Vol. II, chapter 2.1.2);

---

* ALT = L-Alanine: 2-oxoglutarate aminotransferase, EC 2.6.1.2.
** PMS = phenazine methosulphate.

suitable commercial preparations are available. All other reagents should be of the best purity available.

Preparation of solutions (for about 40 determinations): all solutions in re-purified water (cf. Vol. II, chapter 2.1.3.2). The citrate solution (4) is only needed in the fixed-time test.

1. Tris/lactate/MTT-solution (Tris, 0.105 mol/l; pH 8.0; L-(+)-lactate, 42 mmol/l; MTT, 0.21 mmol/l):

   dissolve 0.7 g octylphenylpolyethyleneglycol ether (e.g. Triton X-100®), 0.2 g coconut-pentaethoxymethylammonium methylsulphate (e.g. Steinaquat® CPEM), 403 mg L-(+)-lactate, Li-salt, 1.27 g Tris, 61.8 mg glutamic acid in 80 ml water, adjust to pH 8.0 with HCl, 2 mol/l, dissolve 8.7 mg MTT and dilute to 100 ml with water.

2. Nicotinamide-adenine dinucleotide (β-NAD, 86.8 mmol/l):

   dissolve 144 mg NAD, free acid, with 2.5 ml water.

3. Auxiliary and indicator enzymes (ALT, 88 kU/l, NADH: dye-DH, 10.8 kU/l):

   centrifuge an aliquot of a commercially available ALT suspension in ammonium sulphate and discard the supernatant. Dissolve the sediment in 2.5 ml Tris-citrate buffer, 50 mmol/l, pH 7.6 (0.606 g Tris/100 ml, adjusted to pH 7.6 with citric acid solution, 2 g/100 ml), to obtain 2.75 mg protein (= 80 U/mg, 25 °C). Dissolve 1.8 mg NADH: dye-DH (15 U/mg, available from *Boehringer Mannheim Corp.* or *Toyo Jozo*), in this solution.

4. Citrate buffer (67 mmol/l; pH 2.9):

   dissolve 10 g isotridecyloctaethyleneglycol ether (e.g. Genapol X-080®) in 80 ml water, add 0.141 g citric acid, adjust to pH 2.9 with NaOH and dilute to 100 ml with water.

Stability of solutions: solution (1) is stable for at least 4 weeks at 0 – 4 °C in a dark bottle. Store solutions (2) and (3) at 0 – 4 °C and prepare them freshly every week. Solution (4) is stable for at least one year at room temperature.

Procedure

Collection and treatment of specimen: collect blood from the vein without stasis. Additions of oxalate, 1 mg/ml; citrate, 1 mg/ml; heparin, 0.2 mg/ml; or EDTA, 1 mg/ml are without effect. Centrifuge for 10 min at about 3000 *g* in order to obtain plasma or serum. Use only fresh serum or plasma absolutely free from haemolysis.

Stability of the enzyme in the sample: according to [6], the enzyme in serum loses about 10 – 30% activity in 24 h when kept at room temperature, 4 °C or in the frozen state. However, cf. pp. 123, 130.

Details for measurement in tissues: LDH is a cytoplasmic enzyme [7], therefore it is totally extracted by using a *Potter-Elvehjem* homogenizer. Activity is measured in the supernatant after high speed centrifugation of the homogenate. For measurement in liver tissue 10 mg is enough.

The optimal conditions for human serum do not necessarily apply to animal serum or organs, or other body fluids. Urine has to be dialyzed for 2 hours before measurement.

Assay conditions: wavelength Hg 578 nm; light path 10 mm; final volume 2.12 ml in the continuous monitoring assay and 4.12 ml in the fixed-time assay, respectively; 25, 30 or 37 °C (thermostatted cuvette holder). Measure against air; before starting the assay, adjust temperature of the solutions to the selected temperature.

If another wavelength (520 – 600 nm) is preferred, a calibration curve has to be prepared with known amounts of NADH instead of the sample.

Measurement

Continuous monitoring assay

Pipette successively into the cuvette:			concentration in assay mixture	
Tris/lactate/MTT solution	(1)	2.00 ml	Tris L(+)-lactate glutamate MTT	100 mmol/l 40 mmol/l 4 mmol/l 0.2 mmol/l
NAD solution	(2)	0.05 ml	NAD	2 mmol/l
ALT/NADH: dye-DH	(3)	0.05 ml	ALT NADH: dye-DH	2 U/ml 0.25 U/ml
sample		0.02 ml	volume fraction	0.0094
mix thoroughly with a plastic spatula, wait 1 min and read absorbance after exactly 1, 2, 3 and 4 min or monitor the reaction on a recorder				

The value $\Delta A/\Delta t$ must be ≤ 0.15/min at 578 nm. Otherwise dilute sample 10-fold with saline, 0.9%.

Fixed-time assay

Within each series a reagent blank has to be run with 0.9% saline instead of sample.

Pipette successively into the cuvette:			concentration in assay mixture	
Tris/lactate/MTT solution	(1)	2.00 ml	Tris L-(+)-lactate MTT glutamate	100 mmol/l 40 mmol/l 0.2 mmol/l 4 mmol/l
NAD solution	(2)	0.05 ml	NAD	2 mmol/l
ALT/NADH: dye-DH	(3)	0.05 ml	ALT NADH: dye-DH	2 U/ml 0.25 U/ml
sample		0.02 ml	volume fraction	0.00485
mix thoroughly with a plastic spatula and incubate for exactly (!) 10 min, then pipette				
citrate buffer	(4)	2.00 ml	citrate	33 mmol/l
mix and read absorbance of the sample (A_S) and of the blank (A_R) against air after exactly 20 min. $\Delta A = A_S - A_R$.				

If $\Delta A > 0.8$, the sample has to be diluted 5-fold with 0.9% saline.

Calculation: use calculation formula (k) or (k_1), respectively (cf. "Formulae", Appendix 3). $\varepsilon = 1.72 \; l \times mmol^{-1} \times mm^{-1}$. The following relations are valid at a wavelength of 578 nm (Δt in minutes):

continuous-monitoring assay:	$b = 6163 \times \Delta A/\Delta t$	U/l
	$b = 10.27 \times 10^4 \times \Delta A/\Delta t$	nkat/l
fixed-time assay:	$b = 1198 \times \Delta A/\Delta t$	U/l
	$b = 1.996 \times 10^4 \times \Delta A/\Delta t$	nkat/l

Validation of Method

Precision, detection limits and sensitivity: for values of about 90 U/l in serum an imprecision of ±8 U/l has been found in the continuous-monitoring test. The relative standard deviation within a series was about 1.5% (n = 20) and from day to day it was about 5%, respectively ($n = 20$). The lowest possible value measurable with a spectral-line photometer at Hg 578 nm is approximately 12 U/l with an imprecision < ±20% in the continuous-monitoring test. Sensitivity is found to be $\Delta A/\Delta t$ = 0.01/10 min (*Eppendorf* photometer).

Sources of error: ascorbic acid at the highest concentration present in sera (15 mg/l) [8] increase the absorbance of the sample blank to 0.006. This is normally negligible compared with the value being measured. Other interferences due to therapeutics are unknown.

Specificity: the LDH activity is measured specifically. Nevertheless, if other dehydrogenases and their substrates are present in high concentrations in the sample, a sample blank reaction is possible. This possibility must be noted when other materials than serum are tested.

Reference ranges: in human serum, 46 – 88 U/l at 25 °C and 119 – 238 U/l at 37 °C, respectively, were found.

References

[1] *M. T. Hakala, A. J. Glaid, G. W. Schwert,* Lactic Dehydrogenase II., Variations of Kinetic and Equilibrium Constants with Temperature, J. Biol. Chem. *221*, 191 – 209 (1956).
[2] *N. Hochella, S. Weinhaus,* Automated Assay of Lactate Dehydrogenase in Urine, Anal. Biochem. *13*, 322 – 335 (1965).
[3] *A. L. Babson, G. E. Phillips,* A Rapid Colorimetric Assay for Serum Lactic Dehydrogenase, Clin. Chim. Acta *12*, 210 – 215 (1965).
[4] *A. L. Babson,* The Chemical Differentiation of Tissue Lactic Dehydrogenase, Clin. Chim. Acta *16*, 121 – 125 (1967).
[5] *Boehringer Mannheim*, Temperature Conversion Factors for Enzymes, Mannheim, 1981.
[6] *H. Suedhof, E. Woetzel,* Uber die Haltbarkeit von zur Aktivitätsbestimmung vorgesehenen Serumproben, Klin. Wochenschr. *38*, 1165 – 1167 (1960).
[7] *T. Buecker, P. Braun,* Vortrag, Dtsch. Kongr. f. ärztl. Fortbildung, Berlin 1958.
[8] *J. S. Vobecky, J. Vobecky, D. Shapcott, R. Blauchard:* Vitamin E and C Levels in Infants during the first Year of Life, Am. J. Clin. Nutr. *29*, 766 – 771 (1976).

2.2.4 Isoenzymes of Lactate Dehydrogenase

L-Lactate: NAD^+ oxidoreductase, EC 1.1.1.27

A. Ralph Henderson

General

The total lactate dehydrogenase (LDH) activity of higher animals is found in the cytoplasm of the cell contributed by five isoenzymes [1]. This number derives from the existence of two gene loci which code for two different polypeptides designated A (or

M for muscle) and B (or H for heart). These polypeptides (subunits) combine together in groups of four to produce five different isoenzymes. These are numbered, according to their electrophoretic mobility in a slightly alkaline pH, in order of decreasing anodal mobility as LDH-1 (H_4); LDH-2 (H_3M); LDH-3 (H_2M_2); LDH-4 (HM_3); LDH-5 (M_4). In passing, it should be noted that a third LDH subunit exists designated C (or X) which forms LDH-X (X_4) – this is the sixth isoenzyme only found in post-pubertal testis [1].

LDH isoenzyme patterns are characteristic of a number of tissues: in general, there tend to be three main groups of these patterns which suggest their clinical applications [2]. The most important group is the anodal – where LDH-1 and LDH-2 predominate – found in erythrocytes, cardiac muscle and kidney; a cathodal group – with LDH-4 and LDH-5 predominating – seen in liver, skeletal muscle and many neoplastic (malignant) tissues, and an intermediate group – with LDH-3 predominating – seen in lymphatic tissue, platelets and many malignant tissues.

Application of method: in biochemistry and in clinical chemistry. The isoenzyme pattern in serum is often quite characteristic of certain types of pathological disorders, and is thus of diagnostic importance [2]. The main diagnostic groups are shown in Table 1, although it should be realized that this is not an exhaustive listing: it only covers the more commonly encountered diagnostic categories (cf. also reference [3]).

Table 1. The principal diagnostic categories that produce abnormalities of the serum lactate dehydrogenase isoenzyme pattern

Anodal elevation	Intermediate group elevation	Cathodal elevation
*infarction or surgery of myocardium	*rapid destruction of platelets	liver diseases
*in vivo or in vitro haemolysis	*lymphoid tissue involvement (e.g. infective mononucleosis)	*congestive cardiac failure early stages of acute hepatitis exacerbation of chronic hepatitis
ineffective erythropoiesis	splenic necrosis	
germ cell tumours	malignant pattern	skeletal muscle injury
renal infarction		
muscular dystrophies		prostatic carcinoma *malignant pattern

* the more commonly occuring patterns

Enzyme properties relevant in analysis: the catalytic activity in serum and tissues, is due to the existence of, usually, all five isoenzymes present with sufficient activity to be readily detected. Provided the reaction times are sufficiently short, substrate conversion proceeds linearly with time: certain detection reactions show substrate depletion, if too much enzyme activity is present, leading to non-linearity. The immunological differences between the isoenzymes allow LDH-1 (or LDH-5) to be separately estimated when a anti-LDH-M (or anti-LDH-H) antibody is incubated with the isoenzyme mixture.

Methods of determination: the individual isoenzymes can be separated by electrophoresis or ion-exchange chromatography making use of the catalytic activity of the isoenzyme to estimate their individual contribution to the total activity. Alternatively, estimates of the anodal or cathodal groups can be made by means of heat denaturation or by inactivation in the presence of high concentrations of urea – in both cases catalytic activity is determined before and after the inactivating procedure [6]. The anodal fraction can react preferentially with the substrate 2-oxobutyrate (the so-called "2-hydroxybutyrate dehydrogenase" or "2-HBDH" reaction) and this catalytic assay can therefore be used as a measure of the LDH-1 content of sera [7]. As previously mentioned, the immunological properties of the subunits can be utilized to measure the LDH-1 content of sera. An antiserum against the M-subunit, together with a second antibody, will precipitate all isoenzymes containing one or more of this subunit, leaving only LDH-1 (H_4) in the supernatant which can then be readily assayed by any reaction rate method [8].

Electrophoresis is the most common method because of the highly reliable results that can be obtained by its use. It must be admitted that the successful electrophoretic separation of LDH isoenzymes demands a high level of expertise and that these acquired skills must be maintained. In turn, these restrictions sometimes limit the availability of the assay at nights and during weekends. Thus, the faster methods (mentioned above) requiring less skill are becoming popular. None the less, these assays are not capable of giving the wealth of information available from the electrophoretic separation of the LDH isoenzymes. Clearly there has to be a compromise between the cost (in the widest possible sense) of the particular method with the advantages (again in the widest possible sense) to be derived from its use.

International reference methods and standards: neither standardization at the international level nor reference standard material is currently available.

Enzyme effectors: none is known.

2.2.4.1 UV-method after Separation on DEAE-Sephadex

Assay

Method Design

Principle: serum LDH activity can be separated into two fractions by loading serum on to a mini-column of DEAE-Sephadex and washing the unabsorbed LDH fractions off the column. The eluted, unabsorbed, LDH activity corresponds roughly to the cathodal fraction (i.e., mostly LDH-3, LDH-4 and LDH-5 with a diminution of LDH-1 and -2) of the serum sample (see Figure 1).

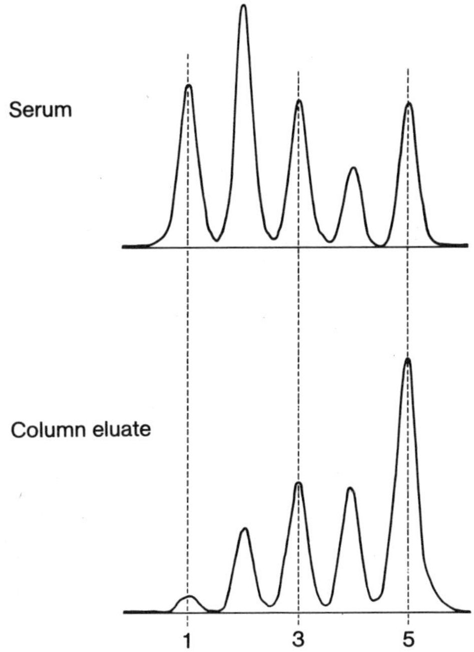

Fig. 1. The cathodal LDH isoenzymes obtained from the one-pass ion-exchange mini-column. Isoenzymes were separated, and detected, by the techniques described in section 2.2.4.2. The positions of isoenzymes LDH-1, LDH-3 and LDH-5 are indicated by the numbers. It is evident that there is an appreciable reduction in the anodal isoenzyme content of the eluate: therefore, the technique gives an approximate indication of the cathodal LDH isoenzyme content of the serum.

The method described here is essentially a column modification of the batch process described by *Bergmeyer & Bernt* [9]. The one-stage column procedure [10] is a simplification of the stepped gradient elution procedure of *Mercer* [11]. The principle of the spectrophotometric assay is described in chapter 2.2.1.

Optimized conditions for chromatography and measurement: the pH of the column buffer is selected to bind the anodal isoenzymes preferentially but allow the cathodal isoenzymes to pass freely through the column. Although the LDH isoenzymes demonstrate different substrate and pH optima – at 37 °C for example, the optimal concentration of pyruvate is about 1.2 mmol/l for LDH-1 and 3.0 mmol/l for LDH-5 with the NADH concentration about 220 μmol/l [12] – it is usual to use the same substrate concentration as is used to measure for the total serum activity (cf. chapter 2.2.1).

Temperature conversion factors: cf. section 2.2.1.

Equipment

5 ml Polypropylene columns (the straight section measuring 4.5 × 0.9 cm contains 3 ml) fitted with a porous (1/32″ thick) polyethylene disc and a 12 ml extension funnel. These are supplied by *Isolab Inc.,* Akron, Ohio 44321, U.S.A. Column flow is started and stopped by means of a bottom closure. It is convenient to mount this column in a rack of sufficient height to allow easy insertion of a test-tube, or graduated cylinder, under the column outlet.

Spectrophotometer or spectral-line photometer suitable for accurate measurements at 339 (Hg 334 or Hg 365) nm, preferably with a constant temperature cuvette holder; water-bath, 37 °C (25 ° or 30 °C); recorder or stopwatch resp.

Reagents and Solutions

Purity of reagents: NADH must be free from inhibitors (see Vol. II, chapter 2.1.2.2); suitable commercial preparations are available. Other reagents should be of a recognized analytical grade.

Preparations of solutions (for about 10 columns): all solutions in re-purified water (see Vol. II, chapter 2.1.3.2).

1. Assay reagents for LDH activity (cf. chapter 2.2.1)

2. Tris/EDTA solution (Tris, 20 mmol/l, EDTA, 2 mmol/l; pH 7.4 at 20 °C):

   dissolve 2.42 g Tris and 0.37 g EDTA-$Na_2H_2 \cdot 2H_2O$ in 900 ml water, adjust to pH 7.4 with HCl, 1 mol/l, and dilute to 1 litre with water.

3. DEAE-Sephadex A-50:

   the bed volume of this ion-exchanger is about 30 ml/g dry powder. Each column has a volume of 3 ml. Allow about 150 mg of dry exchanger for every column to be used.

Stability of solutions: cf. chapter 2.2.1 for stability of LDH assay reagents. Store other reagents at 4 °C. Both the Tris buffer (2) and DEAE-Sephadex (3) are stable as long as no bacterial contamination occurs. A few drops of chloroform can be added to the ion-exchanger to aid preservation.

Procedure

Collection and treatment of specimen: collect blood from the vein with minimal or no stasis. Serum is to be preferred. Cf. section 2.2.4.2 for further details.

Stability of the enzyme in the sample: serum is best stored at room temperature because of the cold lability of LDH-4 and LDH-5. Cf. section 2.2.4.2 for further details.

Details for measurement in tissues: LDH activity is cytoplasmic in origin in the cell but tissues may require to be extracted several times to obtain > 90% of the original activity. As much contaminating blood as possible should be washed out of the tissue before homogenization. The method described here can also be used for the preparation of LDH-5 from tissues [10].

Preparation of column: weigh out 1.2 g DEAE-Sephadex A-50 and mix with 50 ml buffer in a 100 ml measuring cylinder. Stir well and make volume up to 100 ml. Allow the gel to settle and re-suspend it in fresh buffer to the same volume. Repeat this process until the pH of the suspending buffer remains at 7.4: this usually occurs after six washes.

Re-suspend the equilibrated gel in fresh buffer, stir well, and pour 10 ml of gel slurry into each of 10 columns. Allow the gel to settle overnight and adjust the settled bed volume of the gel to 3 ml using a pipette to remove excess gel. The gel buffer should not be above the surface of the gel.

Isoenzyme separation: add 2 ml of serum, remove the bottom closure, and collect 750 μl of eluate (which is discarded) then start collecting the eluate in a second container. As the serum runs into the gel add 1 ml of buffer. The separation is complete when the gel runs dry. The collected fraction should be 3 to 3.25 ml.

Measurement: assay for LDH activity as described in chapter 2.2.1; express result as U/l of eluate.

Validation of Method

Precision, accuracy, detection limits and sensitivity: for the assay of the column eluate of the serum of a healthy person, catalytic activity concentrations about 40 U/l are obtained. At that level an imprecision of about ± 5.0 U/l has been obtained for the manual assay. The relative standard deviation referred to these mean values (within-batch) is about 15%.

As the eluate is a mixture of largely cathodal LDH isoenzymes (i.e. mostly LDH-3, -4 and -5 – see Figure 1) the question of accuracy is not entirely relevant: see however chapter 2.2.4.2 for further discussion of this point.

At 339 nm the lowest possible value to be measured is approximately 15 – 20 U/l with an imprecision of < ±20%. Sensitivity is found to be $\Delta A/\Delta t = 0.005/10$ min for a *Pye Unicam* SP 1700 Spectrophotometer with a recorder scale of 1A = 20 cm.

Sources of error: cf. p. 154.

Specificity: the eluate from the column is a mixture of the cathodal LDH isoenzymes from the source serum (i.e. mostly LDH-3, -4 and -5), although there is a marked reduction in the anodal isoenzymes (i.e. LDH-1 and -2) in the column eluate (Figure 1). Otherwise, the method appears to measure LDH activity specifically.

Reference ranges: in human serum 24 – 49 U/l at 37 °C, measured by the procedure outlined earlier.

2.2.4.2 Measurements after Electrophoretic Separation

Assay

Method Design

Principle and practice of isoenzyme separation by electrophoresis are described by *Wieme* in chapter 1.3.2 of this volume.

Principle: at the pH used (8.6 at 20 °C), LDH-1 to LDH-4 migrate towards the anode and LDH-5 moves towards the cathode of the agarose film. The LDH isoenzymes are detected, after separation, by the reaction

(a) $$\text{L-Lactate} + \text{NAD}^+ \xrightleftharpoons{\text{LDH}} \text{pyruvate} + \text{NADH} + \text{H}^+ .$$

A liquid overlay containing lactate and NAD^+ is applied to the gel, and the gel is incubated at 37 °C. The NADH generated is measured by fluorescence emitted above 410 nm by excitation at 365 nm. The activity of each isoenzyme can be determined by densitometry, measuring either peak height or the area under each peak.

Optimized conditions for measurement: in reaction rate assays, the optimized lactate and NADH concentrations, for LDH-1 and LDH-5 at 37 °C, are 70 and 7 mmol/l, respectively, in 200 mmol/l for DEA or AMPdiol buffer at pH 7.8 [13]. Such assay conditions will determine each isoenzyme with equal efficiency (> 95%). Unfortunately, these conditions are unsifficient to produce detectable amounts of NADH on the agarose gel: while the NAD concentration of 7.5 mmol/l appears adequate, the L-lactate has to be markedly increased to 500 mmol/l. Even at this concentration, it appears that substrate depletion can occur over isoenzyme zones which are present with very high activities: accordingly the amount of LDH activity placed on the agarose gel is limited. Therefore, if the *total* LDH activity exceeds the upper limit of normal serum activity [14] dilute the serum with NaCl, 150 mmol/l, so that the resulting activity is at or less than this upper limit. If a substantial proportion of activity (> 70%) is present as a *single* isoenzyme species (most usually LDH-5) the sample should be diluted, with saline, to the *lower* limit of normal serum activity before applying to the gel [14].

Temperature conversion factors: none is required, or is necessary, because the technique measures the relative proportions of each isoenzyme present.

Equipment

1. Electrophoretic apparatus (e.g. *Shandon* Model U77 electrophoresis tank with safety switch).
2. Power Supply (e.g. *Shandon* VoKam constant voltage/current DC supply).
3. Incubator (37 °C) and Drying (70 °C) oven (e.g. *Corning*/ACI incubator/oven).
4. Recording Fluorescent Densitometer (e.g. *Helena Laboratories* Cliniscan densitometer/analyzer).
5. Gel mould with accessories or commercially available agarose gels (1% agarose *Corning* Universal Film) and GelBond NF polystyrene support film.
6. Oven (50 °C).
7. Levelling table.
8. Hotplate.
9. Long *Pasteur* pipettes.

Reagents and Solutions

Purity of reagents: should be of a recognized analytical grade.

Preparation of solutions (for about 8 determinations): all solutions in re-purified water (cf. Vol. II, chapter 2.1.3.2).

1. Electrophoresis buffer (barbital, 50 mmol/l; EDTA, 0.94 mmol/l):

   dissolve 10.3 g sodium diethylbarbiturate, 0.35 g EDTA-$Na_2H_2 \cdot 2H_2O$ and 7 ml HCl, 1 mmol/l, in 500 ml water and make up to 1 l. The pH (20 °C) should be 8.6.

2. Gel buffer (barbital, 50 mmol/l; EDTA, 0.94 mmol/l; sucrose, 146 mmol/l):

   dissolve 13.4 g sodium diethylbarbiturate, 0.35 g EDTA-$Na_2H_2 \cdot 2H_2O$ and 50 g sucrose in 500 ml water and make up to 1 l after adjusting the pH (20 °C) to 8.6 with HCl, 1 mol/l.

3. L-Lactate (0.5 mol/l):

   dissolve 2.4 g lithium lactate in water, adjust to pH 7.0 (20 °C) with NaOH, 100 mmol/l, (the initial solution is usually slightly alkaline although it depends on the age of the purified water) and make up to 50 ml.

4. Nicotinamide-adenine dinucleotide (β-NAD, 7.5 mmol/l):

   dissolve 5 mg β-NAD, free acid, or β-NAD, lithium salt in 1 ml of lithium lactate solution (3).

Stability of solutions: electrophoresis buffer – store at 4 °C and discard after 10 electrophoretic runs. Stable indefinitely at 4 °C if unused. Gel buffer is best prepared from the electrophoretic buffer when required because the glucose content makes it susceptible to bacterial contamination even at 4 °C. Agarose gels are stable indefinitely at 4 °C. Lithium lactate is stable for one month at 4 °C. β-NAD solution must be freshly prepared immediately before use and any unused solution must be discarded.

Preparation of agarose gel (1%): the technique of making and pouring the gel is important for the preparation of satisfactory gels. This film may be cast using the thin gel mould (fitted with linear walls) described by *Elevitch et al.* [15]. Prepare the agarose [Indubiose A-37, *Pharmindustrie (l'Industrie Biologique Française)*, Clichy, France], at 1% concentration (i.e. 1 g agarose per dl gel buffer).

Mix the agarose powder into a paste with a small amount of gel buffer. Then add the correct amount of buffer to provide a 1% solution and heat the mixture with constant stirring, by *Bunsen* flame or hotplate, to near boiling when the gel particles will be completely dissolved. Place the beaker containing the molten gel in a hot water bath at 45 – 50 °C and allow to cool, with constant stirring (use a thermometer as the stirrer) until the gel temperature is just below 50 °C, then pour it into the pre-heated (to 45 °C) gel mould using pre-warmed long *Pasteur* pipettes to inject the gel.

Elevitch et al. [15] suggest that the gel and gel mould should be chilled before removal of the gel and lid as this allows easier detachment of the gel mould from the gel attached to the lid.

Alternatively, a gel mould obtained from a commercial gel preparation (see below) can be re-used. Clean the mould thoroughly with ethanol to remove the grease along the edges, wash in detergent-containing water, rinse well and air dry. Give the edges of the mould a light greasing with *Dow Corning* High Vacuum Grease. Before pouring the gel (see above) pre-heat the mould to 45 °C and cut to size a piece of GelBond NF (*Marine Colloids Division,* FMC Corporation) film to cover the mould base. After pouring the gel (using a levelling table) roll the GelBond NF (hydrophilic-side facing the gel) onto the molten gel surface so that all air bubbles are excluded. Then lay the mould, GelBond side down, on the bench and place a litre beaker of water on top of the gel mould until the gel is solid. Chilling the gel and gel mould (see above) facilitates removal of the GelBond lid with the gel attached.

It is probably more convenient to use the commercially available agarose gels supplied in the gel mould, the lid of which acts also as a rigid holder for the gel.

Procedure

Collection and treatment of specimen: collect blood from the vein with minimal or no stasis. Serum is to be preferred as plasma can sometimes give spurious results. Blood should be centrifuged for not more than 10 min at about 3000 *g* to obtain serum. Haemolysis should be avoided during the collection of blood or the preparation of the serum as erythrocytes contain about 100 times more LDH activity than normal serum. For the same reason, thoroughly rinse tissues to remove as much blood as possible before homogenization.

Stability of the enzyme in the specimen: LDH-4 and LDH-5 activities are cold-labile, and up to 20% of their activity can be lost within three days at 4 °C [17]. It is suggested [3, 16, 17] that the specimen should be stored at room temperature where it should be stable for up to five days, provided there is no bacterial contamination of the specimen. An alternative storage temperature is −10 °C if the specimen is quickly frozen and thawed. This tends to be less convenient. Tissue samples (and tissue homogenates) are best stored frozen.

Details for measurement in tissues: LDH activity is cytoplasmic in origin in the cell, thus allowing easy extraction of enzyme activity. However, the tissue may require several extractions to obtain > 90% initial yield and the extraction process should always be optimized for the particular tissue being studied. As previously mentioned, blood must be thoroughly washed out of the tissue before homogenization.

Electrophoresis: pour 1 litre of the electrophoresis buffer into the tank and tilt it so that the liquid levels in each compartment run into each other then lay it flat: this ensures that the levels are equal. Remove any traces of buffer along the ledges of the tank to prevent "salt-bridging" between the electrode compartments. The buffer should be at 4 °C and it is sufficiently chilled to complete two sequential runs, although the polarity should be changed with each run. Further runs require the buffer to be re-chilled or another chilled lot to be used.

Place approximately 1 μl of serum in the application well of the gel track using the *Drummond* 1 μl Microcap System. Retain a small amount of serum in the tube to prevent creating a bubble of serum in the application well. Reserve one position on each gel tray for a suitable quality control serum. Add a few crystals of bromophenol blue to one specimen, before it is loaded, to act as a marker of electrophoretic migration.

The bridge pieces in the tank should be about 3 cm apart and should be covered with a well-soaked single piece of suitable filter paper (such as *Whatman* No. 1). Place the loaded gel plate face down on these bridge pieces with the application wells at the cathodal end of the tank. Electrophoresis is performed, with the tank lid in place, at room temperature for about 1 h, using a constant voltage setting of 200 V (i.e. about 15 – 20 V/cm) which gives a current of about 15 – 18 mA. Stop the run when the bromophenol blue marker has migrated about 33 mm anodally.

Incubation: remove the plate, lay it flat on filter paper and apply about 1 ml (20 drops) of the substrate solution along the anode side of the plate and evenly spread the solution backwards and forwards across the plate using the side of a 5 ml pipette in a single sweeping movement. Excess solution is pushed off the edge of the plate at the end of each sweeping motion: it is important not to use any downward pressure on the gel.

During electrophoresis, the incubator should be switched on, allowing equilibration at 37 °C with a humid atmosphere obtained by soaking the paper liner of the incubator tray with water.

Spread the substrate mixture on the gel plate and place it into the pre-equilibrated incubator for 15 min. On completion, wipe the gel mould dry and place the gel and mould in the drying oven (72 °C) for 15 – 20 min. Continue drying until the agarose surface is hard to fingernail pressure. The dried plate has a shiny surface: care should be taken not to dry the plate too much as background fluorescence will increase to such an extent as to swamp the NADH fluorescence.

Evaluation: scan the individual LDH isoenzyme separations in a fluorescent recording densitometer, using an excitation wavelength of 365 nm and recording an emission wavelength signal above 410 nm.

Scanning can either determine peak height or area under each peak, and these measurements can then be converted to percent activity of each peak, by the densitometer processor. If a densitometer-based processor is not available, then calculation of percent activity from the individual peak heights or peak areas is necessary. Some examples of the types of pattern obtained are shown in Figure 2.

Notes on the Procedure

- crystallization of the substrate on the agarose plate after drying will sometimes be seen, particularly around the periphery of the plate: this will not interfere with scanning of the gels. If, however, crystallization occurs over the separation zone then the run must be repeated. Experience in my laboratory suggests that if the substrate solution is initially at pH 7.0, there is considerably less likelihood of crystallization occurring.
- Over-drying will result in excess blank fluorescence: therefore monitor the drying process carefully ensuring that the agarose plate is removed from the drying oven immediately it is dry.
- Quality control can be achieved using any commercial quality control serum containing only human serum with no additives [18]. The serum chosen should have been lyophilized very rapidly after collection because delay in this process appears to lead to asymmetrical isoenzyme peaks [18].

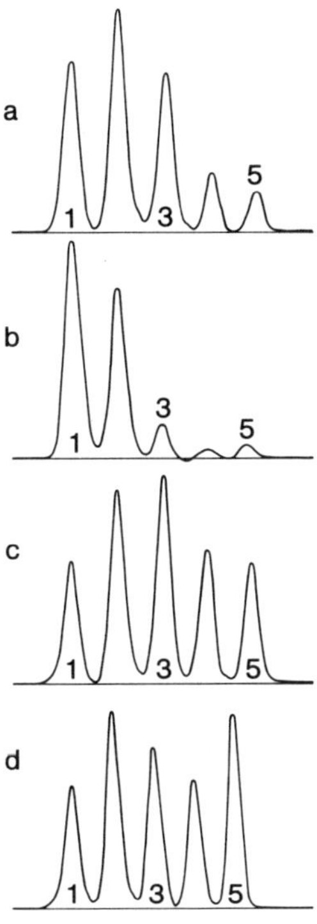

Fig. 2. Examples of serum LDH isoenzyme patterns seen in a General Hospital. The patterns are numbered 1→5 indicating LDH-1, LDH-3 and LDH-5. Pattern *a* is from a healthy individual. Pattern *b* was obtained from a case of myocardial infarction (LDH-1↑). Pattern *c* is typical of a case of infectious mononucleosis (LDH-3↑) and pattern *d* is seen following acute congestive cardiac failure resulting in hepatic anoxia (LDH-5↑).

Validation of Method

Precision, accuracy, detection limits and sensitivity: the precision of the assay is shown in Table 2 for within-day and between-day analyses. Standard deviations of

Table 2. Precision of LDH isoenzyme assay

Sera Used	LDH iso-enzyme	Same-day (%) Mean value	SD*	CV*	*n*	Between-day (%) Mean value	SD*	CV*	*n*
Normal human serum	1	23.26	1.09	4.7	16				
	2	36.80	0.91	2.5	16				
	3	20.59	0.41	2.0	16				
	4	8.53	0.58	6.7	16				
	5	10.76	1.23	11.4	16				
Commercial control serum	1	21.67	0.82	3.8	14	21.71	1.55	7.2	33
	2	35.38	0.68	1.9	14	36.63	1.50	4.1	33
	3	19.99	0.85	4.3	14	20.48	0.97	4.7	33
	4	8.65	0.51	5.9	14	8.42	1.06	12.6	33
	5	14.30	0.64	4.5	14	12.83	1.50	11.7	33

* SD = standard deviation
CV = coefficient of variation, relative standard deviation, in %

less than 1% are usual for most isoenzymes thus giving relative standard deviations of less than 6%. Although no standard reference materials are available, my laboratory has made estimations of accuracy using semi-purified LDH isoenzymes. We measured the relative proportions of the pairs of binary additions [19, 20] – LDH-1 : LDH-5 and LDH-1 : LDH-2 – mixed in known proportions and thus established, in a preliminary fashion, that this separative method has no major analytical bias towards anodal or cathodal isoenzymes. Under optimum conditions the lowest measurable isoenzyme value is 10 – 15 U/l (an average LDH total value might be 275 U/l) with an imprecision of < ±20%. Sensitivity, under optimum conditions, is 5 – 10 U/l.

Sources of error: in this technique there appear to be few. In end-stage renal disease when the patient is on maintenance haemodialysis the albumin, which runs anodally to LDH-1, becomes fluorescent and it appears to be a sixth isoenzyme. It is readily detected by means of a blank assay [21].

Specificity: LDH activity is measured specifically in this technique.

Reference ranges: these are shown in Tab. 3, derived with this assay from a population of 219 healthy subjects. The reference range for the LDH-1/LDH-2 ratio is 0.45 – 0.74 [22].

Table 3. Distribution of LDH isoenzymes in normal serum (n = 219)

	LDH-1	LDH-2	LDH-3	LDH-4	LDH-5
Mean (%)	20.3	34.0	22.8	11.8	11.0
Standard Deviation (%)	2.80	2.26	1.59	1.75	2.49

2.2.4.3 LDH-1 (2-Hydroxybutyrate Dehydrogenase), UV-Assay

Assay

Method Design

Rosalki and *Wilkinson* showed that the affinity of the anodal LDH fractions for 2-oxobutyrate was much greater than that of the cathodal fractions thus suggesting a technique of differentiation of these isoenzyme groups [7].

Therefore, tissues containing predominantly anodal fractions of the LDH isoenzyme (serum, heart muscle and erythrocytes) will show high activities with 2-oxobutyrate as substrate in the LDH reaction, whereas liver and skeletal muscle possess much less activity. LDH-X (found in human testis) also shows considerable activity with 2-oxobutyrate.

The LDH activity thus obtained with 2-oxobutyrate as substrate is called "2-hydroxybutyrate dehydrogenase" or "2-HBDH" activity. The alternative assay, using L or DL-2-hydroxybutyrate, has been described, but the latter substrate is expensive and high concentrations are required for an optimal assay.

Principle

(b) 2-Oxobutyrate + NADH + H^+ $\xrightleftharpoons{\text{LDH}}$ L-2-hydroxybutyrate + NAD^+.

The equilibrium lies to the right as in the LDH reaction with pyruvate. The oxidation of NADH in unit time – determined by the absorbance change at 339 (Hg 334, Hg 365) nm – is a measure of the activity. The method is readily adapted for mechanical analysis. No blanks are necessary for routine assays.

Optimized conditions for measurement: the HBDH reaction is one of the few assays in which the optimum substrate concentration (15 mmol/l at 25 °C) gives less diagnostic information than the use of a sub-optimal concentration (3.3 mmol/l at 25 °C). Despite this well recognized characteristic of the HBDH assay, optimized 30 °C (12 mmol/l) and 37 °C (15 mmol/l) assays have been proposed. Recently, the Association of Clinical Biochemists (UK) have again proposed a sub-optimal assay at 30 °C (3.3 mmol/l) with the intent of retaining the superior diagnostic performance of the original assay [23]. This assay will be described here, as it has been shown that results from it [6, 24] give better correlation with serum LDH-1 values than the 30 °C [25] or 37 °C [26] optimized assays. However, it should be appreciated that LDH-1 determinations by immunoprecipitation [8], or the heat-stable LDH determination [6, 27], may give superior differentiation between patients with cardiac and liver disease than even the sub-optimal HBDH assay [6].

The Association of Clinical Biochemists' (ACB) proposed method uses 2-oxobutyrate (3.3 mmol/l) and NADH (200 µmol/l) in phosphate buffer, 65.4 mmol/l, (pH 7.2 at 30 °C).

Temperature conversion factors: as *Wilkinson* [7] pointed out, measurements at different temperatures usually require different assay conditions so that temperature conversion factors are inappropriate. However, *Wilkinson* had also derived factors [7] by using similar analytical conditions with the warning that they should be applied with extreme caution to individual cases: the main use of the factors is to give an indication of the temperature dependence of a reaction and to allow qualitative comparisons between assay temperatures.

measured at	to convert to 25 °C	30 °C	37 °C
25 °C	1.00	1.19	1.37
30 °C	0.84	1.00	1.15
37 °C	0.73	0.87	1.00

Equipment

Spectrophotometer or spectral-line photometer for measurements at 339 (Hg 334, Hg 365) nm preferably with a thermostatted cuvette holder; water-bath; recorder or stopwatch, respectively.

Reagents and Solutions

Purity of reagents: NADH must be free from inhibitors (see Vol. II, chapter 2.1.2.2); suitable commercial preparations are available. 2-Oxobutyric acid can be transformed by a condensation reaction, after storage for more than two years, to 2-ethyl-3-methyl-tetrahydrofuran-4,5 dione which inhibits the enzyme reaction [7]. It is suggested that the sodium salt of the substrate should be used, as it is more stable. Other reagents should be of a recognized analytical grade.

Preparation of solutions (for about 25 determinations): all solutions in re-purified water (see Vol. II, chapter 2.1.3.2).

1. Phosphate buffer (67 mmol/l; pH 7.2; 30 °C):

   dissolve 9.47 g Na_2HPO_4 in water and make up to 1 l with water. Dissolve 9.08 g KH_2PO_4 in water and make up to 1 l. To 180 ml Na_2HPO_4 solution add 70 ml KH_2PO_4 solution.

2. 2-Oxobutyrate (71 mmol/l):

   dissolve 22 mg 2-oxobutyrate, sodium salt, in 2.5 ml phosphate buffer (1).

3. Reduced nicotinamide-adenine dinucleotide (β-NADH, 220 μmol/l):

   dissolve 7.6 mg NADH, sodium salt, in 50 ml phosphate buffer (1).

Stability of solutions: store all solutions at 0 – 4 °C. The phosphate buffer is stable as long as no bacterial contamination occurs. The 2-oxobutyrate solution is stable for about three months at 0 – 4 °C. The NADH solutions should be prepared freshly each day.

Procedure

Collection and treatment of specimen: collect blood from the vein with minimal or no stasis. The presence of anticoagulants such as oxalate, citrate and fluoride inhibit enzyme activity. EDTA (1 mg/ml) or heparin (0.2 mg/ml) do not inhibit. Centrifuge for no more than 10 min at about 3000 *g* to obtain serum or plasma. Haemolysis should be avoided during the collection of blood or the preparation of serum or plasma, as erythrocytes contain about 100 times more LDH activity than normal serum. For the same reason, thoroughly rinse tissues, to remove as much blood as possible, before homogenization.

Stability of the enzyme in the sample: serum HBDH activity – which corresponds essentially to LDH-1 and LDH-2 activity – is stable for at least a week at 18 °C or 4 °C [9]. Rapid freezing to −10 °C, with subsequent rapid thawing immediately before assaying, will preserve isoenzyme activities for many months. However, as LDH-5 and LDH-4 activity are cold-labile, and as the total LDH activity is also sometimes measured, it is prudent to store this sample under conditions least likely to cause any loss of isoenzyme activity. Ideally, therefore, the specimen should be stored at room temperature where it will be entirely stable for at least five days [16, 17], provided there is no bacterial contamination of the specimen. Tissue samples are best stored frozen: homogenized tissue should also be stored frozen although the enzyme activity can sometimes be much less stable in this form.

Details for measurement in tissues: the total HBDH activity is cytoplasmic in origin in the cell, which allows easy extraction of enzyme activity. The tissue may require several extractions to obtain a >90% yield. As previously mentioned, blood must be thoroughly washed out of the tissue before homogenization.

Assay conditions: wavelength 339 (Hg 334, Hg 365) nm; light path 10 mm; final volume 2.15 ml; 30 °C (constant temperature cuvette holder); read against air. Before starting the assay warm the solutions to 30 °C (allow 10 min as a minimum).

Measurement

Pipette successively into the cuvette:			concentration in assay mixture	
serum phosphate/NADH solution	 (3)	0.05 ml 2.00 ml	volume fraction phosphate NADH	0.023 65.4 mmol/l 0.20 mmol/l
mix thoroughly with plastic spatula; equilibration for 10 – 15 min in a 30 °C water bath to permit non-specific oxidation of NADH to go to completion				
2-oxobutyrate solution (equilibrated to 30 °C)	(2)	0.10 ml	2-oxobutyrate	3.3 mmol/l
mix thoroughly, and pour immediately into a cuvette and monitor the decrease in absorbance				

If $\Delta A/\Delta t$ is > 0.08/min at 339 nm, repeat the measurements; dilute the serum 1 + 4 or 1 + 9 with buffer (1).

Calculation: use calculation formula (k) or (k_1), resp. (cf. "Formulae", Appendix 3). Values for ε: cf. Appendix 4. The following relations are valid:

	Hg 334 nm	339 nm	Hg 365 nm	
	$b = 6958 \times \Delta A/\Delta t$	$6825 \times \Delta A/\Delta t$	$12647 \times \Delta A/\Delta t$	U/l
	$b = 11.60 \times 10^4 \times \Delta A/\Delta t$	$11.38 \times 10^4 \times \Delta A/\Delta t$	$21.08 \times 10^4 \times \Delta A/\Delta t$	nkat/l

Validation of Method

Precision, accuracy, detection limits and sensitivity: for values from 170 – 840 U/l in serum an imprecision of ± 5 to 15 U/l was obtained. The relative standard deviation, referred to the mean value, of a series of within- and between-batch analyses is < 6%. An attempt to assess accuracy indicates that HBDH activity has a bias of about – 8% for LDH-1 and – 24% for LDH-2. At 339 nm the lowest possible value to be measured is approximately 15 – 20 U/l with an imprecision of < ± 20%. Sensitivity is found to be $\Delta A/\Delta t$ = 0.005/10 min for a *Pye Unicam* SP 1700 Spectrophotometer with a recorder scale of 1A = 20 cm.

Sources of error: influences on the measured activity by drug or other therapeutic measures are unknown. The influence on measured activity by enzyme inhibitors in NADH preparations is well recognized – see p. 119.

Specificity: the determination of HBDH activity is selective for LDH-1 activity rather than specific. Other LDH isoenzymes show activity with 2-oxobutyrate and *Wilkinson* gives the following ratios obtained at 25 °C:

	LDH-1	LDH-2	LDH-3	LDH-4	LDH-5
HBDH/LDH:	0.92	0.78	0.55	0.39	0.25

Reference ranges: in human serum a range of 73 – 146 U/l was found (n = 62). The ACB Proposed Method was described as having a range of 111 – 199 U/l from a population of n = 42 [23].

References

[1] *D. W. Moss,* Isoenzyme Analysis, Chemical Society, London 1979, pp. 1 – 35.
[2] *D. W. Moss,* Isoenzymes, Chapman and Hall, London 1982, pp. 1 – 204.
[3] *H. A. Zondag,* Determination and Diagnostic Significance of Lactate Dehydrogenase Isoenzymes, Van Gorcum, Assen 1967, pp. 1 – 118.
[4] *O. Bodansky,* Biochemistry of Human Cancer, Academic Press, New York 1975, pp. 33 – 60.

[5] *U. Bär, S. Ohlendorf,* Studien zur Enzymelimination. I. Halbwertszeiten einiger Zollenzyme beim Menschen, Klin. Wochenschr. *48*, 776 – 780 (1970).

[6] *A. Y. Foo, E. Nemesánszky, S. B. Rosalki,* Comparison of Procedures for the Measurement of the Anodal Isoenzymes of Lactate Dehydrogenase in Serum, Ann. Clin. Biochem. *18*, 232 – 235 (1981).

[7] *J. H. Wilkinson,* Lactate Dehydrogenase, Isoenzymes: LDH_1 ("2-Hydroxybutyrate Dehydrogenase") UV-Assay, in: *H. U. Bergmeyer* (ed.), Methods of Enzymatic Analysis Volume 2, 2nd English Edit., Verlag Chemie, Weinheim 1974, pp. 603 – 607.

[8] *M. Usategui-Gomez, R. W. Wicks, M. Warshaw,* Immunochemical Determination of the Heart Isoenzyme of Lactate Dehydrogenase (LDH_1) in Human Serum. Clin. Chem. *25*, 729 – 734 (1979).

[9] *H. U. Bergmeyer, E. Bernt,* Lactate Dehydrogenase, Isoenzymes: UV-assay after Separation on DEAE-Sephadex, in: *H. U. Bergmeyer* (ed.), Methods of Enzymatic Analysis, Volume 2, 2nd English Edit., Verlag Chemie, Weinheim 1974, pp. 590 – 593.

[10] *P. I. Clark, W. J. Kostuk, A. R. Henderson,* A Simple and Rapid Procedure for the Preparation of Human LDH-1 and LDH-5 using an Ion-exchange Mini-column. Clin. Chim. Acta *69*, 361 – 366 (1976).

[11] *D. W. Mercer,* Improved Column Method for Separating Lactate Dehydrogenase Isoenzymes 1 and 2, Clin. Chem. *24*, 480 – 482 (1978).

[12] *S. N. Buhl, K. Y. Jackson, B. Graffunder,* Optimal Reaction Conditions for Assaying Human Lactate Dehydrogenase Pyruvate-to-lactate at 25, 30, and 37 °C, Clin. Chem. *24*, 261 – 266 (1978).

[13] *S. N. Buhl, K. Y. Jackson, R. Lubinski, R. E. Vanderlinde,* Optimal Conditions for Assaying Human Lactate Dehydrogenase by the Lactate-to-pyruvate Reaction: Arrhenius Relationships for Lactate Dehydrogenase Isoenzymes 1 and 5, Clin. Chem. *23*, 1289 – 1295 (1977).

[14] *D. McKenzie, A. R. Henderson,* The Effect of Dilution and Substrate Repletion on the LDH Isoenzyme Distribution of Sera with High LDH Activities, Clin. Chim. Acta *50*, 147 – 150 (1974).

[15] *F. R. Elevitch, S. B. Aronson, T. V. Feichtmeir, M. L. Enterline,* Thin Gel Electrophoresis in Agarose. Am. J. Clin. Path. *46*, 692 – 697 (1966).

[16] *H. H. Kreutzer, W. H. S. Fennis,* Lactic Dehydrogenase Isoenzymes in Blood Serum after Storage at Different Temperatures, Clin. Chim. Acta *9*, 64 – 68 (1964).

[17] *J. A. Lott, J. M. Stang,* Serum Enzymes and Isoenzymes in the Diagnosis and Differential Diagnosis of Myocardial Ischemia and Necrosis, Clin. Chem. *26*, 1241 – 1250 (1980).

[18] *D. McKenzie, A. R. Henderson,* The Quality Control of Serum Lactate Dehydrogenase Isoenzyme Separations using Commercial Quality Control Sera, Clin. Chim. Acta *51*, 53 – 66 (1974).

[19] *D. McKenzie, P. I. Clark, A. R. Henderson,* How Accurate are Lactate Dehydrogenase Isoenzyme Estimations by the Thin-layer Agarose Fluorescent Technique? Clin. Chem. *22*, 1995 – 1998 (1976).

[20] *F. Y. Leung, A. R. Henderson,* Influence of Hemolysis on the Serum Lactate Dehydrogenase-1/lactate Dehydrogenase-2 Ratio as Determined by an Accurate Thin-layer Agarose Electrophoresis Procedure, Clin. Chem. *27*, 1708 – 1722 (1981).

[21] *D. McKenzie, A. R. Henderson,* An Artefact in Lactate Dehydrogenase Isoenzyme Patterns, Assayed by Fluorescence, Occurring in the Serum of Patients with End-stage Renal Disease requiring Maintenance Haemodialysis, Clin. Chim. Acta *70*, 333 – 336 (1976).

[22] *F. Y. Leung, A. R. Henderson,* Thin-layer Agarose Electrophoresis of Lactate Dehydrogenase Isoenzymes in Serum: a Note on the Method of Reporting and on the Lactate Dehydrogenase Isoenzyme-1/Isoenzyme-2 Ratio in Acute Myocardial Infarction, Clin. Chem. *25*, 209 – 211 (1979).

[23] Association of Clinical Biochemists, Proposed Methods for Determination of some Enzymes in Blood Serum: α-hydroxybutyrate Dehydrogenase (EC 1.1.1.27 HBD) Continuous Monitoring Method, ACB News Sheet Supplement *202*, 27s – 30s (1980).

[24] *F. Y. Leung, A. R. Henderson,* Comparison of Three Serum α-hydroxybutyrate Dehydrogenase Procedures for Lactate Dehydrogenase Isoenzyme Discrimination, Clin. Chim. Acta *115*, 145 – 153 (1981).

[25] *L. M. Shaw, J. Gray,* Optimal α-oxobutyrate Concentration for Assay of α-hydroxybutyrate Dehydrogenase Activity in Normal and Pathological Sera at 30 °C, Clin. Chem. *20*, 494 – 496 (1974).

[26] *G. Ellis, D. M. Goldberg,* Serum α-hydroxybutyrate Dehydrogenase Activity: an Improved Method, Am. J. Clin. Pathol. *56*, 627 – 635 (1971).

[27] *W. G. Brydon, A. F. Smith,* An Appraisal of Routine Methods for the Determination of the Anodal Isoenzymes of Lactate Dehydrogenase, Clin. Chim. Acta *43*, 361 – 369 (1973).

2.3 3-Hydroxy-3-methylglutaryl-coenzyme A Reductase

Mevalonate: NADP oxidoreductase (acylating CoA), E.C. 1.1.1.34

Joachim Siedel

General

3-Hydroxy-3-methylglutaryl-coenzyme A (HMG-CoA) reductase, a key enzyme in the isoprenoid pathway, was discovered independently in the laboratories of *Lynen* [1] and *Rudney* [2] in 1958.

The enzyme occurs in animal and plant tissue as well as in micro-organisms. In liver cells, it is associated with the microsomal membrane [3]; its existence has been demonstrated in the plastids, mitochondria, and microsomes in cells from pea seedlings [4]. The yeast enzyme is localized within the mitochondrial matrix [5].

HMG-CoA reductase has been obtained in the pure state from human [6], rat [7 – 9] and chicken liver [10], baker's yeast [11], *Fusarium oxysporum* [12], and *Pseudomonas* [13].

In contrast to the enzymes from eukaryotes, HMG-CoA reductase from *Pseudomonas* utilizes NADH rather than NADPH as the preferred reductant [13].

Mammalian HMG-CoA reductase, particularly from rat liver and cultured cells, is the subject of intensive investigation because it is widely recognized as the most important regulatory enzyme of sterol biosynthesis [14].

Application of method: in biochemistry.

Enzyme properties relevant in analysis: in the presence of saturating concentrations of HMG-CoA and NADPH, substrate conversion proceeds linearly with time.

Methods of determination: the catalytic activity can be measured by monitoring spectrophotometrically the rate of NADPH consumption during HMG-CoA reduction [15], by the colorimetric [16] or fluorimetric [17] determination of the concomitant release of coenzyme A, by the quantitation of radioactively labelled mevalonate formed from [^{14}C]HMG-CoA in the presence of a NADPH-regenerating system [18, 19], or by multiple-selected ion monitoring [20]. Enzyme mass determination by immunotitration has also been described (cf., e.g. [21]).

Because of its convenience, the direct spectrophotometric assay based on the rate measurement of NADPH oxidation is described here; however, it is applicable only to samples of pure enzyme, or of crude or partially purified HMG-CoA reductase preparations which are largely free from contaminants causing either high background absorbance or non-specific NADPH oxidation. The latter is true for crude yeast autolysates, for example. In cases where the sample to be analyzed contains comparatively low reductase activity, enzymes that rapidly oxidize NADPH even if no other substrate is added, and material causing significant turbidity (e.g., liver microsomes or tissue homogenates), other and more sensitive assays such as those based on the use of ^{14}C-labelled HMG-CoA have to be applied. Here, measurement of NADPH becomes possible only after HMG-CoA reductase has been solubilized from the microsomal membranes [7, 9].

International reference method and standards: standardization of activity measurements or the preparation of reference material at the international level have not yet been reported.

Enzyme effectors: HMG-CoA reductase is strongly inhibited by sulphydryl reagents [15]; the addition of thiol-protecting substances such as β-mercaptoethanol, cysteamine, or dithiothreitol [7, 22, 23] to the assay mixture is therefore essential. The enzyme shows no requirement for free or loosely-bound metal ions [15]. The possibility that the specific activity of hepatic HMG-CoA reductase might be modulated by covalent phosphorylation [24] has been drawn into question more recently [25].

Assay

Method Design

Principle

(a) (3*S*)-3-Hydroxy-3-methylglutaryl-coenzyme A* + 2 NADPH + 2 H^+

$$\xrightarrow{\text{HMG-CoA reductase}} (3R)\text{-mevalonate} + 2\,\text{NADP}^+ + \text{coenzyme A}.$$

The amount of HMG-CoA reacting per unit time, determined by the net decrease in absorbance due to the difference in the rates of NADPH oxidation in the presence and the absence of HMG-CoA, is a measure of the catalytic activity of HMG-CoA reductase. Reversibility of the reaction has been demonstrated; however, mevalonic acid synthesis is strongly favoured [13, 15].

If published data on catalytic enzyme concentrations are being compared, it should be taken into account that some authors define the enzyme activity in terms of the oxidation of NADPH per unit time, which is twice the rate of HMG-CoA turnover referred to by other investigators, cf. eqn. (a). Here, one enzyme unit is defined as the

* For the steric course of the reaction, see [26].

amount of enzyme catalyzing the conversion of 1 μmol HMG-CoA to 1 μmol mevalonate per minute.

Optimized conditions for measurement: the assay outlined here is largely based on the procedure described by *Rétey et al.* [23] and has proved suitable for the determination of yeast HMG-CoA reductase [27]. The *Michaelis* constants of this enzyme have been determined to be 2.4 μmol/l for (3S)-HMG-CoA and 89 μmol/l for NADPH, respectively [22]; maximum activity is observed at a pH of 6.5 [15]. In the following method, (3R, S)-HMG-CoA, 150 μmol/l, NADPH, 250 μmol/l, and phosphate buffer, 70 μmol/l, pH 6.5, are used to approximate to the most satisfactory measurement conditions. The assay might be equally well adapted for the determination of partially or completely purified mammalian HMG-CoA reductase because its kinetic properties are closely similar to those of the enzyme from yeast [8].

Temperature conversion factors: no data have been published so far.

Equipment

Spectrophotometer or spectral-line photometer capable of exact measurement at 339 nm, Hg 334 or Hg 365 nm, with a thermostatted cuvette holder; water-bath; recorder.

Reagents and Solutions

Purity of reagents: HMG-CoA must be free from unbound coenzyme A and 3-hydroxy-3-methylglutaric acid because of the inhibitory action of these substances on the enzyme activity [22, 28]. Suitable preparations are commercially available or may be obtained by the reaction of HMG*-anhydride with coenzyme A and subsequent purification of the HMG-CoA formed [17].

Preparation of solutions (for about 100 determinations): all solutions in re-purified water (see Vol. II, chapter 2.1.3.2).

1. Phosphate buffer (0.1 mol/l; pH 6.5):

   dissolve 1.36 g KH_2PO_4 (anhydrous) in 80 ml water, adjust to pH 6.5 (25 °C) with KOH, 1 mol/l, and dilute to 100 ml with water.

2. EDTA (0.1 mol/l):

   add 292 mg ethylenediamine-tetraacetic acid to 5 ml water, dissolve by addition of KOH, 1 mol/l, until a pH of 7.0 is reached, and dilute to 10 ml with water.

---

* HMG, 3-hydroxy-3-methylglutaric acid.

3. Cysteamine (0.1 mol/l):

   dissolve 22.7 mg cysteamine hydrochloride in 1 ml water, adjust to neutral pH by the addition of KOH, 1 mol/l (about 20 µl), and dilute to 2 ml with water.

4. Serum albumin (1 %):

   dissolve 60 mg of bovine serum albumin in 6 ml water.

5. NADPH (5 mmol/l):

   dissolve 21 mg of NADPH, tetrasodium salt, in 5 ml of a solution of 5 g $KHCO_3$ in 1 l water (50 mmol/l).

6. (3R, S)-HMG-CoA (7.5 mmol/l):

   dissolve 13.7 mg (3R, S)-HMG-CoA (referred to the free acid) in 1 ml water, adjust to pH 5.5 with HCl, 1 mol/l, and dilute to 2 ml with water.

7. Reagent mixture (for about 20 determinations):

   mix 1.0 ml water with 14 ml phosphate buffer (1), 0.4 ml EDTA solution (2), 0.4 ml cysteamine solution (3), 1.2 ml serum albumin solution (4), and 1.0 ml NADPH solution (5).

Stability of solutions: store solutions (1) and (2) at 0 – 4 °C; they are stable as long as no bacterial contamination occurs. Keep solutions (4) and (6) frozen at −20 °C; they are stable for at least three months. Prepare solution (5) freshly every week and keep at −20 °C. Prepare solutions (3) and (7) freshly every day and keep at 0 – 4 °C.

Procedure

Collection and treatment of specimen: preparation of crude yeast HMG-CoA reductase by prolonged autolysis or by sonication of the mitochondrial fraction has been described by *Kirtley & Rudney* [22] and by *Qureshi et al.* [11].

Stability of the enzyme in the specimen: the activity of the enzyme is rapidly lost in buffers of low ionic strength and in the absence of strongly reducing sulphydryl compounds [11]. In the presence of phosphate buffer, 0.1 mol/l, and dithiothreitol, 5 mmol/l, the purified yeast enzyme retains 80 % of its initial activity after two days at 4 °C [11]; with obviously higher salt concentrations (not specified in the reference), the enzyme remains stable for 3 – 4 months at 4 °C [11].

Details for measurement in other samples: liver HMG-CoA reductase may be solubilized from the microsomal membrane by a freeze-thawing procedure, followed by a short-term incubation at 37 °C prior to the assay to restore full enzymic activity [9].

Enzyme preparations from rat liver were found to be stable for at least three months when kept in 50% (v/v) glycerol at −70 °C [8].

Assay conditions: wavelength 339 (Hg 334, Hg 365) nm; light path 10 mm; final volume 1.00 ml; 25 °C (thermostatted cuvette holder). Measure against air. Before starting the assay, adjust temperature of solutions to 25 °C.

Measurement

Pipette sucessively into the cuvette:			concentration in the assay mixture	
reagent mixture	(7)	0.90 ml	phosphate EDTA cysteamine albumin NADPH	70 mmol/l 2 mmol/l 2 mmol/l 0.06% (w/v) 0.25 mmol/l
H_2O sample*		0 to 0.07 ml 0.01 to 0.08 ml	volume fraction	0.01 to 0.08
mix with a plastic spatula, follow the decrease in absorbance until a constant baseline rate is established, measure the decrease in absorbance over a further period of 3 min (ΔA_1);				
(3 R, S)-HMG-CoA	(6)	0.02 ml	(3 S)-HMG-CoA	75 µmol/l
mix and monitor the decrease in absorbance over a period of 3 min after the addition of HMG-CoA (ΔA_2).				

The values $\Delta A_2/\Delta t$ must be $\leq$ 0.06/min at Hg 334 or 339 nm ($\leq$ 0.03/min at Hg 365 nm), and $\Delta A_1/\Delta t$ should not exceed 1/2 of $\Delta A_2/\Delta t$. Otherwise, choose a smaller sample volume fraction or dilute sample appropriately. $\Delta A/\Delta t = \Delta A_2/\Delta t - \Delta A_1/\Delta t$ is used for calculation.

---

* The amount of sample necessary depends on the stage of enzyme purification.

Calculation: the catalytic concentration of the enzyme in the sample (μmol of HMG-CoA reduced per min and litre) is calculated from the decrease in absorbance according to calculation formula (k) or (k_1), respectively (cf. "Formulae", Appendix 3). Values for ε: cf. Appendix 4. The following relations are valid (Δt in minutes):

	Hg 334 nm	339 nm	Hg 365 nm	
$b =$	$\frac{80.9}{v} \times \Delta A/\Delta t$	$\frac{79.4}{v} \times \Delta A/\Delta t$	$\frac{142.9}{v} \times \Delta A/\Delta t$	U/l
$b =$	$\frac{1348}{v} \times \Delta A/\Delta t$	$\frac{1323}{v} \times \Delta A/\Delta t$	$\frac{2381}{v} \times \Delta A/\Delta t$	nkat/l

where v is sample volume in ml.

Validation of Method

Precision, accuracy, detection limits and sensitivity: precision, accuracy and sensitivity of the photometric reductase assay have not been investigated in detail so far. As to the detection limit, the procedure described above has been found reliable for the measurement of HMG-CoA reductase in crude autolysates from fresh baker's yeast prepared according to *Kirtley & Rudney* [22] down to average enzyme concentrations of 50 U/l or 2.5 U/g of protein [27]; below this value, the relative rate of nonspecific NADPH oxidation renders the determination too inaccurate; at least one further purification step (e.g., by acetone precipitation [23]), becomes necessary.

Sources of error: the assay is rather insusceptible to interferences if enzyme preparations from yeast are being determined; on the other hand, if solubilized HMG-CoA reductase from liver microsomes is analyzed, contamination by HMG-CoA deacylase* or by HMG-CoA lyase (E.C. 4.1.3.4) originating from disrupted mitochondria may cause an underestimation of the enzyme activity [29]. Selective inactivation of these interfering enzymes by prolonged treatment of the sample at 37 °C has been reported by *Ness et al.* [8].

Specificity: the HMG-CoA reductase activity is measured specifically.

Reference ranges: no reference values of HMG-CoA reductase in various tissues have been established so far; with liver microsomes from normally treated rats, enzyme activity contents in the range of 0.1 to 1 U/g protein are found [17]**, at an average recovery of 23 mg microsomal protein per g tissue (wet weight) [29].

* No E.C. number assigned yet.

** Hepatic HMG-CoA reductase activity exhibits a circadian variation [30].

References

[1] *F. Lynen, U. Henning, C. Bublitz, B. Sörbo, L. Kröplin-Rueff*, Der chemische Mechanismus der Acetessigsäurebildung in der Leber, Biochem. Z. *330*, 269 – 295 (1958).

[2] *J. J. Ferguson Jr., I. F. Durr, H. Rudney*, Enzymatic Reduction of β-Hydroxy-β-Methyl Glutaryl Coenzyme A (HMGCoA) to Mevalonic Acid in Yeast, Federation Proc. *17*, 219 (1958).

[3] *K. A. Mitropoulos, S. Venkatesan, S. Balasubramaniam, T. J. Peters*, The Submicrosomal Localization of 3-Hydroxy-3-Methylglutaryl-Coenzyme-A Reductase, Cholesterol 7α-Hydroxylase and Cholesterol in Rat Liver, Eur. J. Biochem. *82*, 419 – 429 (1978).

[4] *J. D. Brooker, D. W. Russell*, Subcellular Localization of 3-Hydroxy-3-Methylglutaryl Coenzyme A Reductase in *Pisum sativum* Seedlings, Arch. Biochem. Biophys. *167*, 730 – 737 (1975).

[5] *I. Shimizu, J. Nagai, H. Hatanaka, H. Katsuki*, Mevalonate Synthesis in the Mitochondria of Yeast, Biochem. Biophys. Acta *296*, 310 – 320 (1973).

[6] *R. D. Tanaka, P. A. Edwards, S. F. H. Lan, E. M. Knoeppel, A. M. Fogelman*, Purification of 3-Hydroxy-3-Methylglutaryl Coenzyme A Reductase from Human Liver, J. Lipid Res. *23*, 523 – 530 (1982).

[7] *P. A. Edwards, D. Lemongello, J. Kane, I. Shechter, A. M. Fogelman*, Properties of Purified Rat Hepatic 3-Hydroxy-3-methylglutaryl Coenzyme A Reductase and Regulation of Enzyme Activity, J. Biol. Chem. *255*, 3715 – 3725 (1980).

[8] *G. C. Ness, C. D. Spindler, M. H. Moffler*, Purification of 3-Hydroxy-3-Methylglutaryl Coenzyme A Reductase from Rat Liver, Arch. Biochem. Biophys. *197*, 493 – 499 (1979).

[9] *D. A. Kleinsek, J. W. Porter*, An Alternate Method of Purification and Properties of Rat Liver β-Hydroxy-β-methylglutaryl Coenzyme A Reductase, J. Biol. Chem. *254*, 7591 – 7599 (1979).

[10] *Z. H. Beg, J. A. Stonik, H. B. Brewer, Jr.*, Purification and Characterization of 3-Hydroxy-3-Methylglutaryl Coenzyme A Reductase from Chicken Liver, FEBS Letters *80*, 123 – 129 (1977).

[11] *N. Qureshi, R. E. Dugan, S. Nimmannit, W.-H. Wu, J. W. Porter*, Purification of β-Hydroxy-β-methylglutaryl-coenzyme A Reductase from Yeast, Biochemistry *15*, 4185 – 4190 (1976).

[12] *C. Madhosingh, W. Orr*, Hydroxymethylglutaryl-Coenzyme A Reductase. Purification and Properties of the Enzyme from *Fusarium oxysporum*, Biochim. Biophys. Acta *523*, 283 – 296 (1978).

[13] *W. R. Bensch, V. W. Rodwell*, Purification and Properties of 3-Hydroxy-3-methylglutaryl Coenzyme A Reductase from *Pseudomonas*, J. Biol. Chem. *245*, 3755 – 3762 (1970).

[14] *G. J. Schroepfer, Jr.*, Sterol Biosynthesis, in: *E. E. Snell, P. D. Boyer, A. Meister, C. C. Richardson* (eds.), Ann. Rev. Biochem., Vol. 50, Annual Reviews Inc., Palo Alto, 1981, pp. 585 – 621.

[15] *J. Knappe, E. Ringelmann, F. Lynen*, Über die β-Hydroxy-β-methyl-glutaryl-Reduktase der Hefe, Biochem. Z. *332*, 195 – 213 (1959).

[16] *F. H. Hulcher, W. H. Oleson*, Simplified Spectrophotometric Assay for Microsomal 3-Hydroxy-3-methylglutaryl CoA Reductase by Measurement of Coenzyme A, J. Lipid Res. *14*, 625 – 631 (1973).

[17] *Y. A. Baqir, R. Booth*, A New Method for Assaying Rat Liver Microsomal 3-Hydroxy-3-methylglutaryl-Coenzyme A Reductase Activity and its Application in a Study of the Effect of Dietary Cholesterol on this Enzyme, Biochem. J. *164*, 501 – 508 (1977).

[18] *D. J. Shapiro, J. L. Nordstrom, J. J. Mitschelen, V. W. Rodwell, R. T. Schimke*, Micro Assay for 3-Hydroxy-3-methylglutaryl-CoA Reductase in Rat Liver and in L-Cell Fibroblasts, Biochim. Biophys. Acta *370*, 369 – 377 (1974).

[19] *J. Huber, S. Latzin, B. Hamprecht*, A Simple and Rapid Radiochemical Assay for 3-Hydroxy-3-methylglutaryl-Coenzyme A Reductase, Hoppe-Seyler's Z. Physiol. Chem. *354*, 1645 – 1647 (1973).

[20] *G. Cighetti, E. Santaniello, G. Galli*, Evaluation of 3-Hydroxy-3-Methylglutaryl-CoA Reductase Activity by Multiple-Selected Ion Monitoring, Anal. Biochem. *110*, 153 – 158 (1981).

[21] *J. E. Hardgrave, R. A. Heller, M. G. Herrera, T. J. Scallen*, Immunotitration of 3-Hydroxy-3-methylglutaryl-Coenzyme A Reductase in Various Physiological States, Proc. Natl. Acad. Sci. U.S.A. *76*, 3834 – 3838 (1979).

[22] *M. E. Kirtley, H. Rudney*, Some Properties and Mechanism of Action of the β-Hydroxy-β-methylglutaryl Coenzyme A Reductase of Yeast, Biochemistry *6*, 230 – 238 (1967).

[23] *J. Rétey, E. von Stetten, U. Coy, F. Lynen*, A Probable Intermediate in the Enzymic Reduction of 3-Hydroxy-3-methylglutaryl Coenzyme A, Eur. J. Biochem. *15*, 72–76 (1970).
[24] *Z. H. Beg, D. W. Allman, D. M. Gibson*, Modulation of 3-Hydroxy-3-methylglutaryl Coenzyme A Reductase Activity with cAMP and with Protein Fractions of Rat Liver Cytosol, Biochem. Biophys. Res. Commun. *54*, 1362–1369 (1973).
[25] *G. C. Ness, C. D. Spindler, G. A. Benton*, Characteristics of Magnesium-Adenosine Triphosphate-dependent Inactivators of 3-Hydroxy-3-methylglutaryl Coenzyme A Reductase, J. Biol. Chem. *255*, 9013–9016 (1980).
[26] *P. Blattmann, J. Rétey*, Zur Wirkungsweise und Stereospezifität der Hydroxymethylglutaryl-CoA-Reduktase, Hoppe-Seylers's Z. Physiol. Chem. *352*, 369–376 (1971).
[27] *J. Siedel*, Dissertation, Ludwig-Maximilians-Universität München (1975).
[28] *J. D. Brooker, D. W. Russell*, Properties of Microsomal 3-Hydroxy-3-Methylglutaryl Coenzyme A Reductase From *Pisum sativum* Seedlings, Arch. Biochem. Biophys. *167*, 723–729 (1975).
[29] *N. L. Young, C. D. Saudek, S. A. Crawford, S. L. Zuckerbrod*, Recovery and Activation of Hydroxymethylglutaryl Coenzyme A Reductase from Rat Small Intestine, J. Lipid Res. *23*, 257–265 (1982).
[30] *B. Hamprecht, C. Nüssler, F. Lynen*, Rhythmic Changes of Hydroxymethylglutaryl Coenzyme A Reductase Activity in Livers of Fed and Fasted Rats, FEBS Letters *4*, 257–265 (1969).

2.4 Malate Dehydrogenase

L-Malate : NAD^+ oxidoreductase, EC 1.1.1.37

2.4.1 General

Alistair F. Smith

The enzyme was discovered in 1910 by *Batelli & Stern* [1] and *Thunberg* [2]. The biochemical properties of the enzyme have been reviewed by *Banaszak & Bradshaw* [3].

Malate dehydrogenase occurs in nearly all animal tissues as well as in plants and micro-organisms. Mammalian tissues contain two isoenzymes, one mitochondrial and the other cytoplasmic. These isoenzymes differ from one another in physical and enzymic properties although the differences are generally quantitative rather than qualitative. The cytoplasmic isoenzymes from different tissues are identical to one another, as are the mitochondrial isoenzymes from different tissues. Both isoenzymes are dimers comprising two identical polypeptide chains each of 35000 molecular weight.

Malate dehydrogenase is not completely specific for malate since it will oxidize certain other 2-hydroxydicarboxylic acids [4]. However, there is a specific requirement for NAD, and the enzyme discussed in this section is distinct from the 'malic' enzymes – malate dehydrogenases (oxaloacetate decarboxylating) EC 1.1.1.38 and EC 1.1.1.40, which require NAD and NADP, respectively, as coenzymes.

Malate dehydrogenase activity in serum rises when there is damage to tissues, e.g. heart, liver or skeletal muscle [5]. There is evidence from studies of patients with myocardial infarction [6], liver disease [7] and following trauma [8] that more protracted and severe forms of organ damage are associated with a relatively greater increase in the activity in serum of the mitochondrial than of the cytoplasmic enzyme.

Application of method: in clinical diagnostic enzymology, the serum enzyme activity has been used for detecting heart and liver disease [5, 6, 7]. Measurement of total serum malate dehydrogenase activity is now rarely used since other serum enzymes are more specific and sensitive indices of organ damage. Separate measurement of cytoplasmic and mitochondrial isoenzymes may have a place in clinical diagnosis [6, 7].

The malate dehydrogenase assay is also used in some homogeneous immunoassay methods [9].

The method described here, although of general applicability, has been optimized for measurement of serum malate dehydrogenase activity. There is evidence that the kinetic parameters of malate dehydrogenase in serum differ considerably from those of the enzyme present in tissue extracts [10], so the method is not necessarily optimal for fluids other than serum.

Enzyme properties relevant in analysis: the cytoplasmic and mitochondrial isoenzymes differ quantitatively in their kinetic properties. Available evidence, for pig heart malate dehydrogenase, indicates that, relative to the mitochondrial enzyme, the cytoplasmic enzyme has a higher affinity for malate, a lower affinity for oxaloacetate and is less inhibited by excess oxaloacetate [11, 12]. These differences are unlikely to affect optimal conditions or method selection to any great extent since they are relatively minor. There is some evidence that the kinetic properties of serum malate dehydrogenase may alter with storage of the serum over a period of several days [10]. In devising an assay the principal constraint, apart from the instability of oxaloacetate encountered where the reaction is run in the oxaloacetate to malate direction, is to maintain the substrate concentration sufficiently low to avoid inhibition by substrate excess.

Under the reaction conditions described here the reaction rate remains rectilinear for several minutes. There is no significant lag phase.

Methods of determination: malate dehydrogenase in serum, or other fluids, is measured by determination of catalytic activity. Immunological mass determination has not been used in clinical practice, although the method should be ideal for making separate measurements of the cytoplasmic and mitochondrial isoenzymes.

Apart from assays of purely historical interest, malate dehydrogenase is measured by spectrophotometric methods which monitor absorbance change at 339 nm due to

NAD reduction or NADH oxidation. There are two types of assay – those using oxaloacetate as substrate [13] and those using malate as substrate [14].

With oxaloacetate and NADH as substrates initial reaction velocities are about 4 times greater than with malate and NAD as substrates. However, there is a very serious disadvantage to the former method, as oxaloacetate spontaneously decarboxylates to pyruvate. The pyruvate so formed acts, with the NADH already present in the assay mixture, as a substrate for lactate dehydrogenase which is present in the serum sample. This causes a large and variable blank reaction. Although the problem has been realized in the past, its magnitude has often been ignored. Thus, a 2 mmol/l solution of oxaloacetate used as substrate could give rise to a 100 or 200 μmol/l solution of pyruvate within 30 minutes on standing at room temperature. This substrate concentration is less than an order of magnitude below the optimal pyruvate concentration for serum lactate dehydrogenase, and this enzyme is likely to be present in serum at a considerably higher activity than malate dehydrogenase.

A method has been designed to overcome this problem by generating oxaloacetate immediately prior to its use as substrate, by incubating aspartate and oxoglutarate in presence of aspartate aminotransferase [15]. This is only partially successful because it is difficult to time the various phases of the assay sufficiently rigorously so that no pyruvate can form. Also, the extra reagents add complexities to the method and it is necessary to start the reaction with serum rather than with substrate. For all these reasons it requires great care to assay malate dehydrogenase accurately using oxaloacetate as substrate, the reaction usually used in the past.

If malate and NAD are used as substrates, most of these problems disappear [14]. Malate is more stable than oxaloacetate, a substrate-start is possible and the reaction mixture remains simple. The main disadvantages are that the method is less sensitive since an unfavourable reaction direction is being used, and that side-reactions due to the presence of lactate and lactate dehydrogenase in the sample must be allowed to go to completion before the malate dehydrogenase reaction is started.

Although the malate to oxaloacetate method has only rarely been used in the past, its advantages are such that it has been selected for presentation here.

International reference method and standards: there is currently no international reference method nor are standard enzyme materials available.

Enzyme inhibitors and activators: magnesium ions activate the reaction by about 10%. Although both isoenzymes contain sulphydryl groups, serum malate dehydrogenase activity is not enhanced by sulphydryl agents.

Excess oxaloacetate or malate inhibits the reaction.

References

[1] *F. Batelli, L. Stern,* Oxydation des acides maliques, fumarique et citrique par les tissues animaux, C. R. Soc. Biol., Paris *69,* 552–554 (1910).

[2] *T. Thunberg,* Studien über die Beeinflussung des Gasaustausches des überlebenden Froschmuskels durch verschiedene Stoffe, Scand. Arch. Physiol. *24,* 23–61 (1911).

[3] *L. J. Banaszak, R. A. Bradshaw,* Malate Dehydrogenase, in: *P. D. Boyer* (ed.), The Enzymes, Vol. XI, Academic Press, New York, San Francisco, London 1975, pp. 369–396.
[4] *D. D. Davies, E. Kun,* Isolation and Properties of Malic Dehydrogenase from Ox-Heart Mitochondria, Biochem. J. *66*, 307–316 (1957).
[5] *J. T. Aquilina, W. E. Farnsworth,* Alteration of Serum Enzymes in Clinical Myocardial Infarction, Amer. Heart J. *59*, 166–174 (1960).
[6] *A. F. Smith, P. C.-P. Wong, M. F. Oliver,* Release of Mitochondrial Enzymes in Acute Myocardial Infarction, J. Mol. Med. *2*, 265–269 (1977).
[7] *E. Schmidt, F. W. Schmidt, P. Otto,* Isoenzymes of Malic Dehydrogenase and Glutamic Oxaloacetic Transaminase and Lactic Dehydrogenase in Serum in Diseases of the Liver, Clin. Chim. Acta *15*, 283–289 (1967).
[8] *K. Garbus,* Serum Malate Dehydrogenase Isoenzymes as Indicators of Severe Cellular Injury, Clin. Chim. Acta *35*, 502–504 (1971).
[9] *E. F. Ullman, R. A. Yoshida, J. I. Blakemore, E. Maggio, R. Leute,* Mechanism of Inhibition of Malate Dehydrogenase by Thyroxine Derivatives and Reactivation by Antibodies: Homogenous Enzyme Immunoassay for Thyroxine, Biochem. Biophys. Acta *567*, 66–74 (1979).
[10] *E. Schmidt, F. W. Schmidt,* Über Änderungen der Substrataffinität von Malat-Dehydrogenasen in Serum, Klin. Wochenschr. *38*, 810–811 (1960).
[11] *F. C. Grimm, D. C. Doherty,* Properties of the Two Forms of Malic Dehydrogenase from Beef Heart, J. Biol. Chem. *236*, 1980–1985.
[12] *L. H. Bernstein, M. B. Grisham, K. D. Cole, J. Everse,* Substrate Inhibition of the Mitochondrial and Cytoplasmic Malate Dehydrogenase, J. Biol. Chem. *253*, 8697–8701 (1978).
[13] *A. Siegel, R. J. Bing,* Plasma Enzyme Activity in Myocardial Infarction in Dog and Man, Proc. Soc. Exp. Biol. Med. *91*, 604–607 (1956).
[14] *P. C. P. Wong, A. F. Smith,* Assay of Serum NAD-Dependent Malate Dehydrogenase Using Malate as Substrate, Clin. Chim. Acta *72*, 409–412 (1976).
[15] *H. U. Bergmeyer, E. Bernt,* Malate Dehydrogenase, in: *H. U. Bergmeyer* (ed.), Methods of Enzymatic Analysis, Vol. 2, 2nd. English ed., Verlag Chemie Weinheim/Academic Press, New York 1974, pp. 613–617.

2.4.2 Malate to Oxaloacetate Reaction

Alistair F. Smith

Assay

Method Design

Principle

(a) $$\text{Malate} + \text{NAD}^+ \rightleftharpoons \text{oxaloacetate} + \text{NADH} + \text{H}^+.$$

Malate dehydrogenase reversibly catalyzes the reaction (a). The thermodynamic equilibrium strongly favours the formation of malate from oxaloacetate. This straightforward reaction can be monitored directly by measuring the rate of increase in A_{339} due to NADH formation. The reaction is started by the addition of malate but, prior to this, serum must be allowed to react with NAD^+ so that the endogenous lactate can reduce the NAD to NADH under the action of the lactate dehydrogenase in serum. This reaction may take one hour or sometimes longer and although it might be possible to shorten this period by adding lactate dehydrogenase to the assay mixture, the increased complexity of the reagent and possibility of contamination render this procedure unsatisfactory.

The assay is suitable for performance on both automated reaction rate analyzers and centrifugal analyzers, providing there is the facility for a substrate-start rather than a serum-start.

The reaction rate remains rectilinear for several minutes for samples with malate dehydrogenase activity of the magnitude likely to be encountered in serum. It has been reported that this reaction rapidly halts due to product inhibition and that this can be overcome by addition of glutamate and aspartate aminotransferase [1]. Under the conditions employed in the present assay this additional manoeuvre is unnecessary.

Serum blank reaction rates are normally low and in the interests of convenience have been omitted in the routine assay described here. If desired they can be measured by omitting the addition of malate. The reagent blank is negligible.

Optimized conditions for measurement: the conditions given here have been optimized for serum samples from patients with both liver disease and with myocardial infarction. There is a fairly broad peak of optimal malate concentration lying between 10 and 30 mmol/l; above 30 mmol/l enzyme activity is inhibited. An optimal concentration of NAD has been selected: excess NAD does not inhibit the reaction. The optimal pH for the reaction in the direction malate to oxaloacetate is about pH 9.2 and, of the commonly used buffers effective at this pH, diethanolamine gives rise to the highest activity. Although it has been suggested that the ionic strength of the medium affects malate dehydrogenase activity [2], the ionic strength of the diethanolamine has been found to have little effect on activity when measured under the conditions given here [3].

No studies are available which provide data about the optimal conditions for the measurement of the cytoplasmic and mitochondrial isoenzymes of malate dehydrogenase in human serum after separation from one another; for example, by ion-exchange chromatography [4]. However, it seems unlikely that the two isoenzymes will differ greatly in optimal conditions for measurement since the kinetic differences between them are not great, i.e. they are of the same order of magnitude in those species in which they have been studied in detail [5]. There are only small differences in kinetic parameters between serum from patients with heart disease and those with liver disease [3]. The kinetic parameters have been found to vary with the age of the serum [6] when using oxaloacetate as substrate but no information is available with regard to the assay using malate as substrate.

In order to achieve maximal reaction velocity, 37 °C has been selected for assay temperature. However, the conditions selected are also suitable for assays using 30 °C.

Temperature conversion factors: the following conversion factors have been given [7]:

activity at 30 °C is 0.67 times the activity at 37 °C;
activity at 25 °C is 0.40 times the activity at 37 °C.

These figures have been obtained using an oxaloacetate to malate method – there has been no study of temperature conversion factors for the method described here.

Equipment

The assay may be run on any spectrophotometer capable of monitoring absorbance change at 339 nm and which has facilities for accurate and precise temperature control of the reaction cuvette. Absorbance change may also be monitored at the 334 nm spectral-line of mercury. Most automatic enzyme analyzers and centrifugal analyzers fulfil these conditions.

Reagents and Solutions

Purity of reagents: use reagents of purest available grade.

Preparation of solutions (for about 20 determinations): all solutions in re-purified water (cf. Vol. II, chapter 2.1.3.2).

1. Diethanolamine buffer (DEA, 0.1 mol/l, pH 9.2; containing $MgCl_2$, 5 mmol/l):

   add 9.5 ml diethanolamine to 800 ml water, add 1 g $MgCl_2 \cdot 6H_2O$, mix to dissolve and bring to pH 9.2 by adding HCl, 1 mol/l. Make up to 1 litre with water.

2. NAD (3.6 mmol/l):

   dissolve 120 mg NAD, free acid, in 50 ml DEA buffer (1).

3. L-Malate (250 mmol/l):

   dissolve 0.78 g L-malate (monosodium salt) in 15 ml DEA buffer (1), adjust to pH 9.2 with NaOH, 5 mol/l, and make up to 20 ml.

Stability of solutions: the buffer (1) is stable for at least 1 month at +4 °C. The NAD solution (2) is relatively unstable in alkaline solution so it is advisable to prepare only the amount required for use on the day of analysis. The malate solution (3) is stable for at least a week at +4 °C; alternatively, portions may be stored at −20 °C for several months.

Procedure

Collection and treatment of specimen: collect blood without venous stasis. Fresh serum, rather than plasma, is preferred but blood may be collected into 0.2 mg heparin/ml blood, 1 mg oxalate/ml blood, 1.5 mg EDTA/ml blood or 1 mg citrate/ml blood. Centrifuge blood at about 3000 *g* for 10 minutes to obtain serum or plasma. Haemolysed serum or plasma is unsuitable for analysis for malate dehydrogenase activity.

Stability of the enzyme in the sample: few data are available. Whereas one source [7] states that malate dehydrogenase is stable for more than 48 h at 20 °C and 2 weeks at +4 °C, another finds a 17% loss of activity in 24 h at room temperature, an 11% loss in 24 h at +4 °C and a 2% loss in 24 h at −20 °C [8]. No data are available on the relative stabilities of the cytoplasmic and mitochondrial isoenzymes.

Measurement in fluids other than serum: the method may be applied to tissue extracts but it is likely that the optimal substrate concentration will be lower than for serum.

Assay conditions: wavelength 339 nm (or 334 nm mercury line); light path 10 mm; final volume of reaction mixture 2.5 ml; temperature 37 °C ± 0.05 °C. Incubation time 1 h. Read absorbance against water or air.

For greater accuracy, a serum blank, treated in exactly the same way as the test, including the 1 h incubation at room temperature but omitting the addition of malate, should be measured and subtracted from the test result.

If the measurement are made on certain instruments, e.g. some centrifugal analyzers, the analytical procedure may have to be varied by reducing the reaction volume by a convenient factor e.g. 2-fold, 5-fold or even 10-fold reductions in the volumes of all reagents.

Measurement

Pipette successively into reaction cuvettes:			concentration in assay mixture	
serum		0.25 ml	volume fraction	0.1
NAD^+ solution	(2)	2.00 ml	NAD	2.9 mmol/l
			DEA	90 mmol/l
			$MgCl_2$	4.5 mmol/l
incubate at room temperature for 1 h; bring cuvette temperature to 37 °C (5 to 15 min)				
malate solution	(3)	0.25 ml	malate	25 mmol/l
monitor absorbance continuously (chart recorder) or read at 30 s intervals for 3 min. Record $\Delta A/\Delta t$.				

Calculation: formula (k) or (k_1) (see "Formulae", Appendix 3) is applicable. Values for ε: cf. Appendix 4. Catalytic activity concentration in the serum sample is (Δt in minutes):

	Hg 334 nm	339 nm	
$b =$	$1618 \times \Delta A/\Delta t$	$1587 \times \Delta A/\Delta t$	U/l
$b =$	$26969 \times \Delta A/\Delta t$	$26455 \times \Delta A/\Delta t$	nkat/l

Validation of Method

Precision, accuracy, detection limits and sensitivity: for samples with activities at the upper end of the adult reference range, between-batch imprecision is about 6%. At higher activities the CV is lower, falling to about 3%.

Lactate dehydrogenase can cause inaccuracies in the method, by causing the oxidation of endogenous lactate in the sample, if the period of sample pre-incubation with NADH is omitted. For the greatest accuracy, a sample blank, which omits the addition of malate, should be included with each test: this is unnecessary in routine analysis.

For serum samples the detection limit is about 5 U/l. If samples have previously been partially purified, and larger volumes used for the analysis, the detection limit may be lower – this applies particularly to samples for isoenzyme analysis which have been passed through an ion-exchange column.

Sources of error: drugs and other therapeutic measures are not known to affect serum malate dehydrogenase activity.

Haemolysis of the sample causes a marked increase in malate dehydrogenase activity.

Apart from inadequate blanking of lactate dehydrogenase (discussed above) the method is not known to be subject to systematic error.

Specificity: the rate of the reaction is unlikely to be affected by other enzymes which occur in human serum.

Reference ranges: in healthy male adults, reference values of 5 – 40 U/l were obtained using the present method [3]. Considerably higher values are likely to be obtained if using a method based on oxaloacetate as substrate.

References

[1] *A. Siegel, R. J. Bing,* Plasma Enzyme Activity in Myocardial Infarction in Dog and Man, Proc. Soc. Exp. Biol. Med. *91*, 604 – 607 (1956).

[2] *A. Bracht, A. de P. Campello,* Effect of Ionic Strength on the Kinetic Properties of the Mitochondrial L-Malate dehydrogenase, Experientia *35*, 1559 – 1561 (1979).

[3] *P. C. P. Wong, A. F. Smith,* Assay of Serum NAD-Dependent Malate Dehydrogenase Using Malate as Substrate, Clin. Chim. Acta *72*, 409 – 412 (1976).
[4] *E. Schmidt, F. W. Schmidt, J. Mohr,* An Improved Simple Chromatographic Method for Separating the Isoenzymes of Malate Dehydrogenase and Glutamic Oxaloacetic Transaminase, Clin. Chim. Acta *15*, 337 – 342 (1967).
[5] *F. C. Grimm, D. C. Doherty,* Properties of the Two Forms of Malic Dehydrogenase from Beef Heart, J. Biol. Chem. *236*, 1980 – 1985.
[6] *E. Schmidt, F. W. Schmidt,* Über Änderungen der Substrataffinität von Malat-Dehydrogenasen in Serum, Klin. Wochenschr. *38*, 810 – 811 (1960).
[7] *J. King,* Malate Dehydrogenase, Practical Clinical Enzymology, Van Nostrand, London 1965, pp. 93 – 99.
[8] *H. Südhof, E. Wötzel,* Über die Haltbarkeit von zur Enzymaktivitätsbestimmung vorgesehenen Serumproben, Klin. Wochenschr. *38*, 1165 – 1167 (1960).

2.4.3 Oxaloacetate to Malate Reaction

Hans Ulrich Bergmeyer and Erich Bernt

Assay

Method Design

Principle

(a) L-Aspartate + 2-oxoglutarate $\underset{}{\overset{\text{AST*}}{\rightleftharpoons}}$ L-glutamate + oxaloacetate

(b) Oxaloacetate + NADH + H^+ $\overset{\text{MDH}}{\rightleftharpoons}$ L-malate + NAD^+.

The amount of oxaloacetate reacting per unit time, measured by the decrease in absorbance due to the oxidation of NADH, is a measure of the catalytic activity of MDH. The equilibrium of malate conversion is far in favour of malate. At neutral pH, according to [1], $K' = 2.33 \times 10^{-5}$ at 22 °C. Measurements of the catalytic activity are therefore performed with oxaloacetate as substrate [2]. Since oxaloacetate is unstable in solution (yielding pyruvate by decarboxylation, so that lactate dehydrogenase activity would also be partially measured), it is generated within the assay mixture from 2-oxoglutarate and aspartate by the action of aspartate aminotransferase. Blanks are not necessary for routine assays. The method described below can easily be adapted to automated equipment.

* AST Aspartate aminotransferase, L-Aspartate : 2-oxoglutarate aminotransferase, EC 2.6.1.1.

Optimized conditions for measurement: the most important characteristics of the isoenzymes of MDH, as far as measurement of their activities is concerned, are their different substrate and pH optima. These have been specially studied by *Schmidt* [3]. In the following method, oxaloacetate, 1 mmol/l, NADH, 0.2 mmol/l and pH 7.4 are used, since these conditions approximate to the most satisfactory for measurements of activities in serum both after myocardial infarction and in hepatitis.

Temperature conversion factors: *J. King* [4] found the following factors for measurements at different temperatures (related to the value measured at 25 °C):

°C	18	19	20	21	22	23	24	25	26	27
F	1.80	1.65	1.52	1.39	1.28	1.18	1.09	1.00	0.92	0.86
°C	28	29	30	31	32	33	34	35	36	37
F	0.80	0.74	0.68	0.63	0.58	0.54	0.50	0.46	0.43	0.40

Equipment

Spectrophotometer or spectral-line photometer capable of exact measurement at 339 nm, Hg 334 or Hg 365 nm, preferably with a thermostatted cuvette holder; water-bath; recorder or stopwatch.

Reagents and Solutions

Purity of reagents: the AST preparation must be free from MDH. Contamination with glutamate dehydrogenase, lactate dehydrogenase and oxaloacetate decarboxylase should be < 0.05% (in relation to the AST activity).

NADH and 2-oxoglutarate must be free from inhibitors (cf. Vol. II, chapter 2.1.2.2); suitable commercial preparations are available.

Preparation of solutions (for about 25 determinations): all solutions in re-purified water (cf. Vol. II, chapter 2.1.3.2).

1. Phosphate/aspartate solution (phosphate, 0.1 mol/l, pH 7.4; aspartate, 42 mmol/l):

   dissolve 0.2 g KH_2PO_4, 1.5 g K_2HPO_4, and 0.66 aspartate, sodium salt, or 0.56 g L-aspartic acid in 50 ml water, adjust to pH 7.4 with NaOH, 1 mol/l, and dilute to 100 ml with water.

2. 2-oxoglutarate (60 mmol/l):

   dissolve 17 mg 2-oxoglutarate, sodium salt, in 1.5 ml water.

3. Reduced nicotinamide-adenine dinucleotide (β-NADH, 12 mmol/l):

   dissolve 15 mg NADH, sodium salt, with 1.5 ml $NaHCO_3$ solution, 1% (w/v).

4. Aspartate aminotransferase, AST (180 kU/l):

   AST (≧180 U/mg, 25 °C) from pig heart suspended in ammonium sulphate, 3 mol/l, stabilized with 2-oxoglutarate, 2.5 mmol/l, and maleate, 50 mmol/l (from *Boehringer Mannheim*). If necessary, dilute the stock solution with ammonium sulphate, 3 mol/l.

Stability of solutions: store all solutions at 0 – 4 °C. The phosphate/aspartate solution (1) is stable as long as no bacterial contamination occurs. Prepare the 2-oxoglutarate and NADH solutions freshly every week. Inhibitors of MDH might be formed in the NADH solution during storage.

Procedure

Collection and treatment of specimen: collect blood from the vein without stasis. Additions of oxalate, 1 mg/ml; citrate, 1 mg/ml; fluoride, 2 mg/ml; heparin, 0.2 mg/ml, or EDTA, 1 mg/ml, are unobjectionable. Centrifuge for 10 min at about 3000 *g* in order to obtain plasma or serum. Use only fresh serum or plasma free from haemolysis.

Stability of the enzyme in the sample: according to [5], the enzyme in serum loses about 17% of its activity in 24 h at room temperature, about 11% at 4 °C and about 2% in the frozen state. This does not take into account the change in the optimum substrate concentration due to ageing of the enzyme [3].

According to [4], the enzyme is stable in human serum at 20 °C for at least 48 h and at +4 °C for at least 2 weeks. Samples from other sources may be stabilized with bovine serum albumin, 1% (w/v).

Details for measurement in tissues: the total MDH activity of the cell is quantitatively extracted only after complete homogenization of the tissue [6, 7] because a portion of the activity is due to an isoenzyme specifically located in the mitochondria. By variation of the homogenization technique, it is possible to distinguish between mitochondrial and cytoplasmic MDH. Since the optimum pH and substrate concentration vary considerably for MDH from different tissues, these should be determined in preliminary experiments.

Assay conditions: wavelength 339 (Hg 334, Hg 365) nm; light path 10 mm; final volume 2.75 ml; 30 °C (thermostatted cuvette holder). Measure against air. Before starting the assay, adjust temperature of solutions to 30 °C.

Measurement

Pipette successively in the cuvette:			concentration in assay mixture	
phosphate/aspartate solution	(1)	2.50 ml	phosphate aspartate	91 mmol/l 38 mmol/l
2-oxoglutarate solution	(2)	0.05 ml	2-oxoglutarate	1.1 mmol/l
NADH solution	(3)	0.05 ml	NADH	0.22 mmol/l
AST suspension	(4)	0.05 ml	AST	3270 U/l
mix thoroughly with a plastic spatula, wait for 5 min, until aspartate and 2-oxoglutarate are converted to oxaloacetate				
sample (serum or other sample solution)		0.10 ml	volume fraction	0.036
mix, read absorbance and start stopwatch. Repeat the reading after exactly 1, 2 or 3 min or monitor the reaction on a recorder.				

The values $\Delta A/\Delta t$ must be < 0.030/min at 365 nm (at 334 nm or 339 nm < 0.060/min). Otherwise dilute sample 5 to 10 fold with solution (1) or measure over shorter intervals.

Calculation: use calculation formula (k) or (k_1) from "Formulae", Appendix 3. Values for ε: cf. Appendix 4. The following relations are valid for the catalytic concentration in the sample (Δt in minutes):

	Hg 334 nm	339 nm	Hg 365 nm	
$b =$	$4450 \times \Delta A/\Delta t$	$4365 \times \Delta A/\Delta t$	$8088 \times \Delta A/\Delta t$	U/l
$b =$	$74164 \times \Delta A/\Delta t$	$72748 \times \Delta A/\Delta t$	$134799 \times \Delta A/\Delta t$	nkat/l

Validation of Method

Precision, accuracy, detection limits and sensitivity: for values of about 120 U/l in serum an imprecision of ± 0.7 U/l has been found for manual performance. The relative standard deviation is 0.6%. The relative standard deviation referred to the mean value of a series of interlaboratory measurements was 3%. Data about inaccuracy are not available since standard reference material is not yet established.

For measurements at 339 nm with sufficiently sensitive photometers the lowest possible value to be measured is approximately 4 U/l with an imprecision of 23%. Sensitivity is found to be $\Delta A/\Delta t$ = 0.005/10 min (*Eppendorf* photometer).

Sources of error: influences on the measured activity by therapeutics are unknown. – Interference due to the decarboxylation of the oxaloacetate is reduced to a minimum under the experimental conditions described here. With regard to the influence of enzyme inhibitors in NADH preparations of inadequate quality cf. Vol. II, chapter 2.1.2.2.

Specificity: the MDH activity is measured specifically. Since the added 2-oxoglutarate is converted before the start of the MDH reaction, the presence of glutamate dehydrogenase in the sample does not interfere.

Reference ranges: in human serum 48 – 96 U/l at 25 °C [5]. With oxaloacetate, 0.5 mmol/l, *King & Morris* [8] found activity ranges from 12.5 to 50 U/l in normal adults, but in the neonatal period values up to twice can be considered normal. *Potter* [7] reported the MDH activity in aqueous homogenates of some organs of laboratory animals.

References

[1] *J. R. Stern, S. Ochoa, F. Lynen,* Enzymatic Synthesis of Citric Acid V. Reaction of Acetyl Coenzyme A, J. Biol. Chem. *198*, 313 – 321 (1952).

[2] *A. M. Hehler, A. Kornberg, S. Grisolia, S. Ochoa,* The Enzymatic Mechanism of Oxidation-Reductions between Malate or Isocitrate and Pyruvate, J. Biol. Chem. *174*, 961 – 977 (1948).

[3] *E. Schmidt, F. W. Schmidt,* Über Änderungen der Substrat-Affinität von Malat-Dehydrogenase im Serum, Klin. Wochenschr. *38*, 810 – 811 (1960).

[4] *J. King,* Malate Dehydrogenase, Practical Clinical Enzymology, Van Nostrand London 1965, pp. 93 – 99.

[5] *H. Südhof, E. Wötzel,* Über die Haltbarkeit von zur Enzymaktivitätsbestimmung vorgesehener Serumproben, Klin. Wochenschr. *38*, 1165 – 1167 (1960).

[6] *A. Delbrück, E. Zebe, Th. Bücher,* Über Verteilungsmuster von Enzymen des Energie liefernden Stoffwechsels im Flugmuskel, Sprungmuskel und Fettkörper von Locusta migratoria und ihre cytologische Zuordnung, Biochem. Z. *331*, 273 – 296 (1959).

[7] *V. R. Potter,* The Assay of Animal Tissues for Respiratory Enzymes V. The Malic Dehydrogenase System, J. Biol. Chem. *165*, 311 – 324 (1946).

[8] *J. King, M. Brenda Morris,* Serum Enzyme Activity in the Normal Newborn Infant, Arch. Dis. Child. *36*, 604 – 609 (1961).

2.5 'Malic' Enzyme

L-Malate : NAD$^+$ oxidoreductase (decarboxylating), EC 1.1.1.39

William H. Outlaw Jr. and Stefan A. Springer

General

Three 'malic' enzyme activities catalyze the oxidative decarboxylation of malate to pyruvate, which is linked to the reduction of pyridine nucleotide. These enzymes (EC 1.1.1.38, EC 1.1.1.39 and EC 1.1.1.40) differ in their specificity for pyridine nucleotides and cations, kinetic constants, activators, inhibitors, oxaloacetate-decarboxylating activity, distribution among organisms, intracellular location and biological function. Two major differences among these activities are the basis for their classification. First, ME* (EC 1.1.1.39), unlike ME (EC 1.1.1.38) and ME (EC 1.1.1.40), lacks an intrinsic oxaloacetate-decarboxylating activity. Ordinarily, this distinction is not verified. Second, ME (EC 1.1.1.39) lacks absolute coenzyme specificity, but is more active with NAD than with NADP. The degree of specificity for NAD on the part of ME (EC 1.1.1.38) apparently depends on the biological source [1 – 3]. In general (but see [4]), ME (EC 1.1.1.40) is highly specific for NADP [5 – 7]. This chapter is a description of methods for ME (EC 1.1.1.39) but, as the preceding discussion implies, an assay for activity in raw extracts may not provide sufficient information to allow one to distinguish among the MEs. Therefore, such activities will be referred prudently to as NAD-ME, without regard to EC nomenclature. Methods for ME (EC 1.1.1.38) and ME (EC 1.1.1.40) are published elsewhere [8, 9].

ME (EC 1.1.1.39), first described by *Saz & Hubbard* [10], occurs in some microorganisms [11] and the mitochondria of some plant [12] and animal [13] tissues. This enzyme has diverse metabolic functions. It catalyzes an alternative pathway for malate oxidation (in addition to that via the tricarboxylic acid cycle) [12] and, with pyruvate orthophosphate dikinase (EC 2.7.9.1), forms a gluconeogenic sequence. By

---

* Abbreviations:
BSA, bovine serum albumin
C_3, photosynthetic reductive pentose phosphate pathway
C_4, auxillary photosynthetic pathway involving 4-C acids
CAM, Crassulacean acid metabolism
DTT, dithiothreitol
Hepes, N-2-hydroxyethyl piperazine-N'-2-ethane sulfonic acid
NAD-MDH, NAD$^+$-linked malate dehydrogenase (EC 1.1.1.37)
ME, 'malic' enzyme
PVP-40, polyvinylpyrrolidone (M_r approx. 40000)

regulating the organic acid concentration, ME is part of a pH-stat mechanism [14]. In some plants, the initial fixation of CO_2 is separated spatially (C_4 plants) or temporally (CAM plants) from the carboxylation leading into the reductive pentose phosphate pathway. In certain of these CAM [15] and C_4 [16] plants, NAD-ME [presumably ME (EC 1.1.1.39)] releases the "stored-" or "transport-CO_2" for the latter carboxylation.

Application of method: in biochemistry and, potentially, in clinical chemistry (cf. [17]).

Enzyme properties relevant in analysis: to the extent of present knowledge, the ME (EC 1.1.1.39) activity present in raw extracts of any one tissue is due to one enzyme.

Methods of determination: the catalytic activity may be measured by trapping $^{14}CO_2$ released from (C-4) [^{14}C]malate or, more conveniently, by spectrophotometry as in the present method. Other methods of estimation currently are not in use.

International reference methods and standards: neither standardization at the international level nor the existence of reference standard materials is known.

Enzyme effectors: the catalytic activity from plants is activated by CoA, acetyl CoA, SO_4^{2-} and fructose-1,6-P_2 [15, 16, 18, 19]. The enzyme from some C_4 plants has an absolute requirement for Mn^{2+} [16, 20], but the divalent cation requirement for catalytic activity from C_3 plants [1] may be satisfied by Mg^{2+} or Mn^{2+}. The divalent cation requirement for activity from CAM plants apparently depends on the biological source [15–21]. Thiol reagents are required for full enzyme activity for C_3 [18], C_4 [16] and CAM [21] plants. Reports on the effects of organic acids on the enzyme are conflicting [15–22]. The rabbit heart enzyme [13], unlike the plant enzyme [15], is inhibited by ATP. In contrast to the plant enzyme (see above), the *Ascaris* enzyme [23] is inhibited by $(NH_4)_2\ SO_4$. Because of these variable results on the effects of various compounds, caution is advised.

Assay

Method Design

Principle

$$\text{Malate} + \text{NAD}^+ \xrightarrow{Mn^{2+}} \text{pyruvate} + CO_2 + \text{NADH} + H^+.$$

The amount of malate reacting per unit time, measured as an increase in absorbance due to the reduction of NAD, is a measure of the catalytic activity. Under analytical conditions, the reaction proceeds linearly with time.

Optimized conditions for measurement: important characteristics of NAD-ME, specifically shown to be ME (EC 1.1.1.39), are known for few organisms. Based on findings to date, the properties are sufficiently different to prevent formulation of a generally applicable, ideal assay protocol. In the following method malate, 5 mmol/l; NAD, 2 mmol/l; Mn^{2+}, 4 mmol/l; CoA, 0.1 mmol/l and pH 7.6 are used (see [24]).

Temperature conversion factors: generally-applicable conversion factors are not available.

Equipment

Recording spectrophotometer with a thermostatted cuvette holder, centrifuge, water-bath.

Reagents and Solutions

Purity of reagents: commercially-available reagent-grade preparations are suitable.

Preparation of solutions: (for about 60 determinations): all solutions in re-purified water (cf. Vol. II, chapter 2.1.3.2).

1. KOH (1 mol/l):

   dissolve 0.6 g KOH in 8 ml water, dilute to 10 ml with water.

2. Hepes buffer (100 mmol/l, pH 7.6):

   dissolve 3 g Hepes in 80 ml water, add 6.5 ml KOH solution (1) and dilute to 125 ml with water.

3. $MnCl_2$ (100 mmol/l):

   dissolve 0.1 g $MnCl_2 \cdot 4H_2O$ in 5 ml water.

4. BSA (approx. 10%):

   dissolve 0.25 g BSA in 2.5 ml water.

5. DTT (200 mmol/l):

   dissolve 0.3 g DTT in 10 ml water.

6. Malic acid (100 mmol/l):

   dissolve 160 mg malic acid in 12 ml water.

7. NAD (100 mmol/l):

   dissolve 0.3 g NAD, free acid, in 4 ml water and dilute to 5 ml with water.

8. Carbonate (100 mmol/l; pH approx. 10.6):

   dissolve 80 mg Na_2CO_3 and 20 mg $NaHCO_3$ in 10 ml water.

9. NADH (5 mmol/l):

   dissolve 5.3 mg NADH, disodium salt, in 1.5 ml carbonate solution (8).

10. EDTA (10 mmol/l):

    dissolve 17 mg EDTA-$Na_2 \cdot 2H_2O$, in 5 ml water.

11. CoA (10 mmol/l):

    dissolve 20 mg CoA, free acid, in 2.5 ml water.

12. PVP-40 (approx. 10%):

    dissolve 0.25 g PVP-40 in 2.5 ml water.

13. Extraction cocktail (Hepes, 50 mmol/l, pH 7.6; $MnCl_2$, 2 mmol/l; DTT, 10 mmol/l; BSA, 1%; PVP-40, 1%):

    to 2.5 ml Hepes, solution (2), add 0.1 ml $MnCl_2$, solution (3), 0.5 ml BSA, solution (4), 0.25 ml DTT, solution (5), 0.5 ml PVP-40, solution (12) and dilute to 5 ml with water.

14. Assay cocktail (Hepes, 50 mmol/l, pH 7.6; malate, 5 mmol/l; NAD, 2 mmol/l; NAD-MDH, 1 µg/ml; EDTA, 0.2 mmol/l; CoA, 0.1 mmol/l; NADH, 50 µmol/l; DTT, 5 mmol/l):

    to 25 ml Hepes, solution (2), add 1.25 ml DTT, solution (5), 2.5 ml malic acid, solution (6), 1 ml NAD, solution (7), 0.5 ml NADH, solution (9), 1 ml EDTA, solution (10), 0.5 ml CoA, solution (11), 0.01 ml NAD-MDH (NAD-MDH from pig heart, suspended in ammonium sulphate solution, 3.2 mol/l, $\geqq$500 U/mg protein) and dilute to 50 ml with water.

Stability of solutions: prepare extraction cocktail, solution (13), and assay cocktail, solution (14), within two hours of use. Store NADH, solution (9) at 4 °C and prepare fresh weekly or store at < -50 °C. Store NAD-MDH at 4 °C. Store other solutions at -20 °C.

Procedure

Collection and treatment of specimen: at 0 °C, completely homogenize 0.5 g tissue in 3 ml extraction cocktail, solution (13). Filter homogenate through a 100 μm nylon mesh or other filter. Add 0.015 ml Triton X-100 and mix. After 15 min, centrifuge at 10000 *g* for 1 min. Decant supernatant and bring it to 25 °C for 30 min before assay.

Stability of the enzyme in specimen or sample: according to [24], the activity is stable for several hours if stored at 0 °C.

Details for measurement in tissue: because ME (EC 1.1.1.39) is located in mitochondria, complete homogenization of the tissue is necessary.

Assay conditions: wavelength 339 nm; light path 10 mm; final volume 3.17 ml; 30 °C. Measure against water.

Measurement

Pipette successively into the cuvette:			concentration in assay mixture (rounded)	
assay cocktail solution	(14)	3.00 ml	Hepes	50 mmol/l
			malic acid	5 mmol/l
			NAD	2 mmol/l
			EDTA	0.2 mmol/l
			CoA	0.1 mmol/l
			DTT	5 mmol/l
			NADH	5 μmol/l
			NAD-MDH	500 U/l
tissue extract		0.05 ml	volume fraction	0.016
mix, wait 5 min				
$MnCl_2$ solution	(3)	0.12 ml	$MnCl_2$	4 mmol/l
record absorbance at 339 nm with time.				

Calculations must be based on values obtained when $\Delta A/\Delta t$ remains constant.

Calculation: use calculation formula (k) or (k_1) from "Formulae", Appendix 3. Values for ε: cf. Appendix 4. The following relations are valid for the catalytic concentration:

$$b = 9921 \times \Delta A/\Delta t \quad \text{U/l.}$$

$$b = 16.5 \times 10 \times \Delta A/\Delta t \quad \text{nkat/l}$$

Validation of Method

Precision, accuracy, detection limits and sensitivity: at a moderate level of activity in the extract, precision easily should be within 5%, depending on the spectrophotometer and operator. Because ME (EC 1.1.1.39) levels depend on several non-standardized biological factors, comparisons between different reports are not made. Data on accuracy are not available since a standard reference material does not exist. With standard equipment, a detection limit 500 U/l is easily achieved with no loss of precision. Sensitivity is found to be $\Delta A/\Delta t = 0.005/10$ min.

Sources of error: the presence of abundant NAD-MDH [which has the same substrates as ME (EC 1.1.1.39)] in extracts creates two sources of error in addition to the small burst of NADH as the NAD-MDH equilibrium position is reached. The first source of error arises because NADH is a product of the reactions catalyzed by both NAD-MDH and ME (EC 1.1.1.39). Thus, catalytic activity of the latter will cause a reversal of the NAD-MDH catalyzed reaction; this reversal will cause an underestimate of ME (EC 1.1.1.39) [25]. The second source of error is that Mn^{2+} [a cofactor for ME (EC 1.1.1.39)] stimulates the decarboxylation of oxaloacetate (e.g. 26), a product of NAD-MDH. Oxaloacetate will, then, continue to be produced to maintain the NAD-MDH equilibrium. Therefore, oxalacetate decarboxylation may cause an overestimate of the ME (EC 1.1.1.39) catalytic rate.

The first source of error may be avoided easily by calculation [25]. The second source of error may be avoided also by calculation. (Between 20 °C and 40 °C, the first-order rate constant for Mn^{2+}-stimulated oxaloacetate decarboxylation, k approx. $(-4.47 + 0.26\,T) \times 10^{-2}$ at pH 7 ($R^2 = 0.76$); approx. $(-5.34 + 0.29\,T) \times 10^{-2}$ at pH 7.5 ($R^2 = 0.76$); approx. $(-5.24 + 0.25\,T) \times 10^{-2}$ at pH 8.0 ($R^2 = 0.92$) with T in °C; our unpublished results obtained under assay conditions). Alternatively, the second source of error may be diminished by comparison to control reagents lacking extract, but containing Mn^{2+} and analytical NAD-MDH [27]. Though inappropriate for fluorimetry [27], a simple way to avoid both these errors is to include NADH in the assay cocktail [24], thereby shifting the oxaloacetate concentration of the NAD-MDH equilibrium position to insignificance. Because of its simplicity, this last way is used in this chapter.

It may be necessary to de-salt extracts of some tissues by gel filtration before assay. Whether this de-salting is necessary should be determined by assaying different extract concentrations.

Specificity: ME (EC 1.1.1.39) is not measured with absolute specificity. Calculated activities may be due to ME (EC 1.1.1.38) or ME (EC 1.1.1.40).

Reference ranges: literature values range from non-detectable to 1 U/mg protein.

References

[1] *S. D. Grover, P. F. Canellas, R. T. Wedding,* Purification of NAD malic enzyme from potato and investigation of some physical and kinetic properties, Arch. Biochem. Biophys. *209*, 396–407 (1981).

[2] *E. Weeda,* Some Properties of Mitochondrial NAD^+ Linked Malic Enzyme and Malate Dehydrogenase from the Flight Muscles of *Leptinotarsa decemlineata,* Insect Biochem. *11*, 679–681 (1981).

[3] *K. Takeo,* Existence and Properties of Two Malic Enzymes in *Escherichia coli* – Especially of NAD-linked Enzyme, J. Biochem. *66*, 379–387 (1981).

[4] *P. Pupillo, P. Bossi,* Two Forms of NADP-dependent Malic Enzyme in Expanding Maize Leaves, Planta *144*, 283–289 (1979).

[5] *J. J. Cazzulo, S. M. Juan, E. L. Segura,* The Malic Enzyme of *Trypanosoma cruzi,* J. Gen. Microbiol. *99*, 237–241 (1977).

[6] *B. D. Sanwal, R. Smando,* Malic Enzyme of *Escherichia coli.* Diversity of the Effectors Controlling Enzyme Activity, J. Biol. Chem. *244*, 1817–1823 (1969).

[7] *D. Possner, H. P. Ruffner, D. M. Rast,* Isolation and Biochemical Characterization of Grape Malic Enzyme, Planta *151*, 549–554 (1981).

[8] *S. Ochoa,* "Malic" Enzyme, in: *C. P. Colowick, N. O. Kaplan* (eds.), Methods in Enzymology, Academic Press, New York 1955, Vol. I, pp. 739–753.

[9] *R. Y. Hsu, H. A. Lardy,* Malic Enzyme, in: *S. P. Colowick, N. O. Kaplan, J. M. Lowenstein* (eds.), Methods in Enzymology, Academic Press, New York 1969, Vol. XIII, pp. 230–235.

[10] *J. T. Saz, J. A. Hubbard,* The Oxidative Decarboxylation of Malate by *Ascaris lumbrocoides,* J. Biol. Chem. *225*, 921–933 (1957).

[11] *J. London, E. Y. Meyer,* Malate Utilization by a Group D *Streptococcus:* Physiological Properties and Purification of an Inducible Malic Enzyme, J. Bacteriol. *98*, 705–711 (1969).

[12] *A. R. Macrae, R. Moorhouse,* The Oxidation of Malate by Mitochondria Isolated from Cauliflower Buds, Eur. J. Biochem. *16*, 96–102 (1970).

[13] *R. C. Lin, E. J. Davis,* Malic Enzymes of Rabbit Heart Mitochondria. Separation and Comparison of Some Characteristics of a Nicotinamide Adenine Dinucleotide-Preferring and a Nicotinamide Adenine Dinucleotide Phosphate-Specific Enzyme, J. Biol. Chem. *249*, 3867–3875 (1974).

[14] *D. D. Davis,* The Central Role of Phosphoenolpyruvate in Plant Metabolism, Annu. Rev. Plant Physiol. *30*, 131–158 (1979).

[15] *P. Dittrich,* Nicotinamide Adenine Dinucleotide-Specific "Malic" Enzyme in *Kalanchoe daigremontiana* and Other Plants Exhibiting Crassulacean Acid Metabolism, Plant Physiol. *57*, 310–314 (1976).

[16] *M. D. Hatch, T. Kagawa,* NAD Malic Enzyme in Leaves with C_4-Pathway Photosynthesis and Its role in C_4 Acid Decarboxylation, Arch. Biochem. Biophys. *160*, 346–349 (1974).

[17] *R. Frenkel,* Regulation and Physiological Functions of Malic Enzymes, in: *B. L. Horecker, E. R. Stadtman* (eds.), Current Topics in Cellular Regulation, Vol. 9, Academic Press, New York 1975, pp. 157–181.

[18] *V. Valenti, P. Pupillo,* Activation Kinetics of NAD-Dependent Malic Enzyme of Cauliflower Bud Mitochondria, Plant Physiol. *68*, 1191–1196 (1981).

[19] *K. S. R. Chapman, M. D. Hatch,* Regulation of Mitochondrial NAD-Malic Enzyme Involved in C_4 Pathway Photosynthesis, Arch. Biochem. Biophys. *184*, 298–306 (1977).

[20] *M. D. Hatch, S. L. Mau, T. Kagawa,* Properties of Leaf NAD Malic Enzyme from Plants with C_4 Pathway Photosynthesis, Arch. Biochem. Biophys. *165*, 188–200 (1974).

[21] *R. T. Wedding, P. F. Canellas, M. K. Black,* Slow Transients in the Activity of the NAD Malic Enzyme from *Crassula,* Plant Physiol. *68*, 1416–1423 (1981).

[22] *M. Hirai,* NAD Malic Enzyme from Citrus Fruits, Phytochemistry *17*, 1507–1510 (1978).

[23] *D. W. Fodge, R. W. Gracy, B. G. Harris,* Studies on Enzyme from Parasitic Helminths. I. Purification and Physical Properties of Malic Enzyme from the Muscle Tissue of *Ascaris suum,* Biochim. Biophys. Acta *268*, 271–284 (1972).

[24] *M. D. Hatch, M. Tsuzuki, G. E. Edwards,* Determination of NAD Malic Enzyme in Leaves of C_4 Plants. Effects of Malate Dehydrogenase and Other Factors, Plant Physiol. *69*, 483–491 (1982).
[25] *W. H. Outlaw Jr., J. Manchester,* Conceptual Error in Determination of NAD^+-Malic Enzyme in Extracts Containing NAD^+-Malic Dehydrogenase, Plant Physiol. *65*, 1136–1138 (1980).
[26] *M. Mazelis, B. Vennesland,* Carbon Dioxide Fixation into Oxaloacetate in Higher Plants, Plant Physiol. *32*, 591–600 (1957).
[27] *W. H. Outlaw Jr., J. Manchester, P. H. Brown,* High Levels of Malic Enzyme Activities in *Vicia faba* L. Epidermal Tissue, Plant Physiol. *68*, 1047–1051 (1981).

2.6 Isocitrate Dehydrogenase

Threo-D_s-isocitrate : NAD^+ oxidoreductase (decarboxylating), EC 1.1.1.41
and
Threo-D_s-isocitrate : $NADP^+$ oxidoreductase (decarboxylating), EC 1.1.1.42

David M. Goldberg and Graham Ellis

General

Isocitrate dehydrogenase (ICDH) activity is widely distributed in nature. The enzyme is a component of the Krebs tricarboxylic acid cycle, and has been studied in bacteria, protozoa, fungi, yeasts, higher plants, insects and mammals including man. Most bacteria have only an NADP-dependent enzyme [1]. Cells other than bacteria frequently contain two ICDH enzymes, one of which is specific for NAD and the other for NADP. The activity of these two enzymes is controlled independently and may vary during cell differentiation, even in primitive organisms such as fungi [2]. In higher mammalian tissues the situation is even more complex. For example, in ox heart [3] there are two NADP-specific ICDH isoenzymes located in the mitochondria and cytoplasm respectively. The isoenzymes of these two cell compartments are under independent genetic control. In mammalian cells, the heart and skeletal muscle contain predominantly the mitochondrial NADP-specific isoenzyme, whereas the licer contains mainly the cytoplasmic NADP-specific isoenzyme [3]. Mammalian NADP-specific ICDH enzymes generally have a molecular weight less than 96000 and consist of a single polypeptide chain, or have a dimeric structure. They are not subject to control by modifiers. In contrast, NAD-dependent ICDH of mammalian cells is located exclusively in the mitochondria [4]. This enzyme is much larger, with a molecular weight of 340000; it is composed of multiple subunits, and is considered to be a regulatory enzyme subject to allosteric modifiers.

In man, only the NADP-specific enzyme has been found in serum, and a detailed methodology is described only for this enzyme. References to methods for the assay of NAD-dependent ICDH are presented below.

Application of method: in biochemistry and in clinical chemistry. ICDH has been used as an index of hepatic damage in human subjects. It is particularly sensitive to episodes of minor hepatic cell necrosis occurring as a reaction to drugs [5] and alcohol [6]. Its specificity for the liver is enhanced by the fact that, unlike many other enzymes used as hepatic cell markers, it is rarely raised in patients with myocardial infarction unless congestive cardiac failure is simultaneously present [7]. Unfortunately, it does not discriminate well between patients with primary hepatic cell damage, and those with hepatic parenchymal dysfunction secondary to biliary tract disease [8, 9]. It is, however, very useful as a component of an enzyme profile when mathematical techniques such as discriminant function analysis are applied to the resulting data [10].

Enzyme properties relevant in analysis: two isoenzymes of NADP-specific ICDH have been demonstrated in several bacteria [1, 11, 12] and also in maize mitochondria [13]. The mitochondrial and cytoplasmic NADP-specific ICDH isoenzymes of rat tissues have similar pH optima and K_m values for isocitrate and NADP, but they differ in respect to sensitivity to heat, inhibitors and activators [14]. Two isoenzymes of NADP-dependent ICDH were described in the soluble fraction of rat heart, and two in liver mitochondria and microsomal fractions [15]. Five isoenzymes have been reported in various human tissues after electrophoresis [16]. Further characterization of the isoenzymes was not undertaken in either report. The rate of enzyme activity is linear with respect to time after an initial lag period of 1 – 2 min [17, 18].

Methods of determination: colorimetric methods rely on the formation of 2,4-dinitrophenylhydrazones from the reaction product 2-oxoglutarate [19], or colorimetric estimation of NADPH by reduction of a tetrazolium salt [20]. Most methods use continuous spectrophotometric monitoring of the rate of increase in A_{339} as $NADP^+$ is reduced. No immunoassay methods have been published.

Measurement and properties of NADP-dependent ICDH in mammalian tissues: see the following references: human liver [17], human erythrocytes [22], human heart [21], pig heart [23], ox heart [3, 24], rat heart and liver [14], mouse [25], bovine mammary gland [26].

Measurement and properties of NAD-dependent ICDH in mammalian tissues: see the following references: pig heart [4, 27], ox brain [28].

International reference method and standards: neither are available at the present time.

Enzyme effectors: although the NADP-specific enzymes from many sources are activated by Mg^{2+} or Mn^{2+} [3, 29], the latter is bound with higher affinity [21]. Cu^{2+},

Hg^{2+} [14], oxaloacetate [30, 31] and oxalylglycine [32] are inhibitory to many forms of ICDH. One report [33] described inhibition of a bacterial NADP-specific ICDH by ATP, ADP, CDP, and 2-oxoglutarate. NaCl protects one bacterial ICDH isoenzyme from heat inactivation, and at low concentration enhances the activity of a second isoenzyme in the same organism [12]. In contrast, the human serum and human liver enzymes are inhibited by NaCl in concentrations above 50 mmol/l [17].

Assay

Method Design

Principle

$$threo\text{-}D_s\text{-Isocitrate} + NADP^+ \xrightarrow{Mn^{2+}} \text{2-oxoglutarate} + NADPH + CO_2 + H^+.$$

The enzyme is measured by monitoring the rate of absorbance increase at 339, Hg 334, or Hg 365 nm as NADP is reduced. In the manual method [17], serum is mixed with buffer, Mn^{2+} and NADP, and the reaction is initiated with isocitrate. This procedure has also been used with an *LKB* Reaction Rate Analyzer [34]. With other analyzers it may be necessary to mix all reagents, including isocitrate, and start the reaction by addition of serum. Blank reactions in the presence of serum (measured by omitting isocitrate) range from 0 – 1.1 U/l, with a mean value of 0.6 U/l. Because of this low value, they are not measured routinely. Blanks higher than 1.0 U/l are occasionally seen with highly jaundiced serum and with bilirubin solutions. The mechanism underlying this blank reaction is complex and Mn^{2+}-dependent. Blanks should be run if fluids or homogenates rich in bilirubin and low in ICDH activity are tested.

Optimized conditions for measurement: assay conditions have been optimized for human serum and liver homogenate at 37 °C [17]. In the following method the concentrations in the final assay mixture are: D,L-isocitrate, 6.7 mmol/l; NADP, 0.42 mmol/l; Mn^{2+}, 1.67 mmol/l; triethanolamine hydrochloride buffer, 65 mmol/l, pH 37 °C = 7.3.

Temperature conversion factors: these have not been derived for the assay described here. As a guide only, the following factors have been suggested:

25 °C	30 °C	37 °C	Reference
2.86	1.67	1.0	[19]
2.80	1.87	1.0	[18]

Equipment

A spectrophotometer or spectral-line photometer measuring at 339 nm, Hg 334 or Hg 365 nm, with thermostatted cuvette holder is required. Recorder or stopwatch. With slight modifications the method may be adapted to automated equipment such as the *LKB* Reaction Rate Analyzer [34], or centrifugal analyzers.

Reagents and Solutions

Purity of reagents: use reagents which are of the highest analytical grade. The D,L-isocitrate should be free from possible inhibitors (cf. p. 188).

Preparation of solutions: all solutions should be prepared in re-purified water (see Vol. II, chapter 2.1.3.2).

1. Triethanolamine buffer (0.1 mol/l, pH 7.3):

   add 18.6 g triethanolamine hydrochloride to about 800 ml water. Adjust the pH at 37 °C to 7.3 with NaOH and the volume to 1 litre.

2. Manganese chloride (100 mmol/l):

   dissolve 9.895 g $MnCl_2 \cdot 4H_2O$ in 500 ml water.

3. Nicotinamide-adenine dinucleotide phosphate (NADP, approximately 13 mmol/l):

   daily, dissolve 10 mg NADP, disodium salt, in each 1 ml water.

4. D,L-Isocitrate (67 mmol/l):

   dissolve 1.172 g isocitrate-$Na_3 \cdot 2.5H_2O$ in 60 ml water.

Stability of solutions: store solution (1), (2) and (4) in suitable aliquots at −20 °C for infrequent use up to 3 months. Prepare solution (3) on the day it is required.

Procedure

Collection and treatment of specimen: collect venous blood, allow to clot, and centrifuge at 3000 *g* for 10−15 minutes. Remove the serum promptly because platelets, relative to their volume, contain approximately 700 times the ICDH activity of serum, and erythrocytes 86 times the activity of serum [35].

Stability of the enzyme in the sample: the enzyme in serum is stable for up to 14 days at 4 °C [36 – 38]. Freezing is reported to result in a loss of activity of 10 – 25% [18].

Details for measurements in tissues: tissues are homogenized and processed to yield mitochondria or cytoplasm as required. For preparation of a crude human liver supernatant fraction see [17]; for human heart mitochondrial NADP-specific ICDH see [21]; for rat heart and liver preparations see [14].

Assay conditions: wavelength 339 (Hg 334, Hg 365) nm; light path 10 mm; final volume 3.0 ml; 37 °C (thermostatted cell holder). Measure against air or a suitable blank in a double beam instrument (if indicated). Before starting the assay, adjust the temperature of solutions to 37 °C.

Measurement

Pipette successively into the 10 mm cuvette:			concentration in assay mixture	
triethanolamine buffer	(1)	1.95 ml	TEA	65 mmol/l
manganese chloride solution	(2)	0.05 ml	Mn^{2+}	1.67 mmol/l
NADP solution	(3)	0.10 ml	NADP	0.42 mmol/l
sample (serum or other sample solution)		0.6 ml	volume fraction	0.2
mix, let stand for 20 min at 37 °C.				
D,L-isocitrate solution	(4)	0.3 ml	isocitrate	6.7 mmol/l
mix, read the absorbance and start stopwatch. Repeat reading after exactly 1, 2, 3, 4, and 5 min or monitor the reaction on a recorder.				

The values $\Delta A/\Delta t$ after the 5 – 60 s lag phase must be < 0.100/min at Hg 365 (at Hg 334 or 339 nm < 0.200/min). Otherwise, dilute the sample appropriately and re-assay.

Calculation: use calculation formula (k) or (k_1), cf. "Formulae", Appendix 3. Values for ε: cf. Appendix 4. The catalytic concentration in the sample is (t in minutes):

	Hg 334 nm	339 nm	Hg 365 nm	
b =	$809 \times \Delta A/\Delta t$	$794 \times \Delta A/\Delta t$	$1428 \times \Delta A/\Delta t$	U/l
b =	$1.35 \times 10^4 \times \Delta A/\Delta t$	$1.32 \times 10^4 \times \Delta A/\Delta t$	$2.38 \times 10^4 \times \Delta A/\Delta t$	nkat/l

Validation of the Method

Precision, accuracy, detection limits and sensitivity: the within-batch coefficient of variation for the assay performed on an *LKB* 8600 Reaction Rate Analyzer is 2.9% (serum activity 6.1 U/l), 1.2% (serum activity 20.9 U/l), and 3.5% (liver supernatant activity 73.5 U/l). Accuracy could not be assessed, because no reference method is available. Serum blanks are close to the detection limit; the within-batch coefficient of variation of a serum blank was 36.2% at a value of 0.7 U/l [17].

Sources of error: haemolysis increases the measured serum ICDH activity by approximately 4 U/l for each 1% haemolysis [17]. As mentioned previously, the blank reaction is increased in the presence of gross icterus, or bilirubin solutions, in a complex manner which is Mn^{2+}-dependent, and which is associated with a slow rate of decline of the bilirubin peak at 453 nm. Because sera with gross icterus are usually from patients with liver disease and elevated ICDH, this error can be neglected in routine clinical chemistry practice. However, if ICDH is being studied in tissue homogenates with low ICDH activity, possible Mn^{2+}-dependent spectral changes or Mn^{2+}-associated turbidity effects should be carefully investigated. One author has reported substrate inhibition by isocitrate [18], yet this could not be confirmed in later work [17]. Perhaps some samples of isocitrate contain inhibitors. To test reagents, assay serum with 6.7 or 20 mmol/l D,L-isocitrate and compare the activities at the two concentrations with Fig. 4 of [17]. Storage of the buffer solution at −20 °C reduces the amount of air dissolved in the buffer. Because of the low ICDH activity in serum and the large volume fraction of the serum in the assay, dissolved air released at the time of mixing after substrate addition may cause difficulties with some automated analyzers, e.g., the *LKB* 8600 [34].

Specificity: the assay is specific for NADP-linked ICDH except for the blank reaction, which is minimized by the 20 min pre-incubation of serum, NADP and buffer, prior to the addition of isocitrate.

Reference ranges: in adult human serum 0−7 U/l at 37 °C [17], including serum blanks. Using alternative methodology, the following ranges have been reported at 30 °C: 0.7−2.7 U/l [18] and 1−3.2 U/l [38], and at 25 °C: 0.8−3.8 U/l [36].

There is no sex difference and the enzyme is not increased in pregnancy at term; however, values approximately 2−3 fold higher than those of adults are found in newborn infants [36].

References

[1] *B. Vasquez, H. C. Reeves,* NADP-Specific Isocitrate Dehydrogenase from *Escherichia coli* V. Multiple Forms of the Enzyme, Biochim. Biophys. Acta *660*, 16−22 (1981).

[2] *O. C. Ingebretsen, T. Sanner,* Properties of the Nicotinamide Adenine Dinucleotide-Specific Isocitrate Dehydrogenase from *Blastocladiella emersonii,* Biochim. Biophys. Acta *358*, 25−32 (1974).

[3] *N. MacFarlane, B. Mathews, K. Dalziel,* The Purification and Properties of NADP-Dependent Isocitrate Dehydrogenase from Ox-Heart Mitochondria, Eur. J. Biochem. *74*, 553 – 559 (1977).
[4] *W.-C. Shen, L. Mauck, R. F. Colman,* Physicochemical Properties of the Diphosphopyridine Nucleotide-Specific Isocitrate Dehydrogenase of Pig Heart, J. Biol. Chem. *249*, 7942 – 7949 (1974).
[5] *C. Watts, H. E. D. Griffiths,* Biochemical Studies of Severe Liver Damage Caused by p-Aminosalicylic Acid, Gastroenterology *46*, 317 – 321 (1964).
[6] *D. M. Goldberg, C. Watts,* Serum Enzyme Changes as Evidence of Liver Reaction to Oral Alcohol, Gastroenterology *49*, 256 – 261 (1965).
[7] *J. L. Bell, S. Shaldon, D. N. Baron,* Serum Isocitrate Dehydrogenase in Liver Disease and Some Other Conditions, Clin. Sci. *23*, 57 – 77 (1962).
[8] *G. Ellis, D. M. Goldberg, R. J. Spooner, A. M. Ward,* Serum Enzyme Tests in Diseases of the Liver and Biliary Tree, Amer. J. Clin. Pathol. *70*, 248 – 258 (1978).
[9] *D. M. Goldberg, R. J. Spooner, G. Ellis, A. M. Ward,* Biochemical Features of Intrahepatic Cholestasis, Amer. J. Clin. Pathol. *71*, 557 – 563 (1979).
[10] *G. Ellis, D. M. Goldberg,* Discriminant Function Analysis Applied to Laboratory Tests in Patients with Hepatobiliary Disease, Comput. Biomed. Res. *12*, 483 – 501 (1979).
[11] *C. H. Self, P. D. J. Weitzman,* The Isocitrate Dehydrogenase of *Acinetobacter Lwoffi*, Biochem. J. *130*, 211 – 219 (1972).
[12] *T. Ochiai, N. Fukunaga, S. Sasaki,* Purification and Some Properties of Two $NADP^+$-Specific Isocitrate Dehydrogenases from an Obligately Psychrophilic Marine Bacterium, *Vibrio sp.*, Strain ABE-1, J. Biochem. *86*, 377 – 384 (1979).
[13] *R. A. Curry, I. P. Ting,* Purification, Properties, and Kinetic Observations on the Isoenzymes of NADP Isocitrate Dehydrogenase of Maize, Arch. Biochem. Biophys. *176*, 501 – 509 (1976).
[14] *M. Islam, J. L. Bell, D. N. Baron,* Purification and Comparative Properties of Isoenzymes of Nicotinamide-Adenine Dinucleotide Phosphate-Isocitrate Dehydrogenase from Rat Heart and Liver, Biochem. J. *129*, 1003 – 1011 (1972).
[15] *J. L. Bell, D. N. Baron,* Subcellular Distribution of the Isoenzymes of NADP Isocitrate Dehydrogenase in Rat Liver and Heart, Enzym. Biol. Clin. *9*, 393 – 399 (1968).
[16] *A. W. Opher, J. W. Leonard Jr., J. M. Miller,* Isoenzymes of Isocitrate Dehydrogenase in Cytoplasm of Human Cells, Exp. Med. Surg. *27*, 256 – 266 (1969).
[17] *G. Ellis, D. M. Goldberg,* An Improved Manual and Semi-Automatic Assay for NADP-Dependent Isocitrate Dehydrogenase Activity, with a Description of Some Kinetic Properties of Human Liver and Serum Enzyme, Clin. Biochem. *2*, 175 – 185 (1971).
[18] *G. N. Bowers Jr.,* Measurement of Isocitric Dehydrogenase Activity in Body Fluids, Clin. Chem. *5*, 509 – 518 (1959).
[19] *J. L. Bell, D. N. Baron,* A Colorimetric Method for Determination of Isocitric Dehydrogenase, Clin. Chim. Acta *5*, 740 – 747 (1960).
[20] *M. M. Nachlas, M. B. Davidson, J. D. Goldberg, A. M. Seligman,* Colorimetric Method for the Measurement of Isocitric Dehydrogenase Activity, J. Lab. Clin. Med. *62*, 148 – 158 (1963).
[21] *G. F. Seelig, R. F. Colman,* Characterization of the Physiochemical and Catalytic Properties of Human Heart NADP-Dependent Isocitrate Dehydrogenase, Arch. Biochem. Biophys. *188*, 394 – 409 (1978).
[22] *Y.-M. Wang, J. Van Eye,* Human Erythrocytic NADP-Dependent Isocitrate Dehydrogenase, Int. J. Biochem. *8*, 795 – 799 (1977).
[23] *R. A. Johanson, R. F. Colman,* Cysteine in the Manganous-Isocitrate Binding Site of Pig Heart TPN-Specific Isocitrate Dehydrogenase, Arch. Biochem. Biophys. *207*, 9 – 20 (1981).
[24] *C. C. Fan, L. A. Tomcho, G. W. E. Plaut,* Fluorimetric and Spectrophotometric Studies of DPH-linked Isocitrate Dehydrogenase from Bovine Heart, J. Biol. Chem. *250*, 6197 – 6203 (1975).
[25] *D. J. Fox, R. V. Crockett,* Characterization of Mitochondrial NADP-Dependent Isocitrate Dehydrogenase from Normal and Genetically Obese (OB/OB) Mice, Biochem. Med. *18*, 205 (1977).
[26] *H. M. Farrell Jr.,* Purification and Properties of $NADP^+$: Isocitrate Dehydrogenase from Lactating Bovine Mammary Gland, Arch Biochem. Biophys. *204*, 551 – 559 (1980).
[27] *N. Ramachandran, R. F. Colman,* Chemical Characterization of Distinct Subunits of Pig Heart DPN-Specific Isocitrate Dehydrogenase, J. Biol. Chem. *255*, 8859 – 8864 (1980).

[28] *V. J. Willson, K. F. Tipton,* Purification and Characterization of Ox Brain NAD^+-Dependent Isocitrate Dehydrogenase, J. Neurochem. *33*, 39 – 47 (1979).
[29] *H. C. Reeves, G. O. Daumy, C. C. Lin, M. Houston,* $NADP^+$-Specific Isocitrate Dehydrogenase of *Escherichia coli*. I. Purification and Characterization, Biochim. Biophys. Acta *258*, 27 – 39 (1972).
[30] *S. Nabeshima, S. Ishiyama, A. Tanaka, S. Fukui,* Partial Purification and Some Kinetic Properties of NAD-linked and NADP-linked Isocitrate Dehydrogenases from *Candida tropicalis*, Agric. Biol. Chem. *41*, 509 – 516 (1977).
[31] *T. Saiki, K. Arima,* Studies on $NADP^+$-specific Isocitrate Dehydrogenase from an Extreme Thermophile, *Thermus flavus* AT-62, J. Biochem. *77*, 233 – 240 (1975).
[32] *R. S. Levy, J. J. Villafranca,* Structure-Function Relationships in TPN-Dependent Isocitrate Dehydrogenase. II. Determination of the Paramagnetic Relaxation Rates of Water Protons in Complexes of Enzyme, Mn(II), Substrates, Cofactors, and Inhibitors, Biochemistry *16*, 3301 (1977).
[33] *G. M. Friga, G. L. Farkas,* Isolation and Properties of an Isocitrate Dehydrogenase from *Anacystis nidulans*, Arch. Microbiol. *129*, 331 – 334 (1981).
[34] *D. M. Goldberg, G. Ellis, A. R. Wilcock,* Problems in the Automation of Enzyme Assays with Lag, Accelerated and Blank Reactions, Ann. Clin. Biochem. *8*, 189 – 194 (1971).
[35] *M. Okumura, M. Spellberg,* Serum ICD Activity in the Differential Diagnosis of Liver Disease, Gastroenterology *39*, 305 – 311 (1960).
[36] *S. K. Wolfson Jr., H. G. Williams-Ashman,* Isocitric and 6-Phosphogluconic Dehydrogenases in Human Blood Serum, Proc. Soc. Exp. Biol. Med. *96*, 231 – 234 (1957).
[37] *G. Spano, R. Bragotti, L. Caocci, N. Dioguardi,* Determinazione Quantitativa della Glutammato-Deidrogenasi (GLDH) e della Isocitrato-Deidrogenasi (ISDH) nel Siero in Corso di Epatopatie, Boll. Soc. Ital. Biol. Sper. *43*, 845 – 848 (1967).
[38] *G. A. Lopez, H. P. Dunn,* Serum Isocitric Dehydrogenase. UV Method with NADP, J. Maine Med. Assoc. *59*, 249 – 254 (1968).

2.7 Glucose-6-phosphate Dehydrogenase

D-Glucose-6-phosphate : $NADP^+$ 1-oxidoreductase, EC 1.1.1.49

Jack Deutsch

General

Glucose-6-phosphate dehydrogenase, G6P-DH, the first enzyme in the pentose phosphate pathway, was discovered in red blood cells and yeast by *Warburg* [1 – 3]. It is widely distributed in nature, being found in almost all animal tissues and microorganisms. It is present in large amounts in blood cells, smaller amounts in liver, kidney, and heart and skeletal muscle, and in only trace amounts in serum [4].

Clinically, the red blood cell enzyme is of great importance. Since the erythrocyte lacks the citric acid cycle, it depends on the pentose phosphate pathway for its only supply of NADPH, which is required to maintain the intracellular concentration of reduced glutathione. Deficiency of G6P-DH is an inherited trait, for which more than 80 variants of the enzyme have been described. The common forms of G6P-DH deficiency are manifested by episodes of drug-induced haemolytic anaemia without congenital non-spherocytic disease. Haemolytic anaemia in affected individuals can also be induced by infections or diabetic ketoacidosis, or, in the Mediterranean variant, by exposure to fava beans. The clinical biochemistry of G6P-DH deficiency has been extensively reviewed [5 – 7].

Application of method: in biochemistry and clinical chemistry.

Enzyme properties relevant in analysis: under the conditions described below, enzyme activity is linear for at least 20 minutes following an initial lag period of 2 – 3 minutes.

Methods of determination: G6P-DH is commonly assayed by spectrophotometric measurement of the rate of appearance of NADPH at 339 nm.

International reference method and standards: a standard procedure for determination of erythrocyte G6P-DH activity based on the method of *Zinkham et al.* [8, 9] was proposed by the World Health Organization [10]. A modified procedure, first used by *Glock & McLean* [11] and *Bishop* [12], was proposed by the International Committee for Standardization in Haematology [13].

Enzyme effectors: magnesium ions are required for optimal activity in Tris-buffered assays [4, 12]. Haemolysis of erythrocytes with digitonin or saponin solutions is recommended [12]. Phosphate buffer inhibits enzyme activity [4].

Assay

Method Design

Principle

(a) $\text{Glucose-6-P} + \text{NADP}^+ \xrightleftharpoons{\text{G6P-DH}} \text{6-phosphogluconolactone} + \text{NADPH} + \text{H}^+.$

The rate of increase in absorbance at 339 nm is a measure of G6P-DH activity. Addition of maleimide inhibits further oxidation of reaction products by 6-phosphogluconate dehydrogenase, 6-PGDH* [14]. The method has been automated on a centrifugal analyzer [15], and is adaptable to other analytical systems. A blank of distilled or de-ionized water is required.

* 6-Phosphogluconate dehydrogenase (6-phospho-D-gluconate : NAD(P)$^+$ 2-oxidoreductase, EC 1.1.1.43).

Optimized conditions for measurement: accurate measurement of G6P-DH activity is made problematic by the presence of endogenous 6-PGDH, which produces variable amounts of additonal NADPH by the following reactions:

(b) 6-Phosphogluconolactone + $H_2O \xrightleftharpoons{\text{lactonase}}$ 6-phosphogluconate,

(c) 6-Phosphogluconate + $NADP^+ \xrightleftharpoons{\text{6-PGDH}}$ ribulose-5-P + NADPH + H^+.

Several strategies have been employed to obviate the secondary reactions:

- 6-PGDH activity alone is measured, and is subtracted from combined G6P-DH + 6-PGDH activities; this method has been widely used, but it is cumbersome to perform, and is subject to large analytical error for specimens with low G6P-DH activity.
- Purified 6-PGDH is added to the reaction mixture to drive the secondary reactions to completion; 2 moles of NADPH are produced per mol of G-6-PDH [16].
- Endogenous 6-PGDH activity is inhibited by 2,3-diphosphoglycerate [17] or maleinimide [14]. The use of maleinimide is most effective [15], and is the strategy employed here.

Erythrocyte G6P-DH has a pH optimum of 7.4 – 8.6 [4]. Most assays are performed at pH 7.5. Tris-HCl buffer, 0.05 mol/l, is used here; triethanolamine buffer, 0.05 mol/l, containing EDTA, 5 mmol/l, is equally good. Glycylglycine buffer, used by early investigators, is not recommended [12].

Concentrations of reactants used here are NADP, 0.38 mmol/l, and glucose-6-phosphate, 3.3 mmol/l. The NADP concentration is near optimal. The glucose-6-phosphate concentration is higher than that used by other investigators [4, 10, 12]. $MgCl_2$ is added to a final concentration of 6.3 mmol/l. Maleinimide concentration is approximately 5 mmol/l. Under these conditions, the reaction is linear for at least 20 minutes, and is easily automated using a commercially available reagent kit [15].

Michaelis constants for G6P-DH from various sources are shown in Table 1.

Table 1. *Michaelis* constants of G6P-DH from various sources

Source of enzyme	Michaelis constant (μmol/l)	
	G-6-P	NADP
Erythrocytes, human [6]	35 – 60	2 – 4
Erythrocytes, human [18]	40 – 52	11
Erythrocytes, human [4]	7.4	12
Leukocytes, human [18]	34 – 40	8 – 12
Leukocytes, human [4]	5.6	12
Platelets, human [18]	60 – 70	9.5 – 11
Platelets, human [19]	12	6
Paramyeloblasts, human, acute leukaemia [4]	18	9.1
Lymphocytes, human, chronic lymphadenoma [4]	23	12
Liver, human [4]	33	23
Yeast [4]	69	50

Temperature conversion factors: according to Table 2.

Table 2. Temperature conversion factors [20]

°C	Factor	°C	Factor	°C	Factor
20	1.90	27	1.20	34	0.78
21	1.76	28	1.13	35	0.74
22	1.66	29	1.06	36	0.70
23	1.55	30	1.00	37	0.66
24	1.46	31	0.94	38	0.62
25	1.37	32	0.89	39	0.58
26	1.28	33	0.83		

Equipment

Spectrophotometer or filter photometer capable of accurate measurement of absorbance at 339 nm (bandpass ≤ 8 nm), preferably with thermostatted cuvette; water-bath; recorder or stopwatch; clinical centrifuge; apparatus for counting blood cells.

Reagents and Solutions

Purity of reagents: the G-6-P and NADP should be of the highest purity available. Other chemicals should be reagent grade.

Reagent kits containing various formulations of reagents are available from several different manufacturers.

Preparation of solutions: prepare all solutions with re-purified water (cf. Vol. II, chapter 2.1.3.2).

1. Physiological saline:

   dissolve 9 g NaCl in 1000 ml of water.

2. Lysing solution (0.02% digitonin containing NADP, 15 μmol/l):

   dissolve 16 mg digitonin in 80 ml water. Warm the water to facilitate solution. Cool to room temperature. Filter (Whatman # 1). Add 1 mg NADP (use pre-weighed vial*).

* *Sigma Chemical Co.,* St. Louis, Missouri 63178, USA.

3. NADP (3.8 mmol/l):

   dissolve 0.29 g NADP, sodium salt, in 100 ml water.

4. Tris buffer (0.5 mol/l):

   dissolve 6.05 g Tris base in 70 – 80 ml water. Adjust pH to 7.5 with HCl. Make up to 100 ml with water.

5. $MgCl_2$ (0.63 mol/l):

   dissolve 1.28 g $MgCl_2 \cdot 6H_2O$ in 100 ml water.

6. Glucose-6-phosphate, G-6-P (33 mmol/l):

   dissolve 931 mg glucose-6-phosphate, sodium salt, in 100 ml water.

Stability of solutions: the NADP, $MgCl_2$, and G-6-P solutions are stable for 2 weeks at 0 °C to 4 °C and for 1 year at −20 °C. The lysing solution is stable for 2 months at 0 °C to 4 °C. The saline and Tris buffer solutions are stable at 0 °C to 4 °C as long as no bacterial growth is observed.

Procedure

Collection and treatment of specimen: collect venous blood in tubes containing anticoagulant. ACD* of buffered citrate are preferred; EDTA, heparin, and oxalate are acceptable. Centrifuge at 1000 *g* for 10 minutes. Remove the plasma and buffy coat. Wash the cells twice with saline (solution 1). After the second wash, re-suspend the packed cells in an equal volume of saline. Obtain an accurate cell count using conventional techniques.

Haemolysis: mix 0.05 ml of the washed cell suspension with 0.5 ml of lysing solution (2). Stand for 10 minutes at room temperature. If results are to be expressed as U/g Hb, determine the haemoglobin concentration in the haemolysate by standard methodology.

Stability of enzyme in the specimen: enzyme activity in blood collected in EDTA or citrate is stable for at least 2 weeks at 0 °C to 4 °C, although some unstable variants may exhibit more rapid denaturation. Digitonin haemolysates should be assayed within several hours of their preparation.

Details for measurement in other tissues: G6P-DH has been assayed in many tissues. Reference is made to published methodology for assay in leukocytes, platelets, and liver homogenate [4, 16, 17, 21].

* Acid-citrate-dextrose solution [13].

Assay conditions: wavelength 339 nm; Hg 334 nm or Hg 365 nm can be used. Light-path 10 mm; final volume 1.02 ml; temperature 37 °C; other temperatures may be used with appropriate correction factors; constant temperature must be maintained. Blank: distilled or de-ionized water.

Measurement: prepare the following working reagent immediately before use:

for every sample to be assayed, combine
0.58 ml H_2O,
0.10 ml NADP solution (3),
0.10 ml Tris buffer (4),
0.10 ml $MgCl_2$ solution (5),
0.10 ml G-6-P solution (6).

Add approximately 0.5 mg maleinimide to each ml of working reagent. The amount of maleinimide added is not critical; the assay works equally well in the concentration range of 2 – 8 mmol/l. Mix gently to dissolve the maleinimide.

Pipette successively into the cuvette:		concentration in assay mixture	
working reagent	1.0 ml	Tris NADP $MgCl_2$ G-6-P maleinimide	50 mmol/l 0.38 mmol/l 6.3 mmol/l 3.3 mmol/l 5 mmol/l
mix thoroughly; allow solution to reach reaction temperature;			
haemolysate	0.02 ml	volume fraction	0.02
mix, read absorbance, start recorder or stopwatch. Wait 2 – 3 minutes until reaction becomes linear, then read absorbance at 1 minute intervals for 3 – 5 minutes.			

Normal samples exhibit at 339 nm a change in absorbance of 0.01 – 0.03 per minute. Linearity for each analytical system should be established by assay of serially diluted samples. Samples which exceed the linearity of the system should be diluted with lysing solution and re-assayed.

Calculation: use eqn. (k) or (k_1) for calculation (cf. "Formulae", Appendix 3). Values for ε: cf. Appendix 4. Activity concentration in the haemolysate:

	Hg 334 nm	339 nm	Hg 365 nm	
$b =$	$8252 \times \Delta A/\Delta t$	$8095 \times \Delta A/\Delta t$	$14571 \times \Delta A/\Delta t$	U / l
$b =$	$13.75 \times 10^4 \times \Delta A/\Delta t$	$13.49 \times 10^4 \times \Delta A/\Delta t$	$24.28 \times 10^4 \times \Delta A/\Delta t$	nkat/l

To obtain the activity content in U/g Hb, divide the activity concentration by the haemoglobin concentration (g Hb/l) in the lysate.

To obtain the activity in U/10^9 red blood cells, multiply the activity concentration in the haemolysate by 11 to correct for the dilution made on haemolysis, divide by the number of red blood cells per ml, and multiply by 10^6.

Validation of Method

Precision, accuracy, detection limits and sensitivity: replicate assays of samples containing normal levels of activity are expected to show relative standard deviations of less than 5%. Accuracy of the method has not been determined. The lowest possible value which can be measured is dependent upon the sensitivity of the analytical system used. An instrument capable of accurate measurement of $\Delta A = 0.001$ per min at 339 nm will detect an activity concentration of 8 U/l in the haemolysate at 37 °C. Imprecision of the method at the lowest detection limit has not been determined.

Sources of error: copper (0.1 mmol/l) and sulphate (5 mmol/l) ions are reported to inhibit enzyme activity [20].

Specificity: the method is specific for G6P-DH. Endogenous 6-PGDH does not interfere with the assay.

Reference ranges: a reference range of 7.6 – 16.9 U/g Hb at 37 °C was reported [15]; the calculated reference range at 30 °C is 5.0 – 11.2 U/g Hb. A range of 4.6 – 13.5 U/g Hb at 30 °C is reported by the manufacturer of a reagent kit [20]. Reference ranges for the WHO method [10] and the *Bishop* modification [12] are found in [6]. Reference ranges for other tissues are found in [4].

References

[1] *O. Warburg, W. Christian*, Über Aktivierung der Robisonschen Hexose-Monophosphorsäure in roten Blutzellen und die Gewinnung aktivierender Fermentlösungen, Biochem. Z. *242*, 206 – 227 (1931).

[2] *O. Warburg, W. Christian*, Über ein neues Oxydationsferment und sein Absorptionspektrum, Biochem. Z. *254*, 438 – 458 (1932).

[3] *O. Warburg, W. Christian, A. Griese*, Wasserstoffübertragendes Co-Ferment, seine Zusammensetzung und Wirkungsweise, Biochem. Z. *282*, 157 – 205 (1935).

[4] *G. W. Löhr, H. D. Waller*, Glucose-6-Phosphate Dehydrogenase, in: *H. U. Bergmeyer* (ed.), Methods of Enzymatic Analysis, 2nd English edn., Verlag Chemie, Weinheim and Academic Press, New York 1974, pp. 636 – 643.

[5] *E. Beutler*, Glucose-6-Phosphate Dehydrogenase Deficiency, in: *J. B. Stanbury, J. B. Wyngaarden, D. S. Frederickson* (eds.), The Metabolic Basis of Inherited Disease, 3rd edn., McGraw-Hill, New York 1978, pp. 1430 – 1451.

[6] *E. Beutler*, Hemolytic Anemia in Disorders of Red Cell Metabolism, Plenum, New York 1978, pp. 23 – 167.

[7] *A. Yoshida*, Hemolytic Anemia and G-6-PD Deficiency, Science *179*, 532 – 537 (1973).

[8] *W. H. Zinkham, R. E. Lenhard, Jr., B. Childs*, A Deficiency of Glucose-6-Phosphate Dehydrogenase Activity in Erythrocytes from Patients with Favism. Bull. Johns Hopkins Hosp. *102*, 169 – 175 (1958).

[9] *A. G. Motulsky, A. Yoshida*, Methods for the Study of Red Cell Glucose-6-Phosphate Dehydrogenase, in: *J. J. Yunis* (ed.), Biochemical Methods in Red Cell Genetics, Academic Press, New York 1969, pp. 51 – 93.

[10] *K. Betke, G. J. Brewer, H. N. Kirkman, L. Luzzato, A. G. Motulsky, B. Ramot, M. Siniscalco, E. Beutler, C. C. Standley*, Standardization of Procedures for the Study of Glucose-6-Phosphate Dehydrogenase, WHO Tech. Rep. Ser. No. 366 (1967).

[11] *G. E. Glock, P. McLean*, Further Studies on the Properties and Assay of Glucose-6-Phosphate Dehydrogenase and 6-Phosphogluconate Dehydrogenase of Rat Liver, Biochem. J. *55*, 400 – 408 (1953).

[12] *C. Bishop* ,Assay of Glucose-6-Phosphate Dehydrogenase (EC 1.1.1.49) and 6-Phosphogluconate Dehydrogenase (EC 1.1.1.43) in Red Cells, J. Lab. Clin. Med. *68*, 149 – 155 (1966).

[13] *E. Beutler, K. G. Blume, J. C. Kaplan, G. W. Lohr, B. Ramot, W. N. Valentine*, International Committee for Standardization in Haematology: Recommended Methods for Red Cell Enzyme Analysis, Brit. J. Haematol. *35*, 331 – 340 (1977).

[14] *H. U. Bergmeyer, E. Bernt, F. Schmidt, H. Stork*, D-Glucose: Determination with Hexokinase and Glucose-6-Phosphate Dehydrogenase, in: *H. U. Bergmeyer* (ed.), Methods of Enzymatic Analysis, 2nd English edn., Verlag Chemie, Weinheim and Academic Press, New York 1974, pp. 1196 – 1201.

[15] *J. Deutsch*, Maleimide as an Inhibitor in Measurement of Erythrocyte Glucose-6-Phosphate Dehydrogenase Activity, Clin. Chem. *24*, 885 – 889 (1978).

[16] *Y. Dror, H. F. Sassoon, J. J. Watson, B. C. Johnson*, Glucose-6-Phosphate Dehydrogenase Assay in Liver and Blood, Clin. Chim. Acta *28*, 291 – 298 (1970).

[17] *E. W. Catalano, G. F. Johnson, H. M. Salomon*, Measurement of Erythrocyte Glucose-6-Phosphate Dehydrogenase with a Centrifugal Analyzer, Clin. Chem. *21*, 134 – 138 (1975).

[18] *D. Cottreau, A. Kahn, P. Boivin*, Human Platelet Glucose-6-Phosphate Dehydrogenase, Enzyme *21*, 142 – 151 (1976).

[19] *D. P. Kosow*, Glucose-6-Phosphate Dehydrogenase of Human Blood Platelets: Kinetics and Regulatory Properties, Arch. Biochem. Biophys. *162*, 186 – 193 (1974).

[20] *Sigma Chem. Co.* Tech. Bull. No. 345-UV, Glucose-6-Phosphate Dehydrogenase (G-6-PDH), June, 1980.

[21] *T. K. Chan, D. Todd, C. C. Wong*, Tissue Enzyme Levels in Erythrocyte Glucose-6-Phosphate Dehydrogenase Deficiency, J. Lab. Clin. Med. *66*, 937 – 942 (1965).

3 Oxidoreductases Acting on Groups other than CHOH

3.1 Glutamate Semialdehyde Dehydrogenase

L-Glutamate-γ-semialdehyde: $NADP^+$ oxidoreductase (phosphorylating), EC 1.2.1.41*

James J. Kramer, Jerry G. Henslee, Yasuo Wakabayashi, and Mary Ellen Jones

General

Glutamate semialdehyde dehydrogenase, GSDH (EC 1.2.1.41)* facilitates the ATP- and NADPH-dependent reduction of glutamate to glutamate-5-semialdehyde which spontaneously cyclizes to Δ^1-pyrroline-5-carboxylate (P5C) [1 – 3]. In bacteria, the reduction of glutamate to P5C apparently requires the sequential action of a glutamyl kinase [4] and a γ-glutamyl phosphate dehydrogenase [2]. Since ATP is required, it is believed that γ-glutamyl phosphate is an intermediate in this reaction; however, this has not been directly demonstrated. This conversion of glutamate to P5C has also been exhibited in mammalian cells [5 – 8] and tissues [9]. The enzyme catalyzes the first step in the synthesis of proline and ornithine from glutamate via P5C in mammalian cells. In this laboratory, emphasis has focused on the rat epithelial mucosa, [3] where glutamate semialdehyde dehydrogenase activity has been located in the mitochondria [9 – 11].

Optimum conditions for glutamate-γ-semialdehyde dehydrogenase activity and the subsequent isolation of Δ^1-pyrroline-5-carboxylate are described below. The procedure requires the use of [^{14}C]glutamate which is enzymatically converted to [^{14}C]Δ^1-pyrroline-5-carboxylate which can then be quantitated directly by use of liquid scintillation spectroscopy. Isolation of the radioactive product is conducted under conditions which maintain the stability of P5C [12, 13].

Application of method: in biochemistry.

Enzyme properties relevant in analysis: the enzyme is located in a mitochondrial membrane and is unstable at temperatures above 24 °C [9]; incubations must be at or below 24 °C to obtain a linear conversion of substrate to product.

* This enzyme is supposed to be identical, in part, with L-glutamate-γ-semialdehyde: $NADP^+$ oxidoreductase (phosphorylating) as described in the Recommendations (1978) of the Nomenclature Committee of the International Union of Biochemistry. Recently glutamate semialdehyde dehydrogenase has been purified from *Escherichia coli* (Eur. J. Biochem. *121*, 561, 1982). A spectrophotometric method for the assay of enzymatic activity is described. It involves the measurement of NADPH generated by the oxidation of Δ^1-pyrroline-5-carboxylate in the presence of phosphate. The presumed product formed is γ-glutamyl-phosphate. The mammalian enzyme described here differs from the *E. coli* enzyme in that it has not only the dehydrogenase activity but also the kinase activity that forms γ-glutamyl phosphate from glutamate and ATP; the overall reaction catalyzed is shown in reaction (a) in the text.

Methods of determination: the catalytic activity can be measured by using [^{14}C]glutamate as substrate, followed by isolation of [^{14}C]P5C by ion-exchange chromatography. The amount of the isotope incorporated into P5C is determined by liquid scintillation spectroscopy. An earlier method [14] utilized the isolation of the dihydroquinazolium derivative formed between P5C and o-aminobenzaldehyde (oAB) [1, 13]; however, this method has a very high blank value and does not show saturation with glutamate as the concentration of the substrate is increased.

International reference methods and standards: none is available.

Enzyme effectors: activators are unknown; however, gabaculine indirectly stimulates the enzyme by inhibiting ornithine aminotransferase [10]. Ornithine [10, 15] and orthophosphate [10] inhibit the enzyme.

Assay

Method Design

Principle

(a) $[^{14}C]$ L-Glutamate + ATP + NADPH + H^+ $\xrightleftharpoons{\text{GSDH}}$

$[^{14}C]$ γ-glutamyl semialdehyde + ADP + P_i + $NADP^+$

(b) $[^{14}C]$ γ-Glutamyl semialdehyde $\xrightarrow{\text{spontaneous}}$ $[^{14}C]\Delta^1$-pyrroline-5-carboxylate

Glutamate semialdehyde dehydrogenase activity is measured by formation of Δ^1-pyrroline-5-carboxylate from γ-glutamyl semialdehyde [1 – 2]. Separation of [^{14}C]Δ^1-pyrroline-5-carboxylate from [^{14}C]glutamate is accomplished by use of ion-exchange chromatography [15]. A known quantity of [^{3}H]P5C is added to the assay mixture following termination of the enzymatic reaction with perchloric acid, to determine the recovery of [^{14}C]P5C in the extraction and separation stages.

[^{3}H]P5C is enzymatically synthesized from [^{3}H]ornithine by the action of ornithine aminotransferase [10]; see appendix. The partially-purified ornithine aminotransferase [16], converts [^{3}H]L-ornithine to [^{3}H]P5C with the simultaneous conversion of 2-oxoglutarate to L-glutamate [4].

(c) $[^3H]$L-Ornithine + 2-oxoglutarate $\xrightleftharpoons{\text{ornithine aminotransferase}}$

$[^3H]$L-P5C + L-glutamate.

The radiolabelled product, [^{3}H]P5C, is isolated by ion-exchange chromatography [11].

Optimized conditions for measurement: the optimum conditions for measurement of glutamate semialdehyde dehydrogenase activity include a maximum temperature of 20–24 °C, pH 7.4 and the presence of an ATP regeneration system; K_m values are 2.5 mmol/l for glutamate, 0.19 mmol/l for ATP and 6.5 µmol/l for NADPH. In this assay system, glutamate semialdehyde dehydrogenase, incubated in the presence of glutamate, 1 mmol/l, ATP, 3 mmol/l, and NADPH, 0.2 mmol/l, produces P5C at a rate of 1.3 nmoles per mg protein per min.

Temperature conversion factors: none.

Equipment

Centrifuge; polycarbonate centrifuge tubes; chromatography columns (0.7 cm diameter × 17.0 cm, with sintered glass plate at the bottom); pH indicator paper; liquid scintillation counter.

Reagents and Solutions

Preparation of solutions (for about 25 determinations): all solutions in re-purified water (cf. Vol. II, chapter 2.1.3.2).

1. Hepes buffer (0.3 mol/l; pH 7.4):

   dissolve 7.15 g Hepes in 90 ml water and adjust to pH 7.4 at room temperature with HCl, 1 mol/l. Dilute with water to 100 ml.

2. Hepes buffer (0.1 mol/l; pH 7.4):

   dissolve 2.38 g Hepes in 90 ml water and adjust to pH 7.4 at room temperature with HCl, 1 mol/l. Dilute with water to 100 ml.

3. Creatine kinase (500 kU/l (30 °C)):

   dissolve lyophilized preparation in 0.5 ml water (amount depending on specific activity).

4. Assay reagent (ATP, 40 mmol/l, creatine-P, 201 mmol/l; glutamate, 13.3 mmol/l; NADPH, 2.7 mmol/l):

   dissolve 44.8 mg $ATP\text{-}Na_2 \cdot 3\ H_2O$, 130.2 mg $creatine\text{-}P\text{-}Na_2 \cdot 4\ H_2O$, 4.2 mg glutamic acid, monosodium salt, and 4.6 mg $NADPH\text{-}Na_4$ to 1.875 ml with Hepes buffer (2).

5. [^{14}C]L-Glutamic acid (100 µCi/ml, uniformly labelled):

   re-purify* [11] 100 µC of [^{14}C]L-glutamic acid, ≥240 mCi/mmol, and evaporate to dryness. Re-suspend the amino acid to a concentration of 100 µCi/ml.

* Chromatography (AGI-X8, 200–400 mesh, acetate form, 0.7 cm × 17.0 cm bed volume): apply [^{14}C]glutamate to the column equilibrated with water. Wash the column with 10 ml water, elute [^{14}C]glutamate with acetic acid, 0.2 mol/l. Evaporate the eluate to dryness, re-suspend in water, store at –10 °C.

6. NP-40* (2% v/v):
   dissolve 2.0 ml NP-40 in 8.0 ml water.

7. D,L-gabaculine (0.01 mol/l):
   dissolve 13.9 mg D,L-gabaculine in 10.0 ml water.

8. Potassium hydroxide (10.0 mol/l):
   dissolve 56.11 g KOH in 50 ml water. Once KOH is dissolved, dilute solution to 100 ml.

9. Potassium hydroxide (0.1 mmol/l):
   dissolve 0.56 g KOH in 50 ml water. Once KOH is dissolved, dilute solution to 100 ml.

10. AGl-X8, 200 – 400 mesh, acetate form:
    the resin (strongly basic anion-exchanger) is washed three times with an excess volume of water.

11. Acetic acid (0.025 mol/l):
    to 985.8 ml water add 14.2 ml glacial acetic acid (100% w/w, sp. gr. 1.06).

12. Liquid scintillation cocktail (LSC):
    add 14.67 g PPO (2,5-diphenyloxazole) and 0.265 g POPOP (1,4-bis-2-(5-phenyl-oxazolyl)benzene) to 2.67 l toluene, add 1.33 l Triton X-100 (polyethylene glycol (9 – 10)p-isooctylphenyl ether) [17].

13. $MgCl_2$ (0.4 mol/l):
    dissolve 2.5 g $MgCl_2 \cdot 6\,H_2O$ in water and make up to 10 ml.

14. Perchloric acid (70% w/w):
    use as undiluted stock solution (sp. gr. 1.67).

15. [2,3-^{3}H]Δ^1-Pyrroline-5-carboxylate:
    see appendix.

Stability of solutions: Hepes buffer solutions (1a and 2), NP-40 solution (6), potassium hydroxide solutions (8 and 9), AGl-X8 anion exchange resin (10) and acetic acid solution (11) are stored at 4 °C and are stable indefinitely.

Liquid scintillation cocktail (12) is stable indefinitely at room temperature when stored in a brown bottle.

Re-purified [^{14}C]L-glutamic acid solution (5) is stable for one month when stored at – 20 °C. D,L-gabaculine solution (7) is stable indefinitely at – 20 °C when stored in a dark bottle.

* Nonidet P-40; polyethylene glycol (9)p-isooctylphenyl ether, e.g. from Shell Company

Creatine kinase solution (3) and the assay reagent (4) are stable for several hours when stored at 4 °C.

Procedure

Collection, treatment and stability of specimen: glutamate semialdehyde dehydrogenase is obtained from the mitochondrial fraction of rat epithelial cells and is stable for one week at −70 °C; however, 40−50% of original activity is lost by freezing and thawing of freshly prepared sample [3, 9].

Measurement

Incubation

Pipette into polycarbonate centrifuge tubes (16 mm × 105 mm):			concentration in incubation mixture	
Hepes buffer	(1)	305 µl	Hepes	0.1 mol/l
$MgCl_2$ solution	(13)	50 µl	Mg^{2+}	20.0 mmol/l
creatine kinase solution	(3)	20 µl	CK	10 kU/l
assay reagent	(4)	75 µl	ATP	3 mmol/l
			creatine-P	15 mmol/l
			NADPH	0.2 mmol/l
			$[^{14}C]$L-glutamate	1.0 mmol/l
$[^{14}C]$L-glutamate	(5)	10 µl		1 µCi/µmol
gabaculine	(7)	100 µl	gabaculine	1.0 mmol/l
NP-40	(6)	25 µl	NP-40	0.05% (w/v)
mix and allow to stand in a room-temperature bath for 3 min				
sample + water*		265 µl		1.0 mg/ml
mix, continue incubation for 30 min in the room-temperature bath				
perchloric acid	(14)	50 µl		
cool samples to 4 °C; reaction is terminated;				
$[2,3\text{-}^{3}H]\Delta^{1}$-pyrroline-5-carboxylate	(15)	100 µl		40000 dpm
mix; use this incubation mixture for separation.				

* sample volume depends on protein concentration. 1 mg of protein is added per assay.

Assay conditions: incubation volume 1.0 ml; incubation time 30 min; 20 – 24 °C; blank contains water instead of sample. Volume of enzyme sample used depends on protein concentration with total volume achieved by addition of water. All additions into assay tubes are made at 4 °C.

Assay blanks are routinely used and include all components of the assay system except enzyme sample.

Separation

Centrifuge assay tubes at 21 000 *g* for 10 min to pellet protein. Transfer supernatant solution to clean centrifuge tubes. Allow samples to come to room temperature before proceeding to next step.

Adjust pH of the supernatant to 5.7 – 6.3 by addition of approximately 70.0 µl KOH (8) and a dropwise addition of KOH (9). Verify pH by use of pH indicator paper (place samples on ice).

Re-centrifuge solution at 24000 *g* for 10 min to pellet the $KClO_4$ precipitate.

Apply supernatant to AGl-X8 resin (0.7 cm × 6.0 cm bed volume). Then wash resin with 5 ml H_2O, followed by 3 ml acetic acid (11).

Elute Δ^1-pyrroline-5-carboxylate from the resin by additional acetic acid (11): Add four 1 ml aliquots of acetic acid (11) sequentially to the column. Collect the eluate for each 1 ml acetic acid (11) added. P5C elutes from the column in one or two of these 4 fractions.

Radiometry

Mix eluate with 9 ml liquid scintillation cocktail (12) and count in a liquid scintillation counter for 5 min.

Protein determination

Determine protein according to a modification of the *Lowry* procedure [18], cf. Vol. II, chapter 1.3.

Calculation: the calculation of P5C synthetase activity requires the use of double isotope analysis. This analysis is necessitated by the use of [^{14}C]glutamate for the synthesis of [^{14}C]P5C and the addition of [^{3}H]P5C as an internal standard. Since the energy emission spectra of β particles for ^{3}H and ^{14}C are similar, 100% discrimination in a dual-channel scintillation counter is impossible. Therefore a series of calculations are needed to monitor the efficiency of the scintillation counter for ^{3}H and ^{14}C. The following equations are presented here to clarify this point and to exemplify the mathematics needed to determine P5C synthetase activity (cf. also Appendix 1 and 3 for symbols, quantities, units and equations).

1. Counting efficiency η is used to determine the ability of the scintillation counter to detect a certain β particle emission energy. The dual-channel scintillation counter is specifically preset for ^{3}H and ^{14}C.

$$\eta = \frac{C_{\text{observed}}}{Z} \quad \text{cpm/dpm}$$

C_{observed} is counting rate observed, cpm
Z is decay rate of added labelled material, dpm

2. Calculation of the total dpm from [^{14}C]P5C cpm as detected by the scintillation counter:

$$Z_{\text{total}} = \frac{C_{[^{14}\text{C}]\text{P5C}} - C_{\text{b}[^{14}\text{C}]}}{\eta_{[^{14}\text{C}]}} \quad \text{dpm}$$

Z_{total} is total decay rate detected

3. Correction for ^{14}C spillover into the tritium channel:

$$C_{\text{total}} = (C_{[^{3}\text{H}]\text{P5C}} - C_{\text{b}[^{3}\text{H}]}) - \eta_{[^{3}\text{H}]} \times Z_{\text{total}} \quad \text{cpm}$$

C_{total} is total counting rate for [^{3}H]P5C, corrected for [^{14}C] contribution

4. The recovery of P5C by anion exchange chromatography is monitored by the addition of a known amount of [^{3}H]P5C. To correct for loss of P5C product by this isolation technique, use the ratio of total amount of [^{3}H]P5C cpm detected ($C_{\text{total (detected)}}$) and total amount of [$^{3}H$]P5C cpm added to terminated assay mixture ($C_{\text{total (added)}}$).

5. The total amount of [^{14}C]P5C synthesized is calculated by correcting for loss of P5C product during isolation. For calculation of the total amount of [^{14}C]P5C dpm present in assay mixture use

$$Z_{\text{total (assay)}} = \frac{Z_{\text{total}} \times C_{\text{total (assay)}}}{C_{\text{total (detected)}}} \quad \text{dpm}$$

6. Total amount of nmoles P5C synthesized is

$$n = \frac{Z_{\text{total (assay)}}}{X} \quad \text{nmol}$$

X is specific radioactivity of [^{14}C]glutamate in the assay, nCi/nmol.

The specific radioactivity X is the amount of radioactivity per unit amount of substance. In this assay procedure the specific activity is equal to 2200 dpm/nmol. The resultant value represents the P5C nmoles synthesized per 30 minutes per mg protein.

7. The catalytic activity content of the sample, related to mg protein, is

$$z_c/m_s = \frac{n}{\Delta t \times m} \quad \text{mU/mg}$$

m is mass of protein in the assay.

Example

To illustrate the calculations reviewed above, a representative example is presented below.

a) Standardization of scintillation counter

^{3}H-standard	80000 dpm,
^{14}C-standard	222000 dpm.

Amount of [^{3}H]P5C added to terminated assay 23000 cpm

Isotope added	observed (counting efficiency η) ^{3}H-channel	^{14}C-channel
^{3}H-Standard	23000 cpm (0.29)	20 cpm (0)
^{14}C-Standard	16552 cpm (0.08)	126000 cpm (0.57)

b) Assay results:

	^{3}H-channel	^{14}C-channel
^{14}C counting efficiency η	0.08	0.57
Background counting rate C_b (cpm)	20	10
Assay counting rate $C_{observed}$ (cpm)	18200	37100

c) Calculations:

1. Calculation of total detectable [^{14}C]P5C in dpm (Z_{total}) and [^{3}H]P5C in cpm (C_{total})

$$Z_{total} = \frac{37100 - 10}{0.57} = 65070 \text{ dpm}$$

$$C_{[^3H]P5C} = (18200 - 20) - 0.08 \times 65070 = 12974 \text{ cpm.}$$

2. Calculation of P5C recovery:

$$\frac{12974 \text{ cpm}}{23000 \text{ cpm}} = 0.564$$

3. Calculation of nmoles P5C synthesized:

$$Z_{\text{total (assay)}} = \frac{65\,070}{0.564} = 115\,372 \text{ dpm}$$

$$n = \frac{115\,372}{2\,200} = 52.4 \text{ nmol}$$

If $X = 2200$ dpm/nmol, for $t = 30$ min the catalytic content of the sample is

$$z_c/m_s = \frac{52.4}{30 \times m} \text{ mU/mg protein.}$$

Validation of Method

Precision, accuracy, detection limits and sensitivity: relevant data have not yet been established.

Sources of error: the stability of Δ^1-pyrroline-5-carboxylate depends on temperature, pH and concentration [19]. Care should be taken to keep samples at 4 °C as specified in the assay protocol. When adjusting the pH for ion-exchange chromatography, the procedure is conducted at room temperature. If pH exceeds the value recommended in protocol, re-adjust with $HClO_4$, 0.1 to 1.0 mol/l, *only*. Ion-exchange chromatography should be completed within 1 – 2 hours after centrifugation of the $KClO_4$ precipitate and the sample tube must be kept at 4 °C beforehand.

Specificity: the method described here is specific for the synthesis of Δ^1-pyrroline-5-carboxylate.

Reference ranges: at present, a systematic analysis of P5C synthetase activity in mammalian tissue has not been made. However, a synopsis of current analyses is presented in a recent review [20].

Appendix

Preparation of [2,3-^{3}H]Δ^1-pyrroline-5-carboxylate

[^{3}H]P5C is prepared from [^{3}H]ornithine via ornithine aminotransferase according to the method of *Henslee et al.* [10]:

$$[^3\text{H}]\text{L-Ornithine} + \text{2-oxoglutarate} \longrightarrow [^3\text{H}]\text{P5C} + \text{L-glutamate}.$$

The reaction mixture includes in a final volume of 1.0 ml:

> [^{3}H]L-Ornithine (10 μCi/μmol), 5.0 mmol/l
> 2-Oxoglutaric acid, 10.0 mmol/l
> Pyridoxal-5-phosphate, 55.0 μmol/l
> Tris-HCl buffer, pH 7.4, 0.1 mol/l
> Ornithine aminotransferase, 250 U/l.

Incubate the reaction mixture for 2 h at 37 °C, terminate the reaction by addition of perchloric acid to a final concentration of 3.5% (v/v). Centrifuge the mixture at 24000 *g* for 15 min to pellet denatured protein. Bring to neutrality by addition of KOH, centrifuge at 14000 *g* for 10 min to precipitate $KClO_4$. Re-adjust pH to 1.5 by addition of HCl, 0.1 mol/l.

For isolation of [^{3}H]P5C from unreacted [^{3}H]ornithine use ion-exchange chromatography (cation exchange resin AG5OW-X8, 200 – 400 mesh, hydrogen form, bed volume 0.7 × 17.0 cm; equilibrated in HCl, 0.1 mol/l). Apply the reaction mixture to the column, wash the column with 10 ml HCl, 1 mol/l, followed by 30 ml HCl, 1.5 mol/l; wash the column finally with 10.0 ml HCl, 6 mol/l, to remove untreated [^{3}H]-ornithine. Analysis of the column washes indicates that [^{3}H]P5C is primarily eluted in the HCl fraction of 1.5 mol/l.

Characterization of the radioactivity found in this fraction indicates that approximately 80% is contributed by [^{3}H]P5C. Therefore, aliquots of this solution added to P5C synthetase assay, as an internal standard, should be multiplied by 0.8 to correct for total [^{3}H]P5C cpm used.

References

[1] *H. J. Vogel, B. D. Davis,* Glutamic γ-Semialdehyde and Δ^1-Pyrroline-5-Carboxylic Acid, Intermediates in the Biosynthesis of Proline, J. Amer. Chem. Soc. *74,* 109 – 112 (1952).

[2] *D. J. Hayzer, T. Leisinger,* Proline Synthesis in *Escherichia coli* Stoichiometry and End-Product Identification of the Reaction Catalyzed by Glutamate Semialdehyde Dehydrogenase, Biochem. J. *197,* 269 – 274 (1981).

[3] *Y. Wakabayashi, M. E. Jones,* Pyrroline-5-Carboxylate Synthesis from Glutamate by Rat Intestinal Mucosa Mitochondria, J. Biol. Chem., in press.

[4] *A. Baich,* Proline Synthesis in *Escherichia coli.* A Proline-Inhibitable Glutamic Acid Kinase, Biochim. Biophys. Acta *192,* 462 – 467 (1969).

[5] *L. Levintow, H. Eagle, K. A. Piez,* The Role of Glutamine in Protein Biosynthesis in Tissue Culture, J. Biol. Chem. *227,* 929 – 941 (1957).

[6] *T.-F. Shen, H. J. Strecker,* Synthesis of Proline and Hydroxyproline in Human Lung (WI-38) Fibroblasts, Biochem. J. *150,* 453 – 461 (1975).

[7] *H. Eagle, C. L. Washington, M. Levy,* End Product Control of Amino Acid Synthesis by Cultured Human Cells, J. Biol. Chem. *240,* 3944 – 3950 (1965).

[8] *E. Adams,* Metabolism of Proline and of Hydroxyproline, Int. Rev. Connect. Tissue Res. *5,* 1 – 91 (1970).

[9] *Y. Wakabayashi, J. G. Henslee, M. E. Jones,* Pyrroline-5-Carboxylate synthesis from Glutamate by Rat Intestinal Mucosa Mitochondria: Subcellular Localization and Temperature Stability, J. Biol. Chem., in press.

[10] *J. G. Henslee, Y. Wakabayashi, C. Small, M. E. Jones,* Factors Influencing Pyrroline-5-Carboxylate Synthesis from Glutamate by Rat Intestinal Mucosa Mitochondria. Manuscript being revised for Arch. Biochem. Biophys.
[11] *J. G. Henslee, M. E. Jones,* Ornithine Synthesis from Glutamate in Rat Small Intestinal Mucosa, Arch. Biochem. Biophys. *219*, 186–197 (1982).
[12] *C. Schöpf, F. Oechler,* Zur Frage der Biogenese des Vasicins (Peganins). Die Synthese des Desoxyvasicins unter physiologischen Bedingungen, Ann. Chem. *523*, 1–29 (1936).
[13] *H. J. Strecker,* The Interconversion of Glutamic Acid and Proline. II. The Preparation and Properties of Δ^1-Pyrroline-5-Carboxylic Acid, J. Biol. Chem. *235*, 2045–2050 (1960).
[14] *R. J. Smith, S. J. Downing, J. M. Phang, R. F. Lodato, T. T. Aoki,* Pyrroline-5-Carboxylate Synthase Activity in Mammalian Cells, Proc. Natl. Acad. Sci. USA *77*, 5221–5225 (1980).
[15] *R. F. Lodato, R. J. Smith, D. Valle, J. M. Phang, T. T. Aoki,* Regulation of Proline Biosynthesis: The Inhibition of Pyrroline-5-Carboxylate Synthase Activity by Ornithine, Metabolism *30*, 908–913 (1981).
[16] *R. J. Smith, S. J. Downing, J. M. Phang,* Enzymatic Synthesis and Purification of L-Pyrroline-5-Carboxylic Acid, Anal. Biochem. *82*, 170–176 (1977).
[17] *S. V. Pande,* Liquid Scintillation Counting of Aqueous Samples Using Triton-Containing Scintillants, Anal. Biochem. *74*, 25–34 (1976).
[18] *G. L. Peterson,* A Simplification of the Protein Assay Method of Lowry et al. Which is More Generally Applicable, Anal. Biochem. *83*, 346–356 (1977).
[19] *I. Williams, L. Frank,* Improved Chemical Synthesis and Enzymatic Assay of Δ^1-Pyrroline-5-Carboxylic Acid, Anal. Biochem. *64*, 85–97 (1975).
[20] *M. E. Jones,* Catalysis of the Urea Cycle, New York Academy of Sciences, in press.

3.2 Xanthine Oxidase

Xanthine: oxygen oxidoreductase, EC 1.2.3.2.

Fritz Heinz and Sylvia Reckel

General

The existence of an enzyme which catalyzes the oxidation of hypoxanthine and xanthine to uric acid in mammalian tissues was recognized more than 70 years ago. A second oxidase, which catalyzes the oxidation of a great number of aldehydes, was later isolated from milk by *Schardinger*. In subsequent studies it was demonstrated that the two enzymes were identical. The name xanthine oxidase (XOD) was chosen, because among the various substrates the oxidation rates of xanthine and hypoxanthine to uric acid are the most rapid and the K_m values the lowest. However, the function of the enzyme in purine catabolism is still under discussion. Furthermore, the physiological role of xanthine oxidase as an enzyme which uses oxygen as electron acceptor has become doubtful. It seems that in tissues the precursor, NAD-dependent xanthine dehydrogenase (D-form), is the physiologically effective species, which is converted to the oxygen-dependent enzyme (O-form). Essential factors in the conversion process are thiol groups and proteolysis [1, 2].

Most of the data for xanthine oxidase were derived from the enzyme isolated from cow's milk. Xanthine oxidase is a complex metalloflavoprotein containing one molybdenum, one FAD and two iron-sulphur centres of the ferredoxine type in each of its two independent subunits. Low specificities towards both the substrate and the electron acceptor were reported. More than one hundred different compounds have been described as substrates, e.g. purines and pteridines, which are hydroxylated, and aldehydes, which are oxidized to the corresponding carbonic acids. Apart from molecular oxygen, dyes such as methylene blue, 2,6-dichlorophenol indophenol, etc., can be used as electron acceptors [1 – 3].

In mammals, xanthine oxidase is widely distributed: it can be detected in liver, kidney, jejunal mucosa cells, etc., and is localized in the cell cytosol. In mammary gland and milk, xanthine oxidase is attached to the microsomes or similar particles [1, 2].

Application of method: in biochemistry, clinical chemistry and food chemistry.

Enzyme properties relevant in analysis: the catalytic activity of serum is exclusively dependent on release of the enzyme from the liver. In sera of healthy persons, xanthine oxidase activity is normally absent or only present at very low levels, but it increases markedly in patients with viral hepatitis [4 – 6].

Methods of determination: various methods have been described for the estimation of xanthine oxidase activity.

In general, assay systems which use xanthine as substrate are simpler than those employing hypoxanthine, since the latter undergoes a two-step oxidation, first to xanthine, then to uric acid. The most common assay methods utilize the change in optical density in the UV range. At 265 or 270 nm the enzyme activity may be calculated from the disappearance of xanthine, or at 293 nm from the appearance of uric acid [7]. Other spectrophotometric methods favour electron transfer to dyes [8, 9], which are able to replace oxygen as electron acceptor.

The consumption of oxygen can be measured [10] using manometric techniques or an oxygen-sensitive electrode. The most sensitive methods use ^{14}C-labelled xanthine or hypoxanthine as substrates. The activity of xanthine oxidase is then calculated from the radioactivity of the uric acid produced (assay with xanthine for substrate) [4, 11].

Recently the heat change during the xanthine oxidase reaction has been measured by a microcalorimetric technique [13].

International reference methods and standards: neither standardization at the international level nor reference standard materials are available so far.

Enzyme effectors: xanthine oxidase is inhibited by allopurinol.

Assay

Method Design

Principle

(a) $$\text{Xanthine} + O_2 + H_2O \xrightarrow{\text{XOD}} \text{urate} + H_2O_2$$

(b) $$\text{Ethanol} + H_2O_2 \xrightarrow{\text{Catalase*}} \text{acetaldehyde} + 2\,H_2O$$

(d) $$\text{Acetaldehyde} + \text{NAD(P)}^+ + H_2O \xrightarrow{\text{AlDH**}} \text{urate} + \text{NAD(P)H} + H^+$$

Sum:

$$\text{Xanthine} + O_2 + \text{NAD(P)}^+ + \text{ethanol} \longrightarrow \text{urate} + \text{acetate} + \text{NAD(P)H} + H^+$$

The amount of xanthine oxidized per unit time, measured by the increase in absorbance due to the formation of NAD(P)H, is a measure of the catalytic activity of xanthine oxidase. The equilibrium of xanthine oxidation is far in favour of uric acid. Blanks are necessary for routine assays in sera. The following method can be easily adapted to automated equipment.

Optimized conditions of measurement: the following assay is described for the coenzyme NADP, because in sera the re-oxidation rate of NADPH is smaller than for NADH and aldehyde dehydrogenase from yeast accepts both coenzymes.

In the presence of uricase the change in optical density may be doubled with both coenzymes [14]. However, care must be taken regarding the concentration of xanthine. The cited concentration must be reduced to 1/10; higher concentrations than this may inhibit the uricase reaction. Moreover, in samples containing relatively high concentrations of uric acid, NAD(P)H is formed before the addition of the substrate.

Hypoxanthine, which is also a substrate for xanthine oxidase, should not be used, because at hypoxanthine concentrations higher than 1 mmol/l the reaction may be interrupted at the level of xanthine [14, 15].

The high ethanol concentration is necessary to prevent the normal catalase reaction with two moles of hydrogen peroxide.

K^+ ions are necessary to activate aldehyde dehydrogenase from yeast.

Temperature conversion factors: factors for the conversion of the results determined at 30 °C to other temperatures have not been reported.

---

* Hydrogen peroxide: hydrogen peroxide oxidoreductase, EC 1.11.1.6.

** Aldehyde dehydrogenase, Aldehyde: NAD(P)^+ oxidoreductase, EC 1.2.1.5.

Equipment

Spectrophotometer or spectral-line photometer permitting exact measurement at 339 nm, Hg 334 nm or Hg 366 nm, and with a thermostatted cuvette holder; recorder or stopwatch.

Reagents and Solutions

Purity of reagents: in the commercially available enzymes, aldehyde dehydrogenase from yeast, catalase and if necessary uricase, the NADPH-oxidation activities together should not exceed 0.01% of that of aldehyde dehydrogenase; NADP-dependent alcohol dehydrogenase should be absent. The commercially available quality of NADP and xanthine is sufficient for good results. Other chemicals should be of analytical grade.

Preparation of solutions: all solutions are made in re-purified water (cf. Vol. II, chapter 2.1.3.2).

1. Phosphate buffer (0.1 mol/l) containing EDTA (1 mmol/l, pH 7.8):

   dissolve 22.82 g $K_2HPO_4 \cdot 3\ H_2O$, 1.157 g KH_2PO_4 and 0.372 g EDTA in 1000 ml water; if necessary adjust the pH to 7.8.

2. Buffer mixture (for about 10 determinations):

   to 9 ml phosphate buffer (1) add 0.02 ml catalase (20 mg/ml *Boehringer Mannheim* 106810; ca. 26000 U) and 1 ml ethanol.

3. NADP solution (12.7 mmol/l):

   dissolve 10 mg NADP, Na salt (*Boehringer Mannheim* 128058) in 1 ml water.

4. Aldehyde dehydrogenase solution (50 kU/l):

   dissolve an amount equivalent to 50 U of the lyophilized powder (*Boehringer Mannheim* 171832) in 1 ml of cold water.

5. Xanthine solution (9 mmol/l):

   dissolve 14 mg xanthine (*Sigma* X 0125) in 1 ml NaOH, 0.1 mol/l.

Stability of solutions: store all solutions at 0 – 4 °C. The buffer mixture (2) is stable at least for two days. Prepare the NADP solution (3) and xanthine solution (5) freshly every week. The aldehyde dehydrogenase solution (4) is stable for at least two days.

Procedure

Collection and treatment of specimen: collect blood from the vein. Addition of oxalate, citrate, fluoride and heparin has no effect. Centrifuge for 10 min at about 300 g in order to obtain plasma or serum.

Stability of the enzyme in the sample: the enzyme is stable in human serum at 20 °C 24 h and at 4 °C at least for one week.

Details for measurement in tissues and milk: in tissues xanthine oxidase is localized in the soluble part of the cell and may be extracted as usual.

In milk xanthine oxidase bound to fat particles has to be solubilized by treatment with organic solvents such as butanol, detergents or enzymes [6].

Assay conditions: wavelength 339 (Hg 334, Hg 365) nm; light path 10 mm; final volume 1.61 ml; 30 °C (thermostatted cuvette holder). Before starting the assay, adjust the temperature to 30 °C.

The values of $\Delta A/\Delta t$ must be < 0.05/min at 365 nm (at 334 nm or 339 nm < 0.1/min). Otherwise dilute samples with phosphate buffer, 0.1 mol/l, pH 7.8.

Measurement

Pipette successively into the cuvette:			concentration in assay mixture	
buffer mixture	(2)	1.00 ml	ethanol phosphate EDTA catalase	1.07 mol/l 56 mmol/l 0.56 mmol/l 1.6×10^6 U/l
NADP solution aldehyde dehydrogenase sample	(3) (4)	0.05 ml 0.05 ml 0.50 ml	NADH AlDH volume fraction	0.39 mmol/l 1515 U/l 0.31
mix thoroughly with a plastic spatula. Wait ca. 5 min, until change in absorbance is linear, record (blank)				
xanthine solution	(5)	0.01 ml	xanthine	545 mmol/l
mix, wait 2 min until the change in absorbance becomes linear, then read the absorbance, repeat the reading after exactly 5 min or monitor the reaction on the recorder.				

Calculation: use calculation formula (k) or (k_1), respectively (cf. "Formulae", Appendix 3). For values of ε cf. Appendix 4. The following relations are valid (Δt = 1 min):

	Hg 334 nm	340 nm	Hg 365 nm	
b =	$522 \times \Delta A/\Delta t$	$512 \times \Delta A/\Delta t$	$922 \times \Delta A/\Delta t$	U/l
b =	$8.70 \times 10^4 \times \Delta A/\Delta t$	$8.53 \times 10^4 \times \Delta A/\Delta t$	$15.3 \times 10^4 \times \Delta A/\Delta t$	nkat/l

Validation of Method

Precision, accuracy, detection limits and sensitivity: for an enzyme solution containing a mean of 34.3 U/l the standard deviation was 0.64 and the coefficient of variation 1.87%. For determinations at Hg 334 nm the lowest measurable value is approximately 0.5 U/l. Sensitivity is $\Delta A/\Delta t$ = 0.004/10 min (*Eppendorf* photometer).

Sources of error: when the influence of different drugs was tested Glibonurid in a concentration up to 20 mg/l assay volume did not affect the enzyme activity. Phenylbutazone or disulfiram at normal therapeutic doses do not reach serum levels high enough to inhibit the aldehyde dehydrogenase in the test system. In the case of allopurinol, the serum levels resulting from therapeutic doses are high enough to block xanthine oxidase completely.

Reference ranges: in sera of normal adults the activity of xanthine oxidase is absent or too low for the detection with the method described.

References

[1] *R. C. Bray,* Xanthine oxidase, in: P. D. Boyer, H. Lardy and K. Myrbäck (eds.), The Enzymes, Vol. 7, 2nd edit., Academic Press, New York 1963, pp. 533 – 555.

[2] *R. C. Bray,* Molybdenum Iron-Sulfur Flavin Hydroxylases and Related Enzymes, in: P. D. Boyer (ed.), The Enzymes, Vol. 12, 3rd edit., Academic Press, New York 1975, pp. 300 – 414.

[3] *V. Massey,* Iron-Sulfur Flavoprotein Hydroxylases, in: W. Lovenberg (ed.), Iron-Sulfur Proteins, Vol. 1, Academic Press, New York 1973, pp. 301 – 355.

[4] *T. M. Dougherty,* A Sensitive Assay for Xanthine Oxidase EC 1.2.3.2 Using Commercially Available Carbon-14 Xanthine, Analyt. Biochem. *74,* 604 – 608 (1976).

[5] *S. Giler, D. Sperlig, I. Urca, A. de Vries,* Serum Xanthine Oxidase in Jaundice, Clin. Chim. Acta *63,* 37 – 40 (1975).

[6] *C. H. A. Ramboer, F. Piessens, J. de Grotte,* Serum Xanthine Oxidase and Liver Disease, Digestion *7,* 183 – 195 (1972).

[7] *H. M. Kalckar,* Differential Spectrophotometry of Purine Compounds by Means of Specific Enzymes. I. Determination of Hydroxypurine Compounds, J. Biol. Chem. *167,* 429 – 443 (1947).

[8] *C. Auscher, N. Amdry, C. Pasquier, F. Delbarre,* Localization of Xanthine Oxidase in Hepatic Tissue. A New Histochemical Method, J. Clin. Chem. Clin. Biochem. *14,* 278 (1976).

[9] *R. Fried, L. W. Fried,* Xanthin Oxydase (Xanthin-Dehydrogenase) in: H. U. Bergmeyer (ed.), Methoden der enzymatischen Analyse, 3rd edit., Verlag Chemie, Weinheim/Bergstr., 1974, pp. 682–688.
[10] *W. W. Westerfeld, D. A. Richert,* The Determination of Xanthine Oxidase in Rat Liver and Intestine, J. Biol. Chem. *199,* 393–405 (1952).
[11] *U. A. S. Al-Khalidi, S. Nasrallah, M. H. Shammaa,* A Sensitive Method for the Determination of Xanthine Oxidase Activity. Clin. Chim. Acta *11,* 72–77 (1965).
[12] *H. Kizaki, T. Sakurada,* Simple Micro-Assay Methods for Enzymes of Purine Metabolism, J. Lab. Clin. Med. *89,* 1135–1144 (1977).
[13] *S. Yldiz, B. Pekin,* Determination of Xanthine Oxidase by Microcalorimetric Method, Anal. Lett. *15,* 405–412 (1982).
[14] *F. Heinz, S. Reckel, J. R. Kalden,* A New Spectrophotometric Assay for Enzymes of Purine Metabolism I. Determination of Xanthine Oxidase Activity, Enzyme *24,* 239–246 (1979).
[15] *M. M. Jézweska,* Xanthine Accumulation During Hypoxanthine Oxidation by Milk Xanthine Oxidase, Biochem. J. *36,* 385–390 (1973).

3.3 Glutamate Dehydrogenase

L-Glutamate : NAD oxidoreductase, deaminating (EC 1.4.1.3)

Ellen and Friedrich Werner Schmidt

General

The best characterized of the glutamate dehydrogenases, GlDH, which react with NAD and NADP, is the vertebrate enzyme. It has been purified from the livers of numerous species and from other tissues, e.g. kidney, heart, brain, intestinal mucosa, etc. It consists only of polypeptide chains, whose amino acid composition and sequence are similar in GlDH from different sources [1]. The molecular weight of the chains is between 50000 and 60000. Six chains form the smallest stable active structure with a molecular weight between 310000 and 350000. The hexamers undergo reversible unlimited linear polymerization in most species [2]. GlDH is an enzyme of the mitochondrial matrix. In the liver, it is nearly twice as active in metabolic zone III than in zone I; in kidney tubules it is evenly distributed. The position of GlDH in metabolism, between amino acid metabolism, the citric acid cycle and the urea cycle, and its capacity for transhydrogenation afford a tight regulation, which occurs by the concentration of coenzymes and substrates, by nucleotides, hormones, metal ions and phospholipids. There are different binding sites for the regulatory effectors besides the active site [1, 2, 3]. The highest GlDH activity content is found in liver tissue, followed as a rule by kidney, pancreas, brain, intestinal mucosa, heart, etc. [4, 5]. Between species, GlDH activity content in the liver varies from 396±25 U/g protein in man, increasing through mouse, horse, dog, rat, marmoset, hamster, guinea pig to 834±325 U/g protein in pig.

Application of method: in tissues, particularly in liver, as marker of the mitochondrial matrix (e.g. for the estimation of the purity of cell fractions); for the investigation of amino acid, ammonia and urea metabolism and the interactions between protein and carbohydrate pathways.

In serum for the diagnosis of liver diseases [5, 6] (estimation of the severity of single cell damage, intralobular localization of liver injury, differential diagnosis of very high aminotransferases and of obstructive and non-obstructive jaundice, recognition of fatty liver) and for the determination of plasma volume [7].

Enzyme properties relevant in analysis: several electrophoretic and chromatographic fractions of GlDH have been described. However, these are not true isoenzymes but probably enzyme forms with different molecular masses and, since side chains are buried in the course of polymerization, different charges [8].

GlDH preparations from various sources have slightly different kinetic properties, particularly with respect to coenzyme specificity.

Table 1. Relative catalytic activity of GlDH with its coenzymes

Coenzyme	Human liver homogenate and serum [5]	Human liver purified [9]	Bovine liver purified [9]	Bovine liver purified [10]	Rat brain purified [10]
NADH	100%	100%	100%	100%	100%
NADPH	50%	48%	48%	66%	32%
NAD	20%	7.4%	18%	5.7%	1.8%
NADP	3%	1.2%	2.8%	3.2%	0.4%

Activity with NADH as coenzyme = 100%

Above pH 7 GlDH is activated by ADP, which weakens coenzyme binding to the active site. The degree of activation is more marked with bovine GlDH than with the human enzyme [9].

It is enhanced in the presence of L-leucine. ADP stabilizes GlDH, e.g. during incubation in Tris buffer, and prevents inhibition by excess NADH, competing for the second NADH-binding site, and also inhibition by Ca^{++} ions, excess substrate and pyridoxal-5-phosphate.

No activation by ADP is found with branched chain ketoacids instead of 2-oxoglutarate or with alanine instead of glutamate as substrates. ATP and AMP have much smaller activating and protecting effects. These also depend on pH, coenzyme type and concentration of coenzymes and substrates. Under certain conditions adenosine nucleotides other than ADP can act as inhibitors, as do guanosine nucleotides, which are always inhibitory. NAD^+ in high concentration activates by binding at the allosteric site [1 – 3, 8 – 11].

The function of L-leucine alone as an activator is still under discussion [12].

Native rabbit serum activates purified human and bovine GlDH. Homologous and heterologous antibodies against GlDH activate glutamate formation by human and bovine GlDH. In the reverse reaction no activation is found with anti-human-GlDH and a strong inhibition with anti-bovine-GlDH. Allosteric activation by ADP is reversed partly by anti-human and completely by anti-bovine GlDH antibodies [13].

Methods of determination: the catalytic activity can be measured spectrophotometrically, by the determination of reaction products chromatographically [14] or by a colour reaction (e.g. with tetrazolium salts and phenazine methosulphate) [15] after a fixed incubation period, or, preferably, by following directly the increase or decrease of reduced coenzyme concentrations in the optical test according to *Warburg*. Determination of enzyme mass by immunoprecipitation or titration of active sites are possible but are not in use.

International reference methods and standards: up to now, there are no standardized selected or reference methods at the international level, and no standard reference materials are known.

The German Society for Clinical Chemistry recommended a standard method for the determination of GlDH in body fluids at 25 °C in 1970; the experimental basis for the optimized reaction conditions was published in 1972 and further specifications with respect to the volume fraction of the sample and addition of LDH in 1974 [16]. The standardized optimized method at 37 °C of DAB 7 in the German Democratic Republic with ADP and L-leucine as activators was proposed in 1972 [17] and published officially in 1974 [18].

Assay

Method Design

Principle

$$\text{2-Oxoglutarate} + \text{NADH} + NH_4^+ \xrightleftharpoons{\text{GlDH}} \text{L-glutamate} + NAD^+ + H_2O.$$

The equilibrium of the GlDH reaction is greatly in favour of glutamate formation. $K' = 1.5 \times 10^{-15}$ to 2.5×10^{-13} l/mol, depending on pH, ionic strength and temperature. With NADP(H) instead of NAD(H) it is slightly lower [1 – 3, 8]. Measurements of catalytic activity are thus performed with 2-oxoglutarate and NH_4^+ as substrates and NADH as coenzyme. The decrease in absorbance due to the oxidation of NADH is the measure of catalytic activity of GlDH.

Optimized conditions for measurement

Table 2. Recommended optimized standard conditions for the estimation of GlDH activity in serum

Reference	[16]	[18]	[19]
Temperature (°C)	25	37	37
TEA*, mmol/l	50	50	50
pH	8.0	7.6	7.4
2-Oxoglutarate, mmol/l	7.0	12	10
NH_4^+ acetate, mmol/l	100	125	100**
NADH, mmol/l	0.2	0.3	0.19
ADP, mmol/l	1.0	1.0	0.95
L-Leucine, mmol/l	–	10	–
EDTA, mmol/l	2.5	1.0	–
LDH, U/l	2000	–	–
Volume fraction	0.16	0.13	0.20

* TEA triethanolamine
** $(NH_4)_2SO_4$

Table 2 gives the assay conditions recommended in both parts of Germany for measurement at 25 °C and 37 °C and the optimized conditions at 37 °C as determined by *Ellis & Goldberg* [19]. There is agreement that TEA is the most suitable buffer. Variations between the compositions of the reaction mixtures are not marked. They include differences in pH, 2-oxoglutarate concentration and the addition or omission of EDTA, L-leucine and LDH. As inhibition by excess of NADH, 2-oxoglutarate and NH_4^+ is prevented by the added ADP, the concentrations of these substances above their saturation points are no longer critical. The anion of the ammonium salt can be acetate, sulphate or chloride, but not phosphate or nitrate, which are inhibitory [20].

The activation by L-leucine, 10 mmol/l, alone is only 85% of that by ADP, 1 mmol/l. Additional L-leucine in the presence of optimal ADP was found to increase GlDH activity only by further 4% [19]. LDH is added to remove the endogenous pyruvate more rapidly before starting the specific reaction [16]. Nevertheless, non-specific interference cannot be totally eliminated in about 60% of sera. It averages 0.4 U/l at 25 °C in normal serum [16, 21], but in one quarter of all sera it causes an apparent GlDH activity that is too high by 10 – 30% and in 9% of sera the apparent GlDH activity is between 30 – 100% higher than the true activity [22]. Thus, in view of the frequently low GlDH activity in serum, optimization of conditions requires measurement of the non-specific decrease in absorbance and its subtraction if necessary.

The interactions between pH, temperature and kinetic properties of GlDH demand scrutiny of all other components of the assay system, whenever one of the reaction conditions is modified. In the ADP-activated system the half-saturation constants of NH_4^+ fall slightly more when temperature increases than is the case when ADP is absent, those of 2-oxoglutarate become higher, and those of NADH, which decline with rising temperature without ADP, increase when ADP is added [23].

Measured GlDH activity declines with increasing volume fraction of the sample. Therefore, definition of the latter is a necessary part of optimization. A compromise between too low a signal and diminished apparent activity is found between 0.15 and 0.20 [24].

Temperature conversion factors: the temperature conversion factors determined by *King* [20] are listed in Table 3.

Table 3. Temperature conversion factors for the determination of GlDH activity in serum

Temper-ature (°C)	25	26	27	28	29	30	31	32	33	34	35	36	37
Relative activity	1.00	1.04	1.09	1.14	1.19	1.24	1.28	1.33	1.39	1.47	1.52	1.56	1.61

Jung et al. [23], who investigated the effect of temperature on the reaction conditions, found higher differences when the optimal assay systems for each temperature were used (factor 30°/25°C = 1.3, factor 37°/25°C = 1.7). The pH optimum shifts to lower values with rising temperature. When this is also taken into consideration and ADP and L-leucine are added to the assay system, the *Arrhenius* plot no longer deviates from linearity around 33°C [23]. Therefore, optimized conditions for measurements at 30°C can be extrapolated.

Equipment

A spectrophotometer or spectral-line photometer permitting exact measurements at Hg 334 nm, 339 nm or Hg 365 nm. A thermostatted cuvette holder connected to a water-bath or a similar device, which regulates the desired temperature within ±0.1°C is mandatory. A stopwatch or, better, a recorder or calculator for reading absorbance.

Reagents and Solutions

Purity of reagents: NADH and 2-oxoglutarate must be free from inhibitors. LDH must be free from GlDH. All necessary reagents are commercially available in adequate quality and need no further purification.

Preparation of solutions: (for about 20 macro-determinations: total volume 3.1 ml, or about 100 semi-micro determinations: total volume 0.62 ml). All solutions in re-purified water (cf. Vol. II, chapter 2.1.3.2).

1. TEA/EDTA/CH_3COONH_4 (TEA, 64.5 mmol/l; EDTA, 3.22 mmol/l; CH_3COONH_4, 129 mmol/l; pH 8.0 for determination at 25°C, pH 7.8 for determination at 30°C, pH 7.6 for determination at 37°C):

dissolve 600 mg triethanolamine-HCl, 60 mg EDTA-$Na_2H_2 \cdot 2H_2O$, and 500 mg ammonium acetate in 45 ml water, adjust pH to 8.0 (or 7.8 or 7.6, respectively) with NaOH, 2 mol/l, and fill up to 50 ml with water.

2. NADH/ADP/LDH (NADH, 6.45 mmol/l; ADP, 32.3 mmol/l; LDH 65 kU/l):

   dissolve 9.2 mg NADH, disodium salt, 33 mg ADP monopotassium salt, dihydrate, and 0.2 ml LDH (suspension in ammonium sulphate, 3.2 mol/l; 2 mg/ml, 300 U/mg) in water up to 2 ml.

3. 2-Oxoglutarate (217 mmol/l, pH 6.8 – 7.0):

   dissolve 100 mg 2-oxoglutarate, disodium salt, dihydrate, in 2 ml water.

Stability of the solutions: store all solutions at 0 – 7 °C (do not freeze NADH-containing solutions!). Buffer solution is stable for several months, when contamination by micro-organisms is avoided. 2-oxoglutarate solution is stable for 4 weeks. NADH/ADP/LDH solution must be prepared weekly.

Procedure

Collection and treatment of specimen: collect blood from the vein with as little stasis as possible. Avoid haemolysis (minimal haemolysis, however, does not interfere with the determination). Plasma can be used instead of serum: EDTA, ca. 2 mmol/l, citrate, ca. 10 mmol/l, heparin, ca. 7.5 µmol/l are without effect. Separate serum from the clot as soon as possible, although the few investigations on the effect of clotting on GlDH do not show significant differences. Centrifuge at about 3000 *g* for 10 – 15 min in order to obtain serum or plasma. Determine GlDH activity as soon as possible. Store in the refrigerator until measurement at the same day or in the deep-freezer for longer intervals.

Unlike other enzymes, GlDH activity in rats or mice is virtually the same when blood is obtained by puncture of the retro-orbital plexus, the heart or the aorta, or by venous catheter [25]. Heavily lipaemic sera cause trouble because they must not be cleared by polyanion precipitation with heparin and magnesium chloride for GlDH determination. This leads to a considerable loss of activity, averaging 50% but rather scattered around this value. On the other hand measurement in some photometers is impeded by the turbidity of the relatively large serum portion [26].

Stability of the enzyme in the sample: the stability of GlDH activity in serum samples, stored at different temperatures, is listed in Table 4 [27, 28].

These early findings have been confirmed by several later investigators, who observed stable activity at room temperature for 5 – 6 hours [19, 29], at + 4 °C for 3 days [29] and frozen for at least 7 – 9 days [17, 29] or for several weeks [19]. 10 – 15% loss of activity occurs at room temperature within 1 day [17] or 7 days [19] and at + 4 °C

Table 4. Stability of GlDH in serum

Storage	2nd day	3rd day	5th day	7th day	12th – 16th day
+4 °C	100 ± 5%	95 ± 5%	87 ± 8%	74 ± 11%	70 ± 21%
+25 °C	90 ± 7%	85 ± 6%	76 ± 6%	70 ± 8%	–

Activity at the first day = 100% [27, 28]

within 6 – 7 days [17, 19]. At 37 °C and pH above 8, without ADP and L-leucine, GlDH loses 80% of its activity within 10 min incubation. This is partly reversible by adding ADP before the start of the reaction and can be prevented by addition of ADP before incubation [23]. In frozen liver tissue from man, guinea pig and rat no loss of GlDH activity occurred up to 105 days.

Details for measurements in other specimens

Tissue: as a mitochondrial enzyme, GlDH is only quantitatively extracted from tissues after thorough homogenization (e.g. for at least 2 min in a Waring blendor) in saline or KCl and a delay of at least 30 min (liver homogenate) at +4 °C before centrifugation [27].

Optimal reaction conditions vary between tissues and species, and between crude tissue extracts and purified enzyme from the same source. This is mostly due to conformational changes accompanying polymerization and depolymerization during extraction and purification.

Urine and other body fluids: in order to avoid interference by the high GlDH content of tubular cells, urine must be centrifuged carefully before activity of GlDH is determined. Unlike rat urine, no GlDH is detectable in normal cell-free urine of man. Similarly, GlDH activity is so low in cerebrospinal, synovial, seminal and amniotic fluids and in tears and saliva that it is scarcely measurable by usual methods. These body fluids must be gently concentrated at least ten-fold. GlDH can be determined directly in bile after centrifugation. However, GlDH determination in body fluids other than plasma is so far without clinical significance.

Assay conditions: wavelength Hg 334, 339 or (not favourable) Hg 365 nm; light path 10 mm; final volume 3.10 ml (for semi-micro assay, final volume 0.62 ml, divide all volumes indicated by 5).

Assess correct temperature, 25 °C (30 °C or 37 °C) in a cuvette in the thermostatted cuvette holder by a suitable probe.

Before beginning measurements prepare reaction mixture: mix 25 parts of buffer/EDTA/ammonium acetate solution (1) with 1 part of NADH/ADP/LDH solution (2). The reaction mixture is stable for 5 hours at +4 °C. Immediately before measurement adjust temperature of all solutions and of the sera/plasma to 25 °C (30 °C, 37 °C).

When using a recording device, adjust width and transport speed of paper to the expected – usually low – GlDH activity (e.g. by setting to $A = 0.25$ over the total width). Measure against air.

Measurement

<table>
<tr><td colspan="2">Pipette successively into the cuvette:</td><td colspan="2">concentration in assay mixture</td></tr>
<tr><td>reaction mixture</td><td>2.50 ml</td><td rowspan="3">TEA
EDTA
CH_3COONH_4
NADH
ADP
LDH
volume fraction</td><td rowspan="3">50 mmol/l
2.5 mmol/l
100 mmol/l
0.2 mmol/l
1 mmol/l
2 kU/l
0.16</td></tr>
<tr><td>serum or plasma</td><td>0.50 ml</td></tr>
<tr><td colspan="2">mix thoroughly with a plastic spatula, wait 5 min until endogenous pyruvate is consumed; read adsorbance, start stopwatch and repeat absorbance readings each minute for 5 min ($\Delta A_1/\Delta t$, non-specific interference) or monitor the reaction on a recorder,</td></tr>
<tr><td>2-oxoglutarate (3)</td><td>0.10 ml</td><td>2-oxoglutarate</td><td>7 mmol/l</td></tr>
<tr><td colspan="4">mix again thoroughly. Read absorbance, start stopwatch and repeat absorbance readings each minute for 5 min ($\Delta A_2/\Delta t$) or monitor the reaction on a recorder.</td></tr>
</table>

Instead of determining the non-specific interference before the specific reaction in the same cuvette, in a total volume which is 3.2% lower than that of the specific reaction, it can be measured in a separate cuvette in which 2-oxoglutarate is replaced by water. In about one third of sera $\Delta A_1/\Delta t = 0$.

If $\Delta A/\Delta t$ of the GlDH reaction is >0.040/min at Hg 334 or 339 nm or >0.020/min at Hg 365 nm, the sample must be diluted 5- to 10-fold and the measurement repeated. If the decrease in absorbance is markedly non-linear or comes to a standstill, the sample must be diluted 100-fold and the measurement repeated. It must be ensured that the NADH consumption due to non-specific interference causes a change in absorbance of no more than 0.200 (Hg 334, 339 nm) or 0.100 (Hg 365 nm), when the first and the second 5 min differences are added.

Calculation: the mean value of $\Delta A_1/\Delta t$ is subtracted from the mean value of $\Delta A_2/\Delta t$. The difference $\Delta A/\Delta t$ is used for calculation of GlDH activity.

Use calculation formula (k) and (k_1), resp., cf. "Formulae", Appendix 3. For linear molar absorbance coefficients cf. Appendix 4. For $\Delta t = 1$ min the following relations are valid for the catalytic concentration in the sample:

	Hg 334 nm	339 nm	Hg 365 nm	
$b =$	$1003 \times \Delta A/\Delta t$	$984 \times \Delta A/\Delta t$	$1824 \times \Delta A/\Delta t$	U/l
$b =$	$1.67 \times 10^4 \times \Delta A/\Delta t$	$1.64 \times 10^4 \times \Delta A/\Delta t$	$3.04 \times 10^4 \times \Delta A/\Delta t$	nkat/l

Validation of method

Precision, accuracy, detection limits and sensitivity: imprecision of GlDH determinations in serum is listed in Table 5.

Table 5. Imprecision of the determination of GlDH activity in serum

GlDH activity (U/l)	Temperature °C	CV within day (%)	CV from day to day (%)	Ref.
4.4	37	2.7	3.7	[17]
1.73	37	8.8	–	[19]
5.87	37	2.7	–	[19]
108.2	37	1.0	5.1	[19]
3.8	25	3.2	7.6	own
5.7	25	2.6	6.3	own
15.6	25	2.4	5.3	own

It depends on the activity value. Since the upper limit of normal is twice as high due to optimization, precision is satisfactory even at the decision point between normal and pathological levels. Inter-laboratory comparison is similarly satisfactory [30]. In liver homogenate (about 1500 U/l, volume fraction of the sample $\varphi = 0.032$) the imprecision within day is 2.5%, from day to day 3.8%. Accuracy cannot be assessed since standard reference material is lacking. Data on relative accuracy can be obtained by comparing values assigned by reference laboratories to control sera with actual measurements in routine use. Less than 1% mean inaccuracy was found over a period of 6 months with GlDH activities of 5.75 U/l and 17.5 U/l.

Detection limit is 0.9 U/l (25 °C). Sensitivity: 0.0005 ΔA/min.

Sources of error: disturbance of the analytical procedure arise from very high pyruvate levels in the serum or plasma, which lead to consumption of NADH during the measurement of the non-specific interference. In these cases, when the serum must not be diluted, the lost NADH must be replaced. Difficulties arising in turbid sera have already been mentioned. Drugs which may interfere with the assay are not known at present.

Specificity: when the non-specific interference is accounted for, either by a pre-reaction without 2-oxoglutarate or by a separate blank, GlDH activity is measured specifically.

Reference ranges: table 6 gives some reference values which have been determined so far with optimized methods at 25 °C and 37 °C. No data are available for 30 °C at present.

Table 6. Reference values of GlDH activity in human serum

Age	Sex	Temperature °C	Upper limit of reference range U/l	Reference
Adults	♂	25	4.1	own
Adults	♀	25	3.9	own
Adults	♂	25	4.0	[21]
Adults	♀	25	3.0	[21]
Adults	♂ + ♀	37	7.5	[17]
Adults	♂ + ♀	37	4.0	[19]
Children				
1 – 30 days		25	6.6	[31]
1 – 6 month		25	4.3	[31]
7 – 12 month		25	3.5	[31]
13 – 24 month		25	2.8	[31]
2 – 3 years		25	2.6	[31]
3 – 15 years		25	3.2	[31]

The lowest activities in healthy people are in the range below the detection limit. Thus, only upper limits of the reference range can be determined exactly. Some investigators found slightly higher upper limits in men than in women. This difference may possibly be false, resulting from the higher prevalence of fatty livers in men. Newborns and small infants have higher upper limits of normal; however, after one year, low adult levels are reached. There is no indication of age dependence in older adults.

Reference values in serum or plasma of laboratory animals are given in Table 7.

Table 7. Reference values in serum or plasma of laboratory animals. Mean values ± standard deviation (U/l, 25 °C) (acc. to [32, 33] and unpublished data).

	mouse	rat	marmoset	guinea pig	cat	dog	pig	horse
Mean	11.8	3.8	7.3	4.18	1.28	1.9	1.38	1.3
SD	2.76	1.9	3.5	1.29	0.56	1.4	0.67	0.39

Decline of activity of GlDH in serum with increasing body weight is not without exceptions. Correlation with GlDH in liver tissue of the respective species is absent.

References

[1] *E. L. Smith, B. M. Austen, K. M. Blumenthal, J. F. Nyc*, Glutamate Dehydrogenases, in: *P. D. Boyer* (ed.), The Enzymes, 3rd ed., Vol. XI, part A, Academic Press, New York 1975, pp. 293 – 367.

[2] *M. G. Gore*, L-Glutamic Acid Dehydrogenase. Int. J. Biochem. *13*, 879 – 886 (1981).

[3] *C. Frieden*, The Regulation of Glutamate Dehydrogenase, in: *S. Grisolia* (ed.), The Urea Cycle, *J. Wiley*, New York, 1976, pp. 59 – 71.

[4] *A. Herzfeld*, The Distribution of Glutamate Dehydrogenase in Rat Tissues, Enzyme *13*, 246 – 251 (1972).

[5] *E. Schmidt, F. W. Schmidt*, Methode und Wert der Bestimmung der Glutaminsäure-Dehydrogenase-Aktivität im Serum, Klin. Wochenschr. *40*, 962 – 969 (1962).

[6] *E. Schmidt, F. W. Schmidt*, Enzyme Diagnosis in Diseases of the Liver and the Biliary System, in: *E. Schmidt, F. W. Schmidt, I. Trautschold, R. Friedel* (eds.), Advances in Clinical Enzymology, *S. Karger*, Basel 1979, pp. 239 – 293.

[7] *H. Dornow, R. Friedel, I. Trautschold*, Bestimmung des Plasma- und Blutvolumens durch das Enzym-Aktivitäts-Verdünnungsprinzip, in: *E. Kaiser* (ed.), Fortschritte der Klinischen Chemie, Enzyme und Hormone, Verlag der Wiener Medizinischen Akademie, Wien 1972, pp. 175 – 181.

[8] *F. G. Lehmann, G. Pfleiderer*, Glutamat-Dehydrogenase aus menschlicher Leber: Kristallisation und biochemische Eigenschaften, Hoppe-Seyler's Z. physiol. Chem. *350*, 609 – 616 (1969).

[9] *F. G. Lehmann*, Glutamat-Dehydrogenase aus menschlicher Leber II. Verhalten gegenüber Coenzym-Analogen und allosterische Eigenschaften, Klin. Wochenschr. *47*, 1045 – 1047 (1969).

[10] *P. Gonzáles, E. Ventura, T. Caldes*, Glutamate Dehydrogenase from Rat Brain, in: *S. Grisolia* (ed.), The Urea Cycle, J. Wiley, New York 1976, pp. 73 – 81.

[11] *F. Hucho, I. Rasched, H. Sund*, Studies of Glutamate Dehydrogenase: Analysis of Functional Areas and Functional Groups, Eur. J. Biochem. *52*, 221 – 230 (1975).

[12] *J.-P. Persijn, W. van der Slik, C. J. Timmer, A. Riethorst*, Kinetic Measurement of Serum Glutamate Dehydrogenase Activity in the Presence of Oxamic Acid and L-Leucine, Clin. Chim. Acta *30*, 377 – 386 (1970).

[13] *F. G. Lehmann*, Glutamatdehydrogenase aus menschlicher Leber V, Aktivierung, Hemmung und allosterische Umfaltung des Enzymmoleküls durch seinen Antikörper, Z. exp. Med. *155*, 172 – 186 (1971).

[14] *S. Ohmori, Y. Tanaka, M. Ikeda, K. Hirota*, A Sensitive Determination of α-Keto Acids by Gas-liquid Chromatography and its Application to the Assay of L-Glutamate Dehydrogenase and Aminotransferases, Anal. Biochem. *112*, 204 – 212 (1981).

[15] *B. Lüdtke, K. Jung, E. Egger*, Die Bestimmung der Glutamatdehydrogenaseaktivität im Serum mit einem Farbtest unter optimierten Bedingungen bei 37 °C, Z. med. Labortechnik *16*, 104 – 108 (1975).

[16] Recommendations of the German Society for Clinical Chemistry: Standardisation of methods for the estimation of enzyme activity in biological fluids, J. Clin. Chem. Clin. Biochem. *8*, 659 – 660 (1970), Experimental basis for the optimized standard conditions, ibid. *10*, 281 – 291, Supplement to the Standard Method for the Determination of Glutamate Dehydrogenase (GLDH) Activity, ibid. *12*, 392 (1974).

[17] *K. Jung, A. Sokolowski, E. Egger*, An Optimized Assay of Human Serum Glutamate Dehydrogenase Activity, Enzyme *14*, 44 – 54 (1972/73).

[18] *K. Jung, K. D. Grützmann, E. Egger, A. Sokolowski, H.-G. Mannsfeldt, R. Donath*, Bestimmung der Glutamatdehydrogenase-Aktivität im Serum – DAB 7 D.L.-Standardvorschlag, Zentralbl. Pharm. *113*, 489 – 492 (1974).

[19] *G. Ellis, D. M. Goldberg*, Optimal Conditions for the Kinetic Assay of Serum Glutamate Dehydrogenase Activity at 37 °C, Clin. Chem. *18*, 523 – 527 (1972).

[20] *J. King*, Practical Clinical Enzymology, Von Nostrand, London 1965, pp. 66 – 70.

[21] *H. Schlebusch, W. Rick, H. Lang, M. Knedel*, Normbereiche der Aktivitäten klinisch wichtiger Enzyme, Dtsch. med. Wochenschr. *99*, 765 – 766, (1974).

[22] *E. Schmidt, F. W. Schmidt*, Diagnostische Signifikanz optimierter Enzymaktivitäts-Bestimmungen, in: *H. Lang, W. Rick, L. Róka* (eds.), Optimierung der Diagnostik, Springer, Berlin 1973, pp. 211 – 221.

[23] *K. Jung, E. Egger, R. Neumann, B. Lüdtke*, Zum Einfluß der Temperatur auf Enzymaktivitätsbestimmungen im Serum: Glutamatdehydrogenase, Z. klin. Chem. klin. Biochem. *12*, 159–165 (1974).

[24] *R. Haeckel*, A Simple, Kinetic Procedure for the Optimized Determination of the Activity of Serum Glutamate Dehydrogenase (EC 1.4.1.2) with an Enzyme Analyzer, Z. klin. Chem. klin. Biochem. *11*, 184–188 (1973).

[25] *R. Friedel, I. Trautschold, K. Gärtner, M. Helle-Feldmann, D. Gaudssuhn*, Einfluß verschiedener Methoden zur Blutgewinnung auf Enzym-Aktivitäten im Serum kleiner Laboratoriumstiere, Z. klin. Chem. klin. Biochem. *13*, 499–505 (1975).

[26] *J. Freise, P. Magerstedt, E. Schmidt*, Enzymdiagnostik in lipämischen Seren vor und nach Polyanionenpräzipitation mit Heparin und Magnesiumchlorid, J. Clin. Chem. Clin. Biochem. *15*, 485–488 (1977).

[27] *E. Schmidt, F. W. Schmidt*, Untersuchungen über optimale Reaktions-Bedingungen für Enzym-Aktivitäts-Bestimmungen in Leberextrakten, in Extrakten anderer Organe und im Serum. Die Auswirkungen verschiedener Extraktions-Arten auf Enzym-Aktivitäts-Bestimmungen in Lebergewebe. – Das Verteilungsmuster einiger Enzyme in der menschlichen Leber und seine Veränderungen unter der Zellschädigung, – Untersuchungen über die Alterung von Enzymen in vitro. I.–IV. Mitteilung zur Methodik von Enzym-Bestimmungen in menschlichen Organ-Extrakten und im Serum. Enzymol. Biol. Clin. *2*, 201–232 and *3*, 79–86 (1962/63).

[28] *S. Feissli, G. Forster, G. Laudahn, E. Schmidt, F. W. Schmidt*, Normal-Werte und Alterung von Hauptketten-Enzymen im Serum, Klin. Wochenschr. *44*, 390–396 (1966).

[29] *R. Haschen*, in: Enzymdiagnostik, 2. Aufl. Fischer, Stuttgart 1981, p. 32.

[30] *E. Hansert, D. Stamm*, Determination of Assigned Values in Control Specimens for Internal Accuracy Control and for Interlaboratory Surveys. J. Clin. Chem. Clin. Biochem. *18*, 461–490 (1980).

[31] *St. Veit, F. Sitzmann, H. Prestele*, Normalwerte für Lactat- und Glutamat-Dehydrogenase sowie Leucinarylamidase erstellt mit optimierten Standardansätzen. Klin. Pädiatr. *187*, 244–251 (1975).

[32] *E. Schmidt, F. W. Schmidt, R. Both, S. Ohlendorf, R. Raupach, U. Staar*, Enzyme Patterns in Liver and Serum of Rats during Growth and Ageing, in: *D. Goldberg, M. Werner* (eds.) Progress in Clinical Enzymology, Vol. 2. Masson, New York (in press).

[33] *J. R. Möhr, H. Mattenheimer, A. W. Holmes, F. Deinhardt, F. W. Schmidt*, Enzymology of Experimental Liver Disease in Marmoset Monkeys I. Patterns of Enzyme Activity in Liver, other Organs and Serum of Marmosets compared to Man and other Mammals, Enzyme *12*, 99–116 (1971).

3.4 Monoamine Oxidase

Amine: oxygen oxidoreductase (deaminating) (flavin-containing), EC 1.4.3.4

Jürgen Kusche

General

The first enzyme later to be classified as monoamine oxidase (MAO) was described as tyramine oxidase in rabbit liver by *Hare* (1928) [1]. *Blaschko* and coworkers [2] investigated the deamination (among other substrates) of tyramine, adrenaline and

phenylethylamine in liver, brain and other tissues of mammals [3], other vertebrates, invertebrates and plants [2]. They found very similar properties of the enzymes and suggested one entity "which may be called amine oxidase".
The amine oxidases have been separated by *Zeller* [4] into two types: MAO and diamine oxidase (DAO) (EC 1.4.3.6). This classification was useful but there is a movement to separate the amine oxidases according to the prosthetic groups.

However, since the role of pyridoxal phosphate has been only partially established, *Yasunobu* [5] proposed the terms FAD-amine oxidases (monoamine oxidases) and copper-amine oxidases (diamine oxidases). The latter group included plasma amine oxidases and tissue copper-amine oxidases (aorta amine oxidase, dental-pulp amine oxidase and classical diamine oxidases from placenta, kidney and intestine (see chapter 3.5 "Diamine Oxidase"). Plasma amine oxidase [5 – 7], aorta amine oxidase [8] and dental-pulp amine oxidase [9] were often put in the class of MAO because of their substrate specificities.

With tyramine as substrate, the mitochondrial MAO of rat brain was shown by *Johnston* [10] to be inhibited by clorgyline at much lower concentrations than with benzylamine as substrate. This result indicated the presence of two enzymes called types A and B MAO. However, the discussion of whether the multiple forms of MAO are fact or artefact is not yet finished, since incubation conditions [11], inhibitors [9], separation from the mitochondrial membrane and addition of phospholipids [12] influence the properties of the enzymes. Some of the MAOs seem to fit neither the type A nor B pattern [13, 14].

An important link between the metabolism of monoamines and human disease is provided by the effects of MAO inhibitors. Many attempts have been made to use the activity of the mitochondrial B-type MAO of the platelets for the diagnosis of mental diseases [3, 15]. However, although the results may perhaps indicate an increased "vulnerability for affective disorders" in individuals with low platelet MAO activity, they remain contradictory. For this reason *Sandler* [16] favours the study of metabolic products (e.g. following a tyramine challenge) rather than tests of enzymic activity.

The monoamine-oxidizing activity of the plasma has been used for diagnosis of hepatic fibrosis [6, 7, 17]. The influence of anaesthetic agents on MAO, which was suggested to produce cardiac arrhythmia, may also be of clinical interest [18].

Application of method: in biochemistry and clinical chemistry.

Enzyme properties relevant in analysis: the catalytic activity of tissue is due to the existence of isoenzymes. Benzylamine oxidase is normally present in serum or plasma [6].

Substrate conversion proceeds linearly with time.

Methods of determination: active-site titration with phenylhydrazine has been described for pig plasma amine oxidase [19]. Type A and B isoenzymes can be identified by substrate and inhibitor specificity. *Type A:* preferred substrates: noradrenaline, 5-hydroxytryptamine; inhibitor: clorgyline. *Type B:* preferred substrates: benzylamine, β-phenylethylamine; inhibitor: deprenyl [16].

Numerous methods have been used for the determination of MAO activity [20 – 22]. *Tabor* and later *McEwen* used the photometric measurement of the benzaldehyde derived from benzylamine. However, this method was not appropriate for homogenates and therefore was modified by *Turski* [23], and further by *Christ* [24] and *Deitrich* [25], who used substituted benzylamines. *Ono* [26] and *Suzuki* [27] improved the detection limit of similar methods by introducing naphthol or ninhydrin-coupled aldehydes. Another spectrophotometric principle measures the disappearance of kynuramine (*Weissbach,* cf. [20]) or the appearance of the reaction product, 4-hydroxyquinoline, which is detected fluorimetrically (*Krajl,* cf. [20]). Degradation of serotonin by MAO (following purification by an ion-exchange procedure) was detected fluorimetrically [28], as well as the reaction product, indol acetic acid [29].

Another product of the MAO reaction, H_2O_2, is used in a fluorimetric assay [30] and in a photometric test by the oxidation of leuko-2′-7′-dichlorofluorescein [31] or o-dianisidine [14, 32]. The ammonia released from the substrate can be measured by the coupled NADH test, either fluorimetrically [33] or spectrophotometrically (see chapter 3.5 "Diamine Oxidase").

The most communly used tests, however, are radiometric procedures, the most quoted of which is the assay of *Wurtman & Axelrod* [34].

International reference methods and standards: neither standardization at the international level nor the existence of reference standard materials is known.

Enzyme effectors: when working with cell material, effects of permeability barriers must be eliminated by making sure that cells are broken. Pure substrates must be used since impurities could act as inhibitors of the enzyme and may also give rise to high blanks when extracted together with reaction products. The MAO can be activated by the addition of phosphatidyl ethanolamine [35]. The addition of semicarbazide can improve the extraction of the aldehyde produced [36].

Type A and B MAO can be differentiated by inhibitors and substrates (cf. page 228).

Assay

Method Design

Principle: amine oxidases all catalyze the same general reaction shown in the equation

$$RCH_2-NH_2 + O_2 + H_2O \longrightarrow RCHO + NH_3 + H_2O_2$$

The aldehyde formed from the [^{14}C]labelled substrate is extracted into an organic solvent at low pH values where it will be uncharged. The positively charged amine remains in the aqueous phase.

For routine assay, blanks are neccessary to estimate the small amount of labelled substrate which is extracted.

Optimized conditions for measurement: the method is not greatly affected by high ionic strength. However, it is important that the buffering power of the incubation mixture does not prevent a sufficient lowering of the pH before extraction.

An optimum pH for MAOs between 7.1 and 7.8 is reported; pH 7.6 is chosen for the intestinal enzymes [14].

Many different extraction solvents have been used, since no single solvent is suitable for all the available labelled amines; e.g. toluene, which is suitable for most amines, is poor at extracting products formed from 5-hydroxytryptamine [16].

Temperature conversion factors: most authors use 37 °C as incubation temperature. No conversion factors have been described.

Equipment

Glass-stoppered centrifuge tubes and a thermoblock (37 °C) for the incubation. Horizontal shaker for the extraction. Centrifuge and ethanol bath (−20 °C to −40 °C) for the separation of organic and aqueous phase. Liquid scintillation spectrometer.

Reagents and Solutions

Purity of reagents: all reagents should be of analytical grade if available. The purity of the labelled substrates should be tested, e.g. by thin-layer chromatography.

Preparation of solutions: all solutions in re-purified water (cf. Vol. II, chapter 2.1.3.2).

1. Phosphate buffer (0.15 mol/l, pH 7.6):

   dissolve 5.1 g KH_2PO_4 in 250 ml water (solution 1a); dissolve 6.7 g $Na_2HPO_4 \cdot 2H_2O$ in 250 ml water. Adjust by addition of solution (1a) to pH 7.6.

2. Tryptamine ([side chain-2-^{14}C]tryptamine bisuccinate, specific radioactivity ca. 2.2×10^9 $s^{-1} \times mmol^{-1}$):

   dilute 5 ml with 15 volumes of solution (1).

3. HCl (ca. 2 mol/l):

   dissolve 20 ml of a 32% (w/v) HCl with 60 ml water.

4. PPO/POPOP (0.4%/0.005%):

   to 1000 ml toluene add 4 g PPO and 0.05 g POPOP.

Stability of the solutions: when stored at $-20\,°C$ the radioactive amine is stable for several weeks. The buffer solution should be stored at $0-4\,°C$ and freshly prepared every week. Store the toluene-PPO-POPOP in dark bottles.

Procedure

Collection and treatment of specimen: serum or plasma can be used for the determination of amine oxidases. For the preparation of platelet-rich plasma according to *Tipton* (cf. [16]), 20 ml blood is collected in plastic centrifuge tubes containing 2 ml of sodium citrate (0.19 mol/l). The tubes are centrifuged at 200 *g* for 10 min. The supernatant containing the platelets is then centrifuged at 2000 *g* for 10 min. 1 ml sucrose (0.3 mol/l) is added to the pellet and, after suspension, the platelets are spun down at 2000 *g* for 10 min. This washing step is performed three times. Finally, the platelets are re-suspended in 1 ml sucrose (0.3 mol/l) and disrupted by sonication for 20 sec at low energy.

Homogenization is most often used for tissue preparation. The mitochondrial fraction can be separated by differential centrifugation. For example, frog liver [37] is homogenized with 3 vol. of a sucrose/phosphate buffer (0.25 mol/l sucrose buffered with phosphate buffer, 0.1 mol/l, pH 7.4). The homogenate is suspended in 15 vol. of a phosphate buffer (0.1 mol/l) and centrifuged at 600 *g* for 10 min. The supernatant is re-centrifuged at 8500 *g* for 20 min. The resulting pellet is washed by re-suspension and centrifugation. The pellet is then re-suspended in 15 vol. of the phosphate buffer (0.1 mol/l, pH 7.4) and used as the mitochondrial MAO preparation. Generally, treatment with Triton X-100 and/or sonication are applied in the solubilization of MAO [20].

Stability of the enzyme in the specimen: the stability of MAO has rarely been investigated. *Squires* (cf. [9]) reports that the mitochondrial enzyme is stable for several months when frozen as a mitochondrial pellet in re-purified water. The activity of the soluble fraction of the intestinal MAO of rabbits and dogs decreases considerably during one week [38, 39]. *Gressner* [17] reports a decrease of serum MAO by 30% per day. Therefore, the stability of MAO has to be tested for each single preparation.

Assay conditions: optimized ^{14}C channel of a liquid scintillation counter; automatic subtraction of the instrumental background; quench control (a) by external channel ratio and (b) by internal standard produced by the addition of [^{14}C]toluene (ca. 633 s^{-1}) to the extraction solution (counting efficiency 93 ± 2% in a *Packard* Tricarb 3880 liquid scintillation counter).

For routine assay a reagent blank with buffer (1) instead of sample, and a sample blank with zero min incubation time are neccessary for an estimation of the radioactivity not due to the enzymatic reaction.

Incubation volume 0.3 ml; 37 °C. Triplicates are incubated for various times.

Measurement

Incubation

Pipette succesively into centrifuge tubes:	sample	sample blank	concentration in assay mixture
buffer (1)	0.250 ml	0.250 ml	150 mmol/l
sample	0.025 ml	0.025 ml	volume fraction 0.0833
HCl (3)	–	0.200 ml	
mix, equilibrate temperature at 37 °C			
[^{14}C]tryptamine solution (2)	0.025 ml	0.025 ml	0.01 mmol/l $20 \times 10^6 s^{-1} \times l^{-1}$
mix, incubate for 5, 10, 15 min (sample blank for 15 min)			
HCl (3)	0.200 ml	–	
enzyme reaction is interrupted			
PPO/POPOP (4)	6.0 ml	6.0 ml	
close tubes with glass stoppers, shake for 15 min, centrifuge (3500 *g*; 2 – 5 min), bring tubes into ethanol bath of – 20 to – 40 °C, allow aqueous phase to freeze. Use toluene phase for radiometric measurement.			

Radiometry

Pipette into polyethylene vials:

PPO-POPOP solution (4) 4.0 ml,
organic phase from incubation 6.0 ml.

Close vials and start counting. The counting rate *C* should not exceed 50 s^{-1}, otherwise the sample must be diluted [34].

Calculation: two different blanks are used.

1. Sample blank (used for calculation): inactivation of the enzyme by HCl before addition of the substrate (incubation 0 min).

2. Reagent blank (buffer instead of enzyme):

   there should be no significant difference between the blank readings. The reagent blank may indicate disturbances of components besides the enzyme preparation on the isotope assay.

Correct counting rates of sample for counting rates of the sample blank. The difference of the counting rates ΔC enters the calculation.

The catalytic concentration of the sample is a multiple of $\Delta C/\Delta t$.

From this, U/l can be calculated only when the radiometric assay is calibrated by another assay system (*Tipton,* cf. [16]), since no *true* internal standard is available which provides a calculation of the extraction and counting efficiency (η) of the reaction products.

This procedure is proposed also for the radiometric assay of diamine oxidase (see chapter 3.5 "Diamine Oxidase"). Using the spectrophotometric method and a purified soluble MAO preparation from intestinal mucosa, this calibration can be performed for MAO, also.

Validation of Method

Precision, accuracy, detection limit and sensitivity: according to *Wurtman & Axelrod* [34] the difference between duplicate determinations is less than 2%. The difference of values obtained for the catalytic activity of various volumes of an intestinal MAO preparation corresponds to a coefficient of variation of 3.6% [14].

With the soluble intestinal MAO as enzyme the comparison of three different tests, the [^{14}C]tryptamine assay and spectrophotometric UV- and o-dianisidine methods, resulted in mean values of 1.95, 1.90 and 1.85 U/l, respectively (n = 5 for each method).

Detection limit is 0.02 mU/incubation mixture [34].

Sensitivity is $2 \times \text{cpm}_{\text{blank}}/15$ min.

Sources of error: the linearity of an assay depends on an adequate substrate concentration which should be much greater than the K_m value. For MAOs K_m is of the order of 10^{-4} mol/l. However, the substrate concentration in the assay described is only 0.01 mmol/l. Therefore, the cpm must not exceed 3000 after 15 min incubation. The method can probably be improved by mixing labelled and unlabelled tryptamine.

A further limitation of the radiometric assay is the use of only one substrate. This disadvantage can be overcome by the application of a method independent of substrate, such as the spectrophotometric UV method (see chapter 3.5 "Diamine Oxidase").

No significant difference in extraction of products was obtained whether aldehyde dehydrogenase (EC 1.2.1.3) was present or not [34]. Binding of the reaction products to proteins of the enzyme preparation may affect their extractability but this has been shown to be reversible (*Tipton* cf. [16]).

Specificity: the deamination of tryptamine by MAO is specific since this amine is not metabolized by DAO. The extraction of the reaction products into toluene is not specific; however, specificity is provided by the labelled molecule, which does not exchange the isotope with the surrounding medium.

Table 1. Human amine oxidase activity.

Source of specimen	substrate	enzymic activity concentration (content)		Ref.
Normal adult serum	benzylamine		0.3 μmol $\times$ min^{-1} $\times$ l^{-1}	[6]
Normal adult serum	benzylamine		0.38 μmol $\times$ min^{-1} $\times$ l^{-1}	[7]
Normal adult serum	1[(4-aminomethyl-phenyl)azo]-2-naphthol, 2HCl	(males) (females)	0.38 μmol $\times$ min^{-1} $\times$ l^{-1} 0.39 μmol $\times$ min^{-1} $\times$ l^{-1}	[17]
Normal adult serum	1[(4-aminomethyl-phenyl)azo]-2-naphthol, 2HCl		0.47 μmol $\times$ min^{-1} $\times$ l^{-1}	[26]
Platelets	benzylamine		1.0 nmol $\times$ min^{-1} $\times$ (mg protein)$^{-1}$	[15]
Platelets	serotonin	(males) (females)	0.09 nmol $\times$ min^{-1} $\times$ (mg protein)$^{-1}$ 0.13 nmol $\times$ min^{-1} $\times$ (mg protein)$^{-1}$	[28]
Platelets	tryptamine	(males) (females)	0.03 nmol $\times$ min^{-1} $\times$ (mg protein)$^{-1}$ 0.04 nmol $\times$ min^{-1} $\times$ (mg protein)$^{-1}$	[42]
Platelets	kynuramine		12.2 nmol $\times$ min^{-1} $\times$ (mg protein)$^{-1}$	[43]
Placenta	tryptamine		122 nmol $\times$ min^{-1} $\times$ (g wet weight)$^{-1}$	[44]

Table 2. Monoamine oxidase activity in the rat.

Source of specimen	substrate	enzymic activity content	Ref.
Brain	4-hydroxy-3-methoxybenzylamine	160 nmol $\times$ min^{-1} $\times$ g^{-1}	[24]
Brain	benzylamine	115 nmol $\times$ min^{-1} $\times$ g^{-1}	[45]
Liver	tryptamine	175 nmol $\times$ min^{-1} $\times$ g^{-1}	[34]
Liver	benzylamine	725 nmol $\times$ min^{-1} $\times$ g^{-1}	[45]
Liver	β-phenylethylamine	0.13 nmol $\times$ min^{-1} $\times$ g^{-1}	[47]
Heart	benzylamine	12 nmol $\times$ min^{-1} $\times$ g^{-1}	[45]
Kidney	benzylamine	9.3 nmol $\times$ min^{-1} $\times$ g^{-1}	[45]
Lung	tyramine	0.4 nmol $\times$ min^{-1} $\times$ (mg protein)$^{-1}$	[47]

Reference ranges: very detailed lists of MAO activities with different substrates in various human organs are published by *Owen* [40] and by *Lewinsohn* [41]. Further values from the literature are collected in Tables 1 and 2.

References

[1] *M. C. L. Hare,* Tyramine Oxidase I. A New Enzyme System in Liver, Biochem. J. *22,* 968–979 (1928).

[2] *H. Blaschko, D. Richter, H. Schlossmann,* The Oxidation of Adrenaline and Other Amines, Biochem. J. *31,* 2187–2196 (1937).

[3] *M. B. H. Youdim, M. Holzbauer,* Physiological and Pathological Changes in Tissue Monoamine Oxidase Activity, J. Neurol. Transmission *38,* 193–229 (1976).

[4] *E. A. Zeller,* Oxidation of Amines, in: *J. B. Sumner, K. Myrbäck* (eds.), The Enzymes, 1st ed. Vol. II. Academic Press, New York 1951, pp. 536–538.

[5] *K. T. Yasunoba, H. Ishizuki, N. Minamiura,* The Molecular Mechanistic and Immunological Properties of Amine Oxidases, Mol. Cell. Biochem. *13,* 3–29 (1976).

[6] *J. P. Kirchner, O. T. Nebel, D. O. Castell,* Serum Monoamine Oxidase, an Index of Hepatic Fibrosis: Review of Clinical Studies, Military Med. *139,* 952–954 (1974).

[7] *D. G. Erickson,* Serum Monoamine Oxidase Activity in Experimental Schistosomiasis Japonica, Trans. R. Soc. Trop. Med. Hyg. *67,* 613–614 (1973).

[8] *F. Buffoni, R. Pirisino, P. Marino,* Histamine as Inhibitor of Pig Aorta Amine Oxidase, Agents Actions *5,* 429–430 (1975).

[9] *E. Costa, M. Sandler,* Monoamine Oxidase – New Vistas in: *E. Costa, P. Greengard* (eds.), Advances in Biochemistry and Psychopharmacology, Raven Press, New York 1972.

[10] *J. P. Johnston,* Some Observations upon New Inhibitor of Monoamine Oxidase in Brain Tissue, Biochem. Pharmacol. *17,* 1285–1297 (1968).

[11] *O. Suzuki, H. Hattori, M. Oya et al.,* Oxidation of β-Phenylethylamine by Both Types of Monoamine Oxidase: Effects of Substrate Concentration and pH, Life Sci. *25,* 1843–1850 (1979).

[12] *A. Schurr,* Monoamine Oxidase: To B or not to B? Life Sci. *30,* 1059–1063 (1982).

[13] *T. J. Mantle, M. D. Housley, W. H. Garet et al.,* 5-Hydroxytryptamine is a Substrate for Both Species of Monoamine Oxidase in Beef Heart Mitochondria, J. Pharm. Pharmacol. *28,* 667–671 (1976).

[14] *J. Kusche, K.-D. Feußner, W. Lorenz,* Intestinal Monoamine Oxidase: Does it Have a Role in Histamine Catabolism? Agents Actions *12,* 53–59 (1982).

[15] *D. L. Murphy, C. H. Donnelly, L. Miller, R. J. Wyatt,* Platelet MAO in Chronic Schizophrenia, Arch. Gen. Psychiatry *33,* 1377–1381 (1976).

[16] Monoamine Oxidase and its Inhibition, Ciba Foundation Symposium, Elsevier, Excerpta Medica, North-Holland, Amsterdam, Oxford, New York 1976.

[17] *A. M. Gressner,* Evaluation of the Assay for Serum MAO – an Index of Hepatic Fibrosis, J. Clin. Chem. Clin. Biochem. *18,* 921–927 (1980).

[18] *D. R. Schneider, S. G. Harris, R. W. Gardier et al.,* Increased MAO Activity Produced by General Inhalation Anesthetic Agents, Arch. Int. Pharmacodyn. *211,* 64–73 (1974).

[19] *F. Buffoni, G. Ingesti,* Active-Site Titration of Pig Plasma Benzylamine Oxidase with Phenylhydrazine, Biochem. J. *145,* 369–372 (1975).

[20] *R. Kapeller-Adler,* Amine Oxidases and Methods for their Study, Wiley, Interscience, New York, London, Sydney, Toronto 1970.

[21] *K. F. Tipton,* Monoamine Oxidase, in: *H. K. F. Blaschko, A. D. Smith* (eds.) Handbook of Physiology, Section 7; Endocrinology Vol. 6: Adrenal Gland, American Physiological Soc. Washington, D.C. 1975, pp. 667–691.

[22] *M. B. H. Youdim,* Assay and Purification of Brain Monoamine Oxidase, in: *N. Marks, R. Rodnight* (eds.), Plenum Press, New York 1975, pp. 167–208.

[23] *W. Turski, E. Turska, M. Gross-Bellard,* Modification of the Spectrophotometric Method of the Determination of Monoamine Oxidase, Enzyme *14,* 211–220 (1972/73).

[24] *W. Christ, D. Rakow, M. Fernandes, S. Magour,* A Simple and Sensitive Spectrophotometric Determination of Monoamine Oxidase Activity, Z. Klin. Chem. Klin. Biochem. *11*, 367–370 (1973).
[25] *R. A. Deitrich, V. G. Erwin,* A Convenient Spectrophotometric Assay for Monoamine Oxidase, Anal. Biochem. *30*, 395–402 (1969).
[26] *T. Ono, K. Eto, Y. Sakata, M. Takeda,* A New Colorimetric Assay for Monoamine Oxidase in Serum and its Clinical Application, J. Lab. Clin. Med. *85*, 1022–1031 (1975).
[27] *O. Suzuki, E. Noguchi, K. Yagi,* A Simple Microdetermination of Type B MAO, Biochem. Pharmacol. *25*, 2795–2760 (1976).
[28] *S. Takahashi, T. Karasawa,* A Sensitive Non-Isotopic Assay for Monoamine Oxidase Activity in Human Blood Platelets, Clin. Chim. Acta *62*, 393–400 (1975).
[29] *W. Lovenberg, E. Dixon, H. R. Keiser, A. Sjoerdsma,* A Comparison of Amine Oxidase Activity in Human Skin, Rat Skin and Rat Liver: Relevance to Collagen Cross-Linking., Biochem. Pharmacol. *17*, 1117–1120 (1968).
[30] *G. G. Guilbault, S. S. Kuan, P. J. Brignac,* Fluorometric Determination of Oxidative Enzymes, Anal. Chim. Acta *47*, 503–509 (1969).
[31] *H. Köchl, J. P. von Wartburg,* A Sensitive Photometric Assay for Monoamine Oxidase, Anal. Biochem. *84*, 127–135 (1978).
[32] *G. R. Gordon, J. H. Peters,* Plasma Histaminase Activity in Various Mammalian Species: A Rapid Method of Assay, Proc. Soc. Exp. Biol. Med. *124*, 399–404 (1967).
[33] *M. Harada, T. Nagatsu,* A Sensitive Fluorometric Assay for Monoamine Oxidase Activity, Anal. Biochem. *56*, 283–288 (1973).
[34] *R. J. Wurtman, J. Axelrod,* A Sensitive Specific Assay for the Estimation of Monoamine Oxidase, Biochem. Pharmacol. *12*, 1439–1441 (1969).
[35] *P. H. Yu, A. A. Bouton,* Activation of Platelet MAO by Plasma in the Human, Life Sci. *25*, 31–36 (1979).
[36] *M. Jain,* A Rapid, Sensitive, Radiometric Procedure for the Determination of Human Blood Platelet MAO Activity, Clin. Chim. Acta *47*, 107–109 (1973).
[37] *S. Kobayashi, K. Takahara, K. Kamijo,* Monoamine Oxidase in Frog Liver and Brain, Comp. Biochem. Physiol. *69*, 179–183 (1981).
[38] *J. Kusche, H. Richter, J. Schmidt et al.,* Diamine Oxidase in Rabbit Small Intestine: Separation from a Soluble Monoamine Oxidase, Properties and Pathophysiological Significance in Intestinal Ischemia, Agents Actions *5*, 431–439 (1975).
[39] *J. Kusche, W. Lorenz, J. Schmidt,* Oxidative Deamination of Biogenic Amines by Intestinal Amine Oxidase: Histamine is Specifically Inactivated by Diamine Oxidase, Hoppe-Seyler's Z. Physiol. *356*, 1485–1496 (1975).
[40] *F. Owen, A. J. Cross, R. Lofthouse,* Distribution and Inhibition Characteristics of Human Brain MAO Biochem. Pharmacol. *28*, 1077–1080 (1979).
[41] *R. Lewinsohn, V. Glover, M. Sandler,* Development of Benzylamine Oxidase and Monoamine Oxidase A and B in Man, Biochem. Pharmacol. *29*, 1221–1230 (1980).
[42] *E. Friedman, B. Shopsin, G. Sathananthan, S. Gershon,* Blood Platelet Monoamine Oxidase Activity in Psychiatric Patients, Am. J. Psychiatry *131*, 1392–1394 (1974).
[43] *Z. L. Kruk, A. Moffett, D. F. Scott,* Platelet Monoamine Oxidase Activity in Epilepsy, J. Neurol. Neurosurg. Psychiatry *43*, 68–70 (1980).
[44] *C. H. Chen, D. C. Klein, J. C. Robinson,* Monoamine Oxidase in Rat Placenta, Human Placenta, and Cultured Choriocarcinoma, J. Reprod. Fertil. *46*, 477–479 (1976).
[45] *J. N. Bidard, L. Cronenberger,* Monoamine oxydase dans les homogénats d'organes, Biochimie *54*, 155–159 (1972).
[46] *H. C. Kung, A. G. E. Wilson,* Characterization of Rat Pulmonary Monoamine Oxidase, Life Sci. *24*, 425–432 (1979).
[47] *O. Suzuki, E. Noguchi, K. Yagi,* A Simple Fluorometric Assay for Type B Monoamine Oxidase Activity in Rat Tissues, J. Biochem. *79*, 1297–1299 (1976).

3.5 Diamine Oxidase

Amine: oxygen oxidoreductase (deaminating) (copper-containing), EC 1.4.3.6

Jürgen Kusche and Wilfried Lorenz

General

In 1929 *Best* observed in lung tissue a histamine-destroying principle which was named histaminase [1]. *Zeller,* when studying the enzyme from pig kidney [2], found that diamines such as 1,4-diaminobutane (putrescine) and 1,5-diaminopentane (cadaverine) were oxidatively deaminated and proposed the name diamine oxidase (DAO) for this enzyme. Nowadays the terms histaminase and DAO are used synonymously. However, when histamine degradation forms the basis of the test system "histaminase" is preferred [3, 4].

The enzyme has been detected in micro-organisms [5], leguminous plants [6, 7], fishes [8] and various organs of mammals e.g. kidney, intestinal mucosa, placenta and in the amniotic fluid and plasma of pregnant women [9]. Although DAO was regarded as a soluble, cytoplasmic enzyme [9], this property varies between homologous organs of different species [10]. For instance, in the intestinal mucosa of pigs DAO is completely soluble, whereas the opposite is found in guinea pigs.

The activity of DAO in plasma markedly increases in women during pregnancy. All attempts to use this enzyme activity for the diagnosis of gestational diseases have been unsuccessful until now, because of an uncertain discrimination between the normal and pathological ranges. However, in diseases related to placental insufficiency a significantly elevated activity was observed in the 29th – 34th week of pregnancy [11] and a diminished activity in the 35th – 40th week [11, 12]. The enzyme which is present in the amniotic fluid can be used for the diagnosis of ruptured foetal membranes when demonstrable in the vaginal fluid in the absence of vaginal bleeding [13].

The protective function of DAO in the intestinal mucosa is also of clinical interest [14], as is the alteration of DAO activity in the plasma of patients with medullary carcinoma [4] and in other carcinoma tissue [15, 16]. In inflammatory diseases, especially hepatitis and glomerulonephritis, the DAO activity which normally can be released into the circulation by heparin (200 USP units/kg), reaching about 3 U/l human plasma, is diminished to 5% of this value [17].

Application of method: in biochemistry and clinical chemistry.

Enzyme properties relevant in analysis: the DAO activity in various species is probably due to isoenzymes. This is suggested by the immunological cross-reaction observed between histaminases from different human organs, but not between the human enzymes and those from pig or rat [18]. A further argument is the varying reaction velocities of the enzyme from different sources when acting on the same substrates, e.g. putrescine and histamine. The oxidative deamination of substrates proceeds linearly with time.

Methods of determination: estimation of enzyme mass, active-site titration and iso-enzyme determination are not in use.

The different methods of the determination of the DAO activity have been evaluated [19] and summarized at an international workshop [20] as follows: "methods based on oxidative deamination of p-dimethylaminomethylbenzylamine are most suitable for kinetic work and assay in biological samples, while oxygen consumption is best in kinetic work with other substances". However, because of the lack of a laboratory able to prepare the p-dimethylaminomethylbenzylamine, and because measurement of the oxygen consumption is not suitable for serial measurements, these methods are rarely applied.

The [β^3H]histamine test [3] also needs special techniques for preparation of the substrate. The fluorimetric homovanillic acid method [21] is suspected of producing high blanks, but this test was recently used for an assay for diamines in human urine [7]. The o-dianisidine test [22] should not be used. It is not specific and is sensitive to superoxide radicals.

The most often quoted method is the [^{14}C]putrescine assay of *Okuyama & Kobayashi* [23] and its modifications. For assessing the substrate specificity of DAO the coupled NADH method can be used [24]. Furthermore, this method is needed for the calibration of the [^{14}C]putrescine assay [24].

International reference methods and standards: neither standardization at the international level nor the existence of reference standard materials is known.

Enzyme effectors: DAO can be inhibited by high concentrations of the sodium-potassium phosphate buffer. Optimal concentrations are 0.2 mol/l and 0.1 mol/l for the canine and human intestinal enzymes, respectively. DAO is inhibited by supra-optimal substrate concentrations (> 0.1 mmol/l for histamine and > 1 mmol/l for putrescine).

3.5.1 Radiometric Method

Assay

Method Design

Principle

(a)

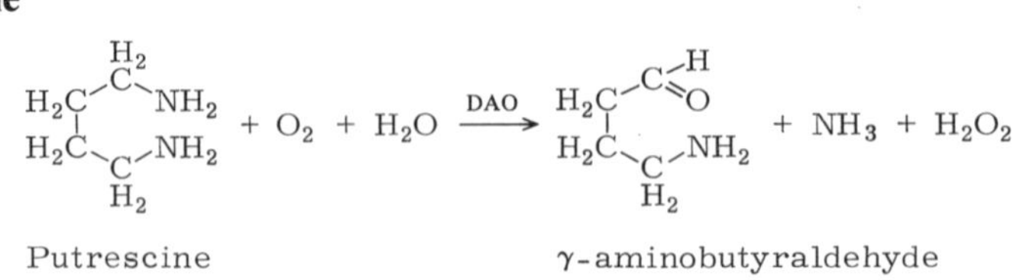

(b)

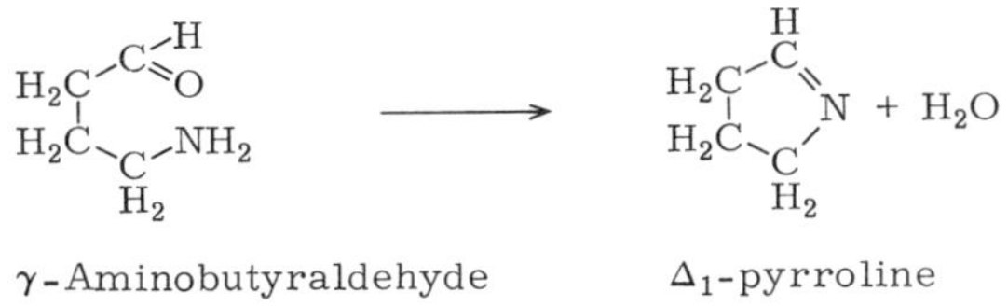

The product of the DAO-catalyzed degradation of putrescine is γ-aminobutyraldehyde, eqn. (a). This compound is converted quantitatively to Δ_1-pyrroline or to its polymers by a spontaneous non-enzymatic cyclization, eqn. (b). Since putrescine is very soluble in water, whereas organic liquids are good solvents for the cyclic compound and its polymers, the radioactive substrate and its products can be separated from each other by extraction into toluene. The amount of radioactivity extracted from the aqueous into the organic phase is then a measure of the DAO activity. The method has not been automated.

Optimized conditions for measurement

Incubation conditions: the optimal conditions for the determination of DAO activity should be determined separately for the enzymes of different origins, such as plasma, kidney or intestine. For the DAO from dog intestine the optimal conditions are pH 7.6 in a sodium-potassium phosphate buffer (0.2 mol/l). The conditions are identical for the NADH test also. With a mixture of unlabelled and ^{14}C-labelled putrescine the optimal concentration is 1 mmol/l. However, when the concentration of unlabelled putrescine is reduced to 0.3 mmol/l the maximum reaction velocity decreases only by 5%, but the specific radioactivity is increased threefold, thus improving the detection limit of the method.

Extraction procedure: according to *Okuyama & Kobayashi* [23] the extraction solution, which is identical with the scintillation cocktail, consists of toluene – 2,5-diphenyloxazol (PPO). For standardization of the extraction procedure, the volume of the extraction solution (6 ml), the extraction time (4 min shaking), and the pH of the aqueous phase (pH 9.6 – 10.2), are optimal conditions. A strong alkaline buffer is necessary in order to achieve the required pH value and to maintain it within ± 0.1 pH unit in the aqueous phase for all samples.

Temperature conversion factors: most authors use 37 °C as the incubation temperature. No conversion factors have been decribed.

Equipment

Glass-stoppered centrifuge tubes and a thermoblock (37 °C) for incubation. Horizontal shaker for extraction. Centrifuge and ethanol-bath (–20 °C to –40 °C) for separation of organic and aqueous phase. Liquid scintillation spectrometer.

Reagents and Solutions

Purity of reagents: all reagents should be of analytical grade if available.

Preparation of solutions: all solutions in re-purified water (cf. Vol. II, chapter 2.1.3.2).

1. Phosphate buffer (0.2 mol/l, pH 7.4):

   dissolve 6.8 g KH_2PO_4 in 250 ml water (solution 1a); dissolve 8.9 g $Na_2HPO_4 \cdot 2H_2O$ in 250 ml water. Adjust by addition of solution (1a) to pH 7.6.

2. Putrescine solution (4.5 mmol/l; 1 mCi/l):

   dissolve 36.2 mg putrescine dihydrochloride in 50 ml of phosphate buffer (1) (i.e. soln. 2a); dissolve [1,4-^{14}C]putrescine dihydrochloride, 1 ampoule (corresponding to about 18.5×10^5 s^{-1}) in the 50 ml solution (2a). This amount of radioactive putrescine increases the final substrate concentration in the incubation mixture only from 0.3 to 0.303 mmol/l, the effect of which on the reaction velocity can be neglected.

   Divide the 50 ml of putrescine solution (2) in 1 ml samples and store at $-20\,°C$.

3. Perchloric acid (ca. 1.7 mol/l):

   dilute 7 ml $HClO_4$ (70% w/w; sp.gr. 1.67) with water to give a volume of 50 ml.

4. Alkaline buffer (NaOH, ca. 0.8 mol/l; $NaHCO_3$, ca. 0.6 mol/l; pH 12.2):

   prepare 1000 ml solution saturated with $NaHCO_3$ at room temperature, centrifuge. Dissolve 40 g NaOH in 500 ml water (2 mol/l). Add NaOH solution to ca. 200 ml of the clear supernatant of $NaHCO_3$ solution until pH 12.2 is reached.

5. Toluene/PPO (0.35%):

   to 1000 ml toluene add 3.5 g PPO.

6. Aminoguanidine* (0.15 mmol/l):

   dissolve 4 mg aminoguanidine sulphate monohydrate in 100 ml buffer (1).

Stability of the solutions: store all buffer, substrate and enzyme solutions at $0-4\,°C$, prepare freshly every monday. Close the NaOH and $NaHCO_3$ solutions tightly. Store the toluene/PPO in dark bottles. The isotope solution is stable for 6 months when stored at $-20\,°C$.

* aminoguanidine is a specific inhibitor of DAO

Procedure

Collection and treatment of specimen: whole blood, plasma or serum can be used. Heparin (3 USP units/ml) or citrate are suitable as anticoagulants *in vitro*.

Tissue samples are homogenized with variable volumes of the phosphate buffer (3 – 100 volumes, depending on the tissue sample [33]), centrifuged (30 min, 48000 *g*) and the supernatant used as the enzyme preparation.

Stability of the enzyme in the sample: the stability of the enzyme preparation depends on its source. We have a preparation from pea seedling which has remained stable since 1964 at –20 °C, whereas a purified enzyme from human intestinal mucosa lost all its activity at –20 °C overnight [10]. Preparations from pig kidney and dog intestine [26] are stable for several weeks.

Measurement

Incubation

Pipette successively into centrifuge tubes:	sample	sample blank	enzyme inhibition	concentration in assay mixture
buffer (1)	0.50 ml	0.50 ml	–	phosphate* 200 mmol/l
aminoguanidine solution (6)	–	–	0.50 ml	aminoguanidine 0.1 mmol/l
sample	0.20 ml	0.20 ml	0.20 ml	volume fraction 0.267
$HClO_4$ (3)	–	0.20 ml	–	
mix, equilibrate temperature at 37 °C				
putrescine solution (2)	0.05 ml	0.05 ml	0.05 ml	putrescine 0.30 mmol 66 μCi/l
mix, incubate for 0, 15, 30, 45 min				
$HClO_4$ (3)	0.20 ml	–	–	
enzyme reaction is stopped				
alkaline buffer (4)	1.00 ml	1.00 ml	1.00 ml	
PPO (5)	6.00 ml	6.00 ml	6.00 ml	
close tubes with glass stoppers, shake for 4 min, centrifuge for 2 min, place tubes in ethanol-bath at –20 °C to –40 °C, allow aqueous phase to freeze. Use toluene phase for radiometric measurement.				

* 147 mmol/l when plasma is used

Assay conditions: optimized ^{14}C-channel of a liquid scintillation counter; automatic subtraction of the instrumental background, quench control (a) by external channel ratio and (b) by internal standard produced by the addition of [^{14}C]toluene (ca. 633 s^{-1}) to the extraction solution (counting efficiency $\eta = 0.93 \pm 0.02$).

For routine assay blanks are necessary to estimate the radioactivity not due to the enzymatic reactions.

Incubation volume 0.75 ml; 37 °C. Triplicates with various incubation times.

Three blanks are used.

- Sample blank: inactivation of the enzyme by $HClO_4$ before addition of substrate (incubation 0 min).
- Enzyme inhibition: aminoguanidine solution (6) instead of buffer (1) is added before the start of the enzyme reaction.
- Reagent blank: buffer instead of enzyme solution.

Radiometry

Pipette into polyethylene vials:

PPO solution (5)	5.0 ml
organic phase from incubation	5.0 ml

close the vials, start counting.

Counting rate C should not exceed 85 s^{-1}, otherwise the sample must be diluted.

Calculation: correct counting rates of sample for counting rates of the sample blank. The difference of the counting rates ΔC enters the calculation.

The catalytic concentration of the sample is a multiple of $\Delta C/\Delta t$.

From this, U/l can be calculated only when the radiometric assay is calibrated by another assay system (e.g. by the spectrophotometric method, p. 243), since no *true* internal standard is available which provides a calculation of the extraction and counting efficiency (η) of the reaction products.

Except of the presence of the indicator system which does not influence the DAO reaction the same incubation conditions can be chosen in both assays. The volume fraction of the sample especially must be identical in both assays.

Example: with substrate solution (2) for conversion of $\Delta C/\Delta t$ to U/l a factor 0.013 is obtained. However, this factor depends on assay and extraction conditions as well as on the specific radioactivity of the [^{14}C]putrescine preparation. Therefore, it can not be generalized nor kept constant when the lot number of the commercial substrate preparation is changed.

3.5.2 Spectrophotometric Method

Assay

Method Design

Principle

(c) NH_4^+ + 2-oxoglutarate + NADH $\xrightleftharpoons{\text{GlDH*}}$ glutamate + NAD^+ + H_2O

Ammonia, another product formed by the DAO reaction (eqn. (a) on p. 238) is determined enzymatically with glutamate dehydrogenase (GlDH), eqn. (c). The equilibrium of this indicator reaction lies predominantly towards the right, since

$$K = \frac{[\text{2-oxoglutarate}] \times [NH_3] \times [NADH] \times [H^+]}{[\text{glutamate}] \times [NAD^+] \times [H_2O]} = 1.8 \times 10^{-13} \quad [25].$$

Thus, the ammonia formed in the DAO reaction is quantitatively transferred to 2-oxoglutarate. The ammonia formation per unit of time (μmol/min) determined by the decrease of absorbance due to an equimolar oxidation of NADH, is a measure of the DAO activity.

Optimized conditions for measurement: the optimal incubation conditions for the spectrophotometric assay of DAO are identical with those worked out for the isotope assay. Applying the microlitre technique [33] the following concentrations are optimal for the indicator reaction, eqn. (c): 24 U GlDH per ml; 0.19 mmoles NADH per litre; 5.7 mmoles 2-oxoglutarate per litre.

Blanks are necessary for routine assay to detect consumption of NADH which is not related to DAO activity. The method can be adapted to automated equipment.

Temperature conversion factors: not reported.

Equipment

Spectrophotometer or spectral-line photometer capable of exact measurement at 339 nm, Hg 334 or Hg 365 nm, thermostatted cuvette holder, preferably with automatic change of cuvettes and recording of the absorbance by plotter or printer.

* L-Glutamate: NAD(P) oxidoreductase (deaminating), EC 1.4.1.3.

Reagents and Solutions

Purity of reagents: the GlDH must be free from ammonia; the content of LDH and MDH should not exceed 0.2% (relative to the specific activity of GlDH).

Preparation of solutions: all solutions in re-purified water (cf. Vol. II, chapter 2.1.3.2).

1. Phosphate buffer (0.2 mol/l; pH 7.4): as described on p. 240.

2. Glutamate dehydrogenase, GlDH (1200 kU/l):

   use commercial preparation with 10 mg/ml, 120 U/mg.

3. 2-Oxoglutarate (285 mmol/l):

   dissolve 322.2 mg 2-oxoglutarate, disodium salt, 2 H_2O, in 5 ml phosphate buffer (1).

4. β-Nicotinamide-adenine dinucleotide (9.5 mmol/l):

   dissolve 33.7 mg NADH, disodium salt, in 5 ml 1% $NaHCO_3$.

5. Putrescine (50 mmol/l):

   dissolve 40.3 mg putrescine dihydrochloride in 5 ml phosphate buffer (1).

Stability of solutions: as described on p. 240. GlDH in glycerol does not lose significant activity within 6 months at 4 °C.

Procedure

Collection and treatment of specimen: only serum, plasma or purified enzyme preparations can be assayed by the NADH method, because of an unspecific consumption of NADH. The minimum requirement for purification is homogenization and centrifugation (cf. p. 241) followed by gel filtration on Sephadex G 50 (2.5 × 30 cm column; all in buffer (1)).

Stability of the enzyme in the sample: as described on p. 241.

Assay conditions: wavelength 339 (Hg 334, Hg 365) nm; light path 10 mm; final volume 1.0 ml; 37 °C (thermostatted cuvette holder); measure against blank (buffer instead of substrate solution). Before starting, adjust temperature of the solution to 37 °C.

Measurement

Pipette successively into the cuvette [27]:			concentration in assay mixture
phosphate buffer*	(1)	420 – 900 µl	ca. 200 mmol/l
GlDH solution	(2)	20 µl	24 kU/l
2-oxoglutarate solution	(3)	20 µl	5.7 mmol/l
NADH solution	(4)	20 µl	0.19 mmol/l
sample		500 – 20 µl	volume fraction 0.5 to 0.02
mix and allow to stand for 10 min, the ammonia already present in the mixture reacts and the temperature reaches 37 °C			
substrate (e.g. putrescine)	(5)	20 µl	1 mmol/l
mix and register the decrease in absorbance with printer or plotter. Automatic change of cuvette every 30 s.			

* when plasma is used, the buffer concentration must be adjusted according to the volume fraction of the sample.

The values $\Delta A/\Delta t$ must be < 0.030/min at Hg 365 nm (at Hg 334 or 339 nm < 0.060/min), otherwise dilute the sample with buffer (1).

Calculation: use eqn. (k) or (k_1), respectively (cf. "Formulae", Appendix 3). Values for ε: cf. Appendix 4. For calculation of the catalytic concentration of the sample the following relationships are valid (t in minutes):

	Hg 334 nm	339 nm	Hg 365 nm	
$b =$	$\frac{162}{v} \times \Delta A/\Delta t$	$\frac{159}{v} \times \Delta A/\Delta t$	$\frac{294}{v} \times \Delta A/\Delta t$	U/l
$b =$	$\frac{2697}{v} \times \Delta A/\Delta t$	$\frac{2645}{v} \times \Delta A/\Delta t$	$\frac{4902}{v} \times \Delta A/\Delta t$	nkat/l

Validation of Methods

Precision, accuracy, detection limits and sensitivity: see Table 1.

Sources of error: introduction of new components into the incubation mixture always requires a full check of the indicator systems, especially for quenching in the isotope assay and for inhibition of GlDH in the NADH test. In some enzyme preparations (e.g. from liver) aldehyde dehydrogenase (EC 1.2.1.3) is present which prevents the formation of Δ1-pyrroline by oxidizing the γ-aminobutyraldehyde. For these organs the isotope assay may be inappropriate [29]. The aldehyde dehydrogenase can be inhibited by chloral hydrate (0.01 mol/l) [30].

Table 1. Precision, accuracy, detection limits and sensitivity

	Radiometric method [24]	Spectrophotometric method [28]
Imprecision	b = 0.18 U/l (c) relative standard deviation 4% 2.2% (within series) 4.4% (between days)	b = 0.29 U/l (a) relative standard deviation 14% 5% (within series) 20% (between days) b = 0.3 U/l (b) relative standard deviation 6.6% b = 21 U/l (c) relative standard deviation 0.3%
Inaccuracy	coefficient of correlation between the two methods r = 0.99 [24]	
Detection limits	0.003 mU in incubation mixture	0.08 mU in incubation sample
Sensitivity	$2 \times \Delta(C_n)_{blank}/30$ min	$\Delta A/\Delta t = 0.05/10$ min (339 nm)

[a] lyophilized pooled serum
[b] native pooled serum
[c] enzyme preparation from dog intestinal mucosa

Interference with the NADH test due to high blank or "creep" can occur with crude samples. This usually results in considerable oxidation of NADH before the start of the actual reaction. The best remedy is to purify the sample on *Sephadex* G 50.

Specificity of methods: the deamination of putrescine by DAO is specific because the amine is not metabolized by monoamine oxidase. Extraction of the reactions products into toluene is not specific; however, specificity is provided by the labelled molecule which does not exchange the isotope with the surroundings.

The enzymatic determination of the liberated ammonia by the GlDH reaction confers a high specificity.

Reference ranges: reference values obtained with various methods from human organs and body fluids are listed in Tables 2 and 3.

Table 2. Diamine oxidase activity in specimens from human subjects

Principle of method	Substrate	Original units of enzymic activity	Source of specimen	Calculated $nmol \times min^{-1} \times g^{-1}$ (mU/g)	Ref.
Inactivation of histamine	histamine	$\mu g \times (3\ h)^{-1} \times g$	duodenum ileum	1.3; 0.7 1.3; 1.8	[31]
^{3}H-Release	[β-^{3}H]histamine	$pmol \times (60\ min)^{-1} \times g^{-1}$	ileum	0.002 – 0.01	[44]
[^{14}C]Putrescine	putrescine	$nmol \times min^{-1} \times g^{-1}$	duodenum	0 – 7.8	[32]
			colon sigmoid rectum	7.2 – 14.8 5.9 – 70.6 2.3 – 36.3	[16]
			rectum (biopsy specimens)	13.7 – 67.2	[33]
^{3}H-Release	[β-^{3}H]histamine	$nmol \times (5 \times 10^6\ cells)^{-1} \times (18\ h)^{-1}$	polymorphonuclear leucocytes	$0.005/10^6$ leucocytes	[34]
[^{14}C]Putrescine	putrescine	nmol/30 min	seminal plasma	6.4×10^{-6}/ml	[35]

Table 3. Diamine oxidase activity during pregnancy

Principles of method	Substrate	Original units of enzymic activity	Source of specimen	Calculated nmol × min^{-1} × g^{-1} (mU/g)	Ref.
Indigo oxidation by H_2O_2	histamine	$\Delta A/\Delta t$ = 1.0/24 h	placenta	2.7	[36]
Direct test of substrate oxidation	p-dimethyl-aminomethyl-benzylamine	μmol/min	placenta	35	[37]
[^{14}C]Putrescine	putrescine	μmol/min	amniotic fluid	1.6	[38]
Guinea pig ileum	histamine	μg × (3 ml)$^{-1}$ × (1.5 h)$^{-1}$	serum	0.44	[39]
	histamine	μg × ml^{-1} × h^{-1}	serum	0.54	[40]
Indigo oxidation by H_2O_2	cadaverine	0.46 μg × ml^{-1} × h^{-1}	serum	2.43	[41]
	putrescine		serum	2.11	
	histamine		serum	1.53	
o-Aminobenz-aldehyde	putrescine	mμ mol × (0.6 ml)$^{-1}$ × (3 h)$^{-1}$	serum	1.30	[42]
[^{14}C]Putrescine	putrescine	0.01 μg × ml^{-1} × (2 h)$^{-1}$	serum	1.33	[43]
NADH test	cadaverine		serum	3.30	[28]
	putrescine	mU/ml	serum	2.42	
	histamine		serum	0.75	

References

[1] *C. H. Best,* The Disappearance of Histamine from Autolysing Lung Tissue, J. Physiol. *67*, 256 – 263 (1929).

[2] *E. A. Zeller,* Über den enzymatischen Abbau von Histamin und Diaminen, Helv. Chim. Acta *21*, 880 – 890 (1938).

[3] *M. A. Beaven, S. Jacobsen,* A New Assay for Histaminase Activity: Measurement of Tritiated Water from β (Side Chain Label)-H^3-Histamine, J. Pharmacol. Exp. Ther. *176*, 1 52 – 64 (1971).

[4] *S. Baylin, G. Mendelsohn, W. R. Weisburger, D. S. Gann, J. C. Eggleston,* Levels of Histaminase and L-Dopa Decarboxylase Activity in the Transition from C-Cell Hyperplasia to Familial Medullary Carcinoma, Cancer *44*, 315 – 321 (1979).

[5] *M. Okada, S. Kawashima, K. Imahori,* Substrate Specificity and Reaction Mechanism of Putrescine Oxidase, J. Biochem. *86*, 97 – 104 (1979).

[6] *E. Werle,* Aminoxydasen, in: *Hoppe-Seyler-Thierfelder* (eds.), Handbuch d. physiol. u. pathol. Chem. Analyse, Vol. VI, A, Springer Verlag, Heidelberg 1964, pp. 653 – 704.

[7] *T. Matsumoto, O. Suzuki et al.,* A New Enzymatic Assay for Total Diamines and Polyamines in Urine of Cancer Patients, J. Cancer Res. Clin. Oncol. *100*, 73 – 84 (1981).

[8] *B. Holstein,* Intestinal Diamine Oxidase of Some Teleostean Fishes, Comp. Biochem. Physiol. *50B*, 291 – 297 (1975).

[9] *R. Kapeller-Adler,* Amine Oxidases and Methods for their Study, Wiley-Interscience New York, London, Sidney, Toronto 1969, pp. 74 – 76.

[10] *T. Biegański, J. Kusche, W. Lorenz, R. Hesterberg, C. D. Stahlknecht, K.-D. Feußner,* Distribution and Properties of Human Intestinal Diamine Oxidase and its Relevance for Histamine Catabolism, Biochem. Biophys. Acta 1983 (in press).

[11] *J. Kusche, U. v. Trotha, G. Mühlenberger, W. Lorenz,* The Clinical-Chemical Application of the NADH Test for the Determination of Diamine Oxidase Activity in Human Pregnancy, Agents Actions *4/3*, 188 – 189 (1974).

[12] *A. L. Southren, A. B. Weingold,* Plasma Diamine Oxidase in Pregnancy Complicated by Age and Weight Factors, Am. J. Obstet. Gynecol. *106* (4) 607 – 616 (1970).

[13] *B. Elmfors, N. Tryding, G. Tufvesson,* The Diagnosis of Ruptured Fetal Membranes by Measurement of the Diamine Oxidase (DAO) Activity in Vaginal Fluid, J. Obstet. Gynecol. *81*, 361 – 362 (1974).

[14] *J. Kusche, W. Lorenz, C.-D. Stahlknecht, H. Richter, R. Hesterberg, A. Schmal, E. Hinterlang, D. Weber, C. Ohmann,* Intestinal Diamine Oxidase and Histamine Release in Rabbit Mesenteric Ischemia, Gastroenterology *80*, 980 – 987 (1981).

[15] *C. W. Lin, N. R. Inglis, A. H. Rule, R. N. Turksoy, C. M. Chapman, S. D. Kirley, L. L. Stolbach,* Histaminase and Other Tumor Markers in Malignant Effusion Fluids, Cancer. Res. *39*, 4894 – 4899 (1979).

[16] *J. Kusche, T. Biegański, R. Hesterberg, C.-D. Stahlknecht, K.-D. Feußner, I. Stahlenberg, W. Lorenz,* The Influence of Carcinoma Growth on DAO in Human Gastrointestinal Tract, Agents Actions *10*, 110 – 113 (1980).

[17] *V. Gäng, G. Gehring, J. Schönborn,* Combined Decrease of Postheparin-DAO (Histaminase) and Postheparin-Lipoproteinlipase in Inflammatory Diseases, Klin. Wochenschr. *55*, 169 – 173 (1977).

[18] *S. B. Baylin,* Histaminase (DAO) Activity in Human Tumors: An Expression of a Mature Genome, Proc. Natl. Acad. Sci. *74*, 883 – 887 (1977).

[19] *M. J. C. Crabbe,* A Study of Indirect Methods of Assay for Diamine Oxidase, Anal. Biochem. *75*, 676 – 681 (1976).

[20] Summary of an International Workshop on Diamine Oxidase Held at the Department of Biogenic Amines, Polish Academy of Sciences, Lodz, Poland, June 7 – 8, 1979. Agents Actions *11*, 3 – 8 (1979).

[21] *G. G. Guilbault, S. S. Kuan, P. J. Brignac,* Fluorimetric Determination of Oxidative Enzymes, Analytical Applications of the Monoamine and Diamine Oxidase Systems, Anal. Chim. Acta *47*, 503 – 509 (1969).

[22] *G. R. Gordon, J. H. Peters,* Plasma Histaminase Activity in Various Mammalian Species, A Rapid Method of Assay, Proc. Soc. Exp. Biol. Med. *124*, 399 – 404 (1967).
[23] *T. Okuyama, Y. Kobayashi,* Determination of Diamine Oxidase Activity by Liquid Scintillation Counting, Arch. Biochem. Biophys. *95*, 242 – 250 (1961).
[24] *J. Kusche, H. Richter, R. Hesterberg, J. Schmidt, W. Lorenz,* Comparison of the ^{14}C-Putrescine Assay with the NADH Test for the Determination of Diamine oxidase: Description of a Standard Procedure with a High Precision and an Improved Accuracy, Agents Actions *3*, 148 – 156 (1973).
[25] *H. J. Strecker,* L-Glutamic Dehydrogenase from Liver, in: *S. P. Colowick, N. O. Kaplan* (eds.), Methods in Enzymology Vol. II, Academic Press, New York 1955, p. 220.
[26] *J. Kusche, W. Lorenz, J. Schmidt,* Oxidative Deamination of Biogenic Amines by Intestinal Amine Oxidases: Histamine is Specifically Inactivated by Diamine Oxidase, Hoppe Seyler's Z. Physiol. Chem. Vol. *356*, 1485 – 1496 (1975).
[27] *J. Kusche, K.-D. Feußner, W. Lorenz,* Intestinal Monoamine Oxidase: Does it Have a Role in Histamine Catabolism? Agents Actions *12*, 53 – 59 (1982).
[28] *J. Kusche, G. Mühlberger, U. v. Trotha, W. Lorenz,* Diamine Oxidase in Human Serum During Pregnancy: Quality Control with Two Substrates and with Native and Lyophilized Mixed Serum, in: *G. A. Anido, E. J. v. Kampen, S. B. Rosalki* (eds.), Progress in Quality Control in Clinical Chemistry, Transactions of the Vth International Symposium Geneva, April 10 – 11, 1973, Hans Huber Publishers, Bern, Stuttgart, Vienna.
[29] *A. C. Andersson, S. Henningsson, L. Persson, E. Rosengren,* Aspects on Diamine Oxidase Activity and its Determination, Acta Physiol. Scand. *102*, 159 – 166 (1978).
[30] *K. D. Rämsch, W. Schmutzler,* Solubilization and Purification of Native Pig Liver Diamine Oxidase, Naunyn-Schmied. Arch. Pharmacol. Suppl. to 287, R63 (1975).
[31] *N. G. Waton,* Studies on Mammalian Histidine Decarboxylase, Br. J. Pharmacol. *11*, 119 – 127 (1956).
[32] *J. Kusche, W. Lorenz, C.-D. Stahlknecht, A. Friedrich, A. Schmidt, K. Boo, G. Reichert,* Diamine Oxidase Activity in Gastric and Duodenal Mucosa of Man and Other Mammals with Special Reference to the Pyloric Junction, Agents Actions *8*, 366 – 371 (1978).
[33] *R. Hesterberg, J. Kusche, C.-D. Stahlknecht, K.-D. Feußner,* The Start of a Programme for Measuring Diamine Oxidase Activity in Biopsy Specimens of Human Rectal Mucosa, Agents Actions *11*, 33 – 37 (1981).
[34] *L. L. Thomas, B. S. Bochner, L. M. Lichtenstein,* Inhibition of Human Polymorphonuclear Leukocyte-Derived Histamine Activity by H-2 Antagonists, Biochem. Pharmacol. *27*, 2562 – 2565 (1978).
[35] *E. Hölttä, P. Pulkkinen, K. Elfving, J. Jänne,* Oxidation of Polyamines by Diamine Oxidase from Human Seminal Plasma, Biochem. J. *145*, 373 – 378 (1975).
[36] *J. K. Smith,* The Purification and Properties of Placental Histaminase, Biochem. J. *103*, 110 – 119 (1967).
[37] *W. G. Bardsley, M. J. C. Crabbe, I. V. Scott,* The Amine Oxidases of Human Placenta and Pregnancy Plasma, Biochem. J. *139*, 169 – 181 (1974).
[38] *G. Tufvesson,* Purification and Properties of Human Amniotic Fluid Diamine Oxidase, Scand. J. Clin. Lab. Invest. *38*, 463 – 472 (1978).
[39] *E. Werle, G. Effkemann,* Über die histaminzerstörende Fähigkeit des Schwangerenblutes, Arch. Gynäkol. *170*, 82 – 89 (1940).
[40] *A. Ahlmark,* Studies on the Histaminolytic Power of Plasma with Special Reference to Pregnancy, Acta Physiol. Scand. *9*, 1 – 107 (1944).
[41] *R. Kapeller-Adler,* Zur Kenntnis der Histaminase, Arch. Exp. Pathol. Pharmakol. *219*, 491 – 501 (1953).
[42] *Ch. McEwen,* Serum Amine Oxidase in Pregnancy, J. Lab. Clin. Med. *64*, 540 – 547 (1964).
[43] *L. Southren, Y. Kobayashi, D. H. Sherman,* Diamine Oxidase in Human Pregnancy, Am. J. Obstet. Gynecol. *89*, 199 – 203 (1964).
[44] *S. Baylin, M. A. Beaven, L. M. Buja, H. R. Keiser,* Histamine Activity: A Biochemical Marker for Medullary Carcinoma of the Thyroid, Am. J. Med. *53*, 723 – 733 (1972).

3.6 Dihydrofolate Reductase

5,6,7,8-Tetrahydrofolate : NADP oxidoreductase, EC 1.5.1.3

Andreas Schalhorn and Wolfgang Wilmanns

General

Dihydrofolate reductase (DHFR) was enriched from mammalian liver by *Futterman, Osborn & Huennekens, Peters & Greenberg,* and *Zakrzewski & Nichol* in 1957 and 1958 [1]. *Bertino et al.* and *Wilmanns* first described DHFR in human leukocytes [1]. They reported high activities in blasts of acute leukaemia, in the cells of chronic myelogenous leukaemia, and in normal bone marrow cells. Activities in chronic lymphatic leukaemia cells and in peripheral normal leukocytes were low [1]. Principally, DHFR is found in micro-organisms with rather high activities in rapidly proliferating bacteria [2]. In vertebrates, highest values of DHFR are found in the liver and kidney, but activity is also detectable in the mucosa of the gastrointestinal tract, the cerebrum, the thymus and muscle [2]. The molecular weight of DHFR of different sources varies between 18000 and 24000 [2 – 4], and improved methods of purification, including affinity chromatography techniques have made purification to homogeneity possible [3 – 5]. Complete amino acid sequences are now available for the enzymes from different antifolate-resistant bacteria and methotrexate-resistant mouse lymphoma L1210 cells [3, 4].

DHFR catalyzes the NADPH-dependent reduction of 7,8-dihydrofolic acid (DHF) to 5,6,7,8-tetrahydrofolic acid (THF), which is the coenzyme in a number of 1-carbon transfer reactions [1, 3, 6 – 9]. In particular, the decrease of the *de-novo* synthesis of deoxythymidine monophosphate (dTMP) from deoxyuridine monophosphate (dUMP) and methylene-THF which leads to inhibition of DNA synthesis and the impairment of purine biosynthesis when DHFR activity is reduced are of great importance as they can cause cell death [1, 2, 3, 7]. This explains the great importance of folate antagonists in the treatment of tumours, lymphomas, and leukaemias [6, 7, 10], of bacterial infections (trimethoprim), and of malaria and toxoplasmosis (daraprim) [4, 11].

Application of method: in biochemistry and clinical chemistry: characterization of proliferating (tumour) tissues and leukaemia cell populations; inhibition of DHFR as measure for the effectiveness of antifolate therapy [9]; determination of methotrexate concentrations in fluids and tissues by a modification of the DHFR assay [9, 12, 13].

Enzyme properties relevant in analysis: DHFR does not occur in extracellular fluids; most of the enzyme is present in the cytoplasmic and microsomal fraction; negligible amounts of DHFR can be demonstrated in the nucleus [2]. Apart from enzymatically active DHFR, a nonfunctional enzyme was detected in human leukaemia cells by immunological techniques [14]. There are several reports on DHFR polymorphism [3]. Multiple forms of DHFR could be demonstrated in bacteria, hamster kidney, human liver and different mammalian liver preparations and in leukocytes [3, 5, 15] which at least in some cases proved to be isoenzymes with different physical and kinetic properties [3]. Increase of DHFR activity due to gene amplification can play a major role in the developement of methotrexate-resistance [6, 16].

Methods of determination: DHFR can be measured spectrophotometrically. The enzyme-catalyzed reduction of DHF to THF with simultaneous oxidation of NADPH to NADP causes a decrease in absorbance at 339 nm (see below). Different methods have been reported involving tritium-labelled DHF to measure the DHFR activity [17, 18]. The method of determining the degradation product, [^{3}H]4-aminobenzoyl glutamic acid, after cleavage of THF proved to be complicated [17]. A new procedure described by *Rothenberg et al.* [18] in which [^{3}H]DHF is generated by adding sodium dithionite in the assay mixture itself, and in which [^{3}H]folic acid and [^{3}H]DHF are separated from the reaction product [^{3}H]THF by acidic zinc precipitation, seems suitable for the estimation of very low DHFR activities.

International reference method and standards: neither standardization at the international level nor the existence of reference standard material is known up to now.

Enzyme effectors: many agents are known to activate DHFR of higher organisms [2]. These include on the one hand, salts (e.g. KCl, RbCl, CsCl, NH_4Cl), urea, and guanidine, and on the other hand, reagents that react with thiol groups (mercurials, DTNB) [2]. Apart from potassium phosphate buffer, the assay described below contains no other activating substance.

Assay

Method Design

Principle

$$\text{DHF} + \text{NADPH} + \text{H}^+ \xrightarrow{\text{DHFR}} \text{THF} + \text{NADP}^+$$

DHFR reduces 7,8-dihydrofolic acid (DHF) to 5,6,7,8-tetrahydrofolic acid (THF) with oxidation of NADPH to $NADP^+$. At 339 nm the specific absorptions of THF and $NADP^+$ are lower than those of the starting materials DHF and NADPH, respectively [1, 2]. Therefore the decrease of absorbance at 339 nm is proportional to the DHFR activity with a linear molar absorption coefficient of 1.127 $l \times mmol^{-1} \times mm^{-1}$ [19].

Optimized conditions for measurement: in contrast to earlier reports [1] we chose a slightly alkaline pH of 7.5 because of the better stability of NADPH at this pH. Moreover the pH optima of DHFR from most sources lie in this pH region [2]. Mercaptoethanol is added to the assay mixture to prevent oxidation of DHF.

Temperature conversion factors: have not been determined and are not available in the literature.

Equipment

DHFR assay: spectrophotometer capable of exact measurement at 339 nm with thermostatted (37 °C) cuvette holder. An automatic cuvette changer and a recorder unit are advisable. Preparation of culture cells and tissues: sonifier and tissue homogenizer.

Reagents and Solutions

Purity of reagents: all reagents must be of analytical grade.

Preparation of solutions (for about 5 determinations): all solutions in re-purified water (cf. Vol. II, chapter 2.1.3.2).

1. Phosphate buffer (0.5 mol/l; pH 7.5):

   a) dissolve 8.7 g K_2HPO_4 ad 100 ml with water,
   b) dissolve 6.5 g KH_2PO_4 ad 100 ml with water,
   mix a and b to achieve pH 7.5 (ca. 80 ml a and 15 ml b).

2. Mercaptoethanol (0.3 mol/l):

   add 46.5 ml water to 1 ml mercaptoethanol.

3. Reduced nicotinamide-adenine dinucleotide phosphate (NADPH, 10 mmol/l):

   dissolve 10 mg NADPH, tetrasodium salt, in 1.2 ml $KHCO_3$ (0.05 mmol/l).

4. Dihydrofolic acid (DHF; 2 mmol/l):

   DHF is prepared by sodium dithionite reduction of folic acid according to *Blakley* [20] and *Futterman* [21]. Dissolve 38.2 mg folic acid in 1.6 ml NaOH (1 mol/l) and add this solution to 10 ml of 10% ascorbate solution (pH 6.0). Add 400 mg sodium dithionite, stir for 5 min at room temperature and cool to below 5 °C in an ice-bath. All further steps are done at 0 °C to 4 °C. Add HCl (1 mol/l) dropwise to pH 2.8 to precipitate the freshly prepared DHF. Centrifuge for 5 min at 1000 *g* (0 °C). Dissolve the precipitate in 10 ml of 10% ascorbate solution by addition of NaOH (1 mol/l) to reach a pH of 6.0. Repeat the acidic precipitation (pH 2.8) of DHF by addition of HCl (1 mol/l) and, after stirring for 5 min, centrifuge the suspension again. Suspend the precipitated DHF in HCl (0.005 mol/l) and store at 0 °C to 4 °C in the dark. Dissolve and dilute DHF to a concentration of 2 mmol/l just before starting the test [1, 20, 21].

Stability of solutions: potassium phosphate buffer [1] and mercaptoethanol [2] are stored at 0 °C to 4 °C and are stable as long as no bacterial or fungal contamination occurs. NADPH [3] is freshly dissolved just before start of the test.

Procedure

Collection and treatment of specimen: for determination of DHFR activity in leukocytes, the cytosol is prepared according to *Wilmanns* [1, 7]. For anticoagulation add 1 volume of EDTA-MgK_2 (0.1 mol/l; pH 7.4) to 10 volumes of blood. After sedimentation for 30 min at 37 °C, carefully remove the supernatant buffy coat containing leukocytes, thrombocytes, and small amounts of erythrocytes with a siliconized pipette and centrifuge (175 *g*, 0 °C, 10 min). All further steps are done at 0 °C to 4 °C. To haemolyze contaminating erythrocytes suspend the sediment for 3 min in NH_4Cl (0.82% w/v) and thereafter centrifuge again. Wash the pellet, containing only leukocytes, 3 times with a calcium free *Ringer* phosphate glucose solution (45 ml *Ringer* solution + 5 ml of potassium phosphate, 67 mmol/l; pH 7.4, + 100 mg glucose). Effect hypotonic lysis of the leukocytes by addition of 1 ml potassium phosphate (1 mmol/l; pH 7.4) per 10^8 cells. After 1 h at 0 °C, centrifuge for 30 min at 38000 *g* and use the supernatant cytosol to assay DHFR activity.

Culture cells are washed 3 times with 0.9% NaCl and sonicated [15]. Liver and tumour tissue is minced into small pieces (diameter 1 – 2 mm) and homogenized mechanically after addition of 3 – 5 volumes of potassium phosphate buffer (0.01 mmol/l; pH 7.5) [15].

Stability of the enzyme in the specimen: in tumour tissue, mammalian liver and human leukocytes, DHFR activity remains stable up to 1 year if stored at or below −40 °C. DHFR preparations lose their activity within 24 h at room temperature.

Details for measurement in other specimens: low DHFR activities only were measured in various tissues at high haemoglobin concentrations. Therefore, special care is needed to wash tumour specimens free from blood. Because of the low DHFR activities in crude tumour homogenates, an ammonium sulphate precipitation of DHFR prior to measurement can be of advantage.

Assay conditions: wavelength 339 nm; light path 10 mm; final volume 2.0 ml; 37 °C (thermostatted cuvette holder). Oxidoreductases present in crude enzyme preparations and spontaneous decomposition of DHF can disturb the assay. Therefore, it is necessary to measure the DHFR assay (M_1) against a sample blank without DHF (M_2) and to run a reagent blank with DHF (B_1) measured against a reference solution without DHF (B_2).

Measurement

Pipette successively into the cuvettes:	main reaction		blank reaction		concentration in assay mixture
	M_1	M_2	B_1	B_2	
phosphate buffer (1)	0.30 ml	0.30 ml	0.30 ml	0.30 ml	75 mmol/l
water	1.64 ml	1.59 ml	1.55 ml	1.60 ml	
mercaptoethanol (2)	0.05 ml	0.05 ml	0.05 ml	0.05 ml	7.5 mmol/l
NADPH (3)	0.05 ml	0.05 ml	0.05 ml	0.05 ml	0.25 mmol/l
sample	0.01 ml	0.01 ml	–	–	volume fraction 0.005
wait 5 min for temperature equilibration					
DHF (4)	0.05 ml	–	0.05 ml	–	0.05 mmol/l
mix and monitor the decrease of absorbance.					

The decrease in absorbance of the main reaction (M_1 *versus* M_2), ΔA_1, and the decrease in absorbance of the blank reaction (B_1 *versus* B_2), ΔA_2, are used for calculation: $\Delta A = \Delta A_1 - \Delta A_2$.

Calculation: use calculation formula (k) or (k_1), respectively (cf. "Formulae", Appendix 3). $\varepsilon = 1.127\ \mathrm{l \times mmol^{-1} \times mm^{-1}}$. The following relation is valid for the catalytic activity concentration of DHFR in the sample:

$$b = 17746 \times \Delta A/\Delta t \qquad \text{U/l}$$

$$b = 2.96 \times 10^6 \times \Delta A/\Delta t \qquad \text{nkat/l}$$

where Δt is the reaction time in min.

The specific activity of DHFR (U/g) or activity per 10^{10} cells are calculated with respect to protein content (g/l) or the cell count (cells/l).

Validation of Method

Precision, accuracy, detection limits and sensitivity: for values greater than 70 U/l the precision of this assay is rather good. In a series of 15 determinations the coefficient of variation was 6.88%. At 10 U/l (10 assays) this coefficient rose to 15%. The lowest values which can be measured are in the range of 1 U/l, but in these cases greater amounts of the enzyme preparation (0.2 – 0.3 ml) must be pipetted into the assay mixture. Sensitivity is found to be $\Delta A/\Delta t = 0.25$/min (*Beckman* model 35 spectrophotometer). Data on accuracy are not available since standard reference material is not yet established.

Sources of error: all antifolate agents induce inhibition of DHFR. In clinical investigations methotrexate has the greatest importance, since concentrations as low as 10^{-9} mol/l cause a 50% inhibition of DHFR. Small molecular-weight inhibitors of DHFR (e.g. trimethoprim) [11] induce false results in enzyme preparations from bacteria. Even treatment with the diuretic triamterene can falsify the results of DHFR measurements [22]. The inhibition of DHFR activity by antifolates can be used to test their activity in biological fluids and tissues [9, 12, 13, 23].

Specificity: the method is very specific. The effects of non-specific oxidoreductases are excluded by the measurement of the main reaction against an assay without DHF and by the blank reaction which is measured against an assay mixture without both DHFR and DHF.

Reference ranges:

	mU/10¹⁰ cells
Normal bone marrow cells	370 ± 120
Normal peripheral leukocytes	0 – 50
Blasts of acute leukaemias (peripheral cells)	480 ± 330
Bone marrow of acute leukaemias	480 ± 300
Chronic myelogenous leukaemia, peripheral cells	300 ± 160
Blast crisis of chronic myelogenous leukaemia, peripheral cells	520 ± 350
Chronic lymphocytic leukaemia, peripheral cells	0 – 70

	U/g
Mouse liver	1.5
Human lymphoma cell line *Ramos*	3.1
Human lymphoblast cell line LS_2	1.5
Different human solid tumours prior to methotrexate	0 – 0.09

Determination of methotrexate by inhibition of DHFR: the MTX concentration of fluids (serum, cerebrospinal fluid) can easily be tested by its inhibitory effects on the DHFR from *Lactobacillus casei* or mouse liver, for example [9, 12, 13]. To determine MTX concentrations in tissues or erythrocytes, the supernatants of the boiled tissue homogenates and haemolysates, respectively, are pipetted into the assay mixture [23]. The inhibitory effect is proportional to the MTX concentration. After preparation of a calibration curve the MTX concentration of the sample can be calculated with respect to the inhibition of DHFR and to the factors of dilution. For more details see refs. [9, 12, 13].

References

[1] *W. Wilmanns,* Bestimmung, Eigenschaften und Bedeutung der Dihydrofolsäure-Reductase in den weißen Blutzellen bei Leukämien, Klin. Wochenschr. *40,* 533 – 540 (1962).

[2] *R. L. Blakley,* The Biochemistry of Folic Acid and Related Pteridines, North-Holland Publishing Comp., Amsterdam-London 1969, pp. 139 – 181.

[3] *J. E. Gready,* Dihydrofolate Reductase: Binding of Substrates and Inhibitors and Catalytic Mechanism, Advances Pharmacol. Chemother. *17,* 37 – 102 (1980).

[4] *G. H. Hitchings, S. L. Smith,* Dihydrofolate Reductases as Targets for Inhibitors, Adv. Enzym. Regul., *18,* 349 – 371 (1980).

[5] *B. T. Kaufman, V. F. Kemerer,* Characterization of Chicken Liver Dihydrofolate Reductase after Purification by Affinity Chromatography and Isoelectric Focusing, Arch. Biochem. Biophys. *179,* 420 – 431 (1977).

[6] *J. R. Bertino,* Resistance of Human Tumors to Cancer Chemotherapeutic Agents: An Important Research Problem, Med. Ped. Oncol. *5,* 105 – 114 (1978).

[7] *H. Sauer, A. Schalhorn, W. Wilmanns,* The Biochemistry of the Citrovorum Factor Rescue Effect in Normal Bone Marrow Cells after High-Dose Methotrexate, Eur. J. Cancer *13,* 1203 – 1209 (1979).

[8] *W. Wilmanns,* Die Thymidinkinase in normalen und leukämischen myeloischen Zellen, Klin. Wochenschr. *45,* 505 – 511 (1967).

[9] *W. Wilmanns, H. Martin,* Wirkung und Verteilung von Methotrexat im menschlichen Organismus bei der Leukämiebehandlung, Klin. Wochenschr. *46,* 281 – 296 (1968).

[10] *W. A. Bleyer,* The Clinical Pharmacology of Methotrexate, Cancer *41,* 36 – 51 (1978).

[11] *J. J. Burchall,* The development of the diaminopyrimidines, J. Antimicrobial Chemother. *5,* (suppl. B), 3 – 14 (1979).

[12] *W. C. Werkheiser, S. F. Zakrzewski, C. A. Nichol,* Assay for 4-Amino Folic Acid Analogues by Inhibition of Folic Acid Reductase, J. Pharmacol. Exp. Ther. *137,* 162 – 166 (1962).

[13] *A. Schalhorn, W. Wilmanns,* Methotrexat-Nachweis im menschlichen Serum – Vergleich zwischen einer enzymatischen und einer enzymimmunologischen Methode, Onkologie *4,* 193 – 196 (1980).

[14] *S. P. Rothenberg, M. P. Iqubal,* Human leukemia and normal leukocytes contain a species of immunoreactive but nonfunctional dihydrofolate reductase, Proc. Natl. Acad. Sci. USA *79,* 645 – 649 (1982).

[15] *A. Schalhorn,* Purification of Dihydrofolate Reductase from Human Leukocytes and from Mouse Liver and Electrofocusing on Thin-Layer Polyacrylamide Gel, Enzyme *25,* 241 – 249 (1980).

[16] *F. W. Alt, R. E. Kellems, J. R. Bertino, R. T. Schimke,* Selective Multiplication of Dihydrofolate Reductase Genes in Methotrexate-resistant Variants of Cultured Murine Cells, J. Biol. Chem. *253,* 1357 – 1370 (1978).

[17] *A. Schalhorn, W. Wilmanns,* Eine neue Bestimmungsmethode der Dihydrofolatreduktase unter Verwendung von (^{3}H)Dihydrofolsäure, Hoppe Seyler's Z. Physiol. Chem. *353,* 1069 – 1074 (1972).

[18] *S. P. Rothenberg, M. P. Iqubal, M. Da Costa,* A Simplified Radioenzymatic Assay for Dihydrofolate Reductase Using (^{3}H)Dihydrofolate, Anal. Biochem. *103,* 152 – 156 (1980).

[19] *D. K. Misra, S. R. Humphreys, M. Friedkin, A. Goldin, E. Crawford,* Increased Dihydrofolate Reductase Activity as a Possible Basis of Drug Resistance in Leukaemia, Nature *189,* 39 – 42 (1961).

[20] *R. L. Blakley,* Crystalline Dihydropteroyl Glutamic Acid, Nature, *40,* 1084 – 1087 (1961).

[21] *S. Futterman,* Enzymatic Reduction of Folic Acid and Dihydrofolic Acid to Tetrahydrofolic Acid, J. Biol. Chem. *228*, 1031 – 1038 (1957).
[22] *A. Schalhorn, W. Siegert, H. Sauer,* Antifolate Effect of Triamterene on Human Leukocytes and on a Human Lymphoma Cell Line, Eur. J. Clin. Pharmacol., *20,* 219 – 224 (1980).
[23] *A. Schalhorn, H. Sauer, W. Wilmanns, G. Stupp-Poutot,* Pharmacokinetics of Erythrocyte Methotrexate after High-Dose Methotrexate, Cancer Chemother. Pharmacol., *9,* 65 – 69 (1982).

3.7 Glutathione Reductase

NAD(P)H: oxidized glutathione oxidoreductase (EC. 1.6.4.2)

David M. Goldberg and Richard J. Spooner

General

Glutathione reductase (GR) was first described in 1951 [1 – 3] and later shown to be a flavoprotein containing one mole of flavin-adenine dinucleotide (FAD) per mole of enzyme [4]. The molecular weight is approximately 110000 daltons and the enzyme has two subunits of equal molecular weight.

Glutathione reductase is found in the soluble fraction of both eukaryotic and prokaryotic cells. The primary role of the enzyme is apparently the maintenance of the intracellular reduced glutathione (GSH) concentration, this being available for the direct reduction of oxidized protein thiol groups (e.g. in enzymes and haemoglobin).

A unique "branching" mechanism, dependent on the substrate concentration, has been proposed [5].

Application of method: the assay of GR has been used in clinical chemistry (detection of hepatic and malignant disease), nutrition (assessment of riboflavin status) and in the detection of genetically determined deficiency states for which a fluorescent spot test has also been employed [6]. Care must be taken to ensure that such apparent enzyme deficiency states are not the result of riboflavin depletion [7].

Enzyme properties relevant in analysis: polymorphism of GR has not been demonstrated. However, under oxidizing conditions molecular aggregates may form. These are associated with a decrease in activity and are dissociable using EDTA or sulphydryl reagents. This phenomenon has been observed in purified enzyme preparations [8, 9] and serum [10]. Purified erythrocyte GR has been shown to exhibit a lag phase of up to one minute at low substrate concentrations [8].

The physical properties of tissue enzymes do not appear to differ from those of the serum enzyme. In general, the *Michaelis* constants for GR in phosphate buffer are of the order of 6×10^{-5} mol/l for K_{GSSG} and 6×10^{-6} mol/l for K_{NADPH}. The value for K_{NADH} is at least an order of magnitude greater [11].

Some species-dependent differences in the *Michaelis* constants have been observed [12].

Methods of determination: the preferred method is continuous monitoring by spectrophotometric analysis. As a consequence of the very low affinity of GR for NADH, saturating concentrations of this cofactor produce high absorbances. A colorimetric assay for GR has been used to circumvent this problem [13].

International reference method and standards: a method for the assay of erythrocyte GR has been recommended by the International Committee for Standardization in Haematology [14]. No reference standard materials are as yet available.

Enzyme effectors: maximal activity of GR is promoted by inclusion of EDTA in the reaction mixture. The inclusion of salts (e.g. NaCl and KCl) has the greatest effect (activation) on purified enzyme preparations. It is uncertain whether this is a specific effect or an ionic strength effect [8].

Assay

Method Design

Principle

$$\text{GSSG} + \text{NADPH} + \text{H}^+ \xrightleftharpoons{\text{GR}} 2\,\text{GSH} + \text{NADP}^+$$

(GSSG and GSH refer to oxidized and reduced forms of glutathione respectively).

The above reaction is shown as reversible, but the reaction forming GSH is strongly favoured. Catalytic activity is measured by following the decrease in absorbance due to the oxidation of NADPH.

Optimized conditions for measurement: maximal GR activity has been obtained using imidazole buffers [8, 15] but it has not been shown that the method described for the serum enzyme [15] is equally applicable to the enzyme from other sources. With a few exceptions, published information relating to kinetic constants comes from studies in which the reaction mixture was buffered by phosphate.

Temperature conversion factors: for the erythrocyte enzyme, the relationship between GR activity at 25 °C and 30 °C to that at 37 °C is 0.557 and 0.725 respectively [14].

Equipment

Glutathione reductase activity may be determined spectrophotometrically at 339 nm or using the mercury lines at 334 nm and 365 nm. The assay procedure described here can be conveniently scaled down to a discrete enzyme analyzer or a centrifugal fast analyzer.

Reagents and Solutions

Purity of reagents: all reagents should be of the highest quality available. NADPH is available at 98% purity and GSSG at 95% purity.

Preparation of solutions: all solutions should be prepared in re-purified water (cf. Vol. II, chapter 2.1.3.2).

1. Imidazole buffer (0.23 mol/l, pH 6.9 (25 °C)):

   for 1 l of buffer, dissolve 15.66 g imidazole in 800 ml water, adjust pH to 6.9 (at 25 °C) with HCl, 1 mol/l, and dilute to 1 l with water.

2. Phosphate buffer (0.12 mol/l, pH 7.2 (37 °C)):

   for 1 l of buffer, dissolve 16.33 g KH_2PO_4 in 800 ml water, adjust pH to 7.2 (at 37 °C) with NaOH, 1 mol/l, and dilute to 1 l with water.

3a. EDTA (8.25 mmol/l):

   dissolve 0.31 g EDTA-$Na_2H_2 \cdot 2H_2O$ in 100 ml water.

3b. EDTA (15 mmol/l):

   dissolve 0.56 g EDTA-$Na_2H_2 \cdot 2H_2O$ in 100 ml water.

4. Flavin-adenine dinucleotide, FAD (155 µmol/l):

   dissolve 12.5 mg FAD, disodium salt, in 100 ml water.

5a. Reduced nicotinamide-adenine dinucleotide phosphate (β-NADPH, 12 mmol/l):

   dissolve 10 mg NADPH, tetrasodium salt, in 1 ml $NaHCO_3$ solution, 1% (w/v).

5b. Reduced nicotinamide-adenine dinucleotide phosphate (β-NADPH, 9.6 mmol/l):

   dissolve 8 mg NADPH, tetrasodium salt, in 1ml $NaHCO_3$ solution, 1% (w/v).

6a. Oxidized glutathione, GSSG (114 mmol/l):

dissolve 70 mg GSSG in 1 ml water.

6b. Oxidized glutathione, GSSG (65.3 mmol/l):

dissolve 40 mg GSSG in 1 ml water.

Stability of solutions: solutions (1), (2), (3a) and (3b) may be divided into aliquots and stored at −20 °C, at which temperature they are stable for at least 6 months. NADPH and GSSG solutions should be prepared freshly as required. The FAD solution is stable at 0−4 °C for two weeks.

Procedure

Collection and treatment of specimen: specimens of serum or heparinized plasma have been found to be suitable for GR assay. Samples containing visible haemolysis should be rejected. Store at 0−4 °C until assayed.

Stability of the enzyme in the sample: serum GR is stable at 0−4 °C for two weeks, and for at least 6 months at −20 °C [15].

The erythrocyte enzyme, stored as packed cells, is stable for 5 days at 25 °C and 20 days at 4 °C [16], and at least three weeks at −20 °C [17].

Haemolysates stored at −20 °C showed slight reductions in activity over a two week period, but the value of the activation coefficient was unaffected [18].

Care must be taken to obtain a homogeneous sample of erythrocytes as GR activity falls by up to 32% during maturation and senescence of erythrocytes [19].

Details for measurement in tissues: glutathione reductase is a cytoplasmic enzyme in mammalian systems (3) and can be quantitatively extracted by conventional homogenization. Care should be taken to remove erythrocyte membranes from haemolysate preparations [16].

Assay conditions: wavelength 339 (Hg 334, Hg 365) nm; light path 10 mm; measure against air; thermostatted cuvette holder.

Assay for serum: final volume 2.75 ml; preferred incubation temperature is 25 °C.

Assay for any source of GR: final volume 3.05 ml (assay with FAD) or 2.95 ml (assay without FAD); preferred incubation temperature is 37 °C.

The order of addition of reagents has been shown to influence the measured enzyme activity [20]. The preferred order of addition is FAD (if assaying red cell enzymes), GSSG and NADPH. The inhibitions found if pre-incubation occurs with NADPH are maximal with highly purified preparations.

Measurement

1. *Assay optimized for serum* (*imidazole buffer* [15])

Pipette successively into the cuvette:			concentration in assay mixture	
imidazole buffer	(1)	2.40 ml	imidazole	200 mmol/l
EDTA	(3a)	0.10 ml	EDTA	0.3 mmol/l
GSSG	(6a)	0.05 ml	GSSG	2 mmol/l
sample (serum)		0.10 ml	volume fraction	0.038
mix contents of cuvette thoroughly, wait 5 min				
NADPH	(5a)	0.10 ml	NADPH	0.43 mmol/l
mix thoroughly and monitor the reaction as convenient.				

2. *Assay optimized for any source of GR* (*phosphate buffer* [21])

Pipette successively into the cuvette:			concentration in assay mixture	
phosphate buffer	(2)	2.60 ml	phosphate	100 mmol/l
EDTA	(3b)	0.10 ml	EDTA	0.5 mmol/l
FAD*/water	(4)	0.10 ml	FAD	5 µmol/l
GSSG	(6b)	0.10 ml	GSSG	2.2 mmol/l
sample		0.10 ml	volume fraction	0.034
mix contents of cuvette thoroughly, wait 5 min				
NADPH	(5b)	0.05 ml	NADPH	0.17 mmol/l
mix thoroughly and monitor the reaction as convenient.				

* FAD should only be included when erythrocyte GR activity is being measured, under which circumstances an assay mixture containing the equivalent volume of water is prepared as a control to provide the baseline GR activity.

In tests of riboflavin status, assay mixtures containing FAD or an equivalent volume of water are used to derive the activation coefficient (see calculation).
The values of $\Delta A/\Delta t$ must be <0.300/min at 339 or Hg 334 nm, or <0.150/min at Hg 365 nm. Dilutions of samples should be made in the appropriate buffer (1) or (2).

Calculation: use calculation formula (k) or (k_1), respectively (cf. "Formulae", Appendix 3). The catalytic concentration in the sample, b, is calculated according to

$$b = F_1 \times \Delta A/\Delta t \qquad U/l$$

or

$$b = F_2 \times \Delta A/\Delta t \qquad nkat/l$$

Factors used in calculating catalytic concentration by various methods

Wave-	Method 1		Method 2		Method 2 + FAD	
length	F_1	F_2	F_1	F_2	F_1	F_2
Hg 334 nm	4.45×10^3	7.42×10^4	4.77×10^3	7.96×10^4	4.94×10^3	8.22×10^4
339 nm	4.37×10^3	7.27×10^4	4.68×10^3	7.80×10^4	4.84×10^3	8.07×10^4
Hg 365 nm	7.99×10^3	13.32×10^4	8.58×10^3	14.29×10^4	8.87×10^3	14.78×10^4

Relating GR activity in erythrocytes to gram haemoglobin, Hb, the catalytic activity concentration z_c/m_s is calculated by use of the same factors, the result being divided by the Hb mass concentration (g Hb/l) of the sample.

When GR assays are used to assess riboflavin status, the results of assays with added FAD (supplemented) and without FAD (unsupplemented) are expressed as an activation coefficient (AC). AC is the quotient of both (AC $>$ 1).

Validation of Method

Precision, accuracy, detection limits and sensitivity: the imprecision of the method expressed as the coefficient of variation, CV (%), can be as low as 3% within the reference range when instruments equipped with automated data acquisition units are used. Manual interpolation of the reaction line from a recorder tracing provides a value of 6% at the same level of activity.

A between-batch CV approximating to 8 – 10% should be expected. Absence of reference material precludes a comment on accuracy. A value of 10 U/l should be considered as the detection limit.

Sources of error: some preparations of FAD may contain inhibitory concentrations of the mononucleotide (FMN) [22]. Administration of chemotherapeutic drugs of the nitroso-urea class, particularly BCNU (1,3-bis(2 chloroethyl)-1-nitroso-urea) has been shown to inhibit erythrocyte GR [23]. No information is available relating to the serum enzyme.

Anomalous behavior (increases in activity) of GR has been shown when the enzyme is pre-incubated in phosphate buffer [15, 24, 25].

Specificity: glutathione reductase is highly specific for GSSG. Only minor activities are seen with other disulphides. NADPH is the preferred cofactor for analytical purposes. The K_m for NADH is so high that saturating concentrations would result in absorbance values greatly exceeding the linear range of most instruments.

Reference ranges

Serum

In an analysis of 528 normal subjects, serum GR activity was observed to have a *Gaussian* distribution, and values were found to increase with age in female subjects [26].

Imidazole buffer (25 °C) [15]	20 – 60 U/l	n = 100
Phosphate buffer (37 °C) [26]	42 – 80 U/l	n = 528
Colorimetric assay (37 °C) [13] (NADPH)	31 – 57 U/l	n = 100
Colorimetric assay [13] (NADH)	26 – 50 U/l	n = 100

Erythrocytes

	– FAD	+ FAD	
Phosphate buffer (37 °C) [26]	5 – 11 U/g Hb	6 – 12.5 U/g Hb	n = 91
[18]	6.7 – 15.5 U/g Hb	10.6 – 15.4 U/g Hb	n = 8
[14]	5 – 9.4 U/g Hb	7.4 – 13.4 U/g Hb	n = 10

Riboflavin status

The literature suggests that, in human subjects, an AC value >1.35 should be considered indicative of marginal riboflavin deficiency. Using optimized assay procedures, higher values of 1.76, based on 15 subjects have been reported by one group [18], and 1.58 based on a population of 91 healthy controls by one of the present authors [26].

References

[1] *L. W. Mapson, D. R. Goddard,* The reduction of glutathione by plant tissues, Biochem. J. *49*, 592 – 601 (1951).
[2] *E. E. Conn, B. Vennesland,* Glutathione reductase of wheat germ, J. Biol. Chem. *192*, 17 – 28 (1951).
[3] *T. W. Rall, A. L. Lehninger,* Glutathione reductase of animal tissues, J. Biol. Chem. *194*, 119 – 130 (1952).
[4] *J. A. Buzard, F. Kopko,* The flavin requirement and some inhibition characteristics of rat tissue glutathione reductase, J. Biol. Chem. *238*, 464 – 468 (1963).

[5] *B. Mannervik,* A branching reaction mechanism of glutathione reductase, Biochem. Biophys. Res. Commun. *53,* 1151 – 1158 (1973).
[6] *E. Beutler,* A series of new screening procedures for pyruvate kinase deficiency, glucose-6-phosphate dehydrogenase deficiency, and glutathione reductase deficiency, Blood *28*, 553 – 562 (1966).
[7] *E. Beutler,* Red cell enzyme defects as nondiseases and as diseases, Blood *54*, 1 – 7 (1979).
[8] *D. J. Worthington, M. A. Rosemeyer,* Glutathione reductase from human erythrocytes. Catalytic properties and aggregation, Eur. J. Biochem. *67*, 231 – 238 (1976).
[9] *I. Ii, H. Sakai,* Glutathione reductase in the sea urchin egg. II cleaving – associated fluctuation of the activity and its possible regulatory mechanism, Biochim. Biophys. Acta *350,* 151 – 161 (1974).
[10] *G. Weidemann, V. Anselstetter,* Multiple Formen der NADPH-abhängigen Glutathion-Reduktase im Serum. Untersuchungen über die NADH-abhängige Glutathion-Reduktase im Serum, J. Clin. Chem. Clin. Biochem. *15*, 411 – 417 (1977).
[11] *A. Icen,* Glutathione reductase of human erythrocytes. Purification and properties, Scand. J. Lab. Clin. Invest. *20*, Suppl. *96,* 1 – 67 (1967).
[12] *R. J. Spooner, A. Delides, D. M. Goldberg,* Heat stability and kinetic properties of human serum glutathione reductase activity in various disease states, Biochem. Med. *26,* 239 – 248 (1981).
[13] *L. Kum-Tatt, I. K. Tan, A. M. Seet,* A new colorimetric method for the determination of NADH/NADPH dependent glutathione reductase in erythrocytes and in plasma, Clin. Chim. Acta *58*, 101 – 108 (1975).
[14] *E. Beutler, K. G. Blume, J. C. Kaplan, G. W. Löhr, B. Ramot, W. N. Valentine,* International Committee for Standardization in Haematology: Recommended methods for red-cell enzyme analysis, Brit. J. Haematol. *35*, 331 – 340 (1977).
[15] *G. Weidemann,* Optimierte Bestimmung und Eigenschaften der NADPH-abhängigen Glutathion-Reduktase im Serum. Untersuchungen über die Glutathion-Reduktase im Serum, I. Mitteilung, Z. Klin. Chem. Klin. Biochem. **13**, 123 – 128 (1975).
[16] *E. Beutler,* in: Red Cell Metabolism, Grune and Stratton, New York 1975.
[17] *G. E. Nicholds, J. D. Lawrence, H. E. Sauberlich,* Assessment of status of riboflavin nutrition by assay of erythrocyte glutathione reductase activity, Clin. Chem. *20*, 624 – 628 (1974).
[18] *R. A. Bayoumi, S. B. Rosalki,* Evaluation of methods of coenzyme activation of erythrocyte enzymes for detection of deficiency of vitamins B_1, B_2 and B_6, Clin. Chem. *22*, 327 – 335 (1976).
[19] *R. J. Spooner, R. A. Percy, A. G. Rumley,* The effect of erythrocyte ageing on some vitamin and mineral dependent enzymes, Clin. Biochem. *12*, 289 – 290 (1979).
[20] *E. Beutler,* Effect of flavin compounds on glutathione reductase activity: in vivo and in vitro studies, J. Clin. Invest. *48*, 1957 – 1966 (1969).
[21] *A. Delides, R. J. Spooner, D. M. Goldberg, F. E. Neal,* An optimised semi-automatic rate method for serum glutathione reductase activity and its application to patients with malignant disease, J. Clin. Pathol. *29*, 73 – 77 (1976).
[22] *C. J. Schorah, D. Messenger,* Flavin nucleotides, glutathione reductase and assessment of riboflavin status, Int. J. Vit. Nutr. Res. *45*, 39 – 50 (1975).
[23] *H. Frischer, T. Ahmad,* Severe generalised glutathione reductase deficiency after antitumor chemotherapy with BCNU [1,3-bis(chloroethyl)-1-nitrosourea], J. Lab. Clin. Med. *89*, 1080 – 1091 (1977).
[24] *R. J. Spooner, A. Delides, D. M. Goldberg,* Anomalous behaviour of glutathione reductase on dilution, Clin. Chem. *22*, 1005 – 1008 (1976).
[25] *D. M. Goldberg, R. J. Spooner,* Clinical and methodological studies on serum glutathione reductase activity, in: *A. Burlina, L. Galzigna* (eds.), Clinical Enzymology Symposia, Vol. 1, Piccin Medical Books, Padua 1977, pp. 259 – 271.
[26] *R. J. Spooner,* Glutathione reductase: Assay and Clinical Applications, *Ph. D. Thesis,* University of Glasgow (1982).

3.8 Cytochrome c Oxidase

Ferrocytochrome c: oxygen oxidoreductase, EC 1.9.3.1

Johannes Rafael

General

Based on *Warburg*'s pioneering work on the Atmungsferment (1926 – 1933) *Keilin & Hartree* [1] defined cytochrome c oxidase and identified it with cytochrome aa_3 in 1938 – 1939. Cytochrome c oxidase, COX*, catalyzes the electron transfer from ferrocytochrome c to dioxygen in the respiratory chain of eukaryotes, as well as in certain prokaryotes. As an integral membrane component, it has been shown to act as proton pump driven by the free energy of the redox process, thus being directly involved in the formation of the electrochemical membrane potential promoting oxidative ATP-production [2]. COX is a complex of polypeptide subunits of different amino acid compositions. In eukaryotes some of the subunits are coded for and produced by the mitochondrial protein-synthesizing system; others are of extramitochondrial origin [3]. The electron transfer from cytochrome c to oxygen occurs via 2 haem a groups of specific characteristics and 2 copper atoms, but little is known about the location of the metal redox centres within the enzyme complex or about the mechanism for pumping protons across the membrane.

The number, structure and function of protein subunits of COX from various sources are still under investigation (for review see [4]).

Application of the method: in biochemical, biophysical and biological research.

Enzyme properties relevant in analysis: determination of COX activity by measuring the rate of electron transport from ferrocytochrome c to oxygen is complicated by the nature of the enzyme as an integral membrane component. It is essential to establish conditions permitting maximal interaction of COX with cytochrome c, i.e. providing optimum accessibility of cytochrome c binding sites without denaturing the protein complex. Sonication or treatment of mitochondrial membranes with detergents may increase the enzyme activity to a variable degree [5]. On the other hand, detachment of the enzyme complex from its membrane environment may lead to dissociation of protein/protein interactions within the enzyme complex or deplete the complex of phospholipids, essential for the catalytic activity [6]. Apart from assay conditions, sample preparation is decisive with respect to the quantity and reproducibility of the measured activity.

* **Abbreviations:** COX: Cytochrome c oxidase EC 1.9.3.1; EDTA: ethylenediaminetetra-acetate disodium salt; TMPD: N,N,N',N'-tetramethyl-p-phenylenediamine dihydrochloride.

Methods of determination: COX is assayed by following the rate of oxygen uptake polarographically by means of an oxygen electrode [7, 8]. The enzyme can also be assayed by measuring the rate of ferrocytochrome c oxidation spectrophotometrically [9, 10], but theoretical and practical reasons appear to favour the polarographic method described here (for review comparing both methods see [11]).

International reference methods and standards: neither standardization at the international level nor the existence of reference standard material is known up to now.

Enzyme effectors: CO, cyanide and azide are inhibitory towards COX. Binding and activity of these agents depend on the state of oxidation of the enzyme. Sulphide, formate and fluoride also react with COX (for details see [4]). No specific activators of COX are known. However, as mentioned before, binding of phospholipids to the protein complex is essential for its catalytic activity. In fact, COX is largely inactive after isolation and purification procedures unless phospholipids and/or suitable detergents are added [6].

Assay

Method Design

Principle

$$2\,\text{Cytochrome c}\,(Fe^{2+}) + 1/2\,O_2 \xrightarrow{\text{COX}} 2\,\text{cytochrome c}\,(Fe^{3+}) + O^{2-}$$

The uptake per unit time of oxygen dissolved in the assay solution is employed as a measure of the catalytic activity of COX. Oxygen concentration in the assay medium is 240 µmol O_2/l at 25 °C when equilibrated with air at 760 mm Hg [11, 12]. The redox equilibrium is far in favour of oxidized cytochrome c and the rate of respiratory activity remains practically independent of oxygen concentration down to 4 µmol O_2/l [13]. Cytochrome c is kept reduced by ascorbate, with TMPD mediating the electron transport [14]. Oxygen concentration in the assay system is measured polarographically by use of a *Clark*-type oxygen electrode. The assay blank, mainly due to low autoxidation rates of ascorbate and TMPD respectively, is subtracted from the measured oxygen uptake.

Optimized conditions for measurement: assay conditions reported so far differ drastically with respect to the nature and ionic concentration of buffers, concentrations of cytochrome c and ascorbate, use of TMPD, temperature and pH. Optimal phosphate concentration (40 – 80 mmol/l), pH optimum (7.0 – 7.4 with the polarographic method) and the effects of ionic strength or cation concentration on COX activity have been studied by many investigators (for review see [15]). Tris buffer (25 mmol/l, pH 7.4) has been used by several workers [14, 16].

Definition of enzyme activity*: the minimum catalytic unit of COX, i.e. the aa_3 monomer, catalyzes the transfer of one electron reducing 1/4 O_2; the standard definition of COX activity is

$$1 \text{ kat} = 1 \text{ mol electrons} \quad s^{-1}$$

(1 mol cytochrome c s^{-1} = 1/2 mol O s^{-1}), for practical use 1 nkat = 1 nmol electrons s^{-1}. Molar catalytic activity of COX can be expressed as mol electrons $s^{-1} \times$ (mol cytochrome $aa_3)^{-1} = s^{-1}$, when cytochrome aa_3 is determined in the sample (by comparing the spectra of the reduced and oxidized forms; see [4, 14–16]). Determination is thus possible without purification of the enzyme.

Temperature conversion factors: the following factors were measured for COX activity at different temperatures (related to the value measured at 25 °C):

°C	5	10	15	20	25	27	29	31	33
F	4.01	2.86	2.04	1.45	1.00	0.87	0.81	0.73	0.65

°C	35	37	39	41
F	0.60	0.55	0.51	0.48

Equipment

Thermostatted incubation cell fitted with *Clark*-type polarographic oxygen electrode and magnetic stirrer. The incubate is protected from the introduction of oxygen during stirring by a plug which permits the use of variable incubation volumes, as well as addition of reagents during the incubation procedure. Recorder, or stopwatch with nano-ampere meter.

Examples of electrode equipment commercially available: *Gilson* oxygraph with *Yellow Springs Instrument* oxygen electrode; oxygen electrode equipment *Hansa Tech.*, King's Lynn, Norfolk, U.K. (distributed in GFR by *Bachofer GmbH + CoKG*, Reutlingen).

Reagents and Solutions

Purity of reagents: cytochrome c is commercially available** or can be prepared from horse or beef heart according to *Brautigan et al.* [11]. Since the assay is very sensitive to changes in the binding of cytochrome c to the oxidase, substantial changes or inhibition of the catalytic activity may occur with products not subjected to careful purification. De-aminated and polymeric forms of cytochrome c are present in even the best commercially available cytochrome c. Whereas no reports are available on the

* Various confusing definitions of COX activity occur in the literature (g atom O min^{-1}, mol O min^{-1}, mol O_2 min^{-1}, mol cytochrome c min^{-1}).

** Horse cytochrome c types III and VI from *Sigma* have been used by many investigators with and without further purification.

influence of de-aminated cytochrome c in the assay, even 5 – 10% cytochrome c polymer can cause non-linearity and low apparent affinity [11]. Purification of cytochrome c [11, 14]: a volume of 0.2 ml containing up to 20 mg of cytochrome c, fully reduced with minimal dithionite, is applied on a column (0.7 × 30 cm) of Sephadex G-75 superfine (*Pharmacia*) and chromatographed in Tris/acetate 50 mmol/l; pH 7.5, at 1 – 2 ml/h. This procedure provides polymer-free cytochrome c which can be used for several weeks if frozen rapidly and kept at – 20 °C. TMPD may result increased assay blank rates by autoxidation in the presence of ascorbate. Base rates may be lowered by recrystallization of TMPD from 85% ethanol containing a small amount of concentrated HCl [14].

Preparation of solutions (solutions 2 and 3 for about 50 determinations): all solutions in re-purified water (cf. Vol. II, chapter 2.1.3.2).

1. Phosphate/EDTA solution (phosphate buffer, 50 mmol/l, pH 7.2; EDTA 1 mmol/l):
   dissolve 3.4 g KH_2PO_4 and 0.186 g EDTA-$Na_2H_2 \cdot 2H_2O$ in 450 ml water, adjust to pH 7.2 with KOH, 2 mol/l, and dilute to 500 ml with water.
2. Ascorbate (0.1 mol/l):
   dissolve 26.4 mg ascorbic acid in 1.5 ml phosphate/EDTA solution (1), adjust to pH 6.0 with KOH, 10 mol/l (higher pH may cause increased assay blanks).
3. Cytochrome c (2 mmol/l):
   dissolve 24.9 mg cytochrome c in 1 ml phosphate/EDTA solution (1), or thaw equivalent amount or purified cytochrome c solution (cf. above).
4. N,N,N',N'-Tetramethyl-p-phenylenediamine, TMPD (50 mmol/l):
   dissolve 11.8 mg TMPD dihydrochloride in 1 ml water.

Stability of solutions: store all solutions at 0 – 4 °C. Phosphate solution (1) is stable as long as no bacterial contamination occurs. Prepare solutions of ascorbate, cytochrome c and TMPD freshly every day.

Procedure

Collection and treatment of specimen: mitochondria are isolated and stored according to standard procedures (1 – 2 mg mitochondrial protein/ml). Tissue specimens (100 – 150 mg) are homogenized in 5 ml ice cold phosphate/EDTA solution (1) in a tightly-fitting glass-glass homogenizer (mechanically driven; 25 – 30 strokes at 200 rpm). Samples are kept at 0 – 4 °C.

Increase of specific COX activity by treating the specimens with certain detergents [5, 16] is not unambiguous (see p. 266). Treatment of specimens with Lubrol* [17] is

* Lubrols are non-ionic detergents consisting of ethylene oxide condensates of fatty-acid alcohols and have been used to dissolve membrane bound enzymes, e.g. adenylate cyclase and Na^+, K^+-ATPase [18, 19].

recommended: A solution of Lubrol (1%; type WX from Sigma) is prepared by dissolving 100 mg detergent in 10 ml phosphate/EDTA solution (1). Dissolution takes 10 – 20 min when the mixture is stirred at room temperature (the solution is slightly opalescent). The Lubrol solution (0.1 ml) is added to 0.9 ml of the stock suspension of mitochondria (cf. above). Tissue specimens (about 100 mg) are homogenized in 2.5 ml phosphate/EDTA solution (1) and diluted 1 + 1 with Lubrol solution, 1%.

Stability of the enzyme in specimen or sample: COX should be assayed immediately after sample preparation. Storage of frozen tissues, tissue homogenates or isolated mitochondria may decrease the enzyme activity significantly. Stable COX activity has been reported [14] from *Keilin-Hartree* particles stored at – 20 °C. Virtually no decrease of COX activity occurs in Lubrol solutions kept at 0 – 4 °C for at least 4 hours.

Assay conditions: final volume 0.7 ml; 25 °C thermostatted electrode vessel stirred magnetically. For equilibration with air, phosphate solution (1) is kept at 25 °C in a gently moving shaker bath for at least 4 h (or over night) prior to the experiment.

Measurement

Calibration of electrode: adjust the electrical zero point of the recorder. Pipette about 1 ml of re-purified water into the electrode chamber, add a few grains of potassium dithionite and adjust the recorder to zero concentration of oxygen. Rinse the cell at least 7 times with re-purified water to remove traces of dithionite.

Pipette successively into the electrode chamber:			concentration in assay mixture	
phosphate/EDTA solution	(1)	0.60 ml	phosphate EDTA	50 mmol/l 1 mmol/l
wait for 1 – 3 min for final equilibration with air; adjust recorder to 100% (oxygen content of air-saturated solution).				
ascorbate solution cytochrome c solution TMPD solution	(2) (3) (4)	0.05 ml 0.03 ml 0.01 ml	ascorbate cytoch. c TMPD	7 mmol/l 85 µmol/l 0.7 mmol/l
wait for 2 – 3 min until the assay blank remains constant.				
sample		0.01 ml	volume fraction	0.014
oxygen consumption is monitored on the recorder.				

Adjust sample concentration to yield an oxygen consumption of 0.070 – 0.150/min (total recorder range = 1). Assay blanks should not exceed 0.020 – 0.030/min. Traces of cytochrome c adsorbed on the glass are removed by washing the cell with HCl, 0.1 mol/l, and rinsing it thoroughly with water.

Calculation: the catalytic activity of the sample (mitochondrial suspension or tissue homogenate; taking account of dilution resulting from treatment with Lubrol) is given by eqn. (k) (cf. "Formulae", Appendix 3):

$$b = \frac{960 \times V}{v} \times \frac{\Delta X}{\Delta t} \quad \mu kat/l$$

960	equivalent to µmol O_2 $l^{-1} \times 4$ equivalent to µmol electrons l^{-1} transferable on oxygen dissolved
ΔX	oxygen consumption recorded (corrected for assay blank)
V	assay volume in ml
v	volume of sample added in ml
Δt	measuring time for oxygen consumption in seconds

Validation of Method

Precision, accuracy, detection limits and sensitivity: for values of about 100 nmol electrons $s^{-1} \times mg^{-1}$ mitochondrial protein, an imprecision of about ± 3 nmol electrons $s^{-1} \times mg^{-1}$ has been found. Data on accuracy are not available since no standard reference material is established. With a sufficiently sensitive oxygen electrode, ΔX of 0.005 min^{-1} (= 2.4 nmol electrons $s^{-1} \times ml^{-1}$) has been measured with an imprecision < ± 20%. Sensitivity is found to be $\Delta X/\Delta t$ = 0.005/5 min (oxygen electrode *Hansa Tech.*, King's Lynn, Norfolk, U.K.).

Sources of error: with the exception of impurities in reagents (see p. 268) no special sources of error are known.

Specificity: COX activity is measured specifically.

Reference ranges: data on COX activity in mitochondria and tissues of laboratory animals reported so far differ extensively because of varying assay conditions and ways of sample preparation; data on *V* and turnover rates are generally lower than those presented in the Table.

Depending on their kind and concentration, detergents may increase or inhibit the catalytic activity of COX [5, 6, 14, 16]. In view of this ambiguity on the one side and the necessity of optimum cytochrome c interaction with the enzyme complex on the other, treatment of the sample with Lubrol is recommended (see p. 269): enzyme

Table 1. COX activity in mitochondria from rat liver, heart and brown adipose tissue treated with 0.1% Lubrol. Molar catalytic activity cf. p. 268. Cytochrome aa_3: liver 0.130 nmol × mg^{-1}, heart 0.308 nmol × mg^{-1}, brown adipose tissue 0.366 nmol × mg^{-1} protein. K_m determined from low affinity binding kinetics.

	V (nmol electr × s^{-1} × mg^{-1})	Molar catalytic activity at V (s^{-1})	K_m (µmol/l)
Liver	122	938	11.3
Heart	187	607	7.7
Brown adipose tissue	230	628	6.5

kinetics with varying cytochrome c concentrations (*Eadie-Hofstee* plot, Fig. 1) show a significant increase of V in the presence of 0.1% Lubrol. Turnover numbers of COX as measured in mitochondria treated with Lubrol (Table 1) are higher than most of the data reported so far. In contrast with other detergents [16], the apparent K_m (low affinity binding site) remains relatively unchanged in liver tissue and mitochondria analyzed with and without the detergent (Fig. 1). COX activity was found to be stable with 0.001 – 0.5% Lubrol present in the assay system.

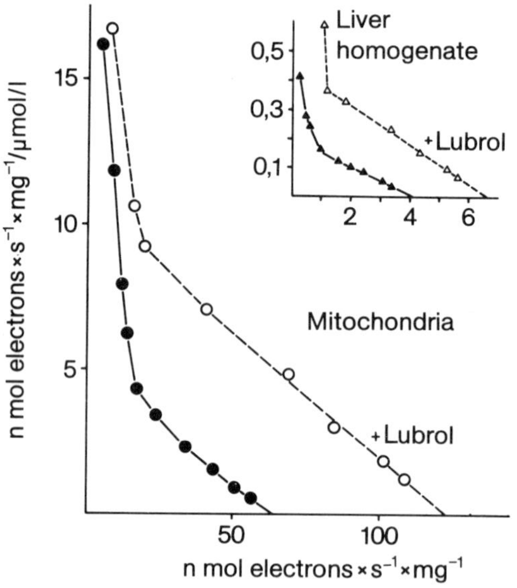

Fig. 1. Effect of Lubrol (0.1%) on COX activity in rat liver and liver homogenate.
Rates of oxygen consumption are measured with 0.7 – 85 µmol/l cytochrome c at 25 °C. The biphasic single reciprocal plot indicates two sites of cytochrome c binding [4, 14].

References

[1] *D. Keilin, E. F. Hartree,* Cytochrome and Cytochrome Oxidase, Proc. R. Soc. Lond. B Biol. Sci. *B 127*, 167 – 191 (1939).

[2] *M. K. F. Wikström, H. T. Saari,* The Mechanism of Energy Conservation and Transduction by Mitochondrial Cytochrome c Oxidase, Biochim. Biophys. Acta *462*, 347 – 361 (1977).

[3] *G. Schatz, L. Mason,* Biosynthesis of Mitochondrial Proteins, Annu. Rev. Biochem. *43*, 51 – 87 (1974).

[4] *M. Wikström, K. Krab, M. Saraste,* Cytochrome Oxidase, Academic Press Inc. (London) Ltd. 1981.
[5] *N. C. Robinson, R. A. Capaldi,* Interaction of Detergents with Cytochrome c Oxidase, Biochemistry *16*, 375 – 380 (1977).
[6] *N. S. Robinson, F. Strey, L. Talbert,* Investigation of the Essential Boundary Layer Phospholipids of Cytochrome c Oxidase Using Triton X-100 Delipidation, Biochemistry *19*, 3656 – 3661 (1980).
[7] *B. Chance,* Spectra and Reaction Kinetics of Respiratory Pigments in Homogenized and Intact Cells, Nature *169*, 215 – 221 (1952).
[8] *T. Yonetani,* Studies on Cytochrome Oxidase. IV. The Cytochrome Oxidase Activity, J. Biol. Chem. *237*, 550 – 559 (1962).
[9] *L. Smith,* Cytochromes a, a_1, a_2 and a_3, in: *S. P. Colowick, N. O. Kaplan* (eds.), Methods in Enzymology, Vol. II, Academic Press, New York 1955, pp. 732 – 740.
[10] *D. C. Wharton, A. Tzagoloff,* Cytochrome Oxidase from Beef Heart Mitochondria, In: *S. P. Colowick, N. O. Kaplan* (eds.), Methods in Enzymology, Vol. X, Academic Press, New York 1967, pp. 245 – 250.
[11] *D. L. Brautigan, S. Ferguson-Miller, E. Margoliash,* Mitochondrial Cytochrome c: Preparation and Activity of Native and Chemically Modified Cytochromes c, in: *S. P. Colowick, N. O. Kaplan* (eds.), Methods in Enzymology, Vol. LIII, Academic Press, New York 1979, pp. 128 – 164.
[12] *J. B. Chappell,* The Oxidation of Citrate, Isocitrate and cis-Aconitate by Isolated Mitochondria, Biochem. J. *90*, 225 – 237 (1964).
[13] *B. Chance,* Cellular Oxygen Requirements, Fed. Proc. *16*, 671 – 680 (1957).
[14] *S. Ferguson-Miller, D. L. Brautigan, E. Margoliash,* Definition of Cytochrome c Binding Domains by Chemical Modification, J. Biol. Chem. *253*, 149 – 159 (1978).
[15] *M. R. Lemberg,* Cytochrome Oxidase, Physiol. Rev. *49*, 48 – 121 (1969).
[16] *P. Merle, B. Kadenbach,* Kinetic and Structural Differences between Cytochrome c Oxidases from Beef Liver and Heart, Eur. J. Biochem. *125*, 239 – 244 (1982).
[17] *W. A. Behrens, J. Himms-Hagen,* Alteration in Skeletal Muscle Mitochondria of Cold-Acclimated Rats: Association with Enhanced Metabolic Response to Noradrenaline. J. Bioenerg. Biomembr. *9*, 41 – 63 (1977).
[18] *P. D. Swanson, H. F. Bradford, H. McIlwain,* Stimulation and Solubilization of the Sodium Ion-Activated Adenosine Triphosphatase of Cerebral Microsomes by Surface-Active Agents, especially Polyoxyethylene Ethers: Action of Phospholipases and a Neuraminidase, Biochem. J. *92*, 235 – 247 (1964).
[19] *G. S. Levey,* Solubilization of Myocardial Adenyl Cyclase, Biochem. Biophys. Res. Commun. *38*, 86 – 92 (1970).

3.9 Catalase

Hydrogen-peroxide: hydrogen-peroxide oxidoreductase E.C. 1.11.1.6.

Hugo E. Aebi

General [1 – 3]

Catalase is widely distributed in nature. It is found in all aerobic microorganisms, in plant and animal cells. The catalase activity of mammalian tissues varies greatly: it is highest in liver and kidney and low in connective tissues. In these cells it is mainly particle-bound (in mitochondria and peroxisomes) whereas in erythrocytes it exists in a soluble state. Human erythrocytes are normally rich in catalase whereas those of the duck are virtually catalase-free. Catalase is absent from the red cells of homozygotes for a rare enzyme-deficiency condition (Japanese-type and Swiss-type acatalasaemia).

Catalase exerts various physiological functions [3]. On one hand the enzyme located in organelles acts as a regulator of H_2O_2 levels and as a specific peroxidase, e.g. in the peroxisomes of the liver it is combined with a number of H_2O_2-producing enzymes, such as D-amino acid oxidase and uricase. On the other hand, in erythrocytes, catalase and glutathione peroxidase jointly exert a protective function for haemoglobin and other SH-proteins (enzymes, membrane, stroma), the relative contributions of each enzyme varying with species and experimental conditions (e.g. the level of the steady-state concentration of H_2O_2). The lower the catalase activity of red cells, the more effective the action of oxidizing agents (e.g. H_2O_2, ascorbate or x-rays).

Catalase is a tetrameric haemin-enzyme consisting of 4 identical tetrahedrally arranged subunits of 60000 g/mol each. Therefore, it contains 4 ferriprotoporphyrin groups per molecule, its molecular mass being approx. 240000. Since, depending on the preparation technique used, the haem component is easily lost, the total proto-haem content is variable (about 1.1%, which corresponds to an iron content of 0.09%).

Application of method: measurements of catalase activity are widely performed in food sciences and industrial food production. Owing to the considerable heat stability of catalase (and other peroxidatic agents), the detection of residual catalase activity in processed and preserved food is important with regard to changes in colour, taste and odour brought about by peroxidatic reactions.

In the dairy industry catalase activity is a sensitive and easily detectable indicator of contamination of milk by neutrophil granulocytes, i.e., it is a reliable test to detect any kind of inflammation or infection process. In biomedical research the study of catalase in liver, leukocytes and erythrocytes has received increased attention owing to its role in oxidative metabolism (e.g. peroxidation of methanol, formate) as well as its protective function by acting as a H_2O_2-scavenger. Furthermore, in genetics the pronounced heterogeneity pattern of plant catalase (e.g. in maize) as well as the rare but puzzling situation in murine and human acatalasaemia has prompted a number of investigations of general relevance.

The use of serum catalase activity has occasionally been recommended as a diagnostic parameter in haematological disorders, but in the author's opinion it is of doubtful value. Finally, a catalase screening test has been proposed for detecting urinary tract infections.

Enzyme properties relevant in analysis [1 – 10]: as heam enzymes purified catalase preparations from any source display a characteristic absorption at 405 nm (Soret band) which can be used to determine the concentration of catalase. The millimolar absorption coefficient is 38 to 40 $1 \times \text{mmol}^{-1} \times \text{mm}^{-1}$. Furthermore, the ratio A_{405}/A_{280} is used as an index of purity of the preparation. The ratio is $A_{405}: A_{280} \sim 1.25$ in practically pure specimens of human erythrocyte catalase.

Catalase is an enzyme showing unusual features in several respects [4, 5]. First, it exerts a dual function because it catalyzes the following reactions:

- decomposition of H_2O_2 to give H_2O and O_2 (catalase activity; see eqn. (a));
- oxidation of H donors, e.g. methanol, ethanol, formic acid, phenols, with the consumption of 1 mole of peroxide ("peroxidase-like" activity, see eqn. (b)).

$$\text{(a)} \qquad 2\,H_2O_2 \xrightarrow{\text{catalase}} 2\,H_2O + O_2$$

$$\text{(b)} \qquad ROOH + AH_2 \xrightarrow{\text{catalase}} H_2O + ROH + A$$

The predominating reaction depends on the concentration of H donor and the steady-state concentration or rate of production of H_2O_2 in the system. In both cases the active catalase-H_2O_2-complex I is formed first. The decomposition of H_2O_2, in which a second molecule of H_2O_2 serves as H donor for complex I, proceeds exceedingly rapidly (rate constant $k \sim 10^7\ l \times mol^{-1} \times s^{-1}$), whereas peroxidative reactions proceed relatively slowly ($k \sim 10^2 - 10^3$) [9].

Second, the kinetics of catalase do not obey the normal pattern. On one hand it is not possible to saturate the enzyme with "substrate" within the feasible concentration range (up to 5 mol/l), and on the other there is a rapid inactivation of catalase at H_2O_2 concentrations above 0.1 mol/l, when the active enzyme-H_2O_2-complex I is converted to the inactive complexes II or III. Measurement of enzyme activity at substrate saturation or determination of the K_s is therefore impossible. In contrast to reactions proceeding at substrate saturation, the enzymatic decomposition of H_2O_2 is a first order reaction, the rate of which is always proportional to the peroxide concentration. Consequently, to avoid a rapid decrease in the initial rate of the reaction, the assay must be carried out with relatively low concentrations of H_2O_2 (ca. 0.01 mol/l). The key to a successful method determining catalase activity is to maintain the enzyme concentration high enough so that the half-time for decomposition of the peroxide is 30 s or less [10]. As the activation energy for the decomposition of H_2O_2 catalyzed by catalase is very low (2500 – 7100 kJ/mol), there is only slight dependence on temperature ($Q_{10} = 1.05 - 1.12$). The decomposition of H_2O_2 initially (ca. 0 – 30 s) follows that of a first order reaction with H_2O_2 concentration between 0.01 and 0.05 mol/l. The rate constant (k) for the overall reaction is given by:

$$\text{(c)} \qquad k = \frac{1}{\Delta t} \times \ln \frac{[S_1]}{[S_2]} = \frac{2.3}{\Delta t} \log \frac{[S_1]}{[S_2]} \quad s^{-1}$$

where $\Delta t = t_2 - t_1$ is measured time interval; $[S_1]$ and $[S_2]$ are H_2O_2 concentrations at times t_1 and t_2. The constant k can be used as a direct measure of the catalase concentration. For the determination of catalase activity in blood or haemolysates the haemoglobin content is usually chosen as the reference unit. In studies with purified enzyme preparations the specific activity (k_1') is obtained by dividing k by the molar concentration of catalase [E].

$$\text{(d)} \qquad k_1' = k/[E] \quad l \times mol^{-1} \times s^{-1}$$

The value for k_1' for pure catalase from human erythrocytes is 3.4×10^7 ($l \times mol^{-1} \times s^{-1}$). This value is used to calculate the absolute content of enzyme in blood and tissues.

Third, there are several mechanisms leading to molecular heterogeneity, thereby complicating the interpretation of quantitative activity data [6–8]. Molecular heterogeneity is either of genetic origin (catalase variants of low stability or low specific activity), or is due to conformational alterations (accompanied by SH-oxidation), or is brought about by the dissociation of the active tetramer into the enzyme dimer, which exerts peroxidatic but not catalatic (i.e. H_2O_2-decomposing) activity.

Methods of determination [2, 4, 10]: in this contribution only the catalatic (not the peroxidatic) activity of catalase is taken into consideration. Catalase activity can be measured by following either the decomposition of H_2O_2 or the liberation of O_2.

Table 1. Other methods of determination

Method/Technique	Author(s)	Material
A. Determination of H_2O_2 removal		
1. Titrimetry		
a) Iodometric	*Bonnichsen et al.* (1947) [11]	Tissues, Blood
b) Permanganometric	*Feinstein* (1949) [12]	Tissues, Blood
2. Spectrophotometry		
a) Substrate: $H_2O_2(A_{240})$	*Bergmeyer* (1955) [13]	Purified preparations
	Cohen et al. (1970) [14]	Tissues, Organelles
b) Substrate: Perborate (A_{220})	*Thomson et al.* (1978) [15]	Tissue fractions
3. Photometry ($A_{405} - A_{415}$)		
a) Vanadic acid	*Warburg & Krippahl* (1963)	Cell cultures
b) Titanium tetrachloride	*Pilz & Johann* (1965)	Blood
4. Fluorimetry		
a) Scopoletin	*Perschke & Broda* (1961)	Aqueous
b) Diacetyldichlorofluorescein	*Keston & Brandt* (1965)	solutions
B. Determination of O_2 production		
1. Oxygen electrode	*Ogata* (1964) [16]	Blood
	Del Rio et al. (1977) [17]	Plant material
	Meerhof & Roos (1980) [18]	Blood
2. Polarography		
a) Catalase	*Jacob* (1964)	Microorganisms
b) Catalase and SOD simultaneously	*Rigo & Rotilio* (1977) [19]	Tissue homogenate Haemolysate
3. Manometry	This technique is considered obsolete	
C. Immunoprecipitation (with anti-catalase)	*Higashi et al.* (1961)	Blood, Tissues
	Ben-Yoseph & Shapira (1973) [20]	Blood
D. Screening Techniques		
Capillary tube	*Fung & Petrishko* (1973)	Microbial cultures
"Siebtest"	*Gross et al.* (1975) [21]	Blood
E. Automated procedure		
Technicon Autoanalyzer®	*Lamy et al.* (1968) [22]	Blood
	Leighton et al. (1968) [23]	Tissue fractions

Accordingly, the remaining substrate concentration at a given moment of the reaction can be determined either by UV-spectrophotometry or – at the end of the reaction period – by simple titration. Alternatively, O_2-production can be followed by means of the oxygen electrode or by polarography. A number of alternative techniques is listed in Table 1 [11–23]. However, the method of choice for biological material is UV-spectrophotometry. Titrimetric methods are suitable for comparative studies whereas for large series of measurements there are simple screening tests which give a quick indication of the approximate catalase activity. Automated methods (e.g. using the *Technicon Autoanalyzer*®) are also available for large series of samples.

International reference method and standards: there is no official reference method at present. However, UV-spectrophotometry has been and still is widely used. Since it is quick, specific and sensitive is has to be regarded as the generally preferred method. In special cases (turbid samples, limited laboratory facilities etc.) a simple titration procedure as presented here may prove to be a useful alternative.

Enzyme effectors: among the many substances affecting catalase activity in vitro are inhibitors, H-donor substances acting as substrates in peroxidatic catalase activity, and catalysts imitating the H_2O_2-decomposing activity of catalase. A selection of substances interfering with catalase activity is listed in Table 2. However, in most instances, particularly if the sample has been adequately diluted before analysis, these effectors do not play a significant role and are only of theoretical interest.

Table 2. Catalase inhibitors (acting in vitro)

Azide	Fluoride
Cyanide	Acetate (acetic acid)
Cyanogenbromide	Formate (formic acid)
Hydroxylamine	Methanol
Ascorbate	Ethanol
Nitrite	

3-Amino-1,2,4-triazole (AT) reacts with Complex I exclusively (therefore acts as an inhibitor in vivo only).

3.9.1 UV-method

Assay

Method Design [4, 5]

Principle: in the ultraviolet range H_2O_2 shows a continual increase in absorption with decreasing wavelength. The decomposition of H_2O_2 can be followed directly by the decrease in absorbance at 240 nm ($\varepsilon_{240} = 0.00394 \pm 0.0002$ l × mmol^{-1} × mm^{-1})

[24]. The difference in absorbance (ΔA_{240}) per unit time is a measure of the catalase activity.

This procedure was originally devised by *B. Chance & D. Herbert* [5], but has been modified by a number of authors for specific purposes.

Optimized conditions for measurement: to avoid inactivation of the enzyme during the assay (usually 30 s) or formation of bubbles in the cuvette due to the liberation of O_2, it is necessary to use a relatively low H_2O_2 concentration (10 mmol/l). The H_2O_2 concentration is critical in that there is direct proportionality between the substrate concentration and the rate of decomposition. As already mentioned, the dependence of H_2O_2-decomposition on temperature is small ($Q_{10} < 1.1$) so that measurement can be carried out between 0–37 °C; however, 20 °C is recommended. The pH-activity curve relative to v_0 has a fairly broad pH optimum (pH 6.8 to 7.5): measurements are made at pH 7.0 [3].

Temperature conversion factors: although temperature dependence is rather small, it is recommended to verify the degree of dependence under specified experimental conditions. In most instances incubation at room temperature (~20 °C) is preferred because there is a relatively rapid inactivation of catalase at higher temperatures, even if the duration of the incubation period is kept as short as possible ($\leq$ 30 s).

Equipment

Spectrophotometer suitable for accurate measurements at 240 nm connected to a recorder (e.g. Hi-speed recorder). Quartz cuvettes (10 mm light path).

Reagents and Solutions

Purity of reagents: glassware, buffer and substrate should be free from heavy metals (catalytic decomposition of H_2O_2).

Preparation of solutions (for about 50 determinations): all solutions in re-purified water (cf. Vol. II, chapter 2.1.3.2).

1. Phosphate buffer (50 mmol/l; pH 7.0):

   a) dissolve 6.81 g KH_2PO_4 in water and make up to 1000 ml,
   b) dissolve 8.90 g $Na_2HPO_4 \cdot 2\,H_2O$ in water and make up to 1000 ml.
   Mix the solutions a) and b) in the proportion 1:1.5 (v/v).

2. Hydrogen peroxide (30 mmol/l):

   dilute 0.34 ml 30% hydrogen peroxide with phosphate buffer (1) to 100 ml.

Stability of solutions: the phosphate buffer is stable as long as bacterial contamination is avoided; store at 2 °C. Prepare the buffered H_2O_2 solution fresh each day.

Procedure

Collection and treatment of specimen: *venous blood:* centrifuge venous blood containing heparin or citrate and take off plasma and leukocyte layer. Wash the erythrocyte sediment 3 times with isotonic NaCl. Prepare a stock haemolysate containing ca. 5 g Hb/100 ml by the addition of 4 parts by volume of water. Prepare a 1:500 dilution of this concentrated haemolysate with phosphate buffer immediately before the assay and determine in duplicate its Hb content (e.g. by the method of *Drabkin*). *Capillary blood:* haemolyze 0.1 or 0.02 ml of blood in 250 or 50 ml water. If the haemoglobin content of the blood is required as reference point it must be determined in a separate sample of blood.

Stability of the enzyme in the sample: although catalase in intact erythrocytes and in concentrated haemolysates is quite stable at 2 °C for several days, there is a relatively rapid loss of activity in dilute haemolysates. This instability is more likely due to decomposition of the enzyme into subunits than to proteolytic changes. No general rules can be made about the stability of the enzyme in dilute haemolysates. *Hübl & Bretschneider* [25] state that haemolysate samples should be analyzed within 5–10 min of dilution. The concentration dependence of this inactivation is shown in Fig. 1.

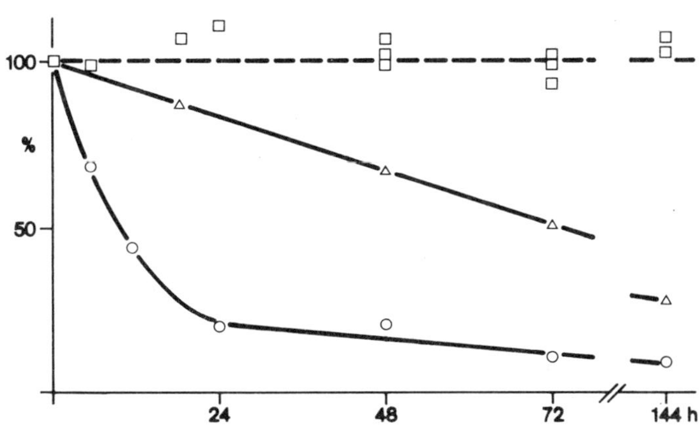

Fig. 1. Effect of haemolysate concentration on the catalase activity in normal human blood during storage (2 °C). Activity determined by the spectrophotometric method.

a) □ 60 mg Hb/ml
b) △ 1.2 mg Hb/ml
c) ○ 0.06 mg Hb/ml

Details for measurements in tissues: catalase in tissues with relatively high activity, such as liver and kidney, can be determined spectrophotometrically if it is possible to obtain complete lysis of all organelles and clear (or only slightly coloured) solutions or extracts. It is necessary to use a detergent (e.g. 1% *Triton* X-100) in the preparation of the stock homogenate (1 + 9 or 1 + 19); otherwise too low values are obtained. Further dilutions can be made with phosphate buffer, pH 7.0 (1 : 100 to 1 : 500 depending on tissue and species). However, if the sample, after lysis of the organelles, cannot be diluted to this extent, the considerable UV-absorption of *Triton* X-100 must be kept in mind. In this case, digitonin (0.01%) or sodium cholate (0.25%) can be used as an alternative. The catalase activity of tissue samples is normally expressed on a mg wet weight or mg total-N basis. A convenient method for the measurement of catalase activity in tissue extracts has been described by *Cohen et al.* [14].

Assay conditions: wavelength 240 nm; light path: 10 mm; final volume 3.00 ml. Read at 20 °C (~ room temperature) against a blank containing enzyme solution or haemolysate but no substrate.

Measurement

Pipette succesively into the cuvette:	blank	sample	concentration in assay mixture
phosphate buffer (1)	1.00 ml	–	phosphate 50 mmol/l
sample (haemolysate)	2.00 mi	2.00 ml	volume fraction 0.67
H_2O_2 solution (2)	–	1.00 ml	H_2O_2 10 mmol/l
start the reaction by addition of H_2O_2. Initial absorbance should be approx. $A = 0.500$. Mix well with a plastic paddle; follow the decrease in absorbance with a recorder for ca. 30 s.			

If a recorder is not available, take readings at 5 or 10 s intervals ("tracking technique"), see *Beers & Sizer* [26]. The values for $\Delta A_{240}/\Delta t = 15\ s^{-1}$ should not be greater than 0.100 and not smaller than 0.020.

Calculation: a definition of catalase units (U) according to the recommendations of the IUB is not feasible because of the abnormal kinetics. Therefore, it is acceptable for this enzyme to use a number of differently defined units and different methods of evaluation (see description of methods listed in Table 1). It is recommended that the rate constant of a first order reaction (k) should be used. The rate constant related to the haemoglobin content (k/g Hb) can serve as a measure of the specific activity of erythrocyte catalase. Eqn. (c) on p. 275 applies in this case. If the decrease in absorbance is recorded, the value of $\log \frac{A_1}{A_2}$ for a measured time interval, or the time required for a certain decrease in absorbance, can be determined.

For a time interval of 15 s the following relationship is obtained according to eqn. (c):

(e) $$k = \frac{2.3}{15} \times \log\frac{A_1}{A_2} = 0.153 \times \log\frac{A_1}{A_2} \quad s^{-1}.$$

To calculate the catalytic concentration b of the sample (unit k/l) or catalytic content z_c/m_s of the sample (unit k/g Hb) proceed as follows:

(f) $$b = \frac{V}{v} \times \frac{2.3}{15} \times \log\frac{A_1}{A_2} \quad s^{-1}/l \quad (k/l)$$

$$b = \frac{3}{2} \times \frac{2.3}{15} \times \log\frac{A_1}{A_2} = 0.23 \times \log\frac{A_1}{A_2} \quad k/l$$

where A_1 is A_{240} at $t = 0$; A_2 is A_{240} at $t = 15$ s; V is total assay volume; v is sample volume in the assay mixture;

(g) $$z_c/m_s = \frac{\alpha \times 2.3}{\rho_{Hb} \times 15} \times \log\frac{A_1}{A_2} = 0.153 \times \log\frac{A_1}{A_2} \quad s^{-1} \times g\,Hb \quad (k/g\,Hb)$$

where α is the quotient of mass concentration of haemoglobin ρ_{Hb} (unit g/l) in blood or erythrocyte sediment, and in the cuvette.

For a difference in absorbance of 0.450 – 0.400 ($\log A_1/A_2 = 0.05115$) the following relation holds:

(h) $$k = \frac{2.3}{\Delta t} \times \log\frac{A_1}{A_2} = \frac{0.1175}{\Delta t} \quad s^{-1}.$$

Validation of Method

Precision, accuracy, detection limits and sensitivity: with values around 300 k/g Hb a precision of ± 10 k/g Hb was found. The coefficient of variation is 2 – 3%.

The limits of this method are set by the turbidity (of tissue homogenates) or by the haemoglobin content (of the haemolysate). Whenever low catalase activity of a sample does not permit dilution of at least 1 : 100 (i.e. activity in blood is less than 20 k/g Hb), the use of the alternative titration technique is recommended.

In assaying the activity of purified catalase preparations by polarography a limit of detection of 5×10^{-4} µmol/l has been found.

Sensitivity has not been established.

Sources of error: in the determination of catalase activity in haemolysates or tissue homogenates the following sources of interference must be considered:

- Lability of catalase in dilute solution. Avoid this by dilution of the stock solution just before the assay.
- Low activity because of restricted permeability due to the binding of catalase to cell structures, e.g. incomplete haemolysis, presence of mitochondria or peroxisomes. Avoid this by addition of saponin (0.2%), *Triton* X-100 (1%) or by freezing and thawing the undiluted samples three times.
- Interference in the assay by strongly absorbing substances or precipitation in the sample. This is particularly troublesome when the sample cannot be diluted greatly because of low catalase activity (e.g. blood samples from homozygotes for the enzyme defect acatalasaemia). In such cases the titrimetric assay is recommended.
- Bacterial contamination: store samples in the frozen state.
- The presence of catalase inhibitors in samples of biological material is less likely (e.g. cyanide, azide, hydroxylamine, 3-amino-1,2,4-triazole).

Specificity of method: any substance which catalyzes the decomposition of H_2O_2 is determined. However, as the activity of other haem compounds is several orders of magnitude smaller ($10^3 - 10^4$), the method can be considered specific for catalase. In doubtful cases, the addition of catalase inhibitors (e.g. sodiumazide, 0.1 mmol/l) provides information as to whether the observed decomposition of H_2O_2 is catalyzed by catalase or other compounds [4].

Reference ranges: the catalase activity in blood of normal subjects was found to be 313 ± 96 k/g Hb ($x \pm 2$ SD).

3.9.2 Titrimetric Method

Assay

Method Design

Principle: the decomposition of H_2O_2 is followed by measurement of the peroxide still present in the mixture after a certain time by back-titration, e.g. with permanganate. Portions of the assay mixture can be taken after 10, 20 and 30 s, or the reaction in the whole mixture can be stopped by addition of acid.

Optimized conditions for measurement: see p. 278. Titrimetric methods with incubation times of longer than 30 s give only comparative values. The preparation of rate curves requires a considerable expenditure of time and effort with a large series of measurements. Perborate, a more stable substrate, is best for routine studies; the inactivation of the enzyme is somewhat slower. The results obtained with H_2O_2 and perborate are of the same order of magnitude. In practice the method of *Feinstein* [12] is suitable, in so far as the assay conditions are modified according to the recommended procedure of *Bonnichsen* [11] (i.e. 30 s incubation time at pH 7.0 and 20 °C). For further information see also references 3 and 9.

Equipment

Thermostatic water bath (20 °C) with rotating shaking attachment.

Reagents and Solutions

Purity of reagents: see remarks on p. 278.

Preparation of solutions (for about 80 determinations): all solutions in re-purified water (cf. Vol. II, chapter 2.1.3.2).

1. Phosphate buffer (50 mmol/l; pH 7.0): see p. 278.

2. Sodium perborate (100 mmol/l; pH 7.0):

   dissolve 7.694 g $NaBO_3 \cdot 4H_2O$ in ca. 20 ml HCl, 2.4 mol/l, add ca. 400 ml water, adjust with HCl to pH 7.0 and make up to 500 ml with water.

3. Potassium permanganate (0.01 mol/l):

   dissolve 1.58 g $KMnO_4$ in water and make up to 1000 ml.

Stability of solutions: Buffer see p. 279; the perborate solution is stable for several days at 2 °C. Check titre with permanganate in every series of measurements.

Procedure

Collection, treatment and stability of specimen: see p. 279.

Assay conditions: incubation at 20 °C; final volume: 5.00 ml.

Measurement

Pipette successively into test tubes:	blank	sample*	concentration in assay mixture
phosphate buffer (1) perborate solution (2)	1.00 ml 3.00 ml	1.00 ml 3.00 ml	phosphate 10 mmol/l perborate 60 mmol/l
pre-incubate for 10 min at 20 °C			
water sample** (haemolysate)	1.00 ml –	– 1.00 ml	volume fraction 0.2
stop the reaction after 30 s by addition of			
H_2SO_4, 1 mol/l	3.00 ml	3.00 ml	
the remaining perborate is back-titrated with permanganate solution.			

* To construct a rate curve, prepare several assay mixtures and stop the reaction at 10, 20 and 30 s.

** The concentration of the enzyme solution should be such that after 30 s not more than 50% and not less than 10% of the perborate is decomposed. If the catalase activity in the sample is low, the incubation time should be increased to 1 min or even to 5 min.

Calculation

a) Titration results, especially those obtained with different incubation times, can be calculated in the same way as in the spectrophotometric method according to eqns. (c) and (d) on p. 275. Because of the progressive inactivation of the enzyme the calculated k values for the various time intervals must be extrapolated to $t = 0$. The values obtained in this way for the rate constant k are somewhat lower.

b) For comparative purposes the results can also be expressed in arbitrarily defined "units", e.g. the number of millimoles of perborate decomposed under the standard experimental conditions related to mg wet weight of tissue (or g haemoglobin, etc.) ("perborate units" [12]).

Validation of Method

Precision, accuracy, detection limits and sensitivity: according to the type and activity of the sample the relative standard deviation is 3 – 5%.

Sources of error: in the titration of concentrated homogenates or blood samples (because of their low catalase activity) the end-point is not sharp since the protein in the sample slowly reacts with permanganate.

If the titration is carried out rapidly and always to the same end-point (relative excess of $KMnO_4$), the resulting error is small. A correction can be applied by titration of an additional control mixture containing no perborate or by deproteinization of the incubation reaction mixtures with trichloroacetic acid. Oxidizable substances (e.g. Tris) must not be used as components of buffers if catalase activity is determined by the titrimetric method.

Specificity: see p. 282.

References

[1] *A. Deisseroth, A. L. Dounce,* Catalase: Physical and Chemical Properties, Mechanism of Catalysis, and Physiological Role, Physiol. Rev. *50,* 319 – 375 (1970).

[2] *G. R. Schonbaum, B. Chance,* Catalase, in: P. D. Boyer (ed.), The Enzymes, Third edition, Vol. 13, Academic Press, New York 1976, pp. 363 – 408.

[3] *B. Chance, H. Sies, A. Boveris,* Hydroperoxide Metabolism in Mammalian Organs, Physiol. Rev. *59,* 527 – 605 (1979).

[4] *A. C. Maehly, B. Chance,* The Assay of Catalases and Peroxidases, in: D. Glick (ed.), Methods of Biochemical Analysis, Vol. 1, Interscience Publishers, Inc., New York 1954, pp. 357 – 424.

[5] *B. Chance, D. Herbert,* The Enzyme-substrate Compounds of Bacterial Catalase and Peroxides, Biochem. J. *46,* 402 – 414 (1950).

[6] *H. Aebi,* La Catalase Erythrocytaire, in: Exposés Annuels de Biochimie Médicale, 29ième série, Masson & Cie, (eds.), Paris 1969, pp. 139 – 166.

[7] *B. Scherz* et al., Heterogeneity of Erythrocyte Catalase, Dissociation, Recombination and Hybridization of Human Erythrocyte Catalases, Eur. J. Biochem. *69,* 603 – 613 (1976).

[8] *H. Aebi, S. R. Wyss, B. Scherz,* Heterogeneity of Human Erythrocyte Catalase, in: C. L. Markert (ed.), Isozymes, Proc. of the 3rd Internatl. Conference on Isozymes, Academic Press, New York 1975, pp. 227 – 244.

[9] *B. Chance,* An Intermediate Compound in the Catalase-hydrogen peroxide Reaction, Acta Chem. Scand. *1,* 236 – 267 (1947).

[10] *R. K. Bonnichsen, B. Chance, H. Theorell,* Catalase Activity, Acta Chem. Scand. *1,* 685 – 709 (1947).

[11] *R. K. Bonnichsen,* Blood Catalase, in: S. P. Colowick and N. O. Kaplan (eds.), Methods in Enzymology, Vol. II, Academic Press, New York 1955, pp. 781 – 784.

[12] *R. N. Feinstein,* Perborate as Substrate in a New Assay of Catalase, J. Biol. Chem. *180,* 1197 – 1202 (1949).

[13] *H. U. Bergmeyer,* Zur Messung von Katalase-Aktivitäten, Biochem. Zeitschr. *327,* 255 – 258 (1955).

[14] *G. Cohen, D. Dembiec, J. Marcus,* Measurement of Catalase Activity in Tissue Extracts, Anal. Biochem. *34,* 30 – 38 (1970).

[15] *J. F. Thomson, S. L. Nance, S. L. Tollaksen,* Spectrophotometric Assay of Catalase with Perborate as Substrate, Proc. Soc. Exper. Biol. Med. *157,* 33 – 35 (1978).

[16] *M. Ogata,* in: The Symposium on Genetics and Biochemistry of Acatalasemia, IX Annual Meeting of the Japan Society of Human Genetics at Wakayama Medical College, 1964.

[17] *L. A. Del Rio* et al., A more sensitive modification of the catalase assay with the Clark oxygen electrode. Application to the kinetic study of the pea-leaf enzyme, Anal. Biochem. *80,* 409 (1977).

[18] *L. J. Meerhof, D. Roos,* An Easy, Specific and Sensitive Assay for the Determination of the Catalase Activity of Human Blood Cells, J. Reticuloendothel. Soc. *28,* 419 – 425 (1980).

[19] *A. Rigo, G. Rotilio,* Simultaneous Determination of Superoxide Dismutase and Catalase in Biological Materials by Polarography, Anal. Biochem. *81,* 157 – 166 (1977).

[20] *Y. Ben-Yoseph, E. Shapira,* Specific immunoassay for quantitative determination of human erythrocyte catalase, J. Lab. Clin. Med. *81,* 133 – 139 (1973).

[21] *J. Gross, A. Hartwig, A. Golding,* Ein verbesserter Siebtest zur Katalasebestimmung im Blut, Z. med. Labortechnik *16*, 336 – 339 (1975).
[22] *J. N. Lamy* et al., Dosage automatique de l'eau oxygénée et application à l'étude cinétique de la catalase, Bull. Soc. Chimie Biol. *49*, 1167 – 1170 (1967).
[23] *F. Leighton* et al., The large-scale separation of peroxisomes, mitochondria, and lysosomes from the livers of rats injected with Triton WR-1339: Improved isolation procedures, automated analysis, biochemical and morphological properties of fractions, J. Cell Biol. *37*, 482 (1968).
[24] *D. P. Nelson, L. A. Kiesow,* Enthalpy of Decomposition of Hydrogen Peroxide by Catalase at 25 °C (with Molar Extinction Coefficients of H_2O_2 Solutions in the UV), Anal. Biochem. *49*, 474 – 478 (1972).
[25] *P. Hübl, R. Bretschneider,* Die Titansulfatmethode zur Bestimmung der Katalase in Blut, Serum und Harn, Hoppe-Seyler's Z. Physiol. Chem. *335*, 146 – 155 (1964).
[26] *R. F. Beers, I. W. Sizer,* Spectrophotometric Method for measuring the Breakdown of Hydrogen Peroxide by Catalase, J. Biol. Chem. *195*, 133 – 140 (1952).

Note: For a more detailed bibliography the reader is referred to the previous edition of "Methods of Enzymatic Analysis" and to references [1 – 10].

3.10 Peroxidases

Donor : hydrogen-peroxide oxidoreductase, EC 1.11.1.7

Johann Pütter und Reinhild Becker

General

The term peroxidase, POD, in its widest meaning comprises a group of specific enzymes such as NAD-POD, fatty acid-POD, cytochrome-POD, glutathione-POD and iodide-POD, as well as a group of non-specific enzymes from different sources which are simply known as POD (donor : H_2O_2-oxidoreductase, EC 1.11.1.7). Only the last group will be considered in this chapter. POD catalyzes the dehydrogenation of a large number of aromatic compounds such as phenols, hydroquinones and hydroquinonoid amines, especially benzidine derivatives. In particular, 2-cresol*, 2-toluidine, guaiacol, pyrogallol, homovanillinic acid, hydroquinone, 1,2- and 1,4-phenylenediamine, leukomalachite green, reduced 2,6-dichlorophenolindophenol, 4,4'-diaminodiphenylamine, propionylpromazine [2], benzidine, o-tolidine, di-o-anisidine and some azo-dyes derived therefrom, should be mentioned.

With few exceptions, the typical PODs included in EC 1.11.1.7 are haemoproteins. Although the normal POD reaction consists of the transfer of hydrogen or electrons

* All references which are cited in the previous edition [1] are not repeated.

from a donor to H_2O_2, there are examples of PODs acting like oxidases ($SH_2 + O_2 \rightarrow S + H_2O_2$) and monooxygenases ($S-H + O_2 + NADPH \rightarrow S-OH + NADP^+ + OH^-$); also, the indirect introduction of an O-atom into a substrate molecule is possible [2]. On the other hand, apart from the true POD, peroxidative reactions may be catalyzed by catalase with certain substrates, e.g. phenols, alcohols and some inorganic compounds; also haemoglobin and some of its breakdown products react with typical POD-substrates. Finally, in some cases oxidases, like POD, can use H_2O_2 as a source of oxygen. Although in the classical POD-reactions the specificity for the hydrogen donor is very low, it is rather high for the peroxide. It appears that, apart from H_2O_2, only compounds bearing the group $-O-OH$, e.g. acetyl-, methyl-, and ethyl-hydroperoxide, can serve as substrates.

Haemoprotein PODs occur in animals, in higher plants such as horse radishes, pineapples, figs, potatoes, legumes, cornroot vegetables and tobacco, in yeasts, in moulds and in bacteria. The longest and best-studied one is horse-radish POD. In mammals PODs occur in leukocytes, milk, liver, spleen, uterus, salivary glands, gut wall, lung, etc. They have been found in mammalian cells in the cytoplasm, the mitrochondria, the microsomes and (perhaps) the lysosomes [3].

Application of method

a) "Pseudoperoxidative" reaction of the blood-pigments, e.g. occult blood [4].
b) Differential staining of leukocytes.
c) As auxiliary enzyme in the determination of H_2O_2 [5], catalase, oxidases, glucose [6], galactose, uric acid [7], cholesterol [8], etc.
d) For characterization of disease states in experimental pathology.
e) In immunochemistry [9, 10].
f) As labelling enzyme in enzyme immuno-assays [11] e.g. of drugs.

Enzyme properties relevant in analysis: the POD reaction consists of 2 successive steps, each involving one H or one electron. A general equation for POD-catalyzed reactions cannot be formulated because the course of the reaction depends on the type of substrate. In the simplest case the same molecule is the hydrogen donor (DH_2)* for both steps. Thus the equation of the overall reaction is usually:

$$H_2O_2 + DH_2 \longrightarrow 2\,H_2O + D$$

The intermediate DH is detectable in many cases.

Often the intermediate radicals can give rise to complicated secondary reactions, e.g. [2]. Also, in the peroxidative oxidation of guaiacol, the POD substrate recommended formerly [1] gives rise to some difficulties, as more than one compound results from its dehydrogenation [12]. In the case of ABTS the subtraction of one electron gives rise to a stable end-product (see below).

* Abbreviations:
ABTS = 2,2′-azino-di-[3-ethyl-benzothiazoline-(6)-sulphonic acid], diammonium salt;
$ABTS^+$ = radical cation derived from ABTS;
D = hydrogen donor.

The kinetics of POD from various origins are not uniform. Generally there is a definable Michaelis constant for H_2O_2 which is dependent on the concentration of the donor. Whether a *Michaelis* constant (K_m) can be obtained for the donor is a matter of controversy. However, in the case of ABTS we found a measurable K_m which was dependent on the concentration of H_2O_2. Under the usual conditions there is a considerable dependence of the reaction rate on the concentration of the donor, whereas the rate is hardly affected by changes in the H_2O_2 concentration.

Methods of determination: principally, POD activity can be determined by the decrease of H_2O_2 or of the hydrogen donor, or by the formation of the oxidized compound. Mostly the third method is applied, and many different substrates have been used. Very accurate values were obtained with di-o-anisidine, but the use of this compound is limited because of its potential carcinogenic potency. Guaiacol was subsequently recommended as substrate; however, some uncertainty exists about its reaction product [1]. A new substance that was originally introduced for the determination of blood glucose, 2,2′-azino-di-[3-ethyl-benzothiazoline-(6)-sulphonic acid] [13, 6], also gives very accurate results in the determination of POD and of H_2O_2 [1, 5, 14].

International reference methods or standards: neither standardisation at the international level nor the existence of internationally recommended standard materials is known up to now.

Enzyme effectors: hormones can sometimes affect POD activity in vivo, especially in hormone-dependent tissues; certain drugs can decrease its activity. There is a decreased activity in some cases of poisoning. Malignant tumours usually have lower POD activity than the original tissue. Cyanide inhibits all POD; this inhibition can sometimes be used to identify "true" POD activity.

Assay

Method Design

Principle

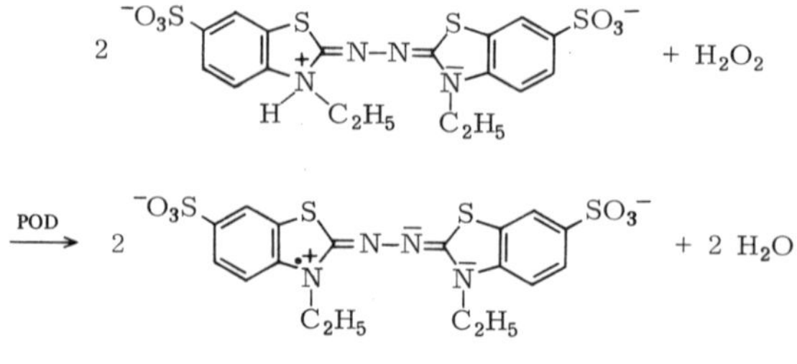

According to this equation, a resonance-stabilized radical cation ($ABTS^+$) is formed [6, 5] if excess of substrate is used. Under other conditions (lower pH, excess of H_2O_2) the reaction may proceed further to form a di-cation. The amount of $ABTS^+$ is proportional to the loss of H_2O_2; thus, with excess POD, the reaction may be used in H_2O_2 determination [5]. Within a certain range the amount of degraded H_2O_2 is proportional to the POD added [5]. It should be mentioned, however, that the reaction rate differs for various donors: PODs from different sources differ with regard to relative reaction rates with different donors. The expression of POD activity in µmol substrate converted per time unit is therefore not generally valid, but applies only for the particular substrate and particular optimum conditions (see below).

Optimized conditions for measurements [5]

Substrate concentration: at pH 6.0 in the presence of 2×10^{-3} mol ABTS per litre, the K_m for H_2O_2 is 3.3×10^{-5} mol/l; in the presence of 10^{-3} mol H_2O_2 per litre, the K_m for ABTS is 3.65×10^{-3} mol/l*. Thus, under our test-conditions saturation with ABTS cannot be achieved. However, with an ABTS concentration of 2×10^{-3} mol/l, the decrease in concentration during the reaction time is small enough to be neglected, and the reaction may be considered to proceed linearly. Higher H_2O_2 concentrations than 10^{-3} mol per l may affect enzyme activity.

pH: horse-radish POD and mammalian PODs such as lacto-POD, differ in their pH optima. pH 6.0 is recommended because at this pH horse-radish POD as well as lacto-POD can be measured with good results.

The temperature of 25 °C is suitable for horse-radish POD as well as for lacto-POD.

Reaction time: during the course of reaction a certain decrease in enzyme activity takes place; this is less than 5% in the recommended time (5 min).

Optimal absorbance of $ABTS^+$ lies at 420 nm. Measurement at 405 nm is recommended in order to allow the use of ordinary spectral-line photometers.

Temperature conversion factors cannot be given generally because PODs of various origins differ highly in this respect.

Equipment

Spectrophotometer or spectrum line photometer capable of exact measurements at 405 nm, preferably with a thermostatted cuvette holder; water bath; recorder or stop watch.

---

* These "substrate-dependent constants" are probably only apparent K_m-values as they are intermediate values between the constants versus the free and the bound enzymes. The K_m value depends not only on the concentration of the other substrate but also on pH, on the purity of the enzyme and on its origin.

Reagents and Solutions

Purity of reagents: ABTS and H_2O_2 have to be free from heavy metal ions. Only highly purified ABTS should be used. POD for calibration should be essentially free from catalase.

Preparation of solutions: all solutions in highly de-ionized water (cf. Vol. II, chapter 2.1.3.2).

1. ABTS solution (2×10^{-2} mol/l):

   dissolve 1.1 g ABTS (diammonium salt) in 100 ml aqueous phosphate buffer, 0.067 mol/l; pH 6.0 (*Sörensen*; 0.144 g $Na_2HPO_4 \cdot 2H_2O$ + 0.798 g KH_2PO_4 in 100 ml water).

2. H_2O_2 (10^{-2} mol/l):

   dilute commercial solutions of about 10 mol/l to a stock solution of 0.1 mol/l. Titrate this stock solution with a $KMnO_4$ solution (0.02 mol/l) and then dilute further to 10^{-2} mol/l. (The $KMnO_4$ solution is titrated once each week with oxalic acid.)

3. POD (for calibration purposes):

   dissolve lyophilized horse radish POD of highest available purity (about 250 U/mg) to various concentrations in phosphate buffer (see under solution 1).

Stability of solutions

ABTS solution: dissolved ABTS is slowly oxidized at room temperature in the presence of oxygen. Essentially no increase in absorption takes place in 10 h.

H_2O_2 solution: the diluted solution should be prepared daily from the stock solution. The concentration of the stock solution should be controlled by titration once a week.

POD solutions: horse-radish POD loses about 10% of its activity per day at room temperature (20 °C) or at 3 °C; therefore, standard solutions should be prepared freshly for each test. With commercially-available lacto-POD, even during 1 hour at 20 °C a considerable loss (about 20%) is found; at 3 °C the stability is somewhat better. Mammalian POD solution should be kept at 0 – 4 °C and tested as soon as possible.

Procedure

Collection and treatment of samples: the collection procedure depends on the type of experimental material. Blood must be removed from organs of experimental animals. Fluids require no special treatment. Tissues are homogenized with a *Potter Elvehjem* homogenizer in phosphate buffer pH 6.0. Refer to the appropriate methods for determination in particular cell fractions (cf. Vol. II, chapter 1.2.2.4, p. 38). Suspension of subcellular particles (cf. Vol. II, chapter 1.2.3, p. 49) can generally be added directly to the cuvette. All samples are appropriately diluted with phosphate buffer pH 6.0.

Stability of the enzyme in the sample: depends on the situation and the kind of POD to be tested. In any case, it is recommended that during the work-up and storage of the sample it should be kept at 0–4 °C. If the stability of the sample has not been specially studied, the assay should be carried out within 1 h of collection of the sample.

Details for measurement in tissues: the total POD activity of the cell is quantitatively extracted only after complete homogenization of the tissue because the enzyme occurs potentially in all cell fractions [3].

Assay conditions: wave-length 405 nm; light path 10.0 mm; final volume 2.4 ml; 25 °C (if possible, thermostatted cuvette holder). Measure against a reagent blank containing ABTS and H_2O_2 but no enzyme. Before starting the assay, adjust temperature of solutions to 25 °C.

Measurement

Pipette into the cuvette:			concentration in assay mixture	
ABTS solution	(1)	0.2 ml	ABTS	1.7 mmol/l
H_2O_2 solution	(2)	0.2 ml	H_2O_2	0.83 mmol/l
			phosphate	67 mmol/l
sample (in phosphate buffer)*		2.0 ml	volume fraction	0.833
mix; read absorbance after 1 min and 6 min				

* sample + buffer volume 2.0 ml

The time between 1 min and 6 min after addition of the sample (i.e. reaction time Δt = 5 min) is used for calculation; the increase in absorption of the blank is subtracted. The enzyme calibration curve is linear up to an absorbance of about 2.0.

Calculation: use formula (k) or (k_1), respectively (cf. "Formulae", Appendix 3). ε = 1.86 $l \times mmol^{-1} \times mm^{-1}$. The following relations are valid for the catalytic concentration in the sample (Δt in minutes):

$$b = 32.26 \times \Delta A/\Delta t \quad \text{U/l}$$

$$b = 537.6 \times \Delta A/\Delta t \quad \text{nkat/l}$$

Note: the degradation of one mole H_2O_2 yields *two* moles oxidized ABTS.

Validation of Method

Precision, accuracy, detection limits and sensitivity: for values of about 1 U/l an imprecision of ±0.08 U/l has been found. The relative standard deviation referred to the mean value of a series tested on different days in the same laboratory was ±2%, when pure POD solution was used. Data about accuracy are not available since standard reference material is not established yet. With H_2O_2, the lowest possible value which can be discriminated from blank values by the ABTS method with a probability of 95% is 0.2 µmol/l, or 14 ng H_2O_2 per cuvette if pure solutions are used [5]. Thus, the detection limit of POD is about 0.04 U/l or 0.08 mU per cuvette. If the enzyme has to be determined in biological materials the detection limit is estimated to be 5 to 10 times as high. Sensitivity* is found to be 0.004 in 5 min (*Eppendorf* photometer).

Sources of error: haemoproteins other than peroxidase and some of their degradation products are known to perform pseudoperoxidative reactions. Also, some organic complexes of heavy metal-ions can simulate peroxidative activity.

Specificity: apart from pseudoperoxidative reactions the described method is specific for POD since other kinds of H_2O_2 degradation (e.g. catalatic) do not react in this test. Whether other enzymes, such as oxidases, are able to form $ABTS^+$ from ABTS in the presence of oxygen remains to be elucidated.

Reference ranges: commercially-available POD with stated activity of 250 U/mg (guaiacol as substrate) exerts an activity of 310 U/mg against ABTS. Literature about peroxidative activity in biological materials with ABTS as substrate is not yet available.

References

[1] *J. Pütter,* Peroxidases, in: *H. U. Bergmeyer* (ed.), Methods of Enzymatic Analysis, Verlag Chemie, Weinheim 1975, pp. 685 – 670.

[2] *J. Pütter,* The participation of peroxidases in the formation of sulfoxides, Eur. J. Drug Metab. Pharmacokin. *2,* 93 – 95 (1977).

[3] *J. Pütter*, Extramicrosomal drug metabolism, Eur. J. Drug Metab. Pharmacokin. *4,* 1 – 7 (1979).

[4] *A. E. Burkhardt, A. M. Tideman,* Devise and method for determining a peroxidatively active substance in a test sample, Eur. Patent Appl. EP 41188 A2 (1981).

[5] *R. Becker,* Die Bestimmung des Wasserstoffperoxids und der Peroxidase mit 2,2′-Azino-di-[3-äthyl-benzthiazolin-sulfonsäure-(6)], Thesis at the University of Gießen (1974), performed in the Institute of Pharmacokinetics of Bayer AG.

[6] *W. Werner, H.-G. Grey, H. Wielinger,* Über die Eigenschaften eines neuen Chromogens für die Blutzuckerbestimmung nach der GOD/POD-Methode, Z. Anal. Chem. *252,* 224 – 228 (1970).

* According to the definition of sensitivity as the quotient of the measured amount (absorbance) divided by the amount to be measured (enzyme content in U/l or in U per cuvette), the sensitivity is 0.155 l/U or 77.5 U^{-1}.

[7] *M. Valickova, V. Chromy,* Enzymic determination of uric acid, Biochem. Clin. Bohemoslov. *9*, 261 – 268 (1980).
[8] *L. W. Siddiqui, J. A. Brochot,* Development of a novel potentiometric assay for peroxidase and peroxidase-coupled reactions based on the fluoride ion-selective electrode, J. Clin. Chem. Clin. Biochem. *19*, 838 (1981).
[9] *N. V. Akzo,* Improved methods for the demonstration and fixing of immunoperoxidase activity, Patent Appl. NL 8001972 A (1981).
[10] *P. W. Alexander, C. Maitra,* Enzyme linked immunoassay of human immunoglobulin G with the fluoride ion selective electrode, Anal. Chem. *54*, 68 – 71 (1982).
[11] *B. K. van Weemen, A. H. W. M. Schuurs,* Immunoassay using antibody-enzyme conjugates, FEBS letters *43*, 215 – 218 (1974).
[12] *H. Booth, B. C. Saunders,* Studies in Peroxidase Action. Part X. The Oxidation of Phenols, J. Chem. Soc. *1956*, 940 – 948.
[13] *K. Gawehn, H. Wielinger, W. Werner,* Screening von Chromogenen für die Blutzuckerbestimmung nach der GOD/POD-Methode, Z. Anal. Chem. *252*, 222 – 224 (1970).
[14] *H. Gallati,* Peroxidase aus Meerrettich: Kinetische Studien sowie Optimierung der Aktivitätsbestimmung mit den Substraten H_2O_2 und 2,2'-Azino-di-(3-ethyl-benzthiazolinsulfonsäure-(6)) (ABTS), J. Clin. Chem. Clin. Biochem. *17*, 1 – 7 (1979).

3.11 Superoxide Dismutase

Superoxide: superoxide oxidoreductase, E.C. 1.15.1.1

Erich F. Elstner, Richard J. Youngman and Wolfgang Oßwald

General

Mc Cord & Fridovich demonstrated in 1968/69 [1] that the copper protein, erythrocuprein, inhibits the xanthine oxidase-mediated reduction of cytochrome c. Reduction of cytochrome c is due to production of the superoxide anion ($O_2^{\dot{}-}$), by xanthine oxidase, XOD (EC 1.2.3.2), in the presence of an electron donor (xanthine, hypoxanthine, acetaldehyde; eqn. a). Inhibition of the reduction occurs because of enzymatic dismutation of the superoxide anion (eqn. b):

(a) $$\text{Xanthine} + O_2 \xrightarrow{\text{XOD}} \text{uric acid} + O_2^{\dot{}-};$$

(b) $$O_2^{\dot{}-} + O_2^{\dot{}-} + 2\,H^+ \xrightarrow{\text{SOD}} H_2O_2 + O_2.$$

Under physiological conditions (neutral pH), the uncatalyzed reaction (b) proceeds with a rate constant of approximately $k = 10^5\,l \times mol^{-1} \times s^{-1}$, but in the presence of erythrocuprein, now called “superoxide dismutase”, SOD [1], reaction (b) is catalytically accelerated by a factor of 10^4 and thus becomes diffusion controlled.

Several different copper, zinc, manganese or iron containing superoxide dismutases have been described [2], isolated from almost all aerobically-living organisms. Algal and mitochondrial enzymes contain manganese, bacterial enzymes manganese or iron, and the plasmatic and plastidic (chloroplasts) enzymes of higher eukaryotes, copper and zinc [3], with certain exceptions [4].

Application of method: in microbiology, cell physiology, biochemistry and clinical chemistry.

Enzyme properties relevant in analysis: crude extracts from eukaryotes may contain different activities measurable as “superoxide dismutation”, due to

- cyanide sensitive SOD (Cu-Zn enzymes)
- cyanide insensitive SOD (Fe and Mn enzymes)
- molecules which mimic SOD, such as metallocomplexes of amino acids or oligopeptides, metal chelates (e.g. iron-EDTA) or certain flavin enzymes [5], and
- superoxide scavengers such as o-diphenols, certain SH-containing compounds or ascorbate.

In order to differentiate between these four superimposed reactions in any test system, it is necessary to measure $O_2^{\cdot -}$ dismutation by comparing

- dialyzed and undialyzed crude extracts
- supernatants of heat treated extracts
- extracts in the presence and absence of cyanide (KCN, about 1 mmol/l).

Methods of determination: the assay of superoxide dismutase is mostly based on the inhibition of a superoxide-driven reaction (oxidation or reduction), by coupling an $O_2^{\cdot -}$ generator with an $O_2^{\cdot -}$ detector [6]:

$$\text{(a)} \quad \text{Xanthine} + O_2 \xrightarrow{\text{XOD}} \text{uric acid} + O_2^{\cdot -}$$

$$\text{(c)} \quad O_2^{\cdot -} + \text{detector} \longrightarrow \text{oxidized/reduced detector}$$

$$\text{(b)} \quad 2\,O_2^{\cdot -} + 2\,H^+ \xrightarrow{\text{SOD}} H_2O_2 + O_2$$

With increasing SOD concentration, the competition between reaction (c) and (b) is measured as a decrease of the rate of the detector reaction. The activity measured in any applied method is related to 50% inhibition (or acceleration) of the detector reaction [1].

A luminometric and a polarographic method have been described [7, 8].

International reference method and standards: not available.

Enzyme effectors: no activators of SOD are known.

Assay

Method Design

Principle: the principle of the method described is based on the enzymatic generation of superoxide by a microbial NADH diaphorase and the detection of $O_2^{\cdot -}$ by oxidation of hydroxylamine, yielding nitrite [9] which is measured by a colorimetric reaction.

The XOD-catalyzed reaction (a) is replaced by the following (simplified) reaction sequence:

$$(a_1) \qquad NADH + H^+ + 2\,AQ^* \xrightarrow{\text{diaphorase}} NAD^+ + 2\,AQH$$

$$(a_2) \qquad 2\,AQH + 2\,O_2 \xrightarrow{\text{autoxidation}} 2\,AQ + 2\,O_2^{\cdot -} + 2\,H^+$$

Conversion of the reaction product, $O_2^{\cdot -}$, is brought about by two competing reactions with comparable reaction rates:

$$(b) \qquad 2\,O_2^{\cdot -} + 2\,H^+ \xrightarrow{\text{SOD}} H_2O_2 + O_2$$

$$(c_1) \qquad 2\,O_2^{\cdot -} + H^+ + NH_2OH \longrightarrow H_2O_2 + H_2O + NO_2^-$$

Reaction (c_1) can be followed by the formation of an azo dye stuff by NO_2^-:

$$(c_2) \qquad NO_2^- + \text{sulphanilamide} + \text{naphthylethylene diamine} \longrightarrow \text{azo dye stuff}$$

The intensity of the red colour correlates with the amount of NO_2^- formed and can be measured at 540 nm.

As a result of the competition between reaction (c_1) and the SOD-catalyzed reaction (b), the decrease of the rate of the detector reaction (c_2) directly reflects thc catalysis of reaction (b) by SOD, i.e. a decrease of the rate of nitrite formation in the presence of SOD (see Fig. 1, p. 301).

Thus, SOD inhibits the $O_2^{\cdot -}$ driven oxidation of hydroxylamine in a concentration-dependent manner. One enzyme unit is defined [1] as 50% inhibition of the detector reaction under the assay conditions. These are: presence of NADH, diaphorase, anthraquinone-2-sulphonate, NH_2OH, and air-saturated buffer. The effect of increasing amounts of SOD (or "sample", respectively) on nitrite formation is measured.

No relation to µmol $O_2^{\cdot -}$ conversion can be established though the indicator colour reaction and the reaction of $O_2^{\cdot -}$ proceed stoichiometrically with nitrite concentration, and with hydroxylamine, resp. The reason is that, due to the complex mechanisms of

* AQ/AQH = anthraquinone sulphonate, oxidized/reduced

autoxidation, eqn. (a_2), the conversion of $O_2^{\cdot-}$ by SOD varies from assay to assay. Thus, zero SOD activity and excess of SOD activity yield differing absorbances for the initial value and for the base line in the colour reaction. Nevertheless, for various SOD calibration curves on the basis of an identical enzyme calibration material, always the half of the absorbance difference indicates the same value on the abscissa. With different enzyme calibration material of fairly good quality the arbitrary units in relation to µg enzyme protein do not vary by more than about ±20%. So far this kind of calibration suits sufficiently the practical needs.

Optimized conditions for measurement

The pH

$O_2^{\cdot-}$ is relatively stable at pH values above 7.5. Since the pK_a of the equilibrium

(d) $$O_2^{\cdot-} + H^+ \rightleftharpoons HO_2^{\cdot} \text{ (perhydroxyl radical)}$$

is at pH 4.88, the following rate constants apply to the spontaneous dismutation of superoxide acc. to eqn. (b) [10]:

(b) $$O_2^{\cdot-} + O_2^{\cdot-} + 2\,H^+ \longrightarrow H_2O_2 + O_2 \quad k = 100\,l \times mol^{-1} \times s^{-1}$$

(b_1) $$O_2^{\cdot-} + HO_2^{\cdot} + H^+ \longrightarrow H_2O_2 + O_2 \quad k = 8.5 \times 10^7\,l \times mol^{-1} \times s^{-1}$$

(b_2) $$HO_2^{\cdot} + HO_2^{\cdot} \longrightarrow H_2O_2 + O_2 \quad k = 7.6 \times 10^5\,l \times mol^{-1} \times s^{-1}$$

At the physiological pH of 7.0 the rate constant of the SOD catalyzed $O_2^{\cdot-}$ dismutation is ca. $2 \times 10^9\,l \times mol^{-1} \times s^{-1}$; for determining SOD activities, a pH of between 7.0 and 8.0 is optimal.

Aerobic conditions

Superoxide is produced from ground state (triplet) dioxygen by one-electron donation according to:

(e) $$^3O_2 + e^- \longrightarrow O_2^{\cdot-}.$$

Therefore, well-aerated buffer solutions are a prerequisite for high rates of enzymatic oxygen reduction. This is achieved by bubbling air for 30 min at room temperature (22 °C) through the phosphate buffer shortly before its use.

In comparison with other methods, for example with xanthine oxidase as $O_2^{\cdot-}$ generator and other detector systems, the method presented here allows the fastest determination of a large number of samples with a minimum of artefacts due to interference from other compounds and also offers a high degree of reproducibility.

Temperature conversion factors: not available.

Equipment

Spectrophotometer, thermostatted cuvette holder not required. Sonicator: *Branson* Sonifier.

Reagents and Solutions

Purity of reagents: all reagents are commercially available (*Boehringer Mannheim; Merck,* Darmstadt; *Sigma Chem. Comp.*) and of the purity required for this SOD test.

Preparation of solutions (for about 10 determinations): all solutions in re-purified water (cf. Vol. II, chapter 2.1.3.2).

1. Phosphate buffer, air-saturated (0.2 mol/l; pH 7.8):

   a) dissolve 7.16 g $Na_2HPO_4 \cdot 12H_2O$ in water and make up to 100 ml.
   b) dissolve 3.56 g $NaH_2PO_4 \cdot 2H_2O$ in water and make up to 100 ml. Titrate solution a) with solution b) until pH 7.8 is reached.

2. Hydroxylammonium chloride (10 mmol/l):

   dissolve 6.9 mg hydroxylammonium chloride in 10 ml water.

3. Reduced nicotinamide-adenine dinucleotide (β-NADH, 10 mmol/l):

   dissolve 7.1 mg NADH, sodium salt, in 1 ml water.

4. Anthraquinone-2-sulphonic acid, AQ (4 mmol/l):

   dissolve 13.2 mg AQ in 100 ml water.

5. Diaphorase (1 mg/ml; ca. 15000 U/l):

   dissolve 1 mg diaphorase from microorganisms (*Boehringer Mannheim*), lyophilized enzyme, ca. 15 U/mg, with 1 ml water.

6. SOD reference solution:

   dilute commercially available SOD (e.g. 15 arb. units/mg) according to 2, 5, 10, 25, 50, 250, 500 units/ml H_2O; if necessary, dilute 1 + 9.

7. Cyanide (31 mmol/l):

   dissolve 2 mg KCN with 1 ml water.

8. Sulphanilamide (1% w/v):
   dissolve 10 g in 1 l HCl, 25% (v/v).

9. Naphthylethylene diamine dihydrochloride (0.02% w/v):
   dissolve 0.2 g in 1 l water.

10. Nitrite standard solution (1 mmol/l):
    dissolve 69 mg $NaNO_2$ in 1 l water. Dilutions from 10 to 1000 μmol/l in water are used for the calibration curve.

Stability of solutions: store all solutions at 0 °C to 4 °C in the dark. NADH, hydroxylamine, sulphanilamide, naphthylethyldiamine dihydrochloride and AQ solutions should be prepared freshly every week, diaphorase and SOD reference solutions should be made up freshly before the experiment.

Procedure

Collection and treatment of specimen: as outlined in detail by *Stein et al.* [11] sample preparation is crucial for cyanide-sensitive, cyanide-insensitive and non-enzymic (artefactual [5, 6]) SOD activities.

In general, we have to envisage the following SOD-active molecules in the samples:

- the centrifuged, sonicated crude extract contains all compounds with potential SOD activity or simulating this activity ($sample_{SOD\ total}$);
- the heat-denatured, centrifuged crude extract may contain non-enzymic SOD-active and SOD-simulating molecules ($sample_{SOD\ non\text{-}enzymic}$);
- the dialyzed crude extract should contain cyanide sensitive and/or cyanide-insensitive SOD, depending on the source ($sample_{true\ SOD}$);
- the heat-denatured, dialyzed and centrifuged crude extract should neither contain SOD nor non-enzymic SOD (-simulating) activities; it thus may be used as an internal blank in addition to the water blank ($blank_{internal}$).

Therefore, the samples should be prepared from biological tissues or cell cultures as follows:

- homogenization of the tissue (2 min *Sorvall* Omnimixer, full speed) with phosphate buffer (1) or buffer of half concentration, depending on the specific tissue weight;
- sonication of the homogenate: 3 min under cooling, *Branson* Sonifier, full power;
- centrifugation for 30 min at 20000 *g*.

After centrifugation divide the supernatant into two aliquots: one is analyzed immediately ($sample_{SOD\ total}$), the other is analyzed after exhaustive dialysis against phosphate buffer (1) ($sample_{true\ SOD}$). Both dialyzed and undialyzed samples are treated as follows:

- heat denaturation: boiling for 2 min in water bath and centrifugation for 30 min at 20000 *g* ($sample_{SOD\ non\text{-}enzymic}$),
- heat denaturation, exhaustive dialysis (see above) and centrifugation for 30 min at 20000 *g* ($blank_{internal}$).

Sample preparation for assaying SOD activities in homogenates: to differentiate between cyanide sensitive, cyanide insensitive, and artefactual SOD activities, the various pretreated samples have to be tested in the presence and absence of KCN, ca. 1.5 mmol/l.

1. $sample_{SOD\ total}$
2. $sample_{SOD\ non\text{-}enzymic}$
3. $sample_{true\ SOD}$
4. $blank_{internal}$

The samples and the internal blank have to be diluted to various concentrations (e.g. 1+99, 1+49, 1+24, 1+9, 1+4, 1+2, 1+1) and these solutions used for the assay.

Since the activity or inhibition, respectively, have to be related to the protein content, a protein determination must be performed on all samples (cf. Vol. II, chapter 1.3).

Stability of the enzyme in the sample: in crude extracts, stable for several h at 4 °C.

Assay conditions

Incubation and measurement: incubation volume 2.0 ml; room temperature. Assay volume in colour reaction 2.0 ml; room temperature; wavelength 540 nm; light path 10 mm. Measure against reagent blank with buffer (1) instead of incubation mixture.

SOD reference curve: use SOD reference solution (6) in at least six various concentrations (see above). It is advisable to run a SOD reference curve for establishing and for checking the performance of the assay system. Furthermore, the activity of any sample with an appropriate content can be roughly estimated by reading the absorbance and relating to that amount of protein which causes an inhibition of 50%. However, to achieve exact results a curve with the various sample concentrations has to be established as for the SOD reference.

Differentiation between the various kinds of SOD activity: for differentiation between total, non-enzymic and true (KCN-sensitive and KCN-insensitive) SOD activity use samples and $blank_{internal}$ treated as described above. In each case run at least duplicates with diluted samples.

Maximal absorbance: this assay is run to yield the initial value for the SOD reference curve as well as for the "sample" curves (no inhibition).

Nitrite calibration curve: it might be established to check the performance of the colour reaction. Use diluted nitrite standard solution (10) instead of incubation mixture in the colour reaction.

Measurement

Incubation

Pipette successively into 10 ml test tubes:	incubation blank ml	maximal absorbance (0% inhibition) ml	sample** ml	concentration in assay mixture
water	0.70	0.60/0.50*	0.50/0.40*	–
buffer, air saturated (1)	1.00	1.00	1.00	phosphate 100 mmol/l
hydroxylamine (2)	0.10	0.10	0.10	hydroxylamine 0.5 mmol/l
AQ (4)	0.10	0.10	0.10	AQ 0.2 mmol/l
diaphorase (5)	–	0.10	0.10	diaphorase 750 U/l
KCN (7)	–	–/0.10	–/0.10	KCN 1.5 mmol/l
mix by shaking				
sample** or	–	–	0.10	dil. 1 + 1 to 1 + 99 e.g. volume fraction 0.05
SOD** resp. (6)	–	–	0.10	SOD 125 – 25 units/l
NADH (3)	0.10	0.10	0.10	NADH 10 mmol/l
mix by shaking, incubate for 15 min at room temperature (approx. 22 °C). 0.5 ml aliquots are used for colour reaction.				

* for the determination of CN^- insensitive SOD maintain the correct total volume by adding 0.40 ml water instead of 0.50 ml

** samples, $blank_{internal}$, or SOD reference resp. diluted as described above

Colour reaction

Pipette successively into 10 ml test tubes:		reagent blank	calibration standards	sample
sulphanilamide	(8)	0.5 ml	0.5 ml	0.5 ml
buffer	(1)	1.0 ml	0.5 ml	0.5 ml
incubation mixture*		–	–	0.5 ml
nitrite standard**	(10)	–	0.5 ml	–
naphthylethylenediamine	(9)	0.5 ml	0.5 ml	0.5 ml
shake; after 20 min read absorbance at 540 nm against reagent blank.				

* apply all incubation mixtures including incubation blank and "maximal absorbance"

** apply the various standard solutions

Plot the absorbance of the nitrite calibration standards (ordinate) against the corresponding nitrite concentration (abscissa). A straight line should be obtained. The red colour is stable for 5 hours. This calibration curve might be established to check the performance of the colour reaction.
Plot the absorbancies of all incubation mixtures, whereby the blank values have to be accounted for (ordinate) against the corresponding protein mass (µg) in the assay mixture (abscissa; cf. Fig. 1).

Calculation: the catalytic activity of the enzyme is defined by the reduction of the colour formation by 50% (cf. Fig. 1).

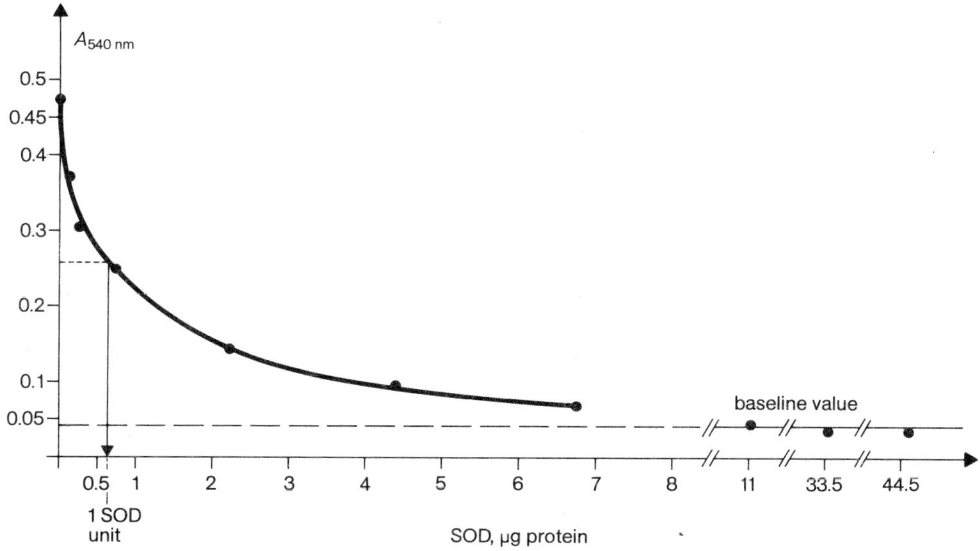

Fig. 1. Inhibition of nitrite formation by superoxide dismutase (SOD).

According to Fig. 1, in this example 50% inhibition of the colour formation (1 arbitrary unit) is caused by 0.65 µg enzyme protein. This corresponds to an absorbance at 540 nm of $A_{1/2} = 0.26$.

$$A_{1/2} = 0.5\,[(A_{540})_{max} + (A_{540})_{base\ line}] = 0.5\,(0.475 + 0.04) = 0.26$$

This arbitrary unit depends on the activity and quality of the commercial SOD preparation used. No link to International Units is possible (cf. pp. 295, 296).

Validation of Method

Precision, accuracy, detection limits and sensitivity: data about accuracy are not available since standard reference material is not yet established. The detectability is ca. ten times higher than in the xanthine oxidase-cytochrome c method [1].

Sources of error: metal chelates or other enzymes may mimic SOD (see above); thus, boiling of crude extracts or purified SOD for longer times (more than 2 min) may not always result in the loss of 100% of dismutase activity.

References

[1] *J. M. Mc Cord, I. Fridovich,* Superoxide Dismutase. An Enzymatic Function for Erythrocuprein (Hemocuprein). J. Biol. Chem. *244*, 6049 – 6055 (1969).
[2] *I. Fridovich,* Superoxide Dismutases. Annu. Rev. Biochem. *44*, 147 – 159 (1975).
[3] *K. Asada, S. Kanematsu,* Distribution of Cuprozinc, Manganic, and Ferric Superoxide Dismutases in Plants and Fungi: An Evolutionary Aspect, in: *H. Matsubara* and *T. Yamanaka* (eds.), Evolution of Protein Molecules, Japan Sci. Soc. Press, Tokyo 1978, pp. 361 – 372.
[4] *E. F. Elstner,* Oxygen Activation and Oxygen Toxicity. Annu. Rev. Plant Physiol. *33*, 73 – 96 (1982).
[5] *J. M. Mc Cord, J. D. Crapo, I. Fridovich,* Superoxide Dismutase Assays: A Review of Methodology, in: *A. M. Michelson, J. M. Mc Cord, I. Fridovich* (eds.), Superoxide and Superoxide Dismutases, Academic Press, London-New York-San Francisco 1977, pp. 11 – 17.
[6] *E. Lengfelder, E. F. Elstner,* Cyanide Insensitive Iron Superoxide Dismutase in *Euglena gracilis.* Comparison of the Reliabilities of Different Test Systems for Superoxide Dismutases, Z. Naturforsch. *34c*, 374 – 380 (1979).
[7] *R. E. Bensinger, Ch. M. Johnson,* Luminal Assay for Superoxide Dismutase, Anal. Biochem. *116*, 142 – 145 (1981).
[8] *A. Rigo, G. Rotilio,* Simultaneous Determination of Superoxide Dismutase and Catalase in Biological Materials by Polarography, Anal. Biochem. *81*, 157 – 166 (1977).
[9] *E. F. Elstner, A. Heupel,* Inhibition of Nitrite Formation from Hydroxylammoniumchloride: A Simple Assay for Superoxide Dismutase, Anal. Biochem. *70*, 616 – 620 (1976).
[10] *D. T. Sawyer, D. T. Richens, E. J. Nanni jr., M. D. Stallings,* Redox Reaction Chemistry of Superoxide Ion, in: *J. V. Bannister, H. A. O. Hill* (eds.), Chemical and Biochemical Aspects of Superoxide and Superoxide Dismutase, Elsevier/North-Holland, Amsterdam-New York-Oxford 1980, pp. 1 – 26.
[11] *T. N. Stein, C. L. Keen, B. Lönnerdal, L. S. Hurley,* Effect of Sample Preparation on Analysis of Superoxide Dismutase Activity and Isoenzymes, J. Inorg. Biochem. *16*, 71 – 77 (1982).

4 Transferases for One-carbon Groups and Ketonic Residues

4.1 Methionine Synthase

5-Methyltetrahydropteroyl-L-glutamate: L-homocysteine *S*-methyltransferase, EC 2.1.1.13

Hansjörg Sauer

General

Methionine synthase together with thymidylate synthase (cf. chapter 4.2) and thymidine kinase (cf. chapter 7.1) and other enzymes are so-called log-phase enzymes [1]. Their characteristic behaviour is an increased activity in the period of logarithmic growth of rapidly proliferating cell populations and a low activity during steady-state growth (the plateau phase). In accordance with this pattern, high enzyme activities were found in E. coli, leukaemia cells and developing brain [2–5]. As shown in Fig. 1 these three enzymes play a key role in the synthesis of DNA precursors and they are directly or indirectly dependent on folic acid derivatives as coenzymes.

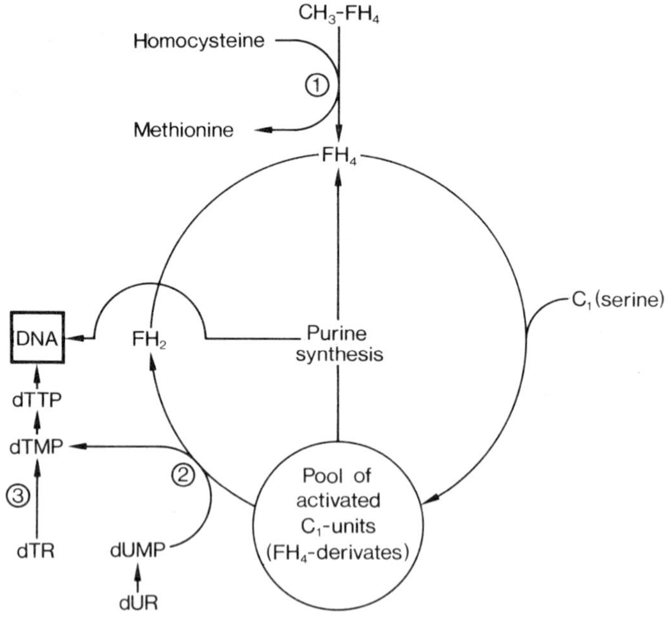

Fig. 1. Folic acid- and vitamin B_{12}-dependent steps in DNA synthesis: 1 = methionine synthase; 2 = thymidylate synthase; 3 = thymidine kinase. CH_3-FH_4 = ^{5}N-methyl-5,6,7,8-tetrahydrofolic acid, FH_4 = 5,6,7,8-tetrahydrofolic acid, C_1-unit = one-carbon unit, FH_2 = 7,8-dihydrofolic acid, dTR = deoxythymidine, dTMP = deoxythymidine monophosphate, dTTP = deoxythymidine triphosphate, dUR = deoxyuridine, dUMP = deoxyuridine monophosphate.

Methionine synthase is activated by a reducing system (flavine nucleotide) and by S-adenosyl L-methionine to produce free tetrahydrofolic acid and methionine from 5-methyl tetrahydrofolic acid and homocysteine. In addition, this enzyme activity is dependent on vitamin B_{12} which is bound in the active centre of the enzyme, generating the holo-enzyme from the apo-enzyme without vitamin B_{12}. The simultaneous dependence on both vitamins – folic acid and B_{12} – explains the disturbances in folate metabolism in the case of vitamin B_{12} deficiency (e.g. pernicious anaemia).

When methionine synthase is inactive, ^{5}N-methyl-tetrahydrofolic acid (which is the metabolically inactive storage form of the tetrahydrofolate coenzymes) cannot be demethylated to the free tetrahydrofolic acid, from which are generated the active coenzyme forms (pool of activated one-carbon units; see Fig. 1). Inactive ^{5}N-methyl-tetrahydrofolate accumulates and this leads to a functional folate deficiency (the methyl-trap-hypothesis) [6–8].

A shown in Fig. 2, deficiency of folic acid as well as of vitamin B_{12} results in a decreased supply of labile methyl groups which explains not only the megaloblastic transformation of DNA synthesis but also the impaired myelin synthesis (subacute combined degeneration of the spinal cord) [6, 9].

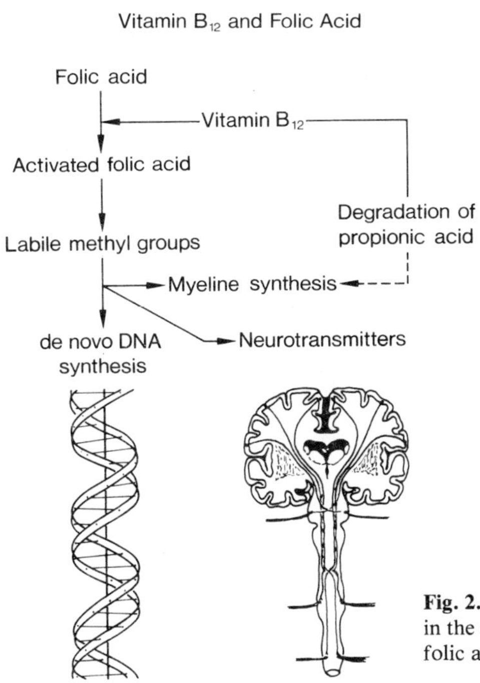

Fig. 2. Dependence of the metabolism of labile methyl groups in the synthesis of DNA and myelin on reactions catalyzed by folic acid and Vitamin B_{12}-dependent enzymes.

By oxidizing the methyl-cobalamin coenzyme of methionine synthase, nitrous-oxide (N_2O) inhibits the enzyme activity completely and can simulate vitamin B_{12} deficiency [10, 11]. The action of other enzyme inhibitors, such as adenosyl-methionine analogues, methyl-B_{12} analogues and other substrate analogues, is more specific [12, 13].

Application of method: in biochemistry and for clinical diagnostic procedures (pernicious anaemia, characterization of proliferating and non-proliferating cell populations).

Enzyme properties relevant in analysis: the enzyme activity is not detectable in extracellular fluids and blood serum. The activity is present in the cytosol of cells which are active in DNA synthesis or with significant metabolism of labile methyl groups (e.g. bone marrow, liver, kidney, brain). In vitro, substrate conversion proceeds linearly with time for at least 3 hours, after a short lag phase (co-operative effect) [14].

Methods of determination: enzyme activity can be assayed by estimating either the production of free tetrahydrofolic acid or the production of methionine [14–16].

International reference methods and standards: not available.

Enzyme effectors: the enzyme activity is initiated by a reducing system: reduced flavine nucleotide in vivo or dithiothreitol in vitro. *S*-adenosyl-L-methionine in catalytic amounts starts the reaction by the primary methylation of the cobalamin coenzyme (methyl-B_{12}).

Assay

Method Design

Principle: the B_{12}-holo-enzyme methionine synthase catalyzes the synthesis of 5,6,7,8-tetrahydrofolic acid (FH_4) from ^{5}N-methyl 5,6,7,8-tetrahydrofolic acid (methyl-FH_4) and homocysteine, in the presence of a reducing system and *S*-adenosyl-L-methionine (SAM):

$$[^{14}C]\text{Methyl-}FH_4 + \text{homocysteine} \longrightarrow FH_4 + [^{14}C]\text{ methionine}.$$

The use of a ^{14}C radioactive label in the transferred methyl group yields the radioactive product methionine which can easily be separated from the radioactive substrate by ion exchange.

Optimized conditions for measurement: for enzyme (cytosol) from mammalian sources as described below. With high concentrations of the reductant, dithiothreitol, no reduced flavine nucleotide is needed for the enzyme reaction. Together with the aquocobalamin in the assay mixture, the enzyme is reduced by the *Peel*-effect [14].

Temperature conversion factors: not tested; not available in the literature.

Equipment

Tissue homogenizers, water-bath, stopwatch, nitrogen gas, ion-exchange columns (see below), facilities to process radioactive material (^{14}C), scintillation counter.

Reagents and Solutions

Purity of reagents: all reagents must be of analytical grade (p. a.).

Preparation of solutions (for about 100 determinations): all solutions in re-purified water (cf. Vol. II, chapter 2.1.3.2).

1. Potassium phosphate buffer (0.5 mol/l; pH 7.5):

   a) dissolve 6.8 g KH_2PO_4 ad 100 ml with water,
   b) dissolve 8.7 g K_2HPO_4 ad 100 ml with water,
   mix a) and b) to pH 7.5 (ca. 15 ml a and 80 ml b). Store at 4 °C.

2. Dithiothreitol, DTT (0.3 mol/l):

   dissolve 46.2 mg DTT ad 1 ml with water (freshly prepared every day).

3. Vitamin B_{12}, aquocobalamin acetate (10.5 mol/l):

   for stock solution, 10^{-4} mol/l, dissolve 6.9 mg vitamin B_{12} ad 5 ml with water (store at −20 °C).
   for assay solution, 10^{-5} mol/l, (freshly prepared every day) dilute stock solution tenfold.

4. Homocysteine (0.05 mol/l):

   preparation from DL-homocysteine thiolactone [14].
   Dissolve 0.384 g DL-homocysteine thiolactone hydrochloride with water. Add quickly 3 ml NaOH, 1 mol/l (makes pH 12). After 30 min under nitrogen atmosphere, add 0.9 ml HCl, 1 mol/l, to neutralize and add water to a volume of 25 ml. Add 25 ml potassium phosphate buffer, 0.1 mol/l, pH 7.5 (fivefold dilution of solution (1)). Store aliquots of 1 ml in glass ampoules under nitrogen at −20 °C in the dark.

5. S-adenosyl L-methionine (0.01 mol/l):

   dissolve 5 mg SAM · H_2SO_4 to 1 ml with water. Store at −20 °C.

6. [^{14}C]Methyl-FH_4 (0.01 mol/l; 7.316 µCi/ml):

   for stock solution dissolve 50 µCi of the radioactive material in 2 ml water and add 20 µl of 2-mercaptoethanol. Store at −20 °C in the dark.

for assay solution dissolve 5 mg non-radioactive methyl-FH_4 in 0.7 ml water, add 0.3 ml radioactive stock solution, 0.01 ml 2-mercaptoethanol, 0.005 ml NaOH, 1 mol/l (to achieve complete solubility). Store at $-20\,°C$ in the dark.

7. Reagent mixture for 12 assays:

   mix 0.455 ml solution (1), 0.13 ml solution (2), 0.13 ml solution (3), 0.39 ml solution (4), 0.065 ml solution (5), and 0.13 ml solution (6).

8. *Bray*'s solution for scintillation counting [17]:

   dissolve 4 g 2.5-diphenyl-oxazole and 200 mg 1.4-bis-2-(4-methyl-5-phenyl-oxazole)benzene in 20 ml ethylene glycol and 100 ml methanol, add 60 g naphthalene and make to 1 litre with dioxane. Store at 4 °C in the dark.

9. Control enzyme solution (acc. to [1]):

   permanently growing human lymphoblasts are cultured in RPMI 1640 medium with 20% foetal calf serum at 37 °C in a humidified atmosphere with 5% CO_2. Harvest cells during the logarithmic growth period of the cultures and wash three times with 0.9% NaCl (centrifugations for 5 min at 800 *g*). After the last centrifugation adjust cell counts to 50000/µl with potassium phosphate buffer, 1 mmol/l, pH 7.5. In this hypotonic buffer the cells are lysed at 0 °C within 1 h. Centrifuge for 5 min at 10000 *g*; the supernatant serves as the control enzyme.
   In principle, every other rapidly growing cell culture can also be used to prepare such a control enzyme. If no cultures are available, preparations from rat or mouse liver are also suitable: mechanically homogenize fresh or frozen liver with 3 volumes of the hypotonic phosphate buffer. After centrifugation for 15 min at 50000 *g* the supernatant can be used as the control enzyme.

Stability of solutions: solutions (2) and (3) must be prepared freshly for each new series of tests. Solution (5) should be replaced every month by a new preparation. All other solutions are stable for several months if stored as indicated.

Ion-exchange columns: small glass columns with an inner diameter of 4 – 5 mm are packed with a 5 mm high bed of *Dowex*-1 × 8(Cl^-) mesh 200 – 400 (*Serva* 41111). It must be assured that none of the resin can escape during the elution procedure! Before every new separation the columns are regenerated with 3 × 0.5 ml HCl, 6 mol/l. Then the exchanger is washed to a neutral pH with water (test with indicator paper!). Columns can be used for several months.

Procedure

Collection and treatment of specimen: cytosol preparations from cell suspensions (cultured cells, leukaemia cells) are prepared according to *Wilmanns* [18] and from solid tissues according to *Sauer* [14].

Stability of the enzyme in specimen or sample: in tissues, cell suspensions or cytosol stored at −40 °C the enzyme activity is stable for up to 2 years. At room temperature the total enzyme activity of tissues or cytosol is lost within 24 hours.

Details for measurement in tissues: tissues and cytosol preparations should not contain more than small amounts of haemoglobin because of its non-specific inhibition of the enzyme activity (possibly due to inactivation of the reducing system) (H. Sauer, unpublished).

Assay conditions: assay mixture containing vials (*Eppendorf* No. 3500924) must have an air-proof cap and must be filled with nitrogen (or argon). Water-bath 37 °C (for mammalian enzyme sources). Incubation in the dark (1 h). A reagent blank with water instead of sample is necessary. Each series is accompanied by controls containing 0.1 ml control enzyme solution (9).

A radioactive standard is made up by mixing 0.03 ml of the blank, 1.5 ml water and 10 ml *Bray*'s solution (8).

Measurement

Incubation

Pipette into reaction vials:		sample	enzyme control	reagent blank	concentration in incubation mixture
reagent mixture	(7)	0.1 ml	0.1 ml	0.1 ml	phosphate 88 mmol/l DTT 14 mmol/l vitamin B_{12} 5×10^{-4} mmol/l homocystein 7.5 mmol/l SAM 2.5×10^{-4} mmol/l methyl-FH_4 5×10^{-4} mmol/l (0.37 mCi/l)
sample		0.1 ml	–	–	volume fraction 0.5
control enzyme solution	(9)	–	0.1 ml	–	
water		–	–	0.1 ml	
incubate for 1 h at 37 °C under nitrogen in the dark; stop reaction by inserting the vials for 2 min into a boiling water-bath. Centrifuge 2 min at 10000 *g*.					

Radiometric measurement: apply 0.1 ml of each supernatant to a *Dowex* column, wash three times with 0.5 ml water, collect eluates in scintillation counter vials and add 10 ml *Bray*'s solution (8). Count for 4 min at 10 °C in a liquid scintillation counter.

Calculation: no correction for counting efficiency according to eqn. (z), chapter 3.9.3, Vol. I, is necessary, since a radioactive standard is used. Therefore, the following simplified equation is valid for calculation of the catalytic activity concentration in the sample. The counting rate of the sample (and of the enzyme control) must be corrected for the blank counting rate, yielding $\Delta(C_n)_{sample}$.

$$b = \frac{\Delta(C_n)_{sample} \times 0.03}{C_{standard} \times \Delta t \times v} \quad U/l$$

where C is the counting rate, $\Delta(C_n)_{sample} = C_{sample} - C_{blank}$, Δt is the reaction time in min, v is the sample volume in l. The factor 0.03 derives from correction for the volume of standard used for counting (× 0.3) and from the content of $[^{14}CH_3]FH_4$ in the assay mixture (× 0.1 µmol).

To obtain the catalytic activity content, z_c/m_s (specific catalytic activity) of the sample, related to the mass of protein, the following equation is valid:

$$z_c/m_s = \frac{b}{\rho_{protein}} \quad U/g$$

where $\rho_{protein}$ is the protein concentration of the sample in g/l.

The catalytic activity content, related to the number of cells, is calculated accordingly.

Validation of Method

Precision, accuracy, detection limits and sensitivity: there are no systematic data available concerning the precision and accuracy of the method described. However, the test seems to have good reproducibility. The detection limits are about 17 mU/g protein in crude extracts of tissues and cell suspensions. Even in this region of low enzyme activity the sensitivity is good because of the very specific measurement of the reaction. Under standard conditions, as described, no radioactive product other than (^{14}C) methionine is found as a reaction product in the eluates from the columns [14].

Sources of error: falsely high activities are measured if the *Dowex*-1 columns are not sufficiently regenerated and neutralized immediately before the separation of a new sample, or if any of the radioactively-loaded resin particles get into the scintillation vials.

Specificity: the reaction is very specific. No radioactive substances other than (^{14}C)-methionine are found in the eluates from the columns [14].

Reference ranges

Rat liver: 83 mU/g protein,
Pig liver: 150 mU/g protein,
Rat kidney: 330 mU/g protein,

Normal human leukocytes: 1 mU/10^{10} cells,
Normal human bone marrow: 5 – 10 mU/10^{10} cells.

References

[1] *H. Sauer, K. Wilms, W. Wilmanns, L. Jaenicke*, Die Aktivität der Methionin-Synthetase (5-Methyl-5,6,7,8-tetrahydrofolsäure: Homocystein Methyltransferase) als Proliferationsparameter in wachsenden Zellen, Acta Haematol. *48*, 200 – 210 (1973).

[2] *A. Paessens, H. Rüdiger*, Purification and chemical characterization of the vitamin-B_{12}-dependent 5-methyltetrahydrofolate: homocysteine methyltransferase from Escherichia coli B, Eur. J. Biochem. *112*, 47 – 51 (1980).

[3] *Y. Kano, S. Sakamoto, K. Hida, A. Tsuboyama, F. Takaku*, Vitamin B_{12}-dependent methyltetrahydrofolate: homocysteine methyltransferase activity in normal and leukemic human hematopoietic cells, Clin. Chim. Acta *109*, 96 – 76 (1981).

[4] *S. A. Suleiman, R. Spector*, Methionine synthetase in mammalian brain: function, development and distribution, Life Sciences *27*, 2427 – 2432 (1980).

[5] *I. Bloos, H. Sauer*, Nachweis und Aktivität der Methioninsynthetase in normalen und pathologischen Zellen des Blutes und des Knochenmarks, Klin. Wochenschr. *50*, 991 – 994 (1972).

[6] *H. Sauer, W. Wilmanns*, Cobalamin dependent methionine synthesis and methyl-folate-trap in human vitamin B_{12} deficiency, Br. J. Haematol. *36*, 189 – 197 (1977).

[7] *J. M. Scott, D. G. Weir*, The methyl folate trap, Lancet *1981/II*, 337 – 340 (1981).

[8] *K. Fujii, T. Nagasaki, F. M. Huennekens*, Vitamin B_{12}-dependent replication of L 1210 mouse leukemia cells, J. Biol. Chem. *256*, 10329 – 10334 (1981).

[9] *J. M. Scott, J. J. Dinn, P. Wilson, D. G. Weir*, Pathogenesis of subacute combined degeneration: a result of methyl group deficiency, Lancet *1981/II*, 334 – 337 (1981).

[10] *R. Deacon, M. J. Lumb, J. Perry*, Vitamin B_{12}, folate and nitrous oxide, Med. Lab. Sci. *39*, 171 – 178 (1982).

[11] *H. Kondo, M. L. Osborne, J. F. Kolhouse, M. J. Binder, E. R. Podell, C. S. Utley, R. S. Abrams, R. H. Allen*, Nitrous oxide has multiple deleterious effects on cobalamin metabolism and causes decreases in activity of both mammalian cobalamin-dependent enzymes in rats, J. Clin. Invest. *67*, 1270 – 1283 (1981).

[12] *T. Ishida, H. Morimoto, M. Inoue, T. Fujiwara, K.-I. Tomita*, Three-dimensional X-ray crystal structure of S-adenosyl-L-homocysteine, a potent inhibitor of S-adenosylmethionine-dependent methyltransferases, J. C. S. Chem. Commun., 671 – 673 (1981).

[13] *F. M. Huennekens, P. M. DiGirolamo, K. Fujii, D. W. Jacobson, K. S. Vitols*, B_{12}-dependent methionine synthetase as a potential target for cancer chemotherapy, in: *G. Weber* (ed.); Advances in Enzyme Regulations, Vol. 14, Pergamon Press, Oxford 1976.

[14] *H. Sauer, L. Jaenicke*, Einfacher Test zur Messung der Methionin-Synthetase-(MS)-Aktivität und seine Anwendungsmöglichkeiten in der Klinik, Klin. Wochenschr. *50*, 986 – 990 (1972).

[15] *H. Rüdiger, L. Jaenicke*, On the role of S-adenosyl-methionine in the vitamin B_{12}-dependent methionine biosynthesis, Eur. J. Biochem. *10*, 557 – 560 (1969).

[16] *H. Weissbach, A. Peterkofsky, B. G. Redfield, H. Dickermann*, Studies on the terminal reaction in the biosynthesis of methionine, J. Biol. Chem. *238*, 3318 – 3324 (1963).

[17] *G. A. Bray*, A simple efficient liquid scintillator for counting aqueous solutions in a liquid scintillation counter, Anal. Biochem. *1*, 179 – 185 (1960).

[18] *W. Wilmanns*, Thymidin-Kinase in normalen und leukämischen myeloischen Zellen, Klin. Wochenschr. *45*, 505 – 511 (1967).

4.2 Thymidylate Synthase

5,10-Methylene-tetrahydrofolate: dUMP *C*-methyltransferase, EC 2.1.1.45

Hansjörg Sauer and Wolfgang Wilmanns

General

Thymidylate synthase is a key enzyme for the *de novo* synthesis of DNA [1 – 3]. By methylation of deoxyuridine monophosphate it produces thymidine monophosphate which, after phosphorylation to thymidine triphosphate, is a substrate for the DNA polymerases (cf. chapter 4.1, page 304). High activities are found in rapidly proliferating cells (log-phase enzyme) [2 – 6]. The physiological methyl donors are the polyglutamate forms of 5,10-methylene-tetrahydrofolate [7 – 9].

5-Fluorouracil (after activation to 5-fluorodeoxyuridine monophosphate) specifically inhibits the enzyme activity. 5-Fluorouracil is an established cytostatic drug in clinical use [10 – 12]. Other substrate analogues are described as inhibitors [13, 14]; pyridoxal 5-phosphate (vitamin B6) protects the enzyme from 5-fluorodeoxyuridylate [15].

Application of method: in biochemistry for clinical diagnostic procedures (folate deficiency, pernicious anaemia, characterization of proliferating and non-proliferating cell populations).

Enzyme properties relevant in analysis: the enzyme activity is not detectable in extracellular fluids and in blood serum. The activity is present in the cytosol of cells which are active in DNA synthesis. *In vitro*, the substrate conversion proceeds linearly with time for at least 3 hours [6].

Methods of determination: besides a spectrophotometric method measuring the oxidation of tetrahydrofolate to dihydrofolate [16] radioactive tracer methods are mostly used [17, 18]. A practicable method makes use of a tritium marker at the 5-C-position of deoxyuridine monophosphate, which generates radioactive water (^{3}HOH) after methylation at this position [6].

International reference methods and standards: not available.

Enzyme effectors: the enzymatic methylation of deoxyuridine monophosphate is magnesium-dependent. During the reaction the substrate, 5,10-methylenetetrahydrofolate is oxidized to 7,8-dehydrofolate. Dihydrofolate reductase activity (cf. chapter 3.6, page 251) must be present in the assay mixture to maintain an adequate substrate concentration.

Assay

Method Design

Principle

1st step: generation of the substrate methylene-tetrahydrofolate (non-enzymatically) with formaldehyde hydrate

(a) 5,6,7,8-Tetrahydrofolate + $CH_2(OH)_2$

$$\longrightarrow \text{5,10-methylene-tetrahydrofolate} + 2\,H_2O$$

2nd step: primary enzymatic reaction

(b) 5,10-Methylene-tetrahydrofolate + [^{3}H]5-deoxyuridine monophosphate

$$\xrightarrow[\text{Mg}^{2+}]{\text{thymidylate synthase}} \text{7,8-dihydrofolate} + \text{deoxythymidine}$$

3rd step: enzymatic regeneration of tetrahydrofolate

(c) 7,8-Dihydrofolate + NADPH + H^+

$$\xrightarrow{\text{dihydrofolate reductase}} \text{5,6,7,8-tetrahydrofolate} + \text{NADP}^+$$

Crude extracts of cells (cytosol) contain enough dihydrofolate reductase activity to make reaction (c) not a limiting step in this cascade of reactions. Reaction (c) regenerates the substrate for reaction (a) to maintain the circle of reactions.

The substrate [^{3}H]5-deoxyuridine monophosphate can easily be separated from the product ^{3}HOH by adsorption to charcoal. The supernatant is assayed for [^{3}H]activity in a scintillation counter.

Optimized conditions for measurement: these are described below for the cytosolic enzyme from mammalian tissues. The assay is dependent on magnesium ions and reducing conditions (maintained by adding 2-mercaptoethanol). Optimal reaction conditions are obtained in Tris buffer, pH 7.2 at 37 °C.

Temperature conversion factors: not tested; not available from the literature.

Equipment

Tissue homogenizers, water-bath, stopwatch, centrifuges, facilities to process radioactive material (^{3}H), scintillation counter.

Reagents and Solutions

Purity of reagents: all reagents must be of analytical purity (p.a.).

Preparation of solutions (6): all solutions in re-purified water (cf. Vol. II, chapter 2.1.3.2).

1. Tris buffer (pH 7.2, 1.0 mol/l):

   dissolve 12.114 g Tris [tris(hydroxymethyl)aminomethane (*Merck* 8382)] in 50 ml water, adjust pH to 7.2 with HCl, 1 mol/l, then add water to 100 ml.

2. 2-Mercaptoethanol (3 mol/l):

   dilute 209 μl = 234 mg 2-mercaptoethanol (*Merck* 805740) with water to 1 ml.

3. K/Mg-EDTA (0.4 mol/l):

   dissolve 171 mg ethylenedinitrilo tetraacetic acid, dipotassium-magnesium salt (*Merck* 8409) to 1 ml with water.

4. Magnesium chloride ($MgCl_2$, 0.4 mol/l):

   dissolve 81 mg $MgCl_2 \cdot 6H_2O$ to 1 ml with water.

5. NADPH (0.06 mol/l):

   dissolve 50 mg NADPH, tetrasodium salt (*Boehringer Mannheim* 107816) to 1 ml with $KHCO_3$ solution (6).

6. Potassium hydrogen carbonate (0.05 mol/l):

   dissolve 5 mg $KHCO_3$ (*Merck* 4854) to 1 ml with water.

7. Formaldehyde (0.1 mol/l):

   dilute 7.94 μl formaldehyde solution (*Merck* 4003) to 1 ml with water.

8. Substrate solution ([^{3}H]dUMP, 0.1 mmol/l, 100 mCi/mmol):

   mix 10 μl [^{3}H]dUMP (10 Ci/mmol, 1 mCi/ml; *Amersham* TRK.287) and 990 μl dUMP (*Serva* 18625; 0.101 mmol/l = 230.5 μg with water to 1 ml).

9. D,L-Tetrahydrofolate* (FH_4, 3.3 mmol/l):

---

* Preparation of FH_4 according to *Jaenicke & Brode* [19]: 30 mg of platinum oxide (PtO_2) and 15 ml acetic acid (100%) are reduced under hydrogen gas in a hydrogenation apparatus (ca. 20 min). Black flocks indicate sufficient reduction. 100 mg of folic acid (*Merck* 3984) are added under hydrogen.

Further hydrogenation for about two hours generates FH_4 (grey-green colour). This solution is centrifuged and the supernatant is added to twice its volume of peroxide-free ether. The resulting precipitate is washed three times with ether and dried under vacuum. The light grey powder (FH_4) is stored at −20 °C in the dark.

10. Control enzyme solution:

    use preparation as described as solution (9) in chapter 4.1 on p. 308.

11. Charcoal-powder suspension:

    suspend 1.25 g Norit A (*Serva* 30890) to 10 ml with water.

Stability of solutions: solutions (1) – (4), (6), (7) and (10) are stored at 4 °C and are stable for several months. Solution (5) is freshly prepared for each series of tests. Solutions (8) and (9) are stored at −20 °C and are stable for several weeks.

Procedure

Collection and treatment of specimens: cytosol preparations from cell suspensions (culture cells, bone marrow cells, leukaemia cells) are prepared according to *Wilmanns* [20] and from solid tissues according to *Sauer* [21].

Stability of the enzyme in specimen or sample: in tissues, cell suspensions and cytosol stored at −40 °C the enzyme activity is stable for up to two years. The total enzyme activity of tissues or cytosol is lost within 24 hours at room temperature.

Details for measurement in tissues: the method is described for measurement in cell extracts (cytosol). The enzyme activity is not detectable in extracellular fluids or in blood serum.

Assay conditions: incubation volume 0.20 ml; temperature 37 °C. To prevent degradation of FH_4 the incubation must be carried out in the dark and under reducing conditions (2-mercaptoethanol). A reagent blank with water instead of sample is necessary. Each series is accompanied by controls containing 0.1 ml of an enzyme preparation from cultured human lymphoblasts [21] with a known thymidylate synthase activity (solution 10). A radioactive standard is made from 10 µl of a 1 : 10 dilution (with water) of solution (8) together with 190 µl water and 5 ml of *Bray*'s solution.

Prior to measurement the following reaction mixture is prepared.

Solution	*for 1 assay*	*for 12 assays (× 13)*
(1) – (5), (7) – (9)	0.010 ml each	0.130 ml each
water	0.020 ml	0.260 ml

Measurement

Enzyme reaction

Pipette successively into reaction vials:	sample	enzyme control	reagent blank	concentration in incubation mixture
reaction mixture	0.10 ml	0.10 ml	0.10 ml	Tris 50 mmol/l mercapto-ethanol 150 mmol/l EDTA 20 mmol/l Mg^{2+} 20 mmol/l HCHO 5 mmol/l NADPH 3 mmol/l dUMP 0.005 mmol/l FH_4 0.165 mmol/l volume fraction 0.5
sample	0.10 ml	–	–	
control enzyme solution (10)	–	0.10 ml	–	
water	–	–	0.10 ml	
incubate in the dark for 1 h				
charcoal suspension (11)	0.20 ml	0.20 ml	0.20 ml	
enzyme reaction is stopped. Mix for 2 min, centrifuge 2 min at 10000 *g*. Use 0.2 ml of supernatant for radioactivity measurement.				

Measurement of tritium activity

For assay of tritium radioactivity mix 0.2 ml supernatant of incubation mixture after centrifugation with 5 ml of *Bray*'s solution (cf. chapter 4.1, page 310) and measure in a scintillation counter.

Calculation: for calculation of the catalytic concentration of the sample (cytosol) the counting rate C of the sample must be corrected:

$$\Delta C = C_{\text{sample}} - C_{\text{blank}} \text{ cpm}.$$

Since

$C_{\text{standard}} \times 10$ corresponds to 1 nmol $[^3H]$dUMP or $[^3H]$dTMP, respectively, for one hour incubation time the catalytic concentration in the sample is

$$b = \Delta C \times 2 \times 10/C_{\text{standard}} \quad \text{nmol} \times 0.1\ \text{ml}^{-1} \times \text{h}^{-1}$$

or

$$b = (20 \times \Delta C/C_{\text{standard}}) \times 10/60 \quad \text{nmol} \times \text{ml}^{-1} \times \text{min}^{-1}$$

or

$$b = (\Delta C/C_{\text{standard}}) \times 3.33 \quad \text{U/l}$$

To obtain specific activity this value has to be divided by the amount of protein or cell-equivalent in the reaction vial.

Validation of Method

Precision, accuracy, detection limits and sensitivity: the precision of the test system is very good. The results of duplicate tests vary less than ± 5%. Accuracy can only be checked with an enzyme preparation from cultured human lymphoblasts which serves as internal standard.

The detection limits are about 1 nmol/min (10^{-3}U) per extract of 10^{10} cells from suspensions or per mg protein of crude extracts from tissues. Even in this region of very low enzyme activity, the sensitivity is good because of the very specific measurement of the reaction. Under standard conditions, as described, more than 99% of the non-transformed radioactive substrate [^{3}H]dUMP is adsorbed to the charcoal and no other radioactive product besides radioactive water (^{3}HOH) can be detected in the supernatant.

Sources of error: falsely high activities are measured if the [^{3}H]dUMP-charcoal is not sedimented completely by centrifugation.

Specificity: the reaction is very specific. No other radioactive product besides ^{3}HOH can be found in the supernatant.

Reference ranges [6]:

Normal human leukocytes	ca. 0.017 mU/10^{10} cells
human bone marrow (normal)	ca. 16 mU/10^{10} cells
human bone marrow (acute leukaemia)	ca. 21 mU/10^{10} cells
human bone marrow (pernicious anaemia)	ca. 29 mU/10^{10} cells

References

[1] *R. L. Blakley,* The Biosynthesis of Thymidylic Acid. IV. Further Studies on Thymidylate Synthetase, J. Biol. Chem. *238*, 2113–2118 (1963).

[2] *A. Fridland, R. J. Langenbach, C. Heidelberger,* Purification of Thymidylate Synthetase from Ehrlich Ascites Carcinoma Cells, J. Biol. Chem. *246*, 7110–7114 (1971).

[3] *J. Marquet, J. Zittoun, C. Weynants, R. Zittoun,* Thymidylate Synthesis in a Folate Deprived Cell Line, Br. J. Haematol. *49,* 87–95 (1981).
[4] *P. A. Bianchi,* Thymidine Phosphorylation and Deoxyribonucleic Acid Synthesis in Human Leukaemic Cells, Biochem. Biophys. Acta *55,* 547–551 (1962).
[5] *V. S. Gupta, J. B. Meldrum,* Purification and Properties of Thymidylate Synthetase from Pig Thymus, Canad. J. Biochem. *50,* 352–362 (1972).
[6] *W. Wilmanns, V. Neef,* Die Thymidylat-Synthetase in weißen Blutzellen und im Knochenmark unter normalen und pathologischen Bedingungen, Klin. Wochenschr. *49,* 755–762 (1971).
[7] *R. L. Kisliuk, Y. Gaumont, E. Lafer, C. M. Baugh, J. A. Montgomery,* Polyglutamyl Derivatives of Tetrahydrofolate as Substrate for Lactobacillus casei Thymidylate Synthase, Biochemistry *20,* 929–934 (1981).
[8] *G. F. Maley, F. Maley, C. M. Baugh,* Studies on Identifying the Folylpolyglutamate Binding Site of Lactobacillus casei Thymidylate Synthetase, Arch. Biochem. Biophys. *216,* 551–558 (1982).
[9] *D. G. Priest, K. K. Happel, M. Mandum, J. M. Bednarek, M. T. Doig, C. M. Baugh,* Tissue Folylpolyglutamate Chain-Length Characterization by Electrophoresis as Thymidylate Synthetase-Fluorodeoxyuridylate Ternary Complexes, Anal. Biochem. *115,* 163–169 (1981).
[10] *J. A. Houghton, C. Schmidt, P. J. Houghton,* The Effect of Derivatives of Folic Acid on the Fluorodeoxyuridylate-Thymidylate Synthetase Covalent Complex in Human Colon Xenografts, Eur. J. Cancer Clin. Oncol. *18,* 347–354 (1982).
[11] *C. A. Lewis, R. B. Dunlap,* Thymidylate Synthetase and Its Interaction with 5′-Fluoro-2′-Deoxyuridylate, Topics in Molecular Pharmacology, Vol. 1, Academic Press, Amsterdam 1981.
[12] *W. L. Wasthien,* Thymidylate Synthetase Levels as a Factor in 5-Fluorodeoxyuridine and Methotrexate Cytotoxicity in Gastrointestinal Tumor Cells, Mol. Pharmacol. *21,* 723–728 (1982).
[13] *E. DeClerq, J. Balzarini, C. T. C. Chang, C. F. Bigge, P. Kalaritis, M. P. Mertes,* 5(E)-(3-Azidostryl)-2′-Deoxyuridine 5′-Phosphate is a Photoactivated inhibitor of Thymidylate Synthetase, Biochem. Biophys. Res. Commun. *97,* 1068–1075 (1980).
[14] *B. J. Dolnick, Y. C. Cheng,* Human Thymidylate Synthetase. II. Derivatives of Pteroylmono- and Polyglutamates as Substrates and Inhibitors, J. Biol. Chem. *253,* 3563–3567 (1978).
[15] *G. P. Tryfiates, H. P. Morris, G. P. Sonidis,* Vitamin B_6 and Cancer (Review), Anticancer Res. *1,* 263–268 (1981).
[16] *A. J. Whaba, M. Friedkin,* Direct Spectrophotometric Evidence for the Oxidation of the Tetrahydrofolate During the Enzymatic Synthesis of Thymidylate, J. Biol. Chem. *236,* PC 11–12 (1961).
[17] *R. D. Armstrong, R. B. Diasie,* Improved Measurement of Thymidylate Synthetase Activity by a Modified Tritium-Release Assay, J. Biochem. Biophys. Meth. *6,* 141–148 (1982).
[18] *P. A. Charlton, D. W. Young,* A Chemical Model for the Thymidylate Synthetase Catalysed Methylation of Deoxyuridine Monophosphate, J. Chem. Soc. No. *6,* 1982, 1363–1370.
[19] *L. Jaenicke, E. Brode,* Untersuchungen über Einkohlenstoffträger. I. Die Tetrahydrofolat-Formylase aus Taubenleber. Reinigung und Mechanismus, Biochem. Z. *334,* 108–132 (1961).
[20] *W. Wilmanns,* Thymidin-Kinase in normalen und leukämischen myeloischen Zellen, Klin. Wochenschr. *45,* 505–511 (1967).
[21] *H. Sauer, L. Jaenicke,* Einfacher Test zur Messung der Methionin-Synthetase-(MS)-Aktivität und seine Anwendungsmöglichkeiten in der Klinik, Klin. Wschr. *50,* 986–990 (1972).

4.3 Ornithine Carbamoyltransferase

Carbamoyl phosphate : L-ornithine carbamoyl transferase, EC 2.1.3.3

4.3.1 Measurement in Serum

Giovanni Ceriotti

General

Ornithine carbamoyltransferase (OCT) was discovered by *Grisolia & Cohen* [1] in 1952 and purified from liver. It catalyzes the transfer of an "active" carbamoyl group to the δ-amino group of ornithine in the urea cycle of *Krebs & Henseleit*, thus forming citrulline. In mammals it is confined almost exclusively to liver mitochondria and is present only in small amounts in the intestine [3]. It has not been found in avian liver, but does occur in many micro-organisms [3] and in plants [4].

Owing to the high OCT concentration in liver and its normally very low level in serum, cytolytic and mitochondrial changes in the hepatocytes due to infections, degenerative or toxic factors, give rise to large increases of activity in serum. The assay of OCT activity in serum is therefore a good diagnostic and prognostic tool in various liver diseases [5 – 7]; it is also useful for testing and monitoring liver toxicity or possible chronic toxicity of new drugs.

Application of method: in clinical chemistry, biochemistry, general microbiology, experimental pathology, toxicology and pharmacology [8].

Enzyme properties relevant in analysis: OCT activity apparently does not require coenzymes [1]. It is inhibited by substances blocking SH groups [2]; it is not influenced by haemolysis. The existence of isoenzymes has not yet been demonstrated. Substrate conversion proceeds linearly with time [9].

Methods of determination: two kinds of assay can be used. The first is based on the arsenolytic activity of the enzyme: in the presence of arsenate, citrulline is converted to ornithine + CO_2 + NH_3. The CO_2 can be measured by the isotope method of *Reichard* [10], or the NH_3 can be determined by diffusion and subsequent colorimetric (*Berthelot*) reaction [11]. The second method measures the transferase reaction by colorimetric determination of citrulline [9, 12, 13]. Estimation of enzyme mass or active-site titration is not in use.

International reference methods and standards: neither standardization at the international level, nor the existence of reference standard materials is known up to now.

Enzyme effectors: no special activators of OCT to be added to the assay mixture are known.

Assay

Method Design

Principle

(a) Carbamoyl-P + ornithine $\xrightleftharpoons{\text{OCT}}$ citrulline + P_i.

The equilibrium of the reaction lies far on the side of citrulline [14]. The amount of citrulline formed per unit time is a measure of the OCT activity.

Citrulline is determined by the highly sensitive diacetylmonoxime-antipyrine reaction [15]. Urea in the sample, which would interfere, is destroyed at the same time by urease. Individual sample blanks are necessary to correct for the citrulline already present (preformed or basal citrulline).

Optimized conditions for measurement: the favourable position of the equilibrium permits the assay to be carried out under optimum experimental conditions. The most important variables to be controlled are the type of buffer, the pH and the respective concentrations of the two substrates, ornithine and carbamoyl phosphate. With some commonly used buffers, namely glycylglycine and Tris, activity increases with increasing pH; however, the blank also increases because carbamoyl phosphate decomposes, forming chromogens non-enzymatically and thus limiting the possibility of increasing the carbamoyl phosphate concentration to its optimum [16–18]. As demonstrated by *Snodgrass* [16, 17] ornithine, even at rather low concentration, exerts an inhibitory action on OCT which varies with the pH, being a function of the "zwitterion" of ornithine. Phosphate buffer used previously [19], with maximum activity at pH 7.0, avoids many of the above cited drawbacks: however, it is somewhat inhibitory. When working with serum, it is also necessary to take into consideration the optimum pH for urease activity. A compromise among all these variables has to be sought to optimize the test conditions. This has been reached [9] by working at pH 7.0 with barbiturate-acetate buffer, increasing the concentration of carbamoyl phosphate to the maximum that does not give rise to unspecific chromogens, and decreasing the ornithine concentration to below the inhibitory level. A slightly higher activity is observed with barbiturate-acetate buffer at pH 6.75. However, pH 7.0 is preferred to favour the activity of urease.

The optimization of the enzymatic reaction conditions allows the use of very small volumes of serum. This facilitates the elimination of protein interference in the chromogenic reaction for citrulline determination. By simply decreasing the sulphuric acid concentration in the chromogenic reagent it is thus possible to perform the reaction directly, without deproteinization.

The use of an incubation temperature of 37 °C is necessary to be able to use sufficiently small volumes of serum.

Under the conditions of the test the OCT activity increases linearity up to 52 °C.

Temperature conversion factors: the factor for conversion from 37 °C to 30 °C is 1.74.

Equipment

Normal laboratory equipment; 37 °C water-bath; boiling water-bath for the colorimetric reaction; micropipettes; specrophotometer capable of accurate measurements at 460 nm.

Reagents and Solutions

Purity of reagents: even in the crystalline state, carbamoyl phosphate must be stored in the cold in dry conditions; moisture is especially dangerous (cf. also chapter 4.3.2 on p. 328).

Preparation of solutions (for about 70 determinations): all solutions in re-purified water (cf. Vol. II, chapter 2.1.3.2).

1. Barbiturate/acetate buffer (70 mmol/l; pH 7.0):

   solution a: dissolve 9.715 g sodium acetate ($3\,H_2O$) and 14.715 g barbiturate, Na salt (sodium veronal) in water and make up to 1 l. Solution b: HCl, 50 mmol/l. Working solution: to 20 ml of solution a add 24 ml of solution b and make up to 100 ml with water. Check the pH with a glass electrode and, if necessary, adjust it to the exact value.

2. Urease (2500 U/l):

   dissolve 5 mg urease (5 U/mg, 30 °C) in 10 ml buffer (1).

3. Substrate/urease solution (carbamoyl-P, 23 mmol/l; ornithine, 2.5 mmol/l; urease, 2500 U/l):

   dissolvc 17.5 mg carbamoyl-P dilithium salt 4.0 mg ornithine hydrochloride in 5 ml urease solution (2).

4. Citrulline standard solution (10 mmol/l):

   dissolve 17.5 mg citrulline in 10 ml water; store in a refrigerator. Immediately before use dilute 1 ml to 10 ml with water (1 mmol/l).

5. Antipyrine/ferric sulphate solution:

   add carefully and slowly with cooling 100 ml conc. H_2SO_4 to about 500 ml water. Dissolve in this solution 50 mg $Fe_2(SO_4)_3 \cdot 9\,H_2O$ and 4 g antipyrine; after cooling, dilute to 1000 ml with water.

6. Acetic acid (5% v/v):

   dilute 50 ml acetic acid with distilled water to 1000 ml.

7. Diacetylmonoxime (0.5% w/v):

   dissolve 5.0 g diacetylmonoxime in 1000 ml acetic acid (6).

8. Chromogenic reagent:

   immediately before use, mix equal volumes of solutions (5) and (7).

Stability of solutions: solutions (2) and (3) are unstable; therefore, they must be prepared immediately before use. Portions may be stored frozen in tightly stoppered tubes and thawed just before use. Solutions (5) and (7) are stable indefinitely, even at room temperature. The chromogenic reagent (8) must be prepared just before use. It is stable at room temperature for about 2 h.

Procedure

Collection and treatment of sample: the assay is usually carried out on serum from venous blood. Plasma from blood to which oxalate, citrate, EDTA or heparin have been added can also be used. Haemolysis does not interfere.

Stability of the enzyme in the sample: according to *Reichard & Reichard* [11] the enzyme is stable in serum for at least a week at 4 °C, and at −15 °C for longer than a year. In our experience, the enzyme is stable for at least 7 days at 25 °C, when stored under sterile conditions.

Assay conditions: incubation temperature 37 °C; incubation volume 75 µl. Colorimetric reaction in boiling water-bath, wavelength 460 nm; light path 10 mm; final volume 3.075 ml; room temperature. Read against blank containing water and substrate (3) instead of sample.

The citrulline already present in the sample before the enzymatic reaction ("basal citrulline" tube) must be determined separately; read against blank containing water and urease (2) instead of sample.

To measure the absorption coefficient or as reference standard, add 25 µl diluted citrulline standard solution (4) (1 mmol/l) equivalent to 25 nmol citrulline instead of sample.

Read against a blank containing water and buffer (1) instead of sample.

Concentrations in incubation mixture

Barbiturate/acetate buffer	9.3 mmol/l
Carbamoyl phosphate	15.3 mmol/l
Ornithine	3.1 mmol/l
Urease	1667 U/l
Volume fraction of sample	0.30

Concentrations in colour reaction

Antipyrine	10.2×10^{-3} mol/l
$Fe_2(SO_4)_3$	43×10^{-6} mol/l
Diacetylmonoxime	24.1×10^{-3} mol/l
Acetic acid	0.427 mol/l
H_2SO_4	0.878 mol/l

Measurement

Incubation

Pipette onto the bottom of test tubes:	total citrulline	blank	basal citrulline	blank basal citrulline	citrulline standard	blank citrulline standard
sample	25 µl	–	25 µl	–	–	–
substrate/ urease (3)	50 µl	50 µl	–	–	–	–
urease solution (2)	–	–	50 µl	50 µl	–	–
citrulline solution (4)	–	–	–	–	25 µl	–
water	–	25 µl	–	25 µl	–	25 µl
buffer (1)	–	–	–	–	–	50 µl
mix carefully and completely by gentle rotation; incubate in a covered water-bath at 37 °C for exactly 30 min.						

Colour reaction

Add to each test tube 3 ml chromogenic reagent (8), mix with a vortex mixer (do not touch the rim of the test tubes with fingers because human skin contains urea), place for 15 min in a boiling water-bath, cool under running water and read the absorbance.

If the absorbance is > 0.800 repeat the procedure with diluted sample.

Calculation: activity is expressed as µmoles of citrulline produced by 1 l of serum per min at 37 °C. Use calculation formula (k) or (k_1), respectively, in combination with eqn. (d_1) (cf. "Formulae", Appendix 3).

Subtract the appropriate blanks from the readings of "total citrulline", "basal citrulline" and "citrulline standard", yielding A_{total}, A_{basic}, $A_{standard}$. For calculation $A_{total} - A_{basic} = \Delta A$ is related to $A_{standard}$. Reaction time is 30 min, concentration of standard is 1000 µmol/l.

The following relations are valid:

$$b = \frac{\Delta A}{A_{standard} \times 30} \times 1000 = 33.33 \times \frac{\Delta A}{A_{standard}} \quad \text{U/l}$$

$$b = 555.5 \times \frac{\Delta A}{A_{standard}} \quad \text{nkat/l}\,.$$

Validation of Method

Precision, accuracy, detection limits and sensitivity: the coefficient of variation ranges from about 6% for low values to about 2% for moderate or high activities. Detection limits are 2 U/l. Sensitivity (discriminative capability) is 0.6 U/l corresponding to $\Delta A = 0.005$, using a high-quality spectrophotometer.

Sources of error: no effects of therapeutically administered drugs have been observed. Contamination of the assay mixture by urea from the skin of the fingers or incomplete destruction of the urea in the serum due to inactive urease can give false values. Under the conditions described here, the urease activity is in excess and can hydrolyze even high urea concentrations in serum, and the ammonia produced is efficiently buffered. The urease activity of the preparation should be checked before use in the assay. No interference has been observed with icteric or with haemolyzed sera. When a very intense lipaemia is present, a complete clarification may not be achieved. However, even in extreme conditions, the determination is possible because the interference is the same for the test and the blank and is therefore eliminated in the calculation of ΔA. The most important possible source of error is instability of carbamoyl phosphate.

Owing to the high stability of the enzyme in serum, it is possible and advisable to run a standard serum with a carefully determined activity, stored frozen in small aliquots, with each set of determinations.

Specificity of the method: the reaction is specific for citrulline in the absence of urea.

Reference ranges: under the conditions described, the range of normal values extends from 5 to 15 U/l.

References

[1] *S. Grisolia, P. P. Cohen,* The Catalytic Role of Carbamyl Glutamate in Citrulline Biosynthesis, J. Biol. Chem. *198*, 561 – 571 (1952).

[2] *H. A. Krebs, K. Henseleit,* Untersuchung über die Harnstoffbildung im Tierkörper, Z. Physiol. Chem. *210*, 33 – 66 (1932).

[3] *H. A. Krebs, L. V. Eggleston, V. A. Knivett,* Arsenolysis and Phosphorolysis of Citrulline in Mammalian Liver, Biochem. J. *59*, 185 – 193 (1955).

[4] *K. Kleczkowski, P. P. Cohen,* Purification of Ornithine Transcarbamylase from Pea Seedlings, Arch. Biochem. Biophys. *107*, 271 – 278 (1964).

[5] *H. Reichard,* Studies on Ornithine Carbamoyl Transferase Activity in Blood Serum, Acta Med. Scand. *172*, Suppl. 390 (1962).

[6] *G. F. Moretti, J. Staeffen, P. Ballan, G. Catanzanox,* L'ornithine carbamoyl transférase du serum (OCT/s) en pathologie hépato-biliaire, Presse Méd. *72*, 325 – 328 (1964).

[7] *G. Ceriotti,* Il significato diagnostico della determinazione della ornitina-carbamil transferasi, Terapia *59*, 94 – 106 (1974).

[8] *L. Bonanomi, A. M. De Paoli, A. Gazzaniga,* Behaviour of Ornithinecarbamoyl Transferase (OCT) in the Toxicological Study of New Drugs, Boll. Chim. Farm. *111*, 443 – 449 (1972).

[9] *G. Ceriotti,* Optimal Conditions for Ornithine Carbamyl Transferase Determination. A Simple Micromethod without Deproteinization, Clin. Chim. Acta *47*, 97 – 105 (1973).

[10] *H. Reichard,* Determination of Ornithine Carbamoyl Transferase in Serum, A Rapid Method, J. Clin. Lab. Med. *63*, 1061 – 1064 (1964).

[11] *H. Reichard, P. Reichard,* Determination of Ornithine Carbamyl Transferase in Serum, J. Lab. Clin. Med. *52*, 707 – 717 (1958).

[12] *R. W. Brown, S. Grisolia,* Ornithine Transcarbamylase Activity in Serum, J. Lab. Clin. Med. *54*, 617 – 620 (1959).

[13] *A. A. Vassef,* Direct Micromethod for Colorimetry of Serum Ornithine Carbamoyl Transferase Activity, with Use of a Linear Standard Curve, Clin. Chem. *24*, 101 – 107 (1978).

[14] *G. H. Burnett, P. P. Cohen,* Study of Carbamyl Phosphate Ornithine Transcarbamylase, J. Biol. Chem. *229*, 337 – 344 (1957).

[15] *G. Ceriotti, L. Spandrio,* A Spectrophotometric Method for Determination of Urea, Clin. Chim. Acta *8*, 295 – 299 (1963).

[16] *P. J. Snodgrass,* The Effects of pH on the Kinetics of Human Liver Ornithine Carbamyl Phosphate Transferase, Biochemistry *7*, 3047 – 3051 (1968).

[17] *P. J. Snodgrass, D. J. Parry,* The Kinetics of Serum Ornithine Carbamoyl Transferase, J. Lab. Clin. Med. *73*, 940 – 950 (1969).

[18] *G. Ceriotti, A. Gazzaniga,* La determinazione della ornitinacarbamil transferasi sierica (OCT) nella dialgnosi della epatite virale. Rapido micro e ultramicrometodo. Epatologia *13*, 1009 – 1019 (1967).

[19] *G. Ceriotti, A. Gazzaniga,* Accelerated Micro and Ultramicro Procedure for Ornithine Carbamyl Transferase (OCT) Determination, Clin. Chim. Acta *16*, 436 – 439 (1967).

4.3.2 Measurement in Liver

Luisa Raijman

General

Ornithine carbamoyltransferase, OCT, catalyzes the second reaction in the pathway of urea synthesis, namely, the synthesis of citrulline from ornithine and carbamoyl phosphate. It is present in liver mitochondria of ureotelic animals and, at much lower activity, in the intestinal mucosa of mammals. OCT has also been detected in the kidney of some uricotelic species in which its function is unknown, and in some plants and micro-organisms. OCT from bovine and rat liver has been purified to apparent homogeneity and has been extensively characterized [1 – 3]; purified human liver OCT has been partially characterized [4 – 6]. While the ornithine carbamoyltransferases from these sources appear to be very similar to one another, some differences have been reported regarding their molecular activities [3, 7] and kinetic behavior with respect to carbamoyl phosphate [2, 3].

In humans, OCT deficiency is the most frequent of the congenital defects of the urea cycle enzymes [8]; it is inherited as an X-linked dominant trait. K_m and pH mutants have been observed (cf. [8]). In three reported studies, the activities of both OCT and carbamoyl phosphate synthetase (ammonia) (EC 6.3.4.16) were markedly decreased [9, 10]. Transient decreases in OCT activity have been observed in patients with Reye's syndrome and in some premature newborns [8].

Application of method: in biochemical, nutritional, and genetic research, and in clinical chemistry.

Enzyme properties relevant in analysis: under the assay conditions described below, the reaction is linear with respect to enzyme concentration provided that not more than about 10% of the carbamoyl phosphate is used during incubation. Beyond this point, the rate of the reaction decreases, most likely as the result of the accumulation of P_i, which is a good competitive inhibitor of OCT [2, 3, 11]. P_i arises not only from the reaction catalyzed by OCT but from the hydrolysis of carbamoyl phosphate by phosphatases present in liver; this non-specific enzymatic hydrolysis could interfere with the assay when, owing to the low activity of OCT in some liver samples (such as in the congenital deficiencies of this enzyme), larger aliquots of tissue homogenate must be used. P_i also arises from the spontaneous decomposition of carbamoyl phosphate, which has a half-life of 42 min at pH 6.8 to 7.9 and 37 °C [12]. For these reasons, the length of incubation is kept relatively short, and the largest sample used for assay is such that less than 0.5 µmol of citrulline is formed in the course of the reaction. The sample sizes given below are selected on the basis of an OCT activity of 40 µmol × min^{-1} × g^{-1} wet weight of tissue, which is in the mid-range of the activity found in human liver (see [8]).

There appear to be no isozymes of OCT in the normal liver of any one species. However, human mutant isoenzymes with pH optimum or K_m values for ornithine and carbamoyl phosphate different from those of the normal human enzyme have been found [8].

Methods of determination: catalytic activity is measured by determining colorimetrically the amount of citrulline formed. For tissues containing extremely low OCT activity, [^{14}C]citrulline synthesized from [^{14}C] carbamoyl phosphate of known specific radioactivity may be measured after decomposing the residual carbamoyl phosphate by heating; alternatively, [^{14}C]ornithine may be used, but in this case this compound must be separated from the [^{14}C]citrulline prior to measuring the latter.

International reference methods and standards: neither standardization at the international level nor the existence of reference standard materials is known up to now.

Enzyme effectors: special activators of ornithine carbamoyltransferase to be added to the assay mixture are not known. P_i, L-norvaline and L-leucine inhibit at relatively low concentrations [2, 3, 11].

Assay

Method Design

Principle

(a) Carbamoyl phosphate + L-ornithine $\xrightarrow{\text{OCT}}$ L-citrulline + P_i.

The amount in µmol of citrulline formed per unit time is a measure of the catalytic activity of OCT. Since the apparent equilibrium constant of reaction (a), with citrulline and P_i as products, is greater than 1×10^5 at pH 7.6 (25 °C) [2], reversal of the reaction is not a source of error under most conditions.

Citrulline is measured colorimetrically by the method of *Ceriotti & Spandrio* [13], which also detects urea. In general, the amounts of urea and citrulline contained in the homogenate samples added to the incubation mixtures are too small to contribute significantly to colour development. Carbamoyl phosphate in solution decomposes upon storage, even at low temperatures; therefore, solutions of this substrate are prepared shortly before use. Enzyme blanks are mixtures containing all reagents and tissue homogenate denatured by heating at 90 °C for 10 min; blank values are generally negligible, but may be significant in some cases (see sp. 4). Reactions are started by addition of carbamoyl phosphate.

Optimized conditions for measurement: ornithine carbamoyl transferases from bovine, rat, and human liver appear to be similar in several respects, including their pH optima and apparent K_m values for carbamoyl phosphate and total ornithine. Despite some reported differences in the numerical values of these constants among the enzymes from different species and within a species, the following generalizations apply to the OCT from the sources mentioned above:

1. the pH optimum ranges between 7.5 and 8.1, depending on the buffer used;
2. over a wide pH range, carbamoyl phosphate, 5 to 10 mmol/l, saturates OCT and does not inhibit it;
3. the K_m for total ornithine (i.e., the sum of all ionic species) decreases with increasing pH;
4. ornithine becomes inhibitory at concentrations equal to or lower than 10 K_m, approximately, depending on the pH.

Nearly optimal conditions for the assay of bovine, rat, and human liver OCT at 37 °C are obtained using triethanolamine-HCl, pH 7.6, 0.2 mol/l; carbamoyl phosphate and ornithine, 5 mmol/l each.

Temperature conversion factors: under the assay conditions described here, the ratio of activity at 37 °C and at 30 °C is 1.31.

Equipment

Spectrophotometer or colorimeter capable of accurate measurement at 460 nm; temperature-controlled incubation-bath, boiling water-bath and cooling-bath with covers; bench-top centrifuge.

Reagents and Solutions

Purity of reagents: ornithine and triethanolamine must be of analytical quality. Dilithium carbamoyl phosphate must not contain significant amounts of P_i, which inhibits OCT, or of urea, which interferes with the colour reaction; suitable commercial preparations are available.

Li_2-carbamoyl phosphate decomposes slowly upon dry storage at 4 °C; old preparations should be assayed and should not be used for assaying OCT if they contain more than 5% of the total phosphate as P_i.

Preparation of solutions: all solutions are made in re-purified water (cf. Vol. II, chapter 2.1.3.2).

1. Triethanolamine buffer (1 mol/l; pH 7.76):

   dissolve 18.56 g triethanolamine hydrochloride in water to 90 ml; adjust the pH to 7.76 with approximately 8 ml KOH, 4.25 mol/l (280.6 g/l, 85% KOH) and dilute to 100 ml with water. Upon dilution to 0.2 mol/l, the final pH is 7.6.

2. Triethanolamine buffer (200 mmol/l, pH 7.6):

   dilute 20 ml solution (1) to 100 ml with water.

3. L-Ornithine (100 mmol/l):

   dissolve 421.5 mg L-ornithine hydrochloride in water to a total volume of 25 ml.

4. Carbamoyl phosphate (50 mmol/l):

   dissolve 38.2 mg carbamoyl phosphate, dilithium salt, in ice-cold water, to a total volume of 5 ml.

5. Perchloric acid (5 mol/l):

   slowly, add 208 ml of 70% perchloric acid to approximately 250 ml water; add water to 500 ml.

6. Citrulline standard (100 µmol/l):

   prepare a stock citrulline solution, 1 mmol/l, by dissolving 17.52 mg citrulline in water to a total volume of 100 ml. Dilute 1 ml of this solution to 10 ml with water to yield citrulline standard, 100 µmol/l.

7. Acetic acid (5%, v/v):

   dilute 25 ml glacial acetic acid to 500 ml with water.

8. Diacetyl monoxime (5 g/l) in acetic acid (5%, v/v):

   dissolve 2.5 g diacetyl monoxime in 500 ml solution (7).

9. Sulphuric acid (40%, v/v):

   to 280 ml water, add slowly, with stirring, 204 ml 98% sulphuric acid. Allow the solution to cool down and add water to 500 ml.

10. Antipyrine (4 g/l in 40% sulphuric acid):

    dissolve 2 g antipyrine in 500 ml solution (9).

Stability of solutions: ornithine and citrulline solutions are stable; they are stored at −20 °C to retard bacterial growth. Triethanolamine buffer is stored at 4 °C for the same purpose. Antipyrine and diacetyl monoxime solutions are stored at room temperature, the latter in the dark. Li_2-carbamoyl phosphate should be dissolved shortly before use.

Procedure

Collection and treatment of specimen: portions of liver are rinsed rapidly in cold saline solution, blotted, and weighed. The samples may be assayed immediately, or may be stored in fragments of known weight, at −70 °C or lower. Fifty mg or more of tissue is homogenized in 19 volumes of ice-cold water, using a cold *Potter-Elvehjem* homogenizer and teflon pestle, making 10 passes at about 1000 rpm. The homogenate should be used within 1 hour. Approximately 15 min before starting the reaction, 0.2 ml of homogenate is diluted with 1.8 ml of cold triethanolamine buffer (0.2 mol/l). Aliquots of the dilute homogenate are placed in tubes containing all reagents except carbamoyl phosphate, while a portion of the dilute homogenate is heated at 90 °C for 10 min.

Stability of the enzyme in specimen or sample: OCT is stable for at least two weeks in samples of liver kept frozen at −70 °C [14] or below [9]; losses of activity are detectable within one week of storage at −20 °C. OCT is unstable in very dilute solutions, but it is stabilized by ornithine [2].

Details for measurement in liver from different sources: the assay system described below uses a range of quantities of liver homogenate adequate to measure OCT in autopsy samples of human liver containing more than 2% of the activity found in control autopsy samples [9, 14]. Human liver specimens obtained by surgical wedge biopsy have higher OCT activity than autopsy specimens [14]; rat and bovine liver have even higher activity. The amounts of homogenate used, therefore, should be adjusted to the material to be assayed.

Extensive homogenization is required by the characteristics of human liver, which is more difficult to disrupt than rat or bovine liver. The method described appears to solubilize OCT from all these tissues completely without loss of activity. Total activity is not increased by further homogenization, by addition of a detergent [14] or by treatment with a sonic disintegrator [15].

When assaying OCT in human liver for diagnostic purposes, the activity in a rat liver is always assayed simultaneously as a control. Special conditions must be used for the detection of K_m or pH mutants (see [9]).

Assay conditions: all measurements in triplicate. Incubation temperature 37 °C; incubation volume 1 ml; final volume after deproteinization 1.5 ml. The colour reaction for standards and unknowns is carried out simultaneously, in a boiling water-bath; final volume 2 ml; readings at 460 nm, room temperature; light path 10 mm. Zero the instrument with the blank containing only the colour reagents. Citrulline standards containing 10, 25, and 50 nmol of this compound are used as reference.

Colour arising from the non-enzymatic carbamoylation of ornithine, and from citrulline, urea, and unidentified chromogens present in the tissue sample is generally negligible. However, colour from endogenous chromogens may be significant when some abnormal samples of human liver are assayed. This is corrected for by subtracting the colour values given by the enzyme blanks (containing heat-treated homogenate) from the respective experimental readings.

Measurement

Enzyme reaction

Pipette successively into the bottom of small tubes:	assay (ml)			blank (ml)			concentration in incubation mixture
water	0.61	0.57	0.49	0.61	0.57	0.49	
triethanolamine buffer (1)	0.19	0.18	0.16	0.19	0.18	0.16	TEA 200 mmol/l
ornithine (3)	0.05	0.05	0.05	0.05	0.05	0.05	ornithine 5 mmol/l
sample	0.05	0.10	0.20	–	–	–	volume fraction: 0.05, 0.10, 0.20
heated sample	–	–	–	0.05	0.10	0.20	
incubate at 37 °C, 5 min							
carbamoyl phosphate (4)	0.10	0.10	0.10	0.10	0.10	0.10	carbamoyl phosphate 5 mmol/l
mix; incubate at 37 °C for exactly 10 min							
perchloric acid (5)	0.50	0.50	0.50	0.50	0.50	0.50	
mix, place on ice for 10 min; centrifuge at 2500 rpm, use supernatant for colour assay.							

Colour reaction

Pipette into 16 mm × 100 mm tubes:	assay (ml)			blank (ml)			citrulline standards (ml)			
supernatant	0.20	0.20	0.20	0.20	0.20	0.20	–	–	–	–
water	0.80	0.80	0.80	0.80	0.80	0.80	0.90	0.75	0.50	1.00
citrulline standard (6)	–	–	–	–	–	–	0.10	0.25	0.50	–
diacetyl monoxime (8)	0.20	0.20	0.20	0.20	0.20	0.20	0.20	0.20	0.20	0.20
antipyrine (10)	0.80	0.80	0.80	0.80	0.80	0.80	0.80	0.80	0.80	0.80
mix thoroughly; cover the tubes with marbles and heat in a covered boiling water-bath for 50 min. Cool to room temperature in a covered water-bath. Mix again and read absorbance at 460 nm.										

When read against water, the absorbance of the colour-reagent blank should be not more than 0.010; higher readings may reflect contamination of the colour reagents. Using a Zeiss PM2 DL spectrophotometer, absorbance is linear with respect to citrulline concentrations as high as 120 nmol/2 ml of colour reaction mixture. Salts affect colour development; if larger aliquots of supernatant are used, the citrulline standards should contain equivalent amounts of incubation mixture without homogenate. To obtain ΔA arising from OCT activity, subtract from the total absorbance the blank absorbance obtained from an equal volume of heated homogenate. If ΔA given by two of the aliquots of homogenate is lower than 0.100 or higher than 1.500, repeat the assay using less active or more diluted homogenate.

Calculation: OCT activity content expressed as µmol of citrulline synthesized per min per g wet weight of tissue is obtained from the following formula (according to eqns. (l_1) in combination with (m_1) in "Formulae", Appendix 3):

$$z_c/m_s = \frac{\Delta A}{\Delta t} \times \frac{c_{\text{standard}}}{\Delta A_{\text{standard}}} \times \frac{V_i \times V_c}{v_i \times v_c \times \rho}$$

$$= \frac{\Delta A}{\Delta t} \times \frac{c_{\text{standard}}}{\Delta A_{\text{standard}}} \times \frac{15}{v_i \times \rho} \quad \text{U/g}$$

where is

$\Delta A_{\text{standard}}$	the absorbance of the particular citrulline standard
c_{standard}	the concentration of that citrulline standard in the colour reaction (5, 12.5, 25 µmol/l)
V_i	volume of incubation mixture after deproteinization (1.5 ml)
v_i	sample volume in incubation mixture
V_c	volume of colour reaction mixture (2 ml)
v_c	volume of supernatant in colour reaction (0.2 ml)
ρ	mass concentration of wet tissue (g/l) in the diluted homogenate (sample).

Validation of Method

Precision, accuracy, detection limits and sensitivity: for values of about 100 U/g wet weight of tissue, an imprecision of less than ±2.5 U/g has been found for manual performance. Data about accuracy are not available since there is no standard reference material. The limit of detection of the enzyme assay is 0.8 U/g of tissue (using 0.1 ml of undiluted homogenate). Using a *Zeiss* PM2 DL spectrophotometer, the sensitivity of the colorimetric method is 1 to 2 nmol/2 ml of colour reaction mixture.

Sources of error: influences of therapeutic agents on the measured activity are not known. Errors are most likely to arise from the use of old or improperly stored carbamoyl phosphate preparations containing inhibitory amounts of P_i, and from excessive conversion of carbamoyl phosphate during incubation. Owing to the high sensitivity of the colorimetric method, contamination of materials with citrulline or urea could cause considerable error. The OCT activity of human liver samples obtained by needle biopsy should not be referred to tissue weight, but to mg of DNA (cf. [15]).

Specificity: the enzymatic reaction is specific for OCT. The colour reaction is specific for citrulline in the absence of urea [13] and carbamoyl aspartate [16].

Reference ranges: in autopsy samples of normal human liver the activity of OCT varies within a broad range of 28 to 72 U/g [8, 9, 14]; in part, the variations are attributable to differences in the number of hours after death at which the tissue samples are obtained. In liver samples obtained by surgical wedge biopsy, OCT activity is higher, 110 ± 26.3 U/g [14]. The OCT activity of adult rat liver is on the order of 200 U/g [15]. In rats [17], monkeys, and presumably also in man [18], the activity of OCT in liver varies significantly as a function of the composition of the diet. In rats, it also varies with age [17].

References

[1] *M. Marshall, P. P. Cohen*, Ornithine Transcarbamylase from *Streptococcus faecalis* and Bovine Liver. I. Isolation and Subunit Structure, J. Biol. Chem. *247*, 1641 – 1653 (1972).

[2] *M. Marshall, P. P. Cohen*, Ornithine Transcarbamylase from *Streptococcus faecalis* and Bovine Liver. II. Multiple Binding Sites for Carbamyl-P and L-Norvaline, Correlation with Steady State Kinetics, J. Biol. Chem. *247*, 1654 – 1668 (1972).

[3] *C. J. Lusty, R. L. Jilka, E. H. Nietsch*, Ornithine Transcarbamylase of Rat Liver. Kinetic, Physical, and Chemical Properties, J. Biol. Chem. *254*, 10030 – 10036 (1979).

[4] *P. J. Snodgrass*, The Effects of pH on the Kinetics of Human Liver Ornithine-Carbamyl Phosphate Transferase, Biochem. *9*, 3047 – 3051 (1968).

[5] *D. L. Pierson, S. L. Cox, B. E. Gilbert*, Human Ornithine Transcarbamylase. Purification and Characterization of the Enzyme from Normal Liver and the Liver of a Reye's Syndrome Patient, J. Biol. Chem. *252*, 6464 – 6469 (1977).

[6] *F. Kalousek, B. François, L. E. Rosenberg*, Isolation and Characterization of Ornithine Transcarbamylase from Normal Human Liver, J. Biol. Chem. *253*, 3939 – 3944 (1978).

[7] *S. Clarke*, The Polypeptides of Rat Liver Mitochondria: Identification of a 36,000 Dalton Polypeptide as the Subunit of Ornithine Transcarbamylase, Biochem. Biophys. Research Commun. *71*, 1118 – 1124 (1976).

[8] *M. Walser*, Urea Cycle Disorders and Other Hereditary Hyperammonemic Syndromes, in: *J. B. Stanbury, J. B. Wyngaarden, D. S. Fredrickson, J. L. Goldstein, M. S. Brown* (eds.), The Metabolic Basis of Inherited Disease, McGraw-Hill Book Co., New York 1982, pp. 402 – 438.

[9] *L. Raijman*, Double Deficiencies of Urea Cycle Enzymes in Human Liver, Biochem. Med. *24*, 226 – 233 (1979).

[10] *A. Matsushima, T. Taga, T. Orii, Y. Matsuda, A. Tsuji, N. Katsunuma*, Hyperammonemia Due to Ornithine Transcarbamylase Deficiency Accompanied by Decreased Activity of Carbamyl-Phosphate Synthetase, Brain Dev. (Japan) *11*, 343 – 346 (1979).

[11] *R. L. Joseph, E. Baldwin, D. C. Watts*, Studies on Carbamoyl Phosphate-L-Ornithine Carbamoyltransferase from Ox Liver, Biochem. J. *87*, 409 – 416 (1963).

[12] *C. M. Allen, M. E. Jones*, Decomposition of Carbamylphosphate in Aqueous Solutions, Biochem. *3*, 1238 – 1247 (1964).
[13] *G. Ceriotti, L. Spandrio*, A Spectrophotometric Method for Determination of Urea, Clin. Chim. Acta *8*, 295 – 299 (1963).
[14] *C. T. Nuzum, P. J. Snodgrass*, Multiple Assays of the Five Urea-Cycle Enzymes in Human Liver Homogenates, in: *S. Grisolia, R. Baguena, F. Mayor* (eds.), The Urea Cycle, John Wiley and Sons, New York 1976, pp. 325 – 349.
[15] *L. Raijman*, Citrulline Synthesis in Rat Tissues and Liver Content of Carbamoyl Phosphate and Ornithine, Biochem. J. *138*, 225 – 232, (1974).
[16] *L. M. Prescott, M. E. Jones*, Modified Methods for the Determination of Carbamyl Aspartate, Anal. Biochem. *32*, 408 – 419 (1969).
[17] *R. T. Schimke*, Adaptive Characteristics of Urea Cycle Enzymes in the Rat, J. Biol. Chem. *237*, 459 – 468 (1962).
[18] *C. T. Nuzum, P. J. Snodgrass*, Urea Cycle Enzyme Adaptation to Dietary Protein in Primates, Science *172*, 1042 – 1045 (1971).

4.4 Transaldolase

Sedoheptulose-7-phosphate : D-glyceraldehyde-3-phosphate dihydroxy-acetone transferase, EC 2.2.1.2

Karl Brand

General

Transaldolase, TA, was first described by *Horecker & Smyrniotis* [1] in 1953. The enzyme was crystallized from *Candida utilis* by *Pontremoli et al.* [2] in 1961. Transaldolase occurs in animal [3 – 6] plant tissues [6] and in micro-organisms [6 – 9]. As an enzyme of the pentose phosphate cycle it is found in the cytoplasmic compartment of the cell. A survey of the distribution of the activities of transaldolase, transketolase, fructose 1,6-bisphosphate aldolase and fructose 1,6-bisphosphatase in rabbit tissues is given in Table 1. The transaldolase activity of normal and pathological human sera is extremely low. *Candida utilis* is particularly suitable for the isolation of the enzyme because it has a high transaldolase activity. In the crystallization of the enzyme from this yeast, it appears that only part of the activity can be obtained in crystalline form; the major part cannot be crystallized. This different behaviour can be explained by the work of *Tsolas & Horecker* [10], who have demonstrated the presence of 3 forms of the enzyme which can be separated by chromatography, and which have been designated types I, II and III on the basis of their elution pattern from a DEAE-Sephadex column. They showed that types I and III are isoenzymes and the type II is a hybrid form of I and III. So far only type III has been obtained in crystalline form by the procedure of *Tchola & Horecker* [11].

Table 1. Activities of transaldolase, transketolase, fructose 1,6-bisphosphate aldolase, fructose 1,6-bisphosphatase in rabbit tissues. (A mU/g fresh weight; B mU/mg extract protein; 25 °C)

Tissue	Transaldolase		Transketolase		Fructose-P_2-aldolase		Fructose-P_2ase	
	A	B	A	B	A	B	A	B
Lung	465	8	–	–	490	6	70	1
Stomach	515	16	295	9	900	27	160	5
Liver	705	6	355	2	1300	22	2000	18
Kidney	645	11	430	7	2200	63	2440	42
Brain	300	10	230	8	3600	115	10	0.3
Bone marrow	480	12	–	–	1080	26	65	0.15
Pancreas	450	7	120	0.2	670	11	475	8
Adrenal	580	32	1160	13	2400	125	–	–
Thyroid	255	12	85	3	3350	136	295	10
Testicle	595	11	435	8	650	12	50	1
Prostate	320	23	125	9	2250	195	215	15
Mammary gland	160	7	–	–	10800	455	255	11
Mammary gland during pregnancy	335	9	195	5	6200	146	280	7
Placenta	225	5	60	1	1000	18	70	1

Frozen tissue was homogenized in a blade homogenizer (3 × 1 min) with the addition of washed glass beads (0.17 – 0.18 mm diameter). Bone marrow and pancreas were homogenized in a Teflon homogenizer. The enzyme activities were measured in the supernatant fluid after centrifugation at 12000 *g*: transketolase by a modification of the method of *Horecker et al.* [13], Fructose-P_2-aldolase by a modification of the method of *Racker* [14], Fructose-P_2ase by a modification of the method of *Pontremoli et al.* [15]. Protein was determined as described by *Bücher* [16].

Application of method: in biochemistry

Enzyme properties relevant in analysis: transaldolase type I has a *Michaelis* constant (K_m) for fructose 6-phosphate of 0.53 mmol/l and for D-erythrose 4-phosphate of 0.02 mmol/l. The corresponding K_m values for type III are 0.71 mmol/l and 0.019 mmol/l respectively [11]. D-Fructose 6-phosphate and D-sedoheptulose 7-phosphate are the most active donor substrates for dihydroxyacetone; D-glyceraldehyde-3-phosphate and D-erythrose 4-phosphate are the most active acceptor aldehydes [7]. Octulose-8-phosphate [16], sorbose-6-phosphate [17] and free fructose [18] are less effective as donor substrates, while free glyceraldehyde [18], L-glyceraldehyde 3-phosphate [17] and ribose 5-phosphate [16] are less effective as acceptor substrates. The molecular weights determined by a sedimentation equilibrium method are 76100 for type I and 65900 for type III [19].

Methods of determination: the catalytic activity can only be measured spectrophotometrically.

International reference methods and standards: neither standardization at the international level nor the existence of reference standard materials is known up to now. now.

Enzyme effectors: the yeast enzyme does not have a prosthetic group and does not require a cofactor for its catalytic activity. Specific activators of transaldolase to be added to the assay mixture are not known.

Assay

Method Design

Principle

(a) $$\text{Fructose-6-P} + \text{erythrose-4-P} \xrightleftharpoons{\text{TA}} \text{sedoheptulose-7-P} + \text{GAP}$$

(b) $$\text{GAP} \xrightleftharpoons{\text{TIM*}} \text{DAP}$$

(c) $$\text{DAP} + \text{NADH} + \text{H}^+ \xrightleftharpoons{\text{GDH**}} \text{glycerol-3-P} + \text{NAD}^+$$

Transaldolase (TA) catalyzes the transfer of dihydroxyacetone, which is liberated by aldol cleavage of fructose 6-phosphate, to erythrose 4-phosphate with the formation of sedoheptulose 7-phosphate and glyceraldehyde 3-phosphate (GAP).

For the determination of transaldolase activity the GAP formed per unit time from Fructose-6-P is measured according to equations (b) and (c). Under the conditions described below the rate of NADH oxidation is proportional to the TA activity. The decrease of absorbance at 339 (Hg 334 or Hg 365) nm per min is measured. According to *Racker* [7], the equilibrium constant K = [sedoheptulose-7-P] × [GAP]/[fructose-6-P] × [erythrose-4-P] = 0.95. The further reactions of GAP and DAP in the auxiliary and indicator reactions give a quantitative conversion of fructose-6-P and erythrose-4-P.

Optimized conditions for measurement: the pH optimum for transaldolase lies between 7.3 and 8.1 with a steep decrease on either side. For routine determinations it is recommended to work at 30 °C, but the enzyme is stable up to 40 °C. Metal ions or other cofactors are not required for the catalytical activity of the yeast enzyme [4]. Transaldolase is inhibited by phosphate, >0.1 mol/l, [20], pyrophosphate or sulphate [6].

* Triosephosphate isomerase, D-Glyceraldehyde-3-phosphate ketolisomerase, EC 5.3.1.1
** Glycerophosphate dehydrogenase, *sn*-Glycerol-3-phosphate: NAD 2-oxido-reductase, EC 1.1.1.8

Equipment

Spectrophotometer or spectrum line photometer suitable for accurate measurements at 339, Hg 334 or Hg 365 nm with constant temperature cuvette holder; water-bath (30 °C); recorder or stopwatch.

Reagents and Solutions

Preparation of solutions (for about 20 determinations): all solutions in re-purified water (cf. Vol. II, chapter 2.1.3.2).

1. Triethanolamine buffer (0.1 mol/l; pH 7.6; EDTA, 10 mmol/l):

   dissolve 1.857 g triethanolamine hydrochloride and 0.372 g EDTA-$Na_2H_2 \cdot 2H_2O$ in 80 ml water, adjust to pH 7.6 with NaOH, 1 mol/l, and dilute to 100 ml with water.

2. Fructose 6-phosphate (0.15 mol/l):

   dissolve 105 mg fructose-6-P, Na salt (75% fructose-6-P) in 2 ml water.

3. Erythrose 4-phosphate (10 mmol/l):

   dissolve and hydrolyze D-erythrose 4-phosphate diethylacetal, barium salt, according to the description of the commercial preparation of *Boehringer Mannheim*. Determine the erythrose-4-P concentration enzymatically as described in Vol. VI and dilute appropriately with water. The pH of the solution should be between 2 and 3.

4. Reduced nicotinamide-adenine dinucleotide (β-NADH, 7.5 mmol/l):

   dissolve 11 mg NADH, sodium salt, with 2 ml $NaHCO_3$ solution, 1% (w/v).

5. Glycerophosphate dehydrogenase/triosephosphate isomerase (GDH, 30 kU/l; TIM, 200 kU/l):

   GDH ($\geqq$ 170 U/mg, 25 °C) and TIM ($\geqq$ 10000 U/mg, 25 °C) from rabbit muscle; 1.8 mg GDH + 0.2 mg TIM suspended in ammonium sulphate, 3 mol/l. Dilute the stock solution 1 + 9 with ammonium sulphate solution, 3 mol/l.

Stability of solutions: store all solutions at 0 to 4 °C. The erythrose-4-P solution (pH 2 – 3) and the fructose-6-P solution are stable for long periods, the NADH solution should be prepared freshly each week. The GDH/TIM mixed suspension is stable for longer than a year.

Procedure

Collection and treatment of specimen: see legend to Table 1.

Stability of enzyme in sample: crystalline transaldolase from *Candida utilis* suspended in 60% saturated ammonium sulphate solution is stable for more than 1 year. Crude extracts of animal and plant tissues and of micro-organisms can be stored at −20 °C for several months without loss of activity.

Assay conditions: wavelength 339 (Hg 334, Hg 365) nm; light path 10 mm; final volume 2.75 ml; 30 °C constant-temperature cuvette holder. Measure against air. Bring the solutions to 30 °C before assay.

Measurement

Pipette successively into the cuvette:			concentration in assay mixture	
triethanolamine buffer	(1)	2.50 ml	TEA EDTA	91 mmol/l 9 mmol/l
fructose-6-P solution	(2)	0.05 ml	fructose-6-P	2.7 mmol/l
erythrose-4-P solution	(3)	0.05 ml	erythrose-4-P	0.18 mmol/l
NADH solution	(4)	0.05 ml	NADH	0.14 mmol/l
GDH/TIM suspension	(5)	0.05 ml	GDH TIM	550 kU/l 3640 kU/l
mix, wait for 2 min, until GAP and DAP in the fructose-6-P and erythrose-4-P preparations are reduced with NADH				
sample		0.05 ml	volume fraction	0.018
mix, read absorbance and start stopwatch. Repeat the reading after exactly 1, 2, 3, 4 and 5 min or monitor the reaction on a recorder.				

The values $\Delta A/\Delta t$ at 339 nm must be < 0.060/min; otherwise dilute the sample with buffer (1) and take the dilution factor into account in the calculation.

Calculation: use calculation formula (k) or (k_1) respectively (cf. "Formulae", Appendix 3). For values of ε cf. Appendix 4. The following relationship holds for the catalytic concentration in the sample solution:

	Hg 334 nm	339 nm	Hg 365 nm	
$b =$	$8900 \times \Delta A/\Delta t$	$8731 \times \Delta A/\Delta t$	$16172 \times \Delta A/\Delta t$	U / l
$b =$	$1.48 \times 10^4 \times \Delta A/\Delta t$	$1.46 \times 10^4 \times \Delta A/\Delta t$	$2.70 \times 10^4 \times \Delta A/\Delta t$	nkat/l

Validation of Method

Precision, accuracy, detection limits and sensitivity: with values around 300 U/l in the sample an imprecision of ± 18 U/l has been found. The coefficient of variation is 3.3%. Data about accuracy are not available since no standard reference material is established yet. For measurements at 339 nm with sufficiently sensitive photometers the lowest possible value to be measured is approximately 5 U/l with an imprecision of $< \pm 20\%$. Sensitivity is found to be $\Delta A/\Delta t = 0.01/10$ min (*Zeiss* photometer).

Sources of error: contamination by GAP and DAP is rapidly and completely removed in the preliminary reaction before addition of sample, so that under the conditions described above the TA activity is measured specifically. In homogenates and extracts from cells and tissues which have been incubated with an excess of glucose, NADH is also oxidized by continuing glycolytic reactions and the added GDH/TIM mixture. This source of error can be avoided by removal of the glycolytic substrates from the sample by dialysis.

Specificity: transaldolase activity is measured specifically.

References

[1] *B. L. Horecker, P. Z. Smyrniotis,* Transaldolase: The Formation of Fructose-6-phosphate from Sedoheptulose-7-phosphate, J. Am. Chem. Soc. *75*, 2021 – 2022 (1953).

[2] *S. Pontremoli, B. D. Prandini, A. Bonsignore, B. L. Horecker,* The Preparation of Crystalline Transaldolase from Candida Utilis, Proc. Natl. Acad. Sci. *47*, 1942 – 1949 (1961).

[3] *Z. Dische,* Some Regulatory Factors in the Interconversion of Glucose-6-phosphate and Ribose-5-phosphate in Human Erythrocytes, Ann. N. Y. Acad. Sci. *75*, 129 – 141 (1958).

[4] *K. Brand, K Decker, J. Musil,* Enzyme Pattern of the Pentose Phosphate Pathway in Ascites Tumor Cells and the Effect of Nucleoside Triphosphate on its Enzyme Activities, Hoppe-Seyler's Z. Physiol. Chem. *351*, 213 – 220 (1970).

[5] *W. Chefurka, Y. Horie, J. R. Robinson,* Contribution of the Pentose Cycle to Glucose Metabolism by Insects, Comp. Biochem. Physiol. *37*, 143 – 165 (1970).

[6] *O. Tsolas, B. L. Horecker,* Transaldolase, in: P. D. Boyer (ed.), The Enzymes, Vol. VII, 3rd. edit., Academic Press, New York 1972, pp. 259 – 280.

[7] *B. L. Horecker, P. Z. Smyrniotis,* Purification and Properties of Yeast Transaldolase, J. Biol. Chem. *212*, 811 – 825 (1955).

[8] *D. Couri, E. Racker,* Oxidative Pentose Phosphate Cycle. V. Complete Oxidation of Glucose-6-phosphate in a Reconstructed System of the Oxidative Pentose Cycle, Arch. Biochem. Biophys. *83*, 195 – 205 (1959).

[9] *P. Srere, J. R. Cooper, M. Tabachnik, E. Racker,* Oxidative Pentose Phosphate Cycle. I. Preparation of Substrates and Enzymes, Arch. Biochem. Biophys. *74*, 295 – 305 (1958).
[10] *O. Tsolas, B. L. Horecker,* Isoenzymes of Transaldolase in Candida Utilis. I. Isolation of Three Isoenzymes from Yeast Extracts, Arch. Biochem. Biophys. *136*, 287 – 302 (1970).
[11] *O. Tchola, B. L. Horecker,* Transaldolase, in: W. A. Wood (ed.), Methods in Enzymology, Vol. IX, Academic Press, New York 1966, pp. 499 – 505.
[12] *B. L. Horecker, P. Z. Smyrniotis, J. Hurwitz,* The Role of Xylulose-5-phosphate in the Transketolase Reaction, J. Biol. Chem. *223*, 1009 – 1019 (1956).
[13] *E. Racker,* Spectrophotometric Measurement of Hexokinase and Phosphohexokinase Activity, J. Biol. Chem. *167*, 843 – 854 (1947).
[14] *S. Pontremoli, S. Traniello, B. Luppis, W.A. Wood,* Fructose Diphosphatase from Rabbit Liver. I. Purification and Properties, J. Biol. Chem. *240*, 3459 – 3463 (1965).
[15] *T. Bücher,* Über ein Phosphatübertragendes Gärungsferment, Biochim. Biophys. Acta *1*, 292 – 314 (1947).
[16] *E. Racker, E. A. R. Schröder,* Formation and Utilization of Octulose-8-phosphate by Transaldolase and Transketolase, Arch. Biochem. Biophys. *66*, 241 – 243 (1957).
[17] *R. Venkataraman, A. G. Datta, E. Racker,* Studies on the Mechanism of Action of Transaldolase, Fed. Proc. *19*, p. 82 (1960).
[18] *A. Bonsignore, S. Pontremoli, E. Grazi, M. Mangiarotti,* The Formation and Cleavage of Fructose Catalyzed by Transaldolase, Biochem. Biophys. Res. Commun. *1*, 79 – 82 (1959).
[19] *O. Tsolas, B. L. Horecker,* Isoenzymes I and III and the Conditions for Hybridization to Form Isoenzyme II, Arch. Biochem. Biophys. *136*, 303 – 319 (1970).
[20] *A. Bonsignore, S. Pontremoli, E. Grazi, B. L. Horecker,* Effect of Orthophosphate on the Transaldolase Reaction, J. Biol. Chem. *235*, 1888 – 1890 (1960).

5 Acyltransferases and Glycosyltransferases

5.1 Lecithin-cholesterol Acyltransferase

Lecithin: cholesterol acyltransferase, EC 2.3.1.43

Gerd Assmann and Hans-Jürgen Menzel

General

Lecithin-cholesterol acyltransferase (LCAT) was described by *Glomset et al.* [1] in 1962 and is an enzyme found in mammalian serum [2]. The cholesteryl esters present in human serum are formed nearly exclusively (approx. 90%) by the action of this enzyme. LCAT is synthesized in the liver, though it is active only in serum [3]. The central role of this enzyme in the metabolism of lipoproteins becomes apparent in the drastic changes observed in the lipoproteins of LCAT-deficient patients (cf. [4]). Reduced LCAT activity is also seen in a number of liver diseases [5].

Application of method: in biochemistry and clinical chemistry.

Enzyme properties relevant in analysis: the enzyme has a number of active isoenzymes [6]. Substrate conversion is linear with time only over a limited range, since both product-saturation and exchange processes involving cholesteryl esters act as limiting factors [7].

Methods of determination: enzymatic activity is assayed by the conversion of radioactive cholesterol to cholesteryl esters. Alternatively, the decrease in cholesterol concentration during incubation of serum may be determined by gas chromatography. As well as the method described here, other methods are in use.

Gas-chromatographic measurement of cholesteryl ester: the content of cholesterol in the sample before and after incubation is determined with the aid of quantitative gas chromatography. The difference corresponds to the activity of the sample [8].

Substitution of artificial substrate for inactivated human serum: micellar complexes of apolipoprotein A-I, phosphatidylcholine and cholesterol in a molar ratio of 1:80:20 are formed in the presence of sodium cholate (PC to cholate ratio 1:1), isolated and used as the substrate for the LCAT reaction [9, 10].

Substitution of endogenous substrate: the LCAT activity in the serum to be assayed is inhibited by adding dithionitrobenzoic acid (final concentration 1.4 mmol/l). [^{14}C]-Cholesterol in albumin solution is then added, incubated for 30 min, and LCAT activity is then restored by adding β-mercaptoethanol. The amount of [^{14}C]-cholesteryl ester formed is measured after incubating again [11].

Enzyme mass measurement: at present, assay of enzyme mass is performed by only one laboratory [12].

International reference methods and standards: to date no standardized methods are known. For standard reference material, see NBS-No. 911a "Cholesterol" (cf. Vol. II, chapter 2.3).

Enzyme effectors: apolipoprotein A-I is a known cofactor for the LCAT reaction. In human serum, apolipoprotein A-I is normally present in amounts sufficient for the enzyme reaction [13]. Apolipoprotein A-I apparently forms a complex with LCAT and apolipoprotein D, and the latter seems to mediate the transfer of cholesteryl esters to other lipoproteins [14]. The reaction is usually run near pH 7 in the presence of a chelating agent, used to inhibit heavy metal-catalyzed peroxidation of lecithin, as well as in the presence of a reducing agent, which is necessary for maximum enzyme activity [15].

Assay

Method Design

Principle

$$\text{Cholesterol} + \text{lecithin} \rightleftharpoons \text{cholesteryl esters} + \text{lysolecithin}$$

Cholesterol in heat-inactivated human serum equilibrated with [^{14}C]cholesterol serves as the substrate. The amount of [^{14}C]cholesteryl esters formed is measured.

Optimized conditions for measurement: the physiological substrate for the LCAT reaction are HDL (high-density lipoprotein) particles newly secreted from the liver, not yet containing cholesteryl esters (nascent HDL [18]). *In vitro* esterification in LCAT-deficient serum to which isolated LCAT has been added proceeds at an initial rate ten times that measured by the standard assay [6]. Since, however, there is no optimum substrate available for routine assays of activity, the assay may be conducted only in the presence of HDL particles already up to 80% esterified [15 – 17] or using an artificial substrate [9, 10].

Temperature conversion factors: the reaction is conducted only at 37 °C.

Equipment

Equipment is required for liquid scintillation counting, freeze-drying, thin-layer chromatography, thermostatted incubations, and photometry.

Reagents and Solutions

Purity of reagents: the reagents used for extraction and chromatography must be of analytical purity. The [^{14}C]cholesterol must be free of labelled cholesteryl esters; human albumin (Cohn fraction V) should be used.

Preparation of solutions (for about 100 determinations): all solutions in re-purified water (cf. Vol. II, chapter 2.1.3.2).

1. Substrate solution (unesterified cholesterol, 300 mg/l, 1.29 mmol/l):

   pool the serum of at least 20 normal human subjects (about 40 ml). Determine unesterified cholesterol enzymatically (cf. Vol. VIII). Adjust cholesterol concentration to 300 mg/l or 1.29 mmol/l (M_r = 386.6) by dilution with 0.9% NaCl/ 0.05% EDTA/NaOH (pH 7.0) and inactivate by heating for 30 min at 56 °C.

2. Labelled substrate solution (cholesterol 0.8 mg/l; 125 μCi/l):

   dissolve 1.85×10^5 s^{-1} of [^{14}C]cholesterol in 6 ml ethanol. Add 1.5 ml of this to 13.5 ml of 5% human albumin solution. Remove the ethanol using N_2.

3. β-mercaptoethanol (0.4 mol/l):

   dilute 28 μl β-mercaptoethanol with water to 1 ml.

4. Chloroform/methanol (2:1, v/v):

   mix two volumes of chloroform with one volume of methanol.

5. Cholesterol/cholesteryl linoleate (2.5 g/l of each):

   dissolve 1.25 mg of each in 0.5 ml chloroform/methanol mixture (4).

6. Petrol ether/diethyl ether/acetic acid (76:20:1):

   mix 76 volumes of petrol ether, 20 volumes of diethyl ether and 1 volume of glacial acetic acid.

7. Scintillation fluid:

   dissolve 4 g 2,5-diphenyloxazole and 0.1 g 2,2′-p-phenylbis-(4-methyl)-5-phenyloxazole in 1 litre toluene.

8. Silica gel plates:

   commercially available, e.g. Polygram Sil 6 from *Macherey and Nagel,* Düren, Federal Republic of Germany.

Stability of solutions: the solutions (1), (2) and (5) should be stored at −20 °C in small aliquots. They remain stable for a period of 2−3 month. Solutions (3) and (6) should be prepared freshly and solutions (4) and (7) can be stored at room temperature in a dark bottle for several months.

Procedure

Collection and treatment of specimen: either fresh serum or plasma may be used. EDTA, heparin or sodium citrate can be used as anticoagulants.

Stability of the enzyme in the specimen: enzyme activity quickly decreases within 12 h at 25 °C as well as at 4 °C. Enzyme stability may be preserved for several days or even weeks at −20 °C.

Assay conditions

Incubation: volume 505 µl; time 6 h; 37 °C.
Each sample is run in duplicate and a blank with 0.9% NaCl instead of serum or plasma is run in parallel.
Thin-layer chromatography: room temperature; about 30 min. Cholesterol/cholesteryl linoleate (5) is run as position reference.
Radiometry: 4 °C; [^{14}C]scintillation counting for 10 min.

Measurement

Incubation

Pipette successively into *Eppendorf* reactions vials:		sample	blank	concentration in incubation mixture
substrate	(1)	350 µl	350 µl	cholesterol 0.89 mmol/l
labelled substrate	(2)	100 µl	100 µl	25 µCi/l
β-mercaptoethanol	(3)	5 µl	5 µl	mercaptoethanol 4 mmol/l
mix				
sample		50 µl	–	cholesterol*
				volume fraction $\varphi = 0.099$
0.9% NaCl		–	50 µl	
mix, incubate for 6 h, lyophilize.				

* to be determined in each individual sample

Extraction

Add to each lyophilizate

500 µl chloroform/methanol (4)
40 µl water,

agitate for 20 min, then centrifuge for 2 min, transfer the organic phase into *Eppendorf* reaction vials. Wash the protein with 250 µl chloroform/methanol (4), centrifuge. Pool the organic phases, dry them at 37 °C. Dissolve in 40 µl chloroform/methanol (4).

Thin-layer chromatography

Apply 40 µl extract in 2-cm-bands. Use solution (5) as a position reference on both sides of the chromatogram. Develop the gel plate with 100 ml chromatography medium (6). Chromatography is complete when the cholesterol and cholesteryl linoleate appear 8 cm apart. Cholesterol and cholesteryl linoleate can be visualized by backlighting as dark spots. Cut out the bands corresponding to cholesterol and cholesteryl linoleate of each sample as well as a TLC piece of equal size for background counting.

Radiometry

Place cut-out TLC pieces into scintillation vials, add 20 ml scintillation fluid (7), and count for 10 min.

Measurement of unesterified cholesterol

Determine unesterified cholesterol of each sample enzymatically (cf. Vol. VII).

Calculation: the ratio of radioactive cholesteryl esters to radioactive total cholesterol multiplied by the total unesterified cholesterol yields catalytic concentration of the enzyme in the sample [15–17]. Subtract background counting rate from the counting rates of unesterified cholesterol and cholesteryl linoleate bands, yielding net counting rates $C_{\text{chol.}}$ and $C_{\text{chol. lin.}}$.

The catalytic concentration of the sample is

$$b = \frac{C_{\text{chol. lin.}}}{C_{\text{chol. lin.}} + C_{\text{chol.}}} \times c_{\text{chol.}} \times \frac{1}{\varphi \times \Delta t} \quad \text{U/l}$$

or

$$b = \frac{C_{\text{chol. lin.}}}{C_{\text{chol. lin.}} + C_{\text{chol.}}} \times c_{\text{chol.}} \times \frac{16.67}{\varphi \times \Delta t} \quad \text{nkat/l}$$

in which

$C_{\text{chol.}}$	is counting rate of cholesterol band
$C_{\text{chol. lin.}}$	is counting rate of cholesteryl linoleate band
$c_{\text{chol.}}$	is concentration of total unesterified cholesterol in incubation mixture (0.89 mmol/l from substrate solution + individual sample cholesterol)
φ	is volume fraction of sample in incubation mixture
Δt	is incubation time in minutes

For conversion to the widely-used units nmol $\times$ h^{-1} $\times$ ml^{-1}, multiply by the factor 60/1000.

The activity of the blank should not exceed 2% of the normal LCAT-activity.

Validation of Method

Precision, accuracy, detection limits and sensitivity: the imprecision of all activity assays using radioactive cholesterol approaches a CV of 4%. The lowest limit of detectable enzyme activity is a value approximately corresponding to a range of 2 standard deviations, with a CV of 20% [8]. Similar values are apparent for enzyme mass assay [9].

Source of error: substrate solution (3) should not be prepared from pathological sera, mainly patients with liver disease. LCAT-activity of the specimen may decrease due to improper storage.

Specificity: the determination of LCAT activity according to the method described here is specific, since there is no other enzyme present in human serum which is responsible for esterifying cholesterol.

Reference ranges: due to the variability of substrates (heat inactivated human serum pool, endogenous substrate, and micellar protein lipid complexes) comparison of measured activities is possible only to a very limited degree. Furthermore, a comparison of activity assay and mass assay shows only partial correspondence [19].

References

[1] *J. A. Glomset, F. Parker, M. Tjaden, R. H. Williams,* The esterification in vitro of free cholesterol in human and rat plasma, Biochim. Biophys. Acta *58,* 398 – 406 (1962).

[2] *K. T. Stokke,* Cholesteryl ester metabolism in liver and blood plasma of various animal species, Atherosclerosis *19,* 393 – 406 (1973).

[3] *R. Osugga, O. W. Portmann,* Origin and disappearance of plasma lecithin: cholesterol acyltransferase, Am. J. Physiol. *220,* 735 – 741 (1971).

[4] *J. A. Glomset, K. R. Norum, E. Gjone,* Familial Lecithin-Cholesterol Acyltransferase Deficiency, in: *J. R. Stanbury, J. H. Wyngaarden, D. S. Fredrickson* (eds.), The Metabolic Basis of Inherited Disease, 5th edit. McGraw-Hill, New York 1982, pp. 643 – 654.

[5] *J. B. Simon,* Lecithin-cholesterol acyltransferase in human liver disease. Scand. J. Clin. Lab. Invest. *33,* Suppl. *137,* 107 – 113 (1974).

[6] *G. Utermann, H. J. Menzel, G. Adler, P. Dieker, W. Weber,* Substitution in vitro of Lecithin: Cholesterol Acyltransferase. Analysis of Changes in plasma Lipoproteins, Eur. J. Biochem. *107,* 225 – 241 (1980).

[7] *K. H. Norum,* The enzymology of cholesterol esterification, Scand. J. Clin. Lab. Invest. *33,* Suppl. *137,* 7 – 11 (1974).

[8] *J. P. Blomhoff,* In vitro determination of lecithin: cholesterol acyltransferase in plasma, Scand. J. Clin. Invest. *33,* Suppl. *137,* 35 – 43 (1974).

[9] *C. E. Matz, A. Jonas,* Micellar Complexes of Human Apolipoprotein A – I with Phosphatidylcholines and Cholesterol Prepared from Cholate-Lipid Dispersions, J. Biol. Chem. *257,* 4535 – 4540 (1982).

[10] *C. E. Matz, A. Jonas,* Reaction of Human Lecithin: Cholesterol Acyltransferase with Synthetic Micellar Complexes of Apolipoprotein A – I, Phosphatidylcholine, and Cholesterol, J. Biol. Chem. *257,* 4541 – 4546 (1982).

[11] *K. T. Stokke, K. R. Norum,* Determination of lecithin: cholesterol acyltransferase in human blood plasma, Scand. J. Clin. Invest. *27,* 21 – 27 (1971).

[12] *J. J. Albers, J. C. Adolphson, Ching-Hong Chen,* Radioimmunoassay of Human Plasma Lecithin-Cholesterol Acyltransferase, J. Clin. Invest. *67,* 141 – 149 (1981).

[13] *C. F. Fielding, V. G. Shore, P. E. Fielding,* A cofactor of lecithin: cholesterol acyltransferase in human plasma, Biochem. Biophys. Res. Commun. *46,* 1493 – 1498 (1972).

[14] *P. E. Fielding, C. J. Fielding,* A cholesteryl ester transfer complex in human plasma, Proc. Natl. Acad. Sci. USA *77,* 3327 – 3330 (1980).

[15] *J. A. Glomset, J. L. Wright,* Some properties of a cholesterol esterifying enzyme in human plasma, Biochim. Biophys. Acta *89,* 266 – 276 (1964).

[16] *R. Kattermann, D. J. Wolfrum,* Cholesterinstoffwechsel und Lezithin-Cholesterin-Acyltransferase im Plasma bei experimenteller Hepatitis and Cholestase in der Ratte, Z. Klin. Chem. Klin. Biochem. *8,* 413 – 419 (1970).

[17] *H. J. Menzel,* Isolierung und Charakterisierung der Lecithin-Cholesterin Acyltransferase, Dissertation, Marburg/Lahn, West Germany, 1978.

[18] *R. L. Hamilton, M. C. Williams, C. J. Fielding, R. J. Havel,* Discoidal Bilayer Structure of Nascent High Density Lipoproteins from Perfused Rat Liver, J. Clin. Invest. *58,* 667 – 680 (1976).

[19] *J. J. Albers, G. Utermann,* Genetic control of lecithin : cholesterol acyltransferase: Measurement of LCAT mass in a large kindred with LCAT-deficiency. Am. J. Hum. Genet. *33,* 702 – 706 (1982).

5.2 γ-Glutamyltransferase

(γ-Glutamyl)-peptide: amino acid γ-glutamyltransferase, EC 2.3.2.2

5.2.1 General

Leslie M. Shaw

γ-Glutamyltransferase (γ-GT) is distributed widely in human tissues. The kidneys, especially the luminal membrane of the proximal tubules, contain the highest quantities, followed by the apical portion of the acinar cells of the pancreas, prostate tissue and the liver. The γ-GT molecule is made up of two subunits, one larger than the other [1 – 3] and has a relative molecular mass of about 86000 [3]. At one end of the larger subunit is a sequence of approximately 40 hydrophobic amino acids that are believed to form the enzyme's anchor to the lipid matrix of the plasma membrane [2]. The rest of the molecule is carbohydrate rich [4], a property that facilitates its interaction with the surrounding plasma. It has been shown that the γ-glutamyl binding site is located on the low molecular weight subunit [5].

γ-GT is located primarily on the outer surface of plasma membranes. This conclusion is based on studies with several tissues including human lymphoid cells, rat proximal tubular epithelial cells, and rat and rabbit liver [6 – 9].

The question of the function of γ-GT has been the subject of many studies. Based on these we believe that γ-GT plays a significant role in the following aspects of glutathione (GSH) metabolism: (a) hydrolysis of GSH; (b) hydrolysis of S-conjugates of GSH to provide the first step in the mercapturic acid pathway; (c) the transfer of the γ-glutamyl moiety to certain amino acids which may be involved in the uptake of amino acids by cells [7, 10 – 13]. We believe that γ-GT in human liver may play a more significant role in total body GSH metabolism than its presumed minimal role in rat liver, since its catalytic activity in normal human liver is twelve-fold higher than in rat liver and catalytic activity in human kidney is approximately 8.5 times lower than that in rat kidney [12].

Application of method: in clinical chemistry. It is generally agreed that clinically significant elevations of γ-GT catalytic activity in patients' sera are associated almost exclusively with hepatobiliary disease [14, 15]. Thus, measurements of total γ-GT activity in patients' sera are regarded as being liver-specific. However γ-GT in sera from normal subjects and liver disease patients is separable into multiple molecular forms which differ in electric charge and relative molecular mass but have similar kinetic properties [15]. The multiple molecular forms of γ-GT in serum are generally regarded as arising by various post-translational modifications (e.g. complex formation with lipoproteins or modifications in the carbohydrate portion of the molecule) [14, 16, 17]. Based on current knowledge of the kinetic, electrophoretic and immunological properties of γ-GT and the fact that the lectin binding properties of γ-GT from four human tissues (liver, pancreas, kidney and duodenum) differ [17, 18], it is likely that the polypeptide portion of γ-GT from different tissues is the same but that the carbohydrate portion differs.

Enzyme properties relevant in analysis: for the reaction mechanism cf. Vol. I, chapter 2.1.3.2, p. 80.

A number of donor substrates have been used for the assay of γ-GT activity. In general, synthetic γ-glutamyl donor substrates are preferred to glutathione, a natural substrate, because of the much greater analytical convenience of the former compared to the latter. L-γ-glutamyl-3-carboxy-4-nitroanilide is preferred to the non-carboxylated substrate because of the much greater solubility of the former and is preferred to another water-soluble substrate, L-γ-glutamyl-3-sulphonic-4-nitroanilide, since the rate of change of absorbance is 20% greater with the carboxylated substrate [19]. Glycylglycine, the preferred γ-glutamyl acceptor substrate, supports high catalytic activity compared to many other acceptor substrates. Several acceptor substrates including L-cystine, L-methionyl glycine, L-glutarinyl glycine, L-alanyl glycine and cystinyl bis-glycine support greater catalytic activities than glycylglycine with rat kidney γ-GT [20]. However, glycylglycine gives the least degree of substrate inhibition compared to these substrates. L-Cystine is of limited practical use due to its low solubility. Glycylglycine is the preferred acceptor substrate since it supports high catalytic activity, relatively low substrate inhibition, a pK_a near the optimum pH, high solubility and low cost. The reference method proposed by the Expert Panel on Enzymes of the IFCC includes the use of L-γ-glutamyl-3-carboxy-4-nitroanilide and glycylglycine as substrates [21].

Methods of determination: the methods described below are based on the principles developed by *Orlowski & Meister* [22] and *Szasz* [23] and later modified by *Persijn & Van der Slik* [19]. The methods use L-γ-glutamyl-3-carboxy-4-nitroanilide, Glucana, as donor substrate. Tris(hydroxymethyl)aminomethane is replaced as the buffer by glycylglycine (the acceptor substrate), and magnesium, formerly included to maintain L-γ-glutamyl-4-nitroanilide in solution, is eliminated.

International reference methods and standards: a provisional IFCC reference method has been published [21]. International reference enzyme material for γ-GT is under development [24]. Standard reference material from National Bureau of Standards (USA) is available: lyophilized human serum, SRM 909 (cf. Vol. II, chapter 2.3).

Enzyme effectors: γ-GT purified from various human tissues is inhibited non-competitively by Glucana at high ratios of concentration of the latter to glycylglycine. Non-competitive inhibition by Glucana has not been observed with serum γ-GT.

Glycine competitively inhibits γ-GT purified from tissues and the γ-GT in human serum. γ-GT in serum and purified from tissues is inhibited by glycylglycine at high concentrations of the latter relative to the donor substrate Glucana.

References

[1] *P. Zelazo, M. Orlowski,* γ-Glutamyltranspeptidase of Sheep Kidney Cortex. Isolation, Catalytic Properties and Dissociation into Two Polypeptide Chains, Eur. J. Biochem. *61*, 147 – 155 (1975).

[2] *R. P. Hughey, N. P. Curthoys,* Comparison of the Size and Physical Properties of γ-Glutamyltranspeptidase Purified from Rat Kidney Following Solubilization with Papain or with Triton X-100, J. Biol. Chem. *251*, 7863 – 7880 (1976).

[3] *S. S. Tate, M. E. Ross,* Human Kidney γ-Glutamyltranspeptidase, J. Biol. Chem. *252*, 6042 – 6045 (1977).

[4] *A. Meister, S. S. Tate, L. L. Ross,* Membrane-Bound γ-Glutamyl Transpeptidase; in: *A. N. Martonosi* (ed.), Enzymes of Biological Membranes, Vol. 3, Plenum, New York 1976, pp. 315 – 347.

[5] *S. S. Tate, A. Meister,* Affinity Labeling of γ-glutamyltranspeptidase and Location of the γ-Glutamyl Binding Site on the Light Subunit. Proc. Natl. Acad. Sci. *74*, 931 – 935 (1977).

[6] *G. V. Marathe, N. S. Damle, R. H. Haschemeyer, S. S. Tate,* Localization of γ-Glutamyl Transpeptidase in Lymphoid Cells, FEBS Lett. *115*, 273 – 277 (1980).

[7] *K. Ormstad, M. Tanaka, S. Orrenius,* Function of γ-Glutamyltransferase in the Kidney; in: *G. Siest, C. Heusghem* (eds.), Gamma-Glutamyltransferases, Masson, Paris 1982, pp. 7 – 14.

[8] *D. Ratanasavanh, A. Tazi, M. Galteau, G. Siest,* Localization of Gamma-Glutamyltransferase in Subcellular Fractions of Rat and Rabbit Liver: Effect of Phenobarbital, Biochem. Pharmacol. *28*, 1363 – 1365 (1979).

[9] *G. D. Smith, T. J. Peters,* Studies on the Subcellular Localization of γ-Glutamyltranspeptidase in Rat Liver, Biochem. Soc. Trans. *6*, 543 – 545 (1978).

[10] *J. S. Elce, B. Broxmeyer,* γ-Glutamyltransferase of Rat Kidney. Simultaneous assay of the hydrolysis and transfer reactions with [glutamate-^{14}C] glutathione, Biochem. J. *153*, 223 – 23 (1976).

[11] *L. M. Shaw, D. A. Newman,* Hydrolysis of Glutathione by Human Liver γ-Glutamyltransferase, Clin. Chem. *25*, 75 – 79 (1979).

[12] *L. M. Shaw, J. W. London,* Kinetic Properties of Human Liver γ-Glutamyltransferase: Comparison with Kidney Enzyme; in: *G. Siest, C. Heusghem* (eds.), Gamma-Glutamyltransferases, Masson, Paris 1982, pp. 23 – 29.

[13] *R. D. Allison, A. Meister,* Evidence that Transpeptidation is a significant Function of γ-Glutamyl Transpeptidase, J. Biol. Chem. *256*, 2988 – 2992 (1982).

[14] *D. M. Goldberg,* Structural, Functional and Clinical Aspects of γ-Glutamyltransferase; in: *J. Batsakis, J. Savory* (eds.), CRC Critical Reviews in Clinical Laboratory Sciences, Vol. 12, CRC Press, Boca Raton, FL 1982, pp. 1 – 58.

[15] *D. W. Moss, Z. O. Echetebu, K. B. Whitaker, O. J. Traynor, C. B. Wood,* Multiple Forms of γ-Glutamyl Transferase and Their Clinical Significance; in: *G. Siest, C. Heusghem* (eds.), Gamma-Glutamyltransferases, Masson, Paris 1982, pp. 41 – 45.

[16] *N. E. Huseby,* Multiple Forms of γ-Glutamyltransferase-biochemical Characterization, in: *G. Siest, C. Heusghem* (eds.), Gamma-Glutamyltransferases, Masson, Paris 1982, pp. 47 – 54.

[17] *L. M. Shaw, L. Petersen-Archer, J. W. London, E. Marsh,* Electrophoretic, Kinetic and Immunoinhibition Properties of γ-Glutamyltransferase from Various Tissues Compared, Clin. Chem. *26*, 1523 – 1527 (1980).

[18] *L. M. Shaw, J. W. London, L. Petersen-Archer,* Isolation of γ-Glutamyltransferase from Human Liver, and Comparison with the Enzyme from Human Kidney, Clin. Chem. *24*, 905 – 915 (1978).

[19] *J. P. Persijn, W. Van der Slik,* A New Method for the Determination of γ-Glutamyltransferase in Serum, J. Clin. Chem. Clin. Biochem. *14*, 421 – 427 (1976).

[20] *G. A. Thompson, A. Meister,* Interrelationships between the Binding Sites for Amino Acids, Dipeptides, and γ-Glutamyl Donors in γ-Glutamyl Transpeptidase, J. Biol. Chem. *252*, 6792 – 6798 (1977).

[21] Expert Panel on Enzymes, Committee on Standards (IFCC), IFCC Methods for Measurement of Enzymes, Part 4. IFCC Method for γ-Glutamyltransferase, Stage 2, 1983.

[22] *M. Orlowski, A. Meister,* γ-Glutamyl-p-nitroanilide: A New Convenient Substrate for Determination and Study of L- and D-γ-Glutamyltranspeptidase Activities, Biochim. Biophys. Acta *73*, 679 – 681 (1963).

[23] *G. Szasz,* A Kinetic Photometric Method for Serum γ-Glutamyl Transpeptidase, Clin. Chem. *15*, 124 – 136 (1969).

[24] *European Economic Community* Bureau of Reference: Working Group on Enzyme Reference Materials, 1983.

5.2.2 Routine Method

August Wilhelm Wahlefeld and Hans Ulrich Bergmeyer

The rigorously-defined provisional IFCC method is also recommended as the routine method. Nevertheless, for continuity, we describe here also the routine method [1] widely used today.

Its modifications with regard to the IFCC method – i.e. lower concentration of γ-glutamyl-3-carboxy-4-nitroanilide (Glucana) and glycylglycine, the addition of Tris as buffer substance and a different pH value – were chosen to obtain identical catalytic activity with the method in which the substrate is unsubstituted γ-glutamyl-4-nitroanilide, according to *Szasz* [2]. All the clinical experience obtained with the older method of *Szasz,* including the reference values, can thus be used with the newer method.

Assay

Method Design

Principle

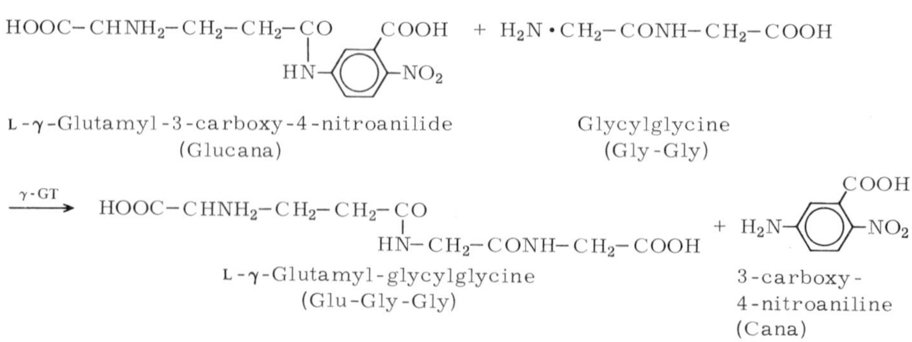

3-Carboxy-4-nitroaniline has an absorption maximum at 380 nm. Since glucana does not absorb significantly above 410 nm, the catalytic activity of γ-GT is measured by monitoring the increase of absorption at Hg 405 nm or higher wavelengths (405 – 410 nm).

Optimum conditions of measurement: as mentioned above, the chosen assay conditions did not aim at the highest catalytic activity possible, but at catalytic activity identical with that given by the *Szasz* method [2]. The variables of the method are described in detail by *Persijn* [3]. The optimal pH value was found to be 8.25 (at 25 °C) with Tris buffer, 100 mmol/l. Tris ions are inhibitory above 200 mmol/l. The concentration of the acceptor substrate, glycylglycine, was set at 100 mmol/l. Under these conditions the concentration of the donor substrate, L-γ-glutamyl-3-carboxy-4-nitroanilide, had to be set at 2.9 mmol/l.

Temperature conversion factors: have been investigated in detail by *Staehler et al.* [4]. The factors showed no dependence on the catalytic activity up to 800 U/l, and no loss of accuracy was seen when measured and calculated catalytic concentrations were compared (for example: measured at 25 °C and 37 °C, but values at 37 °C recalculated to their equivalents at 25 °C). The following temperature conversion factors can be recommended:

	factors for conversion to		
measured	25 °C	30 °C	37 °C
25 °C	1.0	1.35	1.79
30 °C	0.74	1.0	1.32
37 °C	0.56	0.76	1.0

Equipment

As described on p. 361.

Reagents and Solutions

Purity of reagents: as described on p. 361.

Preparation of solutions (for about 50 determinations): all solutions are made with re-purified water (cf. Vol. II, chapter 2.1.3.2).

1. Tris/glycylglycine (Tris, 120 mmol/l; pH 8.25; 25 °C; Gly-Gly, 120 mmol/l):

   dissolve 1.452 g Tris in 90 ml water, while stirring add 1.584 g glycylglycine and adjust to pH 8.25 (25 °C) with diluted hydrochloric acid, 2 mol/l. Sodium azide

(100 mg) should be added to obtain sufficient anti-microbial preservation. Dilute with water to 100 ml.

2. L-γ-Glutamyl-3-carboxy-4-nitroanilide (Glucana, 35.1 mmol/l):

   dissolve 121.6 mg of L-γ-glutamyl-3-carboxy-4-nitroanilide, mono-ammonium salt, in 10 ml water containing 1 mg sodium azide per ml to prevent microbial growth. Check pH and adjust to 7 – 7.5.

3. Tris/Gly-Gly/Glucana (Tris, 110 mmol/l; pH 8.25, 25 °C; Gly-Gly, 110 mmol/l; Glucana, 3.2 mmol/l):

   mix 20 ml solution (1) with 2 ml solution (2).

Stability of solutions: store all solutions at 2 °C to 8 °C. The Tris/glycylglycine solution (1) is stable for 3 months, the Glucana solution (2) for 3 weeks. The reagent Tris/glycylglycine/Glucana solution (3) is stable for 10 days.

Procedure

Collection and treatment of specimens: collect blood from vein without stasis; avoid haemolysis. Only serum or EDTA plasma (up to 1 mg/ml blood) can be used. Citrate-, oxalate- and fluoride plasma give about 10 – 15% lower results, whereas heparin plasma produces turbidity in the reaction mixture. Centrifuge for 10 min at about 3000 *g* to separate serum or EDTA plasma.

Stability of enzyme in the sample: according to our experience, the enzyme in serum is stable for seven days at ambient temperature as well as at 2 °C to 8 °C.

Details for measurement in tissue: no recommendations can be given.

Assay conditions: wavelength Hg 405 nm; light path 10 mm; final volume 2.2 ml; 25 °C, 30 °C, 37 °C (thermostatted cuvette holder). Measure against air. A reagent blank is not necessary. Before starting the assay, adjust the temperature of solutions to the measuring temperature.

When the values $\Delta A/\Delta t$ exceed 0.200 $\min^{-1}$ at Hg 405, dilute sample 1 + 5 with 0.9% sodium chloride solution. ($\Delta A/\Delta t$ value × 6 is used in calculation).

Calculation: use formula (k) or (k_1) (cf. "Formulae", Appendix 3). With $\varepsilon = 0.95\ l \times mmol^{-1} \times mm^{-1}$ [5], and t in minutes, the catalytic concentration in the sample is

$$b = 1158 \times \Delta A/\Delta t \quad U/l$$

$$b = 19297 \times \Delta A/\Delta t \quad nkat/l\,.$$

Measurement

Pipette successively into the cuvette:			concentration in assay mixture	
Tris/glycylglycine/Glucana	(3)	2.0 ml	Glucana Gly-Gly Tris	2.9 mmol/l 100 mmol/l 100 mmol/l
wait for 5 min, until temperature of 25, 30 or 37 °C is reached				
sample		0.2 ml	volume fraction	0.091
mix with plastic spatula and read absorbance, start stopwatch simultaneously and repeat readings after 1, 2 and 3 min or monitor the reaction continuously on a recorder. Use mean increase of absorbance per min during the first three min for calculations.				

When measurement at Hg 405 nm is not possible, it is recommended, especially for filter photometers, to determine the exact absorption coefficient of the product 3-carboxy-4-nitroaniline under the special conditions; for example, the value depends on the bandwidth of the filter used. Pure 3-carboxy-4-nitroaniline (M_r = 182.14) can be obtained commercially. Prepare an aqueous solution, concentration about 0.3 mmol/l (calculate exact concentration according to the weighing protocol). Make a dilution of 2.0 ml with 7.5 ml water. Place 2.5 ml of this solution in the cuvette in the photometer, adjust to $A = 0$, then add the undiluted solution in five different steps, each of 0.2 ml, and measure the absorbances. Calculate from these five measurements the absorption coefficient and take the mean value for calculations.

Validation of Method

Precision, accuracy, detection limit and sensitivity: the following values of imprecision from day-to-day were found for manual performance

$n = 20$;	20 U/l	RSD	3.8%
	93 U/l	RSD	2.3% .

Data about accuracy are not yet available since the reference method is not fully accepted yet. The correlation with the *Szasz* method has been intensively studied and found to be:

$y = 0.43 + 0.985\,x; \qquad r = 0.999,$

where

x = *Szasz* method; y = method described here, for 60 sera from patients, range 9 U/l to 500 U/l.

For measurements at Hg 405 nm with sufficiently sensitive photometers, the lowest value which can be measured is approximately 14 U/l with an imprecision of about 10%. Analytical sensitivity is $\Delta A/\Delta t = 0.010/5$ min with an *Eppendorf* spectral-line photometer.

Sources of error: the glycylglycine should be free of glycine, since glycine is a competitive inhibitor, $K_I = 2.3 \times 10^{-3}$ mol/l. During extended storage at higher temperature glycylglycine solutions may be partially hydrolyzed to glycine. Haemolysis, above 50 mg haemoglobin/100 ml, decreases γ-GT activity, but icteric and lipaemic samples do not influence measurement.

No influence on the assay could be detected in experiments *in vitro* with addition of common drugs (48 different drugs were tested). *In vivo*, it is well known that drugs can increase serum levels of γ-GT, and especially alcohol abuse [6, 7].

Specificity: the assay described measures activity of γ-GT specifically; other tranpeptidases do not use Glucana as donor substrate.

Reference ranges: in human serum, reference values of 6–28 U/l (25 °C) for men and 4–18 U/l (25 °C) for women have been reported. (*G. Szasz*, personal communication).

References

[1] *G. Szasz, G. Weimann, F. Staehler, A. W. Wahlefeld, J. P. Persijn,* New Substrates for Measuring γ-Glutamyltranspeptidase Activity. J. Clin. Chem. Clin. Biochem. *12*, 228 (1974).

[2] *G. Szasz,* A Kinetic Photometric Method for Serum γ-Glutamyl-transpeptidase. Clin. Chem. *15*, 124–136 (1969).

[3] *J. P. Persijn, W. van der Slik,* A New Method for the Determination of γ-Glutamyl-transferase in Serum. J. Clin. Chem. Biochem. *14*, 421–427 (1976).

[4] *F. Staehler,* personal communication (1976). Lecture given at the MEDICA Congress, Düsseldorf, 1976.

[5] *P. Lehmann, H. Lill, E. Schaich, M. Grassl,* The Dependance of the Molar Absorptivity of 4-Nitroaniline, 3-Carboxy-4-nitroaniline and 4-Nitrophenol on Reaction Conditions, Enzymes in Health and Disease, Inaug. Meeting Int. Soc. Clin. Enzymology, London 1977, Karger, Basel, 1978, p. 107–114.

[6] *E. Koettgen,* Die γ-Glutamyltransferase. Pathobiochemie und differentialdiagnostische Wertigkeit. Internist *21*, 213–235 (1980).

[7] *D. Acheampong-Mensah,* Activity of γ-Glutamyl Transpeptidase in Serum of Patients Receiving Anticonvulsant or Anticoagulant Therapy. Clin. Biochem. *9*, 67–70 (1976).

5.2.3 IFCC Reference Method (provisional)

Leslie M. Shaw and Johan H. Strømme

The method we describe for the measurement of the catalytic activity of γ-glutamyltransferase (γ-GT) in serum is the reference method proposed by the Expert Panel on Enzymes of the Commission of Standards of the IFCC [1].

Assay

Method Design

Principle

(a) $\text{Glucana} + \text{Gly-Gly} \xrightarrow{\gamma\text{-GT}} \text{Glu-Gly-Gly} + \text{Cana}$

(b) $\text{Glucana} + \text{Glucana} \xrightarrow{\gamma\text{-GT}} \text{Glu-Glucana} + \text{Cana}$

Glucana (L-γ-glutamyl-3-carboxy-4-nitroanilide) is the donor substrate. Gly-Gly (glycylglycine) is the acceptor substrate.

Under the proposed conditions of measurement, reaction (b) – the autotransfer reaction – accounts for <1 percent of the product formed. The reaction proceeds in the forward direction under the assay conditions and the reverse reaction is not detectable.

The reaction product Cana (3-carboxy-4-nitroanilide or 5-amino-2-nitrobenzoate) absorbs light, with maximum at 380 nm. The γ-GT activity is directly proportional to the amount of Cana liberated per unit time.

Optimized conditions for measurement: these are defined as those conditions which are most favourable for both the kinetic requirements of γ-glutamyltransferases and the technical aspects of the measurement [2]. As is true for the measurement systems for all enzymes, the components of the γ-GT method are interdependent. The reaction conditions described below have been optimized by univariate, initial velocity kinetic and response surface methodology experiments with γ-GT in sera from healthy subjects and patients with liver disorders, and with isolated γ-GT isoenzymes [1].

Buffer and pH

Although tris(hydroxymethyl)aminomethane has been the most widely used buffer in numerous studies of γ-GT it is not included as buffer in the proposed method since it inhibits γ-GT activity. Glycylglycine, the acceptor substrate, has a pK of 8.21 (30 °C) which is close to the optimum pH value of 7.9. Therefore, adequate buffer capacity is provided at the concentration selected for this substrate. A study of the variation of γ-GT catalytic activity as a function of pH and substrate concentrations (1.0 to 11 mmol/l for Glucana and 50 to 250 mmol/l for glycylglycine) has confirmed that a pH in the vicinity of 7.9 is optimal [3].

Specimen type and volume fraction of sample

γ-GT activities measured in plasma from blood samples anticoagulated by heparin with or without fluoride, citrate or EDTA are generally lower than those measured in the corresponding sera. Heparin shows the strongest effect and EDTA the weakest effect. The heparin effect varies markedly from specimen to specimen and the degree of reduction becomes less pronounced as the time before taking the first reading is increased. This effect is present regardless of whether the reaction is initiated with sample, as recommended in the IFCC method, or with the donor substrate.

γ-GT activity is independent of the volume fraction of sample within the range of volume fraction values of 0.009 to 0.228. The volume fraction of 0.091 was selected in order to minimize high initial absorbance values produced by icteric sera and to provide a method with a large analytical range. The selected volume fraction achieves these goals and assures a significant signal-to-noise ratio for sera with γ-GT activities within the reference range.

Selection of wavelength for monitoring catalytic activity

The monitored product, 5-amino-2-nitrobenzoate, has an absorbance maximum at 380 nm at pH 7.9. At this wavelength the donor substrate has an unacceptably high absorbance at the concentration used in the reaction mixture. The wavelength of 410 nm was selected for the IFCC method since the molar linear absorbance of the product is sufficiently high and the blank absorbance due to the donor substrate is reasonably low (about 0.45). It is important to verify the photometric linearity at 410 nm of the instrument used for monitoring catalytic activity over the range of absorbance actually measured.

Substrate concentration

γ-GT catalyzes the transfer of a γ-glutamyl group from a donor substrate (D) to an acceptor substrate (A):

$$\gamma\text{-Glutamyl-D} + \text{A} \longrightarrow \text{D} + \gamma\text{-glutamyl-A}$$

The γ-glutamyl moiety may be transferred to one or more of three essentially different acceptor substrates: water (hydrolytic reaction); the donor substrate itself (autotransfer reaction); or a particular acceptor substrate (transfer reaction) [4, 5]. The enzyme has a low substrate specificity, with a number of L-γ-glutamyl peptides (including glutathione, S-conjugates of glutathione and synthetic γ-glutamyl compounds such as Glucana) serving as donor substrates and a variety of peptides and amino acids that may serve as acceptor substrates. The γ-GT acceptor site has absolute stereospecificity for L-acceptor substrates but the donor-binding site reacts with D-stereoisomers as well as the L-stereoisomers, although the K_m value for D-Glucana (3.61 mmol/l) is 5.6-fold higher than the K_m value for L-Glucana (0.65 mmol/l). Reaction rates obtained with the former are much lower than with the latter [1]. With L-Glucana and glycylglycine as substrates, serum γ-GT catalyzes a transfer (predominant) and a auto-transfer reaction. Under the conditions usually employed for the assay of γ-GT in serum using synthetic substrates the hydrolytic reaction is negligible compared to the transferase reactions [6, 7]. On the other hand, with the physiological substrate, glutathione, the hydrolytic reaction is significant when acceptor substrates are present at physiological concentrations [8, 9]. A "ping-pong bi-bi" mechanism for γ-GT is consistent with numerous kinetic studies [6, 10–13].

At high ratios of acceptor to donor substrate concentration glycylglycine inhibits γ-GT. At high ratios of concentrations of donor substrate to acceptor substrate the overall reaction is reduced due to the displacement of the more efficient acceptor substrate, glycylglycine, from the acceptor site of the enzyme [13]. γ-GT purified from various human tissues is inhibited non-competitively by the donor substrate at high ratios of concentration of the latter to glycylglycine [14, 15]. Non-competitive inhibition by Glucana has not been observed with serum γ-GT [6, 15].

The relationship between γ-GT catalytic activity at pH 7.9 and substrate concentrations has been studied and reported [13]. As is true for any two-substrate enzyme, there are many substrate-concentration pairs (cf. Vol. I, p. 119) that give a particular percentage of the theoretical maximal velocity. Of the factors which must be taken into account in selecting optimal substrate concentrations for γ-GT, the absorbance at 410 nm (or 405 nm) of Glucana is an important one. The initial absorbance of a reaction mixture containing 1 mmol/l in glycylglycine, 150 mmol/l, pH 7.9 (30 °C), is 0.075. Although concentrations of 24 mmol/l of donor substrate and 400 mmol/l of acceptor substrate give the highest catalytic activity ($0.85 \times V$), the initial absorbance value contributed by the donor substrate is 1.8, a value which is unacceptably high. The initial absorbance value will of course be even higher than this due to the absorbance contribution from the specimen, which is often substantial. Thus a lower value of 6 mmol/l was selected for the donor-substrate concentration. This gives an initial absorbance of about 0.45 for the reaction mixture in the absence of the specimen. This absorbance will slowly increase on storage of the reagents due to spontaneous hydrolysis of the donor substrate.

The optimal glycylglycine concentration changes with the concentration of the donor substrate and at high ratios of concentration of glycylglycine to Glucana the inhibition by glycylglycine is greatest, because it is competitive in nature. A glycylglycine concentration of 150 mmol/l was selected since it is in the plateau region of the activity *vs.* concentration curve at 6 mmol/l Glucana. Thus, small changes in glycylglycine concentration from the nominal value of 150 mmol/l will produce minimal change in catalytic activity at a donor-substrate concentration of 6 mmol/l. At concentrations of 6 mmol/l and 150 mmol/l of donor and acceptor substrates respectively, 75% of the theoretical maximal velocity is achieved. No further increase in catalytic activity is obtained by increasing glycylglycine concentration above 150 mmol/l and, beginning with 300 mmol/l, lower catalytic activities are obtained due to competitive substrate inhibition. These conclusions have been verified with the empirical optimization technique of response surface methodology [3].

Table 1. Optimal conditions for IFCC reference method, provisional

Reaction temperature	30 °C
pH	7.9
Volume fraction of serum	0.091
Wavelength for monitoring the reaction	410 nm
L-γ-Glutamyl-3-carboxy-4-nitroanilide	6 mmol/l
N-glycylglycine	150 mmol/l
volume fraction of sample φ	0.75

The reaction is initiated with serum. Equivalent results are obtained when the reaction is initiated with the substrate Glucana and when serum is used.

Temperature conversion factors: the standardized reaction mixture functions equally well at the three most widely used temperatures for initial velocity kinetic assays – 25 °C, 30 °C and 37 °C (Fig. 1). Using the recommended assay conditions we obtained average values of 0.758 (SD 0.021) and 1.34 (SD 0.033) for the ratio of catalytic activity at 25 °C to that at 30 °C and at 37 °C to that at 30 °C, respectively, for 54 patients' sera (16 to 684 U/l with the IFCC method).

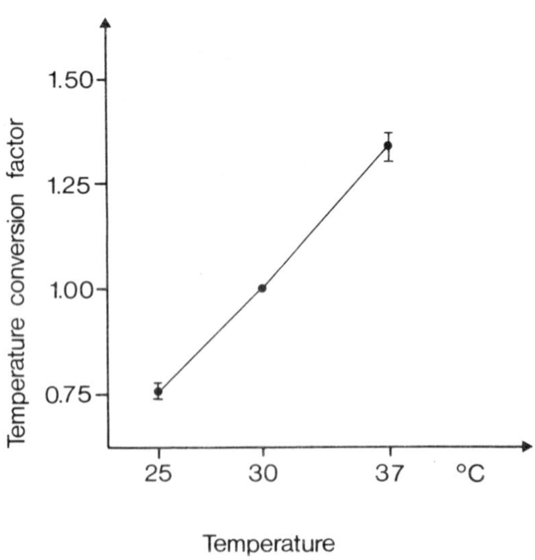

Fig. 1. Mean (± SD) temperature conversion factor as a function of temperature.

Equipment

A precision monochromator, filter- or spectral-line photometer is required, capable of isolating 410 nm, or alternatively 405 nm, with thermostatic control of the cuvette temperature. The specifications for the equipment (i.e. accuracy of the wavelength setting, spectral bandwidth, light path and accuracy of temperature control) should meet those of previous recommendations [2].

Reagents and Solutions

Purity of reagents

(a) Glucana: Pure L-stereoisomer is required since the D-stereoisomer is only 30% as reactive as the L-stereoisomer using the reaction conditions described in Table 1 (30 °C). Other contaminants of this substrate that may be present are L-α-glutamyl-3-carboxy-4-nitroanilide and 5-amino-2-nitrobenzoic acid. These inhibit γ-GT weakly (L-α-glutamyl-3-carboxy-4-nitroanilide) or not at all (5-amino-2-nitrobenzoic acid) but their quantities should be as low as practically possible in order to avoid high reagent-blank absorbance values due to the presence of the latter contaminant, and to avoid weighing out falsely low quantities of Glucana. It is recommended that preparations of Glucana with <0.5% (w/w) of the L-γ-glutamyl isomer and <0.1% (w/w) of 5-amino-2-nitrobenzoic acid be used.

(b) N-glycylglycine: glycine may be present as a contaminant in glycylglycine preparations. Since glycine is an inhibitor of γ-GT it is recommended that glycylglycine with the lowest possible glycine content (i.e. <0.1% w/w) be used.

Preparations of solutions: to prevent the growth of micro-organisms in the solutions, sterilized containers should be used. All solutions should be prepared in calibrated flasks with fresh re-purified water (cf. Vol. II, chapter 2.1.3.2). The pH of the glycylglycine stock solution must be adjusted at 30 °C using a pH-meter standardized with reference buffers (e.g. IUPAC or NBS).

1. N-Glycylglycine solution (183.3 mmol/l; pH 7.90 at 30 °C):

   dissolve 24.22 g glycylglycine in approximately 800 ml water, adjust pH to 7.90 at 30 °C with NaOH, 2 mol/l. Adjust the volume to 1000 ml with water.

2. L-γ-Glutamyl-3-carboxy-4-nitroanilide solution (66.0 mmol/l):

   dissolve 2.286 g L-γ-glutamyl-3-carboxy-4-nitroanilide, monoammonium salt, monohydrate, in water and dilute to 100.0 ml.

3. Reagent solution (Glucana, 6.6 mmol/l, Gly-Gly, 165 mmol/l); pH 7.90 at 30 °C):
   mix 10.00 ml of L-γ-glutamyl-3-carboxy-4-nitroanilide stock stolution (2) with 90.0 ml of glycylglycine solution (1). The absorbance at 410 nm of this solution measured against a water blank should not exceed 0.500.

4. Sodium chloride solution (154 mmol/l):
   dissolve 9.00 g sodium chloride in a total final volume of 1000 ml water.

Stability of solutions: store solutions (1) and (2) at 4 °C or −20 °C. Stability at these temperatures is 1 month and 9 months, respectively [1]. At room temperature bacterial contamination of solutions (1) and (2) is a limiting factor. Moreover, at room temperature spontaneous hydrolysis of Glucana produces a gradual increase in absorbance at Hg 405 nm or 410 nm of solution (2) (after 5 days at room temperature an absorbance increase of about 0.3 at 410 nm is produced). Thus, it is recommended that solution (3) be prepared on each day it is used from solutions (1) and (2) that have been stored at either 4 °C or −20 °C.

Procedure

Collection and treatment of specimen and sample: venous blood is collected under conditions that minimize stasis. For serum collection allow blood to clot and centrifuge for 10 min at a relative centrifugal force of 1000 *g* to obtain cell-free serum. Plasma should not be used.

Stability of the enzyme in the sample: serum can be stored at least 5 days at 4 °C or for 9 months at −80 °C with no loss of catalytic activity of γ-GT when measured by the proposed method.

Assay conditions: wavelength 410 nm (±1 nm); bandwidth ≤2 nm; light path 10.0 mm ± 0.01 mm; volume of reaction mixture 2.20 ml; temperature 30.0 °C ± 0.05 °C (thermostatted cuvette compartment).

Measurement

Pipette successively into cuvettes:	sample	blank	concentration in assay mixture
reagent solution (3)	2.00 ml	2.00 ml	Glucana 6 mmol/l Gly-Gly 150 mmol/l
equilibrate temperature at 30.0 °C			
sample (serum) NaCl solution (4)	0.20 ml –	 0.20 ml	volume fraction φ = 0.09
mix, equilibrate at 30.0 °C; record change in absorbance for up to 5 min.			

The values of $\Delta A/\Delta t$ for the γ-GT reaction rate are constant over an interval of at least 5 min for sera with catalytic concentrations of γ-GT up to about 300 U/l. If the value of $\Delta A/\Delta t$ is greater than about 0.215 per minute dilute the sample 5 to 10 fold with solution (4) and repeat the measurement. The period of observation of the reagent-blank rate should be the same as for the overall reaction.

Calculation: use calculation formula (k) or (k_1), respectively (cf. "Formulae", Appendix 3). The catalytic concentration in the sample with $\varepsilon = 0.7908 \text{ l} \times \text{mmol}^{-1} \times \text{mm}^{-1}$, and t in minutes is given by:

$$b = \frac{2.2 \times 10^3}{7.908 \times 10^2 \times 10 \times 0.2} \times \Delta A/\Delta t = 1.391 \times 10^3 \times \Delta A/\Delta t \quad \text{U/l}$$

or

$$b = 2.318 \times 10^4 \times \Delta A/\Delta t \quad \text{nkat/l}$$

Validation of Method

Precision, accuracy, detection limits and sensitivity: a within-run relative standard deviation of 0.016 was obtained at a catalytic activity of 15.6 U/l (n = 20). Between-run relative standard deviations of 0.029 and 0.021 were obtained when sera of, respectively, 40.2 and 140 U/l were assayed over a period of 9 months (n = 16). In a study of between-laboratories variation using lyophilized human serum (SRM 909 from the National Bureau of Standards) a relative standard deviation of 0.054 was obtained at a catalytic activity of 16.2 U/l (5 laboratories, n = 200) [16].

A study of the accuracy of γ-GT measurement using a well characterized reference standard has not yet been conducted. The lowest value of γ-GT activity that is measurable using the IFCC method and a Cobas Bio centrifugal analyzer has been determined. Under these conditions a detection limit value of 1.84 U/l ± 0.18 U/l (n = 25) found. The sensitivity is estimated to be an increase in absorbance at 410 nm of 0.001/min using the Cobas Bio centrifugal analyzer.

Sources of error: there is little published information on the direct effect of drugs on γ-GT activity. In evaluating elevated γ-GT values, chronic alcohol consumption, gross obesity and therapy with anticonvulsants should be considered.

Specificity: the method is specific for γ-GT.

Reference ranges: the catalytic activity of γ-glutamyltransferase in the serum of healthy adults is presently being determined with the reference method. For the first

173 males (19 – 64 years old) we obtained a reference range of 12 – 64 U/l and for the first 50 females we obtained a range of 9 – 36 U/l. The results of this study will be reported in greater detail when we have accumulated values on a sufficiently large population of healthy men and women.

References

[1] Expert Panel on Enzymes, Committee on Standards (IFCC), IFCC Methods for Measurement of Enzymes, Part 4. IFCC Method for γ-Glutamyltransferase, Stage 2, 1983.

[2] Expert Panel on Enzymes, Committee on Standards (IFCC), Approved Recommendations of IFCC Methods for the Measurement of Catalytic Concentrations of Enzymes, Part I. General Considerations, Clin. Chim. Acta. IFCC Sections 98, 163F – 174F (1979).

[3] *L. W. London, L. M. Shaw, L. Theodorsen, J. H. Strømme,* The Application of Response Surface Methodology to the γ-Glutamyltransferase Assay, Clin. Chem. *28*, 1140 – 1143 (1982).

[4] *M. Orlowski, A. Meister,* γ-Glutamyl-p-nitroanilide: A New Convenient Substrate for Determination and Study of L- and D-γ-Glutamyltranspeptidase Activities, Biochim. Biophys. Acta. *73*, 679 – 681 (1963).

[5] *P. Zelazo, M. Orlowski,* γ-Glutamyltranspeptidase of Sheep Kidney Cortex. Isolation, Catalytic Properties and Dissociation into two Polypeptide Chains, Eur. J. Biochem. *61*, 147 – 155 (1976).

[6] *L. M. Shaw, J. W. London, D. Fetterolf, D. Garfinkel,* γ-Glutamyltransferase: Kinetic Properties and Assay Conditions when γ-Glutamyl-4-nitroanilide and its 3-Carboxy Derivative are used as Donor Substrates, Clin. Chem. *23*, 79 – 85 (1977).

[7] *L. M. Shaw, D. A. Newman,* Hydrolysis of Glutathione by Human Liver γ-Glutamyltransferase, Clin. Chem. *25*, 75 – 79 (1979).

[8] *J. S. Elce, B. Broxmeyer,* γ-Glutamyltransferase of Rat Kidney. Simultaneous assay of the hydrolysis and transfer reactions with [glutamate-^{14}C] glutathione, Biochem. J. *153*, 223 – 232 (1976).

[9] *T. M. McIntyre, N. P. Curthoys,* Comparison of the Hydrolytic and Transfer Activities of Rat Renal γ-Glutamyltranspeptidase, J. Biol. Chem. *254*, 6499 – 6505 (1979).

[10] *J. H. Strømme, L. Theodorsen,* γ-Glutamyltransferase: Substrate Inhibition, Kinetic Mechanism and Assay Conditions, Clin. Chem. *22*, 417 – 421 (1976).

[11] *S. S. Tate, A. Meister,* Interaction of γ-Glutamyltranspeptidase with Amino Acids, Dipeptides and Derivatives and Analogs of Glutathione, J. Biol. Chem. *249*, 7953 – 7602 (1974).

[12] *J. S. Elce, J. Bryson, L. G. McGirr,* γ-Glutamyltranspeptidase of Rat Kidney. Some properties and Kinetic Constants, Can. J. Biochem. *52*, 33 – 41 (1974).

[13] *H. E. Solberg, L. Theodorsen, J. H. Strømme,* γ-Glutamyltransferase in Human Serum: An Analysis of Kinetic Models, Clin. Chem. *27*, 303 – 307 (1981).

[14] *L. M. Shaw, J. W. London, L. E. Petersen,* Isolation of γ-Glutamyltransferase from Human Liver and Comparison with the Enzyme from Human Kidney, Clin. Chem. *24*, 905 – 915 (1978).

[15] *L. M. Shaw, L. Petersen-Archer, J. W. London, E. Marsh,* Electrophoretic, Kinetic and Immunoinhibition Properties of γ-Glutamyltransferase from Various Tissues Compared, Clin. Chem. *26*, 1523 – 1527 (1980).

[16] *G. N. Bowers, J. P. Cali, R. Elser, L. M. Ewen, R. B. McComb, R. Rej, L. M. Shaw,* NBS SRM (909) Study, (1983).

5.3 UDPglucuronosyltransferases

UDPglucuronate β-D-glucuronosyltransferase (acceptor-unspecific), EC 2.4.1.17

Terence Hallinan

General

UDP glucuronyltransferases (GT) are the most versatile and important conjugative enzymes of xenobiotic metabolism. They also conjugate numerous endogenous aglycones and β-glucuronides constitute the principal excretory derivatives of bilirubin, oestrone, oestradiol, androsterone and aetiocholanolone.

Glucuronyltransferases are versatile, being able to couple glucuronic acid with at least 12 separate functional groups, not including C–C glucuronides [1]. GT are important because of their high capacity, UDP-glucuronic acid (UDPGA*) being made from the large glycogen stores. Their location, embedded in endoplasmic reticular membranes (ER) adjacent to the mixed function oxidases which form many highly reactive imtermediates, also contributes to their importance. Most of the GT activity sediments in the microsomal fraction during subcellular fractionation.

Glucuronyltransferases are generally assayed in liver, but gastrointestinal tract, kidney cortex, lung, and skin are also quantitatively important sites and GT are even more widespread. Some GT activities are inducible by sex hormones and many xenobiotics also strongly induce the activity.

Dutton (1980) wrote a most authoritative and comprehensive book on glucuronidation [1]: this is equally valuable to those assaying GT and those writing about its assay.

Applications of method: in membrane biochemistry, studying regulation of latent membrane-embedded enzymes; in toxicology, measuring glucuronidation of xenobiotics and their metabolites; in clinical biochemistry (still poorly developed), mainly with regard to bilirubin glucuronidation at present.

Enzyme properties relevant in analysis: careful purifications, especially in the laboratories of *Burchell, Bock, Tephley & Kaspar* (reviewed in [2]), have demonstrated several GT isoenzymes. Their precise number and the exact boundaries between their

---

* Abbreviations:

UDPGA	uiridine 5′-diphosphoglucuronic acid
ER	endoplasmic reticular membranes
NMR	nuclear magnetic resonance
UDPNAG	uridine 5′-diphospho-N-acetylglucosamine.

aglycone specificities are not yet agreed, but there may be specific enzymes for bilirubin and 3-hydroxysteroids [2]. However, caution may be required over a third possible isoenzyme for 17-β-hydroxysteroids, since this has been monitored with testosterone, which is not extensively glucuronidated, 17-keto-steroids forming the major androgen glucuronides. It has been predicted that specific isoenzymes may exist for each of the major endogenous aglycones, while many xenobiotics could non-specifically cross-react with several GT-isoenzymes [1].

Glucuronyltransferases are phospholipid-dependent [3], apparently requiring phosphatidylcholines which are fluid at the assay temperature fully to express reaction rates and normal regulatory behaviour [4, 5]. For some xenobiotic aglycones, Lubrol $12A_9$ [2] or Brij 58 may substitute well for phospholipids, but they fail to do so for bilirubin or testosterone glucuronidation [2].

The kinetics of glucuronyltransferases are atypical, showing negative co-operativity in microsomes for UDPGA and for several aglycones [6, 7]. This complicates detailed kinetic analysis. In theory, measurement of the maximum velocity of conjugation requires a bi-substrate kinetic analysis. In practice, however, such kinetic analyses are much too tedious and expensive for routine rate measurements. Hence, a method using single fixed substrate concentrations is described below.

Methods of determination: the catalytic activity is generally measured spectrophotometrically or radiochemically [1, 2] but an NMR assay has been described and assays using high-performance liquid chromatography (HPLC) (cf. [8] and [9]) will undoubtedly become more widely used. Neither active-site titration nor enzyme mass determination has yet been applied.

International reference method and standards: neither yet exists but a Working Party chaired by Professor *K. W. Bock*, Department of Pharmacology and Toxicology, University of Göttingen is formulating reference methods and an agreed system of nomenclature [10].

Enzyme effectors: UDP-N-acetylglucosamine (UDPNAG) specifically stimulates *in vitro* glucuronidation of all aglycones examined, without being itself conjugated [1]. Conditions for stimulation by UDPNAG are very specific. It only stimulates the basal glucuronidation rate, measured with latent transferase preparations: it very weakly inhibits the much greater, maximally-expressed rates measured, for example, in detergent-treated homogenate or microsomes, or with purified GT. Reverse GT reactions are also weakly inhibited as is coupled transglucuronidation, which is measured without added UDPGA [11]. Specific UDPNAG-stimulation is abolished when ER-thiol groups are blocked with low concentrations of mercurials or N-ethylmaleimide that are insufficient markedly to depress basal GT rates [1]. Divalent cations are probably needed for optimal UDPNAG-stimulation.

UDP-N-acetylglucosamine typically stimulates basal rates 2 – 5 fold and *Bock and colleagues* have adduced strong evidence that UDPNAG-rates approximate to *in vivo* glucuronidation rates (cf. especially [12, 13]). The mode of action of UDPNAG is still controversial [1].

Other less specific effectors of GT, unlike UDPNAG, also stimulate glucuronidation by purified GT and in detergent-treated preparations: these include the carcinogen diethylnitrosamine (DENA) and non-carcinogenic alkyl ketones (e.g. pentane-3-one) [14]. Most interestingly, these agents also restore defective 2-aminophenol glucuronidation in mutant *Gunn* rat liver preparations [14] and DENA may restore defective hepatic 2-aminophenol conjugation in thyroidectomized or hypophysectomized rats [1]. DENA is thought to alter the conformation of GT, but it is unclear whether its effects are physiologically relevant.

The action of GT on some aglycones (e.g. bilirubin and 2-aminophenol) is strongly dependent on divalent cations (Mg^{++}, Ca^{++}, Mn^{++}, Co^{++}), while glucuronidation of other aglycones, e.g. 4-nitrophenol, is little inhibited by excess EDTA. Divalent cations strongly inhibit reverse GT reactions and coupled transglucuronidation [11].

In most species, nucleotide pyrophosphatase, which hydrolyzes UDPGA to UMP and glucuronate-1-P, is more active than GT. Excess of sacrificial pyrophosphatase substrates, which themselves poorly inhibit GT, e.g. NAD or UDPNAG, will non-specifically alleviate competition from this enzyme, but there is no feasible specific inhibitor of the pyrophosphatase and its effect must always be borne in mind when assaying at low concentrations of UDPGA. The best tactic, where possible, is to assay GT in guinea pig preparations where nucleotide pyrophosphatase (and β-glucuronidase) are poorly active [1], as they may also be in human tissues.

Assay

Method Design

Principle

$$\text{UDPglucuronic acid} + \text{aglycone} \xrightleftharpoons{\text{GT}} \beta\text{-glucuronide} + \text{UDP}$$

Reversibility of reactions *in vitro* varies between aglycones, presumably because of low GT-affinities for such glucuronides as that of 2-aminophenol, whose glucuronidation is therefore irreversible in practice. Even with more readily reversible GT reactions, and in the presence of glucaro-1,4-lactone to inhibit β-glucuronidase, large glucuronide concentrations must be employed to assay reverse reactions *in vitro*. *In vivo*, GT are most unlikely ever to catalyze reverse reactions since ER nucleoside diphosphatase, which hydrolyzes UDP in the presence of divalent cations to UMP + P_i, can exceed GT-activity more than 50-fold*. For this reason, *in vitro* assays of reverse GT-reactions will not be described further.

Since many aglycones can be sulphated as well as glucuronidated, a blank from which UDPGA is ommitted is used with broken-cell preparations to obtain an

* *D. Brown, A. E. R. de Brito, C. Berry, M. Tresize, T. Hallinan,* unpublished.

estimate of the sulphation rate. With intact-cell preparations, β-glucuronide conjugates must be identified by specific hydrolysis with animal β-glucuronidase, which is blocked by glucaro-1,4-lactone (conditions in [8, 15]).

Optimized conditions for measurement: it is recommended that three *in vitro* conjugation rates be assayed:

1. *UDPNAG-stimulated rates,* which are rarely routinely assayed, should be measured in preparations, whose GT-latency is retained and at the closest estimates to the "physiological" concentrations of UDPGA and UDPNAG: these conditions appear to give the best approximation to *in vivo* glucuronidation rates [12, 13]. Routinely assaying with UDPNAG may in future identify "effector-negative" preparations of GT, which fail to respond to UDPNAG. In rodent liver, reported concentrations of UDPGA range from 0.16 – 0.57 mmol/l and of UDPNAG (+ UDP-N-acetylgalactosamine) from 0.32 – 0.41 mmol/l. Excess UDPNAG does not inhibit latent glucuronidation.
2. *Basal rates* are measured in latent GT preparations: they are required to assess stimulation by UDPNAG and by latency-removal, and are less than UDPNAG-stimulated rates at the same UDPGA concentration.
3. *Fully expressible rates* are measured after exposure of the enzyme preparation to a latency removal technique, which is experimentally verified as optimum. The commercially available detergents, Triton X-100 or Brij 35, are recommended for 1-naphthol conjugation described below. Some detergents are unsuitable for removing latency with certain aglycones; e.g. Triton X-100 for 2-aminophenol conjugation, digitonin for 3-hydroxysterol or chloramphenicol conjugation. Fully expressible rates are greater than UDPNAG-stimulated rates.

Next, a suitable aglycone needs to be used. This choice is much more difficult, indeed almost impossible to suit every case. It would obviously be preferable to assay conjugation of a major physiological aglycone but we have found no simple assay which is generally applicable. Thus, although rapid in *Duncan-Hartley* guinea pigs, hepatic oestrogen glucuronidation is sluggish in our *Wistar* and *Sprague-Dawley* rats: high and low rates have also been reported by others. Hepatic aetiocholanolone and androsterone glucuronidation was rapid in guinea pigs but very slow in 5 separate inbred mouse strains and showed marked variation between individual *Wistar* rats, which could either be high or low conjugators, as reported also by others. As expected, hepatic testosterone conjugation was low in all species examined. Present assays of bilirubin glucuronidation are neither simple nor particularly sensitive, though HPLC methods should offer marked improvements in the future [9].

In the absence of a suitable assay for an endogenous aglycone, those for 2-aminophenol or 1-naphthol are probably the most precise and simplest of widely used assays employing xenobiotic aglycones. Conjugation of 4-nitrophenol, measured spectrophotometrically at 400 nm, is less precise despite its popularity, since it is a substrate-disappearance assay and is also quite susceptible to interference from light scattering. The method for 1-naphthol glucuronidation has been chosen for detailed description here because it is the simpler. Naphthol glucuronide is separated from unconjugated

1-naphthol by extraction with $CHCl_3$, unreacted 1-naphthol partitioning into $CHCl_3$ while the glucuronide remains in the aqueous phase [12]. A detailed description of the 2-aminophenol conjugation assay is given in [8], together with a version of the 1-naphthol assay, similar to that described below.

Temperature conversion factors: are inapplicable at least to the membrane-bound enzyme: membrane-bound, latent GT preparations exhibit non-linear temperature-activity profiles, their *Arrhenius* plots showing breaks around 20 °C to 25 °C [16]. Below 13 °C, stimulation by UDP-N-acetylglucosamine is also reportedly abolished (cf. [1]). Most assays are performed at 37 °C.

Equipment

Any modern automatic scintillation spectrometer can be used: quench corrections should not be necessary provided radioactivity is measured in constant aliquots of glucuronide. A thermostatically controlled water-bath, autopipettes, timer, bench centrifuge, disposable screw-cap, glass assay tubes (e.g. 12 × 81 mm GS/BGl/Z) tubes: *Labco*, Marlow Bottom Road, Marlow, Bucks, England).

Reagents and Solutions

Purity of reagents: best grade commercial reagents can be used without further purification.

Preparation of solutions (for about 90 determinations except Triton X-100 and 1-naphthol, nominally about 220 determinations): all solutions in re-purified water (cf. Vol. II, chapter 2.1.3.2).

1. Hepes/$MgCl_2$ (Hepes, 50 mmol/l, pH 7.4; $MgCl_2$, 10 mmol/l):

   dissolve 1.19 g Hepes acid (N-2-hydroxyethylpiperazine-N^1-2-ethanesulphonic acid) and 0.20 g $MgCl_2 \cdot 6H_2O$ in 60 ml water, adjust to pH 7.4 with NaOH, 1 mol/l, and dilute to 100 ml with water.

2. UDPglucuronate (20 mmol/l):

   dissolve 54.6 mg UDPglucuronate trisodium salt (variably hydrated) in 4 ml Hepes/$MgCl_2$ solution (1).

3. UDP-N-acetylglucosamine (10 mmol/l):

   dissolve 27.4 mg UDP-N-acetylglucosamine, disodium salt (variably hydrated) in 4 ml Hepes/$MgCl_2$ solution (1).

4. Triton X-100 (5 g/l, w/v):

   dissolve 50 mg in 10 ml Hepes/$MgCl_2$ solution (1), stirring 1 h.

5. 1-Naphthol (5 mmol/l):

   dissolve 7.2 mg in 200 µl dimethylsulphoxide with warming; gradually make to 10 ml with hot Hepes/$MgCl_2$ solution (1), mixing on a vortex mixer.

6. 1-[^{14}C]naphthol (1.25×10^9 cpm/l):

   take 40 µl stock 1-[^{14}C]naphthol, stored dissolved in ethanol at −20 °C in the dark, at 12.5×10^7 cpm/ml. Gradually add 3.96 ml hot Hepes/$MgCl_2$ solution (1), mixing on a vortex mixer.

7. TCA/glycine (trichloroacetic acid, 0.25 mol/l; glycine, 0.4 mol/l; pH 2.2):

   dissolve 2.04 g trichloroacetic acid in 20 ml water, adjust to pH 2.2 with NaOH, 10 mol/l, and dilute to 25 ml with water. Dissolve 1.50 g glycine in 20 ml water, adjust to pH 2.2 with HCl, 10 mol/l, and dilute to 25 ml with water. Mix TCA and glycine solution 1 : 1, checking pH.

8. Aqueous scintillant:

   use Unisolve, *Koch-Light*, or any compatible scintillant.

Stability of solutions: deep-freeze Hepes/$MgCl_2$ solution (1), UDPGA and UDPNAG solutions (2) and (3) at −20 °C: these are stable for some months. Keep TCA/glycine solution (7), Triton X-100 (4), scintillant and 1-naphthol solutions (8), (5), (6) at room temperature, the last two in the dark. Reject detergent and naphthol solutions if pronounced turbidity or a precipitate develop: 1-[^{14}C]naphthol generally keeps one month.

Procedure

Collection and treatment of specimen: microsomes or homogenates should be prepared promptly from fresh tissue. Some conditions for assaying GT in whole organs, tissue slices, cell suspensions, cultured cells and in "liver snips" are described in [1] and [8].

Stability of the enzyme in the specimen or sample: the ER-permeability barrier, whose integrity is required for UDPNAG specifically to stimulate GT, is much more labile than GT itself. Therefore, if an estimate of *in vivo* GT rates is to be obtained (*UDPNAG-stimulated rates*), assays must be performed on freshly isolated microsomes or those frozen as pellets for up to two weeks. Storage at 0 °C or above generally "spontaneously activates" GT in tissues, homogenates or microsomes; the basal rate may rise up to 8-fold, while latency and the UDPNAG response both decline [1].

Once activated, GT itself becomes mildly unstable [1]. Purified GT is less stable, tending to aggregate and lose activity: phosphatidylcholine and glycerol both stabilize [2].

Details for measurement in tissues: broken cell preparations should be disrupted, using minimum force consistent with acceptable cell disruption. For liver, homogenization in a *Potter-Elvehjem* homogenizer in sucrose, 0.25 mol/l, is recommended. If microsomes are prepared, these are pelleted at an average impulse equivalent to 100000 *g* for 60 min. Sucrose buffered at pH 7.4 may be used and microsomes may be washed. Care must be exercised if media other than sucrose are employed, especially with guinea pig liver. Mannitol or KCl impair the ER-permeability barrier and this is completely disrupted if these microsomes are frozen resuspended in either medium [*D. Brown et al.*, unpublished].

Freshly-disrupted tissue or subcellular fractions should be diluted with sucrose, 0.25 mol/l, so 40 – 200 µl, added to *basal rate* assays, will conjugate about 10 nmol of 1-naphthol in 15 min at 37 °C: this requires about 1 mg liver homogenate protein or 200 µg rat or ginea pig liver microsomal protein. Since GT-activity varies considerably in different tissues, a suitable enzyme concentration must be determined in preliminary experiments, together with the corresponding optimum concentration of Triton X-100. Assay concentrations of UDPGA above 2 mmol/l should also be tested (say up to 6 mmol/l) in case the nucleotide pyrophosphatase activity or the apparent K_{UDPGA} of GT are very high.

Assay conditions: incubation volume 400 µl; 37 °C. Blanks are run simultaneously using buffer (1) instead of UDPGA or UDPNAG.

Measurement

Incubation

Pipette successively into screw-cap tubes:		fully expressible rate	basal rate	stimulated rate	concentration in incubation mixture	
Hepes/$MgCl_2$ solution	(1)	40 µl	80 µl	40 µl	Hepes Mg^{2+}	25 mmol/l 5 mmol/l
UDPGA solution	(2)	40 µl	40 µl	40 µl	UDPGA	2 mmol/l
UDPNGA solution	(3)	–	–	40 µl	UDPNAG	1 mmol/l
Triton X-100	(4)	40 µl	–	–	Triton	0.5 g/l
1-naphthol solution	(5)	40 µl	40 µl	40 µl	1-naphthol	0.5 mmol/l
1-[^{14}C]naphthol solution	(6)	40 µl	40 µl	40 µl		0.07 mCi/l
mix, add water to 400 µl minus sample volume; pre-incubate 5 min						

sample		40 to 200 μl	40 to 200 μl	40 to 200 μl	volume fraction 0.1 – 0.5
mix; incubate		2 min	15 min	5 min	
TCA solution	(7)	500 μl	500 μl	500 μl	
cool on ice to aggregate protein.					

Extraction: add to each tube 3 ml chloroform and shake vigorously 1 min to extract unconjugated 1-naphthol from the aqueous phase. Centrifuge at approx. 2000 *g* for 5 min.

Radiometry: use the aqueous phase from "Extraction". Pipette 0.4 ml into 4 ml aqueous scintillant (8), mix, count radiometry for 10 min or to a counting error of ± 1%.

The *basal rate* should be at least 3 x the blank, which is typically 600 cpm/assay: the *fully expressible rate* assay should not conjugate more than 6000 cpm of 1-[^{14}C]naphthol. Otherwise alter the enzyme concentration and/or the incubation times.

Protein determination cf. Vol. II, chapter 2.4.

Calculation: correct result of liquid scintillation counting rates of sample, C_p, for blank counting rate (without UDPGA), C_n yielding the net counting rate ΔC_n (in cpm). According to eqn. (p_1) (cf. "Formulae", Appendix 3) the catalytic content of the sample is

$$z_c/m_s = \frac{\Delta C_n \times 0.203 \times V}{50000 \times v \times \Delta t \times \rho_{protein}} \quad \text{U/g protein}$$

in this assay it is

- radioactivity of 0.203 μmol 1-naphthol corresponds to 50000 cpm in assay*
- volume of incubation mixture, v = 0.4 ml
- volume of mixture after addition of TCA, V = 0.9 ml
- incubation time Δt = variable
- concentration of protein in g/l used in assay, $\rho_{protein}$ = variable.

* Approx. 3 nmol 1-naphthol is contributed by 50000 cpm of 1-[^{14}C]naphthol, at 10 mci/mmol spec. act.

Validation of Method

Precision, accuracy, detection limits and sensitivity: with properly optimized GT preparations from rodent liver microsomes, the relative standard deviation for replicate assays is ± 2.6%. Data on accuracy are not yet available because of the absence of reference standards, nor have measurements been compared between different laboratories. With freshly prepared 1-[^{14}C]naphthol, which gives a low blank, the limit of detection is about 1 nmol of glucuronide formed.

Sources of error: decomposition of 1-[^{14}C]naphthol (at 1.25×10^9 cpm/l), causing an increased blank and greatly decreased sensitivity, is a possible source of error; make a fresh working solution of 1-[^{14}C]naphthol if this occurs.

Specificity: the solvent partition of unconjugated 1-naphthol from its conjugates is obviously non-specific and any water-soluble conjugate or metabolite will remain in the aqueous phase [8]. Nevertheless, when rodent liver microsomal GT-preparations are assayed, zero time blank values and blanks incubated without UDPGA are virtually identical, suggesting that virtually all of the water soluble product formed in tubes incubated (+UDPGA) is glucuronide. Recent results with rat "liver snips" indicate the over 90% of the water soluble products formed from 1-naphthol is glucuronide [17], while over 94% of those formed by cultured rat lung is glucuronide [15]. In very marked contrast, cultured human lung forms virtually no glucuronide from 1-naphthol, conjugating this instead with sulphate [15].

Reference ranges: the following rates (nmol $\times$ min^{-1} $\times$ mg^{-1} protein) are reported for rodent liver GT preparations: rat microsomes, *basal,* 3; *UDPNAG-stimulated,* 11; *maximally expressible,* 45 [12]: guinea pig microsomes, *basal,* 2.3*: mouse microsomes, *basal,* 1*. Rat hepatocytes, 0.72; perfused rat liver (assuming 180 mg protein/g), 0.55; rat liver snips, 0.53 (cited in [17]).

References

[1] *G. J. Dutton,* Glucuronidation of Drugs and Other Compounds, CRC press, Boca Raton, Florida 1980.

[2] *B. Burchell,* Identification and Purification of Multiple Forms of UDP Glucuronyltransferase, in: *E. Hodgson, J. R. Bend, R. M. Philpot* (eds.), Reviews in Biochemical Toxicology, Vol. 3, Elsevier/North Holland, N.Y. 1981, pp. 1–32.

[3] *T. Hallinan,* Effects upon UDP Glucuronyltransferase of Phospholipase Treatment or of Phospholipid Depletion During Isolation, Biochem. Soc. Trans. *8,* 43–46 (1980).

[4] *O. M. P. Singh, A. B. Graham, G. C. Wood,* The Phospholipid-Dependence of UDP Glucuronyltransferase: Purification, Delipidation and Reconstitution of Microsomal Enzyme from Guinea-Pig Liver, Eur. J. Biochem. *116,* 311–316 (1981).

[5] *D. Zakim, D. A. Vessey,* The Effects of Lipid-Protein Interactions on the Kinetic Parameters of Microsomal UDP Glucuronyltransferase, in: *A. I. Martonosi* (ed.), The Enzymes of Biological Membranes, Vol. 2, Plenum Press, New York 1976, pp. 443–461.

[6] *A. Winsnes,* Kinetic Properties of Different Forms of Hepatic UDP Glucuronyltransferase, Biochim. Biophys. Acta *284,* 394–405 (1972).

---

* *D. Brown, A. E. R. de Brito, C. Berry, M. Trezise, T. Hallinan,* unpublished.

[7] *D. A. Vessey, J. Goldenberg, D. Zakim,* Kinetic Properties of Microsomal UDP-Glucuronyltransferase. Evidence for Cooperative Kinetics and Activation by UDP-N-acetylglucosamine, Biochim. Biophys. Acta *309*, 58 – 63 (1973).
[8] *G. J. Dutton, J. E. A. Leakey, M. R. Pollard,* Assays of UDP Glucuronyltransferase Activities, in: *W. B. Jacoby* (ed.), Methods in Enzymology, Vol. 77, Academic Press, New York 1981, pp. 383 – 391.
[9] *B. H. Billing,* The Role of Conjugating Enzymes in the Biliary Excretion of Bilirubin, in: *K. B. M. Heiwegh, S. B. Brown* (eds.), Bilirubin, Vol. 2, C.R.C. Press, Boca Raton, Florida 1982, pp. 85 – 102.
[10] *K. W. Bock, B. Burchell, G. J. Dutton, O. Hanninen, G. J. Mulder, I. Owens, G. Siest, T. R. Tephley,* UDP glucuronyltransferase Activities: Guidelines for Consistent Interim Nomenclature and Assay Conditions, Biochem. Pharmacol., in press.
[11] *C. Berry, T. Hallinan,* "Coupled Transglucuronidation": a New Tool for Studying the Latency of UDP Glucuronyltransferase, FEBS Lett. *42*, 73 – 76 (1974).
[12] *G. Otani, M. M. Abou-El-Nakarem, K. W. Bock,* UDP glucuronyltransferase in Perfused Rat Liver and in Microsomes III Effect of Galactosamine and Carbon Tetrachloride on the Glucuronidation of 1-Naphthol and Bilirubin, Biochem. Pharmacol. *25*, 1293 – 1297 (1976).
[13] *K. W. Bock, G. Van Ackeren, F. Lorch, F. Birke,* Metabolism of Naphthalene to Naphthalene Dihydrodiol Glucuronide in Isolated Hepatocytes and in Liver Microsomes, Biochem. Pharmacol. *25*, 2351 – 2356 (1976).
[14] *El. N. M. A. Lalani, B. Burchell,* Stimulation of Defective Gunn Rat Liver UDP Glucuronyltransferase Activity *in vitro* by Alkyl Ketones, Biochem. J. *177*, 993 – 995 (1979).
[15] *R. Mehta, G. M. Cohen,* Major Differences in the Extent of Conjugation with Glucuronic Acid and Sulphate in Human Peripheral Lung, Biochem. Pharmacol. *28*, 2479 – 2484 (1979).
[16] *D. T. Pechey, A. B. Graham, G. C. Wood,* Phospholipid-Dependence of UDP-Glucuronyltransferase. Temperature-Dependence of Microsomal Enzyme Activity and Thermotropic Changes in Membrane Structure, Biochem. J. *175*, 115 – 124 (1978).
[17] *M. R. Pollard, G. J. Dutton,* Liver Snips. A Simple and Reproducible Method for Studying Metabolism in Small Fragments of Tissue, as Applied to Glucuronidation in Rat Liver, Biochem. J. *202*, 469 – 473 (1982).

5.4 Galactosyltransferase

UDPgalactose : 2-acetamido-2-deoxy-β-D-glucopyranosyl-glycopeptide galactose β (1 → 4) transferase (EC 2.4.1.38)
UDPgalactose : glucopyranose galactose β (1 → 4) transferase (EC 2.4.1.22)

Bernadette Verdon and Eric G. Berger

General

Galactosyltransferase is a ubiquitous enzyme which exerts two functions: lactose biosynthesis in the mammary gland, and chain elongation of glycoproteins [1].

Both activities are closely related by kinetic [2], isoelectric [3] and immunochemical [4] criteria. Galactosyltransferase is localized to the trans cisternae of the *Golgi*

apparatus [4] in a membrane-bound form, where lactose biosynthesis and chain elongation of glycoproteins take place. In addition, the enzyme is present in soluble form in virtually all body fluids (for review see [5]).
Conversion of the intracellular membrane-bound to the extracellular soluble form involves proteolytic cleavage [6]. During this exocytotic process, the enzyme may be exposed to the cell surface for a variable period of time where it may be detected as "ecto-galactosyltransferase" (for review see [7]).

Application of method: galactosyltransferase is measured

- as a *Golgi* marker enzyme in tissue fractionation ([8] and references cited therein);
- in clinical chemistry as a tumour marker for ovarian [9] and breast [10] carcinomata;
- the purified enzyme is also used as a probe for specific detection of terminal GlcNAc* residues on heteroglycans [11].

The present assay description emphasizes enzyme activity measurements in human sera for clinical purposes.

Enzyme properties relevant in analysis: the formation of a glycosidic linkage is catalyzed by a single glycosyltransferase; thus, donor substrates may be shared by several unrelated glycosyltransferases, all of them forming different glycosidic linkages [5]. In crude enzyme systems, care must be taken that homogeneous acceptor substrates are used and that the glycosidic linkage of transferred galactose is established [12].

Purified soluble galactosyltransferase is heterogeneous with respect to size, due to limited proteolytic degradation [1], and with respect to charge [3].

In addition, a cancer-associated isoenzyme of higher molecular weight has been reported [13].

Under the conditions described here, the reaction is linear with time and irreversible; purified galactosyltransferase has no glycosidase activity [2].

Methods of determination: galactosyltransferase is usually measured by its activity, as described here. Estimation of enzyme protein by immunoassay is being developed in this laboratory [14]. Activity measurements are based either on liberation of free UDP using a coupled spectrophotometric assay [2] or by incorporation of radioactive sugar into the acceptor substrate [5]. Various acceptors for galactosyltransferase have been proposed, including free GlcNAc, ovalbumin, ovomucoid, asialo-agalacto-orosomucoid and asialo-agalacto-fetuin. Except for the latter, all acceptors are homogeneous in exposing GlcNAc residues at the non-reducing end of the glycan moiety.

---

* Abbreviations: GlcNAc N-acetylglucosamine
UDP Uridine 5′-diphosphate
UDPgal Uridine 5′-diphospho-D-galactose
GP Glycoprotein.

International reference methods and standards: not available.

Enzyme effectors: both soluble and membrane-bound galactosyltransferases require Mn^{2+} (for recent data see [15]). Membrane-bound enzyme is detergent-activated, most effectively by Triton-X 100 [16].

Assay

Method Design

Principle:

$$\text{UDPgal} + \text{GlcNAc-GP} \xrightarrow[Mn^{2+}]{\text{galactosyltransferase}} \text{Gal}\beta1{\rightarrow}4\text{GlcNAc-GP} + \text{UDP}$$

Incorporation of radioactive galactose into the acceptor substrate is measured. This procedure requires separation of product from unreacted substrate after a given time and measurement of radioactive product. The assay presented here is based on the transfer of [U-^{14}C]galactose from UDP[U-^{14}C]galactose to ovalbumin, acid precipitation of product and retention of the precipitate on glass fibre filters for heterogeneous liquid scintillation counting.

Optimized conditions for measurement: the following conditions refer to measurement of the human serum enzyme. The assay can, however, easily be adapted to other enzyme preparations if certain precautions are taken.

Buffer: Na-cacodylate or Tris-HCl buffer are suitable as they do not chelate Mn^{2+} and have a good buffering capacity at the pH optimum of 7.0 – 7.5 [17].

Effectors: high Mn^{2+} concentrations decrease stability of UDPgal [18]. Thus, a concentration of 10 mmol/l is preferred, even though activity may be considerably higher at 40 mmol/l [17].

Temperature: maximum activity is at 30 °C [17] and remains linear for at least 4 h under conditions described here.

Substrates: K_m for UDPgal is approximately 15 µmol/l; thus a concentration of 50 µmol/l is saturating. K_m for ovalbumin is approximately 0.9 mmol/l (40 mg/ml), a saturating concentration is therefore not attainable due to limited solubility.

Radioactive substrate: both ^{14}C- and ^{3}H-labelled UDPgal are available from commercial sources. In the present assay the ^{14}C-labelled compound is preferred because of the higher counting efficiency. This can be critical, since enzyme preparations of low specific activity produce dense protein precipitates upon assay termination and this may lead to self-absorption of radiation on filters.

Temperature conversion factors: Q_{10} values for serum galactosyltransferase activity:

2.1 (10 – 20 °C); 1.9 (15 – 25 °C); 2.0 (20 – 30 °C); 1.3 (25 – 35 °C).

Above 30 °C, temperature-dependent activation decreases. At 40 °C, the half-time of inactivation is 9 h, at 45 °C 1 h. At 56 °C, the enzyme is rapidly inactivated [12].

Equipment

Water-bath; stopwatch; 5 ml test tubes (Polystyrene), glass fibre filters (2.4 mm, *Whatman* 934 AH), scintillation vials (20 ml), scintillation counter, filter holder (Micro Filter No 4616 G10, *Thomas Scientific Apparatus,* Philadelphia) and suction device.

Reagents and Solutions

Purity of reagents: analytical grade reagents are required. Ovalbumin and unlabelled UDPgal were obtained from *Sigma,* UDP[U^{14}C]galactose (310 Ci/mol) was from *Amersham,* phosphotungstic acid from *Merck.* Conventional toluene-based scintillant may be used.

Preparation of solutions (for about 100 determinations with respect to solution 6, all other ingredients are prepared as stock solutions): all solutions in re-purified water (see Vol. II, chapter 2.1.3.2).

1. Sodium dimethylarsinate (sodium cacodylate, 0.5 mol/l, pH 7.35):

   dissolve 5.35 g cacodylate, sodium salt in 40 ml water, adjust to pH 7.35 with 1.5 ml HCl, 1 mol/l, and dilute to 50 ml with water.

2. Manganese chloride (0.1 mol/l):

   dissolve 0.99 g $MnCl_2 \cdot 4H_2O$ in 50 ml water.

3. Ovalbumin (80 mg/ml, 1.86 mmol/l):

   dissolve 800 mg ovalbumin in 10 ml water; remove insoluble material by centrifugation; store frozen in aliquots of 1 ml.

4. NaCl (0.154 mol/l):

   dissolve 0.45 g NaCl in 50 ml water.

5. Uridine 5′-diphospho-D-galactose (0.5 mmol/l):

   dissolve 2.83 mg UDPgal in 10 ml water. Store frozen in aliquots of 1.0 ml.

6. Uridine 5′-diphospho-D-[U^{14}C]galactose:

   evaporate 2.5 μCi of UDP[^{14}C]gal solution under nitrogen and dissolve residue in 0.5 ml solution (5).

7. Phosphotungstic acid (5% w/v):

   dissolve 100 g in 2 l of HCl, 2 mol/l.

8. Ethanol (1% v/v):

   either absolute or toluene denatured.

9. Scintillant:

   dissolve 20 g Butyl-PBD (2-(L-*tert*-butylphenyl)-5(4-bisphenylyl)-1,3,4-oxadiazole (*Ciba-Geigy*) in 2.5 l toluene (scintillation grade).

Stability of solutions: ovalbumin (solution 3) and UDPgal (solutions 5 and 6) should be stored frozen. $MnCl_2$ and NaCl should be prepared monthly. The other solutions are stable for months.

Procedure

Collection and treatment of sample: the following preparations are suitable: serum or plasma containing oxalate (0.2 mg/ml), heparin (0.5 mg/ml or 5 drops of Liquemin®/10 ml) or EDTA (1 mg/ml). Citrate must be avoided.

Stability of the enzyme in the sample: 11% of enzyme activity is lost within 24 h at room temperature, 3% at 4 °C, 0.5% at −20 °C. After this initial loss of activity, the enzyme remains fairly stable for 1 week at the temperatures indicated above. After 1 month at room temperature, 50% of activity is lost, 6% at 4 °C. However, if rapidly frozen and stored at −70 °C, an increase of the serum enzyme activity amounting to approximately 20% may be detected.

Details for measurement in samples other than serum or plasma: membrane-bound galactosyltransferase in tissue and cell homogenates should be measured in the presence of 0.1−0.4% (w/v) Triton X-100 [16]. In some cases, ATP (2 mmol/l) or UMP are needed (0.1−0.3 mmol/l) to prevent hydrolysis of UDPgal by nucleotide pyrophosphatase [19].

Assay conditions: set water-bath at 30 °C and place reagents in ice water. Solutions (7) and (8) should be kept at 4 °C in a wash-bottle.
Working solution: just prior to the assay, mix solutions (1)−(3) and (4) in respective volumes of 1 : 1 : 2.5 : 3.5.

Controls: sample containing heat-inactivated enzyme (30 min, 56 °C); zero time incubation by adding solution (7) immediately after labelled substrate, sample without ovalbumin which will indicate the amount of endogenous acceptors (usually slightly above background in human serum).

Measurement

Pipette successively into 5 ml test tubes:			concentration in assay mixture	
working solution		40 µl	cacodylate	50 mmol/l
			$MnCl_2$	10 mmol/l
			ovalbumin	470 µmol/l
			NaCl	54 mmol/l
sample (serum/plasma)*		5 µl	volume fraction	0.1
UDP-[^{14}C]gal	(6)	5 µl	UDPgal	50 µmol/l (25 nCi)
incubate in a thermostatted water-bath (30 °C) for 60 min,				
phosphotungstic acid (chilled)	(7)	1 ml		
reaction is terminated; filter precipitate over pre-wetted glass fibre filter and wash once with solution (7) and twice with chilled ethanol (8); dry filters for 15 min at 60 – 80 °C.				
scintillant solution	(9)	4 ml		
count 10 min/vial in a liquid scintillation counter.				

* add sample carefully without vortexing; fast discharge of sample will insure thorough mixing of reagents.

Calculation: in order to calculate enzyme units (1 U = 1 µmol/min) the following factors have to be defined:

Specific radioactivity (X) of UDPgal used in the assay: 0.025 µCi of UDP[^{14}C]gal is added to 2.5 nmol unlabelled UDPgal; thus, the resulting specific activity is 10 mCi/mmol; therefore, since 2.22×10^6 dpm are defined as 1 µCi, 1 nmol UDPgal corresponds to 22200 dpm.

Counting efficiency (η): ^{14}C counting efficiency is 0.85 under the conditions described here; thus cpm values are converted to dpm by dividing by 0.85.

Subtraction of background counting rate (C_b): control cpm values as obtained by either using heat-inactivated enzyme or by terminating the incubation at zero time, have to be subtracted from the counting rates for samples (C_p). The enzyme concentration b in the sample is then calculated according to the following formula:

$$b = \frac{(C_p - C_b) \times F}{\eta \times X \times \Delta t} \quad U/l$$

F factor to convert 5 µl of added sample volume to one litre (2×10^5)
η counting efficiency (0.85)
X specific radioactivity (22.2×10^6 dpm/µCi)
t incubation time (60 min)
C counting rate, cpm

Validation of Method

Precision, accuracy, detection limits and sensitivity: coefficient of variation within an assay is less than 5% ($n = 6$). Since reference material is not available, accuracy cannot be determined.

Detection limits and sensitivity depend on the specific radioactivity of UDP[U-^{14}C]-galactose used. Theoretically, using labelled substrate with a specific activity of 310 Ci/mol, transfer in the pmol range can be measured (682 dpm = 1 pmol). In practice, sensitivity of this assay is limited by the relatively high cost of UDP[^{14}C]-galactose. Under the conditions described here, the enzyme protein may be detected at the concentration level of 0.05 U/ml serum.

Sources of error: the extent of interference by nucleotide pyrophosphatase in serum samples from diseased patients is unknown. Since the assay described here is based on 2 point measurements, deviations from linearity are not readily detected. The volume of normal human serum may, however, be increased up to 10 µl without affecting linearity. Higher volumes are not recommended as the resulting dense protein precipitate decreases counting efficiency.

Specificity: the assay is specific for this type of galactosyltransferase. Specificity can conveniently be checked by adding free GlcNAc, 50 mmol/l (final concentration) to the assay: presence of free GlcNAc inhibits incorporation of galactose into ovalbumin [12].

Reference ranges: serum galactosyltransferase activity in samples from healthy blood donors was determined according to this protocol. The samples were drawn 2 days prior to the assay and stored at 4 °C. Average values recorded were 1119 ± 219 cpm (n = 20) per 5 µl serum after 1 h incubation. Thus, the average enzyme concentration in normal human serum is 0.20 ± 0.04 U/l. This value corresponds to a concentration of 125 µg enzyme protein per litre assuming an average catalytic activity content of 1.6 U/mg of purified galactosyltransferase as determined in 15 different batches of the purified human milk enzyme. Slightly higher values (0.29 U/l, $n = 10$) have been reported [20] based on asialo-agalacto-fetuin as acceptor substrate (see section "Methods of determination").

References

[1] *R. L. Hill, K. Brew,* Lactose synthetase, Adv. Enzymol *43,* 411 – 490 (1975).

[2] *K. E. Ebner,* Lactose synthetase, in: *P. D. Boyer* (ed.), Enzymes, Academic Press, New York 1973, pp. 363 – 377.

[3] *A. Ch. Gerber, I. Kozdrowski, S. R. Wyss, E. G. Berger,* The Charge Heterogeneity of Soluble Human Galactosyltransferases Isolated from Milk, Amniotic Fluid and Malignant Ascites, Eur. J. Biochem *93,* 453 – 460 (1979).

[4] *J. Roth, E. G. Berger,* Immunocytochemical Localization of Galactosyltransferase in HeLa Cells: Codistribution with Thiamine Pyrophosphatase in *trans*-Golgi Cisternae, J. Cell Biol. *92,* 223 – 229 (1982).

[5] *H. Schachter, S. Roseman,* Mammalian glycosyltransferases, in: *W. J. Lennarz* (ed.), The Biochemistry of Glycoproteins and Proteoglycans, Plenum Press, New York 1980, pp. 85 – 160.

[6] *G. J. A. M. Strous, E. G. Berger,* Biosynthesis, Intracellular Transport, and Release of the Golgi Enzyme Galactosyltransferase (Lactose Synthetase A Protein) in HeLa Cells, J. Biol. Chem. *257,* 7623 – 7628 (1982).

[7] *M. Pierce, E. A. Turley, S. Roth,* Cell Surface Glycosyltransferase Activities, Int. Rev. Cytology *65,* 1 – 47 (1980).

[8] *R. Bretz, H. Bretz, G. E. Palade,* Distribution of Terminal Glycosyltransferases in Hepatic Golgi Fractions, J. Cell Biol. *84,* 87 – 101 (1980).

[9] *S. K. Chatterjea, M. Bhattacharya, J. J. Barlow,* Glycosyltransferase and Glycosidase Activities in Ovarian Cancer Patients, Cancer Res. *39,* 1943 – 1951 (1979).

[10] *J. F. Paone, R. R. Baker, T. P. Waalkes, J. H. Shaper,* Sequential Galactosyltransferase and Carcinoembryonic Antigen Levels in Advanced Breast Carcinoma, J. Surg. Res. *31,* 269 – 273 (1981).

[11] *J. A. Hanover, W. J. Lennarz,* Transmembrane Assembly of Membrane and Secretory Glycoproteins, Arch. Biochem. Biophys. *211,* 1 – 19 (1981).

[12] *E. G. Berger, I. Kozdrowski, M. M. Weiser, D. H. v. d. Eijnden, W. E. C. M. Schiphorst,* Human Serum Galactosyltransferase: Distinction, Separation and Product Identification of Two Galactosyltransferase Activities, Eur. J. Biochem. *90,* 213 – 222 (1978).

[13] *D. K. Podolsky, M. S. McPhee, E. Alpert, A. L. Warshaw, K. J. Isselbacher,* Galactosyltransferase Isoenzyme II in the Detection of Pancreatic Cancer: Comparison with Radiologic, Endoscopic and Serologic Tests, N. Engl. J. Med. *304,* 1313 – 1318 (1981).

[14] *B. Verdon, Th. Mandel, E. G. Berger, H. Fey,* A Solid-Phase ELISA for Human Galactosyltransferase, J. Immunol. Methods *55,* 27 – 33 (1982).

[15] *P. J. Andree, L. J. Berliner,* Metal Ion and Substrate Binding to Bovine Galactosyltransferase, Biochemistry (USA) *19,* 929 – 933 (1980).

[16] *R. Bretz, W. Stäubli,* Detergent Influence on Rat Liver Galactosyltransferase Activities towards Different Acceptors, Eur. J. Biochem. *77,* 181 – 192 (1977).

[17] *Y. S. Kim, J. Perdomo, J. S. Whitehead,* Glycosyltransferases in Human Blood. 1. Galactosyltransferase in Human Serum and Erythrocyte Membranes, J. Clin. Invest. *51,* 2024 – 2032.

[18] *H. A. Nunez, R. Barker,* The Metal Ion Catalyzed Decomposition of Nucleoside Diphosphate Sugars, Biochemistry (USA) *15,* 3843 – 3847 (1976).

[19] *G. Spik, P. Six, J. Montreuil,* Chemical and Enzymic Degradations of Nucleoside Mono- and Diphosphate Sugars. I. Determinations of the Degradation Rate during the Glycosyltransferase Assays, Biochim. Biophys. Acta *584,* 203 – 215 (1979).

[20] *M. Bhattacharya, S. K. Chatterjea, J. J. Barlow,* Uridine 5'-Diphospho-Galactose: Glycoprotein Galactosyltransferase Activity in the Ovarian Cancer Patient, Cancer Res. *36,* 2096 – 2101 (1976).

5.5 Purine-nucleoside Phosphorylase

Purine-nucleoside: orthophosphate ribosyltransferase, EC 2.4.2.1

Alexander Dwenger and Ivar Trautschold

General

Since 1945, when nucleoside phosphorylases were first recognized as a distinct group of enzymes by *Kalckar* [1], the reversible phosphorolytic nitrogen-carbon fission catalyzed by purine nucleoside phosphorylase, PNP, has been demonstrated in a large variety of cells. In general, PNP from mammalian sources catalyzes the readily-reversible phosphorolysis of naturally occurring purine ribonucleosides or 2'-deoxyribonucleosides to form the free purine base and ribose 1-phosphate or 2-deoxyribose 1-phosphate, with the probable exception of adenosine and deoxyadenosine. The natural purine bases, guanine, hypoxanthine, and xanthine, as well as certain related structural analogues such as 8-azaguanine, 6-mercaptopurine, allopurinol and oxypurinol, serve as substrates. Adenosine and deoxyadenosine may become substrates for the enzyme only after being deaminated by adenosine deaminase, which is one of the proteins contaminating PNP during isolation and purification. The equilibrium of the phosphorolytic reaction is quite favourable toward nucleoside synthesis *in vitro*, but PNP is thought to function primarily in nucleoside catabolism *in vivo*. Various PNP's are widely distributed in nature and have been identified and studied in a number of tissues of mammals, chicken and fish, in yeast and in several species of bacteria [2].

PNP plays an important role in the salvage pathway which re-utilizes purines formed from the breakdown of the nucleic acids, and is a useful diagnostic tool since it has been recognized that the absence of PNP is associated with immune dysfunction.

The mammalian enzyme has been isolated to electrophoretic homogeneity from many tissues and has been shown to have a molecular weight in the range of 80000 – 90000. Chemical and genetic evidence suggests that the human enzyme is a trimer with a subunit molecular weight of 30000 [3 – 10].

Application of method: methods for the determination and characterization of PNP activity have been used in biochemistry to investigate the role of PNP in nucleoside metabolism, especially of erythrocytes. Since purine nucleosides penetrate into erythrocytes, it is obvious that this cell must play a major role in the total purine economy of the body, a function that perhaps has not been fully appreciated. The importance of detailed characterization of human PNP and its substrate affinities has increased greatly in clinical biochemistry and genetics since the discovery of a relationship between deficiency of this enzyme in man and serious immunological disease [9]. Inherited deficiency of PNP causes a deficiency in cell-mediated immunity, while humoral immunity appears normal.

Enzyme properties relevant in analysis: *Edwards et al.* [10] have reported the occurrence of several PNP isoenzymes in various human tissues including red blood cells. On the basis of tissue-to-tissue variations in the complexity of PNP, they suggested that the primary isoenzymic form is modified *in vivo* with the generation of several isoenzymes. A survey of 2178 unrelated individuals revealed the presence of three electrophoretically different, genetically determined variants of PNP in red blood cell lysates. Although genetic evidence suggests that there is one pair of alleles for PNP in human subjects and that the enzyme is a trimer composed of subunits, all of which are products of this pair of alleles, the enzyme exhibits a variable degree of heterogeneity, depending on the tissue origin [10]. *Zannis et al.* [4] purified PNP from fresh human erythrocytes to homogeneity. The native enzyme appears to be a trimer with native molecular weight of about 94000. Two-dimensional gel electrophoresis of the purified enzyme under denaturing conditions revealed four major separable subunits with the same molecular weight but different isoelectric points. Besides these four major subunits, which together account for 98 – 99% of the total PNP protein, in all preparations of normal purified PNP from erythrocytes there are also two minor subunits of the same molecular weight which are more alkaline than the major subunits and comprise 1 – 2% of the total PNP protein [5]. The subunits are likely to result from post-translational modifications of basic subunits to the more acidic ones and provide an explanation of the complex native isoelectric focusing pattern of PNP from erythrocytes. *Ghangas & Reem* [6] confirmed the subunit structure of the erythrocyte enzyme reported by *Zannis et al.* [4] by two-dimensional gel electrophoresis following immunoprecipitation analysis.

Although PNP's from various mammalian sources resemble each other in their substrate activities and in many of their molecular properties, such as molecular weight and subunit structure [1 – 8, 11], significant differences may be demonstrated by examination of parameters such as electrophoretic characteristics, response to sulphydryl reagents, and kinetic behaviour, especially with regard to the phenomenon of substrate activation. For example, on isoelectric focusing the crystalline human erythrocytic enzyme yields six peaks of activity ranging from pI = 5.85 to 6.25. Although the crystalline PNP was obtained from the pooled erythrocytes of more than 100 persons, heterogeneity could not be due to the genetic variance in this population because the low frequency of genetic variants of PNP [10] suggests that genetic contribution to the observed heterogeneity is negligible. The phenomenon of substrate activation with both inosine and deoxyinosine was observed with all six electrophoretic variants at higher substrate concentrations [7, 8]. The K_m values for inosine of the variants and the corresponding pI values are given in Table 1 (from [7]).

The K_m values for orthophosphate are relatively similar and were reported to be in the range of 0.3 to 1.5×10^{-3} mol/l [2].

With the crystallized human enzyme from erythrocytes, the temperature dependence of the maximum rate of conversion, V, with inosine as substrate, performed at pH 7.5

Table 1. *Michaelis* constants of the variants of human erythrocytic purine-nucleoside phosphorylase using inosine as substrate [7]

pI	K_m (mol/l) 10^5	pI	K_m (mol/l) 10^5
5.85	2.2	6.08	3.2
5.92	2.9	6.14	4.0
6.02	2.8	6.25	5.3

from 13 °C to 40 °C, yielded a linear *Arrhenius* plot with an activation energy of 10.8 kcal/mol [2]. In general, PNP's exhibit broad pH optima between 6.5 and 8.0. The effect of pH on maximal velocity suggests the occurrence of essential imidazole and sulphydryl groups, whereas the effect of pH on the *Michaelis* constant suggests the existence of an essential ionizing group at about 7.2 in the free enzyme [8].

Methods of determination: a number of assays of PNP have been used which employ different procedures. Some of these procedures are no longer used, such as the colorimetric estimation of pentose formation, or inorganic-orthophosphate formation, and the manometric measurement of oxygen consumption in a coupled assay with xanthine oxidase. Today there are several appropriate assays that allow the measurement of either direction of the enzyme reaction under a variety of test compositions, which can be selected to obtain optimal experimental conditions.

1. *Spectrophotometry*

1.1 Coupled assays

One of the most frequently used assay procedures is based on the method of *Kalckar* [12]. The phosphorolysis of inosine is determined by measuring the increase in absorbance at 293 nm due to the formation of uric acid from hypoxanthine by xanthine oxidase ($\varepsilon = 1.25 \times 10^3 \, l \times mol^{-1} \times mm^{-1}$ at 293 nm).

A similar system was used for 6-mercaptopurine ribonucleoside where the formation of thiouric acid was measured by the increase in absorbance at 348 nm ($\varepsilon = 2.45 \times 10^3 \, l \times mol^{-1} \times mm^{-1}$). This method has proved especially useful when the reaction conditions require the presence of substances with high absorbance in the region of 250 – 300 nm; for example, the competitive inhibitor of PNP, formycin B [13].

Another continuous monitoring spectrophotometric assay is based on the use of aldehyde dehydrogenase to determine H_2O_2-producing reactions [14]. An assay system has been described [15, 16] in which hypoxanthine liberated during the phosphorolytic cleavage of inosine is oxidized to uric acid by xanthine oxidase. The hydrogen peroxide formed is further transformed into water by catalase. In the presence of high concentrations of ethanol, equivalent amounts of acetaldehyde are released which are dehydrogenated to acetic acid with an NAD(P)-dependent aldehyde dehydrogenase. The formation of NAD(P)H can be followed spectrophotometrically (cf. p. 387).

1.2 Direct spectrophotometric measurement

Direct spectrophotometric assay of either phosphorolysis or synthesis is possible with substrates that show significant absorbance differences between the nucleoside and the base (e.g., guanosine ⇌ guanine, $\Delta\varepsilon = 5.1 \times 10^2$ l $\times$ mol^{-1} $\times$ mm at 252 nm, pH 7.5) [17].

2. *Radio-enzymatic procedures*

The use of labelled substrates has made possible a variety of assays that have the advantage of great sensitivity and that may be performed under conditions where spectrophotometric procedures are not practical. Nucleoside synthesis has been measured in a radioactive assay by following the formation of ^{14}C-labelled nucleoside from the ^{14}C-labelled purine [18, 19]. Nucleoside phosphorolysis has been measured by the formation of [^{14}C]ribose-1-phosphate from [U-^{14}C]guanosine or replacing guanosine with inosine and [U-^{14}C]guanosine with [U-^{14}C]inosine. These assays are capable of detecting as little as 0.1 nmol of product in either direction [19].

For both diagnostic and fundamental purposes ultra-microchemical methods offer many advantages. *Uitendaal et al.* [20] scaled down the radioisotopic method for the measurement of PNP activity to the level of single cells by monitoring the reaction of either ^{14}C-labelled hypoxanthine or ^{14}C-labelled inosine as substrates, with descending chromatography on paper strips to separate substrate and products followed by liquid scintillation counting of the separated substances.

3. *Immunological methods*

3.1 Immunoelectrophoresis

Rocket immunoelectrophoresis [3, 21] was used to quantitate the PNP protein over the range of 10 to 100 ng with a limit of detection of about 10 ng of phosphorylase protein. Comparisons of phosphorylase activity with the amount of phosphorylase protein can be useful in defining the nature of biochemical changes that produce phosphorylase deficiency in man. In terms of sensitivity to levels of phosphorylase activity and protein in normal tissues the radiochemical and immunochemical assays appear equally sensitive.

3.2 Enzyme-linked immunosorbent assay

Osborne & Scott [22] reported a sensitive enzyme-linked immunosorbent assay with a minimal detectable concentration of 0.15 ng PNP by which inactive variant nucleoside phosphorylase protein could be quantitated in red cells and cultured skin fibroblasts from enzyme deficient patients.

4. *High-performance liquid chromatography*

HPLC is highly suitable for PNP activity measurement [23]. Both the forward and reverse reactions can be monitored. As both the substrate and the product are determined simultaneously, activity can be calculated from the decrease in substrate and from the increase in product. Any interference would then be seen as a discrepancy in these values. For example, IMP formation by the hypoxanthine-guanine-phosphoribosyl transferase should be detected by an increase in IMP or by noting the unequal changes in products and substrates.

International reference methods and standards: neither standardization at the international level nor the existence of reference standard materials is known up to now.

Enzyme effectors: different influences of effectors have been described according to the source of PNP. For the PNP from *Escherichia coli* and *Salmonella typhimurium,* pyrimidines and pyrimidine nucleosides and purine nucleoside monophosphate and triphosphates had no significant effect on the enzyme reaction when these compounds were present in concentrations of 2 mmol/l in the assay. Similarly, neither chloride salts of the cations (1 mmol/l) Fe^{3+}, Fe^{2+}, Mg^{2+}, Mn^{2+}, Co^{2+}, Ni^{2+}, nor Cl^- or SO_4^{2-} salts of Na^+ and K^+ (200 mmol/l) showed any effect on the enzymatic activity [18]. For the PNP from rabbit liver, product inhibition is observed with guanine ($K_I = 1.25 \times 10^{-5}$ mmol/l) and hypoxanthine ($K_I = 2.5 \times 10^{-5}$ mmol/l) for the enzymatic phosphorolysis of guanosine. Ribose 1-phosphate, another product of the reaction, inhibited non-competitively ($K_I = 3.61 \times 10^{-4}$ mmol/l) with guanosine as the variable substrate. Non-competitive inhibition was also observed with 4-chloromercuribenzoate ($K_I = 5.68 \times 10^{-6}$ mmol/l), the inactivated enzyme being completely reactivated upon the addition of an excess of 2-mercaptoethanol [24].

Many substances have been tested for substrate or inhibitory activity towards PNP from human erythrocytes [25]. Formycin B, an analogue of inosine, is a potent competitive inhibitor of purified erythrocytic PNP, with an inhibitor constant (K_I = 1×10^{-4} mmol/l) of the same order of magnitude as the K_m for inosine (K_m = 5×10^{-5} mmol/l) [13].

The reaction catalyzed by human erythrocytic PNP seems to follow an ordered bi-bi mechanism, with nucleoside being the first substrate to add and base the last product to leave [26]. No cofactor or metal requirements are known. Sulphydryl groups play a key role in the enzymatic behaviour. With labelled pCMB, it is possible to titrate 12-SH groups per mole of enzyme. Activity is lost after the first 3 – 4-SH are reacted but may be fully restored with dithiothreitol. 5,5′-dithiobis(2-nitrobenzoic acid) reacts with only 2 – 4-SH causing partial loss of activity; however, in the presence of SDS, approximately 12-SH groups are titratable [25].

Assay

Method Design

Principle

(a) Inosine + P_i $\xrightleftharpoons{\text{PNP}}$ hypoxanthine + ribose-1-P

(b) Hypoxanthine + $2\,H_2O$ + $2\,O_2$ $\xrightleftharpoons{\text{xanthine oxidase*}}$ uric acid + $2\,H_2O_2$

(c) $2\,H_2O_2$ + 2 ethanol $\xrightleftharpoons{\text{catalase**}}$ 2 acetaldehyde + $4\,H_2O$

(d) 2 Acetaldehyde + $2\,NAD(P)^+$ + $2\,H_2O$ $\xrightleftharpoons{\text{AlDH***}}$

2 acetic acid + 2 NAD(P)H + $2\,H^+$

The amount of inosine reacting per unit time, detected by the increase in absorbance due to the reduction of NAD(P), is a measure of catalytic activity of PNP, if there is an excess of ethanol and of the auxiliary enzymes to ensure that the first reaction is rate-limiting. The direction of the complete reaction is in favour of acetic acid formation. The blank of the assay has to be read before the substrate inosine is added to start the reaction sequence and the blank has to be subtracted.

After the addition of inosine there is a lag period of 1 – 3 min, depending on the activity present in the reaction mixture. The method described below can be adapted to mechanization.

Optimized conditions for measurement: until now PNP measurements have not been optimized as a result, at least in part, of the varying heterogeneity of isoenzymes reported for one type of cell and between different organs and species. In the method described below, assay conditions have been selected that are suitable for the assay of PNP activity in biological materials such as blood cells or tissue extracts, especially for the diagnosis of PNP deficiency in red blood cells. NAD can be replaced by NADP, because aldehyde dehydrogenase from yeast uses either coenzyme. High ethanol concentrations minimize the side reaction between 2 moles of H_2O_2.

Temperature conversion factors: temperature conversion factors themselves have not been reported, but the temperature dependence of V with inosine as substrate has been investigated and described [2].

* EC 1.2.3.2.
** EC 1.11.1.6.
*** Aldehyde dehydrogenase, EC 1.2.1.5.

Equipment

Spectrophotometer or spectral-line photometer capable of exact measurement at 339 nm, Hg 334 or Hg 365 nm, preferably with a thermostatted cuvette holder; water-bath; recorder or stopwatch, respectively.

Reagents and Solutions

Purity of reagents: commercially available preparations of catalase, aldehyde dehydrogenase from yeast and xanthine oxidase of analytical grade can be used. In summary, the auxiliary enzymes should not contain NADH oxidation activities greater than 0.01% of that of aldehyde dehydrogenase activity; NAD-dependent alcohol dehydrogenase must be absent. Since there is an inhibitory effect of ammonium ions in higher concentrations on the PNP reaction, undialyzed enzyme suspensions in ammonium sulphate solution must not give rise to ammonia concentrations of more than about 14 mmol/l [15].

Suitable commercial preparations of NAD and inosine of high purity are available and were used. All other chemicals are of analytical grade.

Preparation of solutions (for about 10 determinations): all solutions in re-purified water (cf. Vol. II, chapter 2.1.3.2).

1. Phosphate buffer (0.1 mol/l; pH 7.5):

   dissolve 22.82 g $K_2HPO_4 \cdot 3\ H_2O$ in water and add water to 1 litre (solution A); dissolve 13.61 g KH_2PO_4 in water and add water to give 1 litre (solution B). By mixing solutions A and B pH 7.5 is adjusted.

2. Inosine (9.3 mmol/l):

   dissolve 2.5 mg inosine (*Boehringer Mannheim* 106658) in 1 ml water.

3. Aldehyde dehydrogenase from yeast (ca. 50 kU/l):

   dissolve an equivalent to 50 U of the lyophilized powder (*Boehringer Mannheim* 171832) in 1 ml water.

4. Xanthine oxidase (ca. 4 kU/l):

   the commercially available enzyme preparation (10 mg/ml, ca. 4 U/ml; *Boehringer Mannheim* 110434) is used without further dilution.

5. Catalase (ca. 1.3×10^9 U/l):

   a commercially available enzyme preparation (20 mg/ml; ca. 1.3×10^6 U/ml; *Boehringer Mannheim* 106810) is used without further dilution.

6. Working solution (for about 10 determinations):

   to 9 ml phosphate buffer (1) add 1 ml ethanol, 0.02 ml catalase (5), and 5 mg NAD (*Boehringer Mannheim* 127990).

Stability of solutions: Store all solutions at 0 – 4 °C. The working solution (6) and the aldehyde dehydrogenase (3) are stable for at least two days. Phosphate buffer (1) and inosine solution (2) are stable for many months if stored frozen.

Procedure

Collection and treatment of specimen

Lymphocytes

Lymphocytes from blood of healthy persons are prepared by means of Ficoll-Metrizoate [27], and diluted to approximately 5×10^6 cells per ml in phosphate buffer, 0.05 mol/l, pH 7.5. The suspension is frozen, then thawed and the cell fragments removed by centrifugation.

Whole blood

0.1 ml blood is obtained from the fingertip with a heparinized capillary and diluted with 0.9 ml phosphate buffer, 0.1 mol/l, pH 7.5, containing EDTA, 0.01 mol/l, saturated with digitonin. The mixture is cleared by centrifugation at 10000 *g* for 2 min.

Erythrocytes

Erythrocytes are prepared on a Ficoll-Paque density gradient from heparinized blood (75000 IU/l) drawn by venepuncture [28]. The isolated erythrocytes are re-suspended with physiological saline to give a concentration of about 4×10^{12}/l; cell counting is performed using a counting chamber and a light microscope with 400× magnification. Alternatively, the haemoglobin concentration can be measured. Cell lysis and further preparation are performed as described above.

Stability of the enzyme in the sample: a simple freezing/thawing as well as storage at – 70 °C for four weeks does not change PNP activity in lymphocyte extracts; at 4 °C PNP activity is stable for at least 8 hours. The erythrocytic PNP activity does not alter within 24 hours in blood samples stored at 4 °C.

Details for measurement in tissues: insufficient details are known about the intracellular distribution of PNP, as well as about suitable extraction procedures from tissues. *Laarhoven et al.* [29] reported large differences of PNP activity between various methods of lymphocyte destruction. Therefore, it is advisable to determine the yield and precision of each extraction procedure employed in preliminary experiments.

Assay conditions: wavelength 339 (Hg 334, Hg 365) nm; light path 10 mm; final volume 1.00 ml; 30 °C (thermostatted cuvette holder). Before starting the assay, adjust temperature of solutions to 30 °C.

The values of $\Delta A/\Delta t$ must be <0.03/min at 365 nm (at 334 nm or 339 nm <0.060/min). Otherwise dilute the sample with solution (1).

Measurement

Pipette successively into the cuvette:			concentration in assay mixture	
working solution	(6)	0.81 ml	phosphate ethanol NAD catalase	72.8 mmol/l 1.39 mol/l 0.61 mmol/l 2.1×10^6 U/l
xanthine oxidase	(4)	0.02 ml	XOD	80 U/l
aldehyde dehydrogenase	(3)	0.05 ml	AlDH	2500 U/l
sample (e.g. whole blood)		0.02 ml	volume fraction	0.10
mix thoroughly with a plastic spatula; wait for ca. 20 min until a linear change in absorbance is observed (blank)				
inosine		0.1 ml	inosine	0.93 mmol/l
mix, wait about 2 min (until the change in absorbance becomes linear), read absorbance, repeat the reading after exactly 5 and 10 min or monitor the reaction on a recorder.				

Calculation: use calculation formula (k) or (k_1), respectively (cf. "Formulae, Appendix" 3). For values of ε cf. Appendix 4. The following relations are valid (Δt in minutes):

	Hg 334 nm	339 nm	Hg 365 nm	
$b =$	$809 \times \Delta A/\Delta t$	$794 \times \Delta A/\Delta t$	$1470 \times \Delta A/\Delta t$	U/l
$b =$	$1.35 \times 10^4 \times \Delta A/\Delta t$	$1.32 \times 10^4 \times \Delta A/\Delta t$	$2.45 \times 10^4 \times \Delta A/\Delta t$	nkat/l

Validation of Method

Precision, accuracy, detection limits and sensitivity: replicate measurements ($n = 10$) of an extract of lymphocytes gave a mean ± SD of 113 ± 1.7 U/l, a relative standard deviation of 1.5% [15, 16]. Data about accuracy are not available since standard reference material is not yet established. For measurements at 339 nm with sufficiently sensitive photometers the lowest measurable value is approximately 0.2 U/l sample, if 0.5 ml of sample is applied.

Sources of error: influences on the measured activity by therapeutics are unknown. The presence of alcohol dehydrogenase must be avoided [14].

Specificity: the PNP activity is measured specifically since the reaction is started with the substrate and interfering NADH-producing reactions can be eliminated with the blank.

Reference ranges: mean values of PNP activity measured by the method described are reported to be 9.4 U/g for rat liver, 0.18 U/g for rat heart, 0.59 U/g for rat muscle, 2.09 mU/10^6 cells for human lymphocytes, 5.00 U/ml blood for human erythrocytes [15].

For healthy children ($n = 30$) (age: 1 – 20 years; $\bar{x} = 8.5$ years) the blood PNP activity ($\bar{x}$ ± SD) has been found to be 3450 ± 358 U/l corresponding to 25.53 ± 2.58 U/g haemoglobin (*F. Heinz,* unpublished results).

References

[1] *M. Friedkin, H. Kalckar,* Nucleoside Phosphorylases, in: *P. D. Boyer, H. Lardy, K. Myrbäck* (eds.), The Enzymes, Vol. 5, 2nd edn., Academic Press, New York 1961, pp. 237 – 255.

[2] *R. E. Parks, Jr., R. P. Agarwal,* Purine Nucleoside Phosphorylase, in: *P. D. Boyer* (ed.), The Enzymes, Vol. 7, 3rd edn., Academic Press, New York 1972, pp. 483 – 514.

[3] *D. A. Wiginton, M. S. Coleman, J. J. Hutton,* Characterization of Purine Nucleoside Phosphorylase from Human Granulocytes and Its Metabolism of Deoxyribonucleosides, J. Biol. Chem. *255*, 6663 – 6669 (1980).

[4] *V. Zannis, D. Doyle, D. W. Martin, Jr.,* Purification and Characterization of Human Erythrocyte Purine Nucleoside Phosphorylase and Its Subunits, J. Biol. Chem. *253*, 504 – 510 (1978).

[5] *J. A. McRoberts, D. W. Martin, Jr.,* Submolecular Characterization of a Mutant Human Purine-Nucleoside Phosphorylase, J. Biol. Chem. *255*, 5605 – 5615 (1980).

[6] *G. Ghangas, G. H. Reem,* Characterization of the Subunit Structure of Human Placental Nucleoside Phosphorylase by Immunochemistry, J. Biol. Chem. *254*, 4233 – 4237 (1979).

[7] *K. C. Agarwal, R. P. Agarwal, J. D. Stoeckler, R. E. Parks, Jr.,* Purine Nucleoside Phosphorylase. Microheterogeneity and Comparison of Kinetic Behavior of the Enzyme from Several Tissues and Species, Biochemistry *14*, 79 – 84 (1975).

[8] *R. P. Agarwal, R. E. Parks, Jr.,* Purine Nucleoside Phosphorylase from Human Erythrocytes, IV. Crystallization and some Properties, J. Biol. Chem. *244*, 644 – 647 (1969).

[9] *E. R. Giblett, A. J. Ammann, D. W. Wara, R. Sandman, L. K. Diamond,* Nucleoside-Phosphorylase Deficiency in a Child With Severely Defective T-Cell Immunity and Normal B-Cell Immunity, Lancet *1*, 1010 – 1013 (1975).

[10] *Y. H. Edwards, D. A. Hopkinson, H. Harris,* Inherited Variants of Human Nucleoside Phosphorylase, Ann. Hum. Genet., Lond. *34,* 395 – 408 (1971).

[11] *Y. H. Edwards, P. A. Edwards, D. A. Hopkinson,* A Trimeric Structure for Mammalian Purine Nucleoside Phosphorylase, FEBS Lett. *32,* 235 – 237 (1973).

[12] *H. M. Kalckar,* Differential Spectrophotometry of Purine Compounds by Means of Specific Enzymes, I. Determination of Hydroxypurine Compounds, J. Biol. Chem. *167,* 429 – 443 (1947).

[13] *M. R. Sheen, B. K. Kim, R. E. Parks, Jr.,* Purine Nucleoside Phosphorylase from Human Erythrocytes, III. Inhibition by the Inosine Analog Formycin B of the Isolated Enzyme and of Nucleoside Metabolism in Intact Erythrocytes and Sarcoma 180 Cells, Mol. Pharmacol. *4,* 293 – 299 (1968).

[14] *R. Haeckel,* The Use of Aldehyde Dehydrogenase to Determine H_2O_2-Producing Reactions, I. The Determination of the Uric Acid Concentration, J. Clin. Chem. Clin. Biochem. *14,* 101 – 107 (1976).

[15] *F. Heinz, S. Reckel, R. Pilz, J. R. Kalden,* A New Spectrophotometric Assay for Enzymes of Purine Metabolism, III. Determination of Purine Nucleoside Phosphorylases, Enzyme *25,* 44 – 49 (1980).

[16] *R. Pilz,* Bestimmung der Aktivität von Purinstoffwechselenzymen in menschlichen peripheren Blutlymphocyten: Entwicklung von Methoden und ihre exemplarische Anwendung, Thesis 1980, Medizinische Hochschule Hannover.

[17] *A. F. Ross, K. C. Agarwal, S.-H. Chu, R. E. Parks, Jr.,* Studies On the Biochemical Actions of 6-Selenoguanine and 6-Selenoguanosine, Biochem. Pharmacol. *22,* 141 – 154 (1973).

[18] *K. F. Jensen, P. Nygaard,* Purine Nucleoside Phosphorylase from Escherichia coli and Salmonella typhimurium, Eur. J. Biochem. 253 – 265 (1975).

[19] *G. Milman, D. L. Anton, J. L. Weber,* Chinese Hamster Purine-Nucleoside Phosphorylase: Purification, Structural, and Catalytic Properties, Biochemistry *15,* 4967 – 4972 (1976).

[20] *M. P. Uitendaal, C. H. M. M. de Bruyn, T. L. Oei, P. Hösli, C. Griscelli,* A New Ultramicrochemical Assay for Purine Nucleoside Phosphorylase, Anal. Biochem. *84,* 147 – 153 (1978).

[21] *W. R. A. Osborne, S.-H. Chen, E. R. Giblett, W. D. Biggar, A. J. Ammann, C. R. Scott,* Purine Nucleoside Phosphorylase Deficiency. Evidence for Genetic Heterogeneity in Two Families with Enzyme Deficient Members, J. Clin. Invest. *60,* 741 – 746 (1977).

[22] *W. R. A. Osborne, C. R. Scott,* Purine Nucleoside Phosphorylase Deficiency. Measurement of Variant Protein in Four Families with Enzyme-Deficient Members by an Enzyme-Linked Immunosorbent Assay, Am. J. Hum. Genet. *32,* 927 – 933 (1980).

[23] *A. P. Halfpenny, P. R. Brown,* Optimized Assay For Purine Nucleoside Phosphorylase by Reversed-Phase High-Performance Liquid Chromatography, J. Chromatogr. *199,* 275 – 282 (1980).

[24] *M. D. Glantz, A. S. Lewis,* Purine Nucleoside Phosphorylase from Rabbit Liver, in: S. P. Colowick, N. O. Kaplan, Methods in Enzymology, Vol. LI, Academic Press, New York 1978, pp. 524 – 530.

[25] *J. D. Stoeckler, R. P. Agarwal, K. C. Agarwal, R. E. Parks, Jr.,* Purine Nucleoside Phosphorylase from Human Erythrocytes, in: S. P. Colowick, N. O. Kaplan, Methods Enzymology, Vol. LI, Academic Press, New York 1978, pp. 530 – 538.

[26] *B. K. Kim, S. Cha, R. E. Parks, Jr.,* Purine Nucleoside Phosphorylase from Human Erythrocytes, I. Purification and Properties, J. Biol. Chem. *243,* 1763 – 1770 (1968).

[27] *J. R. Kalden, H. H. Peter, R. Roubin, J. P. Cesarini,* Human Peripheral Null Lymphocytes. Isolation, Immunological and Functional Characterization, Eur. J. Immunol. *7,* 537 – 542 (1977).

[28] *A. Dwenger, H.-J. Mitzkat, W. Holle, P. Tost, I. Trautschold,* Insulin Binding to Erythrocytes from Pregnant, Postpartum, Follicular and Luteal States, J. Clin. Chem. Clin. Biochem. *20,* 273 – 279 (1982).

[29] *J. P. R. M. van Laarhoven, G. Th. Spierenburg, F. T. J. J. Oerlemans, C. H. M. M. de Bruyn,* Micromethods for the Measurement of Purine Enzymes in Lymphocytes, Adv. Exp. Med. Biol. *122,* 415 – 420 (1979).

5.6 Adenine Phosphoribosyltransferase

AMP: pyrophosphate phosphoribosyltransferase, EC 2.4.2.7

Peter Banholzer and Wolfgang Gröbner

General

Williams & Buchanan (1953) showed that purine bases can be converted to ribonucleotides by a one-step process [1]. The existence of a separate phosphoribosyltransferase for adenine (APRT) was first noted by *Kornberg* (1955) who also discovered phosphoribosyl pyrophosphate (PRPP) to be the source of the ribose phosphate moiety [2]. The direct formation of nucleotides from their bases is also called the "salvage pathway". A complete lack of APRT, first seen by *Cartier* in 1974 [3], is associated with the excretion of excessive amounts of 2,8-dihydroxyadenine which is extremely insoluble and may lead to urolithiasis and/or crystal-induced nephropathy. The defect is inherited in an autosomal recessive manner [4].

Application of method: in biochemistry and clinical chemistry.

Enzyme properties relevant in analysis: substrate conversion is linear with time and protein concentration. The enzyme has an absolute requirement for divalent cations [5]. Magnesium has been shown to be the most effective metal cofactor. In addition, an initial Mg^{2+}-independent "burst synthesis" of AMP at 0 °C has been described [6, 7]. Kinetic analysis and inhibitor profiles have been performed in many tissues. For detailed information see references [8, 9, 10].

The equilibrium constant of the reaction favours synthesis of AMP. However, since the enzyme is inhibited by PP_i, the kinetics of the reaction may influence the apparent equilibrium position of the system under the conditions employed. *Michaelis* constants for adenine and for PRPP for the human enzyme from erythrocytes are approximately 1.4 to 2.7 $\times$ 10^{-6} mol/l, and 6 $\times$ 10^{-6} mol/l respectively [8]. Studies on the molecular weight of a 33000-fold purified APRT from erythrocytes showed that the enzyme exists ordinarily as a trimer, but is catalytically active as hexamer and a dimer [6]. In contrast, no subunit structure was reported for the highly purified APRT from rat liver [7].

Methods of determination: catalytic activity of APRT has been assayed by a variety of spectrophotometric and radiochemical methods. The latter are differentiated largely by the techniques used to separate the radioactive substrate from the product. These techniques include high voltage electrophoresis on paper, ascending chromatography on thin-layer cellulose, filtration over DEAE cellulose filter discs, butanol extraction of [^{14}C]AMP, anion-exchange chromatography, and electrophoresis on polyacrylamide gel and cellulose acetate. For references see [8, 10].

The method described here [11] is a simple, sensitive method which can be used for measurement of all phosphoribosyl transferases by replacing adenine by the appropriate substrate, with no further modifications being necessary in many instances. Paper electrophoresis is the simplest way to prepare the samples for liquid scintillation counting.

International reference method and standards: neither standardization at the international level nor the existence of reference standard materials is known up to date.

Enzyme effectors: the reaction is inhibited by its product AMP, and also by IMP, GMP and all three nucleotide di- and tri-phosphates, with guanylate being the most effective. The enzyme appears to be dependent on both magnesium and PRPP, with an absolute requirement for divalent cations. For detailed information see [8].

Assay

Method Design

Principle

$$\text{Adenine} + \text{PRPP} \underset{Mg^{2+}}{\overset{\text{APRT}}{\rightleftharpoons}} \text{AMP} + PP_i$$

In the micro-radiochemical method described here, the amount of [^{14}C]AMP formed from [8-^{14}C]adenine and PRPP is a measure of APRT activity. Results are expressed as amount of product formed per unit time per amount of protein. When erythrocytes are the enzyme source, results have also been given per amount of haemoglobin.

Optimized conditions for measurement: catalytic activity is measured at pH 7.4, 38 °C. Concentrations in the following method are: $MgCl_2$, 6 mmol/l, Tris buffer, pH 7.4, 55 mmol/l, PRPP, 1 mmol/l, and adenine, 0.25 mmol/l.

Maximum APRT activity in *Ehrlich's* ascites tumour cells or in bacteria is achieved when magnesium is present at approximately twice the concentration of PRPP, and concentrations of Mg^{2+} higher than 1 mmol/l inhibit enzyme activity [5, 12]. The optimum Mg concentration for human tissues is unknown. Most clinical investigations have been performed using the above-mentioned conditions.

Temperature conversion factors: these have not been established so far.

Equipment

Liquid scintillation counter; equipment for high-voltage paper electrophoresis; UV-lamp (wavelength 254 nm), spectrophotometer (for protein determination).

Reagents and Solutions

Purity of reagents: commercially available substances can be used. PRPP shows variations in purity. This has to be considered in calculation.

Preparation of solutions (for about 100 determinations): all solutions in re-purified water (see Vol. II, chapter 2.1.3.2).

1. Tris buffer (pH 7.4, 10 mmol/l):

   dissolve 6.05 g Tris in 5 l water, and adjust to pH 7.4 by addition of HCl, 2 mol/l.

2. $MgCl_2$/Tris ($MgCl_2$, 20 mmol/l; Tris, 200 mmol/l; pH 7.4):

   a) dissolve 20.33 g $MgCl_2 \cdot 6\ H_2O$ in 100 ml water,
   b) dissolve 12.1 g Tris in 200 ml water,
   c) mix 1 ml of (a), 2 ml of (b) and 2 ml water.

3. [8-^{14}C]Adenine (1.7 mmol/l, specific activity 5.6 mCi/mmol):

   a) dissolve 24 mg non-radioactive adenine in 100 ml water,
   b) mix 1.35 ml of (a) and 0.15 ml of the radioactive stock solution (1.7 mmol/l, specific activity 56 mCi/mmol).

4. PRPP (10 mmol/l):

   dissolve 5 mg PRPP, dimagnesium salt, PRPP content 70% in 1.065 ml water.

5. Substrate solution ($MgCl_2$, 12 mmol/l; Tris, 100 mmol/l, pH 7.4; adenine-8-^{14}C, 0.5 mmol/l, specific activity 5.6 mCi/mmol; PRPP, 2 mmol/l):

   mix 2.5 ml of solution (2c), 1.5 ml of solution (3b) and 1.0 ml of solution (4). Prepare this final substrate solution immediately before use and keep on ice until all assays have been started.

6. EDTA (100 mmol/l):

   dissolve 3.72 g EDTA-$Na_2H_2 \cdot 2\ H_2O$ in 10 ml water.

7. Borate buffer (25 mmol/l, pH 8.5):

   dissolve 18.1 g borate-$Na_2 \cdot 10\ H_2O$, 3.16 g boric acid and 1.5 g EDTA-$Na_2H_2 \cdot 2H_2O$ in 4 l water.

8. Carrier solution for electrophoresis (AMP, 2 mg/ml :

   dissolve 20 mg AMP-$Na_2 \cdot 6\ H_2O$ in 10 ml water.

9. Scintillation fluid:

   dissolve 9.45 g 2.5 diphenyloxazole and 0.2 g 2.2′-p-phenylene bis-(5-phenyloxazole) in 2.5 l toluene.

Stability of solutions: solutions (1), (2), (3a), (6), (7) are stable at room temperature or at +4 °C as long as there is no bacterial contamination. Solutions (3b), (4) and (5) have to be prepared freshly for each series of determinations. Solution (8) is stable at +4 °C for 2 weeks. Store the radioactive stock solution according to the manufacturer's advice (usually +4 °C). Solution (9) is stable indefinitely.

Procedure

Collection and treatment of specimen: centrifuge heparinized venous blood 10 min at 600 *g* at room temperature. Remove plasma and wash the sediment two times with cold 0.9% NaCl solution. Freeze-thaw packed erythrocytes two times. Dialyze 1 h against 500 volumes of Tris buffer (1). Dilute enzyme preparation 1 : 50 with this buffer. (In the following text "haemolysate" refers to the 1 : 50 diluted preparation described above, except where indicated.)

Stability of the enzyme in the sample: the enzyme can be stored at 4 °C in the form of crude haemolysate prior to dialysis and dilution at −20 °C for several days without loss of activity. At 4 °C we observed a loss of activity of 10% in 5 days for the normal and defect enzyme. At 55 °C a complete loss of activity occurs within 20 minutes.

Details for measurement in tissues: the enzyme is widely distributed in body tissues with the highest specific activity in nucleated cells.

It is considered to exist free in the cytoplasm except in bacteria, where it is located in the periplasmatic space in loose association with the membrane [8, 10].

Homogenization of tissues in the appropriate buffer directly yields a sample suitable for the determination of cytosolic enzyme activity. To prevent degradation of the product by 5′-nucleotidase, thymidine triphosphate has to be added to the reaction mixture [17].

Assay conditions: reaction volume 0.1 ml; 37 °C (thermostatted gyratory water-bath); reaction time 20 min. Blank: Tris buffer (1) instead of sample.

Electrophoresis: borate buffer (7); separation at 8000 volts for 30 min.

Liquid scintillation counting: volume of scintillation fluid 10 ml; +4 °C; 10 min.

Measurement

Incubation

Pipette successively into reaction vessel:			concentration in incubation mixture	
substrate solution	(5)	0.05 ml	$MgCl_2$ Tris adenine PRPP	6 mmol/l 55 mmol/l 0.25 mmol/l 1 mmol/l
sample (haemolysate or other tissues)		0.05 ml	volume fraction	0.5
mix, incubate for 20 min, stop reaction				
EDTA solution	(6)	0.02 ml	EDTA	17 mmol/l

Electrophoresis: rinse *Whatman* No. 3 MM chromatography paper in sodium borate buffer (7). Place 0.02 ml of the AMP carrier solution (8), then 0.02 ml of the incubation mixtures on the paper. After 30 min electrophoresis dry the paper, locate the AMP spot under ultraviolet light (wavelength 254 nm) and cut out.

Scintillation counting: put paper into the vial, add 10 ml scintillation solution (9) and count for 10 min at 4 °C in a liquid scintillation counter.

Protein determination: determine protein content of the haemolysate after tenfold dilution with 0.9% NaCl according to *Lowry et al.* [13]; for procedure see Vol. II, chapter 1.3.4. Results are calculated as g/l.

Calculation: correct scintillation counting rate of sample, C_p for counting rate of blank C_b. The difference ΔC_n is used in calculation.

A more detailed description of calculation is given on p. 405.

The following equation is valid for expressing the catalytic activity content (specific activity) of APRT related to the mass of protein (cf. "Formulae", Appendix 3, eqn. (p_1)):

$$z_c/m_s = \frac{\Delta C_n \times 10 \times 1000}{\eta \times 2.22 \times 10^{12} \times 10^{-6} \times X_0 \times \rho \times \Delta t \times \varphi \times v_r \times 5/6} \quad \text{U/g}$$

where η is the counting efficiency, X_0 is the specific activity of [8-^{14}C]adenine expressed as Ci/mol, ρ is the protein concentration of the sample (expressed as g/l), Δt is the reaction time in minutes, φ is volume fraction of sample in incubation mixture, v_r is sample volume used in electrophoresis and in radiometry (all volumes in ml).

Counting efficiency depends on the scintillation fluid and the equipment used and normally is of the order of 0.6 (cf. also Vol. IV, chapter 4.1).

The factor 10 in the above equation derives from the dilution of [8-^{14}C]adenine by non-radioactive adenine.

Validation of Method

Precision, accuracy, detection limits and sensitivity: in the range of normal activity in haemolysates (20 – 30 nmol/mg protein per h) maximum imprecision is about ± 2 nmol/mg protein per hour.

Standard reference material is not available: data on accuracy do not exist, therefore. Detection limit is in the range of 10^{-4} to 10^{-3} U/g. Sensitivity is in the same range [14].

Sources of error: interacting drugs are not known up to now. Blood transfusions lead to an elevated APRT activity lasting for some months in patients with enzyme deficiency. Elevated levels of APRT activity have been found in newborns, patients with megaloblastic anaemia and reticulocytosis, as well as in renal failure [15, 16, 8, 10]. In tissues containing 5′-nucleotidase, addition of thymidine triphosphate is necessary to prevent AMP degradation [17].

Specificity: the APRT enzyme reacts with 6-amino purines (adenine, 2- and 8-aza-adenine, 2-fluoroadenine, 2,6-diaminopurine), 6-mercaptopurine and also 4-amino-5-imidazolecarboxamide. For references see [8, 10].

Reference ranges: different laboratories report the mean specific activity of APRT in erythrocyte haemolysates (dialyzed or undialyzed) to be between 0.3 and 0.5 U/g protein. Studies of APRT activity in different tissues have been undertaken. For references see [8].

References

[1] *W. J. Williams, J. M. Buchanan,* Biosynthesis of the Purines, V. Conversion of Hypoxanthine to Inosinic Acid by Liver Enzymes, J. Biol. Chem. *203*, 583 – 593 (1953).

[2] *A. Kornberg, I. Lieberman, E. S. Simms,* Enzymatic syntheses of pyrimidine and purine nucleotides. I. Formation of 5-phosphoribosylpyrophosphate, J. Am. Chem. Soc. *76*, 2027 – 2028 (1954).

[3] *P. Cartier, M. Hamet, J. Hamburger,* Phosphoribosyltransferase and lithiasis of 2,8-dihydroxy-adenine, C. R. Acad. Sci. D. (Paris) Sep *279*, p. 883 – 886 (1974).

[4] *W. N. Kelley, R. J. Rosenbloom, F. M. Henderson, J. E. Seegmiller,* Adenine phosphoribosyltransferase deficiency: a previously undescribed genetic defect in man, J. Clin. Invest. *47*, 2281 – 2289 (1968).

[5] *M. Hori, J. F. Henderson,* Purification and properties of adenylate pyrophosphorylase from Ehrlichs ascites tumor cells, J. Biol. Chem. *241,* 1406 – 1411 (1966).
[6] *G. B. Thomas, W. J. Arnold, W. N. Kelley,* Human adenine phosphoribosyltransferase, J. Biol. Chem. *248,* 2529 – 2535 (1973).
[7] *J. G. Kenimer, L. G. Young, D. P. Groth,* Purification and properties of rat liver adenine phosphoribosyltransferase, Biochem. Biophys. Acta *384,* 87 – 101 (1975).
[8] *H. A. Simmonds, K. J. van Acker,* Adenine phosphoribosyltransferase deficiency: 2,8-dihydroxyadenine lithiasis, in: *J. B. Stanbury, J. B. Wyngarden, D. S. Fredrickson* (eds.), The Metabolic Basis of Inherited Disease, McGraw Hill, New York, 1982 (in press).
[9] *J. F. Henderson, A. R. P. Paterson,* Nucleotide Metabolism, Academic Press, New York and London, 1973, pp. 126 – 127.
[10] *W. C. Arnold,* Purine salvage enzymes, in: *W. N. Kelley, J. M. Weiner* (eds.), Uric Acid, Springer-Verlag Berlin-Heidelberg-New York, 1978, pp. 43 – 73.
[11] *W. S. Kelley, F. M. Rosenbloom, J. F. Henderson, J. E. Seegmiller,* A specific enzyme defect in gout associated with overproduction of uric acid, Proc. Natl. Acad. Sci. – USA *57,* 1735 – 1739 (1976).
[12] *J. Hochstadt-Ozer, E. R. Stadtman,* The regulation of purine utilization in bacteria, J. Biol. Chem. *246,* 5312 – 5320 (1971).
[13] *O. H. Lowry, N. J. Rosebrough, A. L. Farr, R. J. Randall,* Protein measurement with the Folin phenol reagent, J. Biol. Chem. *193,* 265 – 275 (1951).
[14] *W. N. Kelley,* Hypoxanthine-guanine phosphoribosyltransferase in the Lesh-Nyhan-syndrome and gout, Fed. Proc. *27,* 1047 – 1052 (1968).
[15] *H. I. Becher, H. I. Weise, U. Volkermann, P. Schallmeyer,* Enhanced purine nucleotide synthesis in erythrocytes of uremic patients, Klin. Wochenschr. *58,* 1243 – 1249 (1980).
[16] *P. Banholzer, W. Gröbner, W. Löffler, S. Reiter, N. Zöllner,* Adenine phosphoribosyltransferase (APRT)-activity in patients with nephrolithiasis or renal failure, J. Clin. Chem. Clin. Biochem. *20,* 345 (1982).
[17] *W. Gröbner, N. Zöllner,* Eigenschaften der Hypoxanthinguaninphosphoribosyltransferase (HGPRTase) bei einem Gichtpatienten mit verminderter Aktivität dieses Enzymes, Klin. Wochenschr. *57,* 63 – 68 (1979).

5.7 Hypoxanthine Phosphoribosyltransferase

IMP-pyrophosphate phosphoribosyltransferase, EC 2.4.2.8

Werner Löffler and Wolfgang Gröbner

General

In 1953, *Williams & Buchanan* [1] showed that purine bases can be converted to ribonucleotides by a one-step process. The enzymes involved were recognized to constitute a separate class by *Kornberg, Liebermann & Simms* (1955) [2]. Several nucleotides can be formed directly from free bases by this system, which is also called the "salvage pathway". The amount of energy necessary for these reactions is considerably smaller than that for *de novo*-synthesis of nucleotides.

The key enzyme of purine synthesis *de novo*, glutamine phosphoribosylpyrophosphate amidotransferase, is subject to feedback inhibition by nucleotides [3]. Hence, hypoxanthine-phosphoribosyltransferase* plays a major role in the regulation of purine biosynthesis by means of its products, the nucleotides.

In animals and man only liver and placenta are capable of synthesizing nucleotides *de novo* in sufficient amounts. Other tissues will depend solely or partially on salvage enzymes for nucleotide formation [3]. In man, partial or complete deficiency of HPRT occurs as an X-linked inherited disease associated with overproduction of uric acid with the consequence of hyperuricaemia and gout. In addition, the complete enzyme defect is accompanied by a bizarre neurological disorder known as *Lesch-Nyhan*-Syndrom [4].

Human HPRT binds 6-oxo- or 6-thio-purines, but not 6-amino-compounds [5]. According to physicochemical as well as immunological properties there is a wide variety of mutant forms of HPRT in man [6]. In some species different enzymes exist for converting the natural substrates of human HPRT to their respective nucleotides [6]. HPRT exists free in the cytoplasm, but has also been shown to occur in the synaptosomes of rat brain [7].

Application of method: in biochemistry and clinical chemistry.

Enzyme properties relevant in analysis: substrate conversion is linear with protein concentration and time. There have been demonstrated electrophoretic variants of the normal human enzyme, which are considered to be the result of post-transcriptional variations [6]. The enzyme is subject to inhibition by its end-products, purine nucleotides, which results in an inactive di- or trimer form of the subunit of 34500 molecular weight. The active form is regenerated by addition of PRPP [6]. The K_m for hypoxanthine, guanine, and PRPP is approximately 8×10^{-6} mol/l, 1×10^{-5} and 2×10^{-4} mol/l, respectively, with the enzyme from human erythrocytes [8].

Methods of determination: catalytic activity can be measured by micro-radiochemical, autohisto-radiographic, or spectrophotometric methods. In the case of mutant forms there is no correlation between catalytic activity and the amount of cross-reacting material to normal enzyme antibodies. Estimation of enzyme mass by this method will be misleading therefore.

The micro-radiochemical method described here [9] is very sensitive in the detection of the low activity of mutant forms. Paper electrophoresis is the simplest way to prepare the samples for liquid scintillation counting.

International reference methods and standards: these are not known up to now.

Enzyme effectors: no activators are known which must be added to the assay mixture.

---

* Abbreviations:
HPRT = hypoxanthine-phosphoribosyltransferase
PRPP = 5-Phospho-α-D-ribose-1-diphosphate

Assay

Method Design

Principle

$$\text{Hypoxanthine} + \text{PRPP} \xrightleftharpoons{Mg^{2+}} \text{IMP} + \text{PP}_i$$

$$\text{Guanine} + \text{PRPP} \xrightleftharpoons{Mg^{2+}} \text{GMP} + \text{PP}_i$$

The method uses the reaction catalyzed by this enzyme *in vivo*. The equilibrium is far on the side of the nucleotides. The amount of products formed per unit time is a measure of the catalytic activity. With the radiochemical method [^{14}C]hypoxanthine is usually used as a substrate and the amount of IMP formed is calculated from the radioactivity found at the site of IMP after electrophoretic separation. In the blanks necessary to correct for background radioactivity, haemolysate or other enzyme sources are replaced by buffer. The result is expressed as amount of product formed per unit time per amount of protein. In the case of erythrocytes as enzyme source results have also been given per amount of haemoglobin. No applications using automated equipment are known up to now.

Optimized conditions for measurement: irrespective of the source of the enzyme, the only donor of the phosphoribosyl moiety for HPRT is PRPP. Plots of activity versus substrate concentration show hyperbolic curves except under certain conditions, which are not used for measuring absolute catalytic activity. There is an absolute requirement for magnesium, and the enzyme is most active at magnesium concentrations from 5 to 20 mmol/l [6]. The preferred substrate seems to be PRPP, dimagnesium salt [5]. Catalytic activity is measured at pH 7.4, 37 °C. In the following method $MgCl_2$, 5 mmol/l, Tris-HCl buffer, pH 7.4, 55 mmol/l, PRPP, 1 mmol/l, and hypoxanthine, 0.3 mmol/l, are used. In the case of guanine as substrate, guanine hydrochloride (0.3 mmol/l) should be used because of its better solubility. $MgCl_2$, 5 mmol/l, is the minimum content derived from solution (2) (see below); the concentration is 7 mmol/l, if PRPP, dimagnesium salt, is used.

Temperature conversion factors: these have not been established so far.

Equipment

Dialysis tubes; water-bath; stopwatch; high voltage electrophoresis; paper for electrophoresis; ultraviolet light; scissors; liquid scintillation counter; spectrophotometer (for protein determination).

Reagents and Solutions

Purity of reagents: radioactive purine bases more than 99% pure are commercially available, which are all suitable as substrates. PRPP shows variations in purity, to the greater part due to non-enzymatic hydrolysis of PRPP. These impurities do not influence enzyme activity to a significant degree.

Preparation of solutions (for about 100 determinations): all solutions in re-purified water (see Vol. II, chapter 2.1.3.2).

1. Tris buffer (10 mmol/l; pH 7.4):

   dissolve 6.05 g Tris in 5 l water and adjust to pH 7.4 by addition of HCl, 2 mol/l.

2. $MgCl_2$/Tris ($MgCl_2$, 20 mmol/l; Tris, 200 mmol/l, pH 7.4):

   a) dissolve 20.33 g $MgCl_2 \cdot 6H_2O$ in 100 ml water, b) dissolve 12.1 g Tris in 200 ml water, c) mix 1 ml of (a), 2 ml of (b), and 2 ml water.

3. [^{14}C]Hypoxanthine (2 mmol/l, specific radioactivity 5 mCi/mmol):

   a) dissolve 27.2 mg non-radioactive hypoxanthine in 100 ml water, to yield a solution of the same molarity as the stock solution (2 mmol/l); b) mix 1.35 ml solution (a) and 0.15 ml radioactive stock solution (2 mmol/l, specific radioactivity 50 mCi/mmol).

4. PRPP (10 mmol/l):

   dissolve 5 mg PRPP, dimagnesium salt (PRPP content 70%) in 1.065 ml water. (Spontaneous degradation is reported by suppliers to be 0.1% per day with the dry substance).

5. Substrate solution ($MgCl_2$, 10 mmol/l; Tris 100 mmol/l, pH 7.4; hypoxanthine/[^{14}C]hypoxanthine, 0.6 mmol/l, 5 mCi/mmol; PRPP, 2 mmol/l):

   mix 5 volumes (2.5 ml) of solution (2c), 3 volumes (1.5 ml) of solution (3b) and 2 volumes (1.0 ml) of solution (4). Prepare this final substrate solution immediately before use and keep on ice, until all assays have been started.

6. EDTA (100 mmol/l):

   dissolve 3.72 g EDTA-$Na_2 \cdot 2H_2O$ in 10 ml water.

7. Sodium borate buffer (25 mmol/l; pH 8.5):

   dissolve 18.1 g borate-$Na_2 \cdot 10H_2O$, 3.16 g boric acid and 1.5 g EDTA-$Na_2 \cdot 2H_2O$ in 4 l water.

8. Carrier solution for electrophoresis (IMP, 4 mmol/l):

   dissolve 20 mg IMP-$Na_2 \cdot 6H_2O$ in 10 ml water.

9. Scintillation fluid:

   dissolve 9.45 g 2,5-diphenyloxazole and 0.2 g 2,2'-p-phenylene bis-(5-phenyloxazole) in 2.5 l toluol.

10. NaCl solution (0.9% w/v):

    dissolve 900 mg NaCl in 100 ml water.

Stability of solutions: solutions (1), (2), (3a), (6), (7) are stable at room temperature or at +4 °C as long as there is no bacterial contamination. Solutions (3b), (4) and (5) have to be prepared freshly for each series of determinations. Solution (8) is stable at +4 °C for 2 weeks. Store the radioactive stock solution according to the manufacturer's advice (usually +4 °C). Solution (9) and (10) are stable undefinitely.

Procedure

Collection and treatment of specimen: for clinical purposes the most suitable probe for determining HPRT activity is haemolysate. Proper dialysis of samples prior to measurements of the catalytic activity will prevent interfering reactions. Collect heparinized blood from the vein. Centrifuge for 10 min at 600 *g*, remove plasma, and wash the erythrocytes twice with cold NaCl, 0.9% (w/v), solution (10). Prepare haemolysate by freezing (−20 °C) and thawing the erythrocytes twice. Dialyze samples against 500 volumes of Tris-HCl buffer (1). Dilute haemolysate 1:50 with Tris-HCl buffer (1). (In the following text "haemolysate" refers to this diluted sample except where indicated.)

Stability of the enzyme in specimen or sample: if the enzyme is stored at −20 °C in the form of crude haemolysate prior to dialysis and dilution, catalytic activity decreases by about 10% during the first month. Details about storage at −70 °C of the normal enzyme from human red cells, at different stages of purification, have been reported [10]. In these studies stabilization was achieved with various compounds, and bovine serum albumin was necessary under all conditions for expression of maximal catalytic activity. – Though many mutant enzymes can be stored with similar loss of activity as occurs in the normal enzyme, this cannot be applied as a general rule.

Details for measurement in tissues: for the determination of enzyme activity in various tissues, all kinds of homogenization are suitable, which reliably achieve destruction of the cell wall, since HPRT occurs free in the cytoplasm. Only in the case of rat brain significant activity could be demonstrated in cell compartments (synaptosomes) [7].

Homogenization of tissues in the appropriate buffer will directly yield the sample for determination of catalytic activity.

Assay conditions: volume of reaction mixture 0.1 ml; volume after stop 0.12 ml; 37 °C (thermostatted gyratory water-bath); reaction time 20 min. Blank: buffer (1) instead of sample.

Electrophoresis: sodium borate buffer, 25 mmol/l, pH 8.5; 35 mA, 8000 V; separation time 30 min.

Liquid scintillation counting: volume of scintillating agent 10 ml; counting time 10 min at +4 °C.

Measurement

Incubation

Pipette successively into the reaction vessel:			concentration in incubation mixture	
substrate solution	(5)	0.05 ml	$MgCl_2$ Tris hypoxanthine PRPP	5 mmol/l 55 mmol/l 0.3 mmol/l 1.5 mCi/l 1 mmol/l
haemolysate (or other sample)		0.05 ml		
mix thoroughly, incubate for 20 min, stop reaction				
EDTA solution	(6)	0.02 ml	EDTA	16.7 mmol/l

Electrophoresis

Rinse *Whatman* No. 3 MM chromatography paper in sodium borate buffer (7). Transfer 0.02 ml of the reaction mixture onto the paper and add 0.02 ml of the carrier solution (8). After separation dry the paper, locate nucleotide spot under ultraviolett light, and cut out.

Radiometry

For scintillation counting put paper carrying the nucleotide spot into the vial and add 10 ml scintillation fluid (9). Count for 10 min.

Protein determination acc. to Lowry et al. [11]

Use 0.02 ml haemolysate (or other sample) and 0.18 ml NaCl solution (solution 10). Proceed according to [11] or as described in Vol. II, chapter 1.3.4. Results must be expressed in g/l for the 1:50 haemolysate.

Calculation: correct result of liquid scintillation counting rates of samples, C_p, for background counting rate C_b yielding the net counting rates C_n. The same is true for blanks. The difference between C_n for sample and blank is ΔC_n.

According to eqn. (o_1) (cf. "Formulae", Appendix 3) the catalytic concentration of the sample is

$$b = \frac{\Delta C_n \times f_h \times f_i}{\eta \times 2.22 \times 10^6 \times X \times v_h \times \Delta t} \quad \text{U/l}$$

In this assay it is
- specific radioactivity of substrate solution (3), X = 5 mCi/mmol
- volume of 1:50 haemolysate used per incubation, v_h = 0.05 ml
- factor applied to correct the dpm measured in 0.02 ml for total volume of incubation mixture after addition of EDTA, f_i = 6
- factor applied to correct for 1 l 1:50 haemolysate, $f_h = 10^3$
- incubation time, Δt = 20 min
- counting efficiency η must be determined for this assay system (cf. Vol. IV, chapter 4.1).

Consequently, the catalytic concentration in the 1:50 haemolysate is

$$b = \frac{\Delta C_n \times 10^3 \times 6}{\eta \times 2.22 \times 10^6 \times 5 \times 0.05 \times 20} \quad \text{U/l}$$

$$= 0.54 \times 10^{-3} \times \frac{\Delta C_n}{\eta} \quad \text{U/l}$$

For calculation of the catalytic content in the 1:50 haemolysate, the catalytic concentration must be divided by $\rho_{protein}$ (g/l):

$$z_c/m_s = 0.54 \times 10^{-3} \times \frac{\Delta C_n}{\eta \times \rho_{protein}} \quad \text{U/g protein.}$$

Example (normal subject): measured values are C_p = 9200 cpm, C_b = 100 cpm, counting efficiency η = 0.6, $\rho_{protein}$ = 5 g/l.

The catalytic concentration is b = 8.19 U/l 1:50 diluted haemolysate. The catalytic activity content is $z_c/m_s = b/\rho$ = 1.64 U/g protein.

Note: in the literature the catalytic activity content is often expressed in arbitrary units defined as nanomoles of product formed per hour per mg of protein.

Validation of Method

Precision, accuracy, detection limits and sensitivity: in the range of normal activity in 1:50 haemolysate, 1.2–2 U/g (70–120 nmol × mg^{-1} protein × h^{-1}), maximal imprecision is about ±0.05 U/g (±3 nmol × mg^{-1} protein × h^{-1}). Standard reference material is not available: data on accuracy do not exist therefore. Detection limit is reported to be in the range of 1.7×10^{-4} μmol IMP × min^{-1} × g^{-1} protein (0.01 nmol IMP × mg^{-1} protein × h^{-1}) [12]. Sensitivity is in the same range [12].

Sources of error: no influences of drugs on catalytic activity could be demonstrated unequivocally. With respect to clinical diagnosis it is of importance that, with some mutant forms of the enzyme-catalytic activity may be normal in haemolysate with saturating substrate concentrations, while *in vivo,* due to altered K_m values for substrate [13] or increased sensitivity to inhibition by nucleotides [14], catalytic activity apparently is reduced. Careful kinetic studies are necessary in these cases, if a mutant enzyme is suspected on clinical grounds. On the other hand mutant enzymes have been described where no catalytic activity could be demonstrated in haemolysate, while the activity in intact red cells was well in the measurable range [15].

Catalytic activity of enzymes interfering with substrate or product concentrations is very low or absent in haemolysates, but may be a source of error in other tissues. Inhibitors have to be added to the assay mixture in these cases (thymidine 5′-triphosphate, for example, for inhibition of 5′-Nucleotidase).

Specificity: HPRT is measured specifically. No enzyme inhibitors could be demonstrated under the assay conditions described.

Reference ranges: in human erythrocytes (1:50 haemolysate) at 37 °C normal values range from 1.2–2 U/g protein (70–120 nmol IMP × mg^{-1} protein × h^{-1}). Activity in human liver is about half and in human brain about eight times that found in haemolysate [12]. Detailed studies of HPRT activity in the course of both foetal and postnatal development have been published [7, 16, 17]. For references on tissue distribution in animals and man see [3, 6, 18].

References

[1] *W. J. Williams, J. M. Buchanan,* Biosynthesis of the Purines, V. Conversion of Hypoxanthine to Inosinic Acid by Liver Enzymes, J. Biol. Chem. *203*, 583–593 (1953).

[2] *A. Kornberg, J. Lieberman, E. S. Simms,* Enzymatic Synthesis of Purine Nucleotides, J. Biol. Chem. *215*, 417–427 (1955).

[3] *J. B. Wyngaarden, W. N. Kelley,* Gout, in: *J. B. Stanbury, J. B. Wyngaarden, D. S. Fredrickson* (eds.), The Metabolic Basis of Inherited Disease, McGraw Hill, New York 1978, pp. 916–1010.

[4] *J. E. Seegmiller, F. M. Rosenbloom, W. N. Kelley,* Enzyme Defect Associated with a Sex-linked Human Neurological Disorder and Excessive Purine Synthesis, Science *155*, 1682–1684 (1967).

[5] *T. A. Krenitsky, R. Papaioannou, G. B. Elion,* Human Hypoxanthine Phosphoribosyltransferase. I. Purification, Properties, and Specifity, J. Biol. Chem. *244*, 1263–1270 (1969).

[6] *W. N. Kelley, J. B. Wyngaarden,* The Lesch-Nyhan-Syndrome, in: *J. B. Stanbury, J. B. Wyngaarden, D. S. Fredickson* (eds.), The Metabolic Basis of Inherited Disease, McGraw Hill, New York 1978, pp. 1011–1036.
[7] *W. Gutensohn, G. Guroff,* Hypoxanthine-Guanine Phosphoribosyltransferase from Rat Brain: Purification, Kinetic Properties, Development and Distribution, J. Neurochem. *19*, 2139–2150 (1972).
[8] *W. Gröbner, N. Zöllner,* Eigenschaften der Hypoxanthinguaninphosphoribosyltransferase (HGPRTase) bei einem Gichtpatienten mit verminderter Aktivität dieses Enzyms, Klin. Wochenschr. *57*, 63–68 (1979).
[9] *W. N. Kelley, F. M. Rosenbloom, J. F. Henderson, J. E. Seegmiller,* A Specific Enzyme Defect in Gout Associated with Overproduction of Uric Acid, Proc. Natl. Acad. Sci. U.S.A. *57*, 1735–1739 (1967).
[10] *W. J. Arnold, W. N. Kelley,* Human Hypoxanthine-Guanine Phosphoribosyltransferase: Purification and Subunit Structure, J. Biol. Chem. *246*, 7398–7404 (1971).
[11] *O. H. Lowry, N. J. Rosebrough, A. L. Farr, R. J. Randall,* Protein Measurement with the Folin Phenol Reagent, J. Biol. Chem. *193*, 265–275 (1951).
[12] *W. N. Kelley,* Hypoxanthine-Guanine Phosphoribosyl Transferase Deficiency in the Lesh-Nyhan-syndrome and Gout, Fed. Proc. *27*, 1047–1052 (1968).
[13] *P. J. Benke, N. Herrick,* Azaguanine-resistance as a Manifestation of a New Form of Metabolic Overproduction of Uric Acid, Am. J. Med. *52*, 547–555 (1972).
[14] *I. H. Fox, I. L. Dwosh, P. J. Marchant, S. Lacroix, M. R. Moore, S. Omura, V. Wyhofsky,* Hypoxanthine-guanine Phosphoribosyltransferase: Characterization of a Mutant in a Patient with Gout, J. Clin. Invest. *56*, 1239–1249 (1975).
[15] *J. S. Cameron, H. A. Simmonds, D. R. Webster, V. Wass, A. Sahota,* Problems of Diagnosis and *in vitro* Enzyme Instability in an Adolescent with Hypoxanthine-Guanine Phosphoribosyltransferase Deficiency Presenting in Acute Renal Failure, J. Clin. Chem. Clin. Biochem. *20*, 353 (1982).
[16] *C. J. Epstein,* Phosphoribosyltransferase Activity During Early Mammalian Development, J. Biol. Chem. *245*, 3289–3294 (1970).
[17] *J. Allsop, R. W. E. Watts,* Activities of Amidophosphoribosyltransferase and the Purine Phosphoribosyltransferases, and the Phosphoribosylpyrophosphate Content of Rat Central Nervous System at Different Stages of Development, J. Neurol. Sci. *46*, 221–232 (1980).
[18] *J. F. Henderson, A. R. P. Paterson,* Nucleotide Metabolism, Academic Press, New York and London 1973, p. 128.

5.8 Sialyltransferase

CMP-*N*-acetylneuraminate: D-galactosyl-glycoprotein *N*-acetylneuraminyl-transferase, EC 2.4.99.1

Hilko Schriewer, Paul-Gerhardt Berge, Andreas Wilhelm

General

Sialyltransferase, ST, is primarily involved in the synthesis of glycoproteins and was first studied by *Roseman et al.* [1]. The enzyme transfers NANA (N-acetylneuraminic acid) from CMP-NANA ("activated" NANA) to appropriate acceptors. ST occurs in plant and animal tissues and has also been found in most human tissues, with the highest concentration in glycoprotein-releasing tissues [2]. At the cellular level, ST is mainly associated with membranous cellular fractions and the *Golgi* apparatus in particular [3]. Like other glycosyltransferases, ST is also located on cell surfaces [4]. These ecto-glycosyltransferases have been implicated in mechanisms of intercellular adhesion and interaction so that ST has received increasing attention in the study of growth regulation and cell communication [5].

The ST activity in the extracellular space is of about the same order of magnitude as in the intracellular space, suggesting a secretion of the enzyme or a shedding from cell surfaces [2]. The relevance of an elevated plasma ST level as a marker in malignant disease is disputed [6, 7, 8]. Generally, tissue damage cannot be expected to raise plasma ST levels significantly. Major sources of plasma ST are possibly the liver, platelets, and erythrocytes [9]. The function of ST in plasma is unknown. However, it may have a role in the regulation of circulating glycoproteins [5].

Application of method: in biochemistry and clinical research.

Enzyme properties relevant in analysis: the catalytic activity results from a group of enzymes. Individual STs may be distinguished from each other on the basis of their specificities for particular acceptor molecules [5]. Isoenzymes exist in both plasma and tissues [2]. Substrate conversion proceeds linearly with time and is proportional to the enzyme concentration [10].

Methods of determination: the catalytic activity is measured by a radiochemical assay allowing direct determination of the product. Active-site titration or immunological methods have not been employed up to now. Isoenzymes are separated by column chromatography [11].

International reference methods and standards: reference methods and standard materials have not been introduced so far.

Enzyme effectors: special activators are not essential for determination of ST. Enzyme activity may be enhanced by certain bivalent cations (Mg^{2+}, Mn^{2+}) to a variable and generally modest extent, possibly depending upon the ST-isoenzyme composition of the plasma. The effect of EDTA varies from activation to inhibition of enzyme activity. Other anticoagulants (citrate, heparin) always inhibit ST [9, 12].

Assay

Method Design

Principle

$$\text{CMP-[}^{14}\text{C]NANA} + \text{asialogalactoglycoprotein} \xrightarrow{\text{ST}} \text{[}^{14}\text{C]sialogalactoglycoprotein} + \text{CMP}$$

The transfer of radioactively-labelled NANA to a macromolecular acceptor is measured by separation of the acceptor from the incubation mixture and counting the incorporated radioactivity. This is achieved by transferring aliquots of the incubation mixture onto filter paper discs which are dried and then subjected to several washing steps, in order to precipitate the macromolecular acceptor and to remove unbound radioactive NANA [13]. Alternative techniques of separation are affinity chromatography [14] or precipitation with subsequent centrifugation or filtration [15]. – The radioactivity of the paper discs is counted in a liquid scintillation system.

Because of a high background activity, blanks (without enzyme addition) are mandatory.

Optimized conditions for measurement: the group of STs are distinguished not only by their acceptor-specificity, but possibly also by differences in their effector requirements. These are still under investigation and discussion. Several pH-optima exist in an interval from pH 5.0 – 7.5 [10].

The apparent K_m-value of plasma ST is 235 μmol/l for CMP-NANA. The most widely used acceptor is asialofetuin, for which the K_m-value of plasma ST is about 0.1 mmol/l [16]. An assay concentration of 0.5 mmol/l has proved satisfactory.

Temperature conversion factors: temperature conversion factors related to the activity at 37 °C are as follows:

°C:	10	20	22	25	27	30	33	35
	4.71	2.20	2.00	1.73	1.53	1.32	1.15	1.08
°C:	37	38	39	40	42	45		
	1.00	0.97	0.96	0.97	1.12	1.61		

Equipment

Filter paper discs (*Whatman* 3MM); ice-water-bath or refrigerator; dishes for washing solutions; liquid scintillation (β-ray) counter.

Reagents and Solutions

Purity of reagents: ST has a minimal affinity for sialylated glycoproteins. Low molecular-weight-substances such as lactose are unfavourable acceptors. Desialylated glycoproteins with free galactose residues such as asialofetuin are the most appropriate acceptors for NANA. Asialofetuin is prepared from fetuin (from *Sigma*) according to [17]. NANA does not act as an inhibitor of ST [12]; therefore, the presence of free NANA in the asialofetuin preparation is not critical. Other "endogenous" acceptors are invariably introduced into the assay by the addition of the enzyme sample (binding up to 15% of the transferred NANA in the serum).

CMP-NANA should not contain free CMP, as this substance, like most nucleotides [12], is an inhibitor of ST; suitable commercial preparations of radioactive CMP-NANA (from NEN, *Amersham, Buchler*) are available. Non-labelled CMP-NANA is not produced commercially (for synthesis of CMP-NANA cf. [18]).

Reagents for precipitation and washing should meet usual standards of purity.

Preparation of solutions (for 10 determinations in quadruple assays): all aqueous solutions in re-purified water (cf. Vol. II, chapter 2.1.3.2).

1. [^{14}C]CMP-NANA (41 – 57 µmol/l; 4.0 – 5.6 MBq/µmol):

   the radioactive substance is dispensed in NH_4HCO_3, 0.02 mol/l, containing 2% ethanol. Dilute 100 µl of this solution with water in a ratio of 1 + 1.

2. Tris/maleate buffer (Tris, 0.1 mol/l; maleate, 0.1 mol/l; pH 7.3):

   for stock solution a) dissolve 2.42 g Tris and 1.96 g maleinic acid anhydride to 100 ml with water; for stock solution b) dissolve 1.6 g NaOH in 100 ml water. Mix 2 ml solution a) with 2 ml solution b).

3. Asialofetuin (1.1 mmol/l, pH 7.3):

   dissolve 98.6 – 99.6 mg asialofetuin in 2 ml Tris/maleate buffer (2). The pH is adjusted with HCl, 0.2 mol/l, to 7.3.

4. NaCl (150 mmol/l):

   dissolve 2 g NaCl in 18 ml water.

5. $HgCl_2$ (40 mmol/l):

   dissolve 13 mg $HgCl_2$ in 1.2 ml water.

6. Trichloroacetic acid (10%, w/v):

   dissolve 50 g trichloroacetic acid to 500 ml water.

7. Trichloroacetic acid (5%, w/v):

   dissolve 12.5 g trichloroacetic acid to 250 ml water.

8. Ethanol/water:

   mix 125 ml ethanol, reagent grade, with 125 ml water.

9. Ether/ethanol:

   mix 125 ml ethylether, reagent grade, with 125 ml ethanol.

Stability of solutions: CMP-NANA-solution (1) must be stored deep frozen. Thermal degradation takes place at a rate of 2%/h at 37 °C in alkaline solution. Lowering the pH increases the instability. Asialofetuin buffer solution (3) is stable at 4 °C as long as no bacterial contamination occurs. The other solutions may be kept at room temperature.

Procedure

Collection and treatment of specimen: collect blood from vein without stasis. No anticoagulants should be added. After coagulation, blood samples are centrifuged at about 3000 *g* for 10 min and the serum, which should be free from haemolysis, is stored frozen until assayed.

Stability of the enzyme in the sample: the enzyme is stable in human serum at room temperature for at least 48 h and at +4 °C for 3 months. Deep frozen it may be stored for 1 year without loss of activity [10].

Details for measurement in tissues: in deep frozen tissues or homogenates (−20 °C) ST activity remains stable for several months. Crude homogenates may require removal of debris by centrifugation at low acceleration (600 *g*). The membrane-bound ST activity is best solubilized by non-ionic detergents such as Triton X 100. This does not alter enzyme characteristics [19]. The kinetic properties of ST vary considerably depending upon the tissues of origin.

Assay conditions: incubation volume 100 µl; 37 °C; quadruple determinations.

Measurement

Incubation

Pipette successively into incubation tubes:	reagent blank	sample blank	sample	concentration in the incubation mixture
sample	50 µl	–	50 µl	volume fraction 0.5
NaCl (4)	–	50 µl	–	
Tris/maleate buffer (2)	45 µl	–	–	Tris/maleate 45 mmol/l
[^{14}C]CMP-NANA (1)	5 µl	5 µl	5 µl	CMP-NANA 2.0 – 2.8 µmol/l 8.1 – 11.2 MBq/l
asialofetuin (3)	–	45 µl	45 µl	asialofetuin 0.5 mmol/l
mix, incubate for 30 min at 37 °C, add				
$HgCl_2$ (5)	30 µl	30 µl	30 µl	$HgCl_2$ 9.2 mmol/l

Precipitation and washing procedure: transfer aliquots of 100 µl of each sample onto paper discs (*Whatman* 3MM, 2.5 cm in diameter) and dry. Treat the discs at 0–4 °C for 15 min with trichloroacetic acid (6) twice, for 30 min with trichloroacetic acid (7), for 60 min with ethanol/water (8) and for 15 min with ether/ethanol (9), each in a separate disk. Dry the discs subsequently.

Measurement

The radioactivity precipitated on the paper discs is measured in a liquid scintillation counter, each for 2 min.

Calculation: the values measured are given in counts per min. The amount of transferred NANA in nmol per s and l serum is calculated by means of the following formula (cf. eqn. (o_1) in "Formulae", Appendix 3):

$$b = \frac{\Delta C_n}{\eta \times 60 \times X \times \varphi \times v_r \times \Delta t} \quad \text{nkat/l}$$

$$\Delta C_n = C_p - C_b$$

it is

C_p counting rate for sample (cpm)
C_b counting rate for blank (cpm)
η counting efficiency
X specific radioactivity, Bq/nmol
φ volume fraction of sample in total assay mixture (50 µl/130 µl)
v_r volume taken for counting (10^{-4} l)
Δt incubation time (1800 s)

Validation of Method

Precision, accuracy, detection limits and sensitivity: with a coefficient of variation of 20% within-series and 25% from day to day the precision of the method is low. Data about accuracy cannot be given as standard reference material has not been introduced as yet.

At an assay concentration of 4000 pmol CMP-NANA/ml serum the detection limit is 5 fmol $\times$ s^{-1} $\times$ ml^{-1} in serum. The sensitivity amounts to 15 fmol $\times$ s^{-1} $\times$ ml^{-1} in serum in the main measuring range (at around 60 fmol $\times$ s^{-1} $\times$ ml^{-1} in serum).

Sources of error: the serum ST level may be increased by enzyme release from damaged thrombocytes or erythrocytes. Some anti-inflammatory drugs (e.g. phenylbutazone, aspirin) are reported to alter plasma ST activity [9]. Hydrolytic enzyme activity (CMP-NANA hydrolase, neuraminidase) may interfere with the measurement of ST activity [6, 7].

Specificity: parallel to the transfer to asialofetuin, sialylation of endogenous acceptors of the plasma occurs to a varying degree. Unspecific non-enzymatic binding of radioactivity to proteins is observed (high background activity).

Reference ranges: there is no difference in enzyme activity between the sexes. Children under the age of 6 years have significantly lower levels of ST acitivity.

At an assay concentration of 4000 pmol/ml serum the normal range is between 37 and 70 fmol $\times$ s^{-1} $\times$ ml^{-1} in serum [8].

References

[1] *S. Roseman, D. M. Carlson, G. W. Jourdian, E. J. McGuire, B. Kaufmann, S. Basu, B. Bartholomew,* Animal Sialic Acid Transferases (Sialyl-Transferases), in: *S. P. Colowick, N. O. Kaplan* (eds.), Methods in Enzymology, Vol. *VIII,* Academic Press, New York 1966, pp. 354 – 372.

[2] *P. G. Berge, A. Wilhelm, H. Schriewer,* The significance of sialyltransferase as a tumor marker, Klin. Wochenschr., in press.

[3] *H. Schachter, J. Jabbal, R. L. Hudgin, L. Pinteric,* Intracellular localization of liver sugar nucleotide glycoprotein glycosyltransferases in a *Golgi*-rich fraction, J. Biol. Chem. *245,* 1090 – 1100 (1970).

[4] *C. W. Porter, R. J. Bernacki,* Ultrastructural evidence for ectoglycosyltransferase Systems, Nature *256,* 648 – 650 (1975).

[5] *E. Köttgen, Ch. Bauer, W. Reutter, W. Gerok,* Neue Ergebnisse zur biologischen und medizinischen Bedeutung von Glycoproteinen, Klin. Wochenschr. *57,* 157 – 159 (1979).

[6] *M. Henderson, D. Kessel,* Alterations in plasma sialyltransferase levels in patients with neoplastic disease, Cancer *39,* 1129 – 1134 (1977).

[7] *C. Ip, T. Dao,* Alterations in serum glycosyltransferases and 5′-nucleotidase in breast cancer patients, Cancer Res. *38,* 723 – 728 (1978).

[8] *P. G. Berge, A. Wilhelm, H. Schriewer, G. Wüst,* Serum-sialyltransferase activity in cancer patients, Klin. Wochenschr. *60,* 445 – 449 (1982).

[9] *D. Kessel, J. Allen,* Elevated plasma sialyltransferase in the cancer patient, Cancer Res. *35,* 670 – 672 (1975).

[10] *A. Wilhelm, P. G. Berge, H. Schriewer,* Zur Aktivitätsbestimmung der Sialyltransferase im menschlichen Serum, J. Clin. Chem. Clin. Biochem. *18,* 271 – 273 (1980).

[11] *C. K. Liu, R. Schmied, E. M. Greenspan, S. Waxman,* Characterization of sialyltransferase activity from human placenta, Biochim. Biophys. Acta *522,* 375 – 385 (1978).

[12] *W. D. Klohs, R. J. Bernacki, W. Korytnyk,* Effect of nucleotides and nucleotide: analogs on human serum sialyltransferase, Cancer Res. *39,* 1231 – 1238 (1979).

[13] *C. H. Bauer, P. Vischer, H. J. Grünholz, W. Reutter,* Glycosyltransferasen and glycosidases in Morris hepatomas, Cancer Res. *37,* 1513 – 1518 (1977).

[14] *U. Ganzinger, F. Dorner, F. M. Unger, K. Moser, K. Jentzsch,* Erhöhung der Serum-Sialyltransferase bei menschlichen Malignomen, Grundlage für ein neues Diagnostikum? Klin. Wochenschr. *55,* 553 – 555 (1977).

[15] *J. P. Durham, D. Gillies, A. Baxter, R. O. Lopez-Solis,* The application of a simple, sensitive disc assay for sialyl-transferase activity to particulate and soluble enzyme preparations and its use in measuring enzymatic activity in human plasma samples, Clin. Chim. Acta *95,* 425 – 432 (1979).

[16] *D. E. Kessel, T. H. Chou, J. Allen,* Some properties of sialyltransferase in plasma and lymphocytes of patients with lymphocytic leukemia, Eur. J. Biochem. *82,* 535 – 541 (1978).
[17] *R. G. Spiro,* Analysis of sugars found in glycoproteins, in: *S. P. Colowick, N. O. Kaplan* (eds.), Methods in Enzymology, Vol. *VIII,* Academic Press, New York 1966, pp. 3 – 26.
[18] *E. L. Kean, S. Roseman,* The sialic acids X: purification and properties of cytidine 5-monophosphosialic acid synthetase, J. Biol. Chem. *241,* 5643 – 5650 (1966).
[19] *J. A. Alhadeff, R. T. Holzinger,* Solubilization and stabilization of human liver glycoprotein sialyltransferase, Biochim. Biophys. Acta *570,* 56 – 64 (1979).

6 Aminotransferases

6.1 Aspartate Aminotransferase

(Glutamate Oxaloacetate Transaminase)

L-Aspartate: 2-oxoglutarate aminotransferase, EC 2.6.1.1

Robert Rej and Mogens Hørder

6.1.1 General

Although the metabolic conversion of oxo acids into amino acids was recognized as early as 1910, it was not until the late 1930s that *Braunstein & Kritzmann* [1] and *Cohen* [2] demonstrated the existence of aspartate aminotransferase, AspAT (formerly GOT). The enzyme occurs in all animal and plant tissues and has been characterized from a variety of micro-organisms [3]. In mammals it is found in particularly high concentrations in liver, nervous tissue, skeletal and cardiac muscle. It is present in small catalytic amounts in the serum from normal human subjects and its activity is raised in sera from individuals with a variety of diseases and disorders [4].

In most species the enzyme exists in two major forms: a mitochondrial isoenzyme (m-AspAT, cathodal in most electrophoretic separations) and a soluble or cytosolic form (s-AspAT, anodal by electrophoresis). The soluble enzyme is a dimer consisting of identical subunits and possessing a dimer molecular weight of about 93000. The best characterized is porcine heart s-AspAT, whose amino acid sequence has been described [5]. The mitochondrial form is also a dimer of identical or very similar subunits of dimer molecular weight of about 91000 [6]. The isoenzymes from several species have been found to lack tissue specificity and the properties of the enzyme from the same cytological location in different tissues are identical. Preparations of both isoenzymes have been demonstrated to contain several subforms with similar kinetic and immunochemical properties which are probably due to post-translational deamidation of glutamine and asparagine residues [7].

AspAT contains one mole of pyridoxal phosphate per mole of monomer. The coenzyme is required for catalytic activity; the apo-enzyme possesses no catalytic activity.

Application of method: in biochemistry and in clinical chemistry.

Enzyme properties relevant in analysis: the two AspAT isoenzymes possess distinct kinetic properties, with the mitochondrial form usually demonstrating an optimum pH more acidic than that of the soluble form (Fig. 1). In the majority of species, the *Michaelis* constants for the four-carbon substrates (L-aspartate and oxaloacetate) are lower with the mitochondrial isoenzymes than with the cytoplasmic forms; the cytoplasmic isoenzymes exhibit lower K_m values for the five-carbon substrates, L-glutamate and 2-oxoglutarate (Table 1). These differences have been exploited in order to estimate the proportions of each isoenzyme in serum and tissues (cf. p. 434) and may lead to a significant bias towards one of the isoenzymes under some assay conditions. The UV-routine assay described here measures more than 94% of V for each isoenzyme and neither is significantly underestimated. The isoenzymes are differentially inhibited by the product oxaloacetate [8]. This does not affect the UV-routine assays as oxaloacetate is reduced to malate and does not accumulate under the conditions of assay.

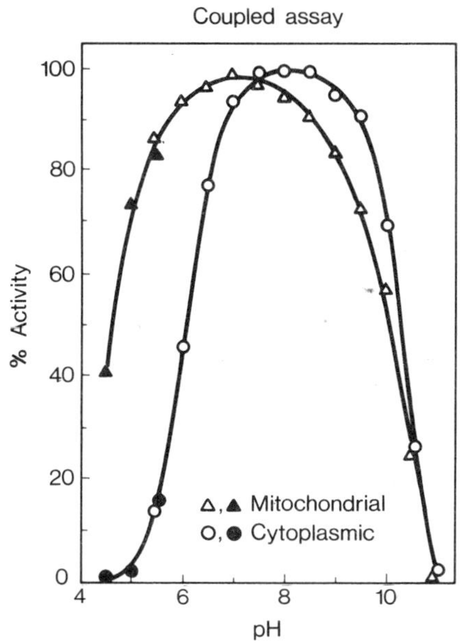

Fig. 1. pH dependence of human AspAT isoenzymes in NADH-MDH coupled assay. Buffers used were acetate (solid symbols) and phosphate (open symbols), each at 50 mmol/l. L-aspartate was at 125 mmol/l, 2-oxoglutarate at 6.7 mmol/l. For m-AspAT 100% was 140 U/l; for s-AspAT 100% was 110 U/l; 0.1 ml of enzyme was used in a total volume of 2.7 ml.

Table 1. *Michaelis* constants of the human aspartate aminotransferase isoenzymes[a]

K_m	m-AspAT	s-AspAT
Oxaloacetate	$9.0 \pm 1.9 \times 10^{-6}$	$3.7 \pm 0.6 \times 10^{-5}$
L-Glutamate	$3.8 \pm 0.2 \times 10^{-2}$	$1.0 \pm 0.1 \times 10^{-2}$
2-Oxoglutarate	$5.6 \pm 0.1 \times 10^{-3}$ [b]	$1.3 \pm 0.1 \times 10^{-4}$
L-Aspartate	$4.0 \pm 0.4 \times 10^{-4}$	$2.0 \pm 0.1 \times 10^{-3}$

[a] All data are in mol/litre.
[b] A value of 3.8×10^{-4} mol/litre was obtained at concentrations of 2-oxoglutarate less than 4 mmol/litre.

The enzyme requires pyridoxal phosphate for catalytic activity and this coenzyme is present in the assay mixture at concentrations sufficient to activate any apo-AspAT present [9]. Tris buffer is used since it possesses a $pK_{30°C}$ near 7.8, the pH used, and it does not interfere with the re-association of pyridoxal phosphate with apo-AspAT [10]. Phosphate buffer slightly inhibits AspAT.

Methods of determination: many techniques have been applied to the measurement of AspAT activity including chromatographic separation of amino acids, manometric determination of CO_2 liberated from oxaloacetate or aspartate, coupling of oxaloacetate with 2,4-dinitrophenylhydrazine or azoene dyes, and coupling of glutamate formation with glutamate dehydrogenase (for review cf. [11]). The preferred technique is that first described by *Karmen* [12] in which oxaloacetate formed is reduced to malate with concomitant oxidation of NADH catalyzed by malate dehydrogenase (MDH, EC 1.1.1.37). This procedure has the advantages of using the well described molar absorptivity of NADH, a high degree of specificity, and removal of oxaloacetate – a point product inhibitor of AspAT – from the reaction.

An alternative technique presented is that based on the colorimetric procedure described by *Babson et al.* [13] using an azoene dye. It is among the more specific colorimetric techniques and is suitable for either batch analysis or for use when available instrumentation cannot accomodate the UV-routine assay. Determination of AspAT on the basis of immunologically-active mass [14, 15] as well as active-site titration techniques [16] have been described, but afford neither the convenience nor the high degree of precision available with the UV-routine assay.

International reference method and standards: a recommended reference method has been developed by the International Federation of Clinical Chemistry [17, 18]. The method includes well documented assays for determining the reliability of auxiliary enzymes and a protocol including the measurement of three separate blank reactions. A standard human enzyme reference preparation from erythrocytes has been described [19] and evaluated with the reference method [20]; cf. also Vol. II, chapter 2.3.

Enzyme effectors: amino-oxyacetate, several dicarboxylic acids (including maleate, adipate, glutarate and succinate), glyceraldehyde 3-phosphate, erythrose 4-phosphate, chloride, phthalate and decaborane inhibit the enzyme. The enzyme exhibits both product and substrate inhibition.

6.1.2 Routine UV-method

Assay

Method Design

Principle

(a) $\text{L-Aspartate} + \text{2-oxoglutarate} \xrightleftharpoons{\text{AspAT}} \text{L-glutamate} + \text{oxaloacetate}$

(b) $\text{Oxaloacetate} + \text{NADH} + \text{H}^+ \xrightleftharpoons{\text{MDH}} \text{L-malate} + \text{NAD}^+ .$

AspAT catalyzes the reaction between L-aspartate and 2-oxoglutarate and the oxaloacetate formed is reduced by NADH catalyzed by MDH. The K_m of MDH for oxaloacetate is low (ca. 1.6×10^{-5} mmol/l), and the equilibrium for reaction (b) is far to the right [21], so that the decrease in absorbance due to NADH oxidation is stoichiometric with AspAT activity.

Pyridoxal phosphate is added to activate apo-AspAT which may be present in the specimen. Lactate dehydrogenase (LDH, EC 1.1.1.27) is added to reduce pyruvate and to shorten the duration of pre-incubation needed in order to obtain a stable initial absorbance. For routine measurements on serum, specimen blanks are not required (cf. "Sources of error", p. 423). A reagent blank should be prepared for each batch of reagents and its value deducted from the overall measured rate.

Optimized conditions for measurement: conditions have been investigated by modelling using the rate equations valid for the aminotransferase-catalyzed reaction and the experimentally-determined K_m values for the human enzyme [18, 22]. These have been confirmed using stepwise optimization of each assay variable [17, 18]. The conditions used in the UV-routine assay support optimal conditions for determining total activity of human AspAT and are similar to those recommended in the International Reference Method.

Temperature conversion factors: the effects of temperature on the measured activity of human AspAT isoenzymes and porcine s-AspAT have been investigated [23] and the following factors are valid for the measurement at different temperatures in order to relate activity to that determined at 30 °C:

°C	20	25	30	32	35	37	40
Factor for							
Human s-AspAT	2.027	1.415	1.000	0.873	0.715	0.626	0.516
Human m-AspAT	2.029	1.416	1.000	0.873	0.715	0.626	0.515
Porcine s-AspAT	1.998	1.405	1.000	0.875	0.719	0.633	0.523

Since there is a measurable effect of temperature on K_m for both substrates [23] these factors are valid only at saturating levels of substrate.

Equipment

Spectrophotometer capable of measurement at 339 nm (preferred) or a spectral-line photometer capable of measurement at Hg 334 or Hg 365 nm. Thermostatted cuvette or cell compartment; attached recorder or stopwatch.

Reagents and Solutions

Purity of reagents: the auxiliary enzymes should be free from AspAT (either apo- or holo-enzyme) and glutamate dehydrogenase (GlDH). Contamination of each should not exceed 0.005% relative to either MDH or LDH catalytic activity. Activity of LDH and MDH may be verified using procedures described on p. 118 and p. 171 respectively. The auxiliary enzymes must be free from ammonium ion in order to reduce the interference from GlDH. Reagent preparation calls for disodium-dihydrate NADH. Adjust amount accordingly if a different salt or hydrated form is used. NADH and 2-oxoglutarate must be free from inhibitors (cf. Vol. II, chapter 2.1.2.2; suitable commercial preparations are available).

Preparations of solutions: all solutions are prepared using re-purified water (cf. Vol. II, chapter 2.1.3.2).

1. Aspartate/Tris solution (L-aspartate, 297 mmol/l; Tris buffer, 100 mmol/l; pH 7.8):

   to about 700 ml water add 39.5 g L-aspartic acid and 12.1 g Tris base. Mix and stir, monitor pH. Slowly add 7 g of solid NaOH (pellets). Constantly monitor temperature, which should be 30 ± 2 °C; place the vessel in warm or cool water to achieve this temperature. Adjust pH to 7.8 ± 0.02 with NaOH solution (10 and 1 mol/l). Dilute to exactly 1 litre with water. Filter using 0.22 µm microporous filter. Refrigerate working solution; freeze remainder.

2. 2-Oxoglutarate/Tris solution (2-oxoglutarate, 144 mmol/l; Tris, 100 mmol/l; pH 7.8):

   to about 70 ml water add 2.10 g 2-oxoglutaric acid and 1.21 g Tris base. Constantly monitor temperature, which should be 30 ± 2 °C; place the vessel in warm or cool water to achieve this temperature. Adjust pH to 7.8 ± 0.02 with NaOH solution (10 and 1 mol/l). Dilute to exactly 100 ml with water. Filter using 0.22 µm microporous filter. Freeze in aliquots of approximately 10 ml.

3. MDH/LDH/glycerol (malate dehydrogenase, 230 kU/l; lactate dehydrogenase, 105 kU/l):

   mix 5.0 ml water with 5.0 ml glycerol; add 0.2 ml porcine muscle LDH (10 mg/ml in glycerol; ca. 550 U/mg at 25 °C), and 0.2 ml of porcine heart MDH (10 mg/ml in glycerol; ca. 1200 U/mg at 25 °C). Mix gently by inversion. Refrigerate – do not freeze.

4. Tris (100 mmol/l; pH 7.8):

   to about 600 ml water add 12.1 g Tris base. Constantly monitor temperature, which should be 30 ± 2 °C; place beaker in warm or cool water to achieve this temperature. Adjust pH to 7.8 ± 0.02 with HCl (1 mol/l). Dilute to exactly 1 litre with water. Refrigerate.

5. Pyridoxal 5-phosphate/Tris (pyridoxal phosphate, 5.44 mmol/l; Tris, 100 mmol/l; pH 7.8):

   weigh 288 mg pyridoxal 5-phosphate monohydrate (most preparations are monohydrate) and dissolve in 200 ml Tris solution [4]. Divide into ca. 10 ml portions; store protected from light at ≤ −20 °C.

6. Working reagent mixture:

   to prepare the working reagent mixture, add the following in the millilitre or milligram quantities shown for the number of determinations desired:

Number of determinations:		10	30	50	100
Aspartate/Tris solution (1)	ml	20.0	60.0	100.0	200.0
MDH/LDH solution (3)	ml	0.1	0.3	0.5	1.0
Pyridoxal phosphate solution (5)	ml	0.5	1.5	2.5	5.0
NADH	mg	3.0	9.0	15.0	30.0

NADH (reduced β-nicotinamide-adenine dinucleotide, disodium salt, dihydrate) is added directly as a dry powder.

Stability of solutions: solution (1) (aspartate/Tris) is stable at 4 °C for at least 1 month and for at least 6 months at ≤ −20 °C. Solutions (2) (2-oxoglutarate) and (5) (pyridoxal phosphate) are stable for at least 2 days at 4 °C and for 6 months at ≤ −20 °C. Solution (3) (MDH/LDH) is stable at 4 °C for at least 1 month – do not freeze this solution. Solution (4) (Tris buffer) is stable at 4 °C for at least 4 months. Solution (6) (working reagent mixture) should be prepared freshly each day of use and remains usable for at least 24 h if stored at 4 °C. Bacterial contamination, in particular of solution (1), may be prevented by the addition of azide (7 mmol/l) without affecting enzyme activity [24].

Procedure

Collection and treatment of specimen: collect blood with a minimum of venous stasis. The preferred specimen is freshly collected serum free from haemolysis. Plasma obtained in the presence of heparin (0.2 mg/ml), EDTA (1 mg/ml), fluoride (2 mg/ml), oxalate (1 mg/ml), or citrate (1 mg/ml) is acceptable.

Stability of the enzyme in the sample: total AspAT activity in serum remains stable at 4 °C for a minimum of 4 days [17]. Storage at < −20 °C for ≤ 1 month is acceptable.

Details for measurement in tissues: the total AspAT activity of tissues can be determined quantitatively only after complete tissue homogenization since a portion (in many tissues such as liver, a preponderance) of the enzyme is located in mitochondria. Use of sonication and detergents such as deoxycholate may be required [25]. Since tissue extracts may also contain glutamate dehydrogenase and other interfering enzyme activities, it may be necessary to perform a specimen blank reaction (see Sources of Error, below). Dilute specimen with Tris buffer (4), saline, or 1% (w/v) albumin solution.

Assay conditions: wavelength 339 nm (alternates with Hg-line photometers only, Hg 334 or Hg 365 nm); light path 10 mm; final volume 2.40 ml; temperature 30.0 °C ± 0.10 °C. Measure absorbance change against air reference. The temperature of solutions [2] and [6] should be at 30 °C.

Measurement

Pipette successively into the cuvette:			concentration in assay mixture
working reagent mixture	(6)	2.0 ml	L-aspartate 240 mmol/l Tris 91 mmol/l NADH 0.16 mmol/l pyridoxal phosphate 0.11 mmol/l MDH ≥ 0.93 kU/l (25 °C) LDH ≥ 0.42 kU/l (25 °C)
sample (serum or diluted tissue homogenate)		0.2 ml	volume fraction 0.083
mix thoroughly without removing any of the solution from the cuvette, incubate at 30 °C for 10 min or until temperature equilibrium has occurred and a stable absorbance reading is observed			
2-oxoglutarate	(2)	0.2 ml	2-oxoglutarate 12 mmol/l
mix thoroughly; monitor change in absorbance with potentiometric recorder or manually record absorbance readings at fixed time intervals using elapsed timer. The reaction should be monitored for about 3 min or until an absorbance change (at 339 nm) of ≥ 0.200 occurs.			

The values of $\Delta A/\Delta t$ should not exceed 0.40 min^{-1}; otherwise dilute sample with solutions recommended under assay of tissues (cf. above, p. 422).

Calculation: the calculation formula (k) or (k_1), respectively, given under "Formulae" (Appendix 3) is applicable. For values of ε cf. Appendix 4. The following relations are valid for the catalytic concentration of the sample (Δt in minutes):

	Hg 334 nm	339 nm	Hg 366 nm	
$b =$	$1942 \times \Delta A/\Delta t$	$1905 \times \Delta A/\Delta t$	$3529 \times \Delta A/\Delta t$	U/l
$b =$	$3.24 \times 10^4 \times \Delta A/\Delta t$	$3.17 \times 10^4 \times \Delta A/\Delta t$	$5.88 \times 10^4 \times \Delta A/\Delta t$	nkat/l

Validation of Method

Precision, accuracy, detection limits and sensitivity: the assay described demonstrated an imprecision of 1.2 U/l for a quality control specimen with a catalytic activity of 80 U/l (relative standard deviation RSD = 1.5%) over a period of six months using different instruments within a single laboratory. At the upper limit of the reference interval for human subjects, the RSD was approximately 2% for within-day measurements and about 3% for long-term imprecision. Reproducibility for interlaboratory measurements (5 laboratories) using the assay conditions described is approximately 4%. With sufficiently sensitive spectrophotometers (e.g. *Varian*, Cary 219), specimens possessing a catalytic concentration of 1.0 U/l can be detected with an overall imprecision of about 0.2 U/l. The magnitude of the reagent blank varies among laboratories and affects the minimum detectable activity (cf. "Sources of error" below). The method shows excellent correlation with the IFCC reference method and has a relative inaccuracy of $< 4\%$ for human serum specimens in comparison with that procedure.

Sources of error

Effects of drugs: several drugs and therapeutic substances have been found to increase the serum catalytic activities of AspAT: for a comprehensive review cf. [26].

Interference by factors influencing blanks: all calculations given above are valid and the numerical apparent AspAT activity determined is deducted from the overall rate found using the routine assay protocol. For human and animal serum specimens this rate is usually $\ll$ 4 U/l, often below the level of imprecision of the technique. However, for tissue extracts, this activity may be significant. For a comprehensive review of blank measurements cf. [17]. For details cf. chapter 6.1.4.

Specificity: with the exceptions noted above under "Sources of error", AspAT activity is measured specifically. With the measurement of blank reactions noted above, no significant non-specific reaction interferes.

AspAT is reasonably specific for L-aspartate and L-glutamate although L-tyrosine, L-cysteinesulphinate, and some aspartate analogues also have been shown to serve as substrates for certain AspAT isoenzymes.

Reference ranges: tentative reference values for serum obtained from normal human subjects are 5 – 34 U/l at 30 °C.

The activity content of tissue extracts is several-fold greater; e.g. 150 U/g for normal human liver tissue [25].

6.1.3 Colorimetric Method

Assay

Method Design

Principle: oxaloacetate formed from equation (a), p. 418 reacts with diazotized 6-benzamido-4-methoxy-3-toluidine chloride (Fast Violet B salt) to form a chromophore absorbing in the 500 – 550 nm range.

Equipment

Photometer suitable for measurements at 530 nm; thermostatted water-bath or dry block heater for 37 °C; stopwatch.

Reagents and Solutions

Purity of reagents: 2-oxoglutarate must be free from inhibitors. Oxaloacetate used for calibration must be verified by using independent techniques such as those described in Vol. VII. Preparations containing < 90% oxaloacetate by weight have been frequently encountered in the laboratory of one of the authors: diazonium salt should be of highest available purity for greatest sensitivity.

Preparation of solutions: all solutions are prepared using re-purified water (cf. Vol. II, chapter 2.1.3.2).

1. Buffer/substrate (phosphate buffer, 200 mmol/l; L-aspartate, 20 mmol/l; 2-oxoglutarate, 5 mmol/l):

   dissolve the following in about 75 ml water: 73 mg 2-oxoglutaric acid, 266 mg L-aspartic acid, 2.84 g Na_2HPO_4 (or 5.36 g of $Na_2HPO_4 \cdot 7\ H_2O$), 1 g polyvinylpyrrolidone, 100 mg EDTA-Na_4. Adjust pH to 7.4 with NaOH or HCl, 0.1 mol/l each, and dilute to exactly 100 ml.

2. Control buffer (phosphate buffer, 200 mmol/l; 2-oxoglutarate, 5 mmol/l):

   dissolve in about 75 ml water the following: 73 mg 2-oxoglutaric acid, 2.84 g Na_2HPO_4 (or 5.36 g $Na_2HPO_4 \cdot 7\ H_2O$), 1 g polyvinylpyrrolidone, 100 mg EDTA-Na_4. Adjust pH to 7.4 with NaOH or HCl, 0.1 mol/l each, and dilute to exactly 100 ml.

3. Diazonium salt (5 mg/ml):

   dissolve 125 mg 6-benzamido-4-methoxy-3-toluidine diazonium chloride (Fast Violet B salt) in 25 ml water.

4. HCl/Lipal (HCl, 0.02 mol/l, 1 % Lipal):

   slowly add 1.7 ml concentrated HCl to about 900 ml water. Add 10 ml ethoxylated tridecylalcohol (Lipal 610) and dilute to exactly 1 litre.

5. Oxaloacetate calibration standards (0 – 4.0 mmol/l in HCl):

   dissolve 66.0 mg oxaloacetic acid in 100.0 ml of cold HCl, 0.01 mol/l. If the purity of oxaloacetic acid is less than 98 % as determined above, correct the amount of oxaloacetic acid added to compensate for the actual amount of this compound. To 0, 1, 2, 4, 6 and 8 ml of this solution add respectively 10, 9, 8, 6, 4, and 2 ml HCl, 0.01 mol/l, and mix. These dilute standard solutions contain 0, 0.5, 1.0, 2.0, 3.0, and 4.0 mmol/l of oxaloacetic acid.

Stability of solutions: the buffer/substrate solution (1) and the control buffer (2) are stable for two months at 4 °C with the addition of several drops of chloroform. Prepare the oxaloacetic acid standard solutions (5) just before use. Fast Violet solution (3) is stable for at least 8 h at room temperature protected from light.

Procedure

Collection, treatment and stability of specimen: see Routine UV-method, p. 418.

Assay conditions: wavelength 530 nm; light path 10 mm; incubator at 37 °C; incubation volume 1.20 ml; final volume 12.2 ml; read against blank containing physiological saline instead of serum. Adjust solutions (1) and (2) to 37 °C before the start of the experiments.

Calibration curve: mix 0.2 ml of each of the diluted oxaloacetic acid standard solutions with 1.0 ml control buffer (2), equilibrate at 37 °C and add 1.0 ml of Fast Violet solution (3). After exactly 10 min dilute with 10.0 ml of HCl/Lipal solution (4) and read the absorbance against the solution without oxaloacetic acid as described below. Plot absorbance against catalytic activity.

Measurement

Pipette into a test tube:			concentration in incubation mixture
buffer/substrate solution	(1)	1.00 ml	phosphate 167 mol/l 2-oxoglutarate 4.2 mmol/l L-aspartate 16.7 mmol/l
serum		0.20 ml	volume fraction 0.17
mix and incubate for exactly 20 min at 37 °C			
Fast Violet solution	(3)	1.00 ml	ca. 6 mmol/l
mix and incubate for exactly 10 min at 37 °C			
HCl	(4)	10.0 ml	
read the absorbance of the sample within 30 min.			

A serum blank must be measured for each specimen. Follow the above protocol substituting control buffer (2) for buffer/substrate (1). The 20 min incubation may be omitted.

Calculation: use the oxaloacetate standard curve. Under the assay conditions the various oxaloacetic acid concentrations correspond to the following AspAT activities:

Standard No.	*1*	*2*	*3*	*4*	*5*	*6*
μmol oxaloacetic acid per assay	0	0.1	0.2	0.4	0.6	0.8
Units described by *Babson* [13]	0	25	52	82	116	157
Catalytic concentration (U/l)*	0	20	42	66	93	126

* U/l refer to approximate catalytic activity concentration of human s-AspAT as determined by the UV-Routine Assay at 30 °C. Formation of oxaloacetate by AspAT is non-linear with time.

Validation of Method

Precision of method: with values around 25 U/l, the relative standard deviation is about 8%.

Sources of error: as oxaloacetate is very unstable in aqueous solution, erroneous values are found with calibration standard solutions that are more than 2 h old. In addition the quality of the Fast Violet chromogen has a marked effect on the final absorbance. It is therefore essential to prepare a new standard curve for each series of measurements. Bilirubin and several drugs interfere.

Specificity: there are many substances in serum which react with the Fast Violet B chromogen [26] including acetoacetate and other oxo-acids.

Reference ranges: range is 5 to 30 *Babson* Units for human serum from normal individuals, corresponding to approximately 4 to 25 U/l.

6.1.4 IFCC Reference Method (provisional)

The IFCC Reference Method has been published [17, 18] and evaluated using human enzyme reference materials [19, 20].

Assay

The method design is based on profound efforts directed towards establishing optimal conditions. These effects have resulted in the assay conditions given in Table 2.

Table 2. Provisional IFCC reference method for AspAT assay: reaction conditions

Temperature	30.0 °C	
pH (30 °C)	7.8	
Volume fraction of sample	φ = 0.0833 (1:12)	
Tris(hydroxymethyl)aminomethane	80 mmol/l	
L-Aspartate	240 mmol/l	
2-Oxoglutarate	12 mmol/l	
Pyridoxal phosphate	0.10 mmol/l	
NADH	0.18 mmol/l	
Malate dehydrogenase	7 µkat/l	(420 U/l)
Lactate dehydrogenase	10 µkat/l	(600 U/l)

Establishing Optimized Conditions

Primary reaction

The assay conditions were optimized theoretically by modelling with the rate equations valid for the reaction catalyzed by the aminotransferase and the experimentally-determined K_m values for the human enzyme. These conditions were confirmed in practice by stepwise optimization of each assay variable.

Coenzyme: pyridoxal phosphate

AspAT is stable after removal of the pyridoxal phosphate coenzyme. The effect of pyridoxal phosphate on AspAT activity has been demonstrated. The catalytic activity in sera of patients with heart or liver diseases increases, on average, by approximately 50% and 15% respectively, after saturation of the enzyme with the coenzyme [22].

Optimal pyridoxal phosphate concentrations have been established experimentally. The time required for the re-combination of the apo-enzyme with pyridoxal phosphate is approximately 7 min; in practice, this is adequately provided by the pre-incubation period required to reach temperature equilibrium before starting the assay.

Phosphate inhibits the re-combination of the apo-enzyme with the coenzyme in all buffers investigated. It can be demonstrated that in the presence of phosphate buffer, 100 mmol/l, the effect of the coenzyme is greatly reduced.

The absorbance at 339 nm of pyridoxal phosphate at the above concentration does not interfere with measurements of change in NADH absorbance.

Buffer, pH

Tris(hydroxymethyl)aminomethane (Tris) buffer does not interfere with the re-association of pyridoxal phosphate with apo-AspAT. The optimal pH lies between approximately pH 7.6 and pH 8.0 for human cytoplasmic enzyme (Fig. 1) and m-AspAT exhibits more than 95% of its maximum activity at this pH. The dependence of the catalytic activity on pH within this interval is of minor importance. The recommended value, pH 7.8, lies close to the pK (30 °C) 7.94 of Tris. Phosphate is excluded because it inhibits the re-combination of the apo-enzyme with pyridoxal phosphate and inhibits the holo-enzyme.

Since aminotransferase activity decreases slightly with increasing buffer concentration, only as much Tris is used as is necessary to keep the pH constant after addition of sample and reagents. At a concentration of about 80 mmol/l an adequate buffering capacity is obtained with only a small decrease in activity.

Substrates

In Tris buffer (80 mmol/l, pH 7.8) optimal rates are obtained with 2-oxoglutarate concentrations of 10 to 17 mmol/l and L-aspartate concentrations of about 240 mmol/l.

The reaction catalyzed by aspartate aminotransferase is a two-substrate, ping-pong bi-bi mechanism with competitive substrate inhibition due to 2-oxoglutarate, of the type demonstrated in Vol. I, chapter 2.1.3.2, Fig. 2, p. 78.

The initial velocity is given by the rate equation

$$v = \frac{V \times [S_1] \times [S_2]}{(K_m)_{S_2} \times [S_1] + (K_m)_{S_1} \times [S_2] \times \left(1 + \frac{[S_2]}{K_I}\right) + [S_1] \times [S_2]}$$

in which $(K_m)_{S_1}$ denotes the *Michaelis* constant for L-aspartate, $(K_m)_{S_2}$ that for 2-oxoglutarate, and K_I the inhibition constant due to excess 2-oxoglutarate. The advantage of using such a technique is that, through an understanding of the reaction mechanism, critical factors – in particular those required for substrate optimization – can be easily identified.

With appropriate values for *Michaelis* constants (Table 2) and inhibitor constants (cf. Vol. I, chapter 2.3.3.3, p. 123), it can be calculated that L-aspartate at 240 mmol/l and 2-oxoglutarate at 12 mmol/l support >94% of V for both human AspAT isoenzymes (cf. Vol. I, chapter 2.3.3.3, p. 116).

Volume fraction of samples in assay mixture

A linear relationship exists between volume fractions of sample from 0.0833 to 0.0167 and measured changes in absorbance, but only if proper corrections are made for blank reactions. It is thus possible to dilute sera with high enzyme concentrations and still obtain an accurate answer. A volume fraction of sample of 0.0833 appears to be the most appropriate for serum samples.

Indicator reaction

The condition that (V/K_m) for aminotransferase should be much smaller than (V/K_m) for malate dehydrogenase is fulfilled by incorporating 10 μkat/l malate dehydrogenase per litre of assay mixture, even for very high catalytic activities of aminotransferase.

Addition of lactate dehydrogenase to the assay system is necessary to achieve complete reduction of the endogenous pyruvate of the serum specimen by NADH more rapidly than if the lactate dehydrogenase of the serum alone is relied upon to catalyze this reaction. A catalytic concentration of 10 μkat/l allows for complete reduction, within ten minutes, of pyruvate concentrations higher than ten times the mean concentration in serum.

The concentration of NADH (0.18 mmol/l) is sufficient to support the optimal reaction rate, as well as to ensure the reduction during the pre-incubation period of concentrations of pyruvate likely to occur in the majority of sera.

Prevention of analytical errors

Interference by factors influencing reagent blank activity

The original development of the UV technique described was aided by the availability of MDH substantially free from AspAT activity [12]. Any such contamination of MDH – and/or LDH – leads to an increased reagent blank activity that must be deducted from the overall measured catalytic rate. Pyridoxal phosphate will activate apo-AspAT in the MDH or LDH preparations.

Another significant reagent blank reaction may result due to the relatively broad substrate specificity of LDH:

$$\text{2-Oxoglutarate} + \text{NADH} + \text{H}^+ \xrightarrow{\text{LDH}} \text{2-hydroxyglutarate} + \text{NAD}^+ .$$

This reaction is catalyzed more efficiently by LDH-1, therefore LDH-5 is used in this procedure. Lability of NADH also contributes to the overall reagent blank reaction. However at pH ≥ 7.4 in the absence of phosphate ion, this reaction is slow and its contribution is usually negligible. The rate of NADH destruction is accelerated as pH decreases and if phosphate buffer is used.

The value of the reagent blank is determined by measurement of the apparent AspAT activity of distilled water analyzed as a specimen. Calculations are identical to those described above.

At the time of writing, available preparations of LDH and MDH are sufficiently free from AspAT and apo-AspAT so that the total reagent-blank rate measured in the described assay is typically equivalent to ≤ 1 U/l at 30 °C. However, the magnitude of this activity varies according to the source, purity, and amount of MDH and LDH used in the assay and the temperature of measurement. The value of the reagent blank is usually quite reproducible for any single batch of reagent and, once determined for a particular preparation, it need only be verified each day of use and should be systematically deducted from all measured activities. Reagent blank activity greater than the equivalent of 3 U/l interferes with the precision of measurement at lower catalytic activity concentrations and alternative sources of auxiliary enzymes should be investigated.

Interference by factors influencing sample blank activity

Although the MDH/NADH coupled system is generally specific for AspAT activity measurements, several competing reactions may give rise to NADH oxidation and therefore to apparent aminotransferase activity. Most notable is activity due to the presence of GlDH (glutamate dehydrogenase, EC 1.4.1.3) and ammonium ion:

$$\text{2-Oxoglutarate} + \text{NADH} + \text{NH}_4^+ \xrightarrow{\text{GlDH}} \text{L-glutamate} + \text{NAD}^+ + \text{H}_2\text{O} .$$

Nearly optimal amounts of both NADH and 2-oxoglutarate are present in the AspAT reaction mixture and an excess of ammonium ion can be introduced by use of suspensions of MDH and LDH in ammonium sulphate. The availability and use of glycerol solutions of the auxiliary enzymes has greatly reduced the error due to this blank reaction. Unless gross contamination of the reagents gives rise to an appreciable NH_4^+ concentration, NH_4^+ in serum specimens – even at elevated levels – supports low GlDH activity for only a short time interval. Measurement of samples containing ammonium ion and/or high activities of glutamate dehydrogenase requires measurement of a sample blank reaction, consisting of the complete reaction mixture lacking only L-aspartate (see below).

Another source of error is due to LDH catalytic activity and the reduction of sample pyruvate and, to a lesser extent to other oxo acids:

(e) $$\text{Pyruvate} + \text{NADH} + \text{H}^+ \xrightarrow{\text{LDH}} \text{L-lactate} + \text{NAD}^+$$

(f) $$\text{Oxo acid} + \text{NADH} + \text{H}^+ \xrightarrow{\text{LDH}} \text{hydroxy acid} + \text{NAD}^+ .$$

To compensate for this LDH reaction, an excess of LDH is added to speed the rate of reduction of pyruvate in order to reach equilibrium in less than 10 min. Errors may be easily introduced, however, if data are taken early, if LDH activity has deteriorated, or if the concentration of pyruvate or another oxo acid is unusually high. If the absorbance reading is observed to be stable before the addition of 2-oxoglutarate solution (2) these errors should make no significant contribution.

For the majority of serum samples, there is evidence that sample-blank reactions represent only a small percentage of measured catalytic activity. However, in the decision range between normal and pathological values the omission of individual sample blanks may in some cases lead to false conclusions [22].

Measurement of an individual sample reagent blank utilizes an alternate Working Reagent Mixture prepared as directed for solution (6) except that Tris buffer (5) is substituted for aspartate/Tris (1). All other additions remain unchanged. The sample is assayed according to the protocol described using this alternative reagent mixture rather than solution (6).

Equipment, Reagents and Solutions

As described in ref. [17, 18]; similar reagent solutions are used in the Routine UV-method on pp. 419–421. Note differences in particular for enzyme concentrations.

Procedure

Collection, treatment, stability of the specimen: as described on pp. 421, 422.

Assay conditions, measurement: as described on pp. 422, 423. However, corrections for blank reactions are mandatory. Four rate measurements are necessary:

A overall AspAT reaction;
B reagent blank reaction in which the sample is sodium chloride, 154 mmol/l;
C overall sample blank which is identical to (A) except for replacement of solution (1) by solution (4) and making up the working solution (6);
D reagent blank for sample blank which is identical to (C) except for replacing serum by sodium chloride, 154 mmol/l.

Before pipetting, the temperature of reagent solutions and sample must be brought to the calibration temperature of the pipettes.

During the pre-incubation period the reaction mixture in the cuvettes must attain a temperature of 30.00 °C.

Calculation: according to p. 423 but taking into account the several blanks. Thus (t in seconds):

$$(\Delta A/\Delta t)_{\text{corrected}} = [(\Delta A/\Delta t)_{\text{A}} - (\Delta A/\Delta t)_{\text{B}}] - [(\Delta A/\Delta t)_{\text{C}} - (\Delta A/\Delta t)_{\text{D}}] \quad \text{s}^{-1}$$

and

$$b = 1.905 \times 10^{-3} \times (\Delta A/\Delta t)_{\text{corrected}} \quad \text{kat/l}$$

Validation of Method

Precision, accuracy, detection limits and sensitivity: as described on p. 423.

Sources of error

Effects of drugs: as described on p. 423.

Interference by factors influencing blanks: the main sources of analytical inaccuracy (bias) are the blank reactions that may occur. It was therefore considered necessary to measure these and to compensate for any blank reaction rate as an integral part of the method.

Specificity: as described on pp. 423, 424.

Reference ranges: as described on p. 424.

References

[1] *A. E. Braunstein, M. G. Kritzman*, Formation and Breakdown of Amino-Acids by Intermolecular Transfer of the Amino Group, Nature *140*, 503 – 504 (1937).
[2] *P. P. Cohen*, Transamination with Purified Enzyme Preparations (Transaminase), J. Biol. Chem. *136*, 565 – 584 (1940).
[3] *A. E. Braunstein*, Amino Group Transfer, The Enzymes *9*, 379 – 481 (1973).
[4] *D. M. Goldberg*, Clinical Enzymology, Prog. Med. Chem. *13*, 1 – 158 (1976).

[5] *Y. A. Ovchinnikov, C. A. Egorov, N. A. Aldanova, M. Y. Feigina, V. M. Lipkin, N. G. Abdulaev, E. V. Grishin, A. P. Kiselev, N. N. Modyanov, A. E. Braunstein, O. L. Polyanovsky, V. V. Nosikov*, The Complete Amino Acid Sequence of Cytoplasmic Aspartate Aminotransferase from Pig Heart, FEBS Lett. *29*, 31 – 34 (1973).

[6] *R. Scandurra, C. Cannella*, Mitochondrial Aspartate Aminotransferase from Beef Kidney, Eur. J. Biochem. *26*, 196 – 206 (1972).

[7] *R. Rej*, Multiple Molecular Forms of Human Cytoplasmic Aspartate Aminotransferase, Clin. Chim. Acta *112*, 1 – 11 (1981).

[8] *J. W. Boyd*, The Intracellular Distribution, Latency and Electrophoretic Mobility of L-Glutamate-Oxaloacetate Transaminase from Rat Liver, Biochem. J. *81*, 434 – 441 (1961).

[9] *M. Hørder, R. E. Moore, G. N. Bowers, Jr.*, Aspartate Aminotransferase Activity in Human Serum. Factors to be Considered in Supplementation with Pyridoxal 5-Phosphate In Vitro, Clin. Chem. *22*, 1876 – 1883 (1976).

[10] *R. Rej, R. E. Vanderlinde*, Effects of Buffers on Aspartate Aminotransferase Activity and Association of the Enzyme with Pyridoxal Phosphate, Clin. Chem. *21*, 1585 – 1591 (1975).

[11] *R. Rej*, Measurement of Aminotransferases: I. Aspartate Aminotransferase, CRC Crit. Rev. Clin. Lab. Sci. *20*, in press (1984).

[12] *A. Karmen*, A Note on the Spectrophotometric Assay of Glutamic-Oxalacetic Transaminase in Human Blood Serum, J. Clin. Invest. *34*, 131 – 133 (1955).

[13] *A. L. Babson, P. O. Shapiro, P. A. R. Williams, G. E. Phillips*, The Use of a Diazonium Salt for the Determination of Glutamic-Oxalacetic Transaminase in Serum, Clin. Chim. Acta *7*, 199 – 205 (1962).

[14] *R. Behra, P. Christen, P. Sonderegger*, Independent Quantitation of the Mitochondrial and the Cytosolic Isoenzyme of Aspartate Aminotransferase in Chicken Tissues by Radioimmunoassays, J. Biol. Chem. *256*, 3381 – 3384 (1981).

[15] *R. Rej, C. R. Keese, I. Giaever*, Direct Immunological Determination of Aspartate Aminotransferase Isoenzymes, Clin. Chem. *27*, 1597 – 1601 (1981).

[16] *M. Roth, L. Selz*, Active Site Titration as a Tool in Enzyme Standardization, Proc. FEBS Meet. *61*, 145 – 153 (1980).

[17] *H. U. Bergmeyer, G. N. Bowers Jr., M. Hørder, D. W. Moss*, Provisional Recommendations on IFCC Methods for the Measurement of Catalytic Concentrations of Enzymes, Part 2. IFCC Method for Aspartate Aminotransferase, Clin. Chim. Acta *70*, F 19 – F 42 following P. 336 (1976).

[18] *H. U. Bergmeyer, M. Hørder, D. W. Moss*, Provisional Recommendations on IFCC Methods for the Measurement of Catalytic Concentrations of Enzymes, Part 2-Revised. Revised IFCC Method for Aspartate Aminotransferase (L-Aspartate : 2-oxoglutarate Aminotransferase, EC 2.6.1.1) (1977), Clin. Chem. *24*, 720 – 721 (1978).

[19] *R. Rej, R. E. Vanderlinde, C. F. Fasce Jr.*, An L-Aspartate : 2-Oxoglutarate Aminotransferase Reference Material from Human Erythrocytes: Preparation and Characterization, Clin. Chem. *18*, 374 – 383 (1972).

[20] *E. J. Sampson, V. S. Whitner, C. A. Burtis, S. S. McKneally, D. M. Fast, D. D. Bayse*, An Interlaboratory Evaluation of the IFCC Method for Aspartate Aminotransferase with Use of Purified Enzyme Materials, Clin. Chem. *26*, 1156 – 1164 (1980).

[21] *H. U. Bergmeyer, G. Rozalskis*, K_m von Malat-dehydrogenase aus Schweineherz in bezug auf Oxalacetat, Z. Klin. Chem. Klin. Biochem. *13*, 509 (1975).

[22] *H. U. Bergmeyer, P. Scheibe, A. W. Wahlefeld*, Optimization of Methods for Aspartate Aminotransferase and Alanine Aminotransferase, Clin. Chem. *24*, 58 – 73 (1978).

[23] *R. Rej, R. E. Vanderlinde*, Effects of Temperature on the Steady-State Kinetics and Measurement of Aspartate Aminotransferases, Clin. Chem. *27*, 213 – 219 (1981).

[24] *R. Rej, R. E. Vanderlinde*, Azide as a Preservative in Assays of Aspartate Aminotransferase Activity, Clin. Chem. *21*, 158 – 161 (1975).

[25] *R. Rej*, Aspartate Aminotransferase Activity and Isoenzyme Proportions in Human Liver Tissues, Clin. Chem. *24*, 1971 – 1979 (1978).

[26] *D. S. Young, L. C. Pestaner, V. Gibberman*, Effects of Drugs on Clinical Laboratory Tests, Clin. Chem. *21*, 1 D – 432 D (1975).

6.1.5 Isoenzymes

Robert Rej, Daniel A. Nealon and Mogens Hørder

General

The existence of isoenzymes of aspartate aminotransferase (AspAT) – or at least the compartmentalization of this enzyme within the cell – was first suggested by the findings of *Hird & Rowsell* [1], who in 1950 reported significant activity of the enzyme in both cytosolic and mitochondrial preparations from rat liver. *Fleisher et al.* [2], using a paper electrophoresis system, observed cathodal and anodal fractions of aspartate aminotransferase activity from heart tissue of porcine, canine, and human origin. The observation of mitochondrial (m-AspAT, cathodal) and soluble (s-AspAT, anodal) isoenzymes has since been extended to a variety of tissues and species [3–5]. These two proteins are true, genetically distinct isoenzymes, which differ in their amino acid compositions and immunochemical properties.

Application of method: in biochemistry and in clinical chemistry. The determination of m-AspAT in serum is useful in detecting mitochondrial damage and estimating the severity of myocardial infarction and liver disease [6–8]. It is of particular value in estimating the extent of liver damage in cases of chronic alcohol abuse [8].

Enzyme properties relevant in analysis: the two AspAT isoenzymes possess distinct kinetic, electrophoretic, and immunochemical properties. Due to these large differences, the two isoenzymes can be separated relatively easily by a variety of techniques. However, bias may be introduced into the determination of the catalytic activities of the two fractions by product inhibition, differing substrate concentrations, and the existence of subforms of each isoenzyme. A suitable procedure that minimizes such bias is given in chapter 6.1.2.

Methods of determination: a number of chromatographic, differential kinetic, electrophoretic, and immunochemical techniques have been applied to separation of the two isoenzymes (for a review see ref. [9]). Procedures for this determination have been evaluated [10], and immunochemical precipitation of s-AspAT is the procedure showing least bias and highest precision. An electrophoretic procedure, minimizing the effects of product inhibition, is also presented.

International reference methods and standards: there are no standardization proposals for measurement of AspAT isoenzymes. A reference method exists for measurement of total AspAT activity (see chapter 6.1.4) and can be used for determining catalytic activity after separation of the two isoenzymes.

Enzyme effectors: see measurement of total AspAT activity, chapter 6.1.2.

6.1.5.1 Determination Using Antibodies Directed Against s-AspAT

Assay

Method Design

Principle: serum samples are incubated with anti-s-AspAT antibodies and centrifuged, and the supernatant fluid is assayed for residual m-AspAT activity [11]. Total AspAT activity is determined in the usual manner (chapter 6.1.2) and s-AspAT activity is estimated by the difference.

Optimized conditions for measurement: sufficient antibody is added to completely precipitate s-AspAT up to 500 U/l at 30 °C. Determination of total and residual AspAT activity should be performed by a technique optimal for measurement of both isoenzymes (chapter 6.1.2).

Temperature conversion factors: factors for temperature conversion of human AspAT isoenzymes are given in chapter 6.1.2.

Equipment

Refrigerated centrifuge; equipment for measurement of total AspAT activity (chapter 6.1.2).

Reagents and Solutions

All reagents for determining total AspAT activity are required. Human s-AspAT antigen can be conveniently prepared from erythrocytes [12]. Porcine s-AspAT, available commercially, may also be used as anti-porcine-s-AspAT cross-reacts with human s-AspAT [5].

Purity of reagents: all reagents should be of the highest purity available.

Production of antisera: adjust purified s-AspAT to 130 µg protein per ml (about 20 U/ml) in Tris buffer (3). Homogenize thoroughly with an equal volume of *Freund*'s complete adjuvant and use to prepare antisera in five- to six-month-old female rabbits. Administer six subcutaneous injections of 2 ml each (130 µg enzyme protein) at one-week intervals. After completion of the course of immunization, draw about

35 ml blood from each animal at two-week intervals. Separate the serum and purify antibodies either by adding ethodin (described below), by ammonium sulphate fractionation or by DEAE-Sephadex chromatography.

Preparation of solutions (for about 5000 determinations): all solutions in re-purified water (cf. Vol. II, chapter 2.1.3.2).

1. PEG/Tris solution (polyethylene glycol, 200 g/l; Tris, 50 mmol/l, pH 7.0):

   dissolve 6.05 g Tris base and 200 g polyethylene glycol 6000 in about 800 ml water. Adjust the pH to 7.0 ± 0.05 at room temperature with HCl, 1 mol/l, and dilute to exactly 1 l with water.

2. Ethodin (Rivanol) solution (4 g/l):

   dissolve 1.4 g ethodin in 350 ml water.

3. Tris buffer (10 mmol/l, pH 7.5):

   dissolve 6.05 g Tris base in about 4 l water, adjust the pH to 7.5 with HCl, 0.1 mol/l, and dilute to 5 l with water.

4. Anti-s-AspAT antibodies:

   to each 100 ml rabbit antiserum (prepared as above), add 350 ml ethodin solution (2). Allow the mixture to stand for 30 min at room temperature. Remove denatured protein by centrifugation az 6000 *g* for 30 min. The supernate contains the IgG serum fraction and ethodin. Dialyze extensively at 4 °C against water, followed by Tris buffer solution (3) until no yellow colour remains. Dispense in 5.0 ml aliquots and lyophilize; store at 4 °C. If lyophilization is not feasible, store at ≤ −40 °C in 5 ml aliquots. Reconstitute (or thaw) one vial of antibody preparation before assay. One 5 ml aliquot suffices for about 500 analyses.

Stability of solutions: store solution (1) at 4 °C; this is stable for at least six months. Lyophilized or frozen antibody solution (4) is stable for longer than six months. Liquid preparations are stable for at least two weeks at 4 °C.

Procedure

Collection and treatment of specimen, stability of the enzyme in the sample, details for measurement and assay conditions of the activity measurement step are described in chapter 6.1.2.

Measurement

Pipette successively into small centrifuge tubes:			concentration in incubation mixture	
sample		0.2 ml	volume fraction	0.87
antibody solution	(4)	0.01 ml	volume fraction	0.043
PEG/Tris solution	(1)	0.02 ml	Tris	4.3 mmol/l
			PEG	17.4 g/l
mix gently but thoroughly, and allow the tube to stand at room temperature with gentle shaking for 1 h. Place at 4 °C for 1 – 16 h, and centrifuge (7000 *g*, 4 °C, 15 min). Use the supernatant for determination of AspAT activity as described in chapter 6.1.2; this is the m-AspAT activity of the specimen.				

Calculation: addition of the antibody preparation and PEG solution results in a slight dilution of sample. Correct for this dilution by a factor of 1.15 in addition to the usual factors for calculating enzyme activity from absorbance changes. s-AspAT activity is calculated from the difference between total AspAT activity and m-AspAT activity according to p. 423.

Validation of Method

Precision, accuracy, detection limits and sensitivity: for a specimen with mean m-AspAT activity of 7.2 U/l, the relative standard deviation was 9.8% (within-day), and 11.5% (day-to-day), respectively. For specimens with high m-AspAT activity (mean 138.2 U/l) the relative standard deviation was 2.6% (within-day) and 2.8% (day-to-day), respectively. Limiting precision for the series was either 0.7 U/l or 2.5%. The combined day-to-day and within-day precision of the procedure was within 1 U/l, or 3.0%, a precision approaching that achievable for measurement of equivalent activities of total AspAT. The largest source of variation is at the stage of measurement of enzyme activity, rather than in the immunoprecipitation step. Detection limits and sensitivity given for the measurement of total AspAT activity (chapter 6.1.2) apply.

Sources of error: measurement of very low m-AspAT activities (< 1.0 U/l) is impeded by the presence of any reagent-blank activity in the activity measurement step. All activities should be corrected for reagent-blank activity (measurement with water as a sample in "Measurement" above; i.e. including antibody). If suitable antibodies are obtained s-AspAT activity ≤ 500 U/l is completely precipitated [11]. Specimens with > 500 U/l total AspAT activity should be diluted.

Specificity: the method is specific for measurement of residual m-AspAT activity. No mitochondrial enzyme is inhibited or precipitated by anti-s-AspAT [11]. Reactions interfering with measurement of total AspAT activity affect determination of m-AspAT activity and appropriate specimen blank determinations should be performed.

Reference ranges: normal human serum contains < 4.0 U/l of m-AspAT at 30 °C. In human liver, m-AspAT comprises over 80% of total activity [5].

6.1.5.2 Measurement after Electrophoretic Separation

Assay

Method Design

Principle: separation of the isoenzymes is achieved by polyacrylamide gel electrophoresis. AspAT is quantitated by measuring glutamate produced, by means of coupled reactions catalyzed by glutamate dehydrogenase (GlDH, L-glutamate: NAD(P) oxidoreductase (deaminating), EC 1.4.1.3) and diaphorase (NADH: lipoamide oxidoreductase, EC 1.6.4.3), leading to the reduction of INT (iodonitrotetrazolium chloride) to a coloured formazan.

(a) $$\text{L-Aspartate} + \text{2-oxoglutarate} \xrightleftharpoons{\text{AspAT}} \text{L-glutamate} + \text{oxaloacetate}$$

(b) $$\text{L-Glutamate} + NAD^+ + H_2O \xrightleftharpoons{\text{GlDH}} \text{2-oxoglutarate} + NADH + NH_4^+$$

(c) $$NADH + INT + H^+ \xrightarrow{\text{diaphorase}} NAD^+ + \text{formazan}$$

The equilibrium for reaction (b) lies well to the left; however, complete oxidation of glutamate is achieved by the continuous removal of NADH by the INT indicator reaction which is irreversible under the assay conditions. Bands of activity appear as pink areas on a clear background which can be visually inspected, or the increase in absorbance can be measured by densitometric scanning near 500 nm.

Optimized conditions for measurement: mitochondrial AspAT is strongly inhibited by oxaloacetate [10]; therefore, the use of the method as a quantitative measure involves the addition of glycine, a suitable decarboxylating agent.

Equipment

Glass plate (125 × 260 × 3 mm); glass plate (125 × 260 × 1 mm); clamps and rubber gasket for moulding 2 mm gels; slot former for uniform 5 µl slots; horizontal electrophoresis cell which can be attached to a circulating cooling system (4 – 10 °C); direct current power supply (1200 V, 50 mA); *Whatman* No. 1 electrode wicks (104 × 253 mm); incubator; densitometer capable of scanning at 492 – 520 nm.

Reagents and Solutions

Purity of reagents: GlDH must be free from glutaminase. Contamination by lactate dehydrogenase, malate dehydrogenase, and alcohol dehydrogenase should not exceed 0.01% of each.

Preparation of solutions: sufficient for about 10 electrophoretic slabs (200 individual specimens): all solutions in re-purified water (cf. Vol. II, chapter 2.1.3.2).

1. Buffer for electrophoresis:

   a) Glycine/Tris stock solution (glycine, 200 mmol/l, pH 7.0; Tris, 5.8 mmol/l):

   dissolve 60.1 g glycine in about 2000 ml water and, with stirring, add 2.81 g tris(hydroxymethyl)aminomethane (Tris base). Adjust pH to 7.0 at 23 °C with HCl, 5 mol/l. Dilute to exactly 4000 ml with water.

   b) Glycine/Tris working solution (glycine, 100 mmol/l, pH 7.0; Tris, 2.9 mmol/l):

   dilute 1000 ml of stock buffer solution (1a) to exactly 2000 ml with water.

2. Ammonium persulphate (66 mmol/l):

   dissolve 150 mg ammonium persulphate in exactly 10 ml water.

3. Polyacrylamide gel

   a) Stock mixture of monomers (acrylamide, 22.2% w/v; N,N′-methylene-bis-acrylamide, 0.6% w/v):

   dissolve 22.2 g acrylamide and 0.6 g N,N′-methylene-bis-acrylamide (Bis) in about 50 ml water. Dilute to exactly 100 ml with water. Filter through a *Whatman* No. 1 filter paper if the solution appears cloudy. Store in a dark bottle. Avoid inhalation of the toxic dust.

   b) Solution for gel formation (acrylamide, 3.5% w/v; N,N′-methylene-bis-acrylamide, 0.09% w/v; ammonium persulphate, 3.2 mmol/l; N,N,N′,N′-tetramethylethylenediamine, 10 mmol/l; glycine buffer, 100 mmol/l, pH 7.0; Tris, 2.9 mmol/l):

mix 19.3 ml water, 33 ml buffer stock solution (1a) and 10.4 ml stock mixture of monomers (3a). De-aerate for a few minutes and immediately before use add 3.2 ml ammonium persulphate solution (2) and 0.1 ml of N,N,N',N'-tetramethylethylenediamine (TEMED).

4. Agarose solution (glycine, 2 mol/l, pH 7.8; agarose, 2% w/v):

   dissolve 15.02 g glycine in approximately 50 ml water, adjust to pH 7.8 at 23 °C with NaOH, 1 mol/l, and dilute to exactly 100 ml with water. Add 2 g low-gel-temperature agarose and dissolve by bringing to the boil. Store in aliquots of exactly 10 ml.

5. Aspartate/Tris solution (L-aspartate, 33.3 mmol/l; Tris, 100 mmol/l, pH 8.1):

   dissolve 0.444 g L-aspartate and 1.211 g Tris base in about 50 ml water, adjust to pH 8.1 at 23 °C with NaOH, 10 mol/l, and dilute to exactly 100 ml with water.

6. 2-Oxoglutarate/Tris solution (2-oxoglutarate, 16.7 mmol/l; Tris, 100 mmol/l, pH 8.1):

   dissolve 0.244 g 2-oxoglutarate and 1.211 g Tris base in about 50 ml water, adjust to pH 8.1 at 23 °C with HCl, 5 mol/l, and dilute to exactly 100 ml with water.

7. β-Nicotinamide-adenine dinucleotide solution (β-NAD, 160 mmol/l):

   dissolve 0.586 g β-NAD, free acid, in 5 ml water.

8. Adenosine 5'-diphosphate solution (ADP, 40 mmol/l; Tris, 100 mmol/l, pH 8.1):

   dissolve 1.01 g ADP, disodium salt, and 0.606 g of Tris base in about 25 ml water, adjust to pH 8.1 at 23 °C with HCl, 5 mol/l, and dilute to exactly 50 ml with water. Freeze in 2 ml aliquots.

9. Glutamate dehydrogenase, GlDH (1200 kU/l at 25 °C):

   use stock solution in 50% glycerol (10 mg/ml; ca. 120 U/mg at 25 °C) without dilution.

10. INT solution (17 mmol/l):

    dissolve 43 mg iodonitrotetrazolium chloride (INT) in exactly 5 ml water.

11. Diaphorase (diaphorase, 39 kU/l; Tris buffer, 100 mmol/l, pH 7.5):

    dissolve 1.211 g Tris base in about 50 ml water, adjust to pH 7.5 at 23 °C with HCl, 1 mol/l, and dilute to exactly 100 ml with water. Dissolve 43 mg diaphorase lyophilizate in 5 ml of this Tris buffer.

Stability of solutions: unless otherwise indicated store all solutions at 0–4 °C. Solutions (1a) and (4) are stable for at least 6 months, solutions (5) and (6) are stable for one week. The INT solution (10) is very sensitive to light and must always be kept in dark bottles; it can be used for about 3 months.

The diaphorase solution (11) and solution (3a) should be prepared freshly every 3 weeks. Solution (3b) should be prepared immediately before use. Solution (4) should be brought to 65 °C, then cooled to 45 °C for use. The GlDH solution (9) is stable for 12 months.

Preparation of gel slabs: solution (3b) which makes a 3.5% (w/v) gel, is poured into the assembled mould, ensuring that no air bubbles remain trapped. The rubber gasket is completely closed and the solution is allowed to polymerize for about 2 h. The clamps are removed from the mould which is then placed in the refrigerator for about 15 minutes to facilitate removal of the slot former. The gasket is removed and a large spatula is twisted between the slot former and the glass plates to allow penetration of air to separate the gel from the slot former.

Procedure

Collection and treatment of blood specimens, stability of the enzyme in the sample and details for measurement in samples other than serum are described in chapter 6.1.2.

Assay conditions: 37 °C during the developer reaction; evaluation by densitometric scanning at 492–520 nm.

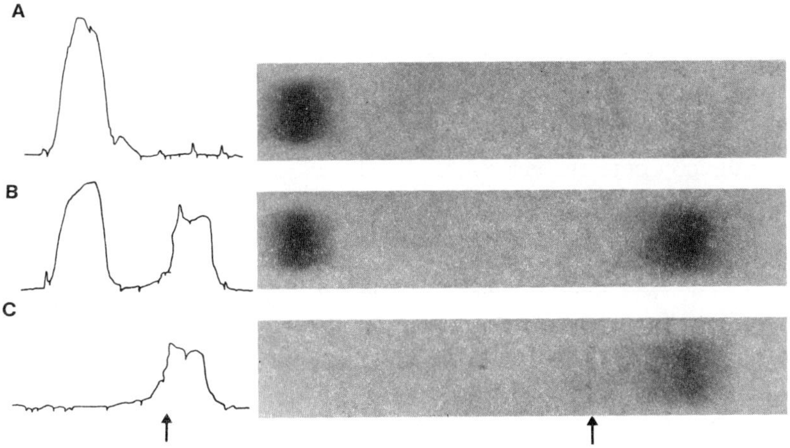

Fig. 1. Electrophoretic separation of AspAT isoenzymes. (A) 250 U/l of s-AspAT, (B) 250 U/l of s-AspAT plus 160 U/l of m-AspAT, and (C) 110 U/l of m-AspAT. Catalytic activities of purified human AspAT isoenzymes were determined by the routine UV-method (chapter 6.1.2).

Electrophoresis: place the gel on the cooling plate. Fill the electrode containers with electrophoresis working buffer (1b). Soak 8 – 10 layers of *Whatman* No. 1 paper electrode wicks in the working buffer and apply the edge of the wet wicks so that they overlap the gel by 10 – 12 mm. Ensure that the two sets of electrode wicks lie exactly parallel to each other and have a uniform contact along the entire length of the gel. Apply the samples (5 μl each) onto the gel and carry out the run for 4 h at 30 V/cm.

Developer reaction

Mix immediately before use:			concentrations in developer solution	
L-aspartate solution 2-oxoglutarate solution NAD solution ADP solution	(5) (6) (7) (8)	3.0 ml 3.0 ml 0.5 ml 0.5 ml	L-aspartate 2-oxoglutarate NAD ADP	5 mmol/l 2.5 mmol/l 4 mmol/l 1 mmol/l
mix				
GlDH solution diaphorase solution	(9) (11)	1.0 ml 0.25 ml	GlDH diaphorase	60 kU/l 490 U/l
bring this mixture to 37 °C by placing it in a water-bath.				
INT solution	(10)	1.75 ml	INT	1.49 mmol/l
immediately after addition of solutions (10) and (11), add at 40 – 45 °C				
agarose solution	(4)	10 ml	agarose glycine	1% (w/v) 1 mol/l
pour this mixture directly onto the gel. Allow the gel to solidify and incubate for at least 60 min at 37 °C. Longer incubation times increase the sensitivity.				

Evaluation: gels can be scanned in a densitometer using wavelengths in the range of 492 – 520 nm. Activity zones appear as an increase in the absorbance; the activity zones appear as pink areas against a clear background (Fig. 1). Quantitation must take into account the background absorbance of a blank sample slot. Electrophoretic separation of the isoenzymes can be performed on any suitable support medium, e.g. cellulose acetate [5], agarose, etc., provided that the electrophoretic buffer has a pH of

7.0 – 8.0 at 23 °C. The developer solution described above can be used to visualize the AspAT activity bands following any such separation.

Calculation: the fractional amount of each isoenzyme is calculated as a percent of the total area scanned. The fractional amount of isoenzyme can then be used, with measurements of total AspAT activity (p. 418), to calculate the catalytic activity of each isoenzyme.

Validation of Method

Precision, detection limits and sensitivity: with overnight incubation of the gel, activities of at least 30 μU of each isoenzyme can be quantitated. With an average of 200 U/l of each isoenzyme, the coefficient of variation for calculating the area under each isoenzyme peak is 3.7% for s-AspAT and 6.8% for m-AspAT.

Sources of error: haemolysis interferes only with the assay of s-AspAT since mature erythrocytes contain no m-AspAT.

Recovery of mitochondrial AspAT is very poor in the absence of a decarboxylating agent since it is strongly inhibited by the oxaloacetate produced in the assay reaction. Thus, in the absence of glycine, recovery of m-AspAT from a mixture of 250 U/l of s-AspAT and 250 U/l of m-AspAT is about 30 – 50%. In the presence of glycine, as recommended in the above procedure, the corresponding recovery of m-AspAT is 80 – 90% (see Fig. 1). Recovery of AspAT isoenzymes can vary among laboratories and source of enzyme, therefore, standards of known catalytic activities of m-AspAT and s-AspAT of the species examined should be included in representative electrophoretic separations.

References

[1] *F. J. R. Hird, E. V. Rowsell,* Additional Transaminations by Insoluble Particle Preparations of Rat Liver, Nature *106*, 517 – 518 (1950).

[2] *G. A. Fleisher, C. S. Potter, K. G. Wakim,* Separation of 2-Glutamic-oxalacetic Transaminase by Paper Electrophoresis, Proc. Soc. Exp. Biol. Med. *103*, 229 – 231 (1960).

[3] *L. H. Bertland, N. O. Kaplan,* Studies on the Conformations of the Multiple Forms of Chicken Heart Aspartate Aminotransferase, Biochemistry *9*, 2653 – 2665 (1970).

[4] *J. W. Boyd,* The Extraction and Purification of Two Isoenzymes of L-Aspartate: 2-oxoglutarate Aminotransferase, Biochem. Biophys. Acta *113*, 302 – 311 (1966).

[5] *R. Rej,* Aspartate Aminotransferase Activity and Isoenzyme Proportions in Human Liver Tissues, Clin. Chem. *24*, 1971 – 1979 (1978).

[6] *A. F. Smith, P. C.-P. Wong, M. F. Oliver,* Release of Mitochondrial Enzymes in Acute Myocardial Infarction, J. Mol. Med. *2*, 265 – 269 (1977).

[7] *E. Schmidt, W. Schmidt, P. Otto,* Isoenzymes of Malic Dehydrogenase, Glutamic Oxaloacetic Transaminase and Lactic Dehydrogenase in Serum in Diseases of the Liver, Clin. Chim. Acta *15*, 283 – 289 (1967).

[8] *H. Ishii, F. Okuno, Y. Shigeta, M. Tsuchiya,* Enhanced Serum Glutamic Oxaloacetic Transaminase Activity of Mitochondrial Origin in Chronic Alcoholics, Current Alcohol *5,* 101 – 108 (1978).
[9] *R. Rej,* Measurement of Aminotransferases. I: Aspartate Aminotransferase, CRC Crit. Rev. Clin. Lab. Sci. *20,* in press (1984).
[10] *R. Rej, J.-P. Bretaudiere, B. Graffunder,* Measurement of Aspartate Aminotransferase Isoenzymes: Six Procedures Compared, Clin. Chem. *27,* 535 – 542 (1981).
[11] *R. Rej,* An Immunochemical Procedure for Determination of Mitochondrial Aspartate Aminotransferase in Human Serum, Clin. Chem. *26,* 1694 – 1700 (1980).
[12] *R. Rej, R. E. Vanderlinde, C. F. Fasce Jr.,* An L-Aspartate:2-Oxoglutarate Aminotransferase Reference Material from Human Erythrocytes: Preparation and Characterization, Clin. Chem. *18,* 374 – 383 (1972).

6.2 Alanine Aminotransferase

(Glutamate Pyruvate Transaminase)

L-Alanine: 2-oxoglutarate aminotransferase, EC 2.6.1.2

Mogens Hørder and Robert Rej

General

Alanine aminotransferase, AlaAT (formerly GPT) is – except for aspartate aminotransferase – the most active and best-characterized aminotransferase from human tissues. It is found in a wide variety of organs (liver, kidney, heart, skeletal muscle, and pancreas) and in serum. It is also found in animal tissues and micro-organisms [1]. In man its concentrations in serum are elevated in diseases of the liver. Multiple molecular forms are known to exist; however the mitochondrial form is very unstable and low in concentration.
AlaAT contains 1 mole of pyridoxal phosphate per mole of monomer. The coenzyme is required for catalytic activity.

Application of method: in biochemistry and clinical chemistry.

Enzyme properties relevant in analysis: the kinetic properties and constants of the enzyme are well enough known to serve as a basis for selecting optimized reaction conditions [2] (cf. chapter 6.2.2 on p. 450).

The coenzyme, pyridoxal phosphate, is included in the assay mixture at concentrations sufficient to activate any apo-AlaAT present [3, 4].

Methods of determination: several techniques can be used to measure AlaAT activity, including chromatographic separation of amino acids, coupling of pyruvate with 2,4-

dinitrophenylhydrazine, and coupling of glutamate formation with glutamate dehydrogenase (GlDH, EC 1.4.1.3). In the preferred technique, the pyruvate formed is reduced to lactate, with concomitant oxidation of NADH catalyzed by lactate dehydrogenase (LDH). The advantages of this procedure are its use of the well-described molar absorptivity of NADH, its high degree of specificity, and its removal of pyruvate, an inhibitory product of the AlaAT reaction, from the reaction mixture.

Enzyme effectors: the enzyme is subject to both product and substrate inhibition (cf. chapter 6.2.2 on p. 450). AlaAT requires pyridoxal phosphate as coenzyme. Phosphate buffer should be avoided since it interferes with the re-association of the co-enzyme with apo-AlaAT.

International reference method and standards: a provisional reference method has been published by the International Federation of Clinical Chemistry [5] (cf. p. 450). The assay described here is based on this method. No certified reference preparation is available yet; however, cf. also Vol. II, chapter 2.3.

6.2.1 Routine UV-method

Assay

Method Design

Principle

(a) $$\text{L-Alanine} + \text{2-oxoglutarate} \xrightleftharpoons{\text{AlaAT}} \text{L-glutamate} + \text{pyruvate}$$

(b) $$\text{Pyruvate} + \text{NADH} + \text{H}^+ \xrightleftharpoons{\text{LDH}} \text{L-lactate} + \text{NAD}^+$$

AlaAT catalyzes the reaction between L-alanine and 2-oxoglutarate, and the pyruvate formed is reduced by NADH in a reaction catalyzed by LDH. After the lag phase the equilibrium of reaction (b) is so far to the right that the decrease in absorbance due to NADH oxidation is stoichiometric with the AlaAT concentration.

Enough pyridoxal phosphate is added to saturate any apo-AlaAT in the specimen. LDH present as indicator enzyme, also reduces pyruvate in the specimen and shortens the required pre-incubation period.

Optimized conditions for measurement: the conditions described are optimal for determining total activity of human AlaAT and are similar to those recommended in the proposed IFCC international reference method [5] (cf. chapter 6.2.2).

Temperature conversion factors: for materials with human serum pools or albumin as the matrix and with added human and animal enzymes, the temperature conversion factors are [6]:

Temperature	25 °C	30 °C	37 °C
Factor	1.00	1.28 – 1.34	1.75 – 1.87

Equipment

A spectrophotometer capable of measurement at 339 nm is preferred. If this is not available a spectral-line photometer capable of measurement at Hg 334 nm or Hg 365 nm can be used. Either instrument should have a thermostatted cuvette or cell compartment and an attached recorder, or stopwatch.

Reagents and Solutions

Purity of reagents: the indicator enzyme should be free from AlaAT (either apo- or holo-enzyme) and GlDH. Contamination by either should not exceed 0.005% relative to LDH catalytic activity. LDH activity can be verified by the procedures described in chapter 2.2. To reduce the interference from trace GlDH the indicator enzyme must also be free from ammonium ion. Reagent preparation calls for disodium NADH. Adjust the amount accordingly if a different salt or hydrated form is used. NADH and 2-oxoglutarate must be free from inhibitors (cf. Vol. II, chapter 2.1.2); suitable commercial preparations are available.

Preparation of solutions: all solutions are prepared with re-purified water (cf. Vol. II, chapter 2.1.3.2).

1. L-Alanine/Tris buffer (L-alanine, 615 mmol/l; Tris, 110 mmol/l; pH 7.3):

   to about 700 ml water add 54.8 g L-alanine and 13.3 g Tris base. Mix and stir. Constantly monitor temperature, which should be maintained at 30 ± 2 °C by placing the vessel in warm or cool water. Monitor pH and adjust in to 7.3 ± 0.02 with HCl, 1 mol/l. Dilute the solution to exactly 1 l with water. Filter through 0.22 μm pore size filters. Refrigate the working solution; freeze the remainder.

2. 2-Oxoglutarate/Tris buffer (2-oxoglutarate, 180 mmol/l; Tris, 110 mmol/l; pH 7.3):

   to about 70 ml water add 2.63 g 2-oxoglutaric acid and 1.33 g Tris base. Maintain temperature at 30 ± 2 °C as described for solution (1). Adjust pH to 7.3 ± 0.02 with NaOH solutions (10 and 1 mol/l). Dilute to exactly 100 ml with water. Filter through 0.22 μm pore size filters. Freeze in aliquots of approximately 10 ml.

3. Lactate dehydrogenase (ca. 3000 kU/l):

   mix 5.0 ml water with 5.0 ml glycerol. Add 0.8 ml of this solution to 1 ml of porcine muscle LDH (10 mg/ml in glycerol; ca. 550 U/mg at 25 °C). Mix gently by inversion. Refrigerate; do not freeze.

4. Tris buffer (110 mmol; pH 7.3):

   to about 600 ml water add 13.32 g Tris base. Maintain temperature at 30 ± 2 °C as described for solution (1). Adjust pH to 7.3 ± 0.02 with HCl (1 mol/l). Dilute to exactly 1 l with water. Refrigerate.

5. Pyridoxal 5-phosphate (500 mmol/l):

   weigh 266 mg of pyridoxal 5-phosphate monohydrate (most preparations are monohydrate), and dissolve in 200 ml Tris buffer (4). Store in ca. 10-ml aliquots, protected from light, at ⩽ −20 °C.

6. Working reagent mixture:

   mix the reagents as shown for the number of determinations desired

Number of determinations:		10	30	50	100
Alanine/Tris	(1) ml	20.0	60.0	100.0	200.0
LDH	(3) ml	0.1	0.3	0.5	1.0
Pyridoxal phosphate	(5) ml	0.5	1.5	2.5	5.0
NADH	mg	3.3	9.9	16.5	33.0

NADH is added directly as a dry powder.

Stability of solutions: solution (1) is stable at 4 °C for at least 1 month and at ⩽ −20 °C for at least 3 months. Solutions (2) and (5) are stable at 4 °C for at least 2 days and at ⩽ −20 °C for 6 months. Solution (3) is stable at 4 °C for at least 1 month. (Do not freeze this solution.) Solution (4) is stable at 4 °C for at least 4 months. Solution (6) should be prepared for use daily; it remains usable for at least 24 h if kept at 4 °C. Bacterial contamination, particularly of solution (1), can be prevented by adding azide (7 mmol/l); this does not affect enzyme activity.

Procedure

Collection and treatment of specimen: collect blood with a minimum of venous stasis. The preferred specimen is freshly prepared serum free from haemolysis. Plasma is not recommended.

Stability of enzyme in specimen: AlaAT activity in serum remains stable at 4 °C for a minimum of 4 days. Storage at ⩽ −20 °C for ⩽ 1 month is acceptable.

Details for measurements in tissues: AlaAT activity of tissue can be determined quantitatively only after the tissue has been completely homogenized. Use of sonication and detergents such as deoxycholate may be required. Since tissue extracts may also contain glutamate dehydrogenase and other interfering enzyme activities, a sample blank reaction may be required (see "Sources of error", below). Dilute the sample with Tris buffer (4), saline, or albumin solution (1%, w/v).

Assay conditions: wavelength 339 nm (alternatively, with a spectral-line photometer only, Hg 334 or Hg 365 nm); light path 10 mm; final volume 2.40 ml; temperature 30.0 ± 0.01 °C.

For routine measurements on serum or measurements of specimens without interfering matrix, sample blanks are not required. A reagent blank with sodium chloride, 154 mmol/l, instead of sample should be prepared for each batch of reagents and its value deducted from the overall measured rate. Measure absorbance change against air as reference. The temperature of solutions (2) and (6) should be 30 °C.

Measurement

Pipette successively into the cuvette:			concentration in assay mixture	
working reagent mixture	(6)	2.0 ml	L-alanine	500 mmol/l
			Tris	100 mmol/l
			NADH	0.18 mmol/l
			pyridoxal phosphate	0.11 mmol/l
			LDH	≥1200 U/l
sample (serum or diluted tissue homogenate)		0.2 ml	volume fraction	0.083
mix thoroughly without removing any of the solution from the cuvette, incubate at 30 °C for 10 min or until temperature equilibrium has been reached and a stable absorbance reading is observed;				
2-oxoglutarate	(2)	0.2 ml	2-oxoglutarate	15 mmol/l
mix thoroughly; monitor change in absorbance with potentiometric recorder, or manually record absorbance readings at fixed time intervals with elapsed timer. A lag phase of up to 1.5 min may occur. The reaction should be monitored for about 3 min or until an absorbance change (at 339 nm) of ≥0.20 occurs.				

The values of $\Delta A/\Delta t$ should not exceed 0.40/min. If a faster change occurs, dilute the specimen with one of the solutions recommended under "Details for measurement in tissues" (above).

Calculation: the calculation formulae (k) and (k_1) are applicable (cf. "Formulae", Appendix 3). For values of ε cf. Appendix 4. The following relations are valid for the catalytic concentration of the sample (Δt in minutes):

	Hg 334 nm	339 nm	Hg 365 nm	
$b =$	$1950 \times \Delta A/\Delta t$	$1912 \times \Delta A/\Delta t$	$3544 \times \Delta A/\Delta t$	U/l
$b =$	$3.25 \times 10^4 \times \Delta A/\Delta t$	$3.19 \times 10^4 \times \Delta A/\Delta t$	$5.91 \times 10^4 \times \Delta A/\Delta t$	nkat/l

Validation of Method

Precision, accuracy, detection limits and sensitivity: the assay described had an intra-laboratory imprecision of <2 U/l for a quality-control specimen with a catalytic activity of 50 U/l (relative standard deviation RSD = 4.0%) over a period of 24 months. The sensitivity (discriminative capability) is approx. 0.5 U/l at catalytic concentrations of 10 U/l and above. At the upper limit of the reference interval for human subjects (35 U/l, see below) the RSD was approximately 2% for within-day measurements (n = 40) and about 4% for long-term imprecision (n = 28). Reproducibility for interlaboratory measurements (5 laboratories) was approximately 5%.

With a sufficiently sensitive spectrophotometer (e.g. *Cary* 219 or *Varian*) the detectability of the method is 1.0 U/l with an overall imprecision of about 0.2 U/l. The magnitude of the reagent blank varies among laboratories and affects the minimum detectable activity (see "Sources of error", below). The method shows excellent correlation with the IFCC reference method and has a relative inaccuracy of <4% for human serum specimens in comparison with that procedure.

Sources of error

Effects of drugs: several drugs and other therapeutic substances increase the catalytic activity of AlaAT. For a comprehensive review cf. ref. [7].

Interference by factors influencing blanks: measurement of samples containing ammonium ion and/or high GlDH activities requires correction for the sample blank reaction, cf. chapter 6.2.2 for details.

For most serum samples there is evidence that sample blank reactions represent only a small percentage of measured catalytic activity. For human and animal sera the sample blank activity is usually ⩽4 U/l, often below the level of imprecision of the technique. It can therefore be neglected. For tissue extracts, however, this activity may be significant.

Specificity: with the exceptions noted in chapter 6.2.2 under "Sources of error", AlaAT activity is measured specifically. Apart from the blank reactions noted there, no significant non-specific reaction interferes.

AlaAT is reasonably specific for L-alanine and L-glutamate. 3-Aminobutyric acid can also serve as a substrate but at a much lower rate of conversion (1).

Reference ranges: the tentative reference interval for sera from healthy adults is 5 – 35 U/l.

6.2.2 IFCC Reference Method (provisional)

An IFCC reference method has been published in a provisional form [5]; it has been modified subsequently by changing the pH of measurement to 7.3.

Assay

The method design is based on profound efforts directed towards establishing optimal conditions. The assay conditions resulting from these efforts are given in Table 1.

Table 1. Provisional IFCC reference method for AlaAT assay: reaction conditions

Temperature	30.0 °C
pH	7.3
Volume fraction of sample	$\varphi = 0.0833$
Tris (hydroxymethyl) aminomethane	100 mmol/l
L-Alanine	500 mmol/l
2-Oxoglutarate	15 mmol/l
Pyridoxal phosphate	0.10 mmol/l
NADH	0.18 mmol/l
Lactate dehydrogenase	20 μkat/l (1200 U/l)

Establishing Optimized Conditions

Primary reaction

The assay conditions were optimized theoretically by modelling with the rate equations valid for the reaction catalyzed by the aminotransferase and the experimentally-determined K_m values for the human enzyme. These conditions were confirmed in practice by stepwise optimization of each assay variable.

Coenzyme: pyridoxal phosphate

AlaAT is unstable after removal of the pyridoxal phosphate coenzyme. It is, however, difficult to dissociate pyridoxal phosphate from the enzyme, since the binding is strong. Nevertheless, an average 20% increase in the catalytic activities of serum specimens results from addition of the coenzyme.

Optimal pyridoxal phosphate concentrations have been established experimentally. The time required for the re-combination of the apo-enzyme with pyridoxal phosphate is approximately 7 min; in practice, this is adequately provided by the pre-incubation period required to reach temperature equilibrium before starting the assay.

Phosphate inhibits the re-combination of the apo-enzyme with the coenzyme in all buffers investigated; e.g. in phosphate buffer, 100 mmol/l, the effect of the coenzyme is virtually reduced to zero.

Buffer, pH

Tris(hydroxymethyl)aminomethane (Tris) buffer does not interfere with the re-association of pyridoxal phosphate with apo-AlaAT. The $pK_{30°C} = 7.9$ of Tris is close enough to the pH of the assay (7.3) to enable Tris to serve as a buffer at 100 mmol/l.

Recent experimental evidence has shown that the pH of 7.5 originally proposed should be lowered to pH 7.3. The optimal pH for alanine aminotransferase in human serum lies close to pH 7.3 (Fig. 1), although the dependence of catalytic activity on pH close to this value is of minor importance. Therefore, pH 7.3 is recommended.

The buffering capacity of Tris at pH 7.3 has been examined both theoretically and experimentally using patients' sera. The complete oxidation of NADH in the assay results in an increase in the proton concentration of $0.18 \text{ mmol} \times l^{-1}$ and a pH change of less than 0.01 pH unit. Addition of an acidified serum sample (pH 5.62) resulted in no significant change in the assay pH.

Substrates

In this two-substrate reaction the concentrations of both substrates had to be optimized (cf. also Vol. I, chapter 2.3.3, pp. 116, 127). AlaAT is subject to both substrate and product inhibition, resulting in a parabolic relationship between the concentrations of the two substrates at a given fraction of V [2]. The IFCC method employs the lowest (most economical) concentrations of substrates that will maintain a steady rate of conversion even if small variations of substrate concentrations occur.

Furthermore, because of its inhibiting effect, the second substrate, 2-oxoglutarate, must be restricted to as low a concentration as possible, provided that the reduced reaction rate that this would produce can be compensated for by a higher alanine concentration. However, the concentration of alanine which is practicable is limited by the solubility and viscosity of the solution.

The assay described here measures 91% of the maximal velocity V for the enzyme [2, 5].

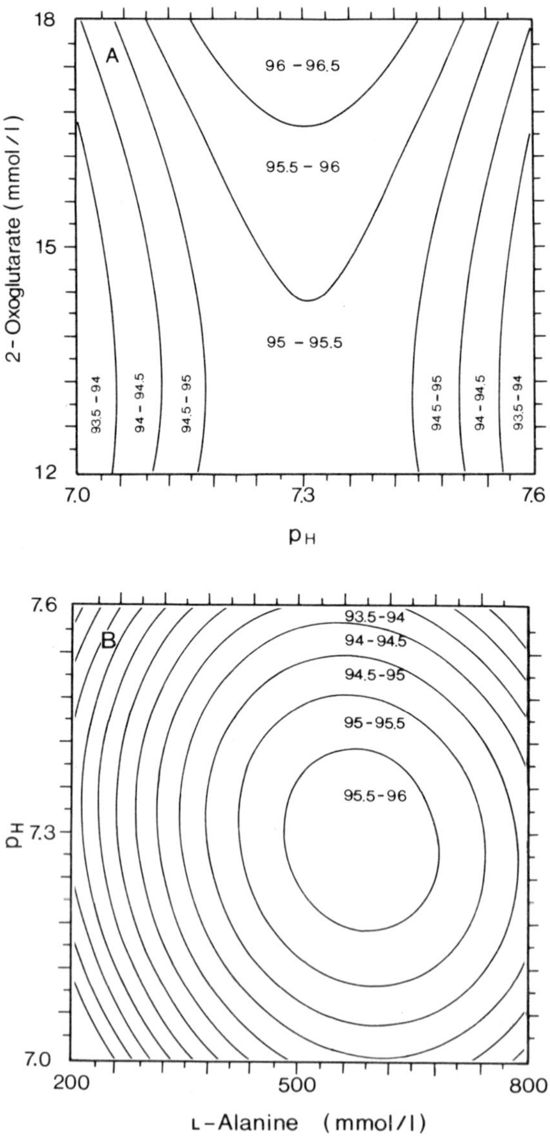

Fig. 1. Confirmation of reaction by response surface optimization: The dependence of the catalytic concentration of alanine aminotransferase on pH and substrate concentration of the reaction mixture.

The sources of enzyme were pools of human serum. 2-Oxoglutarate and L-alanine were varied as shown; L-alanine was present at 500 mmol $\times l^{-1}$ (A), and 2-oxoglutarate at 15 mmol $\times l^{-1}$ (B). Areas of constant catalytic concentration.

Reported by permission from the AlaAT study group of the American Association of Clinical Chemistry).

Volume fraction of sample in assay mixture

A slight dependence of the catalytic activity on the volume fraction exists only for the serum enzyme originating from liver tissue, not that from heart tissue.

Indicator reaction

The catalytic concentration of LDH determines the duration of the lag phase. The catalytic concentrations of LDH required at various steady-state concentrations of pyruvate have been calculated [5]. In this method a lag phase of up to 90 s may be expected at the catalytic concentrations of AlaAT found in normal human serum. At higher concentrations of AlaAT the lag phase will be shorter. However, excess of LDH may cause errors due to contaminant activities in the LDH preparation (cf. "Sources of error").

The NADH concentration of 0.18 mmol/l is sufficient to support the optimal reaction rate. It also ensures the reduction during the pre-incubation period of concentrations of pyruvate likely to occur in the majority of sera.

Prevention of analytical errors

Interference by factors influencing reagent blank activity

The value of the reagent blank is the apparent AlaAT activity of saline analyzed as a sample. Calculations are identical to those described above.

The original development of this UV technique was aided by the availability of LDH substantially free from AlaAT activity. Any such contamination of LDH leads to increased reagent blank activity, which must be deducted from the measured catalytic rate. Pyridoxal phosphate will activate apo-AlaAT in the LDH preparations.

Preparations of LDH available at the time of writing are sufficiently free from AlaAT that the total reagent blank activity is typically equivalent to $\leqslant 1$ U/l at 30 °C. However, the activity varies according to the source, purity and amount of LDH used in the assay and the temperature at the time of measurement. The value is usually quite reproducible for any single batch of reagent and, once determined, need only be verified on each day of use.

Another significant reagent blank reaction may result from the relatively broad substrate specificity of LDH:

$$\text{2-Oxoglutarate} + \text{NADH} + \text{H}^+ \xrightarrow{\text{LDH}} \text{2-hydroxyglutarate} + \text{NAD}^+ .$$

Lability of NADH may also contribute to the reagent blank reaction. However, in the absence of phosphate ion this reaction is slow, and its contribution is usually negligible, although the rate of NADH destruction is accelerated as pH decreases below 7.4.

Reagent blank activity greater than the equivalent of 3 U/l interferes with the precision of measurement at lower catalytic activities. Alternative sources of the indicator enzyme should then be investigated.

Interference by factors influencing sample blank activity

Although the LDH/NADH coupled system is sufficiently specific for AlaAT activity measurements, several competing reactions may give rise to NADH oxidation and therefore to apparent aminotransferase activity. Most notable is activity due to glutamate dehydrogenase and ammonium ion:

$$\text{2-Oxoglutarate} + \text{NADH} + \text{NH}_4^+ \xrightarrow{\text{GlDH}} \text{L-glutamate} + \text{NAD}^+ + \text{H}_2\text{O}.$$

Nearly optimal amounts of NADH and 2-oxoglutarate are present in the AlaAT reaction mixture, and an excess of ammonium ion can be introduced if ammonium sulphate suspensions of LDH are used. The use of solutions of the indicator enzyme in glycerol greatly reduces the error due to this reaction.

Unless gross contamination of the reagents results in an appreciable NH_4^+ concentration, the concentration of NH_4^+ in serum specimens – even at elevated levels – supports low GlDH activity for only a short time.

Equipment, Reagents and Solutions

Equipment, reagents and solutions as described on pp. 446, 447.

Procedure

Collection, treatment, stability of the specimen: as described on pp. 447, 448.

Assay conditions, measurement: as described on pp. 448, 449. However, corrections for blank reactions are mandatory. Four rate measurements are necessary:

A overall AlaAT reaction;

B reagent blank reaction in which the sample is sodium chloride, 154 mmol/l;

C overall sample blank which is identical to (A) except for replacement of solution (1) by solution (4) and making up the working solution (6);

D reagent blank for sample blank which is identical to (C) except for replacing serum by sodium chloride, 154 mmol/l.

Before pipetting, the temperature of reagent solutions and sample must be brought to the calibration temperature of the pipettes.

During the pre-incubation period the reaction mixture in the cuvettes must attain a temperature of 30.00 °C.

Calculation: according to p. 449 but considering the several blanks. Thus (t in seconds):

$$(\Delta A/\Delta t)_{\text{corrected}} = [(\Delta A/\Delta t)_A - (\Delta A/\Delta t)_B] - [(\Delta A/\Delta t)_C - (\Delta A/\Delta t)_D] \qquad s^{-1}$$

and

$$b = 1.905 \times 10^{-3} \times (\Delta A/\Delta t)_{\text{corrected}} \qquad \text{kat/l}$$

Validation of Method

Precision, accuracy, detection limits and sensitivity: as described on p. 449.

Sources of error

Effects of drugs: as described on p. 449.

Interference by factors influencing blanks: the main sources of analytical inaccuracy (bias) are the blank reactions that may occur. It was therefore considered necessary to measure these and to compensate for any blank reaction rate as an integral part of the method.

Specificity: as described on p. 450.

Reference ranges: as described on p. 450.

References

[1] *A. E. Braunstein,* Amino Group Transfer, in: *P. D. Boyer* (ed.), The Enzymes, Vol. IX, Group Transfer, Part B, Academic Press, New York 1973, pp. 462 – 472.

[2] *H. U. Bergmeyer, P. Scheibe, A. W. Wahlefeld,* Optimization of Methods for Aspartate Aminotransferase and Alanine Aminotransferase, Clin. Chem. *24*, 58 – 73 (1978).

[3] *V. Lustig,* Activation of Alanine Aminotransferase by Pyridoxal Phosphate, Clin. Chem. *23*, 175 – 177 (1977).

[4] *M. Plebani, P. Bonvicini, T. DeBesi, A. Giordano, F. Pesarin, G. Ceriotti,* Reference Values for Alanine and Aspartate Aminotransferase Optimized by Addition of Pyridoxal Phosphate, Enzyme *25*, 346 – 352 (1980).

[5] *H. U. Bergmeyer, M. Hørder,* Provisional Recommendations on IFCC Methods for the Measurement of Catalytic Concentrations of Enzymes, Part 3, IFCC Method for Alanine Aminotransferase (L-Alanine: 2 Oxoglutarate Aminotransferase, EC 2.6.1.2), Clin. Chim. Acta *105*, 147F – 172F (1980).
[6] *W. Gruber, D. Hundt, M. Klarwein, H. Möllering,* Comparison of Control Materials Containing Animal and Human Enzymes, J. Clin. Clin. Biochem. *15*, 579 – 582 (1977).
[7] *D. S. Young, L. C. Pestaner, V. Gibberman,* Effects of Drugs on Clinical Laboratory Tests, Clin. Chem. *21*, 1D – 432D (1975).

6.3 Tyrosine Aminotransferase

L-Tyrosine : 2-oxoglutarate aminotransferase, EC 2.6.1.5

Thomas I. Diamondstone

General

Tyrosine aminotransferase, TAT, is a cytosolic enzyme found mainly in liver, although its activity has been demonstrated in other mammalian tissues such as heart, kidney, and thyroid [1, 2]. TAT activity has also been observed in micro-organisms [3, 4, 5] and plants [6]; in these species, the activity appears to play a synthetic, rather than a degradative, role. It is the first mammalian enzyme to have been shown to be inducible [7] and this property has allowed the use of TAT as a marker enzyme in a wide range of studies involving cells in culture [7]. It should be noted that the TAT activity observed in liver mitochondria is generated entirely by mitochondrial aspartate aminotransferase [8].

Application of methods: in biochemistry. See individual methods for more details.

Enzyme properties in relevant analysis: this enzyme has not yet been purified to homogeneity; also enzyme properties and reaction mechanisms have not been studied as extensively as, e.g., for aspartate aminotransferase.

Substrate conversion proceeds linearly with time.

Methods of determination: several methods are available for the determination of TAT activity. These methods fall into two main classes: spectrophotometric, in which a product (usually 4-hydroxyphenylpyruvate) or a derivative is measured either continuously, as in the methods of *Lin et al.* [9] or of *DiCola & Federici* [10], or at fixed time intervals, as in the modified [11] *Briggs* assay [12] or the method of

Diamondstone [13]; and radioisotopic, of which there are several [14–17]. One method of interest relies on separation by means of an amino-acid analyzer of the glutamate formed in the reaction [18]. Of the various methods available, only the modified *Briggs* and the *Diamondstone* assay are widely used, although the other assays, particularly those involving the use of radioisotopes, are useful (and may be preferable) in certain special applications. We will consider only the modified *Briggs* and *Diamondstone* assays.

International reference methods and standards: neither standardization at the international level nor the existence of reference standard materials is known up to now.

Enzyme effectors: the enzyme requires pyridoxal phosphate as coenzyme [19]; this is added to the reaction mixture at a level of 30 to 300 times its K_m value. Sulphydryl reagents have been shown to inhibit the enzyme [20], but chelating agents, such as EDTA, do not.

6.3.1 Briggs Method, Modified

Assay

Method Design

Principle

(a) L-Tyrosine + 2-oxoglutarate $\xrightarrow{\text{TAT}}$ 4-hydroxyphenylpyruvate + L-glutamate

(b) 4-Hydroxyphenylpyruvate + KH_2PO_4 + $(NH_4)_6Mo_7O_{24}$ $\longrightarrow$ blue dye-stuff

In reaction (b), 4-hydroxyphenylpyruvate apparently serves as the reducing agent in a reaction similar to that employed by *Fiske & Subbarow* [22] for the determination of inorganic phosphate. This method is best for measuring TAT activity in homogenates or crude supernatants.

Optimized conditions for measurement: the optimum pH for the reaction is around 7.6 [21, 23], although small variations in pH do not alter the activity appreciably; and because the pK_{a2} of phosphate is around 6.8, the reaction is usually run at around pH 7.2–7.3. The concentrations of 2-oxoglutarate and of pyridoxal phosphate are saturating, in the kinetic sense; that is, they are above the concentration range in which the velocity of the reaction is affected by changes in concentration. However, owing to the slight solubility of tyrosine in aqueous solutions, it is necessary to control carefully the concentration of tyrosine (see below). Although tyrosine is the best amino acid substrate, 3-iodotyrosine [21] and 3-fluorotyrosine [19] are just about as good; and several other analogues of tyrosine can serve as substrates [21]. With crude preparations, a pre-incubation period (see "Assay conditions") of up to 1 h is sometimes necessary before starting the reaction.

The final concentrations of reagents in the incubation mixture are

phosphate buffer, pH 7.3, 190 mmol/l,
tyrosine, 6 mmol/l,
2-oxoglutarate, 9 mmol/l,
pyridoxal phosphate, 0.04 mmol/l.

Temperature conversion factors: no information is available as to variation of activity with temperature; therefore all values are for 37 °C.

Equipment

Spectrophotometer capable of measurement at 850 nm; water-bath (37 °C); stop-watch.

Reagents and Solutions

Purity of reagents: use reagents of the best commercially available grade (reagent grade). Other than this, no special precautions are needed, except as noted in the directions for preparation of the individual solutions.

Preparation of solutions (for about 100 determinations): all solutions in re-purified water (cf. Vol. II, chapter 2.1.3.2).

1. Phosphate buffer (0.2 mol/l, pH 7.3):

   dissolve 6.96 g K_2HPO_4 in 200 ml water; dissolve 2.18 g KH_2PO_4 in 80 ml water. With constant stirring and using a pH-meter, add sufficient KH_2PO_4 solution to the K_2HPO_4 solution to bring the pH to 7.3.

2. 2-oxoglutarate (0.3 mol/l):

   dissolve 554 mg monopotassium salt in 20 ml phosphate buffer (1). The pH may change slightly, but not enough to affect the final pH of the reaction mixture.

3. Pyridoxal phosphate (1.33 mol/l):

   dissolve 3 mg pyridoxal phosphate in 10 ml buffer (1). Solution should be distinctly yellow.

Note: Inasmuch as both the 2-oxoglutarate and the pyridoxal phosphate take some time to dissolve, it is best to prepare these somewhat in advance.

4. Tyrosine (6.85 mmol/l):

   the concentration given is a super-saturated solution; the final mass concentration will be 1.24 g/ml. Therefore, weigh out only enough tyrosine as will be used in a given experiment, rather than for about 100 determinations. Place the tyrosine in the mortar of an all-glass homogenizer (with ground glass walls and pestle) along with a small amount of buffer solution (1). Grind thoroughly until all obvious lumps are broken and a fine uniform suspension is obtained. Add the suspension to some buffer in a test tube; rinse out the mortar with a little more buffer. Add enough buffer to give the indicated final concentration and carefully heat the mixture without boiling until all the tyrosine is dissolved. Place the solution in the 37 °C water-bath, and allow it to reach that temperature before use. Under these conditions, the tyrosine will remain in solution for several hours. (For an alternative method cf. [25].)

5. Trichloroacetic acid, 100% (w/v):

   dissolve 10 g TCA (wet crystal) in enough water (approximately 4.3 ml) to make 10 ml of solution.

6. Potassium phosphate, 1% (w/v):

   dissolve 1 g KH_2PO_4, free of water, in 100 ml water.

7. Ammonium molybdate, 3% (w/v):

   dissolve 3 g $(NH_4)_6Mo_7O_{24} \cdot 4H_2O$ in 100 ml HCl, 5 mol/l.

8. Diethyldithiocarbamate (0.06 mol/l):

   dissolve 200 mg in 200 ml water.
   (Used only for modification, cf. p. 461. Re-crystallization of the diethyldithiocarbamate may be necessary [24]).

Stability of solutions: as described above, the tyrosine solutions (4) must be freshly prepared for each experiment. The phosphate buffer (1), 2-oxoglutarate (2), pyridoxal phosphate (3), KH_2PO_4 (6), and diethyldithiocarbamate (8) should be stored in the refrigerator. The ammonium molybdate solution (7) and the trichloroacetic acid (5) can be kept at room temperature.

Procedure

Collection and treatment of specimen: TAT is a cytosolic enzyme; therefore it can be measured either in tissue homogenates (usually mammalian liver) or in crude or purified cytosolic fractions. Standard procedures for isolation and preparation can be used [25]. A satisfactory procedure for obtaining a preparation for measurement is to homogenize a liver in 2–4 volumes of ice-cold salt solution (KCl, 0.14 mol/l) or in diluted buffer and use the supernatant obtained from high-speed centrifugation (e.g., in the ultracentrifuge).

Stability of the enzyme in specimen or sample: TAT preparations are relatively unstable at refrigerator (0 °C – 4 °C) temperatures. They can be kept for fairly long periods of time at −70 °C.

Assay conditions: wavelength 850 nm; light path 10 mm; final volume (before addition of trichloroacetic acid) 3.2 ml; incubation temperature 37 °C. Since the tyrosine forms the bulk of the reaction mixture, it is not necessary to pre-warm the other solutions. Measure against blank containing buffer in place of sample.

Measurement

Enzyme reaction

Pipette successively into 10 ml conical centrifuge tubes:			concentration in incubation mixture	
tyrosine solution	(4)	2.8 ml	tyrosine	6 mmol/l
pyridoxal phosphate solution	(3)	0.1 ml	pyridoxal-P	0.04 mmol/l
sample (appropriate dilution)		0.2 ml	volume fraction	0.063
mix, allow to stand in water-bath (37 °C) for up to 1 h*				
2-oxoglutarate solution	(2)	0.1 ml	oxoglutarate	9 mmol/l
mix, start stopwatch. Reaction times (usually up to 10 min) must be measured precisely				
TCA solution added at end of desired reaction time (e.g., 2, 4, 6, 8, 10 min)	(5)	0.1 ml	TCA	3.125% (w/v)
mix; centrifuge the tubes to pack the precipitate. Use clear supernatant for colour reaction.				

* Pre-incubation period; may be required for crude preparations (p. 458).

Colour reaction

Pipette successively into 25 ml test tubes:			concentration in reaction mixture
supernatant water KH_2PO_4/ammonium molybdate solution*	 (6)/(7)	1 (or 2) ml 1 ml 2 ml	 KH_2PO_4 0.2 – 0.25% (w/v) ammonium molybdate 0.6 – 0.75% (w/v)
mix; allow to stand at room temperature for 1 h. Pour into cuvette (10 mm light path); read absorbance at 850 nm			

* Mix equal volumes together before pipetting.

Modification: for fresh enzyme preparations, it is sometimes necessary to add 0.2 ml of diethyldithiocarbamate solution (8) in order to inhibit 4-hydroxyphenylpyruvate oxidase. Under these conditions, the amount of tyrosine added to the enzyme reaction mixture will be 2.6 ml, rather than 2.8 ml. This will require a slightly more concentrated tyrosine solution, so that the final concentration of tyrosine in the enzyme reaction mixture is still 6 mmol/l.

Calculation: the absorption coefficient of the coloured complex (as 4-hydroxyphenylpyruvate) is 0.41 $l \times mmol^{-1} \times mm^{-1}$. The catalytic concentration of the sample (diluted supernatant of a homogenate after centrifugation) is given by

$$b = \frac{\Delta A_{850}}{\Delta t} \times \frac{3.3 \times 4 \times 1000}{0.41 \times 10 \times 0.2} = 3.22 \times 10^3 \times \frac{\Delta A_{850}}{\Delta t} \quad \text{U/l}$$

$$b = 5.37 \times 10^4 \times \frac{\Delta A_{850}}{\Delta t} \quad \text{nkat/l}$$

where the factor of 4 accounts for the fact that 1 ml of supernatant was taken from the incubation mixture and diluted to 4 ml for the colour reaction. (If 2 ml had been taken, the corresponding factor would have been 2.5.)

Validation of Method

Precision, accuracy, detection limits and sentivity: it is best to choose concentration of enzyme so that the A_{850} readings fall in the range 0.070 – 0.200. In this range the total amount of 4-hydroxyphenylpyruvate produced per ml of enzyme sample is between about 1.1 and 3.2 μmol/ml. Within this range, the coefficient of variation is about 5%. Below about A_{850} = 0.070, the readings are too variable to be reliable, and above A_{850} = 0.200, the reaction is proceeding so fast that it may not be linear. Therefore, initially it is best to check linearity by setting up various tubes with different amounts of enzyme and/or different time intervals of (for example) 2, 4, 6, ... min or some other convenient range. At A_{850} = 0.020, the concentration of 4-hydroxyphenylpyruvate in the enzyme reaction mixture is about 0.2 mmol/l, approximately 3% of the original tyrosine concentration.

Source of error: the supernatant after precipitation with trichloroacetic acid must be clear, otherwise the mixture for the colour reaction will be cloudy and unreadable. When diethyldithiocarbamate is used, it can produce a colour, so blanks or zero-time controls should always be run. Also, anything that will serve as a reducing agent for the *Fiske-Subbarow* determination for inorganic phosphate will interfere.

Reference ranges: see following *Diamondstone* method.

6.3.2 Diamondstone Method

Assay

Method Design

Principle

1) L-Tyrosine + 2-oxoglutarate $\xrightarrow{\text{TAT}}$ 4-hydroxyphenylpyruvate + L-glutamate

2) 4-Hydroxyphenylpyruvate + OH^- $\longrightarrow$ 4-hydroxybenzaldehyde + oxalate

4-Hydroxybenzaldehyde absorbs strongly at 331 nm. In addition to producing 4-hydroxybenzaldehyde, the OH^- also stops the reaction and dissolves all protein, thus eliminating the need for a separate colour reaction. This method is suitable for measurement of TAT activity in preparations of a wide range of purity.

Optimized conditions for measurement: see modified *Briggs* assay.

Equipment

Spectrophotometer capable of measurement at 331 nm; water-bath (37 °C); stopwatch.

Reagents and Solutions

Same as for modified *Briggs* assay, except that no trichloroacetic acid, KH_2PO_4, or ammonium molybdate is required. However, NaOH is required.

Preparation of solutions (for about 100 determinations): same as for modified *Briggs* assay, except for absence of solutions (5), (6), and (7), and presence of NaOH.

Sodium hydroxide (10 mmol/l):

add *slowly* 4 g NaOH to about 6 ml water. When solution has cooled, add water to give 10 ml final solution. The volume of this preparation can be scaled up for convenience in preparation, since it can be kept at room temperature for periods up to 1 year.

Procedure

Collection and treatment of specimen: as described under modified *Briggs* method.

Stability of the enzyme in specimen or sample: as described under modified *Briggs* methods.

Assay conditions: wavelength 331 nm. Other conditions as for modified *Briggs* method.

Measurement: same as for modified *Briggs* assay, except that 0.1 ml NaOH is added instead of 0.1 ml 100% trichloroacetic acid (final NaOH concentration is 0.1 mol/l); and 12 mm × 75 mm test tubes are used in place of centrifuge tubes.

Shake tube immediately after adding the NaOH, otherwise readings will be less reproducible. Allow to stand at room temperature for about 30 min, then pour directly into cuvette. Read absorbance at 331 nm.

Modification: TES buffer can be used instead of phosphate buffer. (TES, N-tris(hydroxymethyl)methyl-2-aminoethanesulphonic acid, M_r 229).

TES buffer (pH 7.3, 80 mmol/l):

dissolve 9.16 g in ca. 400 ml water; bring pH to 7.3 with concentrated NaOH or KOH. Add water to make 500 ml. The apparent K_m for pyridoxal phosphate is an order of magnitude lower in this* and other non-phosphate buffers [25].

---

* *Diamondstone, T. I.*, unpublished observations.

Calculation: use calculation formula (k) or (k_1), respectively (cf. "Formulae", Appendix 3). The absorption coefficient for 4-hydroxybenzaldehyde under the conditions of the assay is 1.99 l × $mmol^{-1}$ × mm^{-1}. The catalytic concentration of the sample is given by

$$b = \frac{\Delta A_{331}}{\Delta t} \times \frac{3.3 \times 1.03 \times 1000}{1.99 \times 10 \times 0.2} = 854 \times \frac{\Delta A_{331}}{\Delta t} \quad \text{U/l}$$

$$b = 14233 \times \frac{\Delta A_{331}}{\Delta t} \quad \text{nkat/l}$$

where the factor of 1.03 is the ratio of the volume of the reaction mixture after addition of NaOH to that before addition (3.3/3.2 = 1.03).

Validation of Method

Precision, accuracy, detection limits and sensitivity: for this assay, readings in the range of A_{331} = 0.080 – 0.500 are best, although higher readings are acceptable owing to the much lower detection limit of this assay as compared to the modified *Briggs* assay. In the range given, the final amount of 4-hydroxyphenylpyruvate produced per ml of enzyme solution is between about 0.07 and 0.43 μmol/l, about an 18-fold lower concentration of product than for the same absorbancy readings in the modified *Briggs* assay. Again, lower readings result in higher variability (> 5%), but readings up to about A_{331} = 0.900 can be used. Linearity should be checked initially, as described above (p. 462). A measure of the increase in sensitivity can be obtained as follows: A_{850} = 0.200 in the modified *Briggs* assay corresponds to about 3% utilization of substrate (p. 462), whereas A_{331} = 0.200 in the *Diamondstone* assay corresponds to a 4-hydroxyphenylpyruvate concentration (as 4-hydroxybenzaldehyde) of 0.011 mmol/l, about 0.16% utilization.

Sources of error: the figure 1.99 l × $mmol^{-1}$ × mm^{-1} for the absorption coefficient of 4-hydroxybenzaldehyde under the conditions of the assay is derived from the observation that the conversion of 4-hydroxyphenylpyruvate to 4-hydroxybenzaldehyde by the addition of NaOH is about 80% complete; the actual absorption coefficient of 4-hydroxybenzaldehyde in NaOH, 0.1 mol/l, is 2.49 l × $mmol^{-1}$ × mm^{-1} [13]. Therefore, it is suggested that the extent of conversion of 4-hydroxyphenylpyruvate to 4-hydroxybenzaldehyde be checked under actual conditions of use, so that an appropriate absorption coefficient can be calculated. Also, note the caution about immediate shaking under "Procedure".

Reference ranges: the amount of enzyme activity content in normal rat liver is in the range of about 1 – 4 U/g wet weight. Induction of the enzyme produces values about 5 to 10 times higher.

References

[1] *F. Cacioppo, C. Nicotra, G. Tesoriere,* Tyrosine-2-oxoglutaric and Glutamic-4-hydroxyphenylpyruvic Transaminase in Rat Organs, Boll. Soc. Ital. Biol. Sper. *40*, 745 – 748 (1964).

[2] *R. S. Rivlin, C. S. Hollander, S. P. Asper Jr.,* Activity of Tyrosine Transaminase in the Thyroid Gland, Endocrinology *71*, 636 – 638 (1962).

[3] *S. Friedman,* Tyrosine Synthesis in a *Pseudomonas* Species, J. Bacteriol. *96*, 276 – 277 (1968).

[4] *S. Sentheshanmuganathan,* The Purification and Properties of the Tyrosine Aminotransferase of *Saccharomyces Cerevisiae*, Biochem. J. *77*, 619 – 625 (1960).

[5] *C. Mavrides, A. D'Iorio,* The Regulation of Tyrosine Aminotransferase in *Tetrahymena Pyriformis,* Biochem. Biophys. Res. Commun. *35*, 467 – 473 (1969).

[6] *P. Gadal, H. Bouyssou, J. P. Barthe,* Biosynthesis of Aromatic Compounds in Higher Plants. Purification and Properties of the L-Phenylalanine-2-oxoglutarate Aminotransferase from the Leaves of *Quercus Pedunculata,* Physiol. Veg. *1969*, 69 – 80 (1969).

[7] *R. T. Schimke, D. Doyle,* Control of Enzyme Levels in Animal Tissues, Annu. Rev. Biochem. *39*, 929 – 976 (1970).

[8] *G. Litwack, J. E. Miller,* Purification, Properties, and Identity of Liver Mitochondrial Tyrosine Aminotranferase, J. Biol. Chem. *246*, 3234 – 4240 (1971).

[9] *E. C. C. Lin, B. M. Pitt, M. Civen, W. E. Knox,* The Assay of Aromatic Amino Acid Transaminases and Oxo Acid Oxidation by the Enol-borate Tautomerase Method, J. Biol. Chem. *233*, 668 – 673 (1959).

[10] *D. DiCola, G. Federici,* NADH-coupled Spectrophotometric Assay of Tyrosine Aminotransferase, Ital. J. Biochem. *30*, 30 – 39 (1981).

[11] *A. P. Briggs,* A Colourimetric Method for the Determination of Homogentisic Acid in Urine, J. Biol. Chem. *51*, 453 – 454 (1922).

[12] *Z. N. Canellakis, P. P. Cohen,* Purification Studies of Tyrosine-2-oxoglutaric Acid Transaminase, J. Biol. Chem. *223*, 53 – 162 (1956).

[13] *T. I. Diamondstone,* Assay of Tyrosine Transaminase Assay by Conversion of 4-Hydroxyphenylpyruvate to 4-Hydroxybenzaldehyde, Anal. Biochem. *16*, 395 – 401 (1966).

[14] *A. Weinstein, G. Medes, G. Litwack,* Isotopic Method for Tyrosine Transaminase Activity, Anal. Biochem. *21*, 86 – 97 (1967).

[15] *J. Wurtman, F. Larin,* A Sensitive and Specific Isotopic Assay for the Estimation of Tyrosine Transaminase, Biochem. Pharmacol. *17*, 817 – 818 (1968).

[16] *G. Litwack, J. M. Squires,* Use of 2-^{3}H-D,L-Tyrosine in a Rapid and Sensitive Isotopic Assay for Tyrosine Aminotransferase, Anal. Biochem. *24*, 438 – 442 (1968).

[17] *J. V. Miller, E. B. Thompson,* Radioactive Assay for Tyrosine Aminotransferase, Anal. Biochem. *47*, 487 – 494 (1972).

[18] *E. A. Lane, C. Mavrides,* New Assay for Tyrosine Aminotransferase with the Amino Acid Analyzer, Anal. Biochem. *44*, 106 – 110 (1971).

[19] *Z. N. Canellakis, P. P. Cohen,* Kinetic and Substrate Specificity of Tyrosine-2-oxoglutaric Acid Transaminase, J. Biol. Chem. *222*, 63 – 71 (1956).

[20] *J. S. Rosenberg, G. Litwack,* Liver Cytosol Tyrosine Aminotransferase: Product Inhibition and Interaction with Substrates, J. Biol. Chem. *245*, 5677 – 5684 (1970).

[21] *G. A. Jacoby, B. N. LaDu,* Studies on the Specificity of Tyrosine-2-oxoglutarate Transaminase, J. Biol. Chem. *239*, 419 – 424 (1964).

[22] *C. Y. Fiske, Y. Subbarow,* The Colourimetric Determination of Phosphorus, J. Biol. Chem. *66*, 375 – 400 (1925).

[23] *G. Litwack, N. Winicov, J. M. Squires,* The Effects of pH an the Rat Liver Tyrosine Aminotransferase, Biochim. Biophys. Acta *129*, 404 – 406 (1966).

[24] *G. J. van Kolmeschate, R. W. Stern,* Preparation of Sodium Diethyldithiocarbamate, Chem. Weekbl. *45*, 733 – 734 (1949).

[25] *D. K. Granner, G. M. Tomkins,* Tyrosine Aminotransferase (Rat Liver), in: *S. P. Colowick, N. O. Kaplan* (eds.), Methods in Enzymology, Vol. XVII A, Academic Press, New York 1970, pp. 633 – 637.

7 Phosphotransferases

7.1 Thymidine Kinase

ATP: thymidine 5′-phosphotransferase, EC 2.7.1.21

Hansjörg Sauer and Wolfgang Wilmanns

General

Thymidine kinase (TK) is a key enzyme in the generation of substrates for DNA-synthesis. Using pre-formed deoxythymidine (dT) for phosphorylation to deoxythymidine monophosphate (dTMP), deoxythymidine diphosphate (dTDP) and deoxythymidine triphosphate (dTTP), it represents the so-called salvage pathway (in contrast to the *de novo* pathway *via* thymidylate synthase; see figure 1, chapter 4.1, page 304). High activities of TK are found in rapidly proliferating cells (log-phase-enzyme) [1 – 6]. The enzyme activity is not detectable under normal conditions in extracellular fluids and blood plasma, but, in animals and human subjects with neoplastic diseases, significant amounts can be measured [7]. Isoenzymes have been described in tumour cells [8] as well as a mitochondrial-specific enzyme type [9]. TK deficiency and its genetic manipulation with 5-azacytidine has been described for the *herpes simplex* gene in mouse cells and for chinese hamster cells [10, 11].

TK is inhibited physiologically by dTTP (feedback inhibition) [12]. Other inhibiting nucleoside derivatives have been described [13 – 15]. A special TK type from *herpes* viruses very actively phosphorylates acycloguanosine (Acyclovir), which then inhibits viral DNA-polymerase much more effectively than cellular DNA-polymerase. Therefore, Acyclovir is used as a selective virostatic drug in clinical medicine [16]. Mutant-viruses defective in TK are resistant to the drug [17, 18].

Application of method: in biochemistry, for clinical diagnostic procedures (pernicious anaemia, leukaemias, characterization of proliferating and non-proliferating cell populations).

Enzyme properties relevant in analysis: the enzyme activity is present in the cytosol of cells which are active in DNA synthesis (viruses, bacteria, eukaryotic cells). Low activities are found in the blood plasma of animals and patients with malignant diseases. Substrate conversion proceeds linearly with time for at least 3 hours in vitro.

Method of determination: tritium-labelled thymidine is used as a substrate. After phosphorylation the radioactive phosphate, dTMP (and also dTDP and dTTP), are separated from the substrate by ion exchange on DEAE-cellulose paper.

International reference methods and standards: not available.

Enzyme effectors: the enzymatic phosphorylation depends on energy (ATP) and magnesium ions.

Assay

Method Design

Principle

$$\text{[Methyl-}^3\text{H]deoxythymidine} + \text{ATP} \xrightarrow[\text{Mg}^{2+}]{\text{TK}} \text{[methyl-}^3\text{H] dTMP* + ADP.}$$

The radioactive product can easily be separated from the radioactive substrate by ion exchange.

Optimized conditions for measurement: for cytosolic enzyme from mammalian sources as described below. The assay is dependent on ATP and magnesium ions. Optimal reaction conditions are in Tris buffer pH 8.0 at 37 °C.

Temperature conversion factors: not tested, not available from the literature.

Equipment

Tissue homogenizers, water-bath, stopwatch, centrifuges, facilities to process radioactive material [^{3}H], scintillation counter.

Reagents and Solutions

Purity of reagents: all reagents must be of analytical purity.

Preparation of solutions: all solutions in re-purified water (cf. Vol. II, chapter 2.1.3.2).

---

* Abbreviations:

dT	doexythymidine
[^{3}H]dT	[methyl-^{3}H]deoxythymidine
dTMP	deoxythymidine 5′-monophosphate
dTDP	deoxythymidine 5′-diphosphate
dTTP	deoxythimidine 5′-triphosphate
TK	thymidine kinase

1. Tris buffer (1.0 mol/l, pH 8.0):
   dissolve 12.114 g Tris-base in 50 ml water, adjust pH to 8.0 with HCl, 2 mol/l, then add water to 100 ml.

2. EDTA (0.1 mol/l):
   dissolve 43 mg EDTA-MgK to 1 ml with water.

3. Magnesium chloride ($MgCl_2$, 0.2 mol/l):
   dissolve 40.7 mg $MgCl_2 \cdot 6\,H_2O$ to 1 ml with water.

4. Adenosine 5′-triphosphate (ATP, 45 mmol/l):
   dissolve 110 mg ATP-$Na_2H_2 \cdot 3\,H_2O$ in 3.8 ml water, neutralize with ca. 0.06 ml KOH, 5 mol/l, add water to 4 ml. Store aliquots of 0.5 ml at −20 °C.

5. Substrate solution ([methyl-^{3}H]thymidine, 1 mmol/l, 10 mCi/l):
   mix 990 µl [^{3}H]dT (2 mCi/mmol, 1 mCi/ml; *Amersham* TRA.310) and 10 µl dT (non-radioactive, *Fluka* 89270, 0.1 mol/l = 21.8 mg in 1 ml water).

6. Control enzyme solution:
   solution (a) as described in chapter 4.1, p. 308.

7. Ammonium formate (1 mmol/l):
   dissolve 63 mg to 1000 ml with water.

8. Absolute ethanol.

9. Scintillation counting solution:
   dissolve 10 g Omnifluor (*NEN Chemicals* Nef 906) in 2.5 l toluene. Store in the dark at 4 °C.

Ion-exchange paper: DEAE-cellulose (diethylaminoethyl-cellulose, (*Whatman* DE 81), cut square pieces of ca. 2 cm^2.

Stability of solutions: solutions (1) – (3), (6) – (8) are stable for several months at 4 °C. Solutions (4) and (5) are stable for several weeks at −20 °C.

Procedure

Collection and treatment of specimen: cytosol preparations from cell suspensions (culture cells, bone marrow cells, leukaemia cells) are prepared according to *Wilmanns* [1] and from solid tissues according to *Sauer* [6] (cf. chapter 4.1).

Stability of the enzyme in the specimen: in tissues, cell suspensions and cytosol stored at −40 °C the enzyme activity is stable for up to two years. At room temperature the total enzyme activity of tissues or cytosol is lost within 24 hours.

Details for measurement in tissues: the method is described for measurement in cell extracts (cytosol).

Assay conditions: incubation volume 0.2 ml; temperature 37 °C; 60 min. A reagent blank with water instead of sample is necessary. Each series is accompanied by controls containing 0.1 ml enzyme control solution (6) from cultured human lymphoblasts with known thymidine kinase activity. A radioactive standard is made from 10 µl of the blank, which is placed on DEAE-cellulose square and dried without any washing.

Prior to measurement prepare the following reaction mixture:

Solution	*for 1 assay*	*for 12 assays (× 13)*
(1)	0.020 ml	0.260 ml
(2), (3)	0.010 ml each	0.130 ml each
(4), (5) and water	0.020 ml each	0.260 ml each

Measurement

Incubation

Pipette successively into reaction vials:	sample	enzyme standard	reagent blank	concentration in incubation mixture
reaction mixture	0.10 ml	0.10 ml	0.10 ml	Tris 100 mmol/l Mg^{2+} 10 mmol/l EDTA 5 mmol/l ATP 4.5 mmol/l dT 0.1 mmol/l (0.01 Ci/l)
sample	0.10 ml			volume fraction 0.5
control enzyme solution (6)	–	0.10 ml	–	
water	–	–	0.10 ml	
incubate for 1 h. Stop reaction by inserting the vials into a boiling water-bath for two min. Centrifuge 2 min at 10000 *g*. Use supernatant for scintillation countings.				

Radiometric measurement: place a piece of DEAE-cellulose paper in each counting vial. Pipette 20 µl of incubation mixture onto the paper. Wash the paper twice for 10 min with solution (7), once for 5 min with water and once for 5 min with ethanol (8). Dry for 20 min at 80 °C; chill and add 5 ml of the scintillation counting fluid (9). Estimate the radioactivity adsorbed to the DEAE-cellulose ([^{3}H]dTMP) in a scintillation counter.

Calculations: no correction for counting efficiency according to eqn. (z), chapter 3.9.3, Vol. I, is necessary, since a radioactive standard is used. Therefore, the following simplified equation is valid for calculation of the catalytic concentration of the sample (cytosol). The counting rate of the sample (and of the enzyme control) must be corrected for the blank counting rate, yielding $\Delta(C_n)_{\text{sample}}$. The counting rate of the radioactive standard, C_{standard}, corresponds to 1 nmol [^{3}H]dTMP. Therefore,

$$b = \frac{\Delta(C_n)_{\text{sample}} \times 0.200}{C_{\text{standard}} \times 0.5 \times 0.02 \times 60} = 0.333 \times \frac{\Delta(C_n)_{\text{sample}}}{C_{\text{standard}}} \ \text{U/l}\,.$$

To obtain the specific catalytic activity this value has to be divided by the mass concentration of protein, ρ_{protein} (g/l), or cell-equivalent in the reaction vial.

Validation of Method

Precision, accuracy, detection limits and sensitivity: the precision and accuracy of the test system are very good. The results of duplicate tests differ by les than ± 5%. The detection limits are about 0.1 nmol/min (mU) per extract of 10^{10} cells from suspensions, or per mg protein of crude extracts from tissues. Even in this region of low enzyme activity the sensitivity is good because of the specific measurement of the reaction. Under standard conditions, as described, at least 95% of the radioactivity adsorbed to DEAE-cellulose represents dUMP [4]. Traces of dTDP and dTTP account for the remaining 5%.

Sources of error: in crude extracts high enzymatic activities which decompose dTMP can lead to falsely low activities of thymidine kinase.

Specificity: the reaction is very specific. Besides dTMP, only traces of dTDP and dTTP are synthesized.

Reference ranges [1, 19]

Normal human leukocytes	< 1 mU/10^{10} cells
Normal human bone marrow	10 ± 5 mU/10^{10} cells
Human acute leukaemia (peripheral cells)	10 ± 7.5 mU/10^{10} cells
Human acute leukaemia (bone marrow cells)	41 ± 25 mU/10^{10} cells
Human pernicious anaemia (bone marrow)	126 ± 42 mU/10^{10} cells

References

[1] *W. Wilmanns,* Die Thymidin-Kinase in normalen und leukämischen myeloischen Zellen, Klin. Wochenschr. *45*, 505 – 511 (1967).

[2] *S. Schollenberger, D. Taurek, W. Wilmanns,* Enzyme des Thymidin- und Thymidylat-Stoffwechsels in normalen und pathologischen Zellen des Blutes und des Knochenmarks, Blut *25*, 318 – 334 (1972).

[3] *H. A. Hopkins, H. A. Campbell, B. Barbiroli, V. R. Potter,* Thymidine Kinase and Deoxyribonucleic Acid Metabolism in Growing and Regenerating Livers from Rats on Controlled Feeding Schedules, Biochem. J. *136*, 955 – 966 (1973).

[4] *I. Olsen, G. Harris,* Thymidine Kinase of Mouse Spleen Cells *in vivo* and *in vitro*, Biochem. J. *146*, 489 – 496 (1975).

[5] *P. C. Gilette, W. C. Claycomb,* Thymidine Kinase Activity in Cardiac Muscle During Embryonic and Postnatal Development, Biochem. J. *142*, 685 – 690 (1974).

[6] *H. Sauer, K. Wilms, W. Wilmanns, L. Jaenicke,* Die Aktivität der Methionin-Synthetase (5-Methyl-5,6,7,8-Tetrahydrofolsäure: Homocystein Methyltransferase) als Proliferationsparameter in wachsenden Zellen, Acta Haematol. *48*, 200 – 210 (1973).

[7] *W. Kreis, Z. Arlin, A. Yagoda, B. Leyland-Jones, L. Fiori,* Deoxycytidine and Deoxythymidine Kinase Activities in Plasma of Mice and Patients with Neoplastic Disease. Cancer Res. *42*, 2514 – 2517 (1982).

[8] *P. H. Ellmis, T. E. Gan, G. Medley, M. B. Van Der Weiden,* Prognostic Relevance of Thymidine Kinase Isoenzymes in Adult non-Hodgkin's Lymphoma, Blood *58*, 926 – 930 (1981).

[9] *A. J. Berk, B. J. Meyer, D. A. Clayton,* Mitochondrial-specific Thymidine Kinase, Arch. Biochem. Biophys. *154*, 563 – 565 (1973).

[10] *D. W. Clough, L. M. Kunkel, R. L. Davidson,* 5-Azacytidine-induced Reactivation of a Herpes Simplex Thymidine Kinase Gene, Science *216*, 70 – 73 (1982).

[11] *M. Harris,* Induction of Thymidine Kinase in Enzyme-deficient Chinese Hamster Cells, Cell *29*, 483 – 492 (1982).

[12] *W. Wilmanns,* Kompetitive Enzymmechanismen bei der Therapie von Hämoblastosen mit Antimetaboliten. Verh. Dtsch. Ges. Inn. Med. *74*, 822 – 830 (1968).

[13] *A. Hampton, R. R. Chawla, F. Kappler,* Species- or Isoenzyme-specific Enzyme Inhibitors. 5. Differential Effects of Thymidine Substituents on Affinity for Rat Thymidine Kinase Isoenzymes, J. Med. Chem. *25*, 644 – 649 (1982).

[14] *A. Hampton, T. T. Hai, F. Kappler, R. R. Chawla,* Species- or Isoenzyme-specific Enzyme Inhibitors. 6. Synthesis and Evaluation of Two-substrate Condensation Products as Inhibitors of Hexokinases and Thymidine Kinases, J. Med. Chem. *25*, 801 – 805 (1982).

[15] *T. Hamazaki,* Effect of Adenosine-3′,5′-monophosphate on Thymidine Kinase in Tumor Cells. GANN *64*, 219 – 226 (1973).

[16] Editorial: Acyclovir, Lancet *1981/II*, 845 – 846 (1981).

[17] *C. S. Crumpacker, L. E. Schnipper, S. I. Marlowe, P. N. Kowalsky, B. J. Hershey, M. J. Levin,* Resistance to Antiviral Drugs of Herpes Simplex Virus Isolated from a Patient Treated with Acyclovir, N. Engl. J. Med. *306*, 343 – 346 (1982).

[18] *W. H. Burns, R. Saral, G. W. Santos, O. L. Laskin, P. S. Leitman, C. McLaren, D. W. Barry,* Isolation and Characterisation of Resistant Herpes Simplex Virus after Acyclovir Therapy, Lancet *1982/I*, 421 – 423 (1982).

[19] *W. Wilmanns, V. Neef,* Die Thymidylat-Synthetase in weißen Blutzellen und im Knochenmark unter normalen und pathologischen Bedingungen, Klin. Wochenschr. *49*, 755 – 762 (1972).

7.2 Protein Kinases

ATP: protein phosphotransferase, EC 2.7.1.37

7.2.1 Cyclic Nucleotide and Ca^{2+}-independent Protein Kinases

D. Grahame Hardie

General

Protein kinases are a heterogeneous group of enzymes of almost ubiquitous occurrence which are involved in cellular regulation [1, 2]. They have been generally divided into three classes according to their mode of regulation [1]:

a) cyclic nucleotide-dependent protein kinases,
b) Ca^{2+}-dependent protein kinases,
c) non-specified protein kinases.

Class (a) is discussed in chapter 7.2.2; Class (b) in chapter 7.2.3 and Class (c) in this chapter. Since the last is a particularly heterogeneous group, only general guidelines can be given here concerning methods of measurement.

Protein kinases in Class (c) which are reasonably well characterized include casein kinases I and II from rabbit reticulocytes [3] and glycogen synthase kinase-3 from rabbit skeletal muscle [4]. Most of the well-characterized protein kinases phosphorylate serine or threonine residues on their protein substrates. A new class of protein kinases which has recently emerged are those which phosphorylate tyrosine residues [5]. While the methods described in this chapter should be applicable to these protein kinases also, they are not considered in detail here.

Application of method: in biochemistry.

Enzyme properties relevant in analysis: the catalytic activity against a given substrate protein in a cell extract may be a function of several different protein kinases. Purification of the protein kinases will be necessary to examine this possibility. The assay normally proceeds without a lag phase and is linear with time as long as the degree of phosphorylation of the protein substrate is low. (See also below under "Sources of error".)

Method of determination: the catalytic activity is measured in a fixed-time assay by analyzing the incorporation of radioactivity from [γ-^{32}P]ATP into trichloroacetic acid-insoluble protein. Two methods for the collection, washing and counting of the acid precipitate are described below.

International reference methods and standards: none is known.

Enzyme effectors: this possibility must be examined in each individual case. Casein kinase II is inhibited by heparin [6].

Assay

Method Design

Principle

$$[\gamma\text{-}^{32}\text{P}]\text{ATP} + \text{protein} \rightarrow [^{32}\text{P}]\text{phosphoprotein} + \text{ADP}$$

The reaction is stopped by adding trichloroacetic acid to inactivate the protein kinase and precipitate the [^{32}P]phosphoprotein. The [^{32}P]phosphoprotein is washed to remove unreacted [γ-^{32}P]ATP and the radioactivity determined by scintillation counting. Up to three reaction blanks should be performed as discussed below.

Optimized conditions for measurement: the optimal pH and ionic strength must be established empirically in each case. The ATP concentration suggested below (1 mmol/l) should be suitable for most protein kinases: however, it is expensive in radioisotope and many protein kinases which have low K_m values for ATP can be assayed at lower concentrations. The actual substrate for the reaction is Mg-ATP: therefore, Mg^{2+} ions must be included at a concentration in excess of the total concentration of ATP plus any chelating agents present. The protein substrate concentration suggested below (20 mmol/l) is ideal, but such a high concentration will be impractical for many proteins. In such cases the low substrate concentration should be compensated by using ATP of higher specific radioactivity, although this may lead to higher incorporation in reaction blanks.

Temperature conversion factors: must be established empirically in each case.

Equipment

Water-bath, stopwatch, micro-centrifuge, scintillation counter, spectrophotometer (for standardizing ATP solution).

Reagents and Solutions

Purity of reagents: the protein substrate should ideally be free from contaminating protein kinases, although small amounts of contamination can be corrected for by the reaction blanks described below.

Preparation of solutions: all solutions in re-purified water (cf. Vol. II, chapter 2.1.3.2).

1. Protein substrate (50 µmol/l) in suitable buffer:

   the pH, ionic strength and buffer ions used must be determined empirically in each case so that the conditions are optimal for storage of the protein substrate and for the phosphorylation reaction.

2. Buffer solution in which the protein substrate is dissolved (for control reactions).

3. EGTA* (1 mmol/l):

   dissolve 38 mg EGTA (free acid, neutralized with NaOH, 1 mol/l, to pH 7.0) in 100 ml water.

4. Buffer solution in which the protein kinase sample is dissolved (for control reactions).

5. ATP (100 mmol/l):

   dissolve 1.21 g ATP (free acid) in about 10 ml water. Bring to pH 7.0 with NaOH, 1 mol/l, and dilute to 20 ml.

6. Magnesium acetate (100 mmol/l):

   dissolve 2.14 g magnesium acetate (hydrated) in 100 ml water.

7. [γ-^{32}P]ATP/magnesium acetate (ATP, 5 mmol/l, $2-5 \times 10^4$ dpm/nmol; Mg^{2+}, 15 mmol/l):

   add 10 µl solution (5) and 30 µl solution (6) to [γ-^{32}P]ATP ($2-5 \times 10^7$ dpm). Dilute to 0.2 ml with water. The chemical content of the [γ-^{32}P]ATP obtained from the supplier is normally negligible. Note that ATP solutions should always be neutralized before adding Mg^{2+} ions. Attempts to neutralize a mixture of Mg^{2+} and the acid form of ATP will result in the formation of insoluble Mg-ATP complexes.

8. Trichloroacetic acid (25%, w/v):

   dissolve 250 ml trichloroacetic acid in 1 litre water.

* EGTA ethyleneglycol-bis-(β-aminoethyl ether) N,N^1-tetraacetic acid.

9. Bovine serum albumin (10 mg/ml):

   dissolve 100 mg BSA in 10 ml water.

10. Trichloroacetic acid (5%, w/v):

    dissolve 500 g trichloroacetic acid in 10 l water.

Stability of solutions: store solutions (5) and (7) at $-20\,^{\circ}C$. Solution (1) may be stored at either $0-4\,^{\circ}C$ of $-20\,^{\circ}C$ depending on the protein used. Store all other solutions at $0-4\,^{\circ}C$. It may be worth adding an inhibitor of bacterial growth [e.g. sodium azide 0.02% (w/v)] to solution (1). The ^{32}P radioactivity in solution (7) will decay with a half-life of 14 days.

Procedure

Collection and treatment of specimens: will depend on the protein kinase and the tissue. A typical extraction might involve the homogenization of minced tissue in 2.5 ml sucrose, 0.25 mol/l, EDTA, 1 mmol/l, 2-mercaptoethanol, 0.1% (v/v), pH 7.0, per g tissue, followed by centrifugation to separate subcellular fractions. See below under "Details for measurement in tissues" for problems encountered in assaying protein kinases in crude extracts.

Stability of the enzyme in the specimen: will depend on the protein kinase under study.

Details for measurement in tissues: measurement of protein kinase activity in crude tissue extracts is difficult because of three major problems:

(a) the extract may contain other proteins which may be phosphorylated in competition with the protein substrate added.
(b) the extract may contain protein phosphatase activities which will reverse the phosphorylation reaction.
(c) the extract may contain enzymes which hydrolyze ATP.

Problem (a) can be corrected for in favourable cases by performing the blank reactions described below.

Problems (b) and (c) may be diagnosed by conducting a time course. If the incorporation of radioactivity reaches a plateau, and then more radioactivity is incorporated on re-addition of Mg[γ-^{32}P]ATP, enzymes hydrolyzing ATP are present. It may be possible to find an inhibitor of these: if not, keep the reaction period short and use high concentrations of ATP.

If the incorporation of radioactivity rises and then falls, protein phosphatases are present. Most, but not all, protein phosphatases can be inactivated by pre-incubation of the extract (30 min, $30\,^{\circ}C$) with NaF, 50 mmol/l, and EDTA, 1 mmol/l.

Assay conditions: final volume 0.05 ml; 30 °C is suitable for most protein kinases. The reaction is carried out in a 1.5 ml polypropylene micro-centrifuge tube. Pre-warm assay components to the reaction temperature.

Measurement: in addition to the assay of the sample itself, up to three control reactions should be carried out as indicated below.

Incubation

Pipette successively into the reaction tube:	sample	control 1	control 2	control 3	concentration in incubation mixture
protein solution (1)	20 µl		20 µ		protein 20 µmol/l
protein buffer (2)		20 µl		20 µl	variable
EGTA solution (3)	10 µl	10 µl	10 µl	10 µl	EGTA 0.1 mmol/l
sample	10 µl	10 µl			
sample buffer (4)			10 µl	10 µl	variable
mix on a vortex mixer and pre-incubate to assay temperature (30 °C),					
[γ-^{32}P]ATP/Mg solution (7)	10 µl	10 µl	10 µl	10 µl	ATP 1 mmol/l Mg^{2+} 3 mmol/l
mix by vortexing. After a suitable incubation time, stop the reaction by one of the methods below.					

Method 1: add 1 ml of 25% TCA (solution 8) to the reaction tube. Add 50 µl of BSA (solution 9), vortex, and keep tube on ice for 10 min to allow complete precipitation. Centrifuge for 2 min on a micro-centrifuge. Remove supernatant with *Pasteur* pipette. Wash pellet three times by adding 1 ml of cold 25% TCA, vortex-mixing, centrifuging and removing supernatant. Determine the radioactivity remaining in the pellet by *Cerenkov* counting (the micro-centrifuge tube is placed in a scintillation vial and counted without added scintillant using a tritium channel).

Alternatively, add 0.2 ml of 90% formic acid to dissolve the pellet and count in the micro-centrifuge tube using a ^{32}P channel after addition of 1 ml of a scintillation fluid which is compatible with aqueous samples (e.g. Unisolve). This approximately doubles the efficiency of counting when compared with *Cerenkov* counting.

Method 2: stop the reaction by pipetting 40 µl of the mixture onto a 1.5 cm square of paper (*Whatman* ET31) and dropping the paper into 1 l of cold 5% TCA (solution 10). Up to 50 paper squares (numbered in pencil) can be processed simultaneously. Stir the papers gently on a magnetic stirrer for at least 5 min and then carefully pour off the TCA. Wash the papers three times by adding 1 l of cold TCA (solution 10), stirring for 5 min and pouring off the TCA solution. Finally wash in 100 ml of acetone, air dry, and count the papers in 10 ml of toluene-based scintillation fluid using a ^{32}P channel. The scintillation fluid can often be re-used after removal of the paper square.

Comparison of the two methods: method 2 is much more convenient for processing large numbers of samples. However the efficiency of counting on the paper squares cannot be determined with certainty and method 1 is more accurate if absolute activity (i.e. moles of phosphate incorporated) is required.

Calculation: control reaction 1 (sample blank) should always be carried out. Control reactions 2 and 3 (substrate blank and reagent blank) are only necessary if the protein substrate contains contaminating protein kinases.

The various blanks have to be accounted for as follows:

a) Radioactivity incorporated into substrate in presence of added protein kinase

$$C = C_{\text{sample}} - C_{\text{sample blank}} \quad \text{cpm}.$$

b) Radioactivity incorporated into substrate in absence of added protein kinase

$$C_{\text{substrate}} = C_{\text{substrate blank}} - C_{\text{reagent blank}} \quad \text{cpm}.$$

c) Radioactivity incorporated into substrate due to added protein kinase

$$C_{\text{protein kinase}} = C - C_{\text{substrate}} \quad \text{cpm}.$$

The catalytic activity concentration of the sample added is

$$b = \frac{C_{\text{protein kinase}}}{X \times \Delta t} \quad \text{U/10 µl}$$

$$= \frac{C_{\text{protein kinase}} \times 10^5}{X \times \Delta t} \quad \text{U/l}$$

X specific radioactivity of the ATP in cpm/µmol
Δt incubation time in minutes

To determine this specific radioactivity, dilute an aliquot of solution (7) one hundred-fold in water and read its absorbance at 260 nm (for ATP, 5 mmol/l, diluted 1 in 100 the absorbance should be 0.75). Count a 20 μl aliquot of this 1 in 100 diluted solution in the same manner as the experimental samples (i.e. in a micro-centrifuge tube with or without formic acid and scintillation fluid for method 1; on a paper square in scintillation fluid for method 2).

$$X = \frac{(C_{n(\text{recorded})}) \times 50 \times 15}{(A_{260(\text{recorded})})} \quad \text{cpm}/\mu\text{mol}$$

C_n net counting rate (cpm)

If the ATP standard is always counted at the same time as the assay samples, decay of ^{32}P is corrected for automatically. The assumption is made that the efficiencies of counting of the ATP standard and of the assay samples are identical. This is certainly valid for method 1.

Note that for method 2 the activity value should be multiplied by $\frac{5}{4}$ since only 40 μl of the 50 μl reaction is sampled.

Validation of Method

Precision, accuracy, detection limits and sensitivity: no precise data can be given as many factors can affect these parameters (e.g. the substrate concentrations used, the presence of protein substrates in the protein kinase sample).

Sources of error: the presence of contaminating protein phosphatases and ATPases will lead to reaction rates which are non-linear (see above: "Details for measurement in tissues"). The reaction rate may also be non-linear if the degree of phosphorylation of the substrate becomes significant. These possibilities should be checked by conducting time courses.

Specificity: the assay described will detect any protein kinase which phosphorylates the protein substrate, except that Ca^{2+}-dependent protein kinases will be inhibited by the EGTA present. Interference from the catalytic subunit of cyclic AMP-dependent protein kinase (which is active in the absence of cyclic AMP) can be eliminated by titrating out this activity with the heat-stable inhibitor protein first described by *Walsh et al.* [7].

Reference ranges: not applicable.

References

[1] *E. G. Krebs, J. A. Beavo,* Phosphorylation-Dephosphorylation of Proteins, Annu. Rev. Biochem. *48*, 923–959 (1979).

[2] *P. Cohen,* The Role of Protein Phosphorylation in Neural and Hormonal Control of Cellular Activity, Nature *296*, 613–620 (1982).

[3] *G. M. Hathaway, J. A. Traugh,* Cyclic Nucleotide-independent Protein Kinases from Rabbit Reticulocytes, J. Biol. Chem. *254*, 762–768 (1979).
[4] *P. Cohen, D. Yellowlees, A. Aitken, A. Donella-Deana, B. A. Hemmings, P. J. Parker,* Separation and Characterization of Glycogen Synthase Kinase 3, Glycogen Synthase Kinase 4 and Glycogen Synthase Kinase 5 from Rabbit Skeletal Muscle, Eur. J. Biochem. *124*, 21–35 (1982).
[5] *T. Hunter, B. M. Sefton, J. A. Cooper,* Phosphorylation of Tyrosine: Its Importance in Viral Transformation and Normal Cell Metabolism, in: *O. M. Rosen and E. G. Krebs* (eds.), Protein Phosphorylation, Cold Spring Harbor Laboratory 1981, pp. 1189–1202.
[6] *G. M. Hathaway, T. H. Lubben, J. A. Traugh,* Inhibition of Casein Kinase II by Heparin, J. Biol. Chem. *255*, 8038–8041 (1980).
[7] *D. A. Walsh, C. D. Ashby, C. Gonzalez, D. Calkins, E. H. Fischer, E. G. Krebs,* Purification and Characterization of a Protein Inhibitor of Adenosine 3',5'-monophosphate-dependent Protein Kinases, J. Biol. Chem. *246*, 1977–1985 (1971).

7.2.2 Cyclic Nucleotide-dependent Protein Kinase

D. Grahame Hardie

General

Cyclic AMP-dependent protein kinase was discovered by *Krebs* and coworkers in 1968 [1], while cyclic GMP-dependent protein kinase was first described by *Kuo & Greengard* in 1970 [2]. These cyclic nucleotide-dependent protein kinases are widely distributed in all tissues in eukaryotes [3, 4]. The cAMP-dependent protein kinase is the only known intracellular receptor for cyclic AMP, and has a vital role in mediating the effects of hormones and other extracellular signals [5]. The cAMP-dependent enzyme is predominantly located in the cytoplasm, although it has been reported to associate with particulate fractions [6]. The functions of the cGMP-dependent enzyme are still not clear. This enzyme is also largely cytoplasmic although it may associate with the plasma membrane [7]. cAMP-dependent protein kinase consists of catalytic (C) subunits and regulatory (R) subunits, with the (inactive) holo-enzyme being R_2C_2. On addition of cyclic AMP, the R dimers bind cyclic AMP and the (active) C subunits dissociate. cGMP-dependent protein kinase appears to be structurally very similar to the cAMP-dependent enzyme [8], although the R and C domains are fused on a single polypeptide and there is no dissociation on activation by cyclic GMP.

Abbreviations:
cAMP cyclic AMP
cGMP cyclic GMP
EGTA ethyleneglycol-bis-(β-aminoethyl ether) N,N^1-tetraacetic acid.

Application of method: in biochemistry.

Enzyme properties relevant in analysis: in most tissues cAMP-dependent protein kinase is present as two isoenzymes (Types I and II) which differ in their R subunits. Neural tissues contain a specific Type II isoenzyme with an R subunit that is immunologically distinct from the R subunit of the Type II enzyme of heart [9].

Methods of determination: cAMP-dependent and cGMP-dependent protein kinases are generally assayed by measuring the incorporation of radioactivity from [γ-^{32}P] into histone protein (see below). They have also been detected and/or quantitated using photoaffinity labelling with cyclic nucleotide analogues [10] or radioimmunoassay [9] methods which are beyond the scope of this chapter.

International reference methods and standards: none known.

Enzyme effectors: cAMP-dependent and cGMP-dependent protein kinases are activated by low concentrations of cyclic AMP and cyclic GMP respectively. The cAMP-dependent enzyme is potently inhibited by a small heat-stable protein first described in rabbit skeletal muscle by *Walsh et al.* [11]. The cyclic GMP-dependent enzyme is inhibited by polyarginine and certain arginine-rich proteins [12].

Assay

The basic assay described is for cAMP-dependent protein kinase, with those modifications required for the assay of cGMP-dependent protein kinase being mentioned briefly.

Method Design

Principle

$$\text{Histone} + [\gamma\text{-}^{32}\text{P}]\text{ATP} \rightarrow [^{32}\text{P}]\text{phosphohistone} + \text{ADP}$$

The reaction is stopped by pipetting the mixture onto squares of phosphocellulose paper and washing in cold water. Histones, being very basic proteins, bind tightly to phosphocellulose paper, while [γ-^{32}P]ATP is washed off. The cyclic nucleotide-dependent activity can be measured by carrying out assays with and without added cyclic nucleotide. However, this will not detect enzyme which is already partly activated by endogenous cyclic nucleotides (see below under "Specificity").

Optimized conditions for measurement: cAMP-dependent protein kinase has a broad optimum between pH 7.0 – 7.8. Optimal Mg^{2+} concentration is 10 mmol/l. cGMP-dependent protein kinase has a broad optimum between pH 6 – 8 although dependence on cyclic GMP is reduced above pH 6.5 [13]. The optimal Mg^{2+} concentration depends on the substrate used [13].

Temperature conversion factors: not known.

Equipment

Water-bath, stopwatch, scintillation counter, spectrophotometer (for standardizing ATP solutions).

Reagents and Solutions

Purity of reagents: use analytical grade reagents throughout.

Preparation of solutions: all solutions in re-purified water (cf. Vol. II, chapter 2.1.3.2).

1. Calf thymus mixed histone (*Sigma,* Type IIA)/phosphate buffer (phosphate, 75 mmol/l, pH 7.0):

   dissolve 0.23 g NaH_2PO_4 in about 10 ml water, adjust to pH 7.0 with NaOH, 1 mol/l, add 100 mg mixed histone and stir to dissolve, and dilute to 20 ml with water.

2. EGTA (1 mmol/l):

   dissolve 7.6 mg EGTA (free acid) in about 10 ml water, adjust to pH 7.0 with NaOH, 1 mol/l, and dilute to 20 ml with water.

3. cyclic AMP (100 µmol/l):

   dissolve 34.7 mg cyclic AMP in 1 l water.

4. ATP (100 mmol/l):

   dissolve 1.21 g $ATP\text{-}Na_2H_2 \cdot 3H_2O$ in about 10 ml water, adjust to pH 7.0 with NaOH, 1 mol/l, and dilute to 20 ml with water.

5. Magnesium acetate (100 mmol/l):

   dissolve 2.14 g magnesium acetate (hydrated) in 100 ml water.

6. [γ-^{32}P]ATP/magnesium acetate (ATP 1 mmol/l; $1-2 \times 10^4$ cpm/nmol; Mg^{2+}, 50 mmol/l):

   mix [γ-^{32}P]ATP (1.2×10^7cpm), 10 μl solution (4) and 0.5 ml solution (5), and dilute to 1 ml with water. The [γ-^{32}P]ATP obtained from the radiochemical supplier normally has a negligible chemical content.

7. Sample buffer:

   the preparation of this buffer depends on the kind of specimen (cf. "Collection and treatment of specimens").

Stability of solutions: store solutions (1), (4) and (6) at −20 °C. Store all other solutions at 0−4 °C. The ^{32}P radioactivity in solution (6) will decay with a half-life of 14 days.

Procedure

Collection and treatment of specimens: will depend on the tissue involved. A typical extraction would involve homogenization of minced tissue in 2.5 ml sucrose, 0.25 mol/l, EDTA, 1 mmol/l, 2-mercaptoethanol, 0.1% (v/v), pH 7.0, per g tissue followed by centrifugation to separate subcellular fractions. See "Details for measurements in tissues" in chapter 7.2.1 for a discussion of problems encountered in assaying protein kinases in crude extracts.

Stability of the enzyme in the specimen: will depend on the nature of the sample. The enzymes are stabilized in the presence of β-mercaptoethanol, 0.1% (v/v), or dithiothreitol, 1 mmol/l.

Measurement

Pipette successively into the reaction tube:	blank	sample	concentration in the assay mixture	
histone/buffer solution (1)	20 μl	20 μl	histone phosphate	2 mg/ml 30 mmol/l
EGTA solution (2)	5 μl	5 μl	EGTA	0.1 mmol/l
cyclic AMP solution (3)	–	5 μl	cAMP	10 μmol/l (a)
sample	–	10 μl		
sample buffer (7)	10 μl	–		
mix on a vortex mixer and pre-incubate to 30 °C; start the reaction by adding				
[γ-^{32}P]ATP/Mg^{2+} solution (6)	10 μl	10 μl	ATP Mg^{2+}	0.1 mmol/l 10 mmol/l

Details for measurements in tissues: problems caused by the presence of protein phosphatases and/or ATP hydrolases have already been described in chapter 7.2.1.

Assay conditions: final volume 50 µl; 30 °C. Carry out the reactions in 1.5 ml polypropylene micro-centrifuge tubes.

Incubate for, say, 20 min at 30 °C. Stop the reaction by pipetting 40 µl of the mixture onto a 1.5 cm square of phosphocellulose paper (*Whatman* P81) and dropping it into a beaker containing 1 l of ice-cold water. Up to 50 paper squares (numbered in pencil) may be processed simultaneously. Stir the papers gently for 5 min on a magnetic stirrer and carefully pour off the water. Wash six times by stirring in 1 l of cold water for 1 – 2 min and pouring off the water. Finally, wash in 100 ml of acetone, air-dry and count in 10 ml of toluene-based scintillant. The scintillation fluid can often be re-used after the paper is removed.

Modifications for assay of cGMP-dependent protein kinase

This is identical to the assay for cAMP-dependent protein kinase except for the following modifications (based on the method of *Gill et al.* [13]). The pH should be reduced to 6.5; the magnesium concentration should be raised to 50 mmol/l; cyclic AMP should be replaced by cyclic GMP, 1 µmol/l, and mixed histone should be replaced by purified histone H2B, 1 mg/ml. It has recently been found that arginine-rich histones such as H2B produce a time-dependent inactivation of cGMP-dependent protein kinase [12]. Therefore the reaction should be kept short or, alternatively, another substrate such as histone H1 should be used. If a different substrate is used, the optimal assay conditions should be re-examined.

Calculation

Radioactivity incorporated into substrate

$$C = (C_{\text{sample}} - C_{\text{blank}}) \times 5/4 \quad \text{cpm}$$

The catalytic activity concentration of the sample is

$$b = \frac{C}{X \times \Delta t} \times 5/4 \quad \text{U}/10\,\mu\text{l}$$

$$= \frac{C \times 10^5}{X \times \Delta t} \times 5/4 \quad \text{U/l}$$

X specific radioactivity of the [γ-^{32}P]ATP in cpm/µmol. This is determined by pipetting 20 µl solution (6) onto a square of P81 paper and counting in 10 ml toluene-based scintillation fluid (see chapter 7.2.1 for details of the calculation of specific radioactivity).

Δt incubation time in min.

Validation of Method

Precision, accuracy, detection limits and sensitivity: for enzyme solutions of around 2×10^{-3} U/ml the imprecision should be $< \pm 10\%$. Data about accuracy are not available due to the lack of reference standard material. For the assay as described the lowest possible value measurable with an imprecision $< \pm 20\%$ would be about 2×10^{-4} U/ml. Lower values could be measured if the assay time was extended.

Sources of error: as with nucleotide-independent protein kinases, the presence of protein phosphatases and ATP hydrolases can lead to non-linear reactions. Procedures for dealing with these problems are discussed in chapter 7.2.1. An additional problem encountered when assaying cyclic nucleotide-dependent protein kinases may be the presence of phosphodiesterases which hydrolyze the cyclic nucleotide. These can be inhibited by the inclusion of theophylline, 1 mmol/l, in the assay.

Specificity: if the assays are carried out in the presence and absence of the cyclic nucleotide, then the difference will be a specific measure of the cyclic nucleotide-dependent protein kinase. Although cyclic AMP can activate cGMP-dependent protein kinase, and cyclic GMP can activate cAMP-dependent protein kinase, these effects are minimal at the suggested concentrations of cyclic nucleotide.

Some of the histone kinase activity measured in the absence of cyclic AMP will result from the presence of the free catalytic subunit of the cAMP-dependent protein kinase, and can be measured by carrying out the assays in the presence and absence of the heat-stable inhibitor protein [11] which is a potent and specific inhibitor of this enzyme. Use of this inhibitor will also prevent any interference by this latter enzyme in assays of cGMP-dependent protein kinase.

Reference ranges: using the above assay the activity of cAMP-dependent protein kinase in rat liver is $5 - 15 \times 10^{-3}$ U/g wet weight. Activities in other mammalian tissues are similar. The cGMP-dependent protein kinase varies greatly in activity from tissue to tissue.

References

[1] *D. A. Walsh, J. P. Perkins, E. G. Krebs,* An Adenosine 3',5'-monophosphate-dependent Protein kinase from rabbit skeletal muscle, J. Biol. Chem. *243*, 3763 – 3774 (1968).

[2] *J. F. Kuo, P. Greengard,* Cyclic Nucleotide-dependent Protein Kinase VI. Isolation and Partial Purification of a Protein Kinase Activated by Guanosine 3',5'-monophosphate, J. Biol. Chem. *245*, 2493 – 2498 (1970).

[3] *J. F. Kuo, P. Greengard,* Cyclic Nucleotide-dependent Protein Kinase IV. Widespread Occurrence of Adenosine 3',5'-monophosphate-dependent Protein Kinase in Various Tissues and Phyla of the Animal Kingdom, Proc. Natl. Acad. Sci. USA *64*, 1349 – 1355 (1969).

[4] *G. N. Gill, R. W. McCune,* Guanosine 3',5'-monophosphate-dependent Protein Kinase, Curr. Top. Cell. Regul. *15*, 1 – 45 (1979).

[5] *P. Cohen,* The Role of Protein Phosphorylation in Neural and Hormonal Control of Cellular Activity, Nature *296*, 613 – 620 (1982).
[6] *H. Maeno, E. M. Johnson, P. Greengard,* Subcellular Distribution of Adenosine 3′,5′-monophosphate-dependent Protein Kinase in Rat Brain, J. Biol. Chem. *246*, 134 – 142 (1971).
[7] *H. E. Ives, J. E. Casnellie, P. Greengard, J. D. Jamieson,* Subcellular localization of Cyclic GMP-dependent Protein Kinase and its Substrates in Vascular Smooth Muscle, J. Biol. Chem. *255*, 3777 – 3785 (1980).
[8] *T. M. Lincoln, J. D. Corbin,* Adenosine 3′,5′-monophosphate-dependent and Guanosine 3′,5′-monophosphate-dependent Protein Kinases: Possible Homologous Proteins, Proc. Natl. Acad. Sci. USA *74*, 3239 – 3243 (1977).
[9] *C. S. Rubin, R. Rangel-Aldao, D. Sarkar, J. Erlichman, N. Fleischer,* Characterization and Comparison of Membrane-associated and Cytosolic cAMP-dependent Protein Kinases, J. Biol. Chem. *254*, 3797 – 3805 (1979).
[10] *U. Walter, P. Greengard,* Quantitative labelling of the Regulatory Subunit of Type II cAMP-dependent Protein Kinase from Bovine Heart by a Photoaffinity Analog, J. Cyclic Nucleotide Res. *4*, 437 – 444 (1976).
[11] *D. A Walsh, C. D. Ashby, C. Gonzalez, D. Calkins, E. H. Fischer, E. G. Krebs,* Purification and Characterization of a Protein Inhibitor of Adenosine 3′,5′-monophosphate-dependent Protein Kinases, J. Biol. Chem. *246*, 1977 – 1985 (1971).
[12] *G. M. Walton, G. N. Gill,* Regulation of Cyclic Nucleotide-dependent Protein Kinase Activity by Histone and Poly (L-arginine), J. Biol. Chem. *256*, 1681 – 1688 (1981).
[13] *G. N. Gill, K. E. Holdy, G. M. Walton, C. B. Kanstein,* Purification and Characterization of 3′,5′-cyclic GMP-dependent Protein Kinase, Proc. Natl. Acad. Sci. USA *73*, 3918 – 3922 (1976).

7.2.3 Ca^{2+}- and Calmodulin-dependent Protein Kinases

D. Grahame Hardie and Philip Cohen

General

Phosphorylase kinase was the first protein kinase to be discovered [1]. However it was not until 1978 that the activation of the enzyme by calcium ions was found to be due to the presence in the protein of stoichiometric amounts of the Ca^{2+}-binding protein, calmodulin [2]. Since 1978 numerous other examples of Ca^{2+}- and calmodulin-dependent protein kinases have been reported, the best-characterized being myosin light-chain kinase [3]. Unlike phosphorylase kinase, where the calmodulin is tightly

Abbreviations:
NTA nitrilotriacetate
ANSA amino nitro sulphonic acid
EGTA ethyleneglycol-bis-(β-aminoethyl ether) N,N^1-tetraacetic acid

bound, most calmodulin-dependent protein kinases require the addition of exogenous calmodulin in order to exhibit catalytic activity. Ca^{2+}- and calmodulin-dependent protein kinases are widely distributed in the tissues of higher eukaryotes. They are involved in mediating the neural and hormonal signals that use calcium ions as second messenger.

A new class of Ca^{2+}-dependent protein kinase has recently been discovered that is activated by Ca^{2+}, phospholipid and diacylglycerol [4]. Since the regulation of this protein kinase by calcium does not appear to involve calmodulin it is not considered further in this chapter.

Phosphorylase kinase is of clinical importance, and a hereditary defect in this enzyme is responsible for one of the glycogen storage diseases [5].

Application of method: in biochemistry.

Enzyme properties relevant in analysis: phosphorylase kinase [6] and probably also myosin light-chain kinase [7] exist as different isoenzymes in different tissue types. A single organ, e.g. the brain [8], may contain multiple Ca^{2+}- and calmodulin-dependent protein kinases.

The assays described below should be linear with time as long as the degree of phosphorylation of the substrate is small (see below under "Sources of error").

Methods of determination: in general Ca^{2+}- and calmodulin-dependent protein kinases are assayed by following the incorporation of radioactivity from [γ-^{32}P]ATP into substrate protein. However phosphorylase kinase is assayed more conveniently and economically by measuring the activation of phosphorylase *b* (see below).

International reference methods and standards: not available.

Enzyme effectors: all of these enzymes are activated by Ca^{2+} and exogenous calmodulin (although phosphorylase kinase also contains tightly bound endogenous calmodulin). Phosphorylase kinase is also activated by phosphorylation, catalyzed by cyclic AMP-dependent protein kinase. Myosin light-chain kinase is also phosphorylated by cyclic AMP-dependent protein kinase which increases 20-fold the concentration of calmodulin required for half-maximal activation.

Assay

The assay described below is for phosphorylase kinase. However general guidelines are also given for assay of other Ca^{2+}- and calmodulin-dependent protein kinases.

Method Design

Principle

(a) Phosphorylase b* + ATP $\xrightarrow{\text{PhK}}$ phosphorylase a* + ADP

(b) $(\text{Glucose})_n$ + glucose-1-P $\xrightarrow{\text{Ph}\,a}$ $(\text{glucose})_n + P_i$

Phosphorylase kinase (PhK) phosphorylates phosphorylase *b* to form phosphorylase *a* (Ph *a**). Skeletal muscle phosphorylase *b* requires AMP for activity while the *a* form does not. Activation of phosphorylase (Ph *a*) is determined in a second incubation by following the release of inorganic phosphate when the phosphorylase reaction is working in the direction of glycogen synthesis in the absence of AMP.

Optimized conditions for measurement: the activity of skeletal muscle phosphorylase kinase is optimal at the non-physiological pH of 8.2. Optimal free Ca^{2+} concentration is 70 µmol/l although after phosphorylation by cyclic AMP-dependent protein kinase the optimal Ca^{2+} concentration is reduced to about 5 µmol/l.

Temperature conversion factors: not known.

Equipment

UV/visible spectral-line photometer, water-bath, stopwatch.

Reagents and Solutions

Purity of reagents: phosphorylase *b* must be freed from AMP as described below.

Preparation of solutions: all solutions in re-purified water (cf. Vol. II, chapter 2.1.3.2).

1. Tris/glycerophosphate buffer (Tris, 0.25 mol/l; glycerophosphate, 0.25 mol/l; pH 8.6):

   dissolve 0.60 g Tris base and 1.08 g β-glycerol-P-$Na_2 \cdot H_2O$ in about 10 ml water, adjust to pH 8.6 with HCl, 6 mol/l, and dilute to 20 ml with water.

* Phosphorylase *b*, Ph *b*, } 1,4-α-D-Glucan: orthophosphate α-glucosyltransferase, EC 2.4.1.1
Phosphorylase *a*, Ph *a*, }

2. Nitrilotriacetate, NTA (30 mmol/l):

   dissolve 0.57 g nitrilotriacetate in 100 ml water.

3. $CaCl_2$ (225 mmol/l):

   dissolve 3.31 g $CaCl_2$ in 100 ml water.

4. Tris/glycerophosphate/Ca^{2+}/NTA buffer (Tris 0.125 mol/l; glycerophosphate, 0.125 mol/l; $CaCl_2$, 2.25 mmol/l; NTA, 3.0 mmol/l):

   mix 5.0 ml solution (1) with 1.0 ml solution (2), 0.1 ml solution (3), and 3.9 ml water.

5. Mg/ATP solution (ATP, 18 mmol/l; $MgCl_2$, 60 mmol/l):

   dissolve 0.22 g ATP-$Na_2H_2 \cdot 3\,H_2O$ in about 10 ml water and adjust to pH 7.0 with NaOH, 1 mol/l. Dissolve 0.24 g $MgCl_2 \cdot 6\,H_2O$ and dilute to 20 ml with water.

6. Dilution buffer (glycerophosphate, 50 mmol/l; 2-mercaptoethanol, 0.1% v/v):

   dissolve 11 g β-glycerol-P-$Na_2 \cdot H_2O$ in about 500 ml water and adjust to pH 7.0 with NaOH, 1 mol/l. Dilute to 1 l and add 1 ml 2-mercaptoethanol immediately before use.

7. Phosphorylase *b* solution (Ph *b*, approx. 100 mg/ml; glycerophosphate, 50 mmol/l; EDTA, 2 mmol/l; glycerol, 50% v/v; 2-mercaptoethanol, 0.1% v/v):

   dissolve crystalline phosphorylase *b* (80 U/mg) to a final concentration of 30 – 50 mg/ml by warming at 30 °C in solution (6) containing EDTA, 2 mmol/l. Treat several times with Dowex AGI-X4 resin to remove AMP [9] until the absorbance ratio (A_{260}/A_{280}) has decreased to 0.53 – 0.54. Dialyze against solution (6) containing EDTA, 2 mmol/l, and glycerol, 50% (v/v).

8. Phosphorylase *b* solution (Ph *b*, 15 mg/ml; glycerophosphate, 50 mmol/l; 2-mercaptoethanol, 0.1% v/v):

   dialyze solution (7) overnight at 0 – 4 °C against solution (6). Dilute to 15 mg/ml using solution (6). A 1 + 19 dilution should have an absorbance at 280 nm of 0.98.

9. Maleate/BSA solution (maleate, 0.1 mol/l; BSA, 0.01% w/v; 2-mercaptoethanol, 0.1% v/v):

   dissolve 1.8 g sodium maleate and 100 mg bovine serum albumin in about 50 ml water. Adjust to pH 6.5 with NaOH, 1 mol/l, and dilute to 100 ml with water. Add 100 μl 2-mercaptoethanol immediately before use.

10. Glucose-1-P/glycogen/maleate (glucose-1-P, 0.15 mol/l; glycogen, 2% w/v; maleate, 0.1 mol/l):

dissolve 0.4 g rabbit liver glycogen in about 10 ml water. Add a spatula tip full of mixed-bed resin (Amberlite MB-3). Stir for a few minutes, and filter off the resin. Add 1.13 g glucose-1-P, disodium salt, and 0.36 g sodium maleate. Adjust to pH 6.5 with NaOH, 1 mol/l, and dilute to 20 ml with water.

11. Ammonium molybdate solution:

    dissolve 2.5 g ammonium molybdate in 900 ml water and add 14 ml H_2SO_4, 18 mol/l.

12. ANSA reagent:

    dissolve 27.2 g $Na_2S_2O_5$ and 1.0 g anhydrous Na_2SO_3 in about 180 ml water. Dissolve 0.5 g amino nitrosulphonic acid (ANSA) by warming at 70 °C and dilute to 200 ml with water.

13. Phosphate standard (2 mmol/l):

    dissolve 0.31 g NaH_2PO_4 in 1 l water.

Stability of solutions: store solutions (3), (4), (6), (8), (9) at 0 – 4 °C, adding 2-mercaptoethanol to (6) and (9) immediately before use. Solution (8) is stable for several weeks but can be stored as solution (7) for at least 2 years at – 20 °C. The Ca/NTA buffer, solution (4), precipitates slowly on storage and should be prepared weekly. Store solutions (1), (5), (7), (10) at – 20 °C. Solutions (2), (11), (12) and (13) are stable for many months at room temperature.

Procedure

Collection and treatment of specimen: will depend on the tissue. For extraction of phosphorylase kinase from rabbit skeletal muscle, the back and hind-limb muscles are minced and then homogenized in 2.5 ml EDTA, 4 mmol/l, 2-mercaptoethanol, 0.1% (v/v), pH 7.0 per g muscle using a *Waring* blender for 30 s at low speed. The homogenate is centrifuged (6000 *g*, 45 min, 4 °C) and the supernatant filtered through glass wool.

Stability of the enzyme in the sample: this will depend on the source of the enzyme. Phosphorylase kinase [11] and myosin light-chain kinase [12] can be converted by limited proteolysis to forms which have lower requirements for Ca^{2+} or are not activated by exogenous Ca^{2+} and calmodulin.

Details for measurements in tissues: the problems described under this heading in chapter 7.2.1, i.e. the presence of protein phosphatases and ATP hydrolases, may apply to the assay of Ca^{2+}-dependent protein kinases also. However interference from these sources is not serious when assaying phosphorylase kinase in skeletal muscle extracts because of the high dilutions at which this enzyme can be assayed (>100-fold).

Assay conditions

PhK catalyzed reaction: volume of incubation mixture 60 µl; final volume 2.46 ml; 30 °C; 10 × 1.0 cm test-tubes.

Ph a catalyzed reaction: volume of incubation mixture 0.2 ml; 30 °C; 12 × 1.2 cm test-tubes.

Colour reaction: wavelength 620 nm; light path 10 mm; final volume 5.00 ml; 30 °C.

Measurement

PhK catalyzed reaction

Pipette successively into test-tubes:		blank	sample	concentration in assay mixture	
Tris/glycerophosphate/ Ca^{2+}/NTA	(4)	0.02 ml	0.02 ml	Tris glycerophosphate free Ca^{2+} NTA	42 mmol/l 58 mmol/l ~70 µmol/l 10 mmol/l
phosphorylase *b* (in dilution buffer 6)	(8)	0.02 ml	0.02 ml	phosphorylase mercaptoethanol	5 mg/ml 0.05% (v/v)
sample (in dilution buffer 6)			0.01 ml		
dilution buffer	(6)	0.01 ml			
mix on vortex mixer and pre-incubate at 30 °C for 5 min					
Mg^{2+}/ATP solution	(5)	0.01 ml	0.01 ml	Mg^{2+} ATP	10 mmol/l 3 mmol/l
mix on vortex mixer and incubate 5 min					
sodium maleate/ BSA (ice-cold)	(9)	2.4 ml	2.4 ml		
reaction is stopped; mix.					

Now assay for phosphorylase *a* in incubation solution by carrying out Ph *a* catalyzed reaction and colour reaction on both sample and blank.

Ph a catalyzed reaction and colour reaction

Pipette successively into test-tubes:	blank	sample	concentration in assay mixture
glucose-1-P/glycogen (10)	0.1 ml	0.1 ml	glucose-1-P 75 mmol/l glycogen 1% (w/v) maleate 0.1 mol/l
pre-incubate to 30 °C			
incubation solution (blank and sample resp.)	0.1 ml	0.1 ml	
mix on vortex mixer and incubate for 5 min			
ammonium molybdate (11) ANSA reagent (12)	4.55 ml 0.25 ml	4.55 ml 0.25 ml	
reaction is stopped; mix on vortex mixer, incubate for 5 min and read absorbance at 620 nm.			

The phosphorylase kinase should be diluted so that <20% conversion of phosphorylase *b* to *a* occurs. This corresponds to $A_{620} < 0.5$.

Construct a calibration curve using 0.2 ml aliquots of the phosphate standard (solution 13) instead of solution (10) and the incubation solution for the colour reaction.

If necessary determine the specific activity of phosphorylase *b* by carrying out the phosphorylase *a* catalyzed reaction and colour reaction with solution (8) (diluted 1:1000 with solution 9) instead of incubation solution and in the presence of AMP, 1 mmol/l. For crystalline phosphorylase *b* this specific activity should be about 80 U/mg (conversion of 80 μmoles P_i per min and mg).

Calculation: the amount of phosphate formed in each incubation is determined by reference to the calibration curve.

Catalytic activity concentration of phosphorylase kinase in original solution

$$b = \frac{P_{i\,(assay)} - P_{i\,(blank)}}{z_c/m_s} \times 1440 \quad \mu mol/min \times l \quad (U/l)$$

The units obtained are μmoles phosphorylase a formed per min and l of sample. The formula derives from

$$b = \frac{P_{i\,(assay)} - P_{i\,(blank)}}{z_c/m_s \times 97.4 \times 0.7 \times 5 \times 5} \times \frac{2.46}{0.1} \times 100 \times 1000 \quad \mu mol \times min^{-1} \times l^{-1} \quad (U/l)$$

in which

P_i	is expressed in μmoles
z_c/m_s	is specific activity of the phosphorylase *b* used
97.4	is based on the assumption that the molecular weight of Ph *a* is 97400 [13]
0.7	is based on the assumption that the activity of Ph *a* in the absence of AMP is 70% of the activity of Ph *b* in the presence of AMP (1 mmol/l) [14]
5	is incubation time in min (Ph *a* and PhK catalyzed reactions)
2.46	is volume of PhK catalyzed reaction in ml
0.1	is volume of sample
100	is factor relating to 1 ml of sample
1000	is factor relating to 1 l of sample

Alternative assay using [γ-^{32}P]ATP: phosphorylase kinase can also be assayed in an incubation identical with incubation 1, except that [γ-^{32}P]ATP (1 mmol/l; $2-5 \times 10^4$ cpm/ nmole) is used. The reaction is carried out in a 1.5 ml polypropylene micro-centrifuge tube and is stopped and processed as described in chapter 7.2.1. Other Ca^{2+}-and calmodulin-dependent protein kinases can also be assayed by this method using the appropriate protein substrate. Normally, exogenous calmodulin will have to be added. Optimal conditions must be determined either empirically or from the relevant literature.

Validation of Method

Precision, accuracy, detection limits and sensitivity: for phosphorylase kinase at 1.5 U/l the imprecision of the assay described would be $< \pm 5\%$. Data about accuracy are not available. By extending the time for incubation 1 to 60 minutes one could measure a lowest possible value of about 3×10^{-1} U/l with an imprecision of $< \pm 20\%$.

Sources of error: in general incubations should be kept short so that the degree of conversion of substrate is small and interference from protein phosphatases and ATP hydrolases is minimized (see also chapter 7.2.1). In the [γ-^{32}P]ATP assay, the presence of contaminating protein kinases in the protein substrate, or contaminating substrate proteins in the protein kinase sample, should be corrected for as described in chapter 7.2.1.

Specificity: the spectrophotometric assay for phosphorylase kinase assay described is completely specific. The radioactive assay will detect any protein which phosphorylates that substrate. To test for Ca^{2+}-dependence, carry out the reaction in the presence of EGTA (1 mmol/l). To test for calmodulin dependence, assay with and without added calmodulin (e.g. 10 µg/ml). Calmodulin-dependent enzymes can also be completely inhibited using phenothiazine drugs such as trifluoperazine at concentrations of about 50 µmol/l [14]. However trifluoperazine does not inhibit the basal activity of phosphorylase kinase, because it cannot interact with the tightly bound molecule of calmodulin.

Reference ranges: in adult rabbit skeletal muscle extracts after homogenization in 2.5 ml medium/g muscle [10] the activity of phosphorylase kinase is about 1 200 U/l.

References

[1] *E. G. Krebs, E. H. Fischer,* the Phosphorylase *b* to *a* Converting Enzyme of Rabbit Skeletal Muscle, Biochim. Biophys. Acta *20,* 150 – 157 (1955).

[2] *P. Cohen, A. Burchell, J. G. Foulkes, P. T. W. Cohen, T. C. Vanaman, A. C. Nairn,* Identification of the Ca^{2+}-dependent Modulator Protein as the Fourth Subunit of Rabbit Skeletal Muscle Phosphorylase Kinase, FEBS Lett. *92,* 287 – 293 (1978).

[3] *K. Yagi, M. Yazawa, S. Kakiuchi, M. Ohshima, K. Uenishi,* Identification of an Activator Protein for Myosin Light Chain Kinase as the Ca^{2+}-dependent Modulator Protein, J. Biol. Chem. *253,* 1338 – 1340 (1978).

[4] *A. Kishimoto, Y. Takai, T. Mori, U. Kikkawa, Y. Nishizuka,* Activation of Calcium and Phospholipid-dependent Protein Kinase by Diacylglycerol, its Possible Relation to Phosphatidylinositol Turnover, J. Biol. Chem. *255,* 2273 – 2276 (1980).

[5] *F. Huijing,* Phosphorylase Kinase Deficiency, Biochem. Genet. *4,* 187 – 194 (1970).

[6] *A. Burchell, P. T. W. Cohen, P. Cohen,* Distribution of Isoenzymes of the Glycogenolytic Cascade in Different Types of Muscle Fibre, FEBS Lett. *67,* 17 – 22 (1976).

[7] *H. Wolf, F. Hofmann,* Purification of Myosin Light Chain Kinase from Bovine Cardiac Muscle, Proc. Natl. Acad. Sci. USA *77,* 5852 – 5855 (1980).

[8] *T. Yamauchi, H. Fujisawa,* Evidence for Three Distinct forms of Calmodulin-dependent Protein Kinases from Rat Brain, FEBS Lett. *116,* 141 – 144 (1980).

[9] *D. P. Gilboe, K. L. Larson, F. Q. Nuttall,* Radioactive Method for the Assay of Glycogen Phosphorylases, Anal. Biochem. *47,* 20 – 27 (1972).

[10] *P. Cohen,* The Subunit Structure of Rabbit Skeletal Muscle Phosphorylase Kinase, and the Molecular Basis of its Activation Reactions, Eur. J. Biochem. *34,* 1 – 14 (1973).

[11] *P. Cohen,* The Role of Calcium Ions, Calmodulin and Troponin in the Regulation of Phosphorylase Kinase from Rabbit Skeletal Muscle, Eur. J. Biochem. *111,* 563 – 574 (1980).

[12] *T. Tanaka, M. Naka, H. Hidaka,* Activation of Myosin Light Chain Kinase by Trypsin, Biochem. Biophys. Res. Commun. *92,* 313 – 318 (1980).

[13] *K. Titani, A. Koide, J. Hermann, L. H. Ericsson, S. Kumar, R. D. Wade, K. A. Walsh, H. Neurath, E. H. Fischer,* Complete Amino Acid Sequence of Rabbit Skeletal Muscle Glycogen Phosphorylase, Proc. Natl. Acad. Sci. USA *74,* 4762 – 4766 (1977).

[14] *B. Weiss, R. M. Levin,* Mechanism for Selectively Inhibiting the Activation of Cyclic Nucleotide Phosphodiesterase and Adenylate Cyclase by Antipsychotic Agents, Adv. Cyclic Nucleotide Res. *9,* 285 – 314 (1978).

7.3 Pyruvate Kinase

Assay in Serum and Erythrocytes

ATP : pyruvate 2-*O*-phosphotransferase, EC 2.7.1.40

Hisaichi Fujii and Shiro Miwa

7.3.1 UV-method

ICSH Standard Method for Erythrocytes

General

Pyruvate kinase, PK, catalyzes the conversion of phosphoenolpyruvate (PEP) to pyruvate with regeneration of ATP. In human erythrocytes, PK is one of the three postulated rate-controlling enzymes of glycolysis. Deficiency of human erythrocyte PK is the most common enzymatic cause of hereditary haemolytic anaemia [1]. More than 200 cases with erythrocyte PK deficiency have been reported since the first report by *Valentine, Tanaka & Miwa* [2]. Therefore, the measurement of erythrocyte PK activity may aim mainly to detect the patient with enzyme deficiency. Elevation of serum PK has been reported in progressive muscular dystrophy and acute myocardial infarction [3]. However, the assay of PK does not seen to offer advantages in clinical diagnosis compared with widely used enzymes such as aspartate aminotransferase, creatine kinase, lactate dehydrogenase and hydroxybutyrate dehydrogenase.

Application of Method: in biochemistry and clinical chemistry.

Enzyme properties relevant in analysis: the erythrocyte enzyme is unstable after haemolysis of blood.

Methods of determination: PK activity can be measured spectrophotometrically by the consumption of NADH in a coupled optical assay in which LDH is used as an auxiliary enzyme [4, 5], or directly by the decrease in absorbance at 230 nm due to the removal of PEP [6].

International reference methods and standards: a method for erythrocyte enzyme analysis was proposed recently by the International Committee for Standardization in Haematology [4]. Standard reference material is not available.

Assay

Method Design

Principle

(a) $$\text{PEP} + \text{ADP} \xrightarrow[\text{Mg}^{2+},\,\text{K}^+]{\text{PK}} \text{pyruvate} + \text{ATP}$$

(b) $$\text{Pyruvate} + \text{NADH} + \text{H}^+ \xrightleftharpoons{\text{LDH}} \text{L-lactate} + \text{NAD}^+$$

The equilibrium (a) lies on the side of pyruvate and ATP with an equilibrium constant $K = 6 \times 10^3$. Pyruvate is rapidly removed in the indicator reaction (b). The amount of PEP converted to pyruvate per unit time, as determined by the decrease of absorbance due to oxidation of NADH, is a measure of the PK activity.

Optimum conditions for measurement: a variety of buffers and reagent concentrations have been reported. Although triethanolamine-HCl (TEA) buffer gives the highest activity, the use of the standardized method with Tris buffer makes it possible to compare results obtained in different laboratories. The method described here is based on that recommended by the International Committee for Standardization in Haematology [4].

Temperature conversion factors

Serum: activity at 30 °C/activity at 37 °C = 0.563, activity at 25 °C/activity at 37 °C = 0.413.

Erythrocytes: activity at 30 °C/activity at 37 °C = 0.578, activity at 25 °C/activity at 37 °C = 0.433.

Equipment

Spectrophotometer or spectral-line photometer suitable for accurate measurement at 339 nm, preferably with constant temperature cuvette holder; water-bath (37 °C); glass column (2.5 × 10 cm).

Reagents and Solutions

Purity of reagents: LDH must not contain more than 0.01% PK; the preparation of *Boehringer Mannheim* conforms to this requirement.

Preparation of solutions (for about 50 determinations): all solutions in re-purified water (cf. Vol. II, chapter 2.1.3.2).

1. Physiological saline (0.9% NaCl):

   dissolve 9 g NaCl in 1 l water.

2. Microcrystalline cellulose/α-cellulose:

   suspend 20 g of a mixture of equal weights of dry α-cellulose (*Sigma*) and microcrystalline cellulose (Sigmacell, type 50) in 500 ml physiological saline (1).

3. "Stabilizing solution":

   dissolve 100 mg EDTA-$Na_2H_2 \cdot 2H_2O$ and 5 µl 2-mercaptoethanol in 75 ml water, adjust pH to 7.0 with KOH, 10 mol/l, and dilute to 100 ml with water.

4. Glycerol/sorbitol/NaCl:

   dissolve 28 g glycerol and 2.8 g sorbitol in 100 ml physiological saline (1).

5. Tris buffer (1 mol/l; pH 8.0):

   dissolve 12.1 g Tris and 186 mg EDTA-$Na_2H_2 \cdot 2H_2O$ in 75 ml water, adjust pH to 8.0 at 25 °C with conc. HCl and dilute to 100 ml with water.

6. $MgCl_2$ (100 mmol/l):

   dissolve 2.033 g $MgCl_2 \cdot 6H_2O$ in 100 ml water.

7. KCl (1 mol/l):

   dissolve 7.456 g KCl in 100 ml water.

8. NADH (2 mmol/l):

   dissolve 14.2 mg NADH, disodium salt, in 10 ml 1% $NaHCO_3$ solution*.

9. Lactate dehydrogenase, LDH (600 U/l):

   dilute commercially available preparations (e.g. with 300 U/mg) with "stabilizing solution" (3) and check LDH activity concentration at 37 °C using the method of *Beutler* [5] or as described in Vol. II, chapter 2.2.1.

10. ADP (30 mmol/l):

    dissolve 70.7 mg ADP-Na_2 in 5 ml water, adjust pH to 7.0 with KOH, 10 mol/l, and assay enzymatically [5] (cf. also Vol. VII).

* NADH is unstable, not only in solution but even in the dry state, at −20 °C. A solution more concentrated than the desired final concentration should be prepared with 1% $NaHCO_3$ and assayed spectrophotometrically.

11. PEP (50 mmol/l):

dissolve 103.1 mg PEP-K in 10 ml water.

12. Ferricyanide/cyanide (*Drabkin*'s solution):

dissolve 200 mg $K_3[Fe(CN)_6]$, 50 mg KCN and 140 mg KH_2PO_4 in 1 l water.

Stability of solutions: solutions (1), (4), (6), (7), (12), stored in dark bottles, and cellulose suspension (2) are stable for several months at room temperature if growth of micro-organisms is avoided. Solutions (3) and (9) are stable for a week at 4°C. Solutions (10) and (11) are stable for several months at −80°C. Solution (8) should be prepared daily.

Procedure

Collection and treatment of specimen: PK activity in serum and erythrocytes is much lower than that in white blood cells and platelets (leukocytes and platelets are unaffected by the enzyme deficiency in the case of PK deficiency). It is, therefore, essential that the sample should be leukocyte- and platelet-free. Filtration through microcrystalline cellulose/α-cellulose column is satisfactory for this purpose [5].

Serum: draw specimens of venous blood, allow to clot at room temperature, and centrifuge at 3000 *g* for 10 min. The supernatant must be completely free from leukocytes, platelets and haemolysis; it is used for the assay of serum PK activity.

Erythrocytes: the anticoagulants heparin (10 units/ml blood), EDTA (1 mg/ml blood) and acid/citrate/dextrose (ACD) solution (0.15 ml formula A, or 0.25 ml formula B per ml blood) are satisfactory for the assay of erythrocyte PK activity.

Centrifuge anticoagulated blood for 10 min at 3000 *g*; suspend the packed cells with an equal volume of physiological saline (1). Let the erythrocyte suspension flow through an equal volume of swollen cellulose (2). Collect the effluent into a centrifuge tube and wash the erythrocytes with physiological saline (1) three times. (The recovery of erythrocytes exceeds 90% by this method, and there does not seem to be selective retention of reticulocytes.) Re-suspend the packed cells in physiological saline (1) to give an approximately 50% suspension. For haemolysis mix 1 volume of the erythrocyte suspension with 9 volumes of "stabilizing solution" (3). Freeze rapidly in a dry-ice acetone bath and thaw in a water-bath three times. Dialyze haemolysate against 1000 volumes of "stabilizing solution" (3) for 2 h (elimation of intracellular fructose 1.6-bisphosphate, cf. chapter 7.3.2, p. 502).

Stability of the enzyme in the sample: the erythrocyte PK after haemolysis and serum PK are unstable. The assays of activity must therefore be carried out on fresh serum or haemolysate after dialysis. When anticoagulated blood is stored at 4 °C under sterile conditions, erythrocyte PK activity is stable for at least 6 days.

For long-term storage, the erythrocytes are suspended in the same volume of glycerol/sorbitol/saline (4) and, after a 30 min equilibration, the suspension is dropped into liquid nitrogen to freeze the erythrocyte beads. The beads are then stored in liquid nitrogen. After 8 months of storage of normal erythrocytes in liquid nitrogen, erythrocyte PK is essentially unchanged [7].

Assay conditions: wavelength 339 nm; light path 10 mm; final volume 1 ml; 37 °C (constant temperature cuvette holder).

A blank assay is carried out to be certain that the LDH is free of PK activity: use water instead of sample.

For PK assay in serum use 100 µl sample, in erythrocyte haemolysate 20 µl sample, add water ad 1000 µl each. Measure against blank.

Measurement:

Pipette successively into cuvettes:		blank	sample	concentration in assay mixture	
buffer	(5)	100 µl	100 µl	Tris	100 mmol/l
				EDTA	0.5 mmol/l
KCl	(7)	100 µl	100 µl	KCl	100 mmol/l
$MgCl_2$	(6)	100 µl	100 µl	$MgCl_2$	10 mmol/l
NADH	(8)	100 µl	100 µl	NADH	0.2 mmol/l
ADP	(10)	–	50 µl	ADP	1.5 mmol/l
LDH	(9)	100 µl	100 µl	LDH	60 U/l
sample		100 µl	100 µl	volume fraction	0.10
water		300 µl	250 µl		
mix, incubate at 37 °C for 10 min					
PEP	(11)	100 µl	100 µl	PEP	5 mmol/l
mix, monitor absorbance for approximately 10 min.					

A recorder expansion giving a full-scale reading of $A = 0.4$ is suitable for normal serum and haemolysate. Mutations which result in the formation of an enzyme with normal or high PK activity at high substrate concentration are known to occur. Most of variants show low substrate affinity for PEP (increased K_m for PEP). These variants can be detected under the condition of low substrate concentrations. For this purpose, it is important to measure PK activity at PEP, 0.25 mmol/l, (low S system) additionally.

Determination of haemoglobin concentration: add 4.5 ml ferricyanide-cyanide reagent (12) to 0.5 ml haemolysate (to convert the pigment to cyanomethaemoglobin), and read absorbance at 540 nm [8].

Calculation: use calculation formula (k) or (k_1), respectively (cf. "Formulae", Appendix 3). $\varepsilon = 0.63\ \text{l} \times \text{mmol}^{-1} \times \text{mm}^{-1}$. The following relations are valid for the catalytic concentration in the sample (serum or haemolysate):

serum	$b = 1587 \times \Delta A/\Delta t$	U/l	$b = 2.65 \times 10^4 \times \Delta A/\Delta t$	nkat/l
haemolysate	$b = 7937 \times \Delta A/\Delta t$	U/l	$b = 13.2 \times 10^4 \times \Delta A/\Delta t$	nkat/l

Concentration of haemoglobin in the haemolysate is calculated as follows:

$$\rho_{\text{Hb}} = \frac{\Delta A_{540}}{0.0682} \quad \text{g/l}$$

The catalytic activity content of PK (related to gram haemoglobin) follows from

$$z_c/m_s = \frac{b}{\rho_{\text{Hb}}} \quad \text{U/g or nkat/g}$$

Validation of Method

Precision, accuracy, detection limits and sensitivity: in duplicate experiments the imprecision is better than 5% with the proper assay technique. No comments can be given on accuracy due to the lack of standard reference material. Detection limits and sensitivity have not been examined.

Source of error: leukocytes are a rich source of PK, and the activity of this enzyme is not diminished in leukocytes in erythrocyte PK deficiency. It is therefore important that leukocytes must be removed completely for the determination of erythrocyte PK activity. For the determination of serum PK activity, a sample which is completely free from haemolysis must be used. Haemolysis results in a great increase in serum PK activity.

Specificity: ADP supplied by *Boehringer Mannheim* contains less than 0.1% pyruvate. Therefore, the assay is specific for PK activity when this substrate is used.

Reference ranges: in human serum, 49.2 ± 8.0 U/l at 37 °C ($n = 10$, mean $\pm$ standard deviation).

In erythrocytes, 13.8 ± 1.3 U/gHb or 428 ± 46 mU/10^9 erythrocytes at 37 °C ($n = 10$, mean $\pm$ standard deviation). Activity in the low S system is about 15% of that in the regular system at 37 °C.

7.3.2 Isoenzymes of Pyruvate Kinase

ICSH Standard Method for Erythrocytes

General

In mammalian tissues at least three different isoenzymes of PK exist [9, 10]. Type M_1 is found in skeletal muscle, heart, and brain. Type L is the major isoenzyme in liver and the minor one in renal cortex and also occurs in the small intestine. The main PK isoenzyme in the kidney is type M_2 which is also found in most other tissues. The erythrocyte enzyme (type R or L′) is kinetically and immunologically identical with the major isoenzyme found in the liver. The M_1 and M_2 isoenzymes display *Michaelis-Menten* kinetics with respect to PEP. The M_1 isoenzyme is not affected by fructose 1,6-bisphosphate and the M_2 is allosterically activated by this compound. Types L and R exhibit co-operativity in their kinetics towards PEP, and both are allosterically activated by fructose 1,6-bisphosphate. It has been established that both L and R type PK differ from M_1 and M_2 type PK in kinetic, electrophoretic and immunological properties, and that they are probably under the control of different genes [11, 12].

Recent work suggests that the M_1 and M_2 isoenzymes of PK are the products of the same gene [13], and that the L and L′ type isoenzymes are encoded by the same structural gene and are translated from different messenger RNAs [14].

Application of method: this method can be used for the study of the tissue distribution and multimolecular forms of PK isoenzymes. The electrophoretic patterns of the PK isoenzymes from rat tissues change greatly during differentiation from the foetus to the adult. In the liver of cancer-bearing animals, the L type isoenzyme decreases, and the M_2 type becomes predominant with increased growth of the cancer. Furthermore, genetic PK variants in humans can be distinguished from each other by this method.

Methods of determination: these enzyme forms can be differentiated by electrophoresis and ion-exchange chromatography, or ammonium sulphate fractionation [11, 15]. Electrophoretic methods are widely adopted using cellulose acetate strips [16] and polyacrylamide gel [9]. PK activity can be detected by NADH oxidation produced by coupling with the LDH reaction [16]. Thin-layer polyacrylamide gel electrophoresis is described in detail here.

International reference method and standards: the method described below is based on the standardized method for the analysis of PK variants which has recently been established by a working group of the International Committee for Standardization in Haematology Expert Panel on Red Cell Enzymes [17].

Assay

Method Design

Principle: the enzyme is first resolved by electrophoresis on thin layer polyacrylamide gel. The gel is then applied to a thin agar layer containing the assay reagent. After incubation, loss of absorbance of the agar gel due to the PK reaction is photographed with long-wave ultraviolet light.

Equipment

- Horizontal thin-layer electrophoresis apparatus,
- direct current power supply (500 V, 50 mA),
- long-wave ultraviolet light,
- *Whatman* No. 3 filter paper,
- photographic paper,
- soft thin plastic sheet,
- tissue homogenizer.

Reagents and Solutions

Purity of reagents: all reagents must be A.R. or purest grade available. LDH must be free from PK.

Preparation of solutions (for about 50 determinations): all solutions in re-purified water (cf. Vol. II, chapter 2.1.3.2).

1. Electrophoresis buffer (0.01 mol/l; pH 8.3):

   dissolve 1.21 g Tris, 0.6 g $MgSO_4 \cdot 6H_2O$ and 3.5 ml 2-mercaptoethanol in 800 ml water, adjust pH to 8.3 at 25 °C with HCl, 1 mol/l, and dilute to 1 litre with water. Before use add 458 mg F-1,6-P_2, tetracyclohexylammonium salt, $10H_2O$.

2. Reagents for polyacrylamide gel

   a) Solution A:
   dissolve 6.5 g acrylamide and 0.425 g Bis* in 100 ml water.

   b) Solution B:
   dilute 0.5 ml TEMED** to 50 ml with water.

   c) Solution C:
   dissolve 60 mg ammonium persulphate in 50 ml water.

* N,N′-methylene-bis-acrylamide.
** N,N,N′,N′-tetramethylethylenediamine.

3. Buffer for activity stain:

   dissolve 12.1 g Tris, 186 mg EDTA-$Na_2H_2 \cdot 2H_2O$, 7.5 g KCl and 2.03 g $MgCl_2 \cdot 6H_2O$ in 80 ml water, adjust pH to 7.5 with conc. HCl and dilute to 100 ml with water.

4. Agar solution:

   dissolve 300 mg agar in 10 ml water.

Stability of solutions: solutions (1), without F-1,6-P_2, (3) and (4) are stable for several months at room temperature. Solutions (2a) and (2b) are stable for several months at 4 °C in dark bottles. Solution (2c) should be prepared before use as well as complete solution (1).

Preparation of polyacrylamide gel plate: apparatus for preparation of the polyacrylamide gel plate is shown in Fig. 1. 10 ml solution (2a), 5 ml solution (2b) and 5 ml solution (2c) are mixed well and poured onto a glass plate on which a rectangular

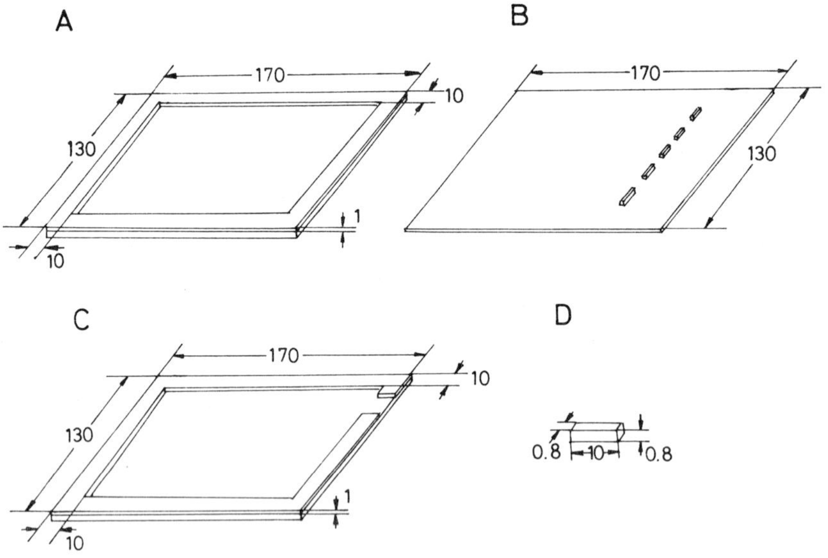

Fig. 1. Apparatus for preparation of polyacrylamide gel plate and activity staining.
A: Glass plate with rectangular acrylic resin frame.
B: Acrylic resin cover plate with slot formers.
C: Glass plate with rectangular acrylic resin frame, leaving a slot of 20 × 10 × 1 mm on one side of the frame.
D: Measurements of slot formers (in mm).

acrylic resin frame is fixed [A]. An acrylic resin cover plate with slot formers [B] is lowered smoothly onto the gel solution taking care to avoid enclosing air bubbles in the gel solution.

After standing for 1 h to allow polymerization, the cover is detached and the polyacrylamide gel slab attached to the glass is dialyzed against re-purified water for 2 h, then against the electrophoresis buffer (1) for 3 h for the gel to be equilibrated with the electrophoresis buffer.

Procedure

Preparation of samples

Tissue extracts: homogenize tissues in several volumes of cold buffer (3) at 0 °C. Centrifuge the homogenates at 30000 *g* for 1 h at 4 °C. Use the clear supernatants for enzyme assay and electrophoresis.

Haemolysate: mix packed cells obtained by the methods described in chapter 7.3.1 with 3 times their volume of the electrophoresis buffer (1). Haemolyse the erythrocyte suspension by freeze-thawing. Add one volume of toluene equal to 20% of the haemolysate, shake vigorously for 10 min, then centrifuge at 4500 *g* for 15 min. Use the clear haemolysate from which the toluene and stroma is removed for enzyme assay and electrophoresis. Adjust the activity of the sample to about 1.5 – 2.0 U/ml with the electrophoresis buffer (1).

Electrophoresis: a control haemolysate prepared from normal cells should be separated in parallel with the unknown samples, to provide reference electrophoretic mobilities and to allow possible genetic variants of the erythrocytic isoenzyme to be identified.

Remove excess buffer (1) from the sample slots of the gel with filter paper. Put the sample (about 5 µl) into the slots in the gel with capillary pipettes. Place the gel slab with the glass plate on a cooling system. Connect the gel slabs and buffer troughs with filter paper bridges (wet with electrophoresis buffer). Use other filter paper bridges to connect the buffer troughs and the electrode troughs.

Fill electrophoresis buffer (1) in both buffer and electrode troughs. Cover the whole gel with a soft thin plastic sheet (e.g. Saran-Wrap) but protect the place where the sample slots are made by a rectangular frame of 13 × 1.5 × 0.1 cm in order to avoid contamination of the samples due to direct contact with the plastic sheet. Perform electrophoresis at 14 V/cm for 5 h.

Detection of PK activity on the gel: prepare the following reaction mixture immediately before use:

dissolve 8 mg PEP-K, 30 mg ADP-Na_2, 20 mg NADH, disodium salt, in 1.5 ml buffer for activity stain (3) (Tris buffer, 1 mmol/l, pH 7.5), add water to 12 ml. Incubate for 10 min at 37 °C. Add 3 ml agar solution (4) incubated at 55 – 60 °C; add 30 U LDH.

After electrophoresis, cover the acrylamide gel plate with another glass plate with rectangular acrylic resin frame [C]. Pour the reaction mixture into the space without air bubbles to make a four-layered sandwich made up to a glass plate, reaction mixture, polyacrylamide gel and another glass plate, and allow to solidify for 10 min.

Incubate 10 – 20 min at 37 °C. Determine photographically the loss of absorbance of the reaction mixture due to the enzyme reaction: Expose photographic paper to long-wave ultra-violet light through the gel at the beginning and at end of the incubation time. After developing the photographic paper, dark bands due to PK activity can be seen.

Evaluation, calculation: photographic paper can be preserved as a permanent record of the results. The electrophoretic mobility of an examined sample can be calculated as a percentage of the control separated simultaneously.

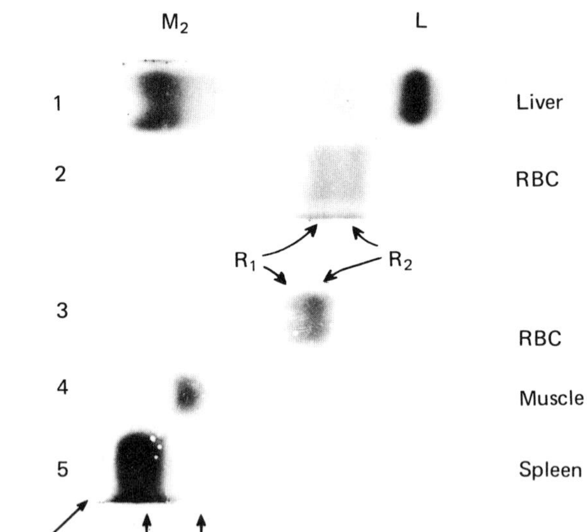

Fig. 2. Pyruvate kinase zymogram of liver, erythrocyte, muscle and spleen. Supporting medium: 3.34% acrylamide gel, 1 mm in thickness.
Buffer: Tris, 10 mmol/l
$MgCl_2$, 5 mmol/l
fructose 1,6-P_2, 0.5 mmol/l
pH 8.3, 25 °C
Voltage: 14 V/cm. Time 5 h.

Normal zymogram: the zymogram of PK in various human tissues is shown in Fig 2. The liver contains at least two forms, i.e. type L and M_2. Type L moves most rapidly towards the anode. Type M_2 as seen in the liver and spleen moves most slowly towards the anode.

Muscle contains a single form of PK (type M_1) which is slightly faster than type M_2. Erythrocyte PK is separated into two bands which have an intermediate mobility between type L and M. The cathodal band is designated as R_1 and the anodal band as R_2.

References

[1] *S. Miwa,* Pyruvate Kinase Deficiency and Other Enzymopathies of the Embden-Meyerhof Pathway, Clin. Haematol. *10,* 57 – 80 (1981).

[2] *W. N. Valentine, K. R. Tanaka, S. Miwa,* A Specific Erythrocyte Glycolytic Enzyme Defect (Pyruvate Kinase) in Three Subjects with Congenital Non-spherocytic Hemolytic Anemia, Trans. Assoc. Am. Phys. *74,* 100 – 110 (1961).

[3] *Y. Harano, R. Adair, P. J. Vignos Jr., M. Miller, J. Kowal,* Pyruvate Kinase Isoenzymes in Progressive Muscular Dystrophy and in Acute Myocardial Infarction, Metabolism *22,* 493 – 501 (1973).

[4] International Committee for Standardization in Haematology, Recommended Methods for Red-Cell Enzyme Analysis, Br. J. Haematol. *35,* 331 – 340 (1977).

[5] *E. Beutler,* Red Cell Metabolism. A Manual of Biochemical Methods, 2nd edit., Grune & Stratton, New York 1975.

[6] *N. G. Pon, R. J. L. Bondar,* A Direct Spectrophotometric Assay for Pyruvate Kinase, Anal. Biochem. *19,* 272 – 279 (1967).

[7] *H. Fujii, S. Miwa, Y. Ishida, E. Oda, T. Nakatsuji, Y. Miura,* Red Cell Enzyme Activities and Properties of Mutant Pyruvate Kinase after Long-Term Storage of Red Cells in Liquid Nitrogen, Vox Sang. *36,* 362 – 366 (1979).

[8] *E. J. van Kampen, W. G. Zijlstra,* Standardization of Hemoglobinometry. II. The Hemiglobincyanide Methode, Clin. Chim. Acta *6,* 538 – 545 (1961).

[9] *K. Imamura, T. Tanaka,* Multimolecular Forms of Pyruvate Kinase from Rat and Other Mammalian Tissues. I. Electrophoretic Studies, J. Biochem. *71,* 1043 – 1051 (1972).

[10] *K. H. Ibsen,* Interrelationship and Functions of the Pyruvate Kinase Isozymes and Their Variant Forms: A Review, Cancer Res. *37,* 341 – 353 (1977).

[11] *K. Imamura, T. Tanaka, T. Nishina, K. Nakashima, S. Miwa,* Studies on Pyruvate Kinase (PK) Deficiency. II. Electrophoretic, Kinetic, and Immunologic Studies on Pyruvate Kinase of Erythrocytes and Other Tissues, J. Biochem. *74,* 1165 – 1175 (1973).

[12] *S. Saheki, K. Saheki, T. Tanaka,* Peptide Structure of Pyruvate Kinase Isoenzymes. 1. Comparison of the Four Pyruvate Kinase Isoenzymes of the Rat, Biochim. Biophys. Acta *704,* 484 – 493 (1982).

[13] *A. J. Hance, J. Lee, M. Feitelson,* The M_1 and M_2 Isoenzymes of Pyruvate Kinase are the Products of the Same Gene, Biochem. Biophys. Res. Commun. *106,* 492 – 499 (1982).

[14] *J. Marie, M.-P. Simon, A. Kahn,* Cotranslation of L and L′ Pyruvate Kinase Messenger RNAs from Human Fetal Liver, Biochim. Biophys. Acta *696,* 340 – 344 (1982).

[15] *J. Carbonell, J. E. Feliu, R. Marco, A. Sols,* Pyruvate Kinase. Classes of Regulatory Isoenzymes in Mammalian Tissues, Eur. J. Biochem. *37,* 148 – 156 (1973).

[16] *W.A. Susor, W. J. Rutter,* Method for the Detection of Pyruvate Kinase, Aldolase, and Other Pyridine Nucleotide Linked Enzyme Activity after Electrophoresis, Anal. Biochem. *43,* 147 – 155 (1971).

[17] International Committee for Standardization in Haematology, Recommended Methods for the Characterization of Red Cell Pyruvate Kinase Variants, Br. J. Haematol. *43,* 275 – 286 (1979).

7.4 Creatine Kinase

ATP : creatine N-phosphotransferase, EC 2.7.3.2

7.4.1 General

Willie Gerhardt

Creatine kinase (CK) catalyzes the reversible N-phosphorylation of creatine by Mg^{2+}-ATP/Mg^{2+}-ADP complexes thus maintaining an immediately accessible energy reservoir in the cells [1].

The dimeric CK molecule consits of two side-by-side, cigar-shaped polypeptide subunits. Two distinct subunits exist, designated B and M, and active CK molecules may be composed of two identical subunits, i.e. the homodimers MM and BB, or of a pair of unlike subunits (CK MB). The M and B subunits are very similar. Each has a molecular mass of about 43000 and contains one catalytic site and one reactive SH group. The reduced state of the latter is a prerequisite for full catalytic activity [2 – 4].

The nomenclature of the subunits and the MM, MB and BB dimeric isoenzymes derives from the main localizations of the isoenzymes: CK BB in brain, CK MM in skeletal muscle [4]. A fourth CK dimer, MiMi, occurs in the mitochondria of myocardium and skeletal muscle and is thought to be composed of a third type of subunit [1, 4]. In man, CK occurs in order of decreasing catalytic concentration in skeletal muscle, myocardium, brain, gastrointestinal tract, thyroid, kidney, and liver [5]. Erythrocytes contain almost no CK activity [4].

The predominant part of the body's CK activity is present in skeletal muscle. About 97% of the CK catalytic activity in this tissue consists of cytoplasmic MM and about 3% of CK MB, part of which is associated with the Z and M bands of myofibrils [4]. CK BB [5] and mitochondrial CK MiMi [1, 4] occur as trace activities. In contrast, in human myocardium about 60% of the CK catalytic activity consists of CK MM, 30 – 40% of CK MB [5] and up to 10% of CK MiMi [4]. The high proportion CK MB in myocardium is the basis of the diagnostic use of CK MB determinations in suspected acute myocardial infarction (see CK isoenzymes). In brain, prostate and most of the gastrointestinal tract CK activity is predominantly due to CK BB [4].

In the sera of healthy persons about 94 – 98% of the CK activity is due to CK MM (see CK isoenzymes). The level of serum CK activity in damage to skeletal muscle or myocardium may depend partially on lymphatic transport of CK [6].

Application of method: in clinical chemistry and in biochemistry.

Enzyme properties relevant in analysis: the catalytic activity in serum is due to the varying catalytic concentrations of creatine kinase isoenzymes with different kinetic properties (see below).

Consequently, reaction conditions must be a compromise. For full catalytic activity, CK must be reactivated for several minutes by a thiol compound such as N-acetyl cysteine (NAC), dithiothreitol, monothioglycerol and others. Following an initial lag phase after initiation of the reaction by addition of creatine phosphate, substrate conversion proceeds linearly with time and amount of enzyme until deceleration occurs due to accumulation of inhibitory NADPH and decrease of the $NADP^+$ concentration. Linearity exists between amount of purified CK MM and reaction rates up to about 80 μkat/l, depending on the measuring conditions (see below p. 523).

In contrast, linearity may not always exist with sera that contain a mixture of different isoenzymes, nor when sera are diluted with NaCl, 150 mmol/l, or heat-inactivated serum. Similarly, a proportionate change in activity may not be observed when the volume fraction of the sample in the assay is varied (cf. p. 516).

Methods of determination: the catalytic activity can be measured spectrophotometrically, fluorimetrically or by bioluminescence (cf. p. 525). The standardized method described below has been developed for routine determination in clinical chemistry. This method can also be used for the determination of CK B-subunit activity in serum after complete immunoinhibition of M-subunit activity by a specific immunoinhibitor, anti-M (cf. Vol. I, chapter 2.7 and pp. 530–539). The method is well suited to automation.

International reference methods and standards: the provisional IFCC CK reference method is based on the European standard CK reagents. This method has been verified and accepted by a number of European societies of clinical chemistry. International reference enzyme standards are under development (cf. Vol. II, chapter 2.3).

Enzyme effectors: Mg^{2+} is required as an activator to form the ATP- and ADP-Mg^{2+} complexes [1].

A number of anions are non-competitive inhibitors, in decreasing order of efficiency: I^-, SO_4^{2-}, NO_2^-, Br^-, SO_3^{2-}, Cl^-, F^- [7], including buffer* anions such as Tris, imidazole, bis-tris, Pipes, Mops [7, 8].

Ca^{2+} competes with Mg^{2+} (K_I = 4.5 mmol/l [9]), as do Fe^{3+} [9, 10], and Cu^{2+} [9]. The endogenous average serum concentration of Ca^{2+}, 2.5 mmol/l, causes up to 10% inhibition of CK reaction rates at a sample volume fraction of 0.04 with Mg^{2+}, 10 mmol/l, in the standardized CK reagent [9, 10]. The inhibition by Ca^{2+} and other polyvalent cations is completely reversed by the inclusion of EDTA, 2 mmol/l, in the CK reagent (see below).

Reactivators are thiols, including N-acetylcysteine (NAC), β-mercaptoethanol, dithiothreitol, monothioglycerol, and glutathione (cf. p. 521).

* bis-tris [bis(2-hydroxyethyl)amino] tris-(hydroxymethyl)methane.
Pipes piperazine-N,N′-bis(2-ethane sulphonate).
Mops 3(N-morpholino)propane sulphonate.

References

[1] *V. A. Saks, G. B. Chernousova, D. E. Gukovsky* et al., Studies of Energy Transport in Heart Cells. Kinetic Properties and Regulatory Action of Mg^{2+} Ions, Eur. J. Biochem. *57*, 273 – 290 (1975).
[2] *Y. Degani, C. Degani,* Enzymes with Assymmetrically Arranged Subunits, T.I.B.S. *5*, 337 – 341 (1980).
[3] *S. B. Rosalki,* An improved Procedure for Serum Creatine Phosphokinase Determination, J. Lab. Clin. Med. *69*, 696 – 705 (1967).
[4] *H. Lang* (ed.), Creatine Kinase Isoenzymes. Pathophysiology and Clinical Application, Springer-Verlag, Berlin 1981.
[5] *W. Gerhardt, S. Hofvendahl, L. Ljungdahl,* Diagnostic Approach in Suspected Acute Myocardial Infarction, in: *A. Burlina, L. Galzigna* (eds.), Clinical Enzymology Symposia 4, Piccin Medical Books, Padua 1983, in press.
[6] *J. Lindena, W. Küpper, I. Trautschold,* Effect of Transient Hypoxia in Skeletal Muscle on Enzyme Activities in Lymph and Plasma, J. Clin. Chem. Clin. Biochem. *20*, 95 – 102 (1982).
[7] *L. G. Morin,* Creatine Kinase: Stability, Inactivation, Reactivation, Clin. Chem. *23*, 646 – 652 (1977).
[8] *G. Szasz, W. Gruber, E. Bernt,* Creatine Kinase in Serum: 1. Determination of Optimum Reaction Conditions, Clin. Chem. *22*, 650 – 656 (1976).
[9] *G. Szasz, J. Waldenström, W. Gruber,* Creatine Kinase in Serum: 6. Inhibition by Endogeneous Polyvalent cations and Effect of Chelator on the Activity and Stability of some Assay Components, Clin. Chem. *25*, 446 – 452 (1979).
[10] *W. Gerhardt, J. Waldenström, W. Gruber,* EDTA Effect on Creatine Kinase (CK) and on the SCE Reagent, Scand. J. Clin. Lab Invest. *39*, 737 – 742 (1979).

7.4.2 Routine UV-method

(European Standard Method)

Willie Gerhardt

Assay

Method Design

Principle

(a) $$\text{Creatine-P} + \text{ADP} \xrightleftharpoons{\text{CK}} \text{creatine} + \text{ATP}$$

(b) $$\text{ATP} + \text{D-glucose} \xrightleftharpoons{\text{HK*}} \text{ADP} + \text{D-glucose-6-P}$$

(c) $$\text{D-Glucose-6-P} + \text{NADP}^+ \xrightleftharpoons{\text{G6P-DH**}} \text{D-gluconate-6-P} + \text{NADPH} + \text{H}^+$$

* Hexokinase, ATP : D-hexose-6-phosphotransferase, EC 2.7.1.1
** Glucose-6-P dehydrogenase, D-glucose-6-phosphate : $NADP^+$ 1-oxidoreductase, EC 1.1.1.49

This method builds on principles proposed by *Oliver* [1], later modified by *Rosalki* [2], *Szasz,* and others (cf. pp. 518 – 523).

The equilibrium in (a) is displaced towards the formation of creatine + ATP (the so-called "reverse reaction") at pH values around 6 – 7 due to the higher energy of creatine phosphate compared to that of ATP, and the lower K_m values for Mg-ADP$^-$ and creatine-P than for Mg-ATP^{2-} and creatine (a). This primary reaction is coupled by the auxiliary hexokinase-reaction (b) to the NADPH-forming indicator reaction (c) catalyzed by G6P-DH. The rate of NADPH formation is monitored continuously at 339 nm or 334 nm. Reagent- and sample blanks may be omitted for the measurement of total CK activity in serum.

Optimized conditions for measurement: these are defined as those conditions which are most favourable both for the kinetic requirements of creatine kinases and the technical aspects of the measurement [3]. The method shows an extremely complex interdependence of most components. The reaction conditions described below have been optimized by univariate and multivariate experiments (cf. p. 518) with CK isoenzymes in sera from healthy individuals, patients with acute myocardial infarction, and patients with skeletal muscle diseases, and with isolated human CK isoenzymes (cf. pp. 530 – 539). The conditions for measurement are identical with those of the provisional IFCC CK reference method (cf. pp. 518 – 524). For measurements at 25 °C pH 6.7 is optimal, at 37 °C pH 6.5.

The assay is initiated by addition of creatine phosphate. It is possible to start the reaction with serum but this is not recommended. During the combined reactivation and lag phases the NADPH concentration will build up and the NADPH + H^+/$NADP^+$ ratio will increase, thus decreasing the range of linearity between catalytic concentration of CK and reaction rate.

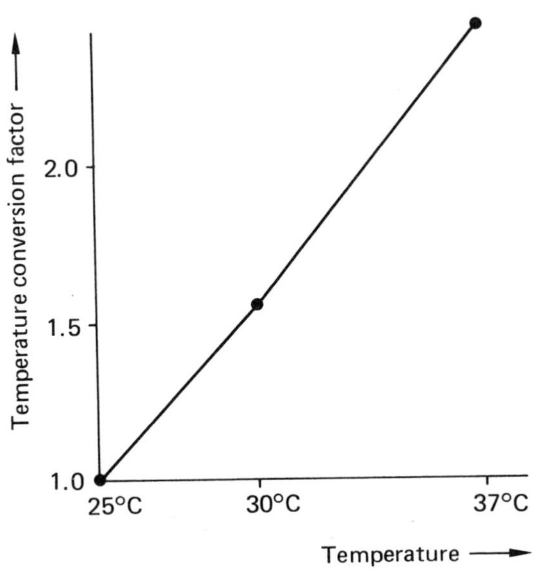

Fig. 1. Average temperature conversion factor as a function of temperature, see conversion table in the text.

Temperature conversion factors: the standardized reaction medium functions equally well at the three currently widely-used measuring temperatures of 25 °C, 30 °C and 37 °C [4]. The imidazole acetate buffer and CK have similar temperature coefficients.

Due to the occurrence of CK isoenzymes temperature conversion factors must be considered approximate (Fig. 1).

Temperature conversion factors

from	25 °C	26 °C	27 °C	28 °C	29 °C	30 °C	31 °C	32 °C	33 °C	34 °C	35 °C	36 °C	37 °C
to 25 °C	1.00	0.91	0.84	0.77	0.70	0.64	0.60	0.56	0.53	0.49	0.46	0.43	0.41
to 30 °C	1.56	1.43	1.31	1.20	1.09	1.00	0.94	0.88	0.82	0.77	0.72	0.67	0.63
to 37 °C	2.44	2.23	2.04	1.87	1.71	1.56	1.46	1.37	1.28	1.20	1.12	1.05	1.00

Equipment

Precision monochromator, filter- or spectral-line photometer capable of isolating 339 nm or the Hg 334 nm emission line, with thermostatic control of cuvette temperature according to IFCC specifications [3]. The instrument should provide the possibility of a) incubating the sample in the reagent without creatine-P for the time required for reactivation of CK by NAC, and b) subsequent start of the reaction by injection of creatine-P [4], and c) frequent or continuous monitoring of NADPH formation rate under steady state conditions 90 s (37 °C) or 120 s (25 °C and 30 °C) after start of the reaction [4, 5].

With sequential analyzers a throughput of 60 samples per hour is possible by having step b) occur 120 s before samples reach the measuring position in the instrument. Parallel and centrifugal analyzers should be provided with facilities to carry out steps a) and b) separately, not starting the reaction by adding serum to the complete reagent.

Reagents and Solutions

Purity of reagents: inhibitors may occur in imidazole, NAC, NADP and creatine-P. Imidazole preparations should be of highest purity, with low fluorescence, and the absorbance at 340 nm of the buffer should be less than 0.050 [4]. A charcoal-filtered imidazole buffer may have an absorbance of less than 0.010 at 339 nm and the creatine-P, 30 mmol/l, of less than 0.050. The absorbance at 339 nm of the complete CK reagent should be less than about 0.250 [4].

Reagents should be free from ATP and contaminating enzymes including CK, AK*, 6-phosphogluconate dehydrogenase**. The reagent blank rate should correspond to an increase in absorbance at 339 nm of less than 0.0007/min (30 °C, [5]) or 0.001/min (37 °C, [4]).

Preparation of solutions: 60 ml stable stock solution for about 30 determinations in 2.3 ml cuvette volume. It is convenient, and contributes to high precision in the weighing-in of the individual components, to prepare 1 l (sufficient for about 450 determinations) of stable stock solution of the CK reagent without NAC, auxiliary and indicator enzymes. This stock solution may be dispensed in, e.g., 60 ml aliquots in glass bottles provided with tight screw caps and stored at − 20 °C for two months or at − 70 °C for 6 months.

All solutions are prepared in re-purified water (see Vol. II, chapter 2.1.3.2).

1. Buffer stock solution (imidazole, 115 mmol/l; Mg acetate, 11.5 mmol/l; EDTA, 2.3 mmol/l; pH about 7.5):

   dissolve 7.83 g imidazole (M_r 68.1), 2.47 g Mg $(CH_3COO)_2 \cdot 4\,H_2O$ (M_r 214.5), and 0.85 g EDTA-$Na_2H_2 \cdot 2H_2O$, (M_r 372.2) in approximately 950 ml water. Adjust pH to approximately 7.5 (25 °C) with acetic acid, 1 mol/l. Add water to a final volume of 1000 ml.

2. CK reagent stock solution (imidazole acetate buffer, 115 mmol/l; EDTA, 2.3 mmol/l; magnesium acetate, 11.5 mmol/l; ADP, 2.3 mmol/l; AMP, 5.8 mmol/l; Ap_5A, 11.5 µmol/l; D-glucose, 23 mmol/l; NADP, 2.3 mmol/l; pH 7.10 to 7.15):

   transfer approximately 950 ml of buffer stock solution (1) to a 1500 ml glass beaker with a magnetic stirrer. Weigh in the following components and dissolve with stirring at room temperature:
   1153 mg ADP-K $\cdot$ $2H_2O$, M_r 501.3),
   2900 mg AMP-$Na_2 \cdot 6H_2O$, M_r 499.2),
   11 mg Ap_5A (p^1, p^5-di(adenosine-5′)-pentaphosphate, trilithium salt, M_r 934.2),
   4140 mg D-glucose (M_r 180.2),
   1810 mg NADP (β-nicotinamide-adenine dinucleotide phosphate, disodium salt, M_r 787.4).
   Adjust pH to 7.10 − 7.15 (25 °C) with acetic acid, 1 mol/l. Dispense this solution into, e.g., 60 ml glass bottles.

3. CK reagent working solution (concs. as in soln. 2, plus NAC, 23 mmol/l; HK, 3500 U/l, 58 µkat/l; G6P-DH, 2300 U/l, 38 µkat/l):

   prepare immediately before use.
   Bring 60 ml stock solution (2) to room temperature. Dissolve 225 mg N-acetyl L-cysteine (M_r 163.2). Adjust pH, if necessary, to 6.6 (30 °C) corresponds to pH 6.7

* Adenylate kinase, ATP : AMP phosphotransferase, EC 2.7.4.3
** 6-Phopho-D-gluconate : $NADP^+$ 2-oxidoreductase, EC 1.1.1.4.4

at 25 °C with acetic acid, 1 mol/l. Add 210 U HK (3.5 μkat, 30 °C) and 138 U G6P-DH (2.3 μkat, 30 °C).

4. Substrate (creatine-P, 345 mmol/l):

   dissolve 1.13 g creatine-P-$Na_2 \cdot 4H_2O$ (M_r 327.2) in 10 ml water.

Stability of solutions: storage of solution (1) at 4 °C. Stability 3 months.
Storage of solution (2) at −20 °C. Stability 2 months.
Storage of solution (3) at 4 °C. Stability 5 days. Stability at room temperature is about 24 hours, at −20 °C 2−3 weeks.
Storage of substrate solution (4) at 4 °C or −20 °C. Stability at these temperatures is 3 months and 12 months, respectively [4].

Procedure

Collection and treatment of specimen: collect blood by venepuncture with minimal manipulation and stasis, to avoid haemolysis which increases adenylate kinase activity and 339 nm absorbance. Use containers without any additives.

Stability of CK in the specimen or sample: the stability of CK isoenzymes decreases in the following order: macro CK forms of BB and MiMi; MM, MB, and BB. Loss of catalytic activity may be caused by thermal inactivation, oxygenation of reactive thiol groups, increasing pH due to loss of CO_2 and energy-rich light.

Consequently, samples should preferably be separated without delay (within 2 h) and the serum stored at low temperatures in tightly closed glass tubes with as little air-space over the serum as possible. The catalytic stability is apparently pH dependent, being maximal around pH 6.5−7.0 and minimal around pH 8.0−8.5 [6, 7]. Stability may thus vary somewhat for individual sera, with the degree of acidosis or alkalosis, and with isoenzyme distribution. About 94−99% of the total CK activity in sera from healthy individuals and patients with skeletal muscle diseases is due to CK MM and 2−6% to CK MB. In patients with uncomplicated acute myocardial infarction these proportions are altered to about 80−95% MM, 5−20% MB (cf. p. 532).

Average stabilities, defined as less than 5% loss under the storage conditions specified are:
CK MM: room temp., 24 h, 4 °C one week, −20 °C 3 months, −70 °C 6 months.
CK B: room temp., 12 h, 4 °C 3 days, −20 °C 1 month, −70 °C 6 months.
CK MB in these studies was determined as CK B activity after immunoinhibition with anti-M (see CK isoenzymes). All isoenzymes are stable in the reaction medium for 20 min (room temperature) and 10 min (37 °C), respectively (cf. p. 535, 536).

Details for measurement in tissues: part of the CK catalytic activity is located in the mitochondria of myocardium and skeletal muscle, cf. p. 508. The mitochondrial CK MiMi shows different K_m values for creatine-P and ADP [8]. A reagent optimized for this enzyme should be used in research.

Assay conditions: wavelength 339 nm or Hg 334 nm; light path 10 mm; cuvette thermostatted at selected reaction temperature; final volume 2.3 ml.

Total CK activity in serum may be measured with or without correction for sample rAK activity (residual adenylate kinase activity). The influence of sample rAK activity is discussed under CK B determination (p. 534). Sample rAK is measured in the CK working reagent, i.e. without creatine-P. With a single-beam instrument total CK (CK + rAK + reagent blank) and rAK (rAK + reagent blank) may be measured in separate cuvettes against air. If very few measurements are to be made, rAK and CK may be measured sequentially in the same cuvette.

Measurement with correction for rAK is most conveniently and precisely done on a double-beam instrument:

sample cuvette: sample total CK + rAK + reagent blank rate
reference cuvette: sample rAK + reagent blank rate.

Measurement

Pipette successively into cuvettes:		sample CK + rAK	reference rAK	concentration in assay mixture	
CK reagent	(3)	2.00 ml	2.00 ml	imidazole	100 mmol/l
				EDTA	2 mmol/l
				Mg^{2+}	10 mmol/l
				ADP	2 mmol/l
				AMP	5 mmol/l
				Ap_5A	10 µmol/l
				NAC	20 mmol/l
				D-glucose	20 mmol/l
				$NADP^+$	2 mmol/l
				hexokinase	50 µkat/l (3000 U/l)
				G6P-DH	33 µkat/l (2000 U/l)
serum		0.10 ml	0.10 ml	volume fraction	0.0435
mix carefully. Wait a minimum of 5 min for full reactivation of CK					
substrate	(4)	0.20 ml	–	creatine-P	30 mmol/l
water		–	0.20 ml		
wait two min (end of lag phase). Monitor increase in absorbance, preferably on a recorder for one or two minutes depending on the reaction rate.					

Samples should be diluted with NaCl, 150 mmol/l, if $\Delta A_{339}/\Delta t$ exceeds 0.6/min, corresponding to about 37 µkat/l or 2200 U/l, giving a final absorbance of about 2 (sum of sample, reagent, and NADPH absorbance) three min after start of the reaction. At a dilution of 1 + 1 the CK activity may be overestimated by about 5 – 10% [4].

Calculation: use calculation formulae (k) or (k_1), resp. (cf. "Formulae", Appendix 3). For values of ε cf. Appendix 4. Calculate $\Delta A/\Delta t$ from linear part of the curve. Δt = 1 min. The following relations are valid.
Catalytic concentration in sample:

	339 nm	Hg 334 nm	
b =	$60.9 \times \Delta A/\Delta t$	$62.0 \times \Delta A/\Delta t$	µkat/l
b =	$3651 \times \Delta A/\Delta t$	$3722 \times \Delta A/\Delta t$	U/l

Validation of Method

Precision, accuracy, detection limits and sensitivity

CK activity		Imprecision, CV (%)			
µkat/l	U/l	within-day	day-to-day	Temp.	Ref.
0.59	35	2.3	2.7	30 °C	[5]
2.69	162	0.9	1.3	30 °C	[5]
5.81	349	0.8	1.6	30 °C	[5]
8.27	496	–	2.5	37 °C	[9]
16.10	966	–	2.8	37 °C	[10]

See also CK B determination, p. 537.

Data on accuracy are not available as no international reference standard has yet been established.

For measurement at 340 nm (339 nm) with sufficiently sensitive photometers the lowest values that is possible to measure may be estimated. At 30 °C, volume fraction of sample 0.03, and three minutes monitoring after the end of the lag phase a detection limit of 24 ± 8 nkat/l (about 2 U/l ± 0.5 U/l) has been reported [5]. At 37 °C, volume fraction of sample 0.0435, two minutes monitoring after the end of the lag phase, the lowest measurable activity is about 60 ± 15 nkat/l, or about 3.6 ± 0.9 U/l, giving a practical detection limit of about 80 nkat/l or 5 U/l with an imprecision corresponding to a CV of about 20% (10) (photometer *LKB* 2086).

The sensitivity may be estimated to lie around an increase of absorbance at 339 nm of 0.002/2 min (photometer *LKB* 2086).

Sources or error: information on the effect of drugs on the measured CK activity is lacking. Physical activity, hypothyroidism and alcoholism should be considered when evaluating increased values.

Specificity: apart from interference by AK as described (cf. pp. 515, 520, 521), the method is specific for CK.

Reference ranges: given at 25 °C, 30 °C and 37 °C in Table 1. Values may not be directly convertible by temperature conversion factors. Differences may be due to varying sample selection criteria or demographic differences.

Table 1. Reference values for CK in serum determined at 30 °C [5] and 37 °C [11]. The values are given as μkat/l, in parenthesis as U/l. The values are not directly convertible between the units by a factor of 60 or 0.017 due to rounding-off of the values.

Sex, age		μkat/l (U/l)		
		25 °C	30 °C	37 °C
Cord blood		–	–	9.50 (<570)
Children	<10 yrs	–	0.30 – 2.10 (18 – 126)	–
Girls	12 – 14 yrs	–	–	4.20 (<250)
Boys	12 – 14 yrs	–	–	6.20 (<370)
Women	10 – 45 yrs	1.17 (<70)	0.30 – 1.50 (18 – 90)	2.50 (<150)
Men	10 – 60 yrs	1.33 (<80)	0.60 – 3.60 (36 – 220)	4.50 (<270)

References

[1] *I. T. Oliver,* A Spectrophotometric Method for the Determination of Creatine Phosphokinase and Myokinase, Biochem. J. *61,* 116 – 122 (1955).

[2] *S. B. Rosalki,* An Improved Procedure for Serum Creatine Phosphokinase Determination, J. Lab. Clin. Med. *69,* 696 – 705 (1967).

[3] Expert Panel on Enzymes, Committee on Standards (IFCC) (1979), Approval Recommendations of IFCC Methods for the Measurement of Catalytic Concentrations of Enzymes, Part 1. General Considerations, Clin. Chim. Acta, IFCC sections. *98,* 163 – 174 (1979).

[4] Scandinavian Committee on Enzymes of The Scandinavian Society for Clinical Chemistry and Clinical Physiology.
a) Recommended Method for the Determination of Creatine Kinase in Blood.
b) Recommended Method for the Determination of Creatine Kinase in Blood Modified by the Inclusion of EDTA, Scand. J. Clin. Lab. Invest. a) *36,* 711 – 723 (1976), b) *39,* 1 – 5 (1979).

[5] Enzymology Commission, Société Française de Biologie Clinique, Recommendations for Measuring the Catalytic Concentration of Creatine Kinase in Human Serum (documents C, 1976, C′ 1980, C, (1980), Ann. Biol. Clin. *40,* 138 – 149 (1982).

[6] *D. A. Nealon, S. M. Pettit, A. R. Henderson,* Effect of Serum pH on Storage Stability and Reaction Lag Phase of Human Creatine Kinase Isoenzymes, Clin. Chem. *26,* 1165 – 1169 (1980).

[7] *J. Bohner, W. Stein, W. Renn, R. Steinhart, M. Eggstein,* Stability of Macro Creatine Kinases and Creatine Kinase Isoenzymes Compared: Heat Inactivation Test for Determination of Thermostable Creatine Kinases, J. Clin. Chem. Clin. Biochem. *19,* 1021 – 1026 (1981).

[8] *H. Lang* (ed.), Creatine Kinase Isoenzymes. Pathophysiology and Clinical Application, Springer-Verlag, Berlin 1981.
[9] *W. Gerhardt, A. Louderback, J. Waldenström,* Stabilized Liquid Control Sera for Creatine Kinase Isoenzymes, Clinical Chemistry. *28*, 719–720 (1982).
[10] *W. Gerhardt, J. Waldenström,* Creatine Kinase B-Subunit Activity in Serum after Immunoinhibition of M-Subunit Activity, Clinical Chemistry, *25*, 1274–1280 (1979).
[11] Scandinavian Committee on Enzymes (SCE), Experience with the Scandinavian Recommended Methods for Determination of Enzymes in Blood, Scand. J. Clin. Lab. Invest. *41*, 107–116 (1981).

7.4.3 IFCC Reference Method (provisional)

Willie Gerhardt

The IFCC reference method* is based on the conditions of the European standard method, described on pp. 510–518. This method has been validated and accepted by a number of professional societies [1–6].

Assay

The method design is the same as described above for the routine method. The assay conditions are summarized in Table 1.

Table 1. Provisional IFCC reference CK method (equivalent to European standard CK method).

Reaction temperature	30 °C
pH	6,6
Volume fraction of serum	0.0435
Reactivation time for CK at 30 °C	5 min
Component	*Concentration*
Imidazole acetate	100 mmol/l
EDTA	2 mmol/l
Magnesium acetate	10 mmol/l
Creatine-P	30 mmol/l
Adenosine 5′-diphosphate (ADP)	2 mmol/l
Adenosine 5′-monophosphate (AMP)	5 mmol/l
P^1, P^5 Di(adenosine 5′-)pentaphosphate (Ap_5A)	10 µmol/l
N-Acetyl cysteine (NAC)	20 mmol/l
D-Glucose	20 mmol/l
Nicotinamide-adenine dinucleotide phosphate ($NADP^+$)	2 mmol/l
Hexokinase	50 µkat/l
Glucose-6-phosphate dehydrogenase	33 µkat/l

* status Sept. 1st, 1982

Establishing Optimized Conditions

Primary reaction

Buffer: imidazole acetate

Imidazole has a pK_a of 7 (25 °C). CK isoenzymes show broad and flat pH optima shifting toward lower values with increasing reaction temperature: 6.7 at 25 °C [1], 6.6 at 30 °C [4], and 6.5 at 37 °C [2]. The imidazole acetate buffer, 100 mmol/l, has a similar temperature dependence, Δ pH/$\Delta T \cong 0.02$, to that of human CK isoenzymes. Consequently, the same imidazole acetate buffer, adjusted to a pH of 6.7 (25 °C) will change pH with increasing temperature and support the reaction equally well at 30 °C and 37 °C, respectively [2].

Buffer anions inhibit CK (cf. p. 509). Consequently, 100 mmol/l is a compromise concentration between sufficient buffer capacity and minimal inhibition [1]. Imidazole may contain or produce inhibitors on ageing in the presence of humidity.

The chelating buffer bis-tris [7, 8] offers no advantage over imidazole with EDTA (see below). The so-called *Good* buffers have been found less suitable [1, 4, 7].

EDTA

Addition of EDTA has two effects: a) stabilization of NAC prevents autoxidation with formation of inhibitors [2, 7, 8]. Stability of the CK reagent at 4 °C is increased from less than 24 hours to 5 days [2, 7, 8], b) CK reaction rates are increased by reversal of the inhibition by endogeneous cations, including Ca^{2+} and Fe^{3+} [2, 7 – 9]. Average increases of apparent CK activity in sera from healthy individuals and from patients with acute myocardial infarction approximate to 1.15 at 25 °C [7], 1.1 to 1.18 at 30 °C [4, 7] and 1.2 at 37 °C [2, 8].

The weakly chelating buffer, bis-tris ± EDTA offers no advantage over imidazole with EDTA [2, 7, 8]. In the pH range 6.5 to 6.7, EDTA has a greater affinity for Ca^{2+} than for Mg^{2+}, leaving sufficient free Mg^{2+} for full CK activity at the selected concentration of 10 mmol/l of Mg^{2+} [2, 7, 8].

Magnesium acetate

Mg^{2+} is required for both CK and HK catalytic activity. Mg^{2+} and ADP concentrations are interdependent [10, 11]. Simultaneous increases of Mg^{2+} and ADP will increase the concentration of Mg^{2+}-ADP complexes required to drive the CK reaction in the direction of formation of ATP. Univariate increases of Mg^{2+} or of ADP beyond the chosen concentration produce no increase in reaction rates [1, 4, 10]. Recent multivariate studies show that the simultaneous increase of both Mg^{2+} and of ADP will increase apparent CK reaction rates ([11], see ADP below).

ADP

ADP is substrate for both CK, and for the interfering side reaction catalyzed by adenylate kinase:

$$ADP + ADP \xrightleftharpoons{AK} ATP + AMP$$

Different K_m values have been reported for the three current measurement temperatures. 25 °C: 95 μmol/l [12], 30 °C: BB 90 μmol/l [12], MM 130 μmol/l [12, 13], 37 °C, sera: 140 μmol/l (own data). Univariate increase of ADP shows a broad plateau at all three temperatures with no increase of reaction rates beyond 2 mmol/l [1, 2, 4, 10]. The multivariate studies referred to above show that simultaneous increases of Mg^{2+} to 20 mmol/l and ADP to 4 mmol/l increase CK reaction rates by 1.08 [11]. However, this change increases the ADP/AMP molar ratio from the current 0.4 [1,14] to 0.8, and may thus increase residual AK activity ([14], see CK B determination).

AMP

AMP is a competitive inhibitor (with ADP) of both AK and CK activities. Consequently, the selected AMP concentration is a compromise between sufficient AK inhibition and minimal CK inhibition. The average K_I of AMP at 25 °C is about 0.3 mmol/l for AK from erythrocytes, muscle, and liver [14]. Inhibition, singly and in combination with Ap_5A, is summarized in Table 2.

The average K_I of AMP for CK at 25 °C is about 10 mmol/l [14]. At the selected ADP/AMP molar ratio of 0.4, inhibition of CK is about 5% at 25° [1, 14] and 37 °C [2].

Table 2. Inhibition of AK and CK (%, average values) by various inhibitors and inhibitor combinations at 25 °C and 37 °C. Modified after refs. [2, 4, 14, 15].

Inhibitor	Adenylate kinases from				Creatine kinase
	Liver	Erythr.	Thromb.	Muscle	
AMP, 5 mmol/l	50	90	80	90	5
Ap_5A, 10 μmol/l	50	97	80	97	0
AMP, 5 mmol/l Ap_5A, 10 μmol/l	75	97	90	99	5
NaF, 25 mmol/l	90	90	85	90	5
NaF, 25 mmol/l AMP, 5 mmol/l	95	98	–	98	10
NaF, 25 mmol/l Ap_5A, 10 μmol/l	95	99	–	99	5

p^1,p^5-Di(adenosine-5′-)pentaphosphate, Ap_5A

Ap_5A is a competitive inhibitor of AK; most efficient against AK from erythrocytes and skeletal muscle with a K_I at 25 °C of 30 nmol/l [14], less effective against AK from liver with a K_I at 25 °C of 3 μmol/l [14].

Fluoride is a non-competitive, weaker AK inhibitor with a K_I at 25 °C of 2.5 mmol/l [15]. Neither singly, nor in combination with AMP [2] or with Ap_5A [2], does fluoride offer any advantage over the selected AMP-Ap_5A inhibitor combination [2, 14, 15]. Fluoride has three side effects: it causes turbidity in the reagent due to precipitation of Mg-fluoride [2, 10, 15]; at 25 mmol/l it causes a 5% inhibition of CK [2, 15]; and inhibition by fluoride requires a lag-phase before maximum inhibition is achieved of 8 min at 25 °C [15], 6 min at 30 °C [15], and 4 min at 37 °C [2].

N-acetylcysteine (NAC)

CK in serum is rapidly inactivated. Incubation with a thiol with a high reduction potential reactivates CK. This thiol must fulfill several requirements, including rapid and complete reactivation of CK, no precipitation of sample proteins, sufficient solubility and stability in solution, lack of obnoxious odour, and possibility of freeze-drying for incorporation into lyophilized reagent kits.

NAC has been found satisfactory in extensive tests. At a volume fraction of sample of 0.02 to 0.04 and with 20 mmol/l of NAC in the CK reagent, reactivation of CK in serum samples stored for one week at 4 °C is 99% complete [1, 2, 4].

In contrast to DTT and DTE at the same molar concentration, NAC does not cause micro-precipitation of sample proteins [2] and is easily soluble at pH 6.5 to 6.7 at the required concentrations.

Thiols in solution undergo reversible oxidation. This has two effects: loss of available -SH groups [16], and formation of potent CK inhibitors [7, 8]. Both processes are accelerated by certain polyvalent cations and are therefore retarded by the presence of chelators. Inclusion of EDTA, 2 mmol/l, stabilizes NAC in the reagent for 24 h at room temperature and 5 days at 4 °C [7, 8]. NAC must be of the highest purity to avoid pre-formed inhibitors [4].

Glutathione has been discarded as a reactivator since 1976 [1, 2, 4]; reactivation is incomplete and requires a longer time, while glutathione reductase in serum causes decreased reaction rates [17].

Creatine phosphate

K_m values of CK in sera for creatine-P in the standardized reagent have been reported to be 1.1 mmol/l at 25 °C [1], 1.2 mmol/l at 30 °C [12], and 1.3 mmol/l at 37 °C [2]. Different values have been reported for CK isoenzymes at 30 °C: BB, MB, and MM, 0.9 mmol/l, 1.2 mmol/l, and 2.5 – 3.0 mmol/l, respectively [12]. Most studies report a broad plateau between 20 mmol/l and 40 mmol/l at all three temperatures. 30 mmol/l has been selected as optimal in univariate and multivariate studies [1, 2, 4, 11, 12].

Auxiliary and indicator reactions

HK and G6P-DH***

The catalytic activity of these enzymes in the reaction medium determine the observed reaction rate of the overall coupled system, and the duration of the lag phase.

The respective activities of HK and G6P-DH may be specified in either of two ways: a) as determined in completely separate assays optimized for measurements of HK, and G6P-DH respectively e.g. at 25 °C, ("manufacturer's specifications"), or, as determined in the actual CK reaction medium at the temperature selected for the CK assay.

The HK measurement is started by ATP [2] or by glucose [4], the G6P-DH by glucose 6-phosphate [2,4]. This procedure also serves as a functional test of the catalytic capacity of the complete reagent. Details are given in ref. [2]. Specifications are given in Table 3.

Table 3. Catalytic activities of HK and G6P-DH in the IFCC reference method (equivalent to European standard CK method)

Enzyme	Manufacturer's specifications, 25 °C [1]		Determined in CK reagent at 30 °C [4]		37 °C [2]	
	μkat/l	U/l	μkat/l	U/l	μkat/l	U/l
HK	33	2000	50	3000	58	3500
G6P-DH	25	1500	33	2000	33	2000

The combination of HK and G6P-DH catalytic activities in Table 3 provides the highest reaction rates [1, 2, 4], the shortest lag phases (120 s at 25 °C [1] and 30 °C [4], less than 90 s at 37 °C [2]), and the lowest reagent blank rates. Further increase of G6P-DH will increase reagent blank rates above 4 U/l at 25 °C [1].

Both enzymes must be as free as possible from contaminating enzymes which will cause a reagent blank rate, including CK and AK. Traces of 6-phosphogluconate dehydrogenase (EC 1.1.1.44) will give an additional molecule of NADPH for each molecule of gluconate-6-P produced in the indicator reaction [18].

HK and G6P-DH must be supplied in glycerol, or lyophilized. Standard enzyme suspensions in $(NH_4)_2SO_4$ cause a loss of about 10% in reaction rates [2, 4]. Comparisons between G6P-DH from yeast and from *Leuconostoc mesenteroides* with NADP as the coenzyme show no differences in apparent CK reaction rates [19]. In contrast, the Leuconostoc G6P-DH with NAD as the coenzyme will give about 6% lower reaction rates with a wide scatter due to interference by other serum enzymes including lactate dehydrogenase (EC 1.1.1.27) [19].

* Hexokinase, ATP: D-hexose-6-phosphotransferase (EC 2.7.1.1)

** Glucose-6-phosphate dehydrogenase, Glucose-6-phosphate: $NADP^+$ 1-oxidoreductase (EC 1.1.1.49)

D-Glucose

Reaction rates are broadly independent of the glucose concentration within 10 mmol/l to 100 mmol/l at all three temperatures [1, 2, 4]. 20 mmol/l provides full catalytic activity with no side reactions.

NADP

The proportionality between CK catalytic activity and reaction rate depends on the $NADP^+/NADPH + H^+$ molar ratio. NADPH in the reaction acts as a competitive inhibitor of G6P-DH [1]. An initial concentration of NADP of 2 mmol/l, i.e. about 20 times the average K_m for NADP at 37 °C (SCE data), will sustain a constant rate until NADPH formation has decreased the $NADP^+/NADPH + H^+$ molar ratio to about 0.13 [1]. At a volume fraction of sample of 0.04 this corresponds to sustaining a constant rate of apparent CK activity of 625 U/l for 10 minutes [1, 2]. Reaction rates monitored 120 to 180 s after the start [2] will be linear with catalytic concentration of CK up to at least 40 µkat/l (about 2500 U/l), or to more than 80 µkat/l (5000 U/l) if monitoring commences immediately after the end of the lag phase.

Equipment, Reagents and Solutions

Equipment as described on p. 512. Reagents and solutions as described on pp. 512 – 514 (status sept. 1982).

Procedure

Assay conditions, measurement: as described on pp. 515, 516 except that a sample blank is mandatory.

Before pipetting, the temperature of reagent solutions and of sample must be brought to the calibration temperature of the pipettes.

During the pre-incubation period the reaction mixture in the cuvettes must attain a temperature of 30 °C. For this final pipetting, use of a pipette that has been calibrated at 30 °C is preferred. However, using a pipette calibrated at a lower temperature will introduce an error that is less than 0.1%.

Calculation: as described on p. 516. Correction for sample blank is necessary.

Validation of Method

Since the IFCC reference method and the European standard (routine UV-)method are identical, except with respect to a sample blank, values given on pp. 516, 517 for imprecision, accuracy, detection limits and sensitivity are comparable for both methods.

References

[1] a) *G. Szazs, W. Gruber, E. Bernt,* Creatine Kinase in Serum: Determination of Optimum Reaction Conditions, Clin. Chem. *22,* 650–656 (1976).
b) Recommendations of the German Society for Clinical Chemistry Standard Method for the Determination of Creatine Kinase Activity Revised Draft of 1976, J. Clin. Chem. Clin. Biochem. *15,* 255–260 (1977).

[2] Scandinavian Committee on Enzymes of The Scandinavian Society for Clinical Chemistry and Clinical Physiology,
a) Recommended Method for the Determination of Creatine Kinase in Blood.
b) Recommended Method for the Determination of Creatine Kinase in Blood Modified by the Inclusion of EDTA, Scand. J. Clin. Lab. Invest. a) *36,* 711–723 (1976), b) *39,* 1–5 (1979).

[3] Aanbevolen methode voor het meten van de activiteit van kreatine kinase (CK) in serum. Enzymcommissie of the Netherlandse Vereniging voor Klinische Chemie. Mededelingen 3, 220 (1978).

[4] Enzymology Commission, Société Française de Biologie Clinique, Recommendations for Measuring the Catalytic Concentration of Creatine Kinase in Human Serum (documents C, 1976, C′ 1980, C, 1980), Ann. Biol. Clin. *40,* 138–149 (1982).

[5] Schweizerische Gesellschaft für Klinische Chemie, Fachkommission: Empfohlene Methoden zur Bestimmung von 3 Enzymen im Blutplasm: ASAT – ALAT – CK. Bulletin Schweiz. Ges. Klin. Chemie supplement *21,* 43–48 (1980).

[6] Association of Clinical Biochemists, Working Party on Enzyme Methods of the Scientific and Technical Committee. "Proposed Methods for Determination of some Enzymes in Blood Serum". Ann. Clin. Biochem (UK) 1980: *17:* (Supplementum) 1s–30s.

[7] *G. Szasz, J. Waldenström, W. Gruber,* Creatine Kinase in Serum: 6. Inhibition by Endogenous Polyvalent Cations and Effect of Chelator on the Activity and Stability of some Assay Components, Clin. Chem. *25,* 446–452 (1979).

[8] *W. Gerhardt, J. Waldenström, W. Gruber,* EDTA Effect on Creatine Kinase (CK) and on the SCE Reagent, Scand. J. Clin. Lab. Invest. *39,* 737–742 (1979).

[9] *D. Nealon, A. R. Henderson,* Effect of Cations on the Human Creatine Kinase Isoenzymes, Clin. Chem. *26,* 1137–1139 (1980).

[10] *L. G. Morin,* Creatine Kinase: Re-examination of Optimum Reaction Conditions, Clin. Chem. *23,* 1569–1575 (1977).

[11] *V. S. Whitner, D. M. Fast, E. J. Sampson,* Reexamination of Reagent Concentrations for Scandinavian Creatine Kinase Method, Clin. Chem. *28,* 1642 (1982).

[12] *G. Szasz, W. Gruber,* Creatine Kinase in Serum: 4. Differences in Substrate Affinity Among the Isoenzymes, Clin. Chem. *24,* 245–249 (1978).

[13] *I. P. Steghens, I. Maire, M. Mathieu,* CK Affinity for Mg^{2+}, ADP, and Creatine Phosphate, (Personal communication).

[14] *G. Szasz, W. Gerhardt, W. Gruber, E. Bernt,* Creatine Kinase in Serum: 2. Interference of Adenylate Kinase with the Assay, Clin. Chem. *22,* 1806–1811 (1976).

[15] *G. Szasz, W. Gerhardt, W. Gruber,* Creatine Kinase in Serum: 3. Further Studies of Adenylate Kinase Inhibitors, Clin. Chem. *23,* 1888–1892 (1977).

[16] *G. Szasz, W. Gerhardt, W. Gruber,* Creatine Kinase in Serum: 5. Effect of Thiols on Isoenzyme Activity during Storage at Various Temperatures, Clin. Chem. *24,* 1557–1563 (1978).

[17] *G. Weidemann,* Störung der mit GSH aktivierten Creatin-Kinase-Bestimmung im Serum durch Glutathion Reduktase, Z. Klin. Chem. Klin. Biochem. *11,* 134–135 (1973).

[18] *W. Gerhardt,* Specific Determination of Glucose-6-Phosphate Dehydrogenase in Erythrocytes Using the LKB 8600: Extraction Procedures, Stability Studies, and Reference Values Calculated Per Mean Erythrocyte, Proceedings of the 4th International Symposium on Clinical Enzymology: *A. Burlina, L. Galzigna* (eds.), Quad. Sclavo Diagn. Clin. Lab. *9* (1), 464–471 (1973).

[19] *G. Szasz, E. Kinne, J. P. Colombo, W. Gruber,* Influence of Indicating Enzyme Reaction on Apparent Creatine Kinase Activity, J. Clin. Chem. Clin. Biochem. *17,* 689–691 (1979).

7.4.4 Luminometric Method

Karl Wulff

The catalytic activity of creatine kinase (EC 2.7.3.2) can be assayed luminometrically in two different ways: either by monitoring ATP formation directly using the firefly system [1 – 9], or by following NADPH formation in the hexokinase/glucose-6-phosphate dehydrogenase reaction using the lucibacteria luminescence system [10]. This contribution will only deal with the creatine kinase assays using firefly bioluminescence.

For basic considerations in luminometry, especially concerning the kinetic requirements for luminometric assays of enzyme activities, see Vol. I, chapter 3.5.

Assay

Method Design

Principle

(a) $$\text{ADP} + \text{creatine-P} \underset{}{\overset{\text{CK}}{\rightleftharpoons}} \text{creatine} + \text{ATP}$$

(b) $$\text{ATP} + \text{luciferin} + O_2 \xrightarrow{\text{luciferase*}} \text{oxyluciferin} + CO_2 + PP_i + \text{AMP} + h\nu$$

The increase in light intensity per unit of time, which is proportional to the amount of ATP formed per unit of time, is a measure of the catalytic activity of CK.

Optimized conditions for measurement: the method described here uses the same substrate concentrations for CK, the same buffer, and the same inhibitor concentrations for adenylate kinase as the European CK standard method (see p. 510).

Although these conditions are not optimal for the luminescence reaction, they allow CK to be assayed by a method which yields the same activity expressed in International Units as the photometric standard assay does.

Temperature conversion factor: at a temperature of 20 to 25 °C the luciferase catalyzed reaction emits light at 562 nm. With increasing temperature there is a shift of the emission wavelength to 615 nm. Since many photomultiplier (PM) tubes are blind at 615 nm or have at least a significantly decreased sensitivity at this wavelength, the corresponding signal is low. Hence the spectral sensitivity differs from one individual instrument of the same kind to another. Therefore, temperature conversion

* Photinus luciferin 4-monooxygenase (ATP-hydrolyzing) (EC 1.13.12.7)

factors can only be given for one individual instrument. Furthermore, they may not be constant over a longer period of time (such as several months).

With some instruments the sensitivity will be decreased by 20% at 30 °C and by a factor of two at 37 °C compared with that at 25 °C; with others the signal at 30 °C will be decreased by 80% and no signal at all will be measured at 37 °C.

Equipment

The procedure described in this article is designed for the luminometer 1250 (*LKB-Wallac*, Turku, Finland). The instrument should be used with a thermostatted cuvette holder, water-bath and recorder.

The procedure can of course also be used with any other luminometer of comparable sensitivity which maintains a constant temperature in the cuvette. However, it should be recognized that other instruments may require a different final volume of assay mixture for optimal results.

Reagents and Solutions

Purity of reagents: luciferase and luciferin must be of the highest purity available. Bovine serum albumin may contain traces of CK, which can be inactivated by heating the solution. ADP always contains variable amounts of ATP as an impurity. The ATP influences the detection limit for CK by giving a constant background emission. If ADP is freshly chromatographed as its lithium salt, it is stable for several weeks as a deep-frozen solution.

P^1,P^5-Di(adenosine 5′-)pentaphosphate (Ap_5A) stimulates light emission from luciferase/luciferin equivalent to 8×10^{-3} mol ATP/mol Ap_5A (11). The chemical nature of this effect is still unknown.

Preparation of solutions (for about 50 determinations): all solutions in re-purified water (see Vol. II, chapter 2.1.3.2).

1. Bovine serum albumin (60 g/l):

   dissolve 6 g bovine serum albumin in water to a final volume of 100 ml. Incubate this solution for 1 h in a water-bath of 65 °C to inactivate contaminating bovine CK.

2. Imidazole/EDTA/acetate buffer (imidazole/acetate, 110 mmol/l, pH 6.7; EDTA, 2.5 mol/l; magnesium acetate, 11 mol/l; albumin, 2.5 g/l):

   dissolve 3.74 g imidazole + 465 mg EDTA-$Na_2H_2 \cdot 2H_2O$ + 1.18 g magnesium acetate tetrahydrate in 300 ml water, adjust to pH 6.7 with acetic acid, 2 mol/l, add 21 ml solution (1), and dilute to 500 ml with water.

3. Luciferin (700 μmol/l):

   dissolve 1 mg D-luciferin (*Photinus pyralis*) in 2.0 ml solution (2). Dilute an aliquot of this solution 20-fold (100 μl luciferin-solution + 2.0 ml solution 2) and measure the absorption at 327 nm. If the stock solution is 700 μmol/l, the absorption should be $A_{327} = 0.595$. Dilute the stock solution according to its A_{327} to get the desired concentration of 700 μmol/l.

4. Tris buffer (500 mmol/l):

   dissolve 6.05 g tris(hydroxymethyl)-aminomethane in 50 ml water, adjust the pH with acetic acid, 2 mol/l, to pH 7.5, and dilute to 100 ml with water.

5. Luciferase (16 mU/l):

   dissolve 1 mg luciferase from *Photinus pyralis* in 2.0 ml ice cold solution (4) without shaking or agitation in a glass vial. This solution can be divided into 10 μl portions and stored frozen for months (but only in glass, polyethylene or polypropylene vials. Other plastic material may cause inactivation of the enzyme). Dilute 10 μl of this stock solution to 2.5 ml with solution 2 to get the final working solution (2 μg/ml).

6. AMP/N-acetylcysteine/Ap_5A (AMP 50 mmol/l; N-acetylcysteine 200 mmol/l; Ap_5A 100 μmol/l):

   dissolve 500 mg $AMP\text{-}Na_2 \cdot 6H_2O$ + 652 mg N-acetylcysteine + 2 mg Ap_5A in 15 ml solution (2), adjust pH to 6.7 by dropwise addition of a KOH-solution, 5 mol/l, and dilute with solution 2 to a final volume of 20 ml.

7. ADP (20 mmol/l):

   dissolve 50 mg $ADP\text{-}K \cdot 2H_2O$ in 5.0 ml solution (2).

8. Creatine phosphate (300 mmol/l):

   dissolve 490 mg $creatine\text{-}P\text{-}Na_2 \cdot 4H_2O$ in 5.0 ml solution (2).

9. ATP (0.5 mmol/l):

   dissolve 6.05 mg $ATP\text{-}Na_2H_2 \cdot 3H_2O$, in 20 ml solution (2).

Stability of solutions

Solution number	*Stability*
(1), (2), (4)	stable as long as no bacterial growth occurs. Store at 0 – 4 °C.
(3)	2 days, if protected from light, at 0 – 4 °C.
(5), (6)	2 days at 0 – 4 °C.
(7), (8), (9)	8 hours at 0 – 4 °C.

Procedure

Concerning specimens and samples see p. 514.

Assay conditions: wavelength of emission 562 nm; final volume 1.02 ml; 25 °C, 30 °C or 37 °C (thermostatted cuvette holder). Before starting the assay, adjust temperature of solutions to measuring temperature.

Measurement

Pipette successively into the cuvette:			concentration in assay mixture	
buffer solution	(2)	0.50 ml	imidazole Mg-acetate EDTA	100 mmol/l 10 mmol/l 2.3 mmol/l
sample		0.10 ml	volume fraction	0.10
incubate 5 min in a water-bath of the desired temperature. Then add				
luciferin solution	(3)	0.05 ml	luciferin	35 μmol/l
luciferase solution	(5)	0.05 ml	luciferase	0.8 mU/l
AMP, N-acetylcysteine, Ap$_5$A solution	(6)	0.10 ml	AMP N-acetylcysteine Ap$_5$A	5 mmol/l 20 mmol/l 10 μmol/l
ADP solution	(7)	0.10 ml	ADP	2 mmol/l
after a further 1 min in the water-bath, place the cuvette in the cuvette holder of the instrument; read for 1 min the background signal $\Delta I_0/\Delta t$*.				
creatine-P solution	(8)	0.10 ml	creatine-P	30 mmol/l
CK reaction is started; follow the linear increase in light intensity $\Delta I_1/\Delta t$, for 2 – 3 min. Then add for calibration				
ATP solution	(9)	0.02 ml	ATP	9.8 μmol/l
after the steplike increase in light intensity (ΔI_{cal})** follow the linear increase in light intensity ($\Delta I_1/\Delta t$) for another 2 min.				

* A slight linear increase in light intensity may occur. It is due to residual adenylate kinase activity.
** This value should be > 50 mV, if the luciferase is sufficiently active.

The optimal range of CK activity in the samples is 1 – 10 U/l. Should a sample have a higher activity, it should be diluted with solution (1).

Calculation: read from the recorder graph the values of

$$\Delta I_0/\Delta t; \quad \Delta I_1/\Delta t; \quad \Delta I_{cal}$$

All values of ΔI have the unit mV.
Calculate the CK activity concentration according to the following equation

$$b = \frac{\Delta I_1 - \Delta I_0}{\Delta t} \times \frac{9.8 \times 10}{\Delta I_{cal}} \quad \text{U/l.}$$

It is

$$\frac{\Delta I_1 - \Delta I_0}{\Delta t} \quad \text{slope in} \quad \text{mV/min}$$

ΔI_{cal} mV	steplike light intensity increase after ATP addition,
9.8 µmol/l	ATP concentration increase after calibration,
10	dilution factor.

Validation of method

Precision, accuracy, detection limits and sensitivity: all these data are dependent on the quality of the reagents (ADP) and of the instrument. For values of 1 – 2 U/l in diluted control serum, a coefficient of variation (CV) of 4 to 5% has been found. The luminometric method correlates well with the standard photometric assay ($R = 0.981$) [8]. The detection limit is 0.2 to 0.5 U/l with ADP of a commercially available quality. It can be improved to 0.05 U/l with freshly purified ADP.

Sources of error: influences of drugs on the measured activity are unknown. A linear increase in light intensity is observed over at least 15 min, but only if the luciferase has the full activity and the luciferin is free from impurities. Otherwise, non-rectilinear curves may be obtained. Constant temperature (±0.2°) is also critical in obtaining a good linearity.

Specificity: the method is specific for measuring CK activity. The method can be used, in combination with known methods of discrimination, for the assay of CK isoenzyme, i.e. after chromatographic separation or after inhibition of one isoenzyme subunit with specific antibodies.

References

[1] *S. A. G. J. Witteveen, B. E. Sobel, M. DeLuca,* Kinetic Properties of the Isoenzymes of Human Creatine Phosphokinase, Proc. Natl. Acad. Sci. USA *71*, 1384 – 1387 (1974).
[2] *P. T. Hwang, P. P. Addis, W. E. Rempel, A. Antonik,* Determination of Porcine Blood Creatine Kinase by Firefly Luciferase, J. Animal Sci. *45*, 1015 – 1022 (1977).
[3] *A. Lundin, I. Styrelius,* Sensitive Assay of Creatine Kinase Isoenzymes in Human Serum using M Subunit inhibiting Antibody and Firefly Luciferase, Clin. Chim. Acta *87*, 199 – 209 (1978).
[4] *P. Tarkkanen, S. Komor, C. Cornely, W. Hacke, H. Greiling,* Bioluminescence Assay of Creatine Kinase and Its Isoenzymes in Serum and Cerebrospinal Fluid, Clin. Chem. *25*, 1644 – 1648 (1979).
[5] *W. Gerhard, J. Waldenström,* Creatine Kinase B-Subunit Activity in Serum after Immunoinhibition of M-Subunit Activity, Clin. Chem. *25*, 1274 – 1280 (1979).
[6] *W. D. Bostick, M. S. Denton, S. R. Dinsmore,* Liquid-Chromatographic Separation and On-Line Bioluminescence Detection of Creatine Kinase Isoenzymes, Clin. Chem. *26*, 712 – 717 (1980).
[7] *A. Lundin, B. Jäderlund, T. Lövgren,* Optimized Bioluminescence Assay of Creatine Kinase and Creatine Kinase B-Subunit Activity, Clin. Chem. *28*, 609 – 614 (1982).
[8] *K. Wulff, W. Döppen, F. Stähler,* Aktivitätsbestimmung der Kreatinkinase unter Standardbedingungen über firefly-Biolumineszenz, Fresenius Z. Anal. Chem. *301*, 173 – 174 (1980).
[9] *K. Wulff, F. Stähler, W. Gruber,* Standard Assay for Total Creatine Kinase and the MB-Isoenzyme in Human Serum with Firefly Luciferase, in: *M. A. DeLuca, W. D. McElroy* (eds.), Bioluminescence and Chemiluminescence, Academic Press, New York 1981, pp. 209 – 214.
[10] *O. Rodriguez, G. G. Guibault,* Immobilized Bacterial Luciferase for Microscale Analysis of Creatine Kinase Activity, Enzyme Microb. Technol. *3*, 69 – 72 (1981).
[11] *G. Momsen,* Firefly Luciferase Reacts with P^1,P^5-Di(adenosine-5′-)pentaphosphate and Adenosine-5′-tetraphosphate, Biochem. Biophys. Res. Commun. *84*, 816 – 822 (1978).

7.4.5 Isoenzymes of Creatine Kinase

Determination of CK B subunit activity (CK MB)

(Photometric measurement after immunoinhibition by anti-M*)

Willie Gerhardt

General

The currently known CK isoenzymes and CK variants that may occur in serum comprise the following (Fig. 1): the three classical dimer isoenzymes numbered in order of decreasing electrophoretic mobility from the anode: CK 1 = BB, CK 2 = MB, and CK 3 = MM [1] and two forms of high molecular mass: macro CK types I and II [1, 2].

* anti-M specific anti-human CK M-subunit inhibiting antibody, from goat

Macro CK (type I) occurs as CK BB-IgG [3 – 5] circulating immune-complexes with different electrophoretic mobilities (Fig. 1 [1 – 6]). Macro CK BB occurs at a prevalence of about 3 percent of hospitalized patients with a slight predominance of elderly women, depending on the material [1, 2, 7]. It is not associated with any specific disease [1 – 7]. It most frequently constitutes 10 – 60% of total CK activity, but may in some cases exceed 98% [8]. The catalytic activity is stabilized in the immune-complex *in vivo* and *in vitro* against thermal inactivation and pH changes [9]. The activity may remain unchanged in patients' serum for months and years [9].

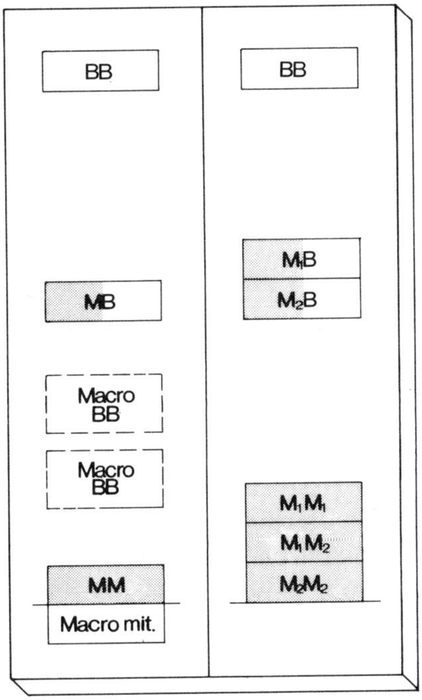

Fig. 1. Schematic representation of CK isoenzyme electrophoresis of the currently identified CK isoenzymes and multiple forms, separated in agarose at pH 8.6 (from ref. [8]).

 M subunits that will be completely inhibited in the presence of the anti-M immunoinhibitor.

Left side: from the anode the three classic, free dimers BB, MB and MM.

Between MB and MM: two macro CK type I (CK BB-IgG complexes) with different electrophoretic mobility. Occurs most frequently in individuals over 60 years, predominantly in women [1, 2, 5, 6, 8]. Most typical is the most cathodical macro BB band shown, but occasional patient sera show both bands in Fig. 1 [6]. In rare cases the electrophoretic mobility may coincide with that of CK MB causing possible misinterpretations.

Most cathodically: macro CK type II (mitochondrial CK, CK MiMi in an unidentified complex): Appears associated with advanced malignancy, may migrate with CK MM mobility. Both forms of macro CK are heat-stable and resistant to immunoinhibition by anti-M [8, 9].

Right side: three MM and MB dimer variants caused by change of normally migrating M_2-subunit into faster migrating M_1-subunit by a heat-labile serum protein. This gives a total of five MB plus MM band as shown.

Macro CK (type II) appears to be an aggregate of CK MiMi* [2, 4, 9], and occurs at a prevalence of 1 – 2 percent depending on the material [2]. In contrast to macro CK BB, macro CK (type II) is most frequently associated with malignancy, and the activity may fluctuate with the state of the disease [4]. It may constitute more than 95 percent of total serum CK activity [4, 8] and, like type I, has a high thermal stability [4, 9].

The electrophoretic mobility of the normal M subunit may be increased by the action of a serum factor producing up to a total of five MB + MM bands ([10] Fig. 1).

Free CK BB may occur in serum in certain cases of malignancy [4, 8, 11] or up to about 20 hours after cardiac arrest [1, 8]. CK MB in serum is increased in acute myocardial infarction in practically all cases within 10 to 20 hours after onset of the acute symptoms [7, 12]. This is the basis of the predominant use of CK MB determinations in clinical chemistry, by electrophoresis, chromatography or as CK B activity after immunoinhibition by anti-M [2, 7, 8, 12 – 14]. The immunoinhibition method has been investigated analytically [15] and clinically [7] by members of the Scandinavian Committee on Enzymes (SCE) and is described in detail in this chapter.

Skeletal muscle contains 2 – 3 percent MB (1 – 1.5 percent CK B) activity, which may appear in serum after resuscitation and precordial thumping after cardiac arrest, after exhaustive muscular efforts [1, 14], extensive skeletal muscle damage [1] including certain intoxications [1, 14], certain muscle diseases, including Duchenne muscular dystrophy and myasthenia gravis, and certain collagenoses e.g. sclerodermatomyositis [1, 7].

Rational application of determinations of CK B (MB) in serum in suspected acute myocardial infarction requires close co-operation between the clinician and the clinical chemist, and a strict diagnostic strategy e.g. sequential application of patient history and clinical symptoms, sampling for determination total CK 10 to 20 hours after onset of the patient's acute symptoms, and determination of CK B(MB) on the samples with CK increased above a sex-related acute myocardial infarction discrimination level [7]. Under these conditions most of the conditions giving rise to extracardial release of CK MB are easily identified, and the frequency of macro creatine kinases as a source of error decreased to about 0.5 percent [2, 7, 8, 14].

Application of method: in clinical chemistry and in biochemistry.

Enzyme properties relevant in analysis: the catalytic activity of serum CK depends on the presence of the various CK isoenzymes and variants described above. In the measurement of CK B activity after immunoinhibition as described below, substrate conversion proceeds linearly with time and amount of enzyme after the end of the lag phase (see under CK, p. 522).

Methods of determination: CK isoenzyme electrophoresis (cf. chapter 1.3.2.3) may be carried out on various media including cellulose acetate and agarose gel. The CK isoenzyme bands may be detected either by coupling a tetrazolium staining reaction to

* CK MiMi mitochondrial form of creatine kinase

the standardized CK reagent or by direct NADPH fluorescence in this medium (for surveys cf. [7, 13]). CK isoenzyme electrophoresis may be considered a referee method for identification of multiple forms of creatine kinases when carried out with proper technique and appropriate human isoenzyme markers. It may be combined with immunoinhibition, immunofixation, and radio-electrophoresis [3, 4, 6]. However, with current technology quantitation of CK isoenzymes by isoenzyme electrophoresis is poor [13].

Chromatography on various ion-exchange media and subsequent elution by increasing ionic strength exists in a great number of commercial variants (cf. also chapter 1.3.2).

In some procedures MB and BB are partially eluted together which may lead to errors [13]. CK BB and macro CK may cause false positives. Suspected macro forms require identification by isoenzyme electrophoresis. Poor recovery may cause false negatives. A U.S. inter-laboratory survey of 350 laboratories using 26 different methods including CK isoenzyme electrophoresis and ion-exchange chromatography documented great inter-laboratory variation and CK BB was falsely detected by 14 percent of the laboratories [16].

Immunoinhibition: determination of S-CK B-subunit activity after inhibition of all CK MM activity by a specific immunoinhibitor, anti-M ([1], [15] described here). The procedure utilizes the standardized CK method with the addition of anti-M [15], has been extensively tested in routine use [1, 7], and is well suited to automation [7, 15]. With the exception of the rare occurrence of macro CK MiMi, the method is specific for CK B subunit activity provided that the residual AK activity of the sample is subtracted from the residual activity measured after immunoinhibition [7, 8, 12, 15, 17]. Discrimination between CK MB and BB, macro BB, and macro MiMi is not possible. Consequently, the result of the immunoinhibition determination should not [7, 12, 15] be multiplied by the theoretical factor two [1] in order to give the impression that the assay is specific for CK MB activity.

International reference methods and standards: determination of CK B-subunit activity after immunoinhibition may be carried out according to the IFCC reference method for CK with anti-M. International standards are under development.

Enzyme effectors: these are the same as described under CK.

Assay

Method Design

Principle

1. Immunoinhibition of CK M subunit activity by anti-M. Out of the total sample creatine kinases only CK B subunits in BB, MB and macro BB, and CK MiMi remain active.

2. CK activity determination as described on p. 515 gives the apparent CK B activity.
3. Determination of sample residual activity of adenylate kinase*, AK (rAK):

(a) $$\text{ADP} + \text{ADP} \xrightleftharpoons{\text{AK}} \text{ATP} + \text{AMP}$$

(b) $$\text{ATP} + \text{D-glucose} \xrightleftharpoons{\text{HK}^{**}} \text{ADP} + \text{D-glucose-6-P}$$

(c) $$\text{D-Glucose-6-P} + \text{NADP}^+ \xrightleftharpoons{\text{G6P-DH}^{***}} \text{D-gluconate-6-P} + \text{NADPH} + \text{H}^+$$

CK B activity is obtained by subtraction of rAK activity from apparent CK B activity. The CK B immunoinhibition reagent functions equally well at 25 °C, 30 °C, and 37 °C as described for CK. As an example, identification of macro CK types I and II by their activation energies may be done in the same reagent at 25 °C, 30 °C, and 37 °C [18].

The application of CK B activity in serum in the investigation of suspected acute myocardial infarction requires high accuracy and precision since the discrimination limit of CK B for this condition is low, 0.2 μkat/l (12 U/l, 37 °C). With the SCE recommended CK methodology and a serum volume fraction of 0.0435 [7,15], this activity corresponds to an increase of 339 nm absorbance of about 0.003/min. Sample rAK rates may be as low as 0.001/min [15]. Any changes in the assay conditions that will result in a lower signal for a given amount of enzyme will reduce accuracy and precision below the values reported for CK B determined under SCE conditions [7, 15]: a) a reaction temperature of 30 °C will reduce reaction rates by more than 35 percent as compared to 37 °C; b) a volume fraction of serum of 0.030 [19] instead of 0.0435 [7, 15] will reduce reaction rates by 30 percent; c) monitoring at 334 nm instead of 340 nm (339 nm) will reduce the signal by about 2 percent. All in all, measurement at 30 °C, with a serum volume fraction of 0.030 [19] at 334 nm wavelength will reduce reaction rates to about 30 percent of those that can be obtained under SCE conditions [15].

Consequently, for highest precision, photometric measurements of CK B activity in serum should be carried out at 37 °C as described below. The determination of apparent CK B activity (CK B + rAK + reagent blank rates) should be corrected by a separate measurement of rAK (rAK + reagent blank rates) [7, 8, 12, 15, 17]. Sample rAK is measured in the CK working reagent (3), i.e. without creatine-P. Serum rAK activities are frequently very low, 0.1 μkat/l – 0.15 μkat/l (6 U/l – 9 U/l, [7, 15]), but may be increased up to, e.g., 2 μkat/l (120 U/l) in liver congestion due to heart failure [7, 8]. The CK reagent will sustain a serum rAK activity of 2 μkat/l at a constant rate for nearly 50 min (see CK). It is only necessary to correct for rAK in samples with increased apparent CK B activity.

* ATP : AMP phosphotransferase, EC 2.7.4.3

** ATP : D-hexose-6-phosphotransferase, EC 2.7.1.1

*** D-glucose-6-phosphate : $NADP^+$ 1-oxidoreductase, EC 1.1.1.49

Optimized conditions for measurement: see for CK, p. 511. Use anti-M as specified below with the IFCC CK reagent at 37 °C.

Temperature conversion factors: as described under CK.

Equipment

as specified for CK, p. 512.

Reagents and Solutions

Purity of reagents: as specified for CK, p. 512.

Quality requirements for anti-M: immunoinhibition of CK M subunit activity in a pure, human CK MM control preparation of 33.3 µkat/l (2000 U/l) should exceed 99.9 percent within 10 min at 25 °C, 30 °C, or 37 °C when the sample is incubated with anti-M in the CK working reagent (3) [15]. At the same anti-M concentration that inhibits CK MM by more than 99.9 percent, residual CK B activity should be within 45 – 50 percent in a MB hybrid, freshly isolated from human serum [15]. At the same anti-M concentration the inhibition of a pure human CK BB control preparation should be less than 5 percent [15].

Preparation of solutions*: as described for the determination of the catalytic activity of CK.

1. Stock solution: cf. p. 513.
2. CK reagent stock solution: cf. p. 513.
3. CK reagent working solution with anti-M:
   prepare solution as described on p. 513, add finally anti-M according to the functional specifications [15].
4. Substrate: cf. p. 514.

Stability of solutions: as described under CK. Stability of anti-M in CK working reagent is the same as for this reagent alone, cf. p. 514.

Procedure

Collection and treatment of sample: as described under CK, p. 514.

Stability of isoenzyme activities in sera: is described under CK. For routine determinations no stabilizers are necessary or desirable ([7], see under CK). For research

* reagent kits are commercially available.

purposes adjustment of sample pH to 6.5 – 7.0 and inclusion of monothioglycerol, 10 mmol/l, will extend the stability of the most labile isoenzyme, CK BB, to at least 7 days at 4 °C [20], 6 months at – 20 °C.

Stability of CK B activity in CK working reagent: 20 min at room temperature, 10 min at 37 °C.

Details for measurement in tissues: the isoenzyme pattern will depend on the completeness of homogenization, e.g. the extent of solubilization of mitochondrial creatine kinases. Results obtained by immunoinhibition or by chromatography should always be verified by isoenzyme electrophoretic identification of isoenzyme, variants, and adenylate kinases, carried out ± creatine phosphate and ± anti-M [4, 6].

Assay conditions: wavelength 339 nm; light path 10 mm; final volume 2.30 ml; 37 °C.

Measurement

Pipette successively into cuvettes:	sample CK B	reference rAK	concentration in assay mixture
working reagent with anti-M (3)	2.00 ml	2.00 ml –	as specified for CK
serum	0.10 ml	0.10 ml	volume fraction 0.0435
mix very carefully; wait at least 10 min for complete reactivation and immunoinhibition			
substrate (4)	0.20 ml	–	as specified for CK
water	–	0.20 ml	
wait 2 min (end of lag phase), monitor the increase of absorbance for at least 1 min preferably on a recorder. Calculate the $\Delta A/\Delta t$ from the linear part of the curve.			

Calculation: calculate $\Delta A/\Delta t$ from the linear part of the conversion curve. Use calculation formulae (k) or (k_1), resp. (cf. "Formulae", Appendix 3). For values of ε cf. Appendix 4. The following relations are valid (t in minutes):

Catalytic concentration in sample is

	339 nm	Hg 334 nm	
b =	$60.8 \times \Delta A/\Delta t$	$62.0 \times \Delta A/\Delta t$	µkat/l
b =	$3649 \times \Delta A/\Delta t$	$3720 \times \Delta A/\Delta t$	U/l

Validation of Method

Precision, accuracy, detection limits and sensitivity: imprecision of the direct, apparent CK B determination of activities greater than about 0.6 µkat/l (36 U/l) lies only slightly higher than that of total CK. The day-to-day imprecision of human purified isoenzyme controls ($n = 23$) using the SCE evaluated CK B method at 37 °C was [15]:

	BB	*B in BB*	*% B in BB*	*MB*	*B in MB*	*% B in MB*
µkat/l	12.45	12.28	98.7	9.20	3.47	37.7
U/l	747	737	98.7	552	208	37.7
CV %	3.8	4.2	3.0	4.5	3.8	3.6

However, CK B in serum is found from two measurements: apparent CK B activity minus sample rAK. Thus, frequently very low values of sample rAK with a relatively high imprecision are subtracted from considerably higher apparent CK B values determined with a better precision.

The within-day imprecision (RSD) of serum rAK, 0.08 µkat/l (5 U/l), is about 15% based on monitoring for two min after the end of the lag phase on an *LKB* 2086 [7, 15].

The within-day imprecision (RSD) of serum CK B values corrected for rAK at the discrimination limit for acute myocardial infarction of 0.20 µkat/l 12 U/l [7] is about 8% and the day-to-day imprecision about 10% [7, 15]. Imprecision as a function of CK activity with and without correction for serum rAK is discussed in ref. [15]. The day-to-day imprecision set in relation to the upper reference limit of serum CK B and to the discrimination limit for acute myocardial infarction is plotted in Fig. 2.

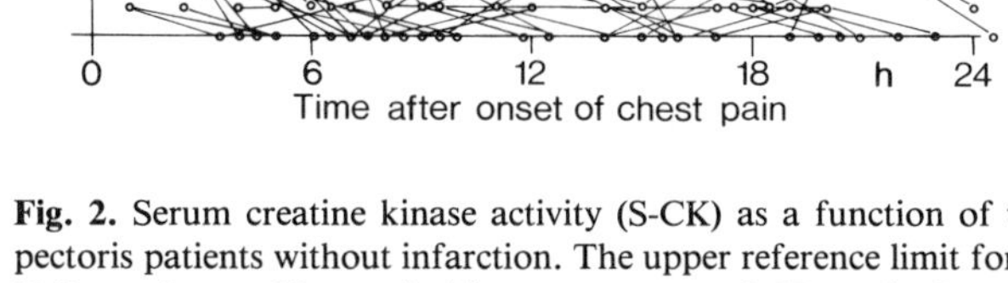

Fig. 2. Serum creatine kinase activity (S-CK) as a function of the time after onset of pain in 36 angina pectoris patients without infarction. The upper reference limit for S-CK B (6 U/l) and the discriminator (12 U/l) are shown. The vertical bar represents ± 2 SD as the "worst case" analytical imprecision of ± 15% (CV) at 9 U/l. Observe that even at + 2 SD the highest value does not reach the discriminator level. With permission from Clin. Chem. 1982 [7].

The detection limit depends on the reagent-blank and rAK rates, instrumentation, and monitoring time. In the absence of rAK activity the detection limit for CK B activity is about the same as for CK, i.e. about 0.08 μkat/l (5 U/l) at an imprecision corresponding to a RSD of 20 percent (see under CK).

The sensitivity at 339 nm, 37 °C, 2 min monitoring time after the end of the lag phase and correction for low reagent and rAK blank rates may be estimated to be around 0.004/2 min (*LKB* 2086). Determination of CK B activity by bioluminescence in the presence of anti-M decreases the detection limit by a factor of around 10 ([15],see luminometric method).

Sources of error: false positive results may be caused by

a) incomplete immunoinhibition due to poor quality of anti-M, too low concentration of anti-M relative to CK M-subunit activity, insufficient mixing of sample with anti-M, or too short an incubation time with the immunoinhibitor [7, 15]. Characteristically, the two latter causes will show decreasing reaction rates with time after start.
b) omission of correction for rAK activity.

The frequency distribution of serum residual AK values in patients suspected of suffered acute myocardial infarction and the significance of AK activity as a source of potential errors have been discussed in detail in refs. [7, 8, 15, 17]. Failure to correct for rAK activity in skeletal muscle extracts has led to gross overestimations of the contribution of MB to the activity of human skeletal muscle in early studies carried out with a now obsolete commercial glutathione-activated CK MB kit with AMP as the only AK inhibitor (discussed in ref. [14]).

False negative results: non-specific inhibition of B-subunit activity by poor quality of anti-M [15]; decrease of CK B activity by incubation of serum in the CK working reagent A for more than 30 min at room temperature; too long storage of samples.

Specificity: with correction for sample rAK activity the assay measures CK B plus (the rarely occurring) CK MiMi activities [4].

Reference ranges: in serum of healthy individuals CK B activity is so low that accurate determinations of CK reference values require a bioluminescence method, calibrated against the routine photometric method performed at 37 °C [15]. By this procedure the upper reference limit of CK B in serum for both sexes has been found to 0.1 μkat/l (6 U/l), 97th percentile, 37 °C [15]. The discrimination limit for acute myocardial infarction, both sexes, lies higher (Fig. 1) at 0.20 μkat/l (12 U/l, 37 °C, [7], [15]). Histograms of reference and patient distributions are given in refs. [7] and [15].

A relative discriminator of 6% CK MB (= 3% CK B activity in serum) is also widely used. For discussions of the respective advantages of the two types of discriminator see refs. [1], [7] and [14].

References

[1] *H. Lang* (ed.), Creatine Kinase Isoenzymes. Pathophysiology and Clinical Application, Springer-Verlag, Berlin 1981.

[2] *W. Stein, J. Bohner, M. Eggstein,* Prevalence of Macro Creatine Kinases, in: *D. M. Goldberg, M. Werner* (eds.), Progress in Clin. Enzymology, Vol. II. Masson, New York 1982 (in press).

[3] *W. Stein, J. Bohner, J. Krais, M. Müller, R. Steinhardt, M. Eggstein,* Macro Creatine Kinase BB: Evidence for Specific Binding between Creatine Kinase BB and Immunoglobulin G, J. Clin. Chem. Clin. Biochem. *19*, 925–930 (1981).

[4] *J. Bohner, W. Stein, R. Steinhardt, U. Würzburg, M. Eggstein,* Macro Creatine Kinases: Results of Isoenzyme Electrophoresis and Differentiation of the Immunoglobulin-Bound Type by Radioassay, Clin. Chem. *28*, 618–623 (1982).

[5] *P. Urdal, J. H. Strömme,* Macro Creatine Kinase BB in Serum is an Antigen-autoantibody Complex, in: *D. M. Goldberg, M. Werner* (eds.), Progress in Clin. Enzymology, Vol. II. Masson, New York 1982 (in press).

[6] *L. Ljungdahl, W. Gerhardt,* Creatine Kinase Isoenzyme Variants in Human Serum, Clin. Chem. *24*, 832–834 (1978).

[7] Scandinavian Committee on Enzymes of the Scandinavian Society for Clinical Chemistry and Clinical Physiology (SCE),
a) Creatine Kinase (EC 2.7.3.2) and Creatine Kinase B-subunit Activity in Serum in Suspect Myocardial Infarction. SCE Evaluation I, Nordic Clinical Chemistry Project (Nordkem) may be requested from: Nordkem Sibeliusgatan 12–14, SF-00260 Helsinki, Finland.
b) Creatine Kinase and Creatine Kinase B-subunit Activity in Cases of Suspected Myocardial Infarction, Clin. Chem. *28*, 277–283 (1982).

[8] *W. Gerhardt, M. Hørder, J. Waldenström,* Immunological Assay of Creatine Kinase and its Isoenzymes, in: *D. M. Goldberg, M. Werner* (eds.), Selected Topics in Clin. Enzymology, Vol. II. W. de Gruyter, Berlin, New York 1983, pp. 275–294.

[9] *J. Bohner, W. Stein, W. Renn, R. Steinhardt, M. Eggstein,* Stability of Macro Creatine Kinases and Creatine Kinase Isoenzymes Compared: Heat Inactivation Test for Determination of Thermostable Creatine Kinases, J. Clin. Chem. Clin. Biochem. *19*, 1021–1026 (1981).

[10] *R. A. Wevers, H. P. Olthius, J. C. C. van Niel, M. G. M. van Wilgenburg, J. B. J. Soons,* A Study on the Dimeric Structure of Creatine Kinase (EC 2.7.3.2), Clin. Chim. Acta *75*, 377–385 (1977).

[11] *L. M. Silverman, G. B. Dermer, M. H. Zweig, A. C. Van Steirteghem, Z. A. Tökés,* Creatine Kinase BB: A New Tumor-Associated Marker, Clin. Chem. *25*, 1432–1435 (1979).

[12] *R. C. Elser, K. McKenna,* Creatine Kinase B-subunit Activity in Human Sera: Temporal Aspects of its Sensitivity Following Myocardial Infarction, Clin. Chem. *27*, 57–60 (1981).

[13] *J. A. Lott, J. M. Stang,* Serum Enzymes and Isoenzymes in the Diagnosis and Differential Diagnosis of Myocardial Ischemia and Necrosis, Clin. Chem. *26*, 1241–1250 (1980).

[14] *W. Gerhardt, S. Hofvendahl, L. Ljungdahl,* Diagnostic Approach in Suspected Acute Myocardial Infarction, in: *A. Burlina, L. Galzigna* (eds.), Clinical Enzymology Symposia 4, Piccin Medical Books, Padova 1983, in press.

[15] *W. Gerhardt, J. Waldenström,* Creatine Kinase B-Subunit Activity in Serum after Immunohibition of M-Subunit Activity. Clin. Chem. *25*, 1274–1280 (1979).

[16] *J. Boone, E. J. Sampson, S. Lewis,* et al., An Interlaboratory Study of Creatine Kinase and Creatine Kinase Isoenzymes, Clin. Chem. *26*, 513–519 (1980).

[17] *P. M. Bayer, M. Boehm, P. Hajdusich,* et al., Immunoinhibition and Automated Column Chromatography Compared for Assay of Creatine Kinase Isoenzyme MB in Serum. Clin. Chem. *28*, 166–169 (1982).

[18] *W. Stein, J. Bohner, R. Steinhard, M. Eggstein,* Macro Creatine Kinase: Determination and Differentiation of Two Types by Their Activation Energies. Clin. Chem. *28*, 19–24 (1982).

[19] Enzymology Commission, Société Française de Biologie Clinique, Recommendations for Measuring the Catalytic Concentration of Creatine Kinase in Human Serum (documents C, 1976, C′ 1980, C, 1980), Ann. Biol. Clin. *40*, 138–149 (1982).

[20] *D. A. Nealon, S. M. Pettit, A. R. Hendersson,* Effect of Serum pH on Storage Stability and Reaction Lag Phase of Human Kinase Isoenzymes, Clin. Chem. *26*, 1165–1170 (1980).

7.5 Adenylate Kinase (Myokinase)

ATP : AMP phosphotransferase, EC 2.7.4.3

Sven E. Brolin

General

Adenylate kinase, AK, occurs in animal and plant tissues and in various microorganisms [1]. Its molecular weight of about 22000 is low compared with enzymes in general, and this is of advantage in studies of leakage from cells with high AK activity. Detecting or excluding such leakage may require analyses at low levels of detectability.

In catalyzing the reaction 2 ADP $\rightleftharpoons$ AMP + ATP, adenylate kinase offers a metabolic short-cut for a rapid supply of ATP. A transient need can thus be met by mobilizing energy stored in ADP. This is facilitated by a high content of the enzyme in heart and skeletal muscle. However, an immediate demand for energy is not limited to muscle movements but concerns also generation of signals, particularly the release of transmitter substances and hormones [2, 3]. Rapid provision of energy may also be needed for changes of membrane potentials. In contrast to muscle cells, cells which exert regulatory functions often occur scattered or in small groups. Studies of the limited quantities which can be isolated require analyses at low levels of detection.

Application of method: the high AK activity of nerve cells is of interest in clinical chemistry since leakage into the cerebral fluid can be utilized for evaluation of cerebral damage and for diagnosis of neurological diseases. Such outflux of AK may reflect also moderate, and to some extent reversible, changes which can be caused by hypoxia [4]. This has occasionally been observed as a sequel to open heart surgery [5]. A survey of AK leakage as a diagnostic tool in neurology has recently been completed [6]. Determinations of AK in red blood corpuscles and in the outflux from them is another relevant application. Like other enzymes with high activities in muscles, AK appears in serum after cardiac infarction, but this has so far been of minor diagnostic value.

The development and rapidly-extending use of micro-techniques [7, 8] (cf. Vol. I, chapter 3.8) permit analyses of small groups of cells and studies of their bioenergetic properties, including their AK activities. This is of considerable interest in physiology and biochemistry. In clinical chemistry AK analyses may be applied to cells from fine needle- and punch-biopsies.

Enzyme properties relevant in analysis: six different tissue-specific human AK isoenzymes are known. Their molecular weights are all between 20000 and 31000 [9]. Variations in organ isoenzyme patterns have been utilized for determining red blood cell phenotypes in genetic studies [9]. It is possible to attain linearity in substrate conversion with respect to time that is satisfactory for analytical purposes. A bivalent metal ion, preferably Mg^{2+}, is required for AK activity.

Methods of determination

Spectrophotometric assay is preferable if the concentration of AK is sufficiently high.

Fluorimetric assay (cf. Vol. I, chapter 3.4 "Fluorimetry") permits measurements of much smaller amounts of AK than absorption spectrophotometry. Coupled reactions are utilized so that either NAD or NADPH is formed. The latter can be measured because of its native fluorescence, whereas NAD has to be converted into a strongly fluorescent compound by treatment with strong alkali.

Luminometric assay (cf. Vol. I, chapter 3.5 "Luminometry") can be employed at low levels of detectability and provides great potentialities in permitting simple analytical performance and fast response. Clinical application of luminescence assay seems to be of considerable interest [10]. Firefly luciferase* can be recommended to those laboratories which are using, or intending to set up, bioluminescence analyses for measurement of AK. In all cases, however, where fluorimetric enzyme assay is in routine use it is not very likely that change of technique will result in any considerable improvement.

International reference methods and standards: there is as yet no standard reference material.

Enzyme effectors: no other activators of AK than Mg^{2+} seem to have been reported.

7.5.1 UV-method

Assay

Method Design

Principle: spectrophotometric and fluorimetric assay by means of ADP formation is accomplished in a system of coupled reactions**

* *Photinus* luciferin : oxygen 4-oxidoreductase (decarboxylating, ATP-hydrolysing), EC 1.13.12.7.

** Abbreviations

AK	Adenylate kinase, EC 2.7.4.3
PK	Pyruvate kinase, EC 2.7.1.40
LDH	Lactate dehydrogenase, EC 1.1.1.27
PEP	Phosphoenol pyruvate

(a) $\text{AMP} + \text{ATP} \xrightleftharpoons{\text{AK}} 2\,\text{ADP}$

(b) $2\,\text{ADP} + 2\,\text{PEP} \xrightleftharpoons{\text{PK}} 2\,\text{ATP} + 2\,\text{pyruvate}$

(c) $2\,\text{Pyruvate} + 2\,\text{NADH} + 2\,\text{H}^+ \xrightleftharpoons{\text{LDH}} 2\,\text{lactate} + 2\,\text{NAD}^+$

Optimized conditions for measurement: the data of the kinetics of phosphoryl transfer from ATP [11] have provided guidance in choosing the concentrations of the various solutions. When suitable concentrations of the reactants were found they have been used, although they may not be optimal. It may be possible to attain some minor improvement but a large number of experiments would be required for a detailed evaluation of the combined reactions.

Temperature conversion factors: the reaction rates depend to some extent on the concentrations of magnesium. The binding constants of the different nucleotides may be influenced by the temperature so that determining useful conversion factors seems to be a rather difficult task.

Equipment

In spectrophotometric analysis the instrumentation should permit reading of small absorption differences. This concerns particularly low levels of detection, since AK occurs often at low concentrations.

Reagents and Solutions

Purity of reagents: auxiliary and indicator enzyme must be free from AK and phosphatases. ATP should not contain more than 0.5% of ADP. All other reagents must be of analytical reagent grade.

Preparation of solutions (for approximately 16 determinations): all solutions in re-purified water (cf. Vol. II, chapter 2.1.3.2).

1. Sodium hydroxide/EDTA (NaOH, 10 mmol/l; EDTA, 5 mmol/l):

   dissolve 41 mg NaOH and 186 mg EDTA-$Na_2H_2 \cdot 2H_2O$ in 100 ml water.

2. Triethanolamine buffer (TEA, 0.07 mol/l, pH 7.6):

   dissolve 1.55 g TEA-HCl in approximately 100 ml water, adjust pH to 7.6 with NaOH, 2 mol/l and make up to 120 ml with water.

3. Working solution (TEA, 0.07 mol/l; $MgSO_4$, 19.5 mmol/l; KCl, 64.5 mmol/l; PEP, 0.45 mmol/l; dithiothreitol, 1.5 mmol/l; PK, 4.1 kU/l; LDH, 22.6 kU/l):

   dissolve in 100 ml TEA buffer (2) 480 mg $MgSO_4 \cdot 7H_2O$ + 481 mg KCl + 20.9 mg PEP (tricyclohexylammonium salt) + 23.1 mg dithiothreitol; (solution 3a). Centrifuge ammonium sulphate suspensions of the ancillary enzymes, dissolve the precipitate in buffer (2) and pipette 410 U PK and 2260 U LDH into 100 ml solution (3a).

4. NADH solution (6.9 mmol/l):

   dissolve 26.1 mg NADH (disodium salt, $2H_2O$) in 5 ml buffer (2).

5. ATP solution (19.5 mmol/l):

   dissolve 118 mg ATP-$Na_2 \cdot 3H_2O$ in 10 ml water.

6. AMP solution (19.5 mmol/l):

   dissolve 34.1 mg AMP-$Na_2 \cdot 6H_2O$ in 3.5 ml water.

Stability of solutions: store buffers frozen in order to avoid disturbing growth of micro-organisms. Use freshly prepared buffer/substrates containing nucleotides and ancillary enzymes and keep them on ice until incubated. Storage below −80 °C may be tried, but do not add the enzymes earlier than a few hours before the assay. The dissolved ATP reagent (6) can be divided in portions and stored below −18 °C for several weeks.

Procedure

Collection and treatment of specimen: various tissues can be assayed after homogenization in 10 to 30 volumes of TEA buffer (2). After lyophilization, conglomerates of cells can alternatively be briefly extracted using NaOH/EDTA solution (1).

Stability of the enzyme in the specimen: AK is regarded as a very resistant enzyme particularly to changes in temperature and pH, but this is not valid for all the isoenzymes [1].

Details for measurements in different kind of specimens: work with homogenates, tissue extracts and body fluids may involve low levels of detection but comparatively large volumes. Volumes larger than those usual in the assay of small cell specimens can accordingly be employed throughout the entire analytical procedure. Certain precautions have to be taken in assay of body fluids, since haemolysis may cause outflux of AK from red corpuscles. Cerebrospinal liquor should be centrifugated twice and the supernatant stored below −20 °C [4].

Assay conditions: blanks without addition of AMP are used to compensate for the formation of ADP which may be caused by phosphatase in the sample. Assay of low AK concentrations are often of interest in clinical chemistry. The assay should therefore be designed for comparatively large sample volumes and for measurements during a sufficiently long period [4]. For the sake of maximum sensitivity at low levels of detectability the absorbance is read at 339 nm.

Light path 10 mm; final volume 3.0 ml; 24 °C (thermostatted cuvette holder). Measure against air. Before starting the assay, adjust temperature of all solutions to 24 °C.

The consumption of NADH is followed after conversion of contaminating ADP and initiated by addition of AMP.

Measurement

Pipette successively into the cuvettes:	reagent blank	sample blank	sample	concentration in assay mixture
working solution (3)	2.00 ml	2.00 ml	2.00 ml	TEA 0.05 mmol/l Mg^{2+} 13 mmol/l KCl 43 mmol/l dithiothreitol 1 mmol/l PK 2.7 kU/l LDH 15 kU/l PEP 0.3 mmol/l
sample (*e.g.* liquor or tissue extract)	–	0.50 ml	0.50 ml	volume fraction 0.167
water	0.50 ml			
NADH solution (4)	0.10 ml	0.10 ml	0.10 ml	NADH 0.23 mmol/l
mix thoroughly with a plastic spatula and read absorption				
ATP solution (5)	0.20 ml	0.20 ml	0.20 ml	ATP 1.3 mmol/l
mix, wait for 5 min, and read absorption. (Decrease of the reagent blank reflects the amount of ADP which contaminates ATP. Phosphatases cause an additional decrease in sample and sample blank.)				
AMP solution (6) (starts the AK reaction)	–	–	0.20 ml	AMP 1.3 mmol/l
water	0.20 ml	0.20 ml	–	
mix and read absorbance per unit time, which can be done for at least 20 min.				

Calculation: correct change in absorbance per unit time of sample and sample blank for blank readings. The sample blank represents the catalytic activity of phosphatases; subtract readings for sample blank from readings for sample, yielding $\Delta A/\Delta t$. Use calculation formula (k) or (k_1) (cf. "Formulae", Appendix 3). Values of ε: cf. Appendix 4. For the catalytic activity concentration of the sample the following relations are valid (Δt in minutes):

$$b = 476 \times \frac{\Delta A}{\Delta t} \quad \text{U/l} \quad (24\,^\circ\text{C})$$

$$b = 7936 \times \frac{\Delta A}{\Delta t} \quad \text{nkat/l} \quad (24\,^\circ\text{C}).$$

If desired, the phosphatase activity concentration in the sample can be estimated by the same formula, using values for $\Delta A/\Delta t$ of the sample blank.

7.5.2 Fluorimetric Methods

7.5.2.1 Determination of ADP Formation

Assay

Method Design

Principle: according to eqns. (a) – (c) of p. 542. NAD serves as the measurable product but its conversion into a strongly fluorescent compound must be preceded by selective destruction of unconsumed NADH.

Optimized conditions, temperature conversion factors: as described on p. 542.

Equipment

In *fluorimetric analyses* the production of about 0.4 nmol NADPH has been measured at concentrations about 1 µmol/l. It is feasible to attain this level of detection with several commercially-available fluorimeters. Conversion of NAD into a strongly fluorescent compound with alkali does not increase the demands on instrumental performance since a much higher intensity is obtained than with the native fluorescence of the reduced form of the nucleotide [7].

At low levels of detectability counting of light quanta is required but measuring the output of a sensitive photomultiplier is generally quite sufficient.

Reagents and Solutions

Purity of reagents: in the ancillary enzymes which are used in the fluorimetric methods, only minimum amounts can be allowed of such contaminants as ATPase or other phosphatases and enzymes interfering with nucleotide conversions.

Preparation of solutions (for approximately 700 determinations; may be scaled down *e.g.* to a fifth if the weighing of the smaller quantities remains convenient).

Solutions of buffers and reagents should be preferably prepared with double quartz-distilled water. When working at low levels of detectability, care should be devoted to minimizing pollution from the air by keeping all solutions covered.

1. Sodium hydroxide/EDTA (NaOH, 10 mmol/l; EDTA, 5 mmol/l):

   dissolve 41 mg NaOH and 186 mg EDTA-$Na_2H_2 \cdot 2H_2O$ in 100 ml water.

2. Triethanolamine buffer (TEA, 0.07 mol/l, pH 7.6):

   dissolve 1.55 g TEA-HCl in approximately 100 ml water, adjust pH ad 7.6 with NaOH, 2 mol/l, and make up to 120 ml with water.

3. Working solution (TEA, 0.07 mol/l; $MgCl_2$, 9.8 mol/l; KCl, 220 mmol/l; AMP, 4.2 mmol/l; ATP, 3.36 mmol/l; PEP, 2.0 mmol/l; NADH, 2.0 mmol/l; BPA*, 0.7 g/l; PK, 0.42 kU/l; LDH, 2.52 kU/l):

   dissolve in 7.14 ml TEA buffer (2) 14.2 mg $MgCl_2 \cdot 6H_2O$ + 117.1 mg KCl + 15.0 mg AMP-$Na_2 \cdot 6H_2O$ + 14.5 mg ATP-$Na_2 \cdot 3H_2O$ + 6.64 mg PEP (tricyclohexylammonium salt) + 10.7 mg NADH (disodium salt, $2H_2O$) + 5 mg BPA; (solution 3a). Centrifuge ammonium sulphate suspensions of the ancillary enzymes, dissolve the precipitate in 100 µl buffer (2) and pipette 3 U PK and 18 U LDH into 7.14 ml solution (3a).

   (3b. Working solution for assay of lyophilized cell samples: pipette 2.86 ml water into 7.14 ml working solution 3).

---

* BPA Bovine plasma albumin.

4. NAD standard solution (1.25 mmol/l):

   dissolve 1.70 mg NAD (free acid) in 2 ml water.

Stability of solutions: as described on p. 543.

Procedure

Collection and treatment of specimen: pieces of organs are quickly cut out from decapitated animals (cf. Vol. II, chapter 1.2.1), attached to cryostat chucks with brain paste as adhesive and plunged into isopentane chilled to −160 °C with liquid nitrogen. Groups of cells are cut out free-hand from lyophilized cryostat sections (25 μm) under a stereomicroscope, see Fig. 1. Specimens in the range of 0.1 to 6 μg dry weight are

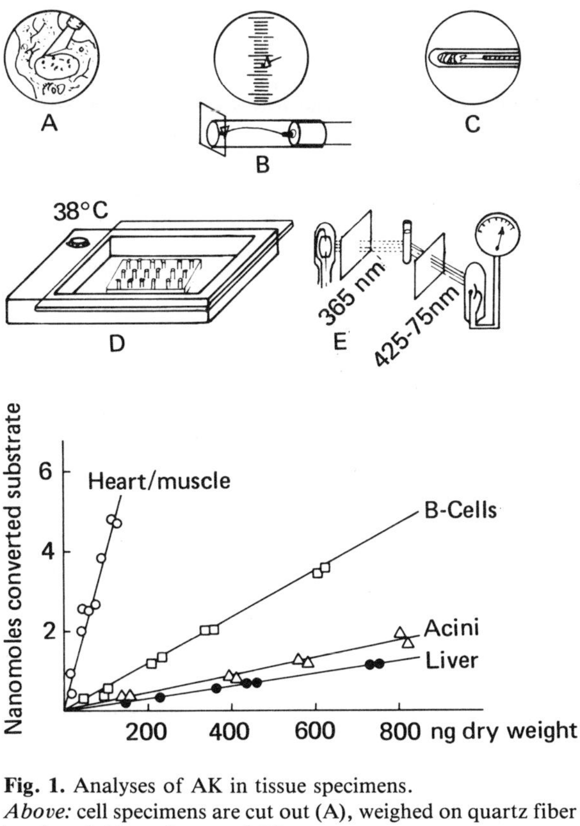

Fig. 1. Analyses of AK in tissue specimens.
Above: cell specimens are cut out (A), weighed on quartz fiber balance (B), inserted into micro-tubes (C), incubated (D) and assayed (E).
Below: conversions related to specimens weights. After 15 min incubation, ATP formation is measured with coupled hexokinase G6P-DH reactions, using native fluorescence of formed NADPH. Each symbol gives the mean of five specimens, which all are prepared from mice.

inserted into microtubes (inner diameter 3 mm; length 50 mm), containing working solution (3b) for production of a measurable nucleotide form. Mixing is not allowed until incubation is completed, since fragments of a freshly inserted specimen may be caused to stick at the wall of the tube. (For assay of liquid samples as homogenates working solution (3) is used.) The tubes are kept on ice until assayed. For micro-techniques cf. also Vol. I, chapter 3.8.

Stability of the enzyme in the specimen: freeze-dried sections can be stored for several weeks at −25 °C in purified argon gas containing less than 1 ppm oxygen. During the isolation of the cell specimens, the relative humidity should be below 40%. In contrast to the sections, the cell specimens should not be stored very long and if possible inserted into the micro-tubes for extraction within two to three hours.

Details for measurements in different kinds of specimens: groups of cells can be isolated from fine needle- or punch-biopsies, either freshly or after lyophilization. If freshly isolated, the specimens should be frozen transiently in order to break cell membranes. Cell specimens can also be prepared by collagenase digestion followed by washing in *Hanks'* balanced salt solution as used for isolation and metabolic studies of the islets of *Langerhans* [12, 13]. Specimens collected in this way are lyophilized, inserted into the sodium hydroxide/EDTA solution (1) and immediately transferred to ice. Thereafter working solution (3) is added as required for production of the nucleotide form used for measurements. The smaller specimens shown in Fig. 1 can be directly inserted into the buffer/substrate. Naturally a few µl of a homogenate or of a preparation of AK can be instilled instead of a solid specimen at step C in Fig. 1.

Assay conditions: incubation of the sample in buffer/substrate is performed at 38 °C, whereas all measurements are carried out at room temperature (24 °C). Primary filters are used for isolating the strong Hg line 365 nm and secondary filters for transmitting the spectral region 425 nm to 475 nm. Incubation volume 14 µl.

Tissue blanks are required if the formation of ADP is erroneously increased as a result of hydrolysis of ATP by phosphatases in the sample. Such blanks are processed without AMP in the incubation medium (3). Generally 20 samples, 4 blanks and 4 standards are run.

Note: Instead of tissue extracts, body fluids or homogenates, small pieces of freeze-dried sections can be inserted directly into 14 µl portions of working solution (3b) which results in the same concentrations in the incubation mixture (cf. Fig. 1). Mixing is not allowed in this kind of assay.

Calculation: the fluorescence intensity F is proportional to the concentration c of the analyte (cf. Vol. I, chapter 3.4). Relating measurements to a standard, the change in substrate or product concentration per unit time ($\mu mol \times l^{-1} \times min^{-1}$) can be calcu-

Measurement

Incubation

Pipette successively into micro-tubes:	sample	NAD standard	tissue blank	concentration in incubation mixture
working solution (3) working solution without AMP (3)	10 µl –	10 µl –	– 10 µl	TEA 50 mmol/l Mg^{2+} 7 mmol/l K^+ 157 mmol/l AMP 3 mmol/l (not in blank) ATP 2.4 mmol/l PEP 1.4 mmol/l NADH 1.4 mmol/l BPA 500 mg/l PK 300 U/l LDH 1800 U/l
NAD standard (4) sample (*e.g.* tissue extract, body fluid or homogenate)	– 4 µl	4 µl –	– 4 µl	NAD 0.36 mmol/l volume fraction 0.29
tap the micro-tubes gently with the forefinger nail close to the bottom (8 – 10 times should be enough for mixing) and incubate 15 min (or 30 min in case of low AK activity) at 38 °C				
HCl (1.4 mol/l)	3 µl	3 µl	3 µl	HCl 247 mmol/l
enzyme reactions are stopped; NADH is destroyed selectively. Use this solution for fluorimetry.				

Fluorimetry

Pipette into fluorimeter tubes:	sample	NAD standard	tissue blank	concentration in assay mixture
NaOH (6.9 mol/l) incubation mixture	120 µl 10 µl	120 µl 10 µl	120 µl 10 µl	NaOH 6.37 mol/l
mix, heat at 60 °C for 5 min. Transfer to ice, after cooling, add 1 ml water, and measure fluorescence at room temperature				

lated. Correct fluorescence of sample for tissue blank reading, yielding ΔF_{sample}. The catalytic concentration b of the enzyme in the cuvette is

$$b = c_{\text{standard}} \times \frac{\Delta F_{\text{sample}}}{\Delta F_{\text{standard}}} / \Delta t \qquad \text{U/l} \quad (38\,°\text{C})$$

ΔF_{sample} is the fluorescence intensity of the sample minus the mean intensity of the blanks,

$\Delta F_{\text{standard}}$ is the fluorescence intensity of the standard nucleotide minus the intensity of the blanks.

c_{standard} is the concentration of the standard nucleotide in the incubation mixture (350 μmol/l).

Δt is the incubation time (15 or 30 min).

In the assay of freeze-dried cell specimens the catalytic content is often related to their dry weight (Fig. 1). For calculation of the catalytic concentration of the sample, the calculated b must be divided by the volume fraction $\varphi = 0.29$.

7.5.2.2 Determination of ATP Formation

Assay

Method Design

Principle

(d) $$2\,\text{ADP} \xrightleftharpoons{\text{AK}} \text{AMP} + \text{ATP}$$

(e) $$\text{ATP} + \text{glucose} \xrightleftharpoons{\text{HK*}} \text{ADP} + \text{G6P}$$

(f) $$\text{G-6-P} + \text{NADP}^+ \xrightleftharpoons{\text{G6P-DH}} \text{6-PG} + \text{NADPH} + \text{H}^+$$

Since the assay is based upon the native fluorescence of the reduced form of the nucleotide, unconsumed NADP does not disturb the measurement and need not be destroyed [2].

Optimized conditions, temperature conversion factors: as described on p. 542.

* Abbreviations:

G6P-DH	Glucose-6-phosphate dehydrogenase, EC 1.1.1.49.
HK	Hexokinase, EC 2.7.1.1.
6-PG	6-phosphogluconate.

Equipment

As described on p. 545.

Reagents and Solutions

Purity of reagents: as described on p. 546. Furthermore, hexokinase and glucose-6-phosphate dehydrogenase should not contain more than 0.01% 6-phosphogluconate dehydrogenase; otherwise additional formation of NADPH may disturb the calculations.

Preparation of solutions: as described on p. 546.

1. Sodium hydroxide/EDTA (NaOH, 10 mmol/l; EDTA, 5 mmol/l):

   dissolve 41 mg NaOH and 186 mg EDTA-$Na_2H_2 \cdot 2\,H_2O$ in 100 ml water.

2. Triethanolamine/EDTA buffer (TEA, 0.07 mol/l, pH 7.6; EDTA, 1.4 mol/l):

   dissolve 1.55 g TEA-HCl and 6.25 mg EDTA-$Na_2H_2 \cdot 2H_2O$ in about 10 ml water, adjust pH ad 7.6 with NaOH, 2 mol/l, and make up to 12 ml.

3. Working solution (TEA, 0.07 mol/l, pH 7.6; EDTA, 1.4 mmol/l, ADP, 4.2 mmol/l; glucose, 2.8 mmol/l; NADP, 2.3 mmol/l; $MgCl_2$, 7 mmol/l; BPA, 700 mg/l; HK 980 U/l; G6P-DH, 420 U/l):

   dissolve in 7.14 ml buffer (5) 10.2 mg $MgCl_2 \cdot 6H_2O$ + 15.04 mg ADP-K $\cdot 2H_2O$ + 3.60 mg D-glucose + 13.8 mg NADP (disodium salt, $3H_2O$) + 5 mg BPA; (solution 3a). Centrifuge ammonium sulphate suspensions of HK (ca. 150 U/mg; 2 mg/ml) and G6P-DH (ca. 350 U/mg; 1 mg/ml), discard supernatants, dissolve the enzyme precipitate in 100 ml buffer/substrate (6a) and add 7 U HK and 3 U G6P-DH to the freshly prepared solution (3a).

   (3b. Complete working solution for assay of lyophilized cell samples. Pipette 2.86 ml water into 7.14 ml working solution 3).

4. NADPH standard solution (1.25 mmol/l):

   dissolve 2.18 mg NADPH (tetrasodium salt, $2H_2O$) in 2 ml of a solution made up by 0.5 ml sodium hydroxide/EDTA solution (1) and 1.5 ml water.

Stability of solutions: as described on p. 543.

Procedure

Collection, treatment, stability of specimen: as described on p. 547.

Assay conditions: incubation of the samples in buffer/substrates is performed at 38 °C, whereas all measurements are carried out at room temperature (24 °C). Primary filters are used for isolating the strong Hg lins at 365 nm and secondary filters for transmitting the spectral region 425 nm to 475 nm. A reagent blank with water instead of sample or standard is needed. Tissue blanks are not required since HK prevents accumulation of ATP and hence any relevant interaction of phosphatases.

Measurement

Incubation

Pipette successively into micro-tubes:	sample	NADPH standard	blank	concentration in incubation mixture	
working solution (3)	10 µl	10 µl	10 µl	TEA	50 mmol/l
				EDTA	1 mmol/l
				ADP	3 mmol/l
				glucose	2 mmol/l
				NADP	1.6 mmol/l
				$MgCl_2$	5 mmol/l
water	–	–	4 µl	BPA	0.5 g/l
				HK	700 U/l
				G6P-DH	300 U/l
NADPH standard (4)	–	4 µl	–	NADPH	0.35 mmol/l
sample (*e.g.* tissue extract, body fluid or homogenate)	4 µl	–	–	volume fraction	0.286
tap the tubes gently with the forefinger nail for mixing and incubate 15 min (or 30 min in case of low AK activity) at 38 °C. Transfer to ice and use for fluorimetry without delay.					

Note: Instead of liquid samples lyophilized cells can be directly inserted into 14 µl portions of working solution (3b), see Fig. 1. This results in the same concentration at the incubation as shown in the right column. These samples may only be mixed after the incubation.

Fluorimetry

pipette successively into fluorimeter tubes:	sample	NADH standard	blank	concentration in assay mixture	
NaOH/EDTA (1)	360 µl	360 µl	360 µl	NaOH EDTA	9.7 mmol/l 5 mmol/l
incubation mixture	10 µ	10 µl	10 µl		
mix well, make up with water to the volume required (0.5 to 1.0 ml), measure fluorescence.					

Calculation: as described on p. 548. However, volume fraction of sample in the assay mixture is $\varphi = 0.286$ and concentration of standard in the incubation mixture is $c = 0.35$ mmol/l. Note that one mole ATP corresponds to two moles ADP.

7.5.3 Luminometric Method

Assay

Method Design

Principle

(g) $$2\,\text{ADP} \xrightleftharpoons{\text{AK}} \text{AMP} + \text{ATP}$$

(h) $$\text{ATP} + \text{luciferin} + O_2 \xrightarrow{\text{luciferase}} \text{ADP} + \text{PP}_i + H_2O + \text{oxyluciferin} + h\nu$$

ATP formation is monitored.

Bioluminescence assay can be used for determining the amount of ATP which is formed in the primary reaction. Disturbing light emission from contaminating adenylates and AK has required measurements using photokinetic technique [14]. Purified firefly luciferase has, however, made the assay much easier and moreover permitted a combination of the incubation and the ATP measurement, resulting in a single step, acceptable analytical performance.

Bioluminescence analyses of AK can be accomplished in a single step performance, using a purified special luciferase preparation. The removal of contaminating AK prevents disturbing emission when ADP is added to the light yielding solution. Addition of ATP yields a flux of light with negligible fading. The intensity level is proportional to the added amount of ATP. The enduring emission is mainly caused by a luciferin analogue serving as inhibitor of the luciferase reaction which yields enough light for the assay. Another requirement for a nearly constant emission is a suitable proportion between concentration of luciferase and luciferin [14, 15].

Optimized conditions for measurement: the assay conditions have not been optimized systematically since crystallized luciferin as well as crystallized firefly luciferase has only recently become available.

The luciferase solution is supplemented with ADP for continuous ATP formation catalyzed by AK in the sample.

Temperature conversion factors: not estimated.

Equipment

The lowest level of detectability is reached by photon counting if the signal-to-noise ratio is increased by cooling the photomultiplier and by minimizing the photosensitive surface [16]. Less elaborate equipment is generally sufficient, for instance instruments which record the photo-current as the voltage across a load resistor. Good-quality instruments of this kind are offered by several manufacturers. They are often equipped with an inbuilt computer which may be useful, but it is more important to have simple outlets for feeding the signals through an interface into a separate microprocessor. In this way programs can be more freely chosen for the processing and storage of analytical data [17]. It may be necessary to avoid high dilution of small samples when recording the light they are induced to emit. Furthermore, maintaining required concentrations of added reactants with increasing volumes decreases the ratio between the amount of sample and the amount of disturbing contaminants. It is therefore advisable to use a bioluminometer which can accomodate small volumes; as a minimum specification, the instrument should be capable of accepting a final volume of 100 µl.

Reagents and Solutions

Purity of reagents: partially purified extracts of firefly luciferase contain ADP and AK but can still be employed analytically. This requires degradation of ADP and ATP with apyrase and a photokinetic technique in order to discriminate between light emission from the sample and from the contaminants [18]. Alternatively, highly purified preparations which now are available simplify the analytical performance and demand less elaborate instrumentation.

For purity requirements of other substances cf. pp. 542 and 546.

Preparation of solutions (for about 500 determinations): solutions of buffers and reagents should be preferably prepared with double quartz-distilled water. When working at low levels of detectability, care should be taken to minimize air pollution by keeping all solutions covered.

1. ADP solution (Tris, 0.1 mol/l; EDTA, 2 mmol/l; ADP, 21 µmol/l; BPA, 1.2 g/l):

   dissolve 1.06 g Trizma-HCl + 0.394 g Trizma base + 74.4 mg EDTA-Na_2H_2 · $2H_2O$ in 100 ml water (adjust pH to 7.8 if required) + 120 mg BPA + 1.05 mg ADP-K · $2H_2O$.

2. Emitting solution (Tris 0.1 mol/l, pH 7.75; EDTA, 2 mmol/l; PP_i, 10 µmol/l; magnesium acetate, 50 mmol/l; BPA, 5 mg/ml; luciferase, 14 mU/l; D-luciferin, 2 mmol/l; L-luciferin 80 µmol/l):

   dissolve the lyophilized content of a vial *LKB-Wallac* 1243 Monitoring Reagent in 10 ml water which will give the aforementioned concentrations (other commercially-available kits containing sufficiently purified luciferase can also be used, as found in tests with *Boehringer* preparation ATP-Bioluminescence CLS (constant light signal).

3. Enzyme standard (Tris, 0.1 mol/l, pH 7.8; EDTA, 2 mmol/l; BPA, 1.2 mg/ml; AK 47.8 U/l):

   dissolve 0.532 g Trizma HCl + 0.197 g Trizma base + 37.2 mg EDTA-Na_2H_2 · $2H_2O$ + 60 mg BPA in 50 ml water, adjust pH to 7.8 if required; (buffer 3a). Centrifuge 20 µl AK suspension in ammonium sulphate solution (e.g. 360 U/mg; 2 mg/ml), dissolve the precipitate in 100 µl buffer (3a). Measure catalytic activity concentration of this standard solution spectrometrically according to p. 544; repeat for each batch after storage. Dilute solution (3) 3000 times with buffer (3a) just before use in the assay.

Stability of solutions: the emitting solution (2) may be stored for a couple of weeks below −20°C, but may not be re-frozen. The ADP solution (1) may be kept covered on ice for several hours. The enzyme standard (3) may be kept on ice a few hours before dilution but should then be used directly.

Procedure

Collection, treatment, stability of specimen: as described on p. 547.

Assay conditions: the assay is carried out at room temperature (24°C). Larger samples can be assayed with larger volumes of reactants. A less concentrated ADP solution (1) of e.g. 7 µmol/l will decrease the background emission from ATP contaminations of ADP, but it is better to use a sufficiently purified ADP preparation. In case of high AK activity the light intensity is measured after 20 s in order to restrict the consumption of ADP. Assay volume 105 µl.

Measurement

Pipette successively into reaction cells:		sample	enzyme standard	blank	concentration in assay mixture	
ADP solution	(1)	–	–	80 µl	ADP	16 µmol/l
					Tris	95 mmol/l
emitting soln.	(2)	–	–	20 µl	EDTA	1.9 mmol/l
water		–	–	5 µl	PP_i	1.9 mmol/l
					Mg^{2+}	9.5 mmol/l
					D-luciferin	380 µmol/l
					L-luciferin	15 µmol/l
					luciferase	2.7 mU/l
					BPA	1.8 g/l
mix and position the blank in front of the photomultiplier; measure in arbitrary units (e.g. mV) the intensity of light emission at 60 s after the mixing						
ADP solution	(1)	–	80 µl	–		
emitting soln.	(2)	–	20 µl	–		
standard	(3)	–	5 µl	–	AK	2.28 U/l
mix the standard and measure light intensity after 60 s						
ADP solution	(1)	80 µl	–	–		
emitting soln.	(2)	20 µl	–	–		
sample (tissue extract, body fluid or homogenate)		5 µl	–	–	volume fraction	0.048
mix the sample and measure light intensity after 60s.						

Calculation: as for fluorimetric measurements, luminometric readings (e.g. in mV) are related to a standard. In this case, the standard is an enzyme standard solution of defined catalytic concentration.

For calculation use eqn. (m_1) (cf. "Formulae", Appendix 3). The following relation is valid for the catalytic concentration in the sample:

$$b = \frac{\Delta I_{\text{sample}}}{\Delta I_{\text{standard}}} \times b_{\text{standard}} \quad \text{U/l} \quad (24\,°\text{C})$$

ΔI are the light emission intensities of sample and standard solution, corrected for intensity of blank.

For calculation of the catalytic content, the result must be divided by the mass concentration of protein ρ (g/l) of the sample. (For protein determinations cf. Vol. II, chapter 1.3.)

Validation of Methods

Precision, accuracy, detection limits and sensitivity: in spectrometric assay according to eqns. (a) – (c), the level of detectability will approach 0.3 to 0.2 U/l if the imprecision is allowed to rise to about 20%. The sensitivity is approximately 35 when estimated as the ratio between units AK/l and the change of absorbance in 25 min.

The total error has been determined for the fluorimetric method for NADPH formation according to eqns. (d) – (f). Calculated as the coefficient of variation and including both method errors and biological variation, it was found to be ± 16%. The total error should not differ much from this value in the assay according to eqns. (a) – (c), provided that the pipetting is adequate and that possible disturbances due to phosphatases are correctly compensated for.

The accuracy has not been specifically determined due to lack of standard reference material. Some information can be derived from the comparisons in Fig. 1, showing the relation between sample weight and conversion for different kinds of cells.

The detection limits in ordinary fluorimetry are also illustrated in Fig. 1. As can be seen smaller samples or shorter incubation time are required for the assay of cells with high AK activity. Concerning the sensitivity, the relationship between conversion rate and time indicates that a difference of about 0.25 nmol of substrate converted could be estimated. This amount would be produced in 0.6 min by 1 µg pancreatic β-cells.

The luminometric assay described with highly purified luciferase will certainly permit a low level of detectability with high precision. This new and convenient single step assay has not yet been applied in experimental series by us. Its applicability has, however, been evident in tests on liver homogenates using samples equivalent to wet weights from 30 ng to 3 µg. A curve demonstrating the increase of ATP concentration with time and a calibration curve for AK assay are shown in Fig. 2.

The maintenance of an emission which is proportional to the rising ATP concentration throughout the entire period of measurements considerably facilities the performance of the analyses. In reality, however, a quite satisfactory control is also obtained by storage of the display in a microprocessor, whether or not the consumption of ATP is restricted.

Sources of error: the detection limits in fluorimetric and luminometric analyses are much lower than in absorption spectrophotometry with the effect that disturbing reactions compete more strongly with the reactions of the analytical system. The main sources of error are phosphatases which may require compensation by special measures as already mentioned. It is naturally also necessary to prevent contamination with ATP (e.g. from water, from finger tips, etc.) and to avoid erronous ATP formation

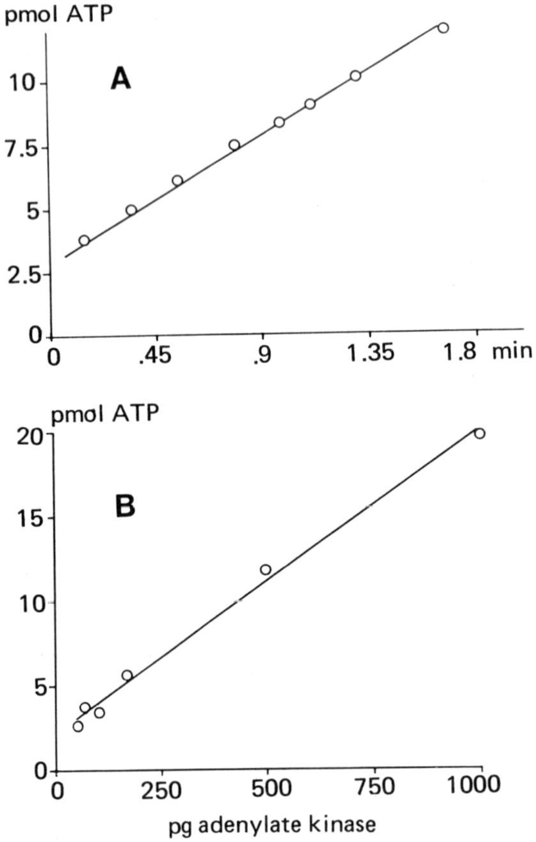

Fig. 2. Analyses of AK by means of a purified firefly luciferase preparation which yields an enduring flux of light.
A: Increase in light emission is elicited by adding 10 ng of AK to a luciferase solution, supplemented with ADP (7 µmol/l in the assay mixture). An initial emission is obtained due to contamination of ADP with ATP but is not subtracted as a blank. The correlation coefficient is 0.996, indicating good linearity between time and the accumulation of ATP.
B: Relationship between conversion rates as reflected by the amount of ATP which is present 60 s after the addition of AK. The concentration of ADP is higher than in *A* (16 µmol/l in the assay mixture). The correlation coefficient is 0.982.

which may occur if the samples contain phosphorylating enzymes together with their substrates. Vigorous mixing may inactivate enzymes, particularly in small volumes, and has to be avoided before incubations.

Specificity: other triphosphates than ATP may serve as phosphate donors but they have higher K_m values. The specificity for the AMP site of the enzyme is more rigorous; substitution by other nucleoside monophosphates leads to much decreased or no activity [1].

Reference ranges: the AK activity of cerebrospinal fluid is below detection or very weak when measured with absorption spectrophotometry. In pathological conditions the activity may just exceed the level of spectrometric detectability but sometimes reaches 1 to 2 U/l [6]. Information about AK activity has been presented for various organs and tissues [3] and can also be derived from Fig. 1.

The correlation coefficients are 0.995 and 0.982, respectively. As an approximate measure of the sensitivity it may be mentioned that the formation of ATP was 20 pmol per min per ng AK.

References

[1] *L. Noda,* Adenylate Kinase, in: *P. D. Boyer, H. A. Lardy, K. Myrbaeck* (eds.), The Enzymes, Vol. VIII, Academic Press, New York 1973, pp. 279 – 305.

[2] *E. Borglund, A. Ågren, S. E. Brolin,* Fluorimetric Assays of Adenylate Kinase, an Enzyme Important in Energy Metabolism, Upsala J. Med. Sci. *83,* 81 – 84 (1978).

[3] *E. Borglund, S. E. Brolin, A. Ågren,* Adenylate Kinase Activity in Various Organs and Tissues of Mice with the Obese-Hyperglycemic Syndrom (Gene Symbol Ob/Ob), J. Histochem. Cytochem. *26,* 127 – 130 (1978).

[4] *G. Ronquist, G. Frithz,* Adenylate Kinase Activity and Glutathione Concentration of Cerebrospinal Fluid in Different Neurological Disorders, Eur. Neurol. *18,* 106 – 110 (1979).

[5] *T. Åberg, G. Ronquist, H. Tydén, P. Åhlund, K. Bergström,* Release of Adenylate Kinase into Cerebrospinal Fluid during Open-heart Surgery and its Relation to Postoperative Intellectual Function, Lancet*θ o. 8282,* 1139 – 1141 (1982).

[6] *G. Ronquist, A. Terént,* Cerebrospinal Fluid Markers of Disturbed Brain Cell Metabolism, Progr. Neurobiol. *18,* 167 – 180 (1982).

[7] *O. H. Lowry, J. V. Passonneau,* A Flexible System of Enzymatic Analysis, Academic Press, New York 1972, pp. 3 – 119.

[8] *V. Neuhoff,* Micromethods in Molecular Biology, Springer Verlag, Berlin, Heidelberg, New York 1973, pp. 85 – 213.

[9] *P. J. Russel jr., J. M. Horenstein, L. Goins, D. Jones, M. Laver,* Adenylate Kinase in Human Tissues, J. Biol. Chem. *249,* 1874 – 1879 (1974).

[10] *A. Thore,* Luminescence in Clinical Analysis, Ann. Clin. Biochem. *16,* 359 – 369 (1979).

[11] *G. Wettermark, E. Borglund, S. E. Brolin,* A Regenerating System for Studies of Phosphoryl Transfer from ATP, Anal. Biochem. *22,* 211 – 218 (1968).

[12] *A. Andersson, C. Hellerström,* Metabolic Characteristics of Isolated Pancreatic Islets in Tissue Culture, Diabetes *21,* Suppl. 2, 546 – 554 (1972).

[13] *S. E. Brolin, J. P. Wersäll, A. Ågren,* Firefly Luciferase Assay of Adenylate Kinase in Insulin and Glucagon Producing Cells, in: *E. Schram, P. Stanley* (eds.), Analytical Applications of Bioluminescence and Chemiluminescence, State Printing & Publising Inc., Westlake Village, California 1979, pp. 458 – 466.

[14] *K. Wulff, H.-P. Haar, G. Michal,* Constant Light Signals in ATP Assay with Firefly Luciferase: A Kinetic Explanation, in: *M. Serio, M. Pazzagli* (eds.), Luminescent Assays in Endocrinology and Clinical Chemistry, Raven Press, New York 1982, pp. 47 – 52.

[15] *A. Lundin,* Applications of Firefly Luciferase, in: *M. Serio, M. Pazzagli* (eds.), Luminescent Assays: Perspectives in Endocrinology and Clinical Chemistry, Raven Press, New York 1982, pp. 29 – 45.

[16] *G. Wettermark, H. Stymne, S. E. Brolin, B. Petersson,* SubstrateAnalyses in Single Cells. I. Determination of ATP, Anal. Biochem. *63,* 293 – 307 (1975).

[17] *A. Ågren, S. E. Brolin, M. Lindfors,* Advantages in Microprocessing Techniques in Bioluminescence Assays, in: *M. A. DeLuca, W. D. McElroy* (eds.), Bioluminescence and Chemiluminescence Basic Chemistry and Analytical Applications, Academic Press, New York 1981, pp. 575 – 582.

[18] *S. E. Brolin, E. Borglund, A. Ågren,* Photokinetic Microassay of Adenylate Kinase Using the Firefly Luciferase Reaction, J. Biochem. Biophys. Meth. *1,* 163 – 169 (1979).

7.6 Galactose-1-Phosphate Uridylyltransferase

UDPglucose: α-D-galactose-1-phosphate uridylyltransferase, EC 2.7.7.12

Graham Ellis and David M. Goldberg

General

Galactose-1-phosphate uridylyltransferase (GALT) is widely distributed in nature, and is found in most organisms that can utilize galactose as a carbohydrate source. It has been studied in bacteria, and purified preparations from these sources have been obtained [1, 2], but because galactose (as lactose) is an important component of milk in many mammalian species, it is in this group that most research has been done (cf. ref. [3] for a description of studies on various mammals). Because GALT is absent from cells and tissues in the "classical" form of the inherited disease galactosaemia, the enzyme has been studied extensively in man. For clinical purposes, the enzyme is measured most frequently in red blood cells (RBC), but it is also present in the cytoplasmic fraction of white blood cells (WBC), liver, spleen, lung, kidney, skeletal and cardiac muscle, brain [4 – 6], cultured amniotic cells [7, 8], placenta [9] and cultured fibroblasts [7, 8, 10]. GALT has been highly purified from calf liver [11], human placenta [9] and human RBC [12, 13], and is thought to be a dimer of molecular weight 88000 – 96000. The enzyme mechanism proceeds by way of a uridylyl-enzyme intermediate [14].

Application of method: in biochemistry, clinical chemistry, and genetic counselling.

Enzyme properties relevant in analysis: GALT activity can be resolved into different isoenzyme fractions by electrophoresis on cellulose acetate, agarose or starch gel. In a study of human autopsy material [5], the following numbers of bands were visualized: RBC (one), liver (one), heart (two), kidney (five). Four bands have been visualized in cultured fibroblasts [10, 15]. When isoelectric focusing was performed in thin-layer acrylamide gel, GALT of RBC was resolved into three bands and that of liver into at least six bands, but the extra bands were thought to be due to post-synthetic modification of the charge of some of the subunits due to ageing [16, 17]. Genetic polymorphism has been studied extensively in RBC, because the commonest (*Duarte*) variant in its homozygous form has activity comparable to that of the heterozygote for galactosaemia. The *Duarte*-variant GALT is best resolved by electrophoresis on starch or agarose gel [18 – 20], because isoelectric focusing does not give good resolution [17]. At least eight other variant forms of the enzyme have been described (cf. ref. [21]). Catalytic activities of these variants in RBC lie between 0 and 141% of that of the normal enzyme, and the variant enzymes usually have abnormal electrophoretic mobility.

GALT may show a lag phase when measured by NADPH-linked continuous monitoring assay. When measured by UDPG consumption methods, substrate utilization by samples with high GALT activity may cause the rate of reaction to slow after the initial part of the incubation. Non-linearity has also been observed in a rare GALT variant which is sensitive to product inhibition by G-1-P [22].

Methods of determination: the reaction catalyzed by GALT is

(a) $$\text{Galactose-1-P} + \text{UDPG} \xrightleftharpoons{\text{GALT}} \text{UDPgal} + \text{glucose-1-P}.$$

GALT activity may be measured as UDPG consumption, cf. eqn. (a). Alternatively [23 – 25], activity may be determined as glucose-1-P formation, monitored by NADP reduction linked with endogenous or exogenous phosphoglucomutase, PGM (EC 2.7.5.1), glucose-6-phosphate dehydrogenase, G6P-DH (EC 1.1.1.49), and 6-phosphogluconate dehydrogenase, 6-PGDH (EC 1.1.1.44) as outlined in eqns. (b), (c), (d).

(b) $$\text{Glucose-1-P} \xrightleftharpoons{\text{PGM}} \text{glucose-6-P}$$

(c) $$\text{Glucose-6-P} + \text{NADP}^+ \xrightleftharpoons{\text{G6P-DH}} \text{gluconate-6-P} + \text{NADPH} + \text{H}^+$$

(d) $$\text{Gluconate-6-P} + \text{NADP}^+ \xrightleftharpoons{\text{6-PGDH}} \text{ribulose-5-P} + \text{NADPH} + \text{H}^+ + CO_2$$

As discussed on pp. 565, 569, this latter principle forms the basis of a simple widely-used manual screening procedure [26], of several semi-automated assays [27, 28], and of a method for isoenzyme localization after electrophoresis [18].

Others have devised a manometric assay in which GALT is measured as the rate of oxygen consumption [29]. The method uses the same series of reactions as that described for G-1-P formation [eqns. (b) – (d)], but incorporates methylene blue to transport electrons from NADPH to molecular oxygen.

Alternatively, under anaerobic conditions, methylene blue is converted to the reduced (colourless) form. This principle was used in an early GALT screening procedure [30].

Radioisotopic methods with exquisite sensitivity have been described, and these are particularly suited to the assay of activity in cultured cells or small samples of tissues. GALT may be monitored as the conversion of [^{14}C]Gal-1-P to [^{14}C]UDPgal, substrate and product being resolved by chromatography on DEAE paper [7, 31] or DEAE columns [4]. Alternatively, alkaline phosphatase may be used to digest the [^{14}C]Gal-1-P and then the [^{14}C]Gal is separated from [^{14}C]UDPgal [32]. Others have used [^{32}P]Gal-1-P or [^{32}P]G-1-P which are converted by GALT to [^{32}P]UDPgal or [^{32}P]UDPG, respectively. The labelled nucleotide is then separated from the sugar phosphate by adsorption on charcoal [33]. The labelled substrates have to be of very high purity for radioisotopic assays, otherwise erroneous results may be generated [34].

International reference methods and standards: neither is currently available.

Enzyme effectors: RBC GALT was shown to be inactivated by 4-chloromercuribenzoate, iodoacetate and diethylpyrocarbonate in studies which concluded that cysteinyl and histidyl residues are close to the active site [12]. Sulphydryl reagents are not normally required in the assay of the RBC enzyme but may restore some activity to the partially denatured enzyme [35, 36] or protect against inactivation [12]. These reagents are essential for the assay of GALT in leukocytes to prevent denaturation. Chloride ions inhibit GALT [37] and the enzyme is subject to product inhibition by UDPgal [35]. One unconfirmed report has suggested that the RBC enzyme is also inhibited by G-1-P, G-6-P, UDPG, ATP and ADP [38]. The action of GALT from calf liver on UDPG is inhibited competitively by UTP, UMP and UDP [39]. In some linked assays, thioglycolic acid, mercaptoethanol, high EDTA concentration [36], Mg^{2+} [11, 24] and various other metals, unfortunately added as their chlorides, were inhibitory [11], although with these particular assays it was not clear whether GALT or the indicator reactions were the site of inhibition. As previously mentioned, a variant RBC enzyme was inhibited by G-1-P [22].

Assay

Method Design

Principle

(a) $$\text{Galactose-1-P} + \text{UDPG} \xrightleftharpoons{\text{GALT}} \text{UDPgal} + \text{glucose-1-P}$$

(e) $$\text{UDPG} + 2\,\text{NAD}^+ + \text{H}_2\text{O} \xrightleftharpoons{\text{UDPG-DH}} \text{UDPglucuronate} + 2\,\text{NADH} + 2\,\text{H}^+$$

GALT of RBC is assayed by a UDPG consumption test [36, 37] in two stages. In the first stage, GALT acts on Gal-1-P and UDPG (eqn. a). The reaction is terminated by dilution and reaction supernatant are assayed for UDPG by UDPG dehydrogenase (UDP glucose: NAD 6-oxidoreductase, EC 1.1.1.22).

The absolute increase in A_{339} is proportional to the UDPG remaining in the reaction mixture after incubation. The difference between the UDPG in blank and test is a measure of the GALT activity. UDPgal-4-epimerase (EC 5.1.3.2) of RBC would interfere with the assay in two ways: it would convert some UDPG to UDPgal in the blank, and it would convert some UDPgal, produced by GALT in the test, back to UDPG. These interconversions would result in lower measured GALT activities.

However, 4-epimerase is NAD dependent and, during a pre-incubation, endogenous NAD nucleosidase (EC 3.2.2.5) destroys the NAD. RBC from infants younger than three months may not contain sufficient NAD nucleosidase to remove the NAD effectively, and a commercial preparation of NAD nucleosidase should be added to prevent activity of the epimerase [35, 36]. In rare instances in older children, for example, where there is a strong clinical indication of galactosaemia but the GALT activity measured by the UDPG consumption test is low but not diagnostic, the NAD-nucleosidase should be added to the haemolysate prior to assay of GALT activity; alternatively, a different assay technique may be used [40]. The consumption test is not applicable to automated analyzers.

Optimized conditions for measurement: the pH optimum is at 8.7; Cl^- is inhibitory; consequently Tris/acetate, or glycine/NaOH buffers are satisfactory. The concentration of UDPG is a compromise. The initial UDPG concentration (560 μmol/l) is approximately $3.7 \times K_m$ [37], and the concentration falls during the reaction. Because UDPG is assayed as NADH, an increase in the initial UDPG concentration would increase the final absorbance of the NADH produced by UDPG dehydrogenase in the indicator reaction. Low activities would then produce small differences between the two large absorbance differences of test and blank. The assay is optimized for low activity samples because a relatively large volume of haemolysate is used. Correction factors have been derived to compensate for non-linearity due to UDPG consumption in high-activity samples [37].

Temperature conversion factors: the enzyme should be assayed at 37 °C. Correction factors for UDPG consumption at high activity are probably inapplicable at other temperatures.

Equipment

Water-bath at 37 °C ± 0.1 °C. Spectrophotometer or spectral-line photometer which can operate at 339 nm or Hg 334 or Hg 365 nm. A pen recorder facilitates the calculation of initial absorbance, and allows easy monitoring of final absorbance. Alternatively, a stopwatch is required.

Reagents and Solutions

Purity of reagents: use reagents of highest analytical quality. UDPG dehydrogenase should be free of contaminants which might oxidize other substrates present in haemolysates [37].

Preparation of solutions: all solutions in re-purified water (cf. Vol. II, chapter 2.1.3.2).

1. Tris/acetate buffer (0.2 mol/l, pH 8.7):

   dissolve 2.4 g tris(hydroxymethyl)aminomethane in about 80 ml water. Adjust pH to 8.7 with glacial acetic acid, and volume to 100 ml with water.

2. UDPglucose (7 mmol/l):

   dissolve 75 mg UDPG-$Na_2 \cdot 4H_2O$ in 10 ml water. Prepare a 100-fold dilution of this solution in water and measure the absorbance at 260 nm. Based on the reading dilute the remainder of the solution so that a 100-fold dilution reads $A = 0.70 \pm 0.01$.

3. Galactose-1-P (8 mmol/l):

   dissolve 3 mg Gal-1-P-$K_2 \cdot 2H_2O$ in 1 ml water.

4. NAD (13.6 mmol/l):

   dissolve 50 g nicotinamide-adenine dinucleotide, free acid, in 4 ml water. Adjust the pH to 7.0 with NaOH, 0.2 mol/l (pH meter), but do not over-titrate because NAD is unstable in alkaline solution. A 300-fold dilution with water should have absorbance at 260 nm of $A = 0.815$. Adjust the concentration of the stock solution if necessary to achieve this.

5. UDPG dehydrogenase, bovine liver (32 U/l):

   dissolve powder in water, or dilute the ammonium sulphate suspension.

6. NaCl solution (0.15 mol/l):

   dissolve 9 g NaCl in 1 l water.

7. NAD nucleosidase solution (630 U/l):

   aqueous solution (required only for blood from umbilical cord or infants younger than 3 months).

8. Buffer/UDPG solution:

   on the day of assay, mix 1 ml of buffer (1) and 0.25 ml of UDPG solution (2).

Stability of solutions: store solutions (1), (2) and (3) in aliquots at −20 °C for up to 3 months. Store solution (6) at 4 °C for up to 1 month. Prepare solutions (4), (5), (7) and (8) freshly on the day they are required.

Procedure

Collection and treatment of specimen: collect 2 – 10 ml of venous blood into a heparinized container, or alternatively, use acetate citrate dextran (ACD) or ethylene diamine tetraacetate (EDTA) as anticoagulants [36, 37]. On the day of assay, centrifuge the sample (10 min at 1400 *g*), aspirate and discard the plasma and buffy coat (leukocytes). Add 1.5 – 2 volumes of NaCl solution, 0.15 mol/l, re-suspend the red blood cells by inverting the tube several times, centrifuge the sample and aspirate saline and any residual buffy coat. Repeat this saline wash procedure once more, centrifuge, and aspirate the saline. Then add a volume of water equal to that of the packed red blood cells and mix. Freeze the haemolysate in a solid CO_2/acetone-bath (or other apparatus). Thaw in a water-bath at room temperature. Repeat the freezing and thawing two more times to ensure complete haemolysis of the sample [25]. Mix well.

Stability of the enzyme in specimen or sample: store the whole blood specimen at 4 °C. Analyze within 7 days. Longer storage at 4 °C may be possible with some specimens [36, 37]. The haemolysate is stable overnight at – 20 °C.

Details for measurement in tissues: the UDPG consumption assay is best suited to red blood cells, which have low activity and high absorbance at 339 nm. Tissue preparations may be more conveniently assayed by NADP-linked systems, as listed on p. 561, and cultured cells by sensitive radioisotope procedures, p. 561. Tissue preparations from foetuses are unstable above 4 °C but these homogenates are stable for several weeks at – 20 °C [6]. Others have noted that tissues obtained at autopsy could be stored at – 20 °C for up to one year with no loss in GALT activity [5]. These authors found it necessary to remove small molecular-weight components from homogenates of heart and skeletal muscle by Sephadex G-25 chromatography. Otherwise these substances interfered with their NADP-linked assay.

Assay conditions

Incubation: volume of incubation mixture 0.25 ml; total volume at reaction termination 0.75 ml; 37 °C (thermostatted water-bath).

UDPG assay: wavelength 339 nm (Hg 334 or Hg 365) nm; light path 10 mm; final volume 1.0 ml; room temperature. Measure the absorbance against air. Before starting the assay, adjust the temperature of the solutions to room temperature.

Assay an aliquot of each haemolysate for haemoglobin concentration by a cyanmethaemoglobin procedure. Prior to assay, incubate the RBC haemolysate at 37 °C for 15 minutes to permit endogenous NAD nucleosidases to destroy the endogenous NAD (cofactor for 4-epimerase). For haemolysates from cord blood and children younger than 3 months, add 0.025 volumes of NAD nucleosidase solution (7), to each volume of haemolysate before it is incubated at 37 °C.

A blank is performed with water replacing Gal-1-P. Assay test samples and blanks in duplicate. Do not assay more than 3 haemolysates (12 tubes) in one batch.

Measurement

Incubation

Pipette successively into Pyrex conical-base centrifuge tubes:	blank	test	concentration in assay mixture
water	0.05 ml	–	
Gal-1-P solution (3)	–	0.05 ml	Gal-1-P 1.6 mmol/l
buffer/UDPG solution (8)	0.10 ml	0.10 ml	Tris 64 mmol/l UDPG 0.56 mmol/l
place all tubes in a water-bath at 37 °C for 5 – 10 min; at 1 min intervals, add sample			
haemolysate	0.10 ml	0.10 ml	volume fraction $\varphi_1 = 0.4$
mix, note time; incubate each sample exactly for 15 min			
NaCl solution +4 °C (6)	0.50 ml	0.50 ml	
immediately transfer the tube to a boiling water-bath for 2 min to terminate the reaction. When the batch has been completed, stir the contents of the tubes with glass rods to break up the coagulum. Centrifuge at 1400 *g* for 15 min. Transfer the clear supernatants to other tubes. Cap and store at 4 °C for up to 5 h, or at – 20 °C overnight (although up to 5% of UDPG may be lost). Warm solutions to room temperature prior to UDPG assay.			

Calculation: extrapolate the initial linear portion of the recording to obtain the absorbance for each cuvette at zero time (the moment of addition and mixing of the UDPG dehydrogenase). Subtract this value from the final stable absorbance, for test and blank respectively, to obtain ΔA_{test} and ΔA_{blank}. The ΔA values are proportional to the UDPG present in the supernatants.

The UDPG consumption per unit time, related to the mass concentration of haemoglobin in the haemolysate (ρ_{Hb} in g/l) represents the GALT activity content of the haemolysate. According to eqns. (l) or (l_1), respectively, of "Formulae", Appendix 3, it is (t in minutes):

$$z_c/m_s = \frac{1\,000 \times f}{0.63 \times 2 \times 10 \times \varphi_1 \times \varphi_2 \times \rho_{Hb}} \times 1\,000 \times F \times (\Delta A_{blank} - \Delta A_{test})/\Delta t$$

$$= \frac{1.98 \times 10^5}{\rho_{Hb}} \times F \times (\Delta A_{blanks} - \Delta A_{test})/\Delta t \quad U/kg\ Hb$$

or

$$z_c/m_s = \frac{3.31}{\rho_{Hb}} \times F \times (\Delta A_{blank} - \Delta A_{test})/\Delta t \quad mkat/kg\ Hb$$

where is

φ_1 volume fraction of sample in incubation mixture
φ_2 volume fraction of sample in UDPG assay
f factor 3 considering dilution of assay mixture by NaCl solution
F factor to correct for non-linearity of the UDPG consumption as listed below.

Assay of UDPG

Pipette successively into cuvettes:			concentration in assay mixture	
water		0.45 ml		
Tris buffer	(1)	0.15 ml	Tris	30 mmol/l
NAD solution	(4)	0.10 ml	NAD	1.36 mmol/l
sample (supernatant from the previous procedure)		0.20 ml	volume fraction	φ_2= 0.2
mix, start the chart recorder* at 2.5 cm/min; at intervals of 1 – 2 min add solution (5)				
UDPG dehydrogenase solution	(5)	0.10 ml	UDPG-DH	3.2 U/l
mix immediately, and mark the chart to indicate the instant of mixing. Place the cuvette in the light path and monitor the absorbance for 1 – 2 min. Repeat this procedure for all other samples. When the reactions have been started in all cuvettes, reduce the chart speed to 0.5 cm/min and monitor the absorbances until all have become stable (20 – 40 min).				

* If a recording spectrophotometer is not available, read the absorbances at exactly 0.5, 1.0, 1.5, and 2 min after addition of UDPG dehydrogenase. When the reactions in all cuvettes have been started, read cuvettes at 5 min intervals until the absorbances become stable.

Table 1. Correction factors [36, 37]

$(A_{blank} - A_{sample})$	F	$(A_{blank} - A_{sample})$	F
<0.200	1.000	0.265	1.132
0.200	1.010	0.270	1.144
0.205	1.020	0.275	1.160
0.210	1.029	0.280	1.175
0.215	1.037	0.285	1.189
0.220	1.045	0.290	1.203
0.225	1.053	0.295	1.220
0.230	1.061	0.300	1.243
0.235	1.072	0.305	1.269
0.240	1.083	0.310	1.294
0.245	1.094	0.315	1.317
0.250	1.100	0.320	1.353
0.255	1.110	0.325	1.412
0.260	1.119	0.330	1.479

For measurements at Hg 334 nm and Hg 365 nm calculate with the appropriate ε values (cf. Appendix 4) instead of $\varepsilon = 0.63\ l \times mmol^{-1} \times mm^{-1}$.

Validation of Method

Precision, accuracy, detection limits and sensitivity: the within-run coefficient of variation was 5.4% at a mean activity of 318 U/kg Hb [36]. Accuracy cannot be assessed, because no reference method is available. Values for individuals with classical galactosaemia are in the range −8 to +8 U/kg Hb [36, 37]. In our experience the detection limit is probably 10−20 U/kg Hb, depending upon the haemoglobin concentration of the prepared haemolysates. Sensitivity cannot be accurately defined with this system.

Sources of error: potential interference by 4-epimerase has previously been referred to (p. 563). This enzyme interferes with the assay in two ways: it converts UDPG in the blank to UDPgal, and it converts some of the UDPGal produced in the test by GALT to UDPG, resulting in falsely low apparent activity [36]. 4-Epimerase may also result in apparent activity in unusual galactosaemic samples, by a mechanism which is not entirely clear [40]. These authors observed apparent measurable GALT activity in RBC from a patient with galactosaemia. Activity was undetectable after incubation of the haemolysate with added NAD nucleosidase. Non-linearity of the UDPG consumption assay causes error which is reduced by use of the correction factors listed in the table. Some samples of UDPG dehydrogenase reportedly contain enzyme contaminants which act on low molecular weight substances in the supernatants and reduce NAD [37]. In this event, obtain UDPG dehydrogenase from an alternative batch or supplier.

Reference ranges: RBC from patients with classical galactosaemia contain no GALT activity (< 8 U/kg Hb). The reference range for normal subjects is difficult to define because of the high prevalence of *Duarte* variant heterozygotes, the gene frequency being 0.07 – 0.08 [41, 42]. Subjects with this phenotype have low normal activity and can only be distinguished from normal by electrophoresis. On average, values are higher in black Americans than whites [43]. Approximate reference values for the procedure are 304 – 468 U/kg Hb [36]. Individuals who are heterozygotes for classical galactosaemia give results between 140 and 222 U/kg Hb. Similar ranges have been found by other groups using this method [44]. For reference values in various laboratory and domestic animals see [3], and in various species of lions and tigers [45].

Screening methods: the UDPG consumption test presented above has proved reliable for the quantitation of GALT activity over many years. However it is labour-intensive and is best suited to small numbers of analyses. A simple screening procedure [26] based on NADPH reduction (p. 561) is more appropriate for larger batches (5 – 80 samples). In this procedure the NADPH content of the reaction mixture is assessed visually after aliquots have been transferred to filter paper and viewed under UV light. The procedure will detect homozygous subjects and may be adapted to detect most heterozygotes [36, 46, 47]. Some subjects with deficiencies of RBC enzymes (e.g., severe G-6-P dehydrogenase deficiency) give false positive results by this test; consequently, all abnormal results should be confirmed with the quantitative assay.

References

[1] *S. Saito, M. Ozutsumi, K. Kurahashi,* Galactose-1-Phosphate Uridylyltransferase of *Escherichia coli,* J. Biol. Chem. *242,* 2362 – 2368 (1967).

[2] *R. Raychowdhury, D. H. Schlesinger, D. B. Wilson,* E. coli Galactose-1-Phosphate Uridyl Transferase: N-Terminal and C-Terminal Sequences, Mol. Cell. Biochem. **23**, 167 – 175 (1979).

[3] *S. Kelly, P. Sweeney, L. Schedlbauer,* Galactose-1-Phosphate Uridyl Transferase Activity in Red Cells of Various Animal Species, Experientia *37,* 550 – 552 (1981).

[4] *Y. S. Shin-Buehring, T. Beier, A. Tan, M. Osang, J. Schaub,* The Activity of Galactose-1-Phosphate Uridyltransferase and Galactokinase in Human Fetal Organs, Pediat. Res. *11,* 1003 – 1009 (1977).

[5] *J. M. Connellan, V. Buchanan,* The Presence of Isozymes of Hexose-1-Phosphate Uridylyltransferase from Postmortem Human Liver, Heart and Kidney, Biochim. Biophys. Acta *364,* 68 – 77 (1974).

[6] *Y. S. Shin-Buehring, L. Stuempfig, E. Pouget, P. Rahm, J. Schaub,* Characterization of Galactose-1-Phosphate Uridyl-Transferase and Galactokinase in Human Organs from the Fetus and Adult, Clin. Chim. Acta *112,* 257 – 265 (1981).

[7] *W. G. Ng, G. N. Donnell, W. R. Bergren, O. Alfi, M. S. Golbus,* Prenatal Diagnosis of Galactosemia, Clin. Chim. Acta *74,* 227 – 235 (1977).

[8] *Y. Shin-Buehring, H. Leitner, H. Henseleit, A. Wirtz, B. Haas, J. Schaub,* Characteristics of Galactokinase and Galactose-1-Phosphate Uridyltransferase in Cultivated Fibroblasts and Amniotic Fluid Cells, Hum. Genet. *48,* 31 – 37 (1979).

[9] *V. P. Williams, G. R. Helmer, Jr., C. Fried,* Human Galactose-1-Phosphate Uridylyltransferase: Purification and Comparison of the Red Blood Cell and Placental Enzymes, Arch. Biochem. Biophys. *216,* 503 – 511 (1982).

[10] *G. Hammersen, R. Mandell, H. L. Levy,* Galactose-1-Phosphate Uridyl Transferase in Fibroblasts: Isozymes in Normal and Variant States, Ann. Hum. Genet. *39*, 147 – 150 (1975).

[11] *J. S. Mayes,* Purification, Properties, and Isozyme Pattern of Galactose-1-Phosphate Uridyl Transferase from Calf Liver, Arch. Biochem. Biophys. *172*, 715 – 720 (1976).

[12] *V. P. Williams, C. Fried, G. Popják,* Human Red Cell Galactose 1-Phosphate Uridylyltransferase: Effects of Site-Specific Reagents on Catalytic Activity, Arch. Biochem. Biophys. *206*, 353 – 361 (1981).

[13] *J. Banroques, C. Gregori, F. Schapira,* Purification and Characterization of Human Erythrocyte Uridylyl Transferase, Biochim. Biophys. Acta *657*, 374 – 382 (1981).

[14] *P. A. Frey, L.-J. Wong, K.-F. Sheu, S.-L. Yang,* Galactose-1-Phosphate Uridylyltransferase: Detection, Isolation, and Characterization of the Uridylyl Enzyme, in: *S. P. Colowick, N. O. Kaplan* (eds.), Methods in Enzymology, Vol. 87, Academic Press, New York 1982, pp. 20 – 36.

[15] *G. Hammersen, H. L. Levy, F. Frigoletto, R. Mandell,* Comparison of Galactose-1-Phosphate Uridyl Transferase in Fetal and Adult Tissues, Clin. Chim. Acta *60*, 281 – 284 (1975).

[16] *F. Schapira, C. Gregori, J. Banroques,* Microheterogeneity of Human Galactose-1-Phosphate Uridylyl Transferase, Isoelectrofocusing Results, Biochem. Biophys. Res. Commun. *80*, 291 – 297 (1978).

[17] *F. Schapira, C. Gregori, J. Banroques, M. Vidailhet, S. Despoisses, C. Vigneron,* Isoelectrofocusing of Erythrocyte Galactose 1 Phospho Uridyl Transferase in a Family with Both Galactosemia and Duarte Variants, Hum. Genet. *46*, 89 – 96 (1979).

[18] *W. G. Ng, W. R. Bergren, M. Fields, G. N. Donnell,* An Improved Electrophoretic Procedure for Galactose-1-Phosphate Uridyl Transferase: Demonstration of Multiple Activity Bands with the Duarte Variant, Biochem. Biophys. Res. Commun. *37*, 354 – 362 (1969).

[19] *M. C. Sparkes, M. Crist, R. S. Sparkes,* Improved Technique for Electrophoresis of Human Galactose-1-P Uridyl Transferase (EC 2.7.7.12), Hum. Genet. *40*, 93 – 97 (1977).

[20] *J. E. S. Lee, W. G. Ng,* Semi-micro Techniques for the Genotyping of Galactokinase and Galactose-1-Phosphate Uridyltransferase, Clin. Chim. Acta *124*, 351 – 356 (1982).

[21] *B. Ibarra, G. Vaca, J. Sánchez-Corona, A. Hernández, M. L. Ramirez, J. M. Cantú,* Los Angeles Variant of Galactose-1-Phosphate Uridyltransferase (EC 2.7.7.12) in a Mexican Family, Hum. Genet. *48*, 121 – 124 (1979).

[22] *A. Lang, H. Groebe, B. Hellkuhl, K. von Figura,* A New Variant of Galactosemia: Galactose-1-Phosphate Uridylyltransferase Sensitive to Product Inhibition by Glucose-1-Phosphate, Pediat. Res. *14*, 729 – 734 (1980).

[23] *K. J. Isselbacher,* Uridyl Transferase, in: *H. U. Bergmeyer* (ed.), Methods of Enzymatic Analysis, Vol. 2, Verlag Chemie, Weinheim, Academic Press, New York 1974, pp. 802 – 805.

[24] *M. A. Pesce, S. H. Bodourian, R. C. Harris, J. F. Nicholson,* Enzymatic Micromethod for Measuring Galactose-1-Phosphate Uridylyltransferase Activity in Human Erythrocytes, Clin. Chem. *23*, 1711 – 1717 (1977).

[25] *R. B. H. Schutgens, W. J. M. Berntssen, L. Pool,* An Improved Quantitative Assay of Galactose-1-Phosphate Uridyltransferase Activity in Erythrocytes Based on the Determination of Glucose-1-Phosphate Generation, Clin. Chim. Acta *86*, 301 – 305 (1978).

[26] *E. Beutler, M. C. Baluda,* A Simple Spot Screening Test for Galactosemia, J. Lab. Clin. Med. *68*, 137 – 141 (1966).

[27] *P. Sturgeon, E. Beutler, D. McQuiston,* Automated Method for Screening for Galactosemia, in: Automation in Analytical Chemistry, Technicon Symposia 1966, Mediad, New York 1966, pp. 75 – 77.

[28] *P. D. M. Frazier, G. K. Summer,* Automated Fluorometric Micromethod for Detection of Transferase-Deficiency Galactosemia, J. Lab. Clin. Med. *83*, 334 – 338 (1974).

[29] *A. Westwood, D. N. Raine,* Galactose-1-Phosphate Uridyl Transferase: A New Manometric Assay Specially Suited to Erythrocyte Studies in Galactosaemia, Clin. Chim. Acta *49*, 435 – 443 (1973).

[30] *E. Beutler, M. Baluda, G. N. Donnell,* A New Method for the Detection of Galactosemia and its Carrier State, J. Lab. Clin. Med. *64*, 694 – 705 (1964).

[31] *Y. Shin-Bühring, M. Osang, R. Ziegler, J. Schaub,* A Method for Galactose-1-Phosphate Uridyltransferase Assay and the Separation of its Isozymes by DEAE-Cellulose Column Chromatography, Clin. Chim. Acta *70*, 371 – 377 (1976).

[32] *A. H. Fensom, P. F. Benson,* Assay of Galactose-1-Phosphate Uridyl Transferase in Cultured Amniotic Cells for Prenatal Diagnosis of Galactosaemia, Clin. Chim. Acta *62*, 189 – 194 (1975).
[33] *Z. Grodzka, T. Chojnacki,* The Use of Glucose-1-[^{32}P] Phosphate for the Determination of Hexose-1-Phosphate Uridyl Transferase Activity in Red Blood Cells, Biochem. Med. *16*, 182 – 186 (1976).
[34] *Y. S. Shin-Buehring, J. Schaub,* Pitfalls in the Radioactive Method of Galactose-1-Phosphate Uridyltransferase Activity Measurement, Clin. Chim. Acta *106*, 231 – 234 (1980).
[35] *E. Beutler, M. Mitchell,* UDPGlu Consumption Methods, in: *D. Y.-Y. Hsia* (ed.), Galactosemia, Charles C. Thomas, Springfield, Illinois 1969, pp. 72 – 82.
[36] *F. A. Ibbott,* A Screening Procedure and a Specific Quantitative Method for UDPglucose: α-D-Galactose-1-Phosphate Uridylyltransferase (EC 2.7.7.12), Clin. Chem. *23*, 1348 – 1355 (1977).
[37] *E. Beutler, M. C. Baluda,* Improved Method for Measuring Galactose-1-Phosphate Uridyl Transferase Activity of Erythrocytes, Clin. Chim. Acta *13*, 369 – 379 (1966).
[38] *H. Z. Hill,* Enzyme Kinetics in Mammalian Cells. III. Regulation of Activities of Galactokinase, Galactose-1-Phosphate Uridyl Transferase and Uridine Diphosphogalactose-4-Epimerase in Human Erythrocytes, J. Cell. Physiol. *78*, 419 – 430 (1971).
[39] *S. Segal, S. Rogers,* Nucleotide Inhibition of Mammalian Liver Galactose-1-Phosphate Uridylyltransferase, Biochim. Biophys. Acta *250*, 351 – 360 (1971).
[40] *C. H. M. M. De Bruyn, C. Raymakers, A. Wensing, T. L. Oei,* Galactose-1-Phosphate Uridyltransferase Activities in Erythrocytes from a Patient with Galactosemia: Discrepancy between Two Methods, Clin. Chim. Acta *78*, 145 – 150 (1977).
[41] *R. Pflugshaupt, R. Scherz, R. Bütler,* Polymorphism of Human Red Cell Adenosine Deaminase, Esterase D, Glutamate Pyruvate Transaminase and Galactose-1-Phosphate-Uridyltransferase in the Swiss Population, Hum. Hered. *26*, 161 – 166 (1976).
[42] *B. Eriksen, J. Dissing,* Human Red Cell Galactose-1-Phosphate Uridylyltransferase (EC 2.7.7.12), Hum. Hered. *30*, 27 – 32 (1980).
[43] *T. A. Tedesco, K. L. Miller, B. E. Rawnsley, M. T. Mennuti, R. S. Spielman, W. J. Mellman,* Human Erythrocyte Galactokinase and Galactose-1-Phosphate Uridylyltransferase: A Population Survey, Am. J. Hum. Genet. *27*, 737 – 747 (1975).
[44] *G. Ellis, D. M. Goldberg,* The Enzymological Diagnosis of Galactosaemia, Ann. Clin. Biochem. *6*, 70 – 73 (1969).
[45] *B. E. Rabinow, P. W. K. Wong, E. R. Maschgan, S. Natelson,* Screening for Errors in Galactose Metabolism with the Erythrocyte, Clin. Chem. *22*, 2010 – 2017 (1976).
[46] *G. Ellis, A. R. Wilcock, D. M. Goldberg,* Experience of Routine Live-Birth Screening for Galactosaemia in a British Hospital Population with Emphasis on Heterozygote Detection, Arch. Dis. Child. *47*, 34 – 40 (1972).
[47] *R. Gitzelmann, I. Schneller,* Fluorescent Spot Screening Test for Galactosemia: Increased Sensitivity, Z. Klin. Chem. Klin. Biochem. *11*, 46 – 47 (1973).

Appendix

1 Symbols, Quantities, Units and Constants

Units and symbols in ⟨ ⟩ should not be used.

m	Metre, m	g	Gram, g
mm	Millimetre, 10^{-3} m	mg	Milligram, 10^{-3} g
µm	Micrometre, 10^{-6} m	µg	Microgram, 10^{-6} g
nm	Nanometre, 10^{-9} m	ng	Nanogram, 10^{-9} g
		pg	Picogram, 10^{-12} g
l	Litre, 10^{-3} m^3	fg	Femtogram, 10^{-15} g
ml	Millilitre, 10^{-3} l	ag	Attogram, 10^{-18} g
µl	Microlitre, 10^{-6} l		
h	Hour	kg	Kilogram, 10^3 g
min	Minute	Mg	Megagram, 10^6 g
s	Second		

t	Time, s (h, min)
Δt	Interval between measurements, s (h, min)
t	Temperature, °C
T	Temperature, K
K	Kelvin
V	Volume (usually volume of assay mixture), ml, l
v	Volume (usually volume of sample in assay mixture), ml, l
φ	Volume fraction of sample in assay mixture
m_s	Mass of substance, g
c	Substance concentration, mol/l
ρ	Mass concentration, g/l
n_c/m_s	Substance content, mol/kg
w	Mass fraction
⟨ppm	Parts per million⟩
⟨ppb	Parts per billion⟩
%	Percentage
% (v/v)	Percentage, volume related to volume
% (v/w)	Percentage, volume related to weight
% (w/v)	Percentage, weight related to volume
% (w/w)	Percentage, weight related to weight
M_r	Molecular weight, relative molecular mass

v	Rate of reaction, mol × l^{-1} × s^{-1} (μmol × ml^{-1} × min^{-1})
V	Maximum rate of reaction, mol × l^{-1} × s^{-1}
$\dot{\xi}$	Rate of conversion, mol × s^{-1} (μmol × min^{-1})
ν_i	Stoichiometric coefficient
kat	Katal, mol × s^{-1}
U	International unit (for enzymes), μmol × min^{-1}
mU	International milliunit, 10^{-3} U
kU	International kilounit, 10^3 U
MU	International megaunit, 10^6 U
Inh.U	Inhibitor unit, μmol × min^{-1}
z	Catalytic activity, U, kat
b	Catalytic activity concentration, U/l, kat/l
z_c/m_s	Catalytic activity content (specific catalytic activity), U/g, kat/kg
ε	Linear molar absorption coefficient, l × mol^{-1} × mm^{-1}
d	Light path, mm
A	Absorbance
F	Fluorescence intensity
I	Light intensity
T	Transmission
Bq	Becquerel, s^{-1}
⟨Ci	Curie, s^{-1}⟩
cpm	Counts per minute, min^{-1}
dpm	Disintegrations per minute, min^{-1}
C	Counting rate (cpm) of the radioactive product
η	Counting efficiency
X	Specific radioactivity, *Bq* × mol^{-1} ⟨Ci × mol^{-1}⟩
Y	Number of counts measured
Z	Decay rate, disintegrations per minute, dpm
k	Reaction constant
K	Equilibrium constant
K'	Apparent equilibrium constant
K_m	Michaelis constant, mol/l
K_I	Inhibitor constant, mol/l
pH	Negative logarithm of the hydrogen ion concentration
pK	Negative logarithm of the dissociation constant, $-\log K$
$[\alpha]_D^{20}$	Specific rotation (sodium D-line at 20 °C)
sp.gr.	Specific gravity at 20 °C relative to water at 4 °C
g	Acceleration due to gravity; 9.81 m × s^{-2}
rpm	Revolutions per minute, min^{-1}

J	Joule, $m^2 \times kg \times s^{-2}$
R	Gas constant, 8.312 $J \times mol^{-1} \times K^{-1}$
h	*Planck's* constant
ν	Frequence of emitted light
$\bar{x}$	Mean value
s, SD	Standard deviation
RSD	Relative standard deviation
⟨CV	Coefficient of variation⟩

2 Abbreviations for Chemical and Biochemical Compounds

It is unavoidable with the numerous abbreviations in use that one abbreviation occasionally is used for different compounds. In such cases the correct meaning can be obtained from the text. Only unequivocal abbreviations are used in the book without further explanation.

A	Adenosine
ABTS®	2,2′-Azino-bis(3-ethylbenzthiazoline) sulphonate
Acetoacetyl-CoA	Acetoacetyl coenzyme A
Acetyl-CoA	Acetyl coenzyme A
AChE	Acetylcholinesterase
AcP	Acid phosphatase
ADA	Adenosine deaminase
ADH	Alcohol dehydrogenase
ADP	Adenosine 5′-diphosphate
ADPglucose	Adenosine 5′-diphosphoglucose
ALAT (ALT, AlaAT)	Alanine aminotransferase
Alcohol-OD	Alcohol oxidase
ALD	Aldolase
AK	Acetate kinase
AK	Adenylate kinase
Ammediol	2-Amino-2-methyl propane-1,3-diol
AMP	Adenosine 5′-monophosphate
A-2-MP	Adenosine 2′-monophosphate
A-3-MP	Adenosine 3′-monophosphate
A-3,(2)-MP	Adenosine 3′(2′)-monophosphate
A-3 : 5-MP (cAMP)	Adenosine 3′ : 5′-monophosphate, cyclic
AOD	Amino acid oxidase
AP	Alkaline phosphatase
Ap_5A	Diadenosine 5′-pentaphosphate
APAD	Acetylpyridine-adenine dinucleotide
APADH	Acetylpyridine-adenine dinucleotide, reduced
ASAT (AST; AspAT)	Aspartate aminotransferase
Ascorbate-OD	Ascorbate oxidase
ATP	Adenosine 5′-triphosphate
ATPase	Adenosine 5′-triphosphatase
BAEE	Benzoyl-L-arginine ethyl ester
BAPNA	N-Benzoyl-arginine-4-nitroanilide
Benzoyl-CoA	Benzoyl coenzyme A
BES	Bis(2-hydroxyethylamino)ethanesulphonic acid
Bicine	N,N-Bis(2-hydroxyethyl)glycine

BMTD	6-Benzamido-4-methoxy-3-toluidinediazonium chloride
BSA	Bovine serum albumin
C	Cytidine
CDP	Cytidine 5′-diphosphate
CDPglucose	Cytidine 5′-diphosphoglucose
CE	Cholesterol esterase
Cellosolve	Ethylene glycol monomethyl ether
CHA	Cyclohexylammonium
ChE	Cholinesterase
ChOD	Cholesterol oxidase
CK	Creatine kinase
CL	Citrate lyase
CMP	Cytidine 5′-monophosphate
C-2-MP	Cytidine 2′-monophosphate
C-2:3-MP	Cytidine 2′ : 3′-monophosphate, cyclic
C-3-MP	Cytidine 3′-monophosphate
C-3,(2)-MP	Cytidine 3′(2′)-monophosphate
CoA	Coenzyme A
COX	Cytochrome c oxidase
CS	Citrate synthase
CTP	Cytidine 5′-triphosphate
Cyt-c	Cytochrome c
d	deoxy (prefix)
DAP	Dihydroxyacetone phosphate
DEA	Diethanolamine
DEAE	Diethylaminoethyl
DFP	Diisopropyl fluorophosphate
DM-POPOP	2,2′-(1,4-Phenylene)bis(4-methyl-5-phenyloxazole)
DMSO	Dimethylsulphoxide
DNase	Deoxyribonuclease
DNP	Dinitrophenylhydrazine
DTNB	5,5′-Dithiobis(2-nitrobenzoic acid)
DTT	Dithiothreitol
EDTA	Ethylenediamine tetraacetate
EGTA	Ethyleneglycol-bis(β-aminoethyl ether) N,N′-tetraacetic acid
FAD	Flavin-adenine dinucleotide
FDH	Formate dehydrogenase
FDNB	1-Fluoro-2,4-dinitrobenzene
FH_2	Dihydrofolate
FH_4	Tetrahydrofolate

FiGlu	N-Formimino-L-glutamate
FMN	Flavin mononucleotide
F-1-P	Fructose 1-phosphate
F-1,6-P_2	Fructose 1,6-bisphosphate
F-2,6-P_2	Fructose 2,6-bisphosphate
F-6-P	Fructose 6-phosphate
F-6-PK	Fructose-6-phosphate kinase
GABAse	Mixture of GAB-AT and SS-DH
GAB-AT	γ-Aminobutyrate aminotransferase
Gal-DH	Galactose dehydrogenase
Gal-OD	Galactose oxidase
GAP	Glyceraldehyde phosphate
GAPDH	D-Glyceraldehyde-3-phosphate dehydrogenase
GDH	L-Glycerol-3-phosphate dehydrogenase (glycerol-1-phosphate dehydrogenase; α-glycerophosphate dehydrogenase)
GDP	Guanosine 5′-diphosphate
GlDH	L-Glutamate dehydrogenase
GMP	Guanosine 5′-monophosphate
G-2-MP	Guanosine 2′-monophosphate
G-3-MP	Guanosine 3′-monophosphate
G-3,(2)-MP	Guanosine 3′(2′)-monophosphate
G-3 : 5-MP (cGMP)	Guanosine 3 : 5-monophosphate, cyclic
GMPK	GMP kinase
GOD	Glucose oxidase
GOT	Glutamate-oxaloacetate transaminase/Aspartate aminotransferase
G-1-P	Glucose 1-phosphate
G-1,6-P_2	Glucose 1,6-bisphosphate
G-6-P	Glucose 6-phosphate
G6P-DH	Glucose-6-phosphate dehydrogenase
GPT	Glutamate-pyruvate transaminase/Alanine aminotransferase
GR	Glutathione reductase
GSH	Glutathione, reduced
GSSG	Glutathione, oxidized
γ-GT	γ-Glutamyltransferase
GTP	Guanosine 5′-triphosphate
Hb	Haemoglobin
Hepes	N-2-Hydroxyethyl piperazine-N′-2-ethanesulphonic acid
HK	Hexokinase
HMG-CoA	3-Hydroxy-3-methylglutaryl coenzyme A
I	Inosine
IDP	Inosine 5′-diphosphate

Ig	Immunoglobulin
IMP	Inosine 5′-monophosphate
INT	2-(4-Iodophenyl)-3-(4-nitrophenyl)-5-phenyltetrazolium chloride
ITP	Inosine 5′-triphosphate
LDH (L-LDH)	L-Lactate dehydrogenase
D-LDH	D-Lactate dehydrogenase
MDH	L-Malate dehydrogenase
MES	2-(N-Morpholino)ethanesulphonic acid
4-Met-um	4-Methylumbelliferon (4-methylumbelliferyl)
MK	Myokinase/adenylate kinase
Mops	3-(N-Morpholino)propanesulphonic acid
4-NA	4-Nitroaniline
NAC	N-Acetylcysteine
NAD	Nicotinamide-adenine dinucleotide
NADH	Nicotinamide-adenine dinucleotide, reduced
NADP	Nicotinamide-adenine dinucleotide phosphate
NADPH	Nicotinamide-adenine dinucleotide phosphate, reduced
NBT	Nitro-BT-tetrazolium salt; 2,2′-bis(4-nitrophenyl)-5,5′-diphenyl-3,3′-(dimethoxy-4,4′-diphenylene)ditetrazolium chloride
NBTH	N-Methyl-2-benzothiazolone hydrazone
NDPK	Nucleoside diphosphate kinase
NMN	Nicotinamide mononucleotide
NMPK	Nucleoside monophosphate kinase
NP	Nucleoside phosphorylase
4-NP	4-Nitrophenol (4-Nitrophenyl)
4-NP-G_1	4-Nitrophenyl α-D-glucopyranoside
4-NP-G_2	4-Nitrophenyl [α-D-glucopyranosyl-(1 → 4)-α-D-glucopyranoside]/4-Nitrophenyl α-D-maltoside
4-NP-G_3	4-Nitrophenyl {di[α-D-glucopyranosyl-(1 → 4)]-α-D-glucopyranoside}/4-Nitrophenyl α-D-maltotrioside
4-NP-G_4	4-Nitrophenyl {tri[α-D-glucopyranosyl-(1 → 4)]-α-D-glucopyranoside}/4-Nitrophenyl α-D-maltotetraoside
4-NP-G_5	4-Nitrophenyl {tetra[α-D-glucopyranosyl-(1 → 4)]-α-D-glucopyranoside}/4-Nitrophenyl α-D-maltopentaoside
4-NP-G_6	4-Nitrophenyl {penta[α-D-glucopyranosyl-(1 → 4)]-α-D-glucopyranoside}/4-Nitrophenyl α-D-maltohexaoside
4-NP-G_7	4-Nitrophenyl {hexa[α-D-glucopyranosyl-(1 → 4)]-α-D-glucopyranoside}/4-Nitrophenyl α-D-maltoheptaoside
4-NPgal	4-Nitrophenyl galactoside
4-NPgluc	4-Nitrophenyl glucoside

4-NPman	4-Nitrophenyl mannoside
4-NPP	4-Nitrophenyl phosphate
NTA	Nitriloacetate
OMP	Orotidine 5′-monophosphate
PALP	Pyridoxal 5′-phosphate
PAMP	Pyridoxamine 5′-phosphate
PDE	Phosphodiesterase
PEP	Phosphoenolpyruvate
6-PGDH	6-Phosphogluconate dehydrogenase
PGK	3-Phosphoglycerate kinase
PGluM	Phosphoglucomutase
PGM	Phosphoglycerate mutase
P_i	Inorganic phosphate
Pipes	Piperazine-N,N′-bis(2-ethanesulphonic acid)
PK	Pyruvate kinase
PL-A	Phospholipase A
PL-C	Phospholipase C
PL-D	Phospholipase D
PMS	Phenazine methosulphate
POD	Peroxidase
POPOP	2,2′-(1,4-Phenylene)bis(5-phenyloxazole)
PPase	Pyrophosphatase, inorganic
PP_i	Inorganic pyrophosphate
PPO	2,5-Diphenyloxazole
PRPP	5-Phospho-α-D-ribose 1-diphosphate
PTA	Phosphotransacetylase
PyDC	Pyruvate decarboxylase
RNA	Ribonucleic acid
rRNA	Ribosomal ribonucleic acid
tRNA	Transfer ribonucleic acid
RNase	Ribonuclease
SOD	Superoxide dismutase
SS-DH	Succinate-semialdehyde dehydrogenase
ST	Sialyltransferase
SUPHEPA	N-Succinyl-L-phenylalanine-4-nitroanilide
T	Ribosylthymidine
dT	Thymidine
TA	Transaldolase
TAT (TyrAT)	Tyrosine aminotransferase
dTDP	Thymidine 5′-diphosphate

dTDPglucose	Thymidine 5′-diphosphoglucose
TCA	Trichloroacetic acid
TEA	Triethanolamine
THF	Tetrahydrofolic acid
TIM	Triosephosphate isomerase
TPP	Thiamine pyrophosphate
Tricine	N,N,N-Tris(hydroxymethyl)methylglycine
Tris	Tris(hydroxymethyl)aminomethane
dTTP	Thymidine 5′-triphosphate
U	Uridine
UDP	Uridine 5′-diphosphate
UDPG (UDPglucose)	Uridine 5′-diphosphoglucose
UDPAG	Uridine 5′-diphospho-N-acetylglucosamine
UDPGA	Uridine 5′-diphosphoglucuronate
UDPGal	Uridine 5′-diphosphogalactose
UDPG-DH	UDPglucose dehydrogenase
UDPG-PP	UDPglucose pyrophosphorylase
UMP	Uridine 5′-monophosphate
U-2-MP	Uridine 2′-monophosphate
U-2:3-MP	Uridine 2′ : 3′-monophosphate, cyclic
U-3-MP	Uridine 3′-monophosphate
U-3,(2)-MP	Uridine 3′(2′)-monophosphate
UTP	Uridine 5′-triphosphate
X	Xanthosine
XOD	Xanthine oxidase

3 Formulae

Experimental data must be converted to concentrations or contents; only then one can speak of a result (cf. Vol. II, chapter 3.1). According to international recommendations [1, 2] the following symbols and units are equivalent:

Metabolites		*Enzymes*	
Substance concentration c	mol/l	Catalytic activity concentration b	kat/l; U/l
	mmol/l		mkat/l
Mass concentration ρ	g/l		μkat/l
	mg/l		nkat/l
Substance content n_c/m_s	mol/kg	Catalytic activity content z_c/m_s	kat/kg
	mmol/g		(U/g, U/mg)
Mass fraction w	1		mkat/g
	0.001		μkat/g

The use of the basic kind of quantities of the SI (cf. Vol. I, pp. 10, 22) involves some changes in the symbols, quantities and units formerly used. For example, the base unit of length is the metre (m) and thousands or thousandths of it. The path length of a photometer cuvette is in general 10 mm (not 1 cm). Accordingly, the unit of the absorption coefficient also changes ($1 \times mol^{-1} \times mm^{-1}$, instead of $1 \times mol^{-1} \times cm^{-1}$).

For example, for NADH at 339 nm $\varepsilon = 6.3 \times 10^2\ 1 \times mol^{-1} \times mm^{-1}$ instead of $6.3 \times 10^3\ 1 \times mol^{-1} \times cm^{-1}$. For practical purposes it is proposed to use the path length $d = 10$ mm and the concentration c in mmol/l (μmol/ml). Then for NADH at 339 nm $\varepsilon \times d = 6.3\ 1 \times mmol^{-1}$ applies.

The following symbols, kind of quantities and units are customary in this context. Throughout this book we use v, V for reaction rates and v, V for volumes (deviating from international recommendations). We also write T for transmission of light and T for thermodynamic temperature.

Symbols in radiometry are still not completely coherent with those in other laboratory sciences; also for various commonly used terms apparently no official symbols exist (cf. [3]).

General

c	substance concentration, mol/l
ρ	mass concentration, g/l
n_c/m_s	substance content, mol/kg
w	mass fraction, 1

MW	mass of one millimole or mole of substrate, mg/mmol, g/mol
V	assay volume, l
v	volume of sample used in assay, l
φ	volume fraction of sample in assay (incubation) mixture, v/V
t	time, s (min, h)
z	catalytic activity, kat, U
b	catalytic activity concentration, kat/l, U/l
z_c/m_s	catalytic activity content (specific catalytic activity), kat/kg, U/g
ν_i	stoichiometric coefficient

Photometry, fluorimetry, luminometry

A	absorbance
ε	linear millimolar absorption coefficient, $l \times mmol^{-1} \times mm^{-1}$
I	light intensity
F	fluorescence intensity
d	light path, mm

For experimental data obtained by fluorimetric methods, F and ΔF are used instead of A and ΔA. In luminometry I and I_0 are used.

Radiometry

Y	number of counts measured
Z	decay rate, disintegrations per minute, dpm
C_b	background counting rate, cpm
C_p	sample (probe) counting rate, cpm
C_n	net counting rate, cpm
η	counting efficiency, cpm/dpm
X	specific radioactivity, Bq/mol (Ci/mol)
v_r	volume taken for scintillation counting, l
cpm	counts per minute
dpm	disintegrations per minute
2.22×10^{12}	factor for conversion of decay rate Ci to dpm (1 Ci = 2.22×10^{12} dpm)
3.7×10^{10}	factor for conversion of Ci to Bq (1 Ci = 3.7×10^{10} Bq)
2.7×10^{-11}	factor for conversion of Bq to Ci (1 Bq = 2.7×10^{-11} Ci)

Metabolites

Photometry, fluorimetry, luminometry

From *Lambert-Beer*'s law (cf. Vol. I, p. 283) follows

(a) $$c = \frac{\log I_0/I}{\varepsilon \times d} = \frac{A}{\varepsilon \times d} \quad \text{mmol/l}.$$

For chemical reactions this gives

$$c_1 - c_2 = \frac{A_1 - A_2}{\varepsilon \times d}; \qquad \Delta c = \frac{\Delta A}{\varepsilon \times d} \quad \text{mmol/l}$$

and for complete conversion ($c_2 = 0$)

$$\text{(a}_1\text{)} \qquad c = \frac{\Delta A}{\varepsilon \times d} \quad \text{mmol/l} \qquad \text{(in the cuvette)}\,.$$

For the determination of the concentration of the analyte in the sample the ratio of assay volume : sample volume (V: v) or the volume fraction of sample in the assay (v/V) = φ), respectively, must be considered:

$$\text{(b)} \qquad c = \frac{\Delta A \times \mathrm{V}}{\varepsilon \times d \times \mathrm{v}} = \frac{\Delta A}{\varepsilon \times d \times \varphi} \quad \text{mmol/l} \qquad \text{(in the sample)}$$

$$\text{(b}_1\text{)} \qquad \rho = \frac{\Delta A \times \mathrm{V} \times \mathrm{MW}}{\varepsilon \times d \times \mathrm{v}} = \frac{\Delta A \times \mathrm{MW}}{\varepsilon \times d \times \varphi} \quad \text{mg/l} \qquad \text{(in the sample)}\,.$$

The substance content of the sample n_c/m_s (i.e. of the analyte in the material under investigation) is

$$\text{(c)} \qquad n_c/m_s = \frac{\Delta A \times \mathrm{V}}{\varepsilon \times d \times \mathrm{v} \times \rho_{\text{sample}}} = \frac{\Delta A}{\varepsilon \times d \times \varphi \times \rho_{\text{sample}}} \quad \text{mmol/g}\,.$$

The mass fraction of the analyte in the sample (g/g) is

$$\text{(c}_1\text{)} \qquad w = \frac{\Delta A \times \mathrm{V} \times \mathrm{MW}}{\varepsilon \times d \times \mathrm{v} \times \rho_{\text{sample}}} = \frac{\Delta A \times \mathrm{MW}}{\varepsilon \times d \times \varphi \times \rho_{\text{sample}}}\,.$$

If two or more moles of light-absorbing reaction products (e.g. NADH) are formed or consumed per mole substrate that reacts, the corresponding stoichiometric coefficient ν_i (cf. Vol. I, p. 10) also appears in the denominator of eqns. (b) and (c). If several moles of one substrate go to form the unit of substance on which the measurement is based, the corresponding factor appears in the numerator.

The result for the sample in relation to a standard is

$$\text{(d)} \qquad c = \frac{c_{\text{sample (measured)}}}{c_{\text{standard (measured)}}} \times c_{\text{standard (weighed out)}} \quad \text{mmol/l}$$

or

$$\text{(d}_1\text{)} \qquad c = \frac{A_{\text{sample (measured)}}}{A_{\text{standard (measured)}}} \times c_{\text{standard (weighed out)}} \quad \text{mmol/l}\,.$$

From the concentration of the substance in the sample solution (e.g. tissue extract), the content of the substance in the material under investigation is calculated by relating to its mass concentration in the standard solution:

(e) $$n_c/m_s = \frac{c_{\text{sample (measured)}}}{\rho_{\text{standard (weighed out)}}} \quad \frac{\text{mmol/l}}{\text{mg/l}} = \text{mmol/mg}\,,$$

(f) $$w = \frac{\rho_{\text{sample (measured)}}}{\rho_{\text{standard (weighed out)}}} \quad \frac{\text{mg/l}}{\text{mg/l}} = 1\,.$$

Example

Determination of fructose 1,6-bisphosphate (molecular weight: 340) in rat liver. Enzymatic analysis with aldolase/triosephosphate isomerase/glycerophosphate dehydrogenase. Two moles of NADH are transformed per mole of fructose-1,6-P_2. To prepare an "extract", 1 g of fresh liver was homogenized in 7.25 ml of $HClO_4$. With a value of 75% (w/w) for the fluid content of the liver, the volume of the extract is 7.25 + 0.75 = 8.00 ml. To neutralize and remove perchlorate, 0.2 ml K_2CO_3 solution was added to 6 ml of the extract. The volume of the perchlorate-free extract is thus 6.2 ml. The dilution factor for the extract is 6.2/6.0 = 1.033, and that for the tissue is 8 × 6.2/6.0 = 8.267. The experimental data must be multiplied by these values to express the results per 1 ml of acid extract or per 1 g of tissue.

The measured change in absorbance at 339 nm (ε = 0.63 l × mmol^{-1} × mm^{-1}) was ΔA = 0.120; the volume of the assay solution was 3 × 10^{-3} l, the volume of sample was 1.5 × 10^{-3} l, and the light path was 10 mm. The concentration in the perchlorate-free sample used for the assay was:

according to eqn. (b)

$$c = \frac{0.120 \times 3 \times 10^{-3}}{0.63 \times 10 \times 1.5 \times 10^{-3} \times 2} = 0.0190\ \text{mmol/l}$$

or according to eqn. (c)

$$\rho = \frac{0.120 \times 3 \times 10^{-3} \times 340}{0.63 \times 10 \times 1.5 \times 10^{-3} \times 2} = 6.46\ \text{mg/l}\,.$$

Multiplication by the dilution factor gives the concentration in the acid extract:

$$c = 0.0190 \times 1.033 = 0.0196\ \text{mmol/l}$$

$$\rho = 6.46 \times 1.033 = 6.67\ \text{mg/l}\,.$$

The substance content of fructose-1,6-P_2 in the tissue is

$$n_c/m_s = 0.0190 \times 8.267 = 0.157 \text{ mmol/kg}$$

or its mass fraction (g/g)

$$w = 6.46 \times 10^{-6} \times 8.267 = 5.4 \times 10^{-5}.$$

Radiometry

For calculation of the analyte concentration and content from the net counting rate C_n of sample and blank, the following relationships are valid.

Substance concentration

(g) $$c = \frac{\Delta C_n}{\eta \times 2.22 \times 10^{12} \times X \times \varphi \times v_r} \text{ mol/l (in the sample)}$$

(units: C_n in cpm; X in Ci/mol or Bq/mol; φ = v/V in l/l; v_r in l).

Mass concentration

(g_1) $$\rho = \frac{\Delta C_n \times MW}{\eta \times 2.22 \times 10^{12} \times X \times \varphi \times v_r} \text{ g/l (in the sample)}$$

(units as for eqn. (g); MW in g/mol).

Substance content

(h) $$n_c/m_s = \frac{\Delta C_n}{\eta \times 2.22 \times 10^{12} \times X \times \varphi \times v_r \times \rho_{sample}} \text{ mol/g (in the sample)}$$

(units as for eqn. (g); ρ_{sample} in g/l).

Mass fraction

(h_1) $$w = \frac{\Delta C_n \times MW}{\eta \times 2.22 \times 10^{12} \times X \times \varphi \times v_r \times \rho_{sample}} \text{ (in the sample)}$$

(units as for eqn. (g_1); ρ_{sample} in g/l).

Enzymes

Photometry, fluorimetry, luminometry

For measurement of the catalytic activity z of enzymes the rate of the catalyzed substrate conversion per time unit is used, µmol/min or mol/s.

Concerning the stoichiometric coefficient ν_i cf. p. 585.

According to eqn. (a)

catalytic activity (conversion in mol* per time unit) is

(i) $$z = \frac{\Delta c \times V}{\Delta t} = \frac{\Delta A \times V}{1000 \times \varepsilon \times d \times \Delta t} \quad \text{mol/s (kat)}$$

(units: ε in $l \times mmol^{-1} \times mm^{-1}$; V of the assay volume in l; d in mm; t in s)

(i_1) $$z = \frac{\Delta c \times V}{\Delta t} = \frac{\Delta A \times V \times 1000}{\varepsilon \times d \times \Delta t} \quad \mu\text{mol/min (U)}$$

(units as for eqn. (i); t in min).

The catalytic activity concentration in the sample is

(k) $$b = \frac{\Delta A \times V}{1000 \times \varepsilon \times d \times \Delta t \times v} = \frac{\Delta A}{1000 \times \varepsilon \times d \times \Delta t \times \varphi} \quad \text{mol} \times s^{-1} \times l^{-1} \text{ (kat/l))}$$

(units as for eqn. (i); v in l; $\varphi = v/V$ in l/l).

(k_1) $$b = \frac{\Delta A \times V \times 1000}{\varepsilon \times d \times \Delta t \times v} = \frac{\Delta A \times 1000}{\varepsilon \times d \times \Delta t \times \varphi} \quad \mu\text{mol} \times \text{min}^{-1} \times l^{-1} \text{ (U/l)}$$

(units as for eqn. (i_1); v in l; $\varphi = v/V$ in l/l).

The catalytic activity related to the mass of protein, catalytic activity content z_c/m_s (specific activity) is

(l) $$z_c/m_s = \frac{\Delta A \times V}{1000 \times \varepsilon \times d \times \Delta t \times v \times \rho_{protein}} = \frac{\Delta A}{1000 \times \varepsilon \times d \times \Delta t \times \varphi \times \rho_{sample}} \quad \text{kat/g}$$

(units as for eqn. (k); $\rho_{protein}$ in g/l)

(l_1) $$z_c/m_s = \frac{\Delta A \times V \times 1000}{\varepsilon \times d \times \Delta t \times v \times \rho_{protein}} = \frac{\Delta A \times 1000}{\varepsilon \times d \times \Delta t \times \varphi \times \rho_{protein}} \quad \text{U/g}$$

(units as for eqn. (k_1); $\rho_{protein}$ in g/l).

---

* $c \times V$ (mol/l) $\times$ l = mol.

Example

Determination of the (specific) catalytic activity of an enzyme. The measurements were made at 339 nm; $\Delta A/\Delta t = 0.063/60$ s in a 3 ml assay mixture ($V = 3 \times 10^{-3}$ l). The volume of the sample was 2×10^{-4} l. The sample diluted over thousandfold for measurement contained 10 g of enzyme protein per litre.

In the assay mixture the catalytic activity is according to eqn. (i) or (i_1), respectively

$$z = \frac{0.063 \times 3 \times 10^{-3}}{10^3 \times 0.63 \times 10 \times 60} = 5 \times 10^{-10}\,\text{kat} = 0.5\,\text{nkat}$$

or

$$z = \frac{0.063 \times 3 \times 10^{-3} \times 10^3}{0.63 \times 10 \times 1} = 0.03\,\text{U}.$$

The catalytic activity concentration in the sample solution according to eqn. (k) or (k_1), respectively, is

$$b = \frac{0.063 \times 3 \times 10^{-3}}{10^3 \times 0.63 \times 10 \times 60 \times 2 \times 10^{-4}} = 2500\,\text{nkat/l}$$

or

$$b = \frac{0.063 \times 3 \times 10^{-3} \times 10^3}{0.63 \times 10 \times 1 \times 2 \times 10^{-4}} = 150\,\text{U/l}\,.$$

Related to the mass of protein (dilution factor 1000), the catalytic activity content is

$$z_c/m_s = \frac{2500 \times 1000}{10} \frac{\text{nkat/l}}{\text{g/l}} = 250\,\mu\text{kat/g}$$

or

$$z_c/m_s = \frac{150 \times 1000}{10} \frac{\text{U/l}}{\text{g/l}} = 15\,\text{U/g}.$$

The result for the sample in relation to an enzyme standard solution is

$$\text{(m)} \qquad b = \frac{b_{\text{sample (measured)}}}{b_{\text{standard (measured)}}} \times b_{\text{standard (indicated)}} \quad \text{nkat/l (U/l)}$$

or simply

$$(\text{m}_1) \qquad b = \frac{A_{\text{sample (measured)}}}{A_{\text{standard (measured)}}} \times b_{\text{standard (indicated)}} \quad \text{nkat/l (U/l)}$$

Radiometry

According to eqns. (k) and (k_1), respectively, and using eqns. (g) and (h) catalytic activity is

(n) $$z = \frac{\Delta C_n \times V}{\eta \times 2.22 \times 10^{12} \times 10^{-9} \times X \times v_r \times \Delta t} \text{ nmol/s (nkat)}$$

(units: C_n in cpm; X in Ci/mol or Bq/mol; v_r in l; t in s),

(n_1) $$z = \frac{\Delta C_n \times V}{\eta \times 2.22 \times 10^{12} \times 10^{-6} \times X \times v_r \times \Delta t} \text{ µmol/min (U)}$$

(units as for eqn. (n); t in min).

The catalytic activity concentration in the sample is

(o) $$b = \frac{\Delta C_n}{\eta \times 2.22 \times 10^{12} \times 10^{-9} \times X \times \varphi \times v_r \times \Delta t} \text{ nkat/l}$$

(units as in eqns. (n), (n_1); φ = v/V in l/l),

(o_1) $$b = \frac{\Delta C_n}{\eta \times 2.22 \times 10^{12} \times 10^{-6} \times X \times \varphi \times v_r \times \Delta t} \text{ U/l}$$

(units as in eqn. (o); t in min).

The catalytic activity related to the mass of protein, catalytic activity content z_c/m_s (specific activity), is

(p) $$z_c/m_s = \frac{\Delta C_n}{\eta \times 2.22 \times 10^{12} \times 10^{-9} \times X \times \varphi \times v_r \times \Delta t \times \rho_{protein}} \text{ nkat/g}$$

(units as for eqn. (o); $\rho_{protein}$ in g/l),

(p_1) $$z_c/m_s = \frac{\Delta C_n}{\eta \times 2.22 \times 10^{12} \times 10^{-6} \times X \times \varphi \times v_r \times \Delta t \times \rho_{protein}} \text{ U/g}$$

(units as for eqn. (o_1); $\rho_{protein}$ in g/l).

Statistics

For parameters and sample statistics, for mathematical evaluation of experiments (mean value, standard deviation, imprecision, inaccuracy) and for statistical evaluation of results, especially for regression analysis, cf. Vol. II, chapter 3.2.

Refering the Experimental Results to Biological Material

In the analysis of organ extracts, blood, serum, etc., the dilution resulting from deproteinization of samples must also be taken into account along with the fluid content of the sample. The following values give reasonable accuracy: blood 80% (w/w), tissue (liver, kidney, muscle, heart) 75% (w/w). Tissue samples are weighed out; a sp. gr. of 1.06 is used for the conversion of blood volumes into mass.

Apart from the volume (in the case of serum, plasma, blood, urine, etc.), other reference quantities that may be used for biological material are the fresh weight, dry weight, total nitrogen, protein content, protein nitrogen; cell count, e.g. erythrocyte count; haemoglobin content, cytochrome c content, and dry weight of the cell-free sample solution.

Examples

Determination of the dilution factor for blood.

The specific gravity of blood is taken as 1.06, and its fluid content is taken as 80% (w/w). 2 ml of blood are deproteinized with 3 ml of perchloric acid and centrifuged, and 2.5 ml of the supernatant are neutralized with 1 ml K_2CO_3 solution. The blood is therefore diluted by the following factor:

$$F = \frac{2 \times 1.06 \times 0.8 + 3}{2} \times \frac{2.5 + 1}{2.5} = \frac{1.696 + 3}{2} \times \frac{3.5}{2.5} = 3.29\,.$$

The experimental result obtained with the neutralized blood extract must be multiplied by this factor to obtain the content of the metabolite in the blood.

Calculation of the dilution factor for multi-stage assay mixtures.
In the determination of maltose in biological fluids, the various reaction steps have different pH-optima. The assay begins with an incubation: 0.5 ml sample with 0.5 ml acetate buffer and 0.02 ml α-glucosidase. After inactivation of the enzyme, the second step is the determination of the glucose in 0.2 ml of the incubation solution ($V = 3.42$ ml). The dilution factor is thus:

$$F = \frac{0.5 + 0.5 + 0.02}{0.5} \times \frac{3.42}{0.2} = 2.04 \times 17.1 = 34.88\,.$$

The experimental result, after division by 2 (1 maltose ≙ 2 glucose), must be multiplied by this factor to obtain the maltose content in 1 ml sample.

To calculate the metabolite content of the cells of a tissue, the metabolite content of the blood in this tissue must be taken into account.

The mass fraction of blood in the tissue is determined according to *Bücher et al.* [4] from absorbance measurements at 578, 560, and 540 nm. Assuming that the proportion of oxyhaemoglobin (HbO_2) in the circulating blood and the tissue is approximately the same, if follows that the fraction of blood w in the tissue is

$$w = \frac{\Delta A_{HbO_2} \times F_1 \times d_1}{\Delta A'_{HbO_2} \times F_2 \times d_2}$$

where

ΔA_{HbO_2} is absorbance difference for tissue extract
$\Delta A'_{HbO_2}$ is absorbance difference for blood dilution
F_1 and F_2 are dilution factors
d_1 and d_2 are light paths of the cuvettes

ΔA_{HbO_2} and $\Delta A'_{HbO_2}$ are calculated [4] from the absorbance measurements at 578, 560, and 540 nm, according to the formula:

$$\Delta A_{HbO_2} \text{ or } \Delta A'_{HbO_2} = (A_{578} - A_{560}) + [(A_{540} - A_{578}) \times 0.47].$$

If the metabolite concentration in the tissue sample is to be referred to the true volume of cellular fluid in order to give the physiological concentration, the result (in mmol/l of tissue extract) is multiplied by the following factor (*G. Michal,* unpublished):

$$F = \frac{V_{\text{after neutralization}}}{V_{\text{before neutralization}}} \times \frac{\dfrac{m_{\text{tissue}}}{\text{spec. gr.}} \times \varphi + V_d}{\dfrac{m_{\text{tissue}}}{\text{spec. gr.}} \times \varphi}$$

φ is the volume fraction of liquid in the tissue volume; m_{tissue} is tissue wet weight; specific gravity relates to the sample. The latter quantity may be taken as unity in most cases. V_d is volume of reagent solution (e.g. $HClO_4$) used for deproteinization.

A similar calculation for erythrocytes has been published by *Bürgi* [5] (cf. Vol. II, chapter 1.1.1.3, p. 13).

References

[1] International Union of Pure and Applied Chemistry (IUPAC) and International Federation of Clinical Chemistry (IFCC): Approved Recommendation (1978), Quantities and Units in Clinical Chemistry, Clin. Chim. Acta *96*, 157F–183F (1979).

[2] Expert Panel on Nomenclature and Principles of Quality Control in Clinical Chemistry; Committee on Standards (IFCC): Approved Recommendation (1978) on Quality Control in Clinical Chemistry, Part 1: General Principles and Terminology, J. Clin. Chem. Clin. Biochem. *18*, 69 – 77 (1980).

[3] International Commission on Radiation Units and Measurements: ICRU Report 33 "Radiation Quantities and Units", 1980, 7910 Woodmont Ave., Washington D.C. 20014, USA.

[4] *H. J. Hohorst, F. H. Kreutz, Th. Bücher,* Über Metabolitgehalte und Metabolit-Konzentrationen in der Leber der Ratte, Biochem. Z. *332*, 18 – 46 (1950).

[5] *W. Bürgi,* The Volume Displacement Effect in Quantitative Analysis of Red Blood Cell Constituents, J. Clin. Chem. Clin. Biochem. *7*, 458 – 460 (1969).

4 Absorption Coefficients of NAD(P)H

The absorption curves and absorption coefficients depend on the temperature, pH, and the ionic strength of the solution. The best-investigated cases [1 – 4] are NADH and NADPH. At λ = 334 nm the temperature dependence of ε is approximately zero. The value of ε falls with rising temperature at wavelengths $\lambda > 334$ nm (including the maximum of the absorption curve) and increases at $\lambda < 334$ nm; the absorption maximum is shifted accordingly.

Table 1. Molar decadic absorption coefficients ($l \times mol^{-1} \times mm^{-1}$) for β-NADH and β-NADPH (measured in triethanolamine/HCl buffer, 0.1 mol/l; pH = 7.6) [1].

	°C	Hg 334 nm	339 nm	340 nm	339.85 nm^{+}	Hg 365 nm
β-NADH	25	6.176×10^2	no measurement	$6.317 \times 10^{2++}$	6.292×10^2	3.441×10^2
		6.182×10^2*	cf. Table 2	–	6.298×10^2*	3.444×10^2*
β-NADH	30	6.187×10^2*	–	–	–	3.427×10^2*
β-NADPH	25	6.178×10^2*	–	–	–	3.532×10^2*
β-NADPH	30	6.186×10^2*	–	–	–	3.515×10^2*

$^{+}$ Checking the photometer revealed 339.85 nm to be correct, instead of 340 nm.
$^{++}$ Acc. to [2] in Tris buffer, 0.1 mol/l; pH 7.8.
* Values were not corrected for beam convergence or intrinsic absorbance of the oxidized coenzyme.

The above mentioned investigations [1 – 4] on NAD(P)H show:

- ε is different for NADH and NADPH,
- ε is temperature-dependent,
- ε depends on the pH and the ionic strength of the solution to be measured,
- ε cannot be determined sufficiently accurately at 37 °C because of the instability of the coenzyme,
- the absorption maximum of NAD(P)H is not located at exactly 340 nm; 339 nm can be taken as a first approximation,
- the absorption maximum is temperature-dependent,
- the differences in the value of ε due to the factors mentioned above are smallest at Hg 334 nm.

All of the influences mentioned above lead to deviations of the value of ε, not exceeding 0.5% at Hg 334 nm. Consequently, it is best, since practically independent of the conditions of measurement, to perform measurements at this wavelength. Moreover, the values of ε are identical here for both coenzymes.

However, the values of ε at 25 °C and 30 °C at 340 nm (or 339 nm) and Hg 365 nm are also sufficiently close for practical purposes (<1% error at 340 (339) nm; about 2% at Hg 365 nm) and are independent of the other conditions of measurement. For measurement at Hg 365 nm, however, one must distinguish between the values of ε for NADH and NADPH.

In a routine laboratory it would not be practical to use the exact values of the absorption coefficients given above. It would then be necessary, under certain circumstances, to determine the exact value of ε for each experiment. The figures recommended for practical purposes in the routine laboratory vary, according to temperature and other measurement conditions, by less than 1 to 2%, which is within the limits of error attainable for routine enzymatic analyses.

Molar decadic absorption coefficients ($l \times mol^{-1} \times mm^{-1}$) for NADH and NADPH at temperatures of 25 °C and 30 °C [3] are for practical use:

	Hg 334 nm (334.15 nm)	*340 nm (339 nm)*	*Hg 365 nm (365.3 nm)*
NADH	6.18×10^2	6.3×10^2	3.4×10^2
NADPH	6.18×10^2	6.3×10^2	3.5×10^2

References

[1] *J. Ziegenhorn, M. Senn, T. Bücher,* Molar Absorptivities of β-NADH and β-NADPH, Clin. Chem. *22*, 151 – 160 (1976).

[2] *R. B. McComb, L. W. Bond, R. W. Burnett, R. C. Keech, G. N. Bowers, jr.,* Determination of the Molar Absorptivity of NADH, Clin. Chem. *22*, 141 – 150 (1976).

[3] *H. U. Bergmeyer,* Neue Werte für die molaren Extinktions-Koeffizienten von NADH und NADPH zum Gebrauch im Routine-Laboratorium, J. Clin. Chem. Clin. Biochem. *13*, 507 – 508 (1975).

[4] *Th. Bücher, G. Lüsch, H. Krell* in: *G. Anido, E. J. van Kampen, S. B. Rosalki, M. Rubin* (eds.), Temperature Dependence of Difference-Absorption Coefficients of NADH Minus NAD^+ and NADPH Minus $NADP^+$ in the Near Ultraviolet, Quality Control in Clinical Chemistry, Walter de Gruyter, Berlin, New York 1975, pp. 301 – 310.

Index